An Introduction to Drug Synthesis

An Introduction to
Drug
Synthesis

Graham L. Patrick

OXFORD
UNIVERSITY PRESS

Great Clarendon Street, Oxford, OX2 6DP,
United Kingdom

Oxford University Press is a department of the University of Oxford.
It furthers the University's objective of excellence in research, scholarship,
and education by publishing worldwide. Oxford is a registered trade mark of
Oxford University Press in the UK and in certain other countries

Published in the United States of America by Oxford University Press
198 Madison Avenue, New York, NY 10016, United States of America

British Library Cataloguing in Publication Data

Data available

Library of Congress Control Number: 2014945280

ISBN 978–0–19–870843–8

Printed in Italy by
L.E.G.O. S.p.A.

Links to third party websites are provided by Oxford in good faith and
for information only. Oxford disclaims any responsibility for the materials
contained in any third party website referenced in this work.

Preface

This text is written for undergraduates and postgraduates who have a basic grounding in organic chemistry and are studying a module or degree in a chemistry-related field such as medicinal chemistry. It attempts to convey, in a readable and interesting style, an understanding about some of the issues and strategies relating to drug synthesis. In particular, the book aims to show the importance of organic synthesis to various aspects of the drug discovery process. For example, the kind of synthetic route chosen for a particular drug has an important bearing on what kind of analogues can be synthesized. These analogues can be used to study structure–activity relationships, as well as helping us to identify structures with improved activities and properties. Synthesis is also crucial to the economic and practical feasibility of manufacturing drugs on a commercial scale. Consequently, the book is of particular interest to students who might be considering a future career in research and development in the pharmaceutical industry.

The book is divided into three parts.

Part A contains six chapters that provide some general background on medicinal chemistry and organic synthesis. The first chapter gives an overview of the process involved in getting a drug to market and the impact that organic synthesis has at various stages of that journey. It also defines various medicinal chemistry terms that are used throughout the other chapters.

Chapter 2 identifies the different types of reactions that are involved in a drug synthesis, highlighting the importance of five general categories—coupling reactions, functional group transformations, functionalizations, functional group removals, and the use of protecting groups. Throughout the chapter, simple examples of drug syntheses are provided to illustrate these different reaction categories, and how the structure of a target compound has a crucial impact on the complexity of the overall synthesis.

Chapters 3–5 are overviews of retrosynthesis, cyclization reactions, and the synthesis of chiral compounds, where the emphasis is on explaining the key principles of these topics and relating them to the synthesis of important drug structures.

Finally, Chapter 6 describes the role of combinatorial and parallel synthesis in drug synthesis.

There are also three case studies. Two of these look at reactions that are particularly important in drug synthesis. Case study 1 considers the role of protecting groups and coupling agents in peptide synthesis, while case study 2 provides an overview of palladium-catalysed coupling reactions. Case study 3 provides an example of how retrosynthesis is used in designing a synthesis. In this case, the target structure is a natural product called huperzine A, which has interesting pharmacological properties.

Part B contains five chapters that describe how synthesis impacts on various stages of the drug design and development process. Chapter 7 focuses on the synthesis of novel structures as potential lead compounds in medicinal chemistry, whereas Chapter 8 looks at synthetic approaches to the analogues of known active compounds.

Chapter 9 covers synthetic and semi-synthetic approaches to the synthesis of medicinally important natural products and their analogues, and also describes how biosynthesis and genetic engineering has been used to generate such compounds.

Chapter 10 describes chemical and process development, and identifies many of the key issues that have to be considered in the synthesis of drugs on a commercial scale.

Finally, Chapter 11 describes the synthesis of isotopically labelled drugs, and the uses of such drugs in therapy, diagnosis, and scientific study.

A case study on gliotoxin provides an example of the use of radiolabelling studies to determine the biosynthesis of an important natural product.

Part C contains three chapters that focus on the design, synthesis, and activities of particular antibacterial agents. Chapters 12 and 13 describe tetracyclines and macrolides, respectively, while Chapter 14 describes different synthetic approaches to quinolones and fluoroquinolones.

In addition to the three main parts of the textbook, there are several appendices that summarize many of the most commonly used reactions in drug synthesis. Further information about these reactions is provided in the Online Resource Centre, as explained in the guide to the book that follows.

G. L. P.
June 2014

About the book

An Introduction to Drug Synthesis and its Online Resource Centre contain many learning features which will help you to understand this fascinating subject. This section explains how to get the most out of these features.

Emboldened key words

Terminology is emboldened within the main text and defined in a glossary at the end of the book, helping you to become familiar with the language of drug synthesis.

target to the one being tested in the *in vitro* test. Physiological effects are often the result of a variety of different biological mechanisms, and carrying out specific *in vitro* tests alone may miss an important new lead compound. Furthermore, it may not be known what role a newly discovered protein has in the body. An *in vitro* test will show whether a compound interacts with that novel target, while an *in vivo* test will identify the overall effect of that interaction on the organism.

Secondly, *in vitro* tests are excellent at establishing whether a drug interacts with its target to produce a pharmacological effect (pharmacodynamics), but they

1.5.3 Pharmacokinetics

If a drug is to be effective, it must not only interact with a particular molecular target, but must also reach that target. However, there are many different factors which can prevent that happening. The main ones are **absorption**, **distribution**, **metabolism**, and **excretion** (commonly referred to as **ADME**). Another factor which is commonly considered is **toxicity** (**ADMET**).

Absorption

Boxes

Boxes are used to present in-depth material and to explore how the concepts of drug synthesis are applied in practice.

BOX 6.2 Dynamic combinatorial synthesis of vancomycin dimers

Vancomycin is an antibiotic that works because it masks the building blocks required for bacterial cell wall synthesis. Binding takes place specifically between the antibiotic and a peptide sequence (*L*-Lys-*D*-Ala-*D*-Ala) which is present in the building block. It is also known that this binding promotes dimerization of the vancomycin–target complex, which suggests that covalently linked vancomycin dimers might be more effective antibacterial agents than vancomycin itself. A dynamic combinatorial synthesis was carried out to synthesize a variety of different vancomycin dimers covalently linked by bridges of different

The tripeptide target was present to accelerate the rate of bridge formation and to promote formation of vancomycin dimers having the ideal bridge length. As shown in Figure 2, the vancomycin monomers bind the tripeptide, which encourages the self-assembly of non-covalently linked dimers. Once formed, those dimers having the correct length of substituent are more likely to react together to form the covalent bridge (Fig. 2).

Having established the optimum length of bridge, another experiment was carried out on eight vancomycin monomers which had the correct length of 'tether' but varied slightly in

Key points

Summaries at the end of major sections within chapters highlight and summarize key concepts, and provide a basis for revision.

KEY POINTS

- Tagging involves the construction of a tagging molecule on the same solid support as the target molecule. Tagging molecules are normally peptides or oligonucleotides. After each stage of the target synthesis, the peptide or oligonucleotide is extended and the amino acid or nucleotide used defines the reactant or reagent used in that stage.

- Dynamic combinatorial chemistry involves the equilibrium formation of a mixture of compounds in the presence of a target. Binding of a product with the target amplifies that product in the equilibrium mixture.
- Diversity-orientated synthesis aims to produce compounds with as wide a diversity as possible in order to fully explore the conformational space around a molecule when it interacts with a target binding site.

Questions

End-of-chapter questions allow you to test your understanding and apply concepts presented in the chapter to solve the problems presented to you.

QUESTIONS

1. Carry out a retrosynthetic analysis of the muscle relaxant **pirindol** and propose a possible synthesis.

2. **Proparacaine** (**proxymetacaine**) is a local anaesthetic that is used in ophthalmology and is applied in eye drops. Carry out a retrosynthetic analysis of its structure and propose a possible synthesis.

Further reading

Selected references allow you to easily research those topics that are of particular interest to you.

FURTHER READING

General references

Patrick, G.L. (2013) *An introduction to medicinal chemistry* (5th edn). Oxford University Press, Oxford (Chapter 24, 'The opioid analgesics'; Chapter 23, 'Drugs acting on the adrenergic nervous system'; Section 19.5.1, 'Penicillins'; Section 21.6.2, 'Protein kinase inhibitors').

Specific syntheses

analgesic', *Journal of Medicinal Chemistry*, **29**, 2290–7 (alfentanil).

Lawrence, H.R., *et al.* (2005) 'Novel and potent 17β-hydroxysteroid dehydrogenase type I inhibitors⊖, *Journal of Medicinal Chemistry*, **48**, 2759–62 (estrone analogues).

Lipkowski, A.W., *et al.* (1986) Peptides as receptor selectivity modulators of opiate pharmacophores, *Journal of Medicinal*

Case studies

Case studies within several chapters and at the end of Parts A and B demonstrate the practical application of drug synthesis by exploring the synthesis of a number of drugs in detail.

CS1.1 Introduction

Peptide synthesis has been an important area of organic synthesis for many years. Many of the body's neurotransmitters and hormones are peptides or proteins, and the ability to carry out peptide synthesis has allowed the medicinal chemist to prepare these structures, as well as their analogues. This provided an understanding of structure–activity relationships and led to useful drugs. The same holds true for peptides and proteins that have

there are several examples where peptide-like drugs have proved clinically useful.

CS1.2 Amino acids—the building blocks for peptide synthesis

Amino acids are the building blocks used for the biosynthesis and synthesis of peptides and proteins. They all contain an amine and a carboxylic acid functional group,

Appendix

There are seven appendices which summarize many of the most commonly used reactions in drug synthesis.

Appendix 1

Functional group transformations

There are a large number of possible functional group transformations (FGTs) in organic synthesis. The following are examples of the most commonly used FGTs in

drug synthesis. Further details on each reaction are available in the book's Online Resource Centre.

About the Online Resource Centre

Online Resource Centres provide students and lecturers with ready-to-use teaching and learning resources to augment the printed book.

You will find the material to accompany *An Introduction to Drug Synthesis* at: www.oxfordtextbooks.co.uk/orc/patrick_synth/

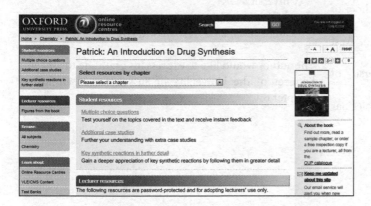

Student resources

Multiple-choice questions
Test yourself on the topics covered in the text and receive instant feedback.

Additional case studies
Further your understanding with extra case studies.

Key synthetic reactions in further detail
Gain a deeper appreciation of key synthetic reactions by following them in greater detail.

Lecturer resources

For registered adopters of the book

Figures from the book
All of the figures from the textbook are available to download electronically for use in lectures and handouts.

Acknowledgements

The author and Oxford University Press would like to thank the following people who have given advice on the textbook.

Dr John Spencer, Department of Chemistry, School of Life Sciences, University of Sussex, UK

Dr Ciaran Ewins, School of Science, University of West of Scotland, UK

Dr Callum McHugh, School of Science, University of West of Scotland, UK

Dr Chris Rostron, School of Pharmacy and Biomolecular Sciences, Liverpool John Moores University, UK

Dr Klaus Pors, School of Life Sciences, University of Bradford, UK

Dr Michael John Hall, School of Chemistry, Newcastle University, UK

Dr Neil Press, Novartis Institutes for Biomedical Research, Horsham, West Sussex, UK

The author would like to express his gratitude to Dr John Spencer of the University of Sussex for co-authoring Chapter 6.

Brief contents

Detailed contents

Abbreviations and acronyms

aa	amino acid
Ac	acetyl
7-ACA	7-aminocephalosporinic acid
AcCl	acetyl chloride
ACE	angiotensin-converting enzyme
ACh	acetylcholine
AChE	acetylcholinesterase
Ac$_2$O	acetic anhydride
ACP	acyl carrier protein
AD	Alzheimer's disease
ADH	aldehyde dehydrogenase
ADHD	attention deficit hyperactivity disorder
ADME	absorption, distribution, metabolism, excretion
ADMET	absorption, distribution, metabolism, excretion, toxicity
ADP	adenosine 5′-diphosphate
AIDS	acquired immune deficiency syndrome
AMD	amorphadiene synthase
AMP	adenosine 5′-monophosphate
cAMP	cyclic adenosine 5′-monophosphate
amu	atomic mass unit
6-APA	6-aminopenicillanic acid
AT	acyltransferase
ATP	adenosine 5′-triphosphate
BBB	blood–brain barrier
BINAP	2,2′-bis(diphenylphosphino-1, 1′-binaphthyl)
BnBr	Benzyl bromide
Boc	*tert*-butyloxycarbonyl
Boc anhydride or (Boc)$_2$O	di-*t*-butyl dicarbonate
BOP	(benzotriazol-1-yloxy) tris(dimethylamino)phosphonium hexafluorophosphate
_n_-Bu	*n*-butyl
_t_-Bu	*tertiary*-butyl
cbz	benzyloxycarbonyl or carboxybenzyl
CDI	N,N′-carbonyldiimidazole
CHIRAPHOS	bis(diphenylphosphino)butane
Clog_P_	calculated logarithm of the partition coefficient
CNS	central nervous system
CoA	coenzyme A
COD	cyclooctadiene
COX	cyclooxygenase
cyclic AMP	cyclic adenosine 5′-monophosphate
CYP	cytochrome P450
DAGO or DAMGO	[D-Ala^2,MePhe^4,Glyol^5]enkephalin
DAST	diethylaminosulphur trifluoride
dba	dibenzylidene acetone
DBTA	dibenzoyl tartaric acid monohydrate
DBU	1,8-diazobicyclo[5.4.0]undec-7-ene
DCC	dicyclohexylcarbodiimide
DCM	dichloromethane
DCU	dicyclohexylurea
DEAD	diethyl azodicarboxylate
DEBS	6-deoxyerythronolide B synthase
DET	diethyl tartrate
DH	dehydratase
DIBAL or DIBAL-H	diisobutylaluminium hydride
DIC	N,N′- diisopropylcarbodiimide
DIOP	O-isopropylidene-2,3-dihydroxy-1,4-bis(diphenylphosphino)butane
DIPAMP	ethane-1,2-diylbis[(2-methoxyphenyl) phenylphosphane]
DIPC	N,N′-diisopropylcarbodiimide
DIPEA	N,N-diisopropylethylamine
DIU	diisopropylurea
DMA	dimethylacetamide
DMAP	4-dimethylaminopyridine
DMF	dimethylformamide
DMSO	dimethylsulphoxide
DNA	deoxyribonucleic acid
DOR	delta opioid receptor
DPDPE	tyr-c(D-Pen-Gly-Phe-D-Pen)
DPP-4	dipeptidyl peptidase-4
EC$_{50}$	concentration of drug required to produce 50% of the maximum possible effect
EDC or EDCI	1-ethyl-3-(3-dimethylaminopropyl) carbodiimide
EDU	1-ethyl-3-(3-dimethylaminopropyl)urea
ee	enantiomeric excess

EGF	epidermal growth factor
EGF-R	epidermal growth factor receptor
EMEA	European Agency for the Evaluation of Medicinal Products
ER	enoyl reductase
Et	ethyl
F-SPE	fluorous solid phase extraction
FDA	US Food and Drug Administration
^{18}F-FDG	[^{18}F]fluorodeoxyglucose
FG	functional group
FGI	functional group interconversion
FGT	functional group transformation
Fmoc	fluorenylmethyloxycarbonyl
Fmoc-Cl	fluorenylmethyloxycarbonyl chloride
FPP	farnesyl pyrophosphate
F-SPE	fluorous solid phase extraction
FT	farnesyl transferase
G-protein	guanine nucleotide binding protein
GABA	γ-aminobutyric acid
GABA-R	benzodiazepine receptor
GCP	Good Clinical Practice
GDP	guanosine 5′-diphosphate
GIT	gastrointestinal tract
GLP	Good Laboratory Practice
GMP	Good Manufacturing Practice
GMP	guanosine 5′-monophosphate
GTP	guanosine 5′-triphosphate
H-R	histamine receptor
HATU	N-[(dimethylamino)-1H-1,2,3-triazolo[4,5-b]pyridin-1-ylmethylene]-N-methylmethanaminium hexafluorophosphate
HBA	hydrogen bond acceptor
HBD	hydrogen bond donor
HFC-134a	1,1,1,2-tetrafluoroethane
HIV	human immunodeficiency virus
HMG-CoA	3-hydroxy-3-methylglutaryl-coenzyme A
HMPA	hexamethylphosphoramide
HOBt	1-hydroxybenzotriazole
HOMO	highest occupied molecular orbital
HPLC	high performance liquid chromatography
17β-HSD1	17β-dehydroxysteroid dehydrogenase type 1
HTS	high throughput screening
IC_{50}	concentration of drug required to inhibit a target by 50%
^{d}Ipc$_2$BCl	d-enantiomer of diisopinocampheylchloroborane
^{l}Ipc$_2$BCl	l-enantiomer of diisopinocampheylchloroborane
IpcBH$_2$	diisopinocampheylborane
K_d	binding affinity or dissociation binding constant
K_i	inhibition constant
KIE	kinetic isotope effect
KN(TMS)$_2$	potassium bis(trimethylsilyl)amide
KOR	kappa opioid receptor
KR	ketoreductase
KS	ketosynthase enzyme
LDA	lithium diisopropylamide
LDH	lactate dehydrogenase
LiHMDS or LiN(TMS)$_2$	lithium bis(trimethylsilyl)amide
Log$P$	logarithm of the partition coefficient
LUMO	lowest unoccupied molecular orbital
M-receptor	muscarinic receptor
MAA	Marketing Authorization Application
MAOS	microwave-assisted organic synthesis
mcpba	*meta*-chloroperbenzoic acid
Me	methyl
MIBK	methyl isobutyl ketone (4-methylpentan-2-one)
MOR	mu opioid receptor
mRNA	messenger RNA
Ms	mesyl
MsCl	methanesulphonyl chloride
MWt	molecular weight
N-Receptor	nicotinic receptor
NAD or NADH	nicotinamide adenine dinucleotide
NADP or NADPH	nicotinamide adenine dinucleotide phosphate
NaN(TMS)$_2$	sodium bis(trimethylsilyl)amide
NBS	N-bromosuccinimide
NCE	new chemical entity
NDA	New Drug Application
NH(TMS)$_2$	bis(trimethylsilyl)amine
Ni(cod)$_2$	bis(cyclooctadiene)nickel(0)
Ni(dppp)Cl$_2$	dichloro(1,3-bis(diphenylphosphino)propane)nickel
NIS	N-iodosuccinimide
NME	new molecular entity
NMP	N-methylpyrrolidinone
NMR	nuclear magnetic resonance

NNRTI	non-nucleoside reverse transcriptase inhibitor	**RNA**	ribonucleic acid
nor-BNI	norbinaltorphimine	**rRNA**	ribosomal RNA
NRPS	non-ribosomal peptide synthase	**SAR**	structure–activity relationships
NRTI	nucleoside reverse transcriptase inhibitor	**SCAL**	safety catch acid-labile linker
		SOP	standard operating procedure
NSAID	non-steroidal anti-inflammatory drug	**SPA**	scintillation proximity assay
NVOC	nitroveratryloxycarbonyl	**SPE**	solid phase extraction
P	partition coefficient	**SPECT**	single photon emission computer tomography
PBS	phosphate-buffered saline	**SSRI**	selective serotonin reuptake inhibitor
Pd/C	palladium charcoal catalyst	**TBAF**	tetrabutylammonium fluoride
Pd$_2$(dba)$_3$	tris(dibenzylideneacetone) dipalladium(0)	**TBDMS or TBS**	*tert*-butyldimethylsilyl
PEG	polyethylene glycol	**TCA**	tricyclic antidepressant
PET	positron emission tomography	**TFA**	trifluoroacetic acid
Ph	phenyl	**TfOH**	triflic acid or trifluorosulphonic acid
PI	protease inhibitor	**THF**	tetrahydrofuran
PKS	polyketide synthase	**TIPS**	triisopropylsilyl
PLP	pyridoxal phosphate	**TLC**	thin layer chromatography
PMP	1,2,2,6,6-pentamethylpiperidine	**TMEDA**	tetramethylethylenediamine
PPA	polyphosphoric acid	**TMSCN**	trimethylsilyl cyanide
PPE	polyphosphoric ethyl ester	**(TMS)$_2$NLi**	lithium bis(trimethylsilyl)amide
PPI	proton pump inhibitor	**TMSOMe**	methoxytrimethylsilane
PPts	pyridinium *para*-toluenesulphonate or pyridinium 4-toluenesulphonate	**T$_2$O**	tritiated water
		o-Tol	*ortho*-tolyl
PTFE	polytetrafluoroethylene	**Tris**	tris(hydroxymethyl)aminomethane
P(o-Tol)$_3$	tri(*o*-tolyl)phosphine	**Tris-HCl**	tris hydrochloride
ptsa	*para*-toluenesulphonic acid	**tRNA**	transfer RNA
PyBOP	benzotriazol-1-yloxytripyrrolidinophosphonium hexafluorophosphate	**TsCN**	*para*-toluenesulphonyl cyanide or 4-toluenesulphonyl cyanide
PyBrOP	bromotripyrrolidinophosphonium hexafluorophosphate	**TsDAEN**	*N*-[2-amino-1,2-bis(4-methoxyphenyl) ethyl]-4-methylbenzenesulphonamide
Q-phos	pentaphenyl(di-*tert*-butylphosphino) ferrocene	**UTI**	urinary tract infection
		Vdw	van der Waals
R	symbol used to represent the rest of the molecule	**Voc-Cl**	vinyloxycarbonyl chloride
		X	halogen or leaving group
Rapid	random peptide integrated discovery	**Z**	benzyloxycarbonyl
RedAl or Red-Al	sodium bis(2-methoxyethoxy) aluminiumhydride		

PART A

Concepts

There are various stages involved in discovering a drug and getting it to market. Chapter 1 provides an overview of that process, clearly identifying the stages where organic synthesis plays a crucial role. The chapter also defines a number of important terms used in drug design and medicinal chemistry.

Chapter 2 provides examples of drug syntheses and categorizes the various reactions involved such that the reader can better understand the overall rationale of any synthetic scheme. For example, coupling reactions are one category that involves linking simple molecular building blocks to create a more complex structure. However, it is rarely possible to create complex structures from coupling reactions alone, and other categories of reaction in a synthetic route will be defined.

Chapter 3 covers the topic of retrosynthesis. This is a topic which many students find 'tricky', and the chapter aims to provide a relatively simple overview with several examples involving simple drug structures. Case Study 3 provides a more complex example of a retrosynthesis and the corresponding synthesis.

Chapter 4 looks at how cyclic systems are introduced into drugs. This is a vast topic due to the variety of different ring systems that are possible, and so there is a focus on the general principles and strategies used to introduce ring systems into clinically important drugs.

Chapter 5 covers the preparation of chiral drugs—structures that can exist as two non-superimposable mirror images. A large number of clinically important drugs are chiral in nature and there are several reasons why it may be preferable to prepare such drugs as a single mirror image or enantiomer. The chapter describes different approaches to producing single enantiomers of chiral drugs.

Finally, Chapter 6 covers combinatorial and parallel syntheses. These are procedures that allow large numbers of structures to be synthesized in a swift and efficient manner. This is particularly crucial in the early stages of a drug discovery programme when looking for new 'hits' or preparing analogues of a lead compound.

1 The drug discovery process

1.1 Introduction

Synthesis has played a crucial role in the discovery of drugs for well over a century. To begin with, chemical reactions were used to create analogues of clinically important **natural products**. Such analogues are defined as **semi-synthetic agents** as they are derived from a natural compound rather than from simple starting materials through a full synthesis. Nevertheless, it was not long before totally synthetic drugs were being generated in the chemistry laboratory. For much of the twentieth century, pharmaceutical companies employed synthetic chemists to generate as many novel synthetic compounds as they could in the quest for new drugs. Usually, this was carried out on a trial and error process and there was little understanding of how these drugs worked. However, from the 1960s to the present day there has been an increasing understanding of the mechanisms by which drugs interact with molecular targets in the body. This knowledge has been fuelled by a number of scientific revolutions that occurred in the latter half of the twentieth century involving both chemistry and biology. As a result, it has become feasible to design drugs in an increasingly rational manner. This has led to the growth of the scientific discipline known as **medicinal chemistry**, which involves the design and synthesis of new drugs. Medicinal chemists are now key players in the pharmaceutical industry, and are required to have an understanding of both the biological and chemical aspects of drug design, as well as a thorough practical and theoretical knowledge of organic synthesis. The two skills go hand in hand. For example, there is no point in designing a drug that is difficult or impossible to synthesize. Similarly, it is wasteful to generate thousands of novel compounds if many of these structures have little chance of being active drugs. In this chapter, we shall look at some key concepts relating to drug design and define a number of terms that are commonly used by practising medicinal chemists.

1.2 The pathfinder years

The birth and development of the pharmaceutical industry is still a relatively modern event. For the vast majority of human history, different cultures around the world depended on herbs and extracts from natural sources to treat the many ailments that afflict the human body. Witch doctors, shamans, medicine men, and herbalists experimented with what was available in the natural world around them, whether that was plants, trees, insects, reptiles, or animals. Numerous potions and lotions were prepared and applied, often accompanied with spells or supernatural rites designed to impress the patient with their power. No doubt, a powerful **placebo effect** was involved as most ancient therapies have subsequently been shown to have little beneficial effect. However, some of these concoctions *are* effective. Examples include the sedative effects of various opium preparations, and the strength, stamina, and hunger suppression obtained by chewing coca leaves—a habit which is still common in some South American communities.

The birth of the pharmaceutical industry in the nineteenth century resulted from chemists investigating known herbs and extracts in order to isolate and purify the chemical components that were present. Their aim was to isolate the pure chemical that was mainly responsible for the pharmacological effects of the extract—the **active principle**. Various successes ensued. For example, **morphine** was isolated and purified in 1816 and shown to be the active principle responsible for the sedative properties of opium, while **cocaine** was isolated from coca leaves in 1860 and shown to be the active principle responsible for hunger suppression and increased stamina. Other active principles isolated in the nineteenth century included the emetic **emetine** from ipecacuanha and **quinine** from cinchona bark. All of these compounds contain basic amine groups, and so it was possible to form water-soluble salts by treating them with acids. Consequently, such compounds were classed as

alkaloids. Other alkaloid active principles isolated in the nineteenth century included **colchicine** (1820), **caffeine** (1821), **atropine** (1833), **physostigmine** (1864), **hyoscine** (1881), and **theophylline** (1888).

Several of these active principles were marketed and used in medicine, but although they proved potent they were by no means ideal. For example, morphine had to be administered by injection for effective analgesia, and it suffered from a range of side effects, some of which could be life-threatening. Therefore, it was not long before chemists were synthesizing analogues of active principles to try and discover agents which might have improved clinical properties. This led to the concept of the **lead compound**—a compound with a useful pharmacological activity that could act as the starting point for further research aimed at developing an improved drug.

Fully synthetic compounds were also tested as drugs during the nineteenth century. Success stories included the gaseous and volatile general anaesthetics, as well as the barbiturates, demonstrating that the natural world did not have exclusive rights on compounds with pharmacological activity.

Throughout the first half of the twentieth century, work continued apace identifying both synthetic and natural products as lead compounds, and then synthesizing as many analogues as possible. During this time, good progress was made in the development of local anaesthetics and opioid analgesics. Important neurotransmitters and hormones involved in human biochemistry were also identified, such as **adrenaline**, **thyroxine**, **estrone**, **estradiol**, **progesterone**, **cortisol**, **histamine**, and **insulin**. Several of these **endogenous compounds** were used as lead compounds for further research, which would eventually result in the development of effective anti-asthmatic agents, contraceptives, anti-inflammatory agents, and antihistamines. But perhaps the greatest advance in the first half of the century was in the field of antimicrobial agents. At the beginning of the century, Paul Ehrlich developed arsenic-containing drugs such as **arsphenamine (Salvarsan)** which proved effective against syphilis, while early antimalarial agents were discovered in the 1920s. The **sulphonamides** were the first important antibacterial agents to be discovered in the 1930s—the lead compound was a commercial synthetic dye called **prontosil**. In the 1940s, the fungal metabolite **penicillin** was successfully purified, and proved to be even more effective than the sulphonamides. With the realization that fungi could be a source of useful antibiotics, there was a massive worldwide study of fungal cultures in the post-war years which led to the identification of many of the antibiotics used in medicine today. The middle part of the twentieth century was a golden age of antibacterial research, and marked one of the most important breakthroughs in medical history. Before the antibiotic revolution, even simple wounds could prove life-threatening and many of the surgical operations carried out routinely today were totally impractical.

1.3 The development of rational drug design

The 1960s can be viewed as the birth of rational drug design. During that period there were important advances in the design of effective anti-ulcer agents, beta-blockers and anti-asthmatics. Much of this was based on trying to understand how drugs worked at the molecular level and proposing theories about why some compounds were active and some were not. A more targeted approach to drug synthesis started to take hold.

However, the story of rational drug design was boosted enormously towards the end of the century by important advances in both biology and chemistry. Advances in molecular genetics led to the sequencing of the human genome, which led in turn to the identification of proteins that had not previously been isolated and could serve as potential novel drug targets. For example, kinase enzymes have proved important targets for novel anticancer agents in recent years. Similarly, the mapping of genomes in viruses led to the identification of viral-specific proteins, which could serve as novel targets for new antiviral agents. Advances in automated small-scale testing procedures (**high throughput screening**) also allowed the rapid testing of potential drugs.

In chemistry, advances have been made in X-ray crystallography and NMR spectroscopy, allowing scientists to study the structure of drugs and their mechanisms of action. Powerful molecular modelling software packages have been developed that allow researchers to study how a drug binds to a protein binding site. Important advances have also been made in the development of novel synthetic methods that have vastly increased the chemical arsenal available to the synthetic chemist. The development of combinatorial and parallel syntheses has also opened the way for the synthesis of vastly increased numbers of compounds using robotic systems.

As a result of these advances, massive progress has been made over the last half century in the design and development of drugs that are effective in virtually every area of medicine.

There has also been a significant change in the way that pharmaceutical research is tackled. For most of the twentieth century, drug research depended on the discovery of a lead compound with a useful pharmacological activity, either an active principle from the natural world or a purely synthetic compound produced in the laboratory. Thousands of analogues were then synthesized in an effort to find an improved compound. Years later, the molecular target for an active compound might be discovered, allowing a better understanding of the biological mechanisms affected by these agents. In this approach, progress was dictated by whatever lead compound was discovered.

The situation has now changed. The genomics revolution has led to the identification of large numbers of proteins, many of which could be potential drug targets for future therapies (**proteomics**), and so most research projects are initiated by choosing a potential drug target to study. Therefore, it is crucial that a lead compound is found for that protein target as quickly as possible. Of course, there are still research projects that are determined by the chance discovery of a pharmacologically active compound, but the scientific approach towards the design of novel drugs now follows the pathway shown in Figure 1.1. In the next few sections, we shall look at some of these stages in more detail.

KEY POINTS

- Medicinal chemistry involves the design and synthesis of novel pharmaceuticals.

- Organic synthesis plays a role in many parts of the drug discovery process.

- An active principle is the chemical component within a natural extract that is chiefly responsible for the extract's pharmacological activity.

- Alkaloids are naturally occurring compounds that contain a basic amine group. Many of the early active principles isolated from natural extracts were alkaloids.

- A lead compound is a compound that binds to a particular target and/or has a potentially useful pharmacological activity. It acts as a starting point in the design and synthesis of analogues having improved properties.

- Rational drug design generally involves understanding how a drug binds to its target and designing improved analogues based on that knowledge.

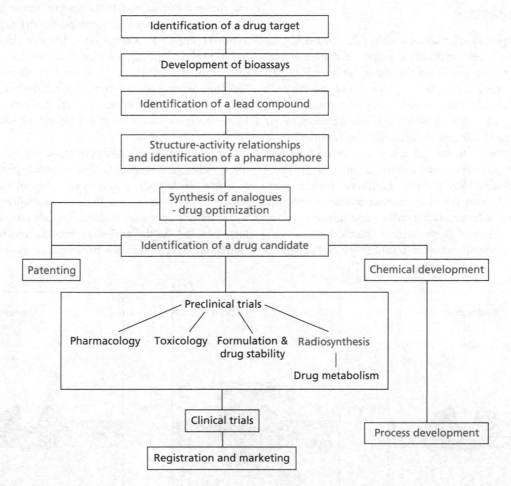

FIGURE 1.1 A typical approach to the development of a drug. Headings in blue indicate stages where the practice or theory of organic synthesis plays a crucial role. Note that various toxicological and pharmacokinetic studies traditionally associated with preclinical trials may be carried out as part of the drug optimization stage to identify potential problems at an earlier stage—an approach known as 'fail early, fail cheap'.

- The typical approach to a drug discovery process is to identify a target of interest, and then identify compounds (hits) that interact with that target. A lead compound is then identified to act as the basis for further research.
- During the drug optimization stage, a large number of analogues are synthesized in order to identify a clinical candidate.

1.4 Identification of a drug target

Drugs have to interact with molecular targets within the body if they are to have a pharmacological and clinical effect. These molecular targets include the nucleic acids DNA and RNA. However, the vast majority of clinically useful drugs interact with proteins. Proteins have a range of different functions, but the most important proteins in terms of drug action are receptors, enzymes, and transport proteins.

1.4.1 Receptors

Most receptors are embedded within the cell membrane with part of their structure exposed on the outer surface and part exposed on the inner surface (Fig. 1.2). Therefore, they are perfectly placed to act as the cell's 'letter boxes'. In other words, they can interact with chemical messengers outside the cell and transmit their message to the inside of the cell. The chemical messengers involved in this process are **neurotransmitters** released from nerve cells known as **neurons**, or **hormones** released from glands. Examples of such messengers include **acetylcholine**, **noradrenaline**, **serotonin**, **dopamine**, **adrenaline**, **insulin**, and **growth factor hormone**. Each of these natural chemical messengers binds to a specific protein receptor which recognizes that particular messenger. In the normal scheme of things, the natural messenger (the **endogenous ligand**) binds to a binding site that is exposed on the outer surface of the cell. The binding process causes the protein to change shape (an **induced fit**), which triggers a series of further events affecting other proteins within the cell membrane and the cell cytoplasm—a process known as **signal transduction**. We do not have the space here to go into this process in detail. Suffice it to say that the chemical messenger does not need to cross the cell membrane in order to deliver its message. The protein receptor is responsible for carrying out that task. Moreover, no chemical reaction takes place between the messenger and the receptor. The message is received and transmitted into the cell as a result of the receptor protein changing shape (the induced fit). Once the message has been received, the chemical messenger departs the binding site unchanged, and the protein returns to its inactive conformation or shape.

Drugs can be designed to bind to a receptor's binding site in a similar way to the natural messenger, such that they, too, activate the receptor. Such drugs are called **agonists** (Fig. 1.3a). Alternatively, drugs can be designed to bind to the binding site in such a way that they produce a different induced fit which fails to activate the receptor. As long as that drug is bound, the natural messenger cannot bind and so its message is not received by the receptor. Such drugs are known as **antagonists** (Fig. 1.3b).

There are a large number of drugs which act as receptor agonists or antagonists. For example, **propranolol** is a beta-blocker which acts as an antagonist at adrenergic receptors (receptors that are normally activated by adrenaline or noradrenaline). Other examples of antagonists include the anti-ulcer agents **cimetidine** and **ranitidine** which block the histamine receptors present

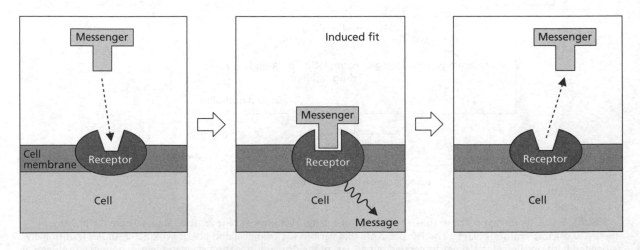

FIGURE 1.2 Binding of a chemical messenger to a protein receptor.

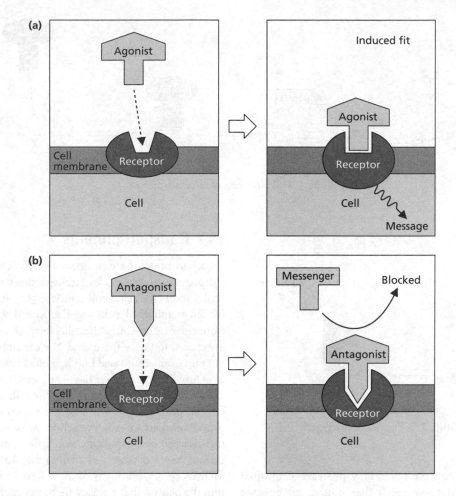

FIGURE 1.3 (a) Binding of an agonist to a receptor. (b) Binding of an antagonist to a receptor.

in cells lining the stomach wall. Examples of receptor agonists include the anti-asthmatic drug **salbutamol**, which triggers adrenergic receptors in the airways, and the analgesic **morphine**, which triggers opioid receptors that normally interact with endogenous peptides such as enkephalins and endorphins.

Most of the receptors that serve as targets for clinically important drugs exist in the cell membrane, but there are some which are situated within the cell itself, and so the natural ligand for those receptors has to cross the cell membrane in order to reach them. Important examples are the different steroid receptors which interact with steroid hormones. This means that steroid hormones such as **estradiol**, **testosterone**, and **cortisol** have to cross cell membranes in order to reach their receptors. Any drugs designed to interact with these receptors will also have to cross cell membranes. This is an important consideration in drug design, as the cell membrane acts as a hydrophobic (water-hating) barrier. Therefore, drugs designed to act within the cell must be sufficiently hydrophobic (or lipophilic) to negotiate that barrier.

1.4.2 Enzymes

Enzymes act as the cell's catalysts and are mostly situated within the cell. Like receptors, they contain a binding site (known as the **active site**) which is responsible for binding a substrate and then catalysing a particular reaction (Fig. 1.4). Many enzymes require small molecules called **cofactors** if they are to be effective. Cofactors include **adenosine triphosphate** (ATP), **nicotinamide adenine dinucleotide** (NADH), and **nicotinamide adenine dinucleotide phosphate** (NADPH). These molecules can be considered as the body's 'reagents'. For example, NADH and NADPH both act as reducing agents and are the body's equivalent of synthetic reagents such as sodium borohydride, lithium aluminium hydride, or hydrogen gas.

Drugs can be designed to bind to an enzyme's active site and be stable to the reaction that is normally catalysed. As long as they remain bound, the enzyme cannot accept its normal substrate. Such drugs are known as **enzyme inhibitors** (Fig. 1.5). **Saquinavir**, which is one of several drugs now used in AIDS therapy,

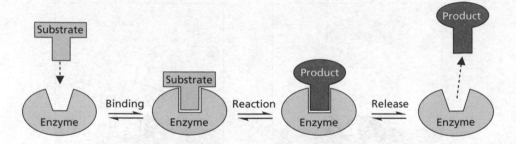

FIGURE 1.4 The process of enzyme catalysis.

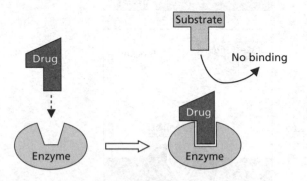

FIGURE 1.5 Enzyme inhibition.

1.4.3 **Transport proteins**

Transport proteins are embedded within cell membranes and play an important role in transporting small polar molecules across the cell membrane (Fig. 1.6). The hydrophobic cell membrane blocks water, ions, and polar molecules from entering or exiting the cell. However, such molecules are crucial to the cell's survival. For example, proteins are built up from amino acid building blocks, which are polar in nature. Therefore, there has to be a mechanism by which amino acids can be smuggled into the cell.

Transport proteins also play an important role in the overall mechanism of nerve action. Most neurotransmitters released by neurons are small polar molecules. They do not have to cross cell membranes to interact with their receptors, but many of them have to be taken back into the neuron that released them in order to limit their period of activity. Transport proteins are responsible for this salvaging operation.

Transport proteins contain a binding site which is capable of binding a specific polar molecule. This results in an induced fit which causes the protein to transport the polar molecule through the cell membrane such that it does not actually interact with the membrane. In other words, it essentially travels through the protein.

inhibits a viral enzyme called **HIV protease**. **Captopril** is an antihypertensive agent that inhibits an enzyme called **angiotensin-converting enzyme** (ACE). This enzyme is involved in the generation of a potent vasoconstrictor called **angiotensin II**. Inhibiting the enzyme lowers the level of angiotensin II, resulting in dilation of blood vessels and reduced blood pressure. Enzymes are located inside cells, and so it is important that enzyme inhibitors are designed such that they can cross the cell membrane.

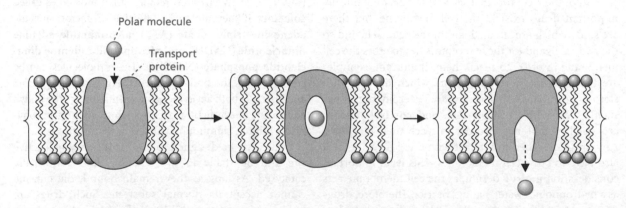

FIGURE 1.6 Transport proteins.

Drugs can be designed to bind to these transport proteins such that they prevent the natural ligand from binding. For example, a group of antidepressants known as **selective serotonin reuptake inhibitors** (SSRIs) bind to the transport protein that normally transports the neurotransmitter **serotonin** back into the neuron from which it was released. By blocking serotonin reuptake, serotonin levels increase at target cells resulting in prolonged activation of serotonin receptors. This results in the observed antidepressant effect.

KEY POINTS

- Proteins are the most common targets for drugs. Typical targets are receptors, enzymes, and transport proteins.

- Many drugs target receptors, most of which are embedded in cell membranes and normally interact with neurotransmitters and/or hormones.

- Agonists are drugs that mimic the effects of the normal ligand when they bind to a receptor.

- Antagonists bind to a target receptor but fail to activate it. By binding to the receptor, they prevent the endogenous ligand from binding and activating it.

- Some receptors are present within the cell. Drugs that are designed to interact with these receptors must be sufficiently hydrophobic to cross the cell membrane.

- Enzymes act as the body's catalysts. Substrates are compounds which bind to the active site of an enzyme and undergo a reaction catalysed by the enzyme.

- Enzyme cofactors are small molecules which bind to enzymes and act as the 'reagents' for an enzyme-catalysed reaction.

- Enzyme inhibitors bind to a target enzyme and prevent the natural substrate from binding, thus inhibiting the enzyme-catalysed reaction.

- Transport proteins are embedded in cell membranes and transport important polar molecules across the membrane.

- Several drugs act as reuptake inhibitors and block the reuptake of important neurotransmitters, such as dopamine, noradrenaline, and serotonin, into the neurons that released them. As a result, the activity of these neurotransmitters is enhanced.

1.5 Drug testing and bioassays

In order to identify whether a drug interacts with a protein target, it is important to develop a suitable test or bioassay. This is normally the province of the pharmacologist rather than the medicinal chemist. However, the availability of such tests is crucial if the medicinal chemist is to establish whether a particular compound is active or not. Tests can be defined as *in vitro* or *in vivo*, and it is important that medicinal chemists appreciate what kind of information can be gleaned from them.

1.5.1 *In vitro* tests

In vitro tests are carried out on target molecules or cell cultures. For example, enzyme inhibitors can be tested on an isolated enzyme in the test tube to see whether they prevent the enzyme catalysing a particular reaction. In order to do this, it is important that the substrate or the product can be easily detected in a quantitative manner in order to measure the rate at which product is formed or substrate is consumed. The effectiveness of an inhibitor can then be determined by how effectively it inhibits the enzyme-catalysed reaction.

In vitro tests can also be used to test receptor agonists and antagonists. However, it is difficult to carry out these tests on isolated receptor proteins. For a start, most receptor proteins are embedded within cell membranes, and removing them from that environment is likely to result in the protein forming a different 3D shape which would be incapable of activation. Moreover, the measurable effects of activating receptors are often 'downstream' of the receptor within the cell itself. Therefore, *in vitro* tests on receptors are normally carried out on whole cells or isolated tissue samples.

In vitro tests can be automated and carried out on a small scale, allowing the rapid and efficient testing of thousands of compounds in a very small time period. This is known as **high throuput screening** (HTS). These tests are ideal for identifying whether drugs interact with a molecular target to produce a particular pharmacological effect. The ability of a drug to bind to its target and produce such an effect is known as **pharmacodynamics**.

1.5.2 *In vivo* bioassays

In vivo bioassays are carried out on living organisms, either humans or laboratory animals. They tend to be slower, more costly, and more controversial than *in vitro* tests. Therefore, one might ask why they should be done at all? There are several reasons.

First of all, *in vivo* tests are complementary to *in vitro* tests. *In vitro* tests are designed to establish whether a particular drug interacts with a molecular target to produce a measurable pharmacological effect such as enzyme inhibition or receptor activation, whereas *in vivo* tests establish whether the drug produces a physiological effect such as analgesia or the lowering of blood pressure. If the agent proves active in both the *in vitro* and *in vivo* tests, then it is probable that the observed physiological effect is a result of the observed pharmacological effect.

On the other hand, if an agent proves to be active in the *in vivo* test, but is inactive in the *in vitro* test, then the agent may well be interacting with a different molecular target to the one being tested in the *in vitro* test. Physiological effects are often the result of a variety of different biological mechanisms, and carrying out specific *in vitro* tests alone may miss an important new lead compound. Furthermore, it may not be known what role a newly discovered protein has in the body. An *in vitro* test will show whether a compound interacts with that novel target, while an *in vivo* test will identify the overall effect of that interaction on the organism.

Secondly, *in vitro* tests are excellent at establishing whether a drug interacts with its target to produce a pharmacological effect (pharmacodynamics), but they give no information about the ability of a drug to reach that target in a living system. *In vivo* tests will establish whether a drug can both reach its target and interact with it. The range of factors affecting a drug's ability to reach its target is known as **pharmacokinetics** and is covered in the next section.

Thirdly, *in vivo* tests are often crucial for identifying unexpected activity that would not be picked up by *in vitro* tests. For example, a dye called **prontosil** was shown to have antibacterial activity *in vivo*, but was inactive *in vitro*. This was because prontosil itself was inactive, but was metabolized to an active sulphonamide within the body. A compound such as prontosil is known as a **prodrug**.

Finally, *in vivo* tests can be useful in detecting side effects which would not be observed in *in vitro* tests. The appearance of side effects can often sound the death knell for a particular drug, but, on occasions, it can identify unexpected applications for the drug. For example, the anti-impotence drug **sildenafil** was originally tested as an antihypertensive drug, and its anti-impotence effects were only identified during early clinical trials.

1.5.3 **Pharmacokinetics**

If a drug is to be effective, it must not only interact with a particular molecular target, but must also reach that target. However, there are many different factors which can prevent that happening. The main ones are **absorption**, **distribution**, **metabolism**, and **excretion** (commonly referred to as **ADME**). Another factor which is commonly considered is **toxicity** (**ADMET**).

Absorption

Orally administered drugs are convenient and easy to take, but they face the most challenges in their journey to reach their target. An orally administered drug has to survive various hazards as it proceeds through the gastrointestinal tract (GIT), such as the acid contents of the stomach and digestive enzymes in the intestines. The stomach is a particular problem for drugs that are acid-sensitive, such as penicillins. Fortunately, specially coated tablets or capsules can be designed which protect the drug in its journey through the stomach and then dissolve in the less acidic environment of the intestines to release the drug. Such a preparation is described as being **enterosoluble**.

Once a drug has reached the intestines, it has to cross through the cells lining the gut wall in order to reach the blood supply. This means that the drug has to cross a cell membrane on two occasions—first to enter the cell, and then to exit the cell (Fig. 1.7).

For that to take place, the drug must be sufficiently fatty (hydrophobic) to dissolve through the cell membranes.

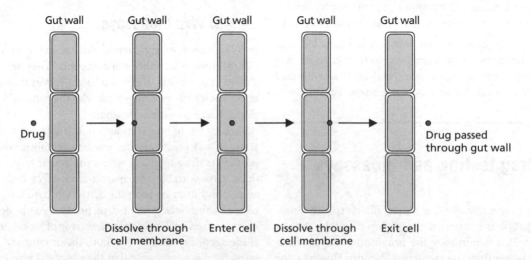

FIGURE 1.7 Absorption from the gastrointestinal tract into the blood supply.

However, the drug must also be sufficiently polar (hydrophilic) to be soluble in the aqueous conditions found in the gastrointestinal tract and the blood supply. Therefore, orally active drugs have to have a good balance of hydrophobic and hydrophilic properties.

Having said that, a number of drugs can cross the gut wall without having to cross cell membranes.

- Some drugs can be transported across the cells lining the gut wall by transport proteins present within the cell membranes. These proteins normally transport the highly polar building blocks required for various biosynthetic pathways (e.g. amino acids and nucleic acid bases) across cell membranes, but if a drug bears a structural resemblance to one of these building blocks, then it too may be smuggled across. For example, **levodopa** is transported by the transport protein for the amino acid phenylalanine, while **fluorouracil** is transported by transport proteins for the nucleic acid bases thymine and uracil. The antihypertensive agent **lisinopril** is transported by transport proteins for dipeptides, and the anticancer agent **methotrexate** and the antibiotic **erythromycin** are also absorbed by means of transport proteins.
- Small polar drugs with a molecular weight less than 200 amu can cross through pores between the cells lining the gut wall.
- A certain number of large drugs can cross by a process known as **pinocytosis** where the cell membrane engulfs the drug to form a membrane-bound vesicle. This carries the drug across the cell and then fuses with the membrane to release the drug on the other side (Fig. 1.8).

Drugs vary in how effectively they are absorbed from the gut. In general, drugs which are either too polar or too hydrophobic are poorly absorbed and are better administered by a different method, such as injection or transdermal absorption.

There have been various studies over the years to identify which structural features are associated with good or bad absorption.

The best known of these studies was carried out by Lipinski, who analysed the structures of a large number of drugs from the World Drugs Index database. From this study, he found that orally active drugs generally had structures which obeyed the following characteristics:

- a molecular weight less than 500
- no more than five hydrogen bond donor groups (HBDs)
- no more than ten hydrogen bond acceptor groups (HBAs)
- a logP value less than + 5 (logP is a measure of a drug's hydrophobicity).

This has become known as **Lipinski's rule of five**, so called because the numbers quoted are multiples of five. However, it is a bit misleading to view these characteristics as 'rules', since there are several orally active drugs which break the 'rules'. Therefore, it is better to consider them as guidelines. Lipinski himself stated that it was possible for structures to be orally active as long as they did not break more than one of his 'rules'. If a structure broke two or more of the rules, there was significantly less chance of it being orally active.

In essence, Lipinski's rule of five identifies that orally active drugs need to have a good balance of hydrophilic and hydrophobic properties. LogP is a measure of a drug's hydrophobicity. A value greater than 5 indicates that the drug is likely to be too hydrophobic for good absorption. Such drugs tend to become trapped within fat globules derived from food and fail to interact with the gut wall.

The number of hydrogen bond donors and acceptors indicates the number of polar functional groups that are present, such as amines, carboxylic acids, alcohols, and phenols. If there are too many of these groups, the drug will be too polar to cross the cell membranes.

The fact that most orally active compounds have a molecular weight less than 500 might indicate that size has an effect on absorption. However, further analysis has established that a high molecular weight is not, in itself, detrimental to oral activity. It just so happens that most structures having a molecular weight greater than 500 amu have too many polar functional groups, and that is why they have poor absorption.

Consequently, another set of guidelines called **Veber's parameters** have been proposed which avoid molecular weight as a condition for oral activity. These parameters have demonstrated the rather surprising finding that molecular flexibility plays an important

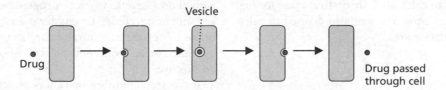

FIGURE 1.8 Pinocytosis.

role in absorption—the more flexible the molecule, the less likely it is to be orally active. In order to measure flexibility, one can count the number of freely rotatable bonds that result in significantly different conformations. Bonds to simple substituents such as methyl or alcohol groups are not included in this analysis as their rotation does not result in significantly different conformations.

Veber's studies also demonstrated that the polar surface area of the molecule could be used as a factor to determine the hydrophilic/hydrophobic character of the molecule, instead of the number of hydrogen bonding groups. These findings led to the following parameters for predicting acceptable oral activity:

- either a polar surface area ≤ 140 Å and ≤ 10 rotatable bonds, or
- ≤ 12 hydrogen bond donors and acceptors in total and ≤ 10 rotatable bonds.

Some researchers set the limit of rotatable bonds to ≤ 7 since the analysis shows a marked improvement in absorption for such molecules.

Therefore, orally active drugs should have a good balance of hydrophobic and hydrophilic properties and be reasonably rigid. These properties also play a role in how effectively a drug is distributed around the body, as well as its ability to cross other cell membranes. It is also the case that molecules with a large number of rotatable bonds generally have a lower activity at target binding sites than more rigid analogues. This is because they are less likely to be in the active conformation when they enter the target binding site (section 1.7).

Lipinski's rule of five and Veber's parameters are good guidelines for prioritizing which analogues might be worth synthesizing in a research project. The molecular weight, as well as the number of HBDs, HBAs, and rotatable bonds, can be determined from the structure of each intended analogue. The P of log$P$ is the partition coefficient which can be measured experimentally by dissolving a compound in a two-phase system consisting of octanol and water. The more polar the compound, the more of it dissolves in the water layer. The more hydrophobic the compound, the more of it dissolves in the octanol layer. However, this method of determining the partition coefficient is not much use if you want to decide whether to synthesize an analogue or not. Therefore, it is more common to use software programs to calculate a theoretical value for log$P$ (clog$P$). Software programs can also be used to calculate the polar surface area.

Metabolism

Once an orally administered drug has reached the blood supply, it is carried immediately to the liver. The purpose of the liver is to 'vet' foreign chemicals and detoxify any potential poisons that have been absorbed. This is carried out by an arsenal of metabolic enzymes that catalyse reactions capable of modifying the structure of the foreign chemical. As drugs are foreign chemicals, they are potentially susceptible to these enzymes. The most important class of metabolic enzymes present in the liver are the **cytochrome P450 enzymes**, which are responsible for catalysing a variety of mainly oxidative reactions. In general, the reactions introduce a polar functional group such as an alcohol or a phenol. This is known as **phase I metabolism**. The functional groups that are introduced serve as polar 'handles' which are recognized by another battery of enzymes which catalyse the addition of a highly polar molecule (e.g. a sugar molecule) to the polar handle. This is known as **phase II metabolism**. The reactions involved are typically conjugation reactions and serve to produce a highly polar metabolite. The increase in polarity is important, as this plays a crucial role in how quickly these metabolites are excreted by the kidneys.

A certain percentage of an orally administered drug will be metabolized as it passes through the liver, and so not all the drug absorbed is capable of reaching its target. This is known as the **first-pass effect**.

Distribution

Once a drug has passed through the liver, it is rapidly transported round the body in the blood supply. However, this is an uneven process. The proportion of drug reaching different parts of the body is proportional to the level of blood supply to those parts. Therefore, organs with a rich supply of blood vessels will be supplied with a greater percentage of the drug than organs with a less rich supply.

It is relatively easy for drugs to exit the blood vessels and reach the cells of organs and tissues. This is because the cells lining the blood vessels are relatively loose fitting, which means that there are pores between the cells that are sufficiently large for drug-sized molecules to pass through. Therefore, drugs exit the blood vessels by travelling through these pores, rather than having to cross cell membranes through the cells themselves.

Drugs that are designed to interact with membrane-bound receptors or transport proteins can now interact with those targets, whereas drugs designed to interact with intracellular targets have another cell membrane to negotiate. Naturally, the latter drugs have to have the correct polar/non-polar balance if they are to be effective.

There are a number of factors which can divert a drug from this journey. For example, polar drugs can become tightly bound to **plasma proteins** within the blood supply. The percentage of drug that is bound in this way is trapped within the blood vessels and is unable to reach target tissues. In contrast, a certain percentage of hydrophobic drugs is likely to be taken up into fatty tissue and stored there. Once again the percentage of drug stored in fatty tissue is not available to interact with its intended target, although it will be slowly released from fatty tissue over time.

Drugs designed to interact with targets in the central nervous system (CNS) have to negotiate an extra barrier known as the **blood–brain barrier** (BBB). The blood vessels supplying the brain have tighter-fitting cells than the blood vessels supplying the rest of the body. As a result, drugs are unable to escape these vessels unless they pass through the cells lining the blood vessels. Moreover, there is a fatty barrier surrounding the blood vessels in the CNS which also has to be negotiated. Therefore, the BBB can act as an effective barrier to polar drugs and hinder their access.

The placental barrier is also selective in the drugs that are allowed to cross. The placental membranes separate a mother's blood from the blood of her fetus. The mother's blood provides the fetus with essential nutrients and carries away waste products, but these chemicals must pass through the placental barrier. As food and waste products can pass through the placental barrier, it is perfectly feasible for drugs to pass through as well. Drugs such as **alcohol**, **nicotine**, and **cocaine** can all pass into the fetal blood supply. Fat-soluble drugs will cross the barrier most easily, and drugs such as **barbiturates** will reach the same levels in fetal blood as in maternal blood. Such levels may have unpredictable effects on fetal development. They may also prove hazardous once the baby is born. Drugs and other toxins can be removed from fetal blood by the maternal blood and detoxified. Once the baby is born, it may have the same levels of drugs in its blood as the mother, but it does not have the same ability to detoxify or eliminate them. As a result, drugs will have a longer lifetime and may have fatal effects.

Excretion

The kidneys are chiefly responsible for the excretion of drugs, drug metabolites, and other waste chemicals. They are particularly efficient at excreting polar molecules, which is why the reactions catalysed by the liver are so important.

Blood enters the kidneys by means of the **renal artery**. This divides into a large number of capillaries, each of which forms a knotted structure called a **glomerulus** that fits into the opening of a duct called a **nephron** (Fig. 1.9). The blood entering these glomeruli is under pressure, and so plasma is forced through the pores in the capillary

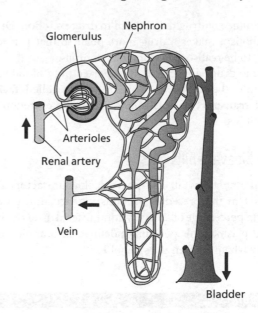

FIGURE 1.9 Excretion by the kidneys.

walls into the nephron, carrying with it any drugs and metabolites that might be present. Any compounds that are too big to pass through the pores, such as plasma proteins and red blood cells, stay in the capillaries with the remaining plasma. Note that this is a filtration process, so it does not matter whether the drug is polar or hydrophobic—all drugs and drug metabolites will be passed equally efficiently into the nephron. However, this does not mean that every compound will be *excreted* equally efficiently, because there is more to the process than simple filtration.

The filtered plasma and chemicals now pass through the nephron on their route to the bladder. However, only a small proportion of what starts that journey actually finishes it. This is because the nephron is surrounded by a rich network of blood vessels carrying the filtered blood away from the glomerulus, permitting much of the contents of the nephron to be reabsorbed into the blood supply. Most of the water that was filtered into the nephron is quickly reabsorbed through pores in the nephron cell membrane which are specific for water molecules and bar the passage of ions or other molecules. These pores are made up of protein molecules called **aquaporins**. As water is reabsorbed, drugs and other agents are concentrated in the nephron and a concentration gradient is set up. There is now a driving force for compounds to move back into the blood supply down the concentration gradient. However, this can only happen if the drug is sufficiently hydrophobic to pass through the cell membranes of the nephron. This means that hydrophobic compounds are efficiently reabsorbed back into the blood, whereas polar compounds remain in the nephron and are excreted. This process of excretion explains the importance of drug metabolism to drug excretion. Drug metabolism makes a drug more polar so that it is less likely to be reabsorbed from the nephrons.

Some drugs are actively transported from blood vessels into the nephrons. This process is called **facilitated transport**, and is important in the excretion of penicillins.

Oral bioavailabilty

When one takes all the pharmacokinetic factors described in the preceding sections into account, only a certain percentage of an orally administered drug reaches blood plasma. This percentage defines the oral bioavailability which is given the symbol F.

KEY POINTS

- Tests or bioassays are necessary to determine whether a drug binds to a target protein, and to identify what effect that binding has.

- *In vitro* tests are carried out on purified enzymes, whole cells, or isolated tissues. Such tests can be automated in a process known as high throughput screening. They are used to study binding affinities and pharmacological properties.

- *In vivo* tests are carried out on living organisms and can identify the physiological effects of a drug, as well as its pharmacokinetic properties.

- Prodrugs are inactive compounds which are converted to active compounds by metabolic reactions. They are useful in solving pharmacokinetic problems such as poor solubility and absorption.

- Pharmacodynamics refers to the effects caused by a drug interacting with its target, whereas pharmacokinetics refers to the factors that affect a drug's survival and ability to reach its target in the body.

- ADME refers to absorption, distribution, metabolism, and excretion.

- Lipinski's rule of five and Veber's parameters are guidelines which are used to predict whether a drug is likely to be orally active.

- Drug metabolism involves phase I and phase II metabolism which result in more polar drug metabolites.

- Cytochrome P450 enzymes in the liver are responsible for many phase I metabolic reactions.

- The first-pass effect is the percentage of orally absorbed drug that is metabolized by metabolic enzymes in the liver before the drug is distributed round the body.

- The amount of drug distributed to different tissues is affected by the amount of blood supply to the tissue concerned and the physical properties of the drug itself.

- The kidneys are responsible for drug excretion and excrete polar compounds more efficiently than non-polar compounds.

- The oral bioavailability of a drug defines how effectively an orally administered drug reaches the blood plasma.

1.6 Identification of lead compounds

Having identified suitable assays for a specified molecular target (sections 1.5.1–1.5.2), the next stage is to identify a lead compound which will interact with that target (section 1.2). A lead compound is a chemical structure which has some binding affinity for the target binding site, or produces an observed pharmacological effect as a result of that binding interaction. The desired affinity or activity may not be particularly high, and there might be a range of problems which prevent the compound being of any clinical use. For example, it might be difficult to synthesize, have unacceptable side effects, and have poor pharmacokinetic properties. However, that is not important at this stage. The main priority is to find a compound that interacts with the target binding site. Further drug design and development can then be used to find synthetically accessible analogues that have stronger binding interactions and increased selectivity and activity, as well as improved pharmacokinetic properties.

Lead compounds have been identified from both the natural world and the laboratory. Historically, the natural world has been a rich source of important novel lead compounds and remains so today. However, there are distinct disadvantages to natural lead compounds. Finding such compounds is usually slow, inefficient, and expensive. Moreover, there is no guarantee of success. Nowadays, there is a much greater emphasis on generating lead compounds by synthesis or rational design (section 6.1 and Chapter 7).

1.7 Structure–activity relationships and pharmacophores

Having identified a lead compound, it is important to establish which features of the compound are important to its activity. This, in turn, can give a better understanding of how the compound might bind to its molecular target.

Most drugs are small molecules compared with their molecular target. For example, a typical orally active

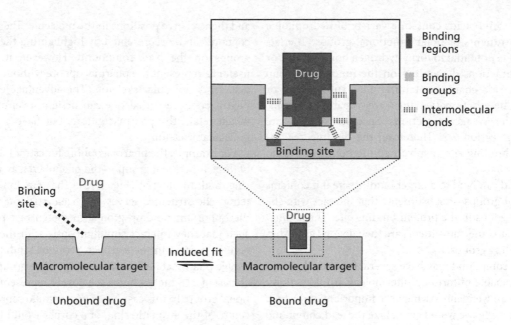

FIGURE 1.10 The equilibrium between a drug being bound and unbound to its target.

drug has a molecular weight between 200 and 500 amu, whereas proteins have molecular weights measuring several thousands of atomic mass units and can be defined as **macromolecules**. This means that a protein's binding site has quite a small area compared with the rest of the protein, often consisting of a hollow or cleft within the protein surface (Fig. 1.10). When one views the relative size of a drug with its protein target (e.g. an agonist and a protein receptor), it can sometimes be difficult to see why the interaction between a drug and a protein should have such an important influence. However, the binding process and the resulting induced fit are key to this. When the binding site alters shape, it can trigger a kind of domino effect which results in conformational changes throughout the whole protein. This in turn may influence how the receptor protein interacts with other proteins.

So how does the binding process cause an induced fit? The answer lies in a variety of **binding regions** within the binding site that are capable of forming different types of intermolecular interactions such as van der Waals interactions, hydrogen bonds, ionic interactions, dipole–dipole interactions, induced dipole interactions,

pi–cation interactions, etc. If a drug (or ligand) has functional groups and substituents capable of interacting with these binding regions, then binding can take place. In order to maximize the different interactions, the binding site alters shape such that the different interactions are as strong as possible. This is the induced fit. So what are these binding regions and binding groups?

As far as a protein binding site is concerned, we should be aware that a protein is a macromolecule made up of amino acid building blocks linked together through peptide bonds (Fig. 1.11). Each amino acid has a particular side chain (R^1–R^4 in Fig. 1.11). Some side chains are simple alkyl substituents. Other side chains contain a functional group, such as an alcohol, phenol, aromatic ring, carboxylic acid, amide, thioether, amine, or heterocyclic ring.

This means that the binding site is lined with a variety of amino acids, some of which form important binding interactions with a ligand. The peptide backbone itself is capable of forming hydrogen bonding interactions, with the carbonyl oxygen serving as a potential hydrogen bond acceptor and the NH proton as a potential hydrogen bond donor (Fig. 1.11). The side chains of

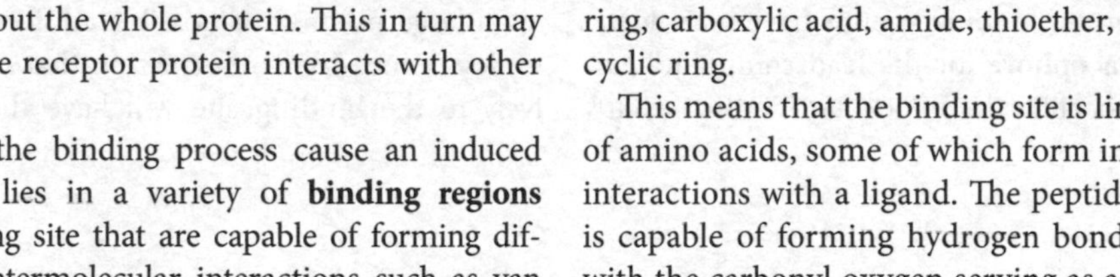

FIGURE 1.11 General structure for a protein. Peptide bonds are in blue. Side chains are represented by R^1–R^4.

different amino acids contain a variety of hydrophobic alkyl substituents or polar functional groups. The latter have the potential to form hydrogen bonding and/or ionic interactions depending on the functional group involved. Side chains containing alkyl substituents or aromatic/heteroaromatic rings are capable of forming van der Waals interactions. Specific examples are shown in section 7.2. Therefore, the binding regions within a binding site involve peptide bonds or amino acid side chains.

A ligand can bind to a target binding site if it contains functional groups or substituents that interact with the binding regions in the protein binding site. Those functional groups and substituents are then defined as the ligand's binding groups.

A lead compound may have several functional groups that are capable of forming intermolecular interactions, but how can we identify which are important and which are not? Ideally, we would crystallize the lead compound with its target protein, and then identify the structure of the ligand–protein complex by X-ray crystallography. Molecular modelling can then be used to study how the ligand binds to the binding site and identify the important interactions. However, it is not always possible to carry out such a study because of the difficulty in crystallizing many target proteins.

Therefore, it is usually the case that a number of analogues are synthesized where a particular functional group is modified or removed. Comparing the activity of an analogue with the lead compound can then determine the importance or otherwise of a particular functional group. If there is little change in activity, the functional group is not essential to binding or activity. If the activity drops significantly, the functional group concerned is likely to be an important binding group. This will be covered in more detail in section 8.2.

Once the important binding groups have been identified, the **pharmacophore** for the lead compound can be defined. This specifies the important binding groups and their relative positions in the molecule. The pharmacophore can be represented by highlighting the binding groups on the lead compound. However, it is more useful to represent the pharmacophore without showing a specific molecular skeleton. The advantage of such a pharmacophore is that one can design novel molecules which retain the pharmacophore, but have a different molecular skeleton.

For example, the pharmacophore for estradiol consists of three functional groups—the phenol OH, an aromatic ring, and an alcohol (Fig. 1.12). The remainder of the tetracyclic structure serves as a rigid structure to hold the important binding groups in the correct positions, such that they interact simultaneously with the binding regions present in the estrogen receptor binding site. A simple pharmacophore representing these groups would consist of a triangle where each corner represents a functional group. In this case, one corner would represent the centre of the aromatic ring, one corner would represent the oxygen of the phenol group, and one corner would represent the oxygen of the alcohol group (structures II and III). Another way of representing the pharmacophore is to identify the type of binding interaction that is taking place (structure IV).

Identifying the pharmacophore of a rigid molecule such as estradiol is relatively straightforward, but there is an added complication when we consider flexible lead compounds. We may know what the important binding groups are, but identifying their relative positions is not so straightforward. For example, the binding groups for dopamine are defined in Figure 1.13 as the two phenol groups, the aromatic ring and the charged amine group. The relative positions of the phenol groups and the aromatic ring are easy to define as this is a rigid part of the molecule, but where is the charged amine positioned? As the bonds in the side chain are freely rotatable, there are a large number of possible conformations available to dopamine. The conformation that binds most effectively to the binding site will have the charged amine

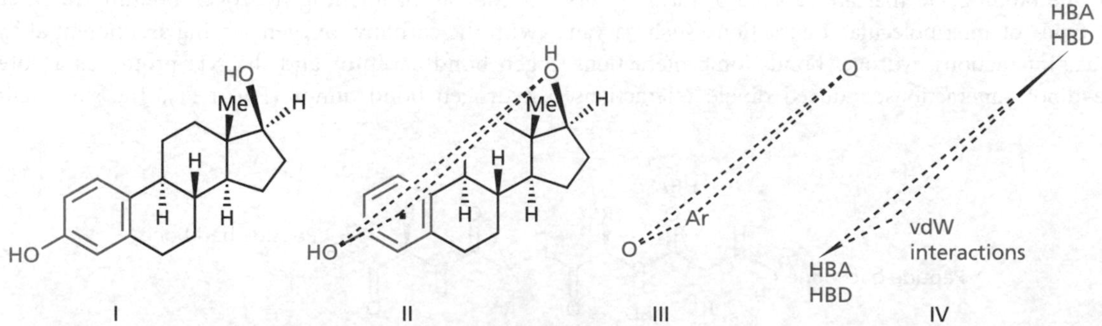

FIGURE 1.12 The pharmacophore for estradiol. The important binding groups are shown in blue (structure I).

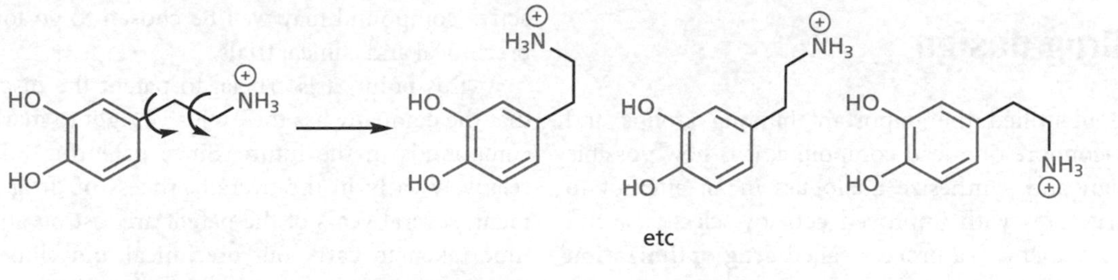

FIGURE 1.13 Different conformations of dopamine (important binding groups shown in blue).

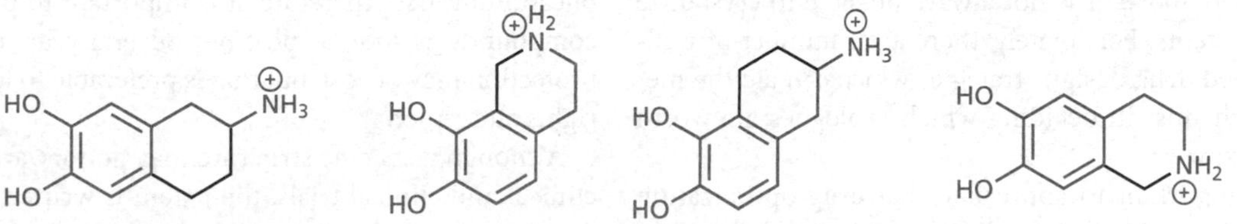

FIGURE 1.14 Rigid analogues of dopamine (the skeleton of dopamine is defined in blue).

group oriented in a particular way, and this is known as the **active conformation**. However, there are many other conformations for dopamine which will fail to bind effectively.

One way of identifying the active conformation of a flexible lead compound is to synthesize rigid analogues of the lead compound where the binding groups have been locked into defined positions. This is known as **rigidification** or **conformational restriction**. The pharmacophore will then be represented by the most active analogue. For example, the structures shown in Figure 1.14 are rigid analogues of dopamine which have particular conformations of dopamine trapped within their structure. If one of these proves significantly more active than the others, it can be used to define the active conformation and the pharmacophore.

Clearly, organic synthesis plays an important role in preparing these analogues, and this is discussed in more detail in Chapter 8.

Finally, in section 1.5.3 we mentioned that the presence of a large number of rotatable bonds in a molecule is likely to have an adverse effect on activity. This is because a flexible molecule can adopt a large number of conformations (shapes), and only one of these shapes corresponds to the active conformation where all the binding groups are properly positioned for effective binding. If the molecule enters the binding site in an inactive conformation, it will depart again without binding. Indeed, a flexible molecule may have to enter and depart a binding site several times before it is in the correct active conformation for binding. In contrast, a totally rigid molecule containing the

required pharmacophore will bind the first time it enters the binding site, resulting in greater activity.

KEY POINTS

- A lead compound is a compound that is known to interact with a specific target. It acts as the starting point for further research in order to improve pharmacodynamic and pharmacokinetic properties.

- Drugs are small molecules that bind to the binding sites of much larger target molecules within the body—typically proteins.

- Drugs form intermolecular bonds, such as hydrogen bonds, van der Waals interactions, and ionic bonds, with target binding sites.

- The functional groups or substituents responsible for these interactions are called binding groups on the drug and binding regions in the target binding site.

- The binding of a drug to a target binding site normally results in an induced fit which alters the shape of the binding site.

- Analogues of a lead compound can be synthesized to study structure–activity relationships in order to identify which functional groups and substituents in a drug are important for binding and activity.

- A pharmacophore identifies the functional groups and substituents that are important for a drug's activity, along with their relative position in the molecule.

- The active conformation defines the shape or conformation that a drug adopts when it is bound to a target binding site.

1.8 **Drug design**

Having identified the important binding groups and pharmacophore of a lead compound, it is now possible to design and synthesize analogues in an attempt to find structures with improved activity, selectivity, and pharmacokinetics—a process called **drug optimization** (section 8.4).

A crystal structure of the lead compound bound to the target protein will help enormously in this quest—a process known as **structure-based drug design**—but, as mentioned above, it is not always possible to crystallize target proteins. Fortunately, there are a number of well-established drug design strategies which can aid the medicinal chemist in deciding which analogues are worth synthesizing.

It is important to appreciate that drug optimization can involve tackling several different issues at the same time, and it is not just a case of optimizing a drug's ability to interact with the target binding site. For example, it is possible to design drugs which interact extremely well with the target binding site, and yet find that these molecules are inactive when given to patients. The pharmacokinetic properties of a drug (i.e. its ability to reach the target) are just as important as its pharmacodynamic properties (its ability to interact with the target). Moreover, it is important to design drugs which are selective for the target protein over similar proteins in order to minimize the possibility of side effects. More information on the synthesis of such analogues is provided in Chapter 8.

1.9 **Identifying a drug candidate and patenting**

Following the drug optimization process, a large number of compounds will have been synthesized, several of which could be considered as drug candidates. Deciding which one actually goes forward to preclinical trials is not always as straightforward as identifying which compound has the greatest activity. Several other factors have to be considered. For example, the drug must have useful activity and minimal side effects, show selectivity for its target, demonstrate good pharmacokinetic properties, lack toxicity, and have no interactions with other drugs that might be taken by a patient. In addition, it is important that the drug is synthesized as cheaply and efficiently as possible in order to make the maximum profit. Therefore, if one has to choose between two compounds where one of the compounds is slightly less active than the other, but considerably cheaper to synthesize, the less active compound may well be chosen to go forward to preclinical and clinical trials.

At this point, it is crucial to patent the discovery so that the company has the exclusive right to market these compounds in the future. Since patenting takes place relatively early in the overall process of drug development, several years of the patent are lost because of the time taken to carry out preclinical and clinical trials. One might ask why, in that case, the patent is not submitted at a later date. The reason is that the longer one delays a patent, the greater the chance that a competitor will make a similar discovery and get their patent application in first. Therefore, it is important to patent the compounds as soon as possible. Several years of patent protection may be lost, but that is preferable to losing the rights altogether.

Although a specific structure may go forward to preclinical and clinical trials, the patent is written to cover compounds of a similar structure, as well as the synthetic route used to prepare them. This guards against a rival company 'piggybacking' on the work that has been carried out and producing a closely related analogue. Moreover, it is possible that the chosen drug candidate may give unsatisfactory results during preclinical or clinical trials. Although the original structure has been 'knocked back', it may be possible to progress a related structure.

1.10 **Chemical and process development**

Once a candidate drug has been identified, work starts immediately on devising a large-scale synthesis such that sufficient quantities of the drug are available for the preclinical and clinical trials to follow. This is known as chemical development. The development chemist has a demanding role as it is important to produce large quantities of the drug as quickly as possible, whilst maintaining the quality of each batch produced. This is crucial since it is important that the various tests used in preclinical and clinical trials are carried out on batches having a consistent purity. Otherwise, the tests are not comparing like with like, and some of the tests may be invalidated. The importance of defining a drug's purity and characteristics early on is particularly critical since the chemical development process is more than just scaling up the original synthesis. Normally, reactions need to be modified or completely altered to optimize yields. Indeed, the final production synthesis may be totally different from the original research synthesis. Chapter 10 looks into chemical and process development in more detail.

1.11 Preclinical trials

Preclinical trials involve testing the drug candidate for a variety of properties. Toxicology tests are carried out to detect any toxic side effects such as carcinogenicity or teratogenicity. Further pharmacological tests determine how selective the drug is for its intended target. Most of this work is carried out by toxicologists, pharmacologists, and biochemists. However, organic synthesis still has a role to play during preclinical trials. Several tests have to be carried out to check what kind of metabolites are formed when the drug is administered to test animals and human beings. In order to detect these metabolites, it is important to synthesize the drug with an isotopic label—usually a radioisotope such as ^{14}C. Incorporating such a radiolabel may require a fundamental change in the synthetic route used. This is covered in more detail in Chapter 11.

1.12 Formulation and stability tests

A large quantity of drug is required in order to carry out formulation and stability tests, such that pharmacists and pharmaceutical chemists can identify how best to store and administer the drug. For example, if the drug is to be administered orally, it is usually provided as a capsule or a tablet. If it is to be administered in tablet form, work has to be carried out to identify what other components should be present in the tablet, and to ensure that the drug is compatible with those components. Stability tests have to be carried out to assess the stability of the drug under varying conditions of temperature, pH, light, humidity, etc.

1.13 Clinical trials

Clinical trials are the province of the clinician, and so chemists have no role to play in the process other than producing the drug. There are four phases of clinical trials. Phase I involves a small group of healthy volunteers who are administered the drug at levels considered to be safe, based on previous preclinical trials carried out on animals. The aim is to identify what kind of doses can be tolerated and to identify any minor side effects. A certain number of volunteers will take a radiolabelled drug in order to identify drug metabolites.

Phase II trials are carried out on patients to see whether the drug has a significant beneficial effect. The levels and frequency of drug administration will also be varied to assess the best dosing regimen. Half the patients involved in the study will be given the drug, and half will be given a **placebo**. A placebo is a compound which should have no significant effect on a patient. This is important because there is a well-known placebo effect where a patient's health can improve based on a belief that he or she has been given an effective medication. When analysing the results, there should be a significantly greater improvement in the patients taking the drug over those taking the placebo. These tests are carried out as **double-blind studies** such that neither the patient nor the clinicians involved in the study are aware of which patients are receiving the drug and which patients are receiving the placebo.

Phase III trials are a continuation of phase II trials and are carried out on a much larger number of patients. If the drug proves effective during this phase, and has no serious adverse effects, it can be registered and marketed.

Phase IV trials continue after the drug has been marketed and are useful in picking up extremely rare side effects that would be missed by earlier studies.

1.14 Regulatory affairs and marketing

1.14.1 The regulatory process

Regulatory bodies such as the **Food and Drug Administration (FDA)** in the USA and the **European Agency for the Evaluation of Medicinal Products (EMEA)** in Europe come into play as soon as a pharmaceutical company believes that it has a useful drug. Before clinical trials can begin, the company has to submit the results of its scientific and preclinical studies to the relevant regulatory authority for approval. This includes information about the chemistry, manufacture, and quality control of the drug, as well as information on its pharmacology, pharmacokinetics, and toxicology.

If the clinical trials proceed smoothly, the company then applies for marketing approval. In the USA, this involves the submission of a **New Drug Application (NDA)** to the FDA; in Europe, the equivalent submission is called a **Marketing Authorization Application (MAA)**. The application has to state what the drug is intended to do, along with scientific and clinical evidence for its efficacy and safety. It should also give details of the chemistry and manufacture of the drug, as well as the controls and analysis which will be in place to ensure that the drug has a consistent quality. Any advertising and marketing material must be submitted to ensure that it makes accurate claims and that the drug is being promoted for its intended use. The labelling of a drug preparation must also be approved to ensure that it instructs physicians about the mechanism of action of the drug, the medical situations for which it should be used, and the correct dosing levels and frequency. Possible side effects, toxicity, or addictive effects should be detailed, as well as special precautions which might need to be taken (e.g. avoiding drugs that interact with the preparation).

Once an application has been approved, any modifications to a drug's manufacturing synthesis or analysis must also be approved. In practice, this means that the manufacturer will generally perfect the route described in the application, rather than consider alternative routes.

Abbreviated applications can be filed by manufacturers who wish to market a generic variation of an approved drug whose patent life has expired. The manufacturer is only required to submit chemistry and manufacturing information, and demonstrate that the product is comparable with the product already approved.

The term **new chemical entity** (**NCE**) or **new molecular entity** (**NME**) refers to a novel drug structure. In the 1960s about 70 NCEs reached the market each year, but this had dropped to less than 30 per year by 1971. In 2002, only 18 NCEs were approved by the FDA and 13 by the EMEA.

1.14.2 **Fast-tracking and orphan drugs**

The regulations of many regulatory bodies include the possibility of **fast-tracking** certain types of drug, so that they reach the market as quickly as possible. Fast-tracking is made possible by requiring a smaller number of phase II and phase III clinical trials before the drug is put forward for approval. Fast-tracking is carried out for drugs that show promise for diseases where no current therapy exists, and for drugs that show distinct advantages over existing ones in the treatment of life-threatening diseases. An example of a fast-tracked drug is **oseltamivir** (Tamiflu) for the treatment of flu.

Orphan drugs are drugs that are effective against relatively rare medical problems. In the USA, an orphan drug is defined as one that is used for less than 200 000 people. Because there is a smaller market for such drugs, pharmaceutical companies may be less likely to reap huge financial benefits and may decide not to develop and market an orphan drug. Therefore, financial and commercial incentives are given to firms in order to encourage the development and marketing of such drugs.

1.14.3 **Good laboratory, manufacturing, and clinical practice**

Good Laboratory Practice (**GLP**) and **Good Manufacturing Practice** (**GMP**) are scientific codes of practice for a pharmaceutical company's laboratories and production plants. They detail the scientific standards that are necessary, and the company must prove to regulatory bodies that it is adhering to these standards.

GLP regulations apply to the various research laboratories involved in pharmacology, drug metabolism, and toxicology studies. GMP regulations apply to the production plant and chemical development laboratories. They encompass the various manufacturing procedures used in the production of the drug, as well as the procedures used to ensure that the product is of a consistently high quality. As part of GMP regulations, the pharmaceutical company is required to set up an independent quality control unit which monitors a wide range of factors including employee training, the working environment, operational procedures, instrument calibration, batch storage, labelling, and the quality control of all solvents, intermediates, and reagents used in the process. The analytical procedures which are used to test the final product must be defined, as well as the specifications that have to be met. Each batch of drug that is produced must be sampled to ensure that it passes those specifications. Written operational instructions must be in place for all special equipment (e.g. freeze-dryers), and standard operating procedures (SOPs) must be written for the use, calibration, and maintenance of equipment. Detailed and accurate paperwork on the above procedures must be available for inspection by the regulatory bodies. This includes calibration and maintenance records, production reviews, batch records, master production records, inventories, analytical reports, equipment cleaning logs, batch recalls, and customer complaints. Although record-keeping is crucial, it is possible that the extra paperwork involved can stifle innovations in the production process.

Investigators involved in clinical research must demonstrate that they can carry out the work according to **Good Clinical Practice** (**GCP**) regulations. The regulations require proper staffing, facilities, and equipment for the required work, and each test site involved must be approved. There must also be evidence that a patient's rights and well-being are properly protected.

1.15 **Conclusion**

In this chapter we have considered the overall drug design and development process, and highlighted those areas where organic synthesis plays a crucial role. In the following chapters, we shall look more closely at the role of organic synthesis in producing pharmacologically active compounds at different stages of the drug development process.

- Drug optimization involves developing a structure that has an optimum balance of properties.

- Following drug optimization a drug candidate is chosen to go forward for preclinical trials.

- Novel drugs are patented before preclinical and clinical trials.

- Chemical development involves developing large-scale syntheses of drug candidates for preclinical and clinical trials. Process development develops an efficient production process for an approved drug.

- Preclinical trials include studies into toxicology, drug metabolism, pharmacology, formulation, and stability.

- Clinical trials involve four phases. Phase I trials are normally carried out on healthy volunteers to assess dosing regimes and safety. Phase II and III trials are carried out on patients to establish whether the drug is effective in treating a target disease. Phase IV trials continue after the drug has been approved and marketed in order to identify any rare side effects.

- Regulatory authorities approve whether clinical trials can be carried out and whether a drug can be marketed.

- Regulatory authorities monitor pharmaceutical companies to ensure that they abide by good laboratory practice and good manufacturing practice.

FURTHER READING

Brunton, L., Chabner, B., and Knollman, B. (eds) (2010) *Goodman and Gilman's The pharmacological basis of therapeutics* (12th edn). McGraw-Hill, New York.

Drews, J. (2003) *In quest of tomorrow's medicines*. Springer-Verlag, New York.

Ganellin, C.R. and Roberts, S.M. (eds) (1994) *Medicinal chemistry—the role of organic research in drug research* (2nd edn). Academic Press, London.

King, F.D. (ed.) (2002) *Medicinal chemistry: principles and practice* (2nd edn). Royal Society of Chemistry, Cambridge.

Krogsgaard-Larsen, P. (ed.) (2009) *Textbook of drug design and development* (4th edn). CRC Press, Boca Raton, FL.

Le Fanu, J. (2011) *The rise and fall of modern medicine*. Abacus, London.

Patrick, G. (2001) *Instant notes in medicinal chemistry*. Bios Scientific, Oxford.

Patrick, G.L. (2013) *An introduction to medicinal chemistry* (5th edn). Oxford University Press, Oxford.

Silverman, R.B. (2004) *The organic chemistry of drug design and action* (2nd edn). Academic Press, San Diego, CA.

Sneader, W. (2005) *Drug discovery: a history*. Wiley, Chichester.

Thomas, G. (2007) *Medicinal chemistry: an introduction* (2nd edn). Wiley, Chichester.

Wermuth, C.G. (ed.) (2008) *The practice of medicinal chemistry* (3rd edn). Academic Press, London.

Williams, D.A. and Lemke, T.L. (eds) (2012) *Foye's Principles of medicinal chemistry* (6th edn). Lippincott–Williams & Wilkins, Philadelphia, PA.

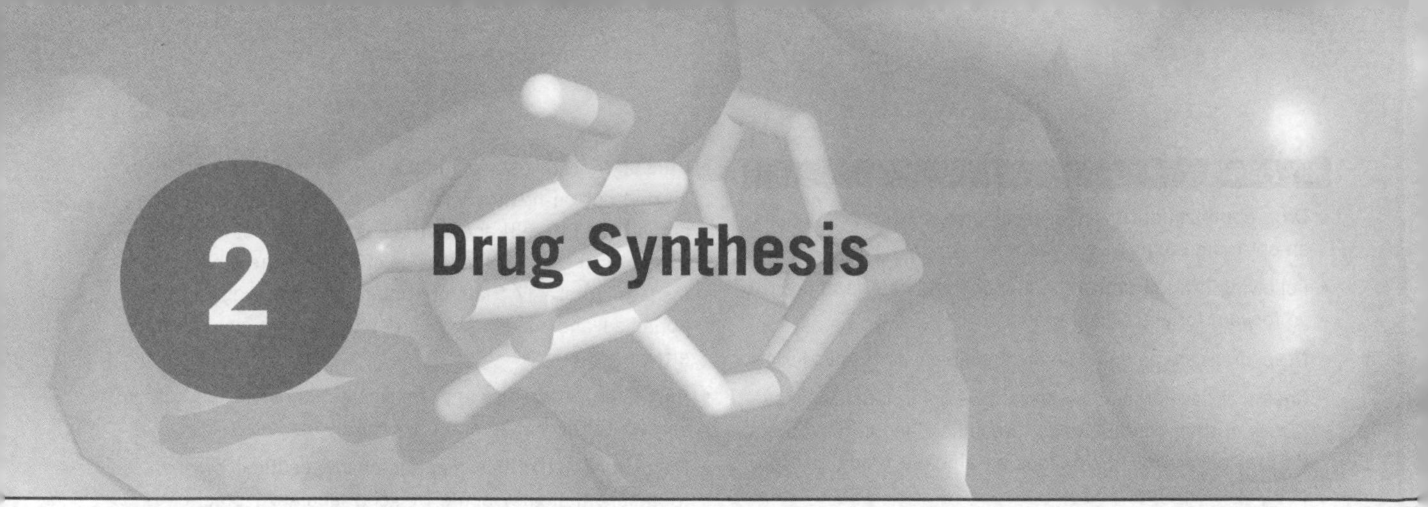

2 Drug Synthesis

2.1 The role of organic synthesis in the drug design and development process

Organic synthesis plays a crucial role in several of the key stages involved in the development of a novel pharmaceutical agent (Fig. 1.1). Large numbers of compounds need to be synthesized to create chemical libraries that can be screened to identify hits for a relevant target (Chapter 6). From these hits, a suitable lead compound can be identified as a basis for further study (Chapter 7). Further synthesis is required to produce analogues of the lead compound in order to identify structure–activity relationships, and for drug optimization (Chapter 8). Once a structure that has the optimum balance of desired activity, minimum side effects, and suitable pharmacokinetic properties has been identified, further synthetic studies are required to identify the cheapest, safest, and most environmentally friendly method of producing the compound in the production plant (Chapter 10). Other studies are required to identify a synthesis that can efficiently incorporate radioisotopes into the structure to allow drug metabolism studies to be carried out (Chapter 11). It may also be necessary to synthesize potential drug metabolites to establish their structure. In this chapter, we will consider a number of drug syntheses and highlight some general principles involved. We shall first consider what kind of structural features may affect how easily a drug can be synthesized.

2.2 Structural features that affect the ease of synthesis

There are several structural features that need to be considered when synthesizing a drug, all of which need to be correctly constructed or incorporated. The number of such features often gives an indication of how complex a synthesis is likely to be.

2.2.1 The molecular skeleton

The molecular skeleton or framework of a drug determines its overall size and shape which, in turn, determines how well it fits into its target binding site. The skeleton of some drugs involves a multicyclic ring system that results in a rigid structure with a clearly defined shape. For example, **morphine** contains a pentacyclic ring system where two of the rings are oriented roughly at right angles to the other three to produce a T-shaped molecule (Fig. 2.1). Rigid molecules are likely to have

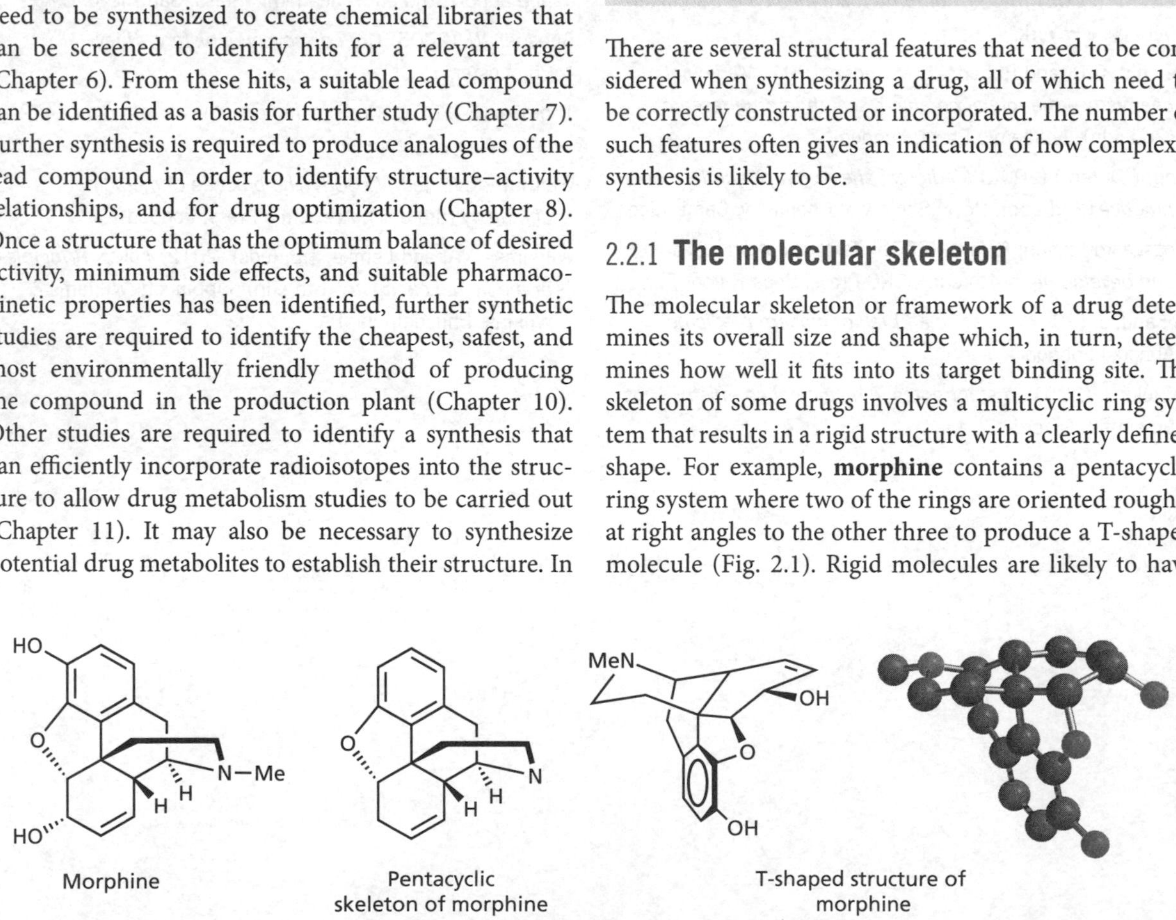

| Morphine | Pentacyclic skeleton of morphine | T-shaped structure of morphine |

FIGURE 2.1 The T-shape defined by the pentacyclic ring system of morphine.

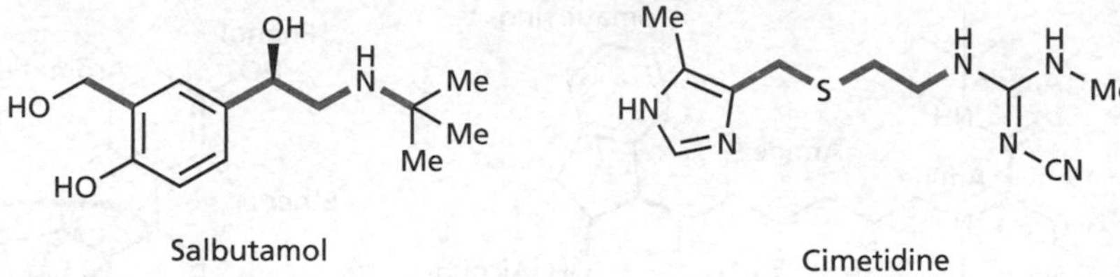

FIGURE 2.2 Examples of 'flexible' drugs containing several rotatable bonds (shown in blue). Salbutamol is an anti-asthmatic agent and cimetidine is an anti-ulcer agent.

very good activity if they are the ideal shape for the target binding site and contain the necessary pharmacophore for activity (section 1.7). However, the drawback with multicyclic ring systems is that they can be very difficult to synthesize. In some cases it is not economically viable to construct the molecular framework of such molecules, and it is more convenient to find a relevant natural product that already contains the required framework (Chapter 9).

In contrast, other drugs have a linear skeleton that includes several rotatable single bonds (Fig. 2.2) (Box 2.1). As a result, these drugs are flexible and can adopt a number of different shapes or conformations. Drugs of this nature are generally easier to synthesize, but their activity might not be as good as that of more rigid molecules (see also section 4.1).

Note that a drug may contain several rings within its structure, yet still be considered linear and flexible. For example, both morphine and the antiviral agent **indinavir** (Fig. 2.3) contain five rings. However, indinavir is much easier to synthesize because the rings are not part of a multicyclic ring system. Instead, they are incorporated within a linear flexible skeleton. Indinavir is an example of a class of drugs known as **protease inhibitors**, which have proved very effective in the treatment of AIDS. The drugs act as enzyme inhibitors by inhibiting a viral protease enzyme.

The nature of the molecular skeleton also plays an important role in a drug's pharmacokinetic properties. For example, drugs containing planar aromatic or heteroaromatic ring systems often suffer from poor solubility compared with drugs containing saturated non-planar ring systems (see also section 7.3). Drugs containing a large number of rotatable bonds are often found to be poorly absorbed from the gastrointestinal tract (section 1.5.3).

2.2.2 Functional groups

Functional groups are important to a drug's activity and mechanism of action. There are a large number of functional groups in chemistry, but some crop up more commonly than others in drug structures. For example,

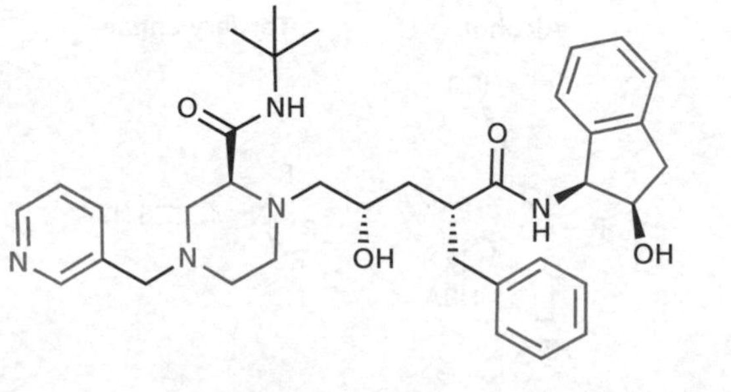

BOX 2.1 Identifying rotatable bonds in a drug structure

In medicinal chemistry, the number of rotatable bonds present in a drug is defined by those bonds that result in a significant change in shape. For example, indinavir contains 12 such rotatable bonds (coloured blue in Fig. 1). The single bonds to the two hydroxyl groups and the three methyl groups certainly rotate, but have no significant effect on the overall shape of the molecule. Note also that the single bond between the NH and the CO of the two amide groups is not counted as a rotatable bond. This is because there is significant double-bond character in an amide bond which prevents bond rotation.

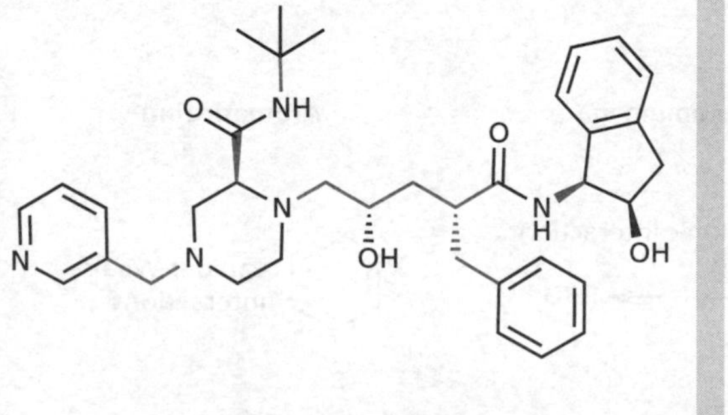

FIGURE 1 Indinavir.

FIGURE 2.3 Indinavir.

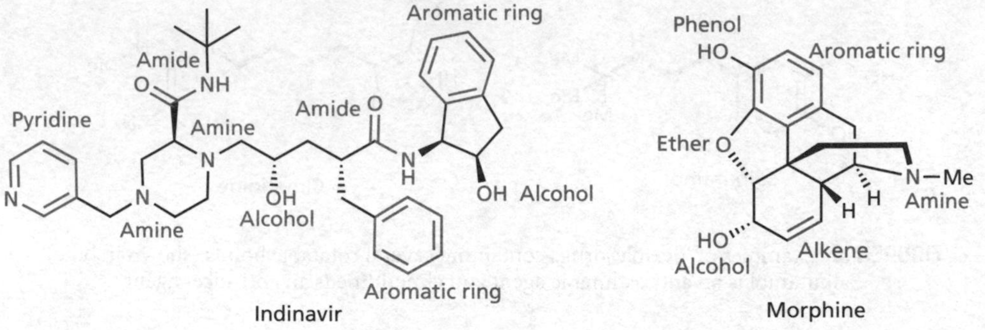

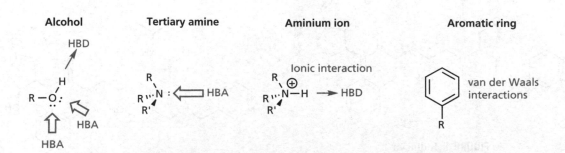

FIGURE 2.4 Functional groups present in indinavir and morphine.

alcohols, phenols, amines, amides, and aromatic rings are particularly common. The functional groups present in morphine and indinavir are identified in Figure 2.4.

Functional groups are often present in drugs because they are involved in the binding interactions between the drug and its target binding site. These binding interactions involve intermolecular bonds such as van der Waals interactions, hydrogen bonds, and ionic bonds (Fig. 2.5) (see also section 8.2.2).

However, not all functional groups within a drug are necessarily involved in binding interactions. Some functional groups might have an indirect effect on binding, while others may play a role in determining a drug's stability and pharmacokinetic properties. For example, functional groups such as nitrile, fluoro, or chloro groups have an electron-withdrawing effect, which may affect how well a different functional group in the structure interacts with the binding site. Alternatively, they may have been incorporated into the structure to block drug metabolism at a particular position such that the lifetime of the drug is increased. Electron-withdrawing substituents have also been added to aromatic rings to affect the basicity of aromatic amines, resulting in different pharmacokinetic properties. Other functional groups, such as alkenes, alkynes, and amides, may have been incorporated into a structure to introduce a measure of rigidity

(conformational restriction) so that there is more chance of the drug being in its active conformation.

Different drugs differ in the number of functional groups present but, in general, the more functional groups that are present in a drug, the more challenging the synthesis is likely to be.

2.2.3 Substituents

The number of substituents attached to a drug's molecular skeleton is another factor that influences how easily a synthesis might be carried out. Substituents include functional groups that are peripheral to the overall molecular skeleton (alcohols, phenols, nitriles, halogens, etc.), as well as alkyl groups and side chains (Fig. 2.6). We have already seen that functional groups can play an important role in the activity and pharmacokinetic properties of a drug (section 2.2.2). Alkyl substituents can also be important in binding interactions if they fit into hydrophobic pockets within a target binding site and form van der Waals interactions. Alkyl chains that include aromatic and heteroaromatic rings are another type of substituent which is found in several drugs. In general, the more substituents that are present, the more challenging the synthesis is likely to be.

FIGURE 2.5 Possible intermolecular binding interactions for selected functional groups.

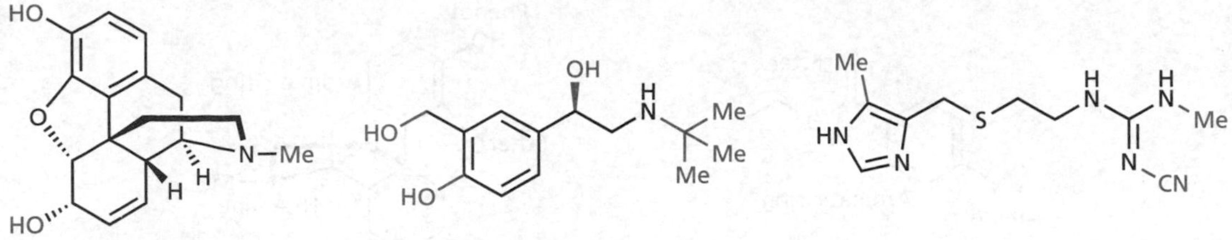

Morphine Salbutamol Cimetidine

FIGURE 2.6 Substituents on different drugs.

2.2.4 Chirality and asymmetric centres

About 30% of drugs on the market are chiral. A chiral molecule is asymmetric in nature, which means that the molecule can exist as two non-superimposable mirror images called **enantiomers**. Because drug targets within the body are also chiral, they usually distinguish between the two enantiomers of a drug, which means that one enantiomer is more active than the other. Moreover, the two enantiomers may differ in their side effects, which can result in one enantiomer being more toxic than the other. For example, one enantiomer of the sedative **thalidomide** proved to be teratogenic, whereas the other enantiomer did not.

Some chiral drugs such as the beta-blocker **propranolol** are sold as a mixture of the two enantiomers (a **racemate**), while others are marketed as the single enantiomer. Synthesizing one enantiomer in preference to the other adds an extra dimension to the synthetic challenge (Chapter 5).

Chiral molecules can often be identified by the fact that they contain asymmetric centres. In general, the more asymmetric centres that are present in a target drug, the more complex the synthesis is likely to be. Asymmetric centres are normally carbon atoms containing four different substituents (Fig. 2.7). However, it is important to appreciate that not all chiral molecules contain an asymmetric centre. For example, the antiviral agent **nevirapine** is a chiral molecule, but lacks an asymmetric centre (Fig. 2.8). Nevirapine inhibits a viral enzyme called **reverse transcriptase** and

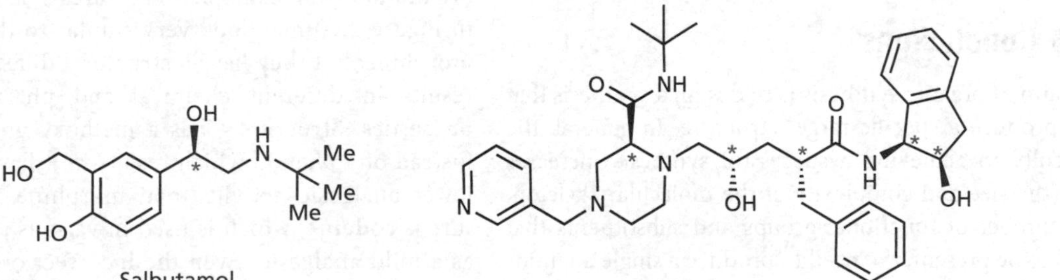

Salbutamol Indinavir

FIGURE 2.7 Asymmetric centres in salbutamol and indinavir. Asymmetric centres are marked with a blue asterisk.

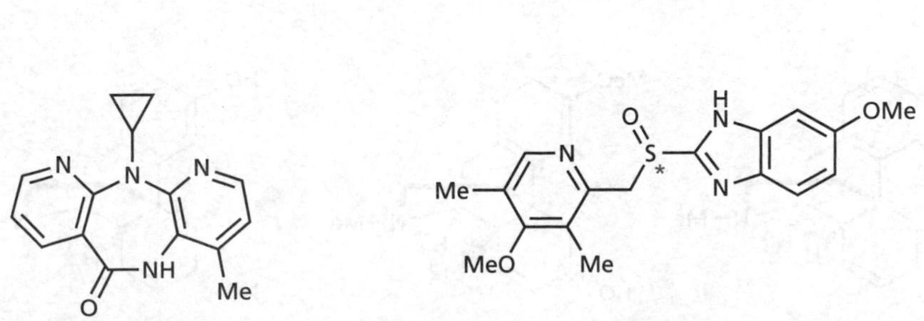

Nevirapine Omeprazole

FIGURE 2.8 Nevirapine and omeprazole.

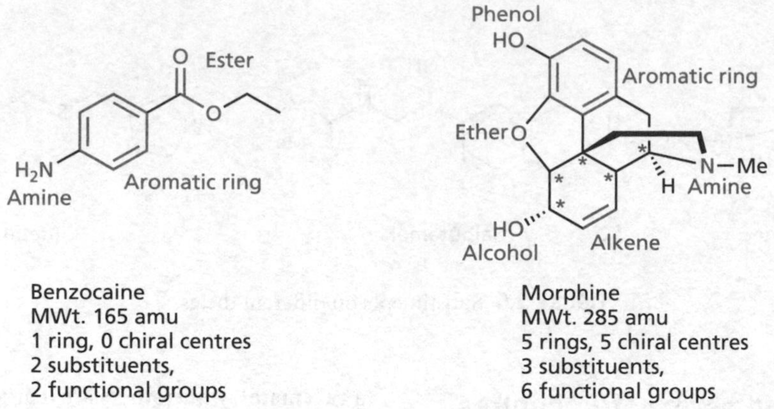

FIGURE 2.9 Comparison of benzocaine and morphine.

is classed as a **non-nucleoside reverse transcriptase inhibitor (NNRTI)**.

It is also possible for atoms other than carbon to act as asymmetric centres. For example, the **proton pump inhibitors** (PPIs) used in the treatment of ulcers contain a sulphur atom that acts as an asymmetric centre. One example is **omeprazole** (Fig. 2.8; see also Box 5.8). The proton pump is an enzyme complex that is present in the parietal cells that line the stomach wall. The pump transports protons from the parietal cells into the stomach, thus generating hydrochloric acid.

2.2.5 Conclusions

The aim of organic synthesis is to design a synthesis that will produce a specific target structure. In general, the difficulty in achieving a successful synthesis increases with the size and complexity of the molecular skeleton, the number of functional groups and substituents that need to be present, the need to produce a single enantiomer for a chiral compound, and the number of asymmetric centres that are present in such a molecule. For example, **benzocaine** (Fig. 2.9) is a relatively simple

achiral molecule with weak local anaesthetic activity which can be synthesized very easily. It contains one ring, two substituents, and three functional groups. In contrast, morphine contains a complex pentacyclic ring system, five asymmetric centres, three substituents, and six functional groups (Fig. 2.9). The synthesis is far more challenging, and the first synthesis of morphine required 29 steps (section 8.3).

There are no prizes for synthesizing a compound which is almost correct. Every feature in the final product must be present at the correct position and orientation. For example, structures I and II shown in Figure 2.10 may look very similar to the analgesic morphine, but they have a structural difference which results in different chemical and pharmacological properties. Structure I has a methoxy group present instead of a phenol, which results in it having a much lower analgesic activity than morphine. This structure is **codeine**, which is used in various preparations as a mild analgesic. Even the incorrect orientation of a single proton results in a totally different structure (structure II in Figure 2.10) with a drastically reduced analgesic activity.

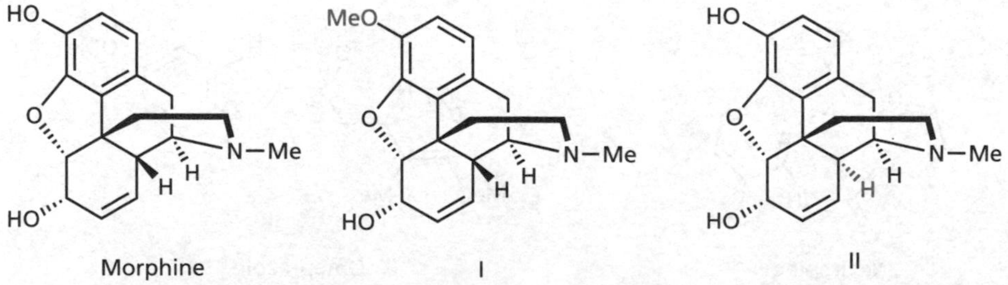

FIGURE 2.10 Comparison of morphine with other opioid structures.

2.2.6 Exceptions to the rule

One of the golden rules in chemistry is that there is invariably an exception to every rule, and that holds true for the rules provided here. Some compounds can be made very easily despite having a large molecular skeleton and a large number of functional groups, substituents, and asymmetric centres. Peptides are a case in point. Peptide synthesis is relatively straightforward and it is possible to prepare large peptides using automated processes. This is possible because there is a ready supply of amino acid building blocks which already contain the necessary functional groups, substituents, and asymmetric centres. It is merely a case of linking the amino acid building blocks through the formation of peptide bonds (see Case Study 1).

2.3 Synthetic approaches to drugs

2.3.1 Introduction

A large number of synthetic routes to drugs have been reported in the literature. Some are very simple, involving a handful of reactions, and some are very complex. It is also worth mentioning that it is usually possible to synthesize a specific drug by a number of different synthetic routes, and each route will have its own advantages and disadvantages. A vast number of reactions and reagents have been used in order to carry out these various drug syntheses. Examples are given in Appendices 1–6 and in the Online Resource Centre, but it is not possible to cover these in any detail in a textbook of this nature. Interested readers should consult one of the excellent textbooks that are available on organic synthesis, particularly the textbook by Clayden, Greeves, and Warren. In this chapter, we will consider some general principles that have been applied to drug syntheses.

First of all, we have to appreciate that functional groups play a crucial role in drug synthesis. We have already seen that they play a crucial role in the activity and properties of a drug, but they are also required if the drug is to be synthesized in the first place. That is because reactions take place at or near functional groups. Functional groups undergo particular types of reaction and the functional group involved is often converted to a different functional group as a result of that reaction.

Secondly, the aim of drug synthesis is to synthesize a target molecule from one or more molecular building blocks. In general, these are simpler molecules that are commercially available and affordable, and have the necessary functional groups present to allow suitable reactions to be carried out. Clearly, they should also bear a structural similarity to the target molecule, or to some part of the target molecule.

Thirdly, the reactions that are carried out at each stage of the synthesis should ideally give a single product and not a mixture of products. This can be quite challenging to achieve if several functional groups are present in the building blocks or the synthetic intermediates. In an ideal situation, a reaction should take place at one specific functional group in preference to any others that might be present. This is called **chemoselectivity**. Examples of chemoselectivity will be highlighted in various syntheses described in later sections, as well as in section 2.13. Another form of reaction selectivity is called **regioselectivity**. This is where a reaction at a functional group occurs selectively at one of several possible positions. Examples of functional groups where this can occur include aromatic and heteroaromatic rings, epoxides, and α,β–unsaturated ketones. One example of regioselectivity is the reaction of an amine with an epoxide during the synthesis of the anti-asthmatic agent (**R**)-**salmeterol** (Fig. 2.11). The reaction is regioselective because the amine reacts with the less

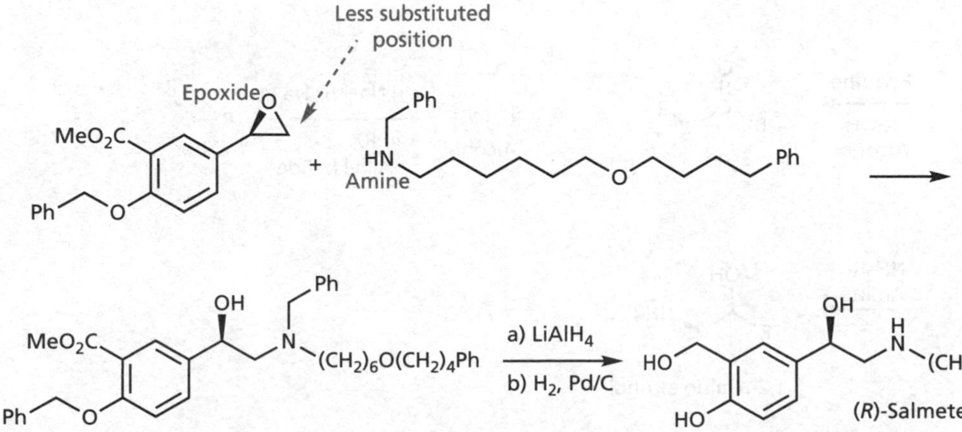

FIGURE 2.11 Synthesis of salmeterol.

substituted position of the epoxide. The reaction is also chemoselective since the amine reacts with the epoxide and not the methyl ester.

There are other forms of reaction selectivity, such as **stereoselectivity, enantioselectivity**, and **diastereoselectivity**, which will be discussed in Chapter 5.

2.3.2 Types of reaction

For the purposes of this book, reactions are classified as coupling reactions if two building blocks are linked together. These are the crucial reactions required to create a target structure, and efficient synthetic routes will be predominantly made up of such reactions. However, a range of other functional group reactions may be necessary such as functionalization, functional group removal, functional group activation, functional group transformation, protection, and deprotection.

Coupling reactions are reactions which are used to link together two molecular building blocks to create a larger molecule. Most drug syntheses involve at least one coupling reaction, with some coupling reactions being particularly popular. For example, reactions involving the formation of ethers, amines, esters, amides, imines, and 1,2- amino alcohols are all commonly employed in a range of different drug syntheses (Fig. 2.12). In all

these reactions, the building blocks are linked together through the formation of a new O–C or N–C bond to give the functional groups indicated.

Coupling reactions which link building blocks by forming a new C–C bond are also common in drug synthesis. Some of the most common reactions are the Friedel–Crafts alkylation, Friedel–Crafts acylation, the Wittig, Suzuki, and Sonogashira reactions, and nucleophilic additions to carbonyl groups involving Grignard or organolithium reagents (Fig. 2.13).

The Grignard reaction can be carried out on the carbonyl group of aldehydes and ketones. A carbon–carbon bond is formed which links the two building blocks, and the original functional group is transformed into an alcohol group. The same kind of reaction can be carried out using an organolithium reagent. Similarly, the Wittig reaction is carried out on the carbonyl group of an aldehyde or ketone. This time the building blocks are linked through a carbon–carbon double bond to form an alkene, and the carbonyl group is no longer present. The Friedel–Crafts alkylation and acylation are useful reactions for adding substituents to aromatic rings, while the Suzuki reaction is effective in linking two aromatic or heteroaromatic rings together.

In addition to coupling reactions, most drug syntheses will include a range of other reactions which involve 'fine

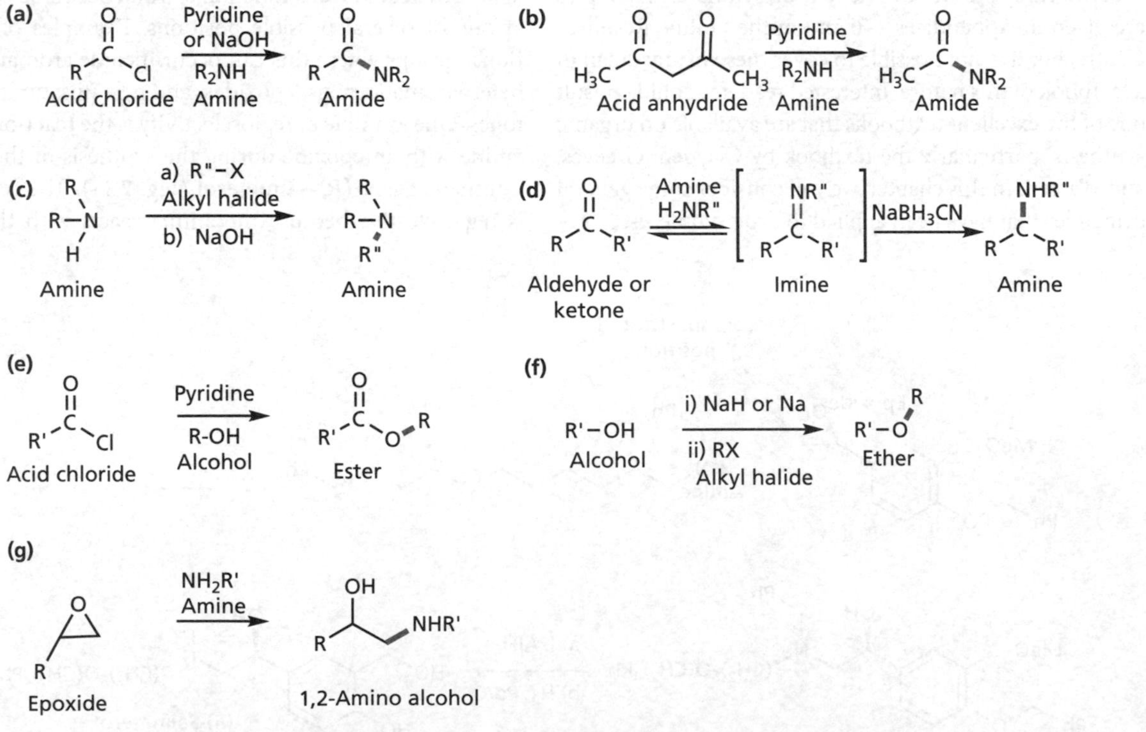

FIGURE 2.12 N–C and O–C coupling reactions that are commonly used to link molecular 'building blocks' to form more complex molecules. The new O–C or N–C bond is coloured blue.

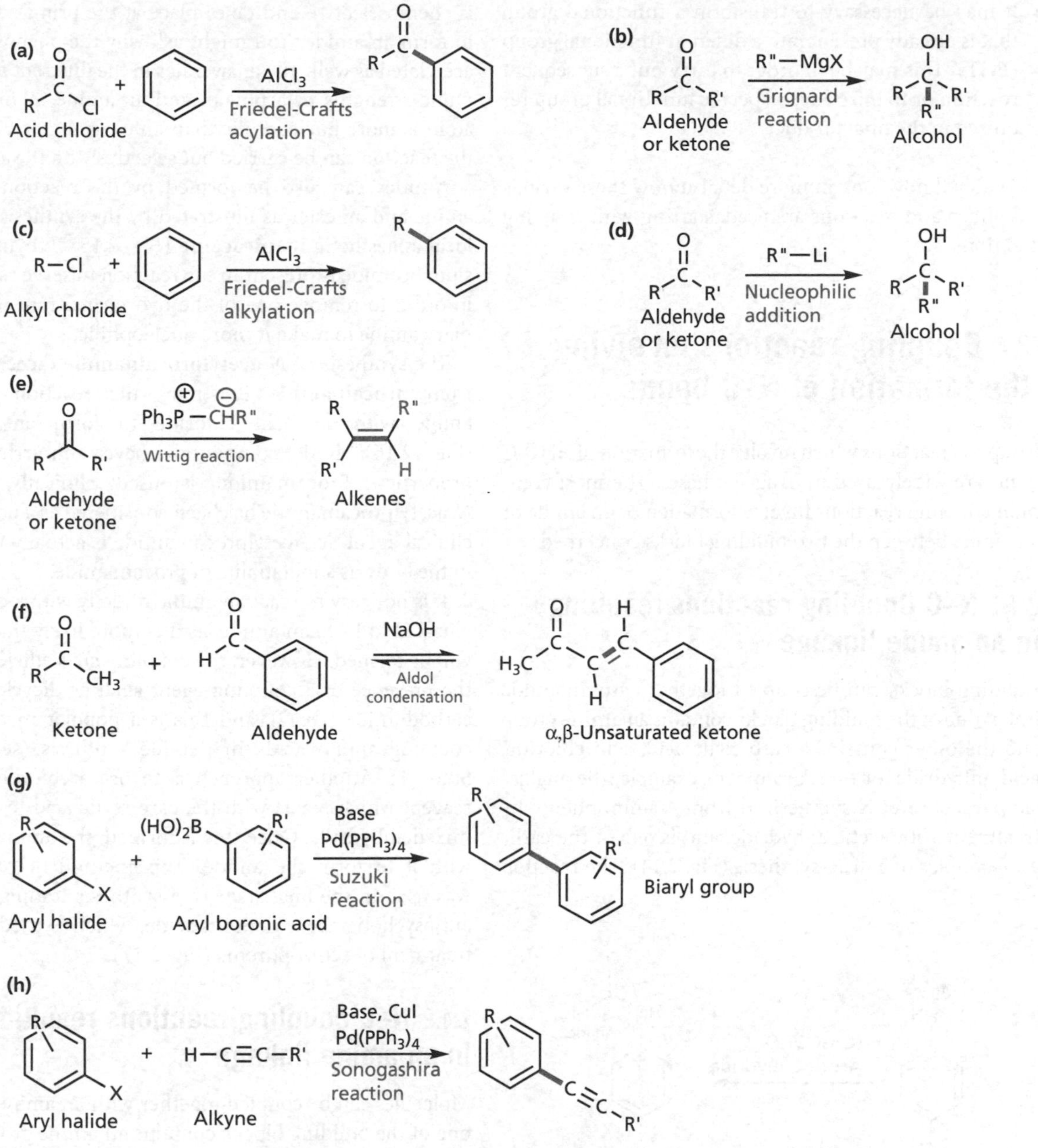

FIGURE 2.13 C–C coupling reactions that are commonly used to link molecular 'building blocks' to form more complex molecules. The new C–C bond is coloured blue.

tinkering' with the functional groups that are present in the building blocks or the synthetic intermediates. There are several reasons for carrying out these reactions. Examples include the following

- It may be necessary to introduce a functional group at a position that does not already have one. This may be necessary in order to carry out a subsequent reaction at that position or to introduce a functional group that is required in the final structure (*functionalization*).

- A functional group that is not required in the final product may have to be removed (*FG removal*).

- A functional group might have to be converted into a more reactive functional group to allow a subsequent reaction to be carried out more effectively (*FG activation*).

- A protecting group may need to be added to a functional group to prevent it undergoing an unwanted reaction (*protection*). Afterwards the protecting group will need to be removed (*deprotection*).

- It may be necessary to transform a functional group that is already present into a different functional group (FGT). This may be in order to carry out a subsequent reaction or to introduce a specific functional group required in the final product.

We shall now look in more detail at how these various couplings and reactions are used, starting with coupling reactions.

2.4 Coupling reactions involving the formation of N–C bonds

Coupling reactions which involve the formation of an N–C bond are widely used in drug syntheses. The most common coupling reactions involve formation of an amide or an amine between the two building blocks concerned.

2.4.1 N–C Coupling reactions resulting in an amide linkage

Building blocks can be coupled together with an amide link if one of the building blocks contains an amine group and the other contains a carboxylic acid, acid chloride, acid anhydride, or ester group. For example, the analgesic **paracetamol** is synthesized from 4-aminophenol by treatment with acetic anhydride, and is one of the earliest examples of a drug synthesis (Fig. 2.14). The reaction is chemoselective and takes place at the primary amine to form an amide. You might ask why the phenol is not acetylated as well. The answer lies in the different nucleophilic strengths of a phenol and an amine. A nitrogen atom is more nucleophilic than an oxygen atom, and so the reaction can be carried out selectively on the amine.

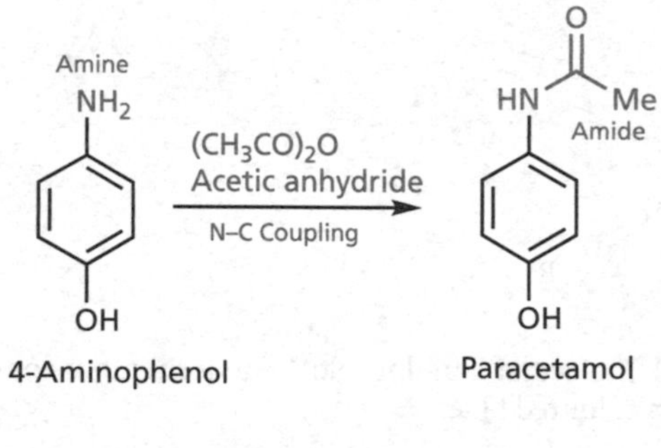

Amides can also be formed by the reaction of an amine and an ester as illustrated by the synthesis of the local anaesthetic **mepivacaine** (Fig. 2.15). Ethylmagnesium bromide is present in the reaction to serve as a base in order to remove one of the hydrogens from the primary amine to make it more nucleophilic.

The synthesis of *N*-acetylprocainamide (**accecainide**) from **procainamide** illustrates the reaction of an amine with an acid chloride to form an amide (Fig. 2.16). Both compounds have anti-arrhythmic properties. Procainamide is used clinically, while *N*-acetylprocainamide has been considered as a potential clinical agent. *N*-Acetylprocainamide is actually formed in the body as a metabolite of procainamide.

It is not easy to react an amine directly with a carboxylic acid to form an amide, as it is more likely that a salt will be formed. However, the coupling goes efficiently in the presence of a coupling agent such as dicyclohexylcarbodiimide (DCC), and this is a popular method of coupling amino acids in peptide synthesis (see Case Study 1). Another approach is to use a chloroformate reagent which reacts with the carboxylic acid to form a mixed anhydride. Once this is formed, the amine reacts with it to form the amide. This method of coupling was used in the final stage of a synthesis leading to the antipsychotic agent **nemonapride**, which is used in the treatment of schizophrenia (Fig. 2.17).

2.4.2 N–C coupling reactions resulting in an amine linkage

Molecules can be coupled together with an amine link if one of the building blocks contains an amine group and the other contains an alkyl halide or an epoxide. For example, **azaperone** was synthesized by reacting an amine with an alkyl chloride (Fig. 2.18). The reaction is chemoselective since the amine reacts with the alkyl halide rather than the ketone. Azaperone is a sedative which is

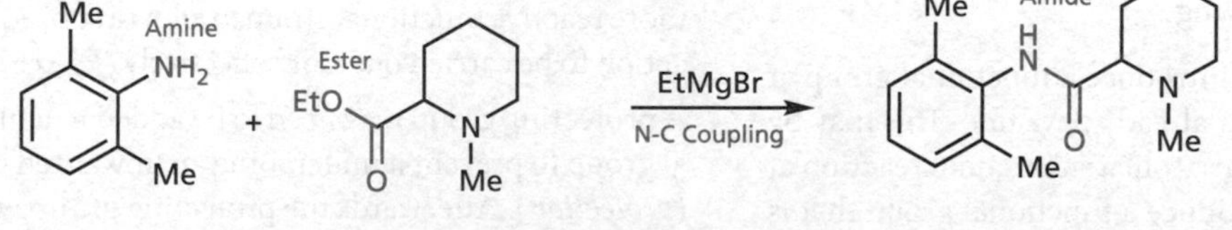

FIGURE 2.14 Synthesis of paracetamol from 4-aminophenol.

FIGURE 2.15 Synthesis of mepivacaine.

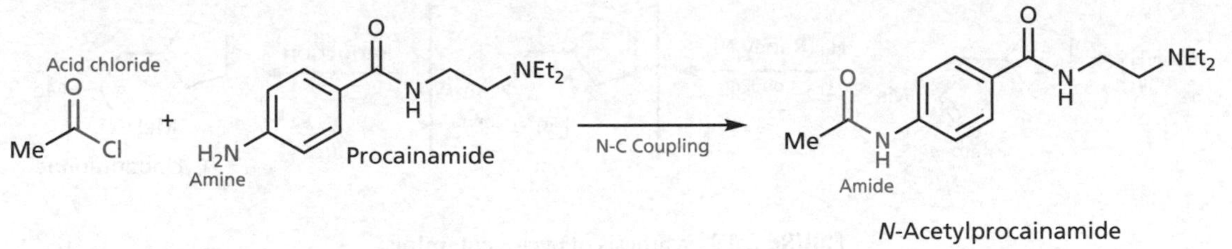

FIGURE 2.16 Synthesis of N-acetylprocainamide.

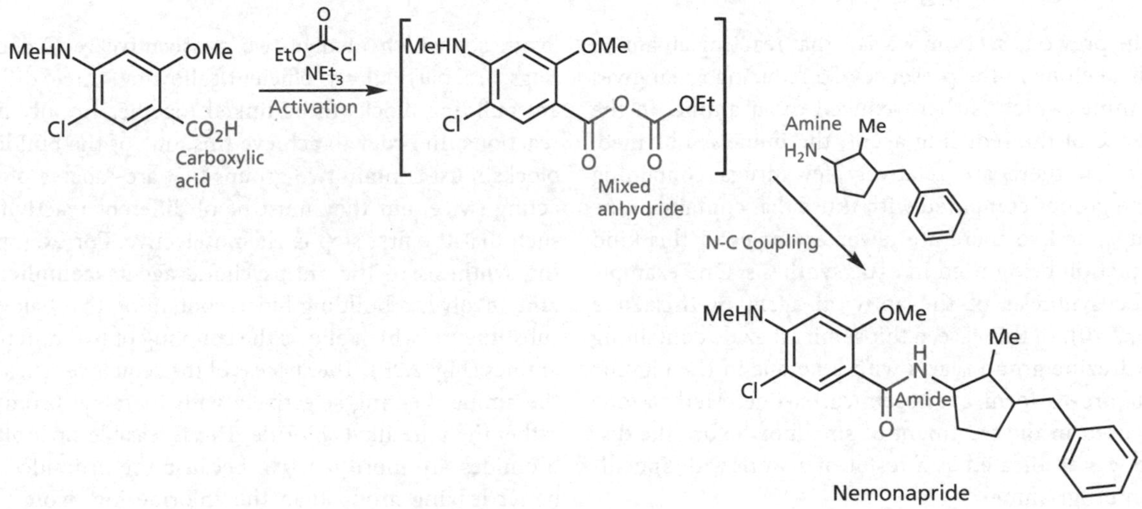

FIGURE 2.17 Final coupling stage in the synthesis of nemonapride.

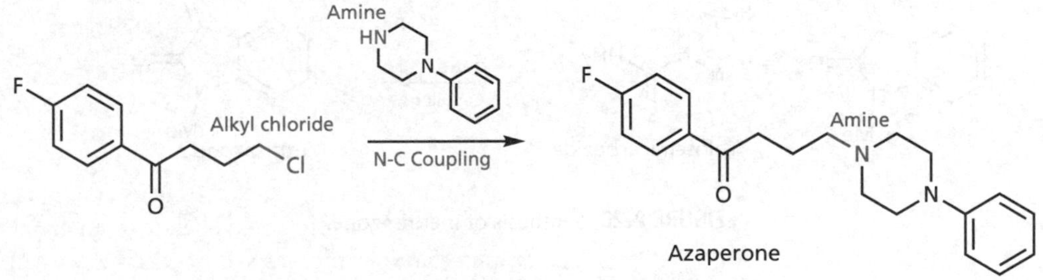

FIGURE 2.18 Synthesis of azaperone.

used in veterinary medicine. It has also been used with the opioid sedative **etorphine** to tranquillize large animals such as elephants. In exceptional circumstances, it has been used in human medicine as an antipsychotic agent.

Another method of coupling building blocks to form an amine linkage is to react an amine with a ketone under reducing conditions. The reaction initially forms an imine which is reduced to the amine in situ. The reducing agent can be hydrogen gas in the presence of a catalyst, or sodium cyanoborohydride. For example, the vaso-constrictor **cyclopentamine** was synthesized by reacting an amine with a ketone in the presence of hydrogen gas and Raney nickel (Fig. 2.19). Cyclopentamine used to be available as a nasal decongestant, but has now been replaced by more modern agents.

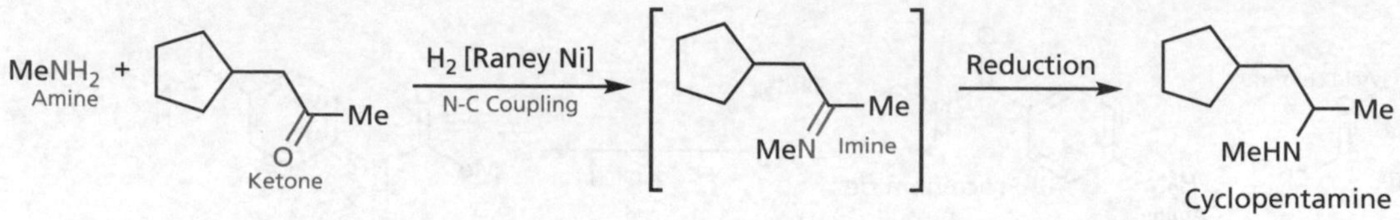

FIGURE 2.19 Synthesis of cyclopentamine.

2.4.3 N–C coupling reactions resulting in an imine linkage

In the previous section, we saw that reacting an amine with a ketone in the presence of a reducing agent gives an imine, which is then reduced to an amine. In the absence of the reducing agent, the imine is obtained. However, there are relatively few drugs containing imine groups compared with those that contain amine groups, and so there are fewer examples of this kind of reaction being used in drug synthesis. One example is the synthesis of the antiviral agent **methisazone** (Fig. 2.20). In this case, a thiosemicarbazide containing a hydrazine group reacts with a ketone in the bicyclic structure to form a thiosemicarbazone. Methisazone was used in the treatment of smallpox before the disease was eradicated as a result of a worldwide inoculation programme.

2.4.4 Syntheses involving consecutive N–C couplings

Syntheses which involve two consecutive N–C couplings are particularly efficient, allowing three different building blocks to be linked together in only two reactions. In order to achieve this, one of the building blocks must contain two groups that are capable of reacting twice, but they must be of different reactivities such that the first step is chemoselective. For example, the synthesis of the antipsychotic agent **acetophenazine** involves a building block containing two halogen substituents, which allows the coupling of two different amines (Fig. 2.21). The success of the sequence relies on the amine reacting selectively with the alkyl bromide rather than the alkyl chloride. This is feasible since alkyl bromides are more reactive because the bromide is a better leaving group than the chloride ion. Note that

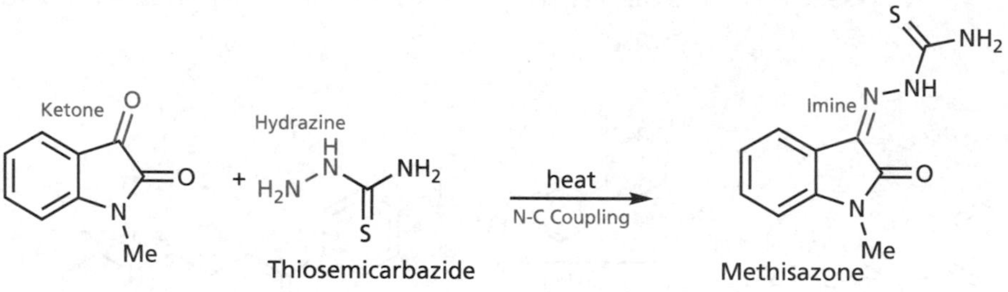

FIGURE 2.20 Synthesis of methisazone.

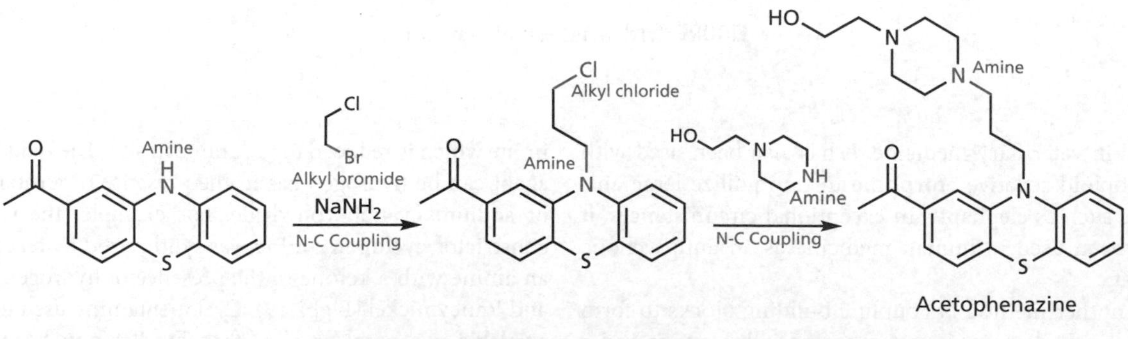

FIGURE 2.21 Synthesis of acetophenazine.

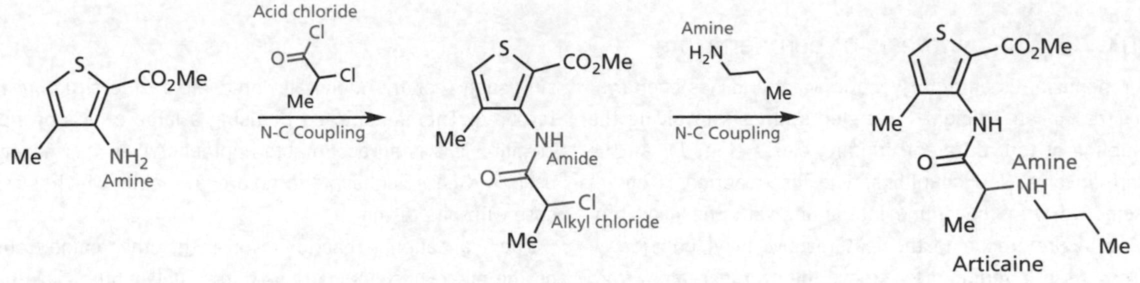

FIGURE 2.22 Synthesis of articaine.

sodium amide is used in this reaction as the amine in the tricyclic ring system is not very nucleophilic. This is because the lone pair of electrons on nitrogen interacts with the pi systems of the neighbouring aromatic rings. The sodium amide base removes the proton from the amine nitrogen to form an anion which is more reactive.

The second coupling reaction can now be carried out whereby a piperazine structure reacts with the alkyl chloride. In this reaction, a strong base is not required because the piperazine nitrogen is more nucleophilic. This reaction is also chemoselective with respect to the piperazine structure since the secondary amine reacts with the alkyl chloride rather than the alcohol. This is because the nitrogen atom of an amine group is more nucleophilic than the oxygen atom of an alcohol group.

Another example involving a building block containing two functional groups of different reactivities is the synthesis of **articaine** (carticaine), which is a local anaesthetic used in dentistry (Fig. 2.22). In this case, the building block contains an alkyl chloride group and an acid chloride. The latter is a more reactive functional group, and so it is possible to carry out the first coupling reaction such that it occurs selectively at the acid chloride to form an amide link. The second amine building block is then introduced to react with the alkyl chloride and

displace the chloride ion in a nucleophilic substitution reaction.

In some syntheses, it has proved possible to couple two different building blocks to a primary amine group in the third building block. For example, the synthesis of the antihistamine agent **chloropyramine** involved the reaction of a primary amine with an aldehyde under reducing conditions to form a secondary amine (Fig. 2.23). The resulting secondary amine was then used to substitute a bromine substituent from a pyridine ring. Chloropyramine acts as an antagonist at H1 histamine receptors and has been used for the treatment of allergies and asthma.

The more consecutive coupling reactions that can be included in a synthesis, the more efficient the synthesis is likely to be. An example of a synthesis involving three consecutive N–C couplings is given in Box 2.2.

2.5 Coupling reactions involving the formation of O–C bonds

Although less common than N–C coupling, O–C coupling is still frequently used in drug syntheses. The most important reactions involve the formation of ethers and esters.

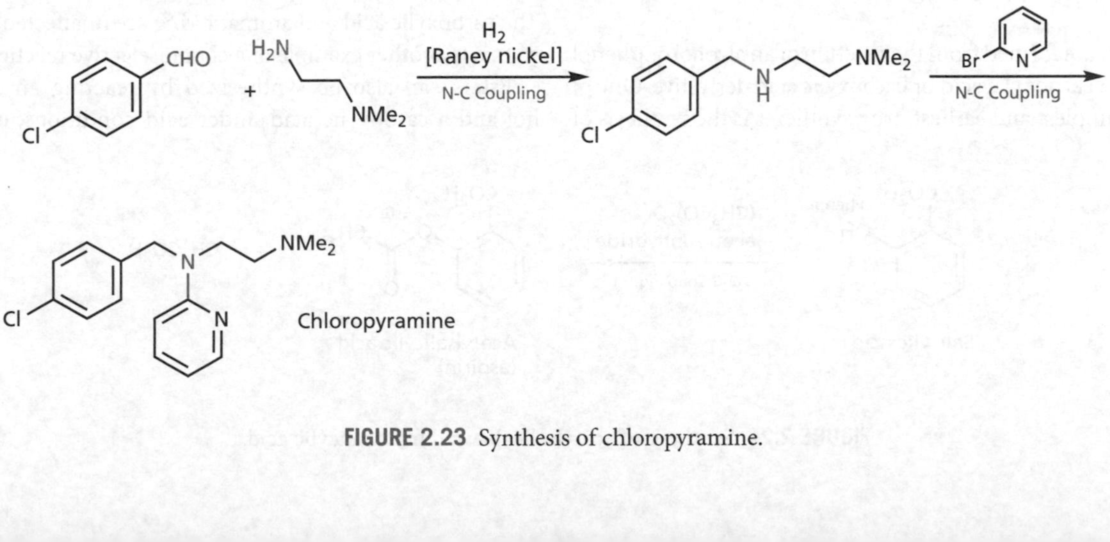

FIGURE 2.23 Synthesis of chloropyramine.

BOX 2.2 The synthesis of perphenazine

Perphenazine is an antipsychotic agent and has been synthesized in an efficient three-step synthesis involving the coupling of four different building blocks (Fig. 1). All the steps involve N–C couplings. The first reaction is chemoselective with the amine interacting with the alkyl bromide in preference to the less reactive alkyl chloride. A strong base is required to remove the hydrogen atom from the amine group in the tricyclic ring system, such that it is a strong enough nucleophile. The second reaction involves nucleophilic substitution of the alkyl chloride with pipera-

zine such that the mono-alkylated piperazine structure is favoured. This is achieved by using a large excess of piperazine. This is an acceptable approach when the reagent used in excess is cheap and readily available, which is the case with piperazine.

The final coupling reaction involves the other amine group on the piperazine ring carrying out a nucleophilic substitution of an alkyl bromide present in the final building block. The reaction is chemoselective for the alkyl bromide over the alcohol group.

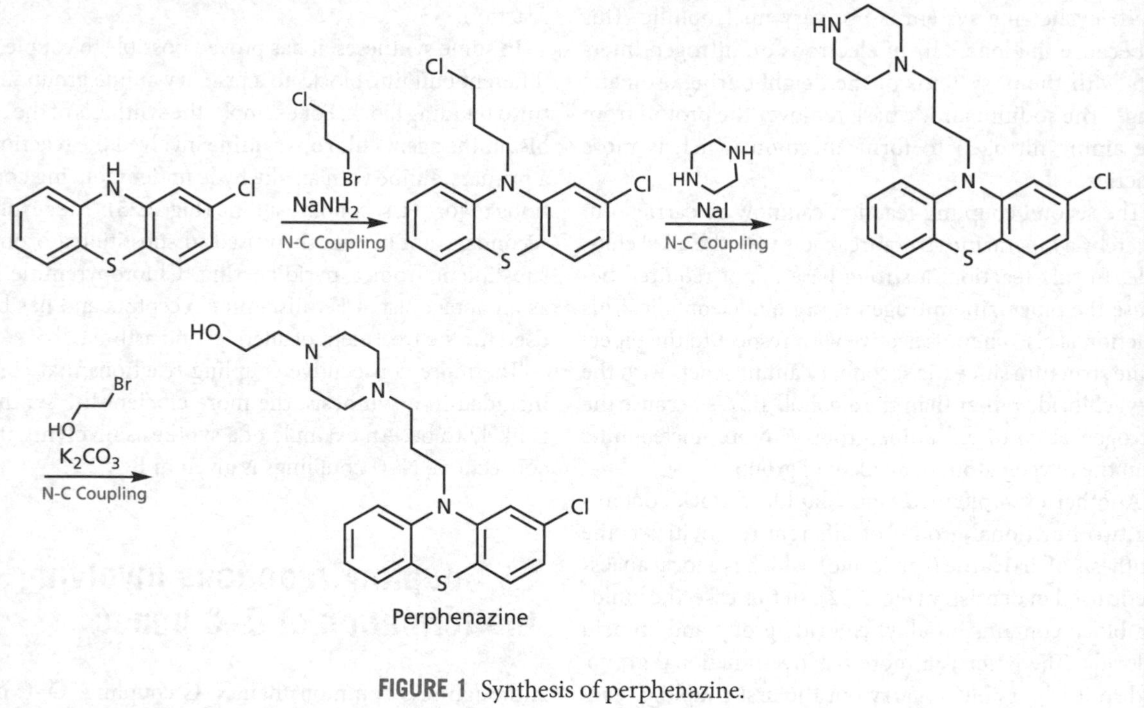

FIGURE 1 Synthesis of perphenazine.

2.5.1 O–C coupling reactions resulting in an ester linkage

Esters are formed from the reaction of an alcohol or phenol with a carboxylic acid or carboxylic acid derivative. One of the simplest and earliest drug syntheses is the synthesis of

acetylsalicylic acid (aspirin), which is easily carried out by acetylating the phenol group of salicylic acid (Fig. 2.24). The carboxylic acid and aromatic ring are unaffected, and so this is another example of a chemoselective reaction.

Esters can also be synthesized by reacting an alcohol and a carboxylic acid under acid conditions, or by

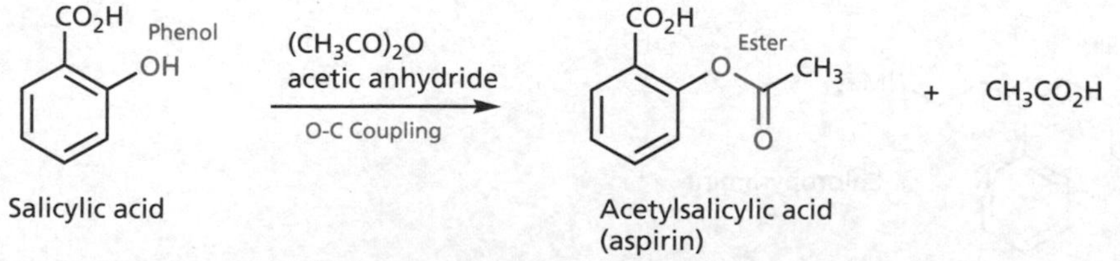

FIGURE 2.24 Synthesis of acetylsalicylic acid from salicylic acid.

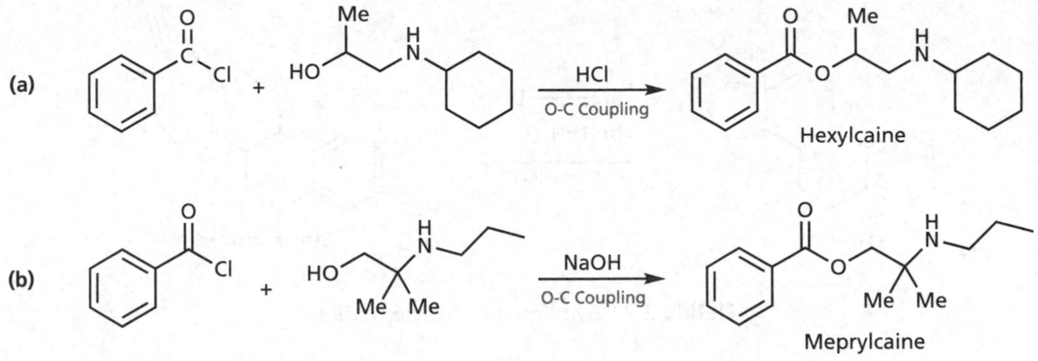

FIGURE 2.25 Synthesis of (a) hexylcaine and (b) meprylcaine.

reacting an alcohol with an acid chloride as illustrated in the syntheses of the local anaesthetic agents **hexylcaine** and **meprylcaine** (Fig. 2.25).

2.5.2 O–C coupling reactions resulting in an ether linkage

Ethers can be formed by reacting an alcohol or a phenol with an alkyl halide. For example, the anticancer agent **etoglucid** was synthesized by reacting a diol with two equivalents of an alkyl chloride (Fig. 2.26). The reaction is chemoselective for the alkyl chloride over the epoxide rings. The latter are important in the drug's mechanism of action as they act as alkylating groups of nucleic acids.

2.6 Coupling reactions involving the formation of C–C bonds

In sections 2.4 and 2.5, we saw how building blocks can be linked together by forming N–C or O–C bonds through the formation of amides, amines, esters, and ethers (see also Appendix 4). These couplings have the advantage of being generally favourable reactions that can be achieved in good yield using relatively mild reaction conditions. However, such reactions are limited in scope since it is not possible to create the more complex carbon skeletons that are present in many drugs. Thus, reactions that allow molecular building blocks to be combined by carbon–carbon bond formations are also important in drug synthesis. Examples of reactions that are commonly used include the Wittig reaction, Friedel–Crafts alkylation, Friedel–Crafts acylation, crossed aldol condensation and palladium-catalysed couplings (Case Study 2), as well as nucleophilic additions to carbonyl groups using Grignard or organolithium reagents (Fig. 2.13). Many other examples are provided in Appendix 5 and the Online Resource Centre.

A simple example of a drug synthesis involving nucleophilic addition to a ketone is the synthesis of the contraceptive **ethynylestradiol** from the naturally occurring female sex hormone **estrone**. The reaction involves the linkage of two building blocks—estrone and an acetylene moiety in the form of an organolithium reagent. The ketone group of estrone reacts with the organolithium agent to give the new C–C bond and formation of a tertiary alcohol group (Fig. 2.27). The presence of a tertiary alcohol group in ethynylestradiol is a 'signature' that is characteristic of the nucleophilic addition of a carbanion to a ketone. Therefore, a C–C coupling reaction involving a ketone with an organolithium reagent or Grignard reagent is always worth considering if there is a tertiary alcohol present in the structure of a drug.

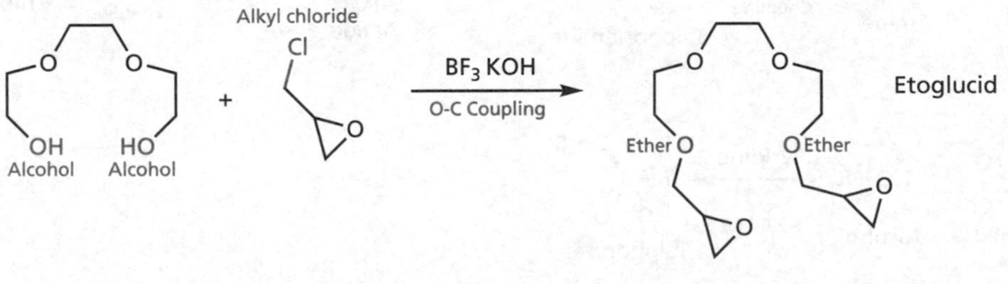

FIGURE 2.26 Synthesis of etoglucid.

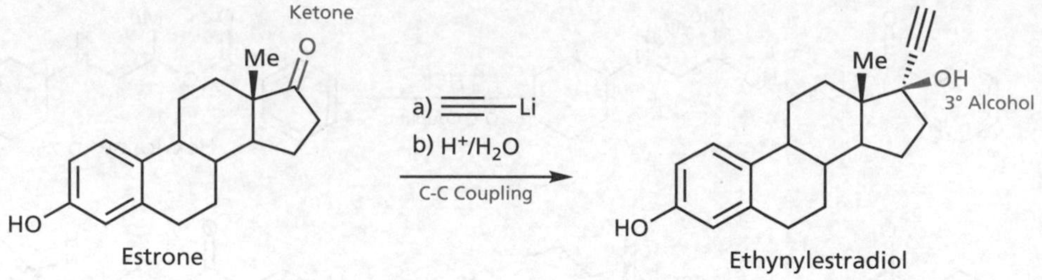

FIGURE 2.27 Synthesis of ethynylestradiol.

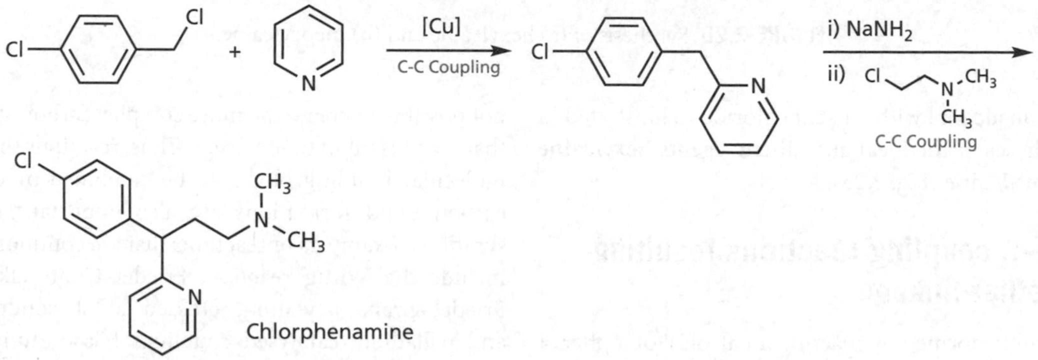

FIGURE 2.28 Synthesis of chlorphenamine.

The synthesis of the antihistamine **chlorphenamine** involves three building blocks which are linked together through two C–C coupling reactions (Fig. 2.28). Chlorphenamine is used in the treatment of allergies and hay fever.

C–C coupling reactions may or may not result in a distinctive functional group in the product. We saw that the reaction of an organolithium reagent with a ketone resulted in a characteristic tertiary alcohol group in ethynylestradiol. However, the C–C couplings involved in the synthesis of chlorphenamine do not produce any recognizable 'signatures' to indicate that these reactions

were used. Thus, it is often difficult to determine from the structure of a drug whether a C–C coupling can be used in its synthesis.

2.7 **Other types of coupling reaction**

N–C, O–C, and C–C couplings are the most common types of couplings carried out in drug synthesis, but there are instances of other types of coupling reaction such as N–S, O–S, and S–C couplings (Fig. 2.29). These result in

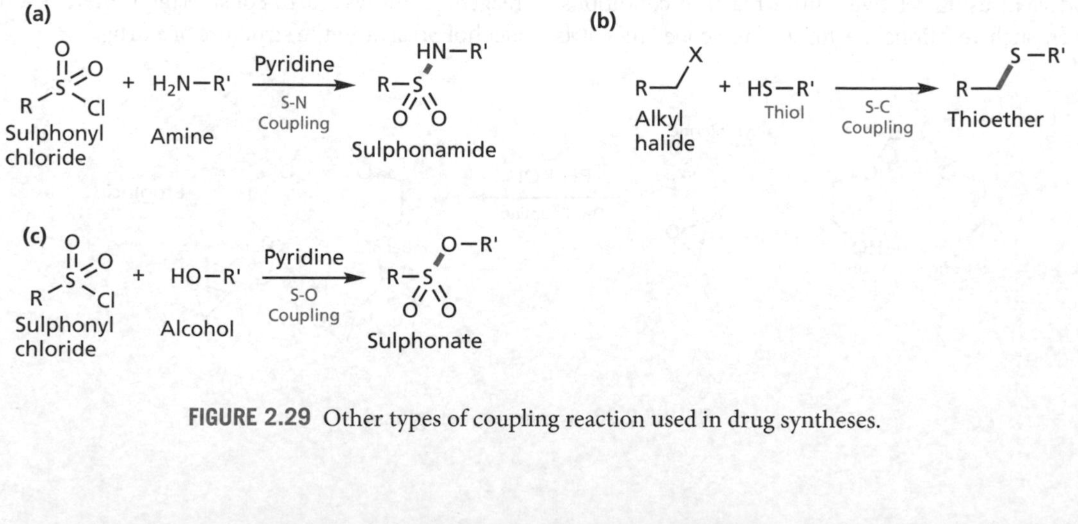

FIGURE 2.29 Other types of coupling reaction used in drug syntheses.

functional groups such as sulphonamides, sulphonates, and sulphides, respectively.

For example, the synthesis of the anticancer agent **busulfan** involves O–S couplings between an alcohol and methanesulphonyl chloride to produce two sulphonate groups (Fig. 2.30). These act as good leaving groups, allowing busulfan to act as an alkylating agent with DNA in tumour cells. An example of an N–S coupling to produce a sulphonamide group can be seen in the synthesis of the cardiovascular drug **ibutilide** (Box 2.3).

2.8 Syntheses involving different consecutive coupling reactions

We have already seen examples of efficient syntheses where consecutive N–C or C–C couplings have been used. It is also possible to carry out syntheses where different types of couplings are used consecutively. For example, the synthesis of the beta-blocker **propranolol** involves three molecular building blocks which are

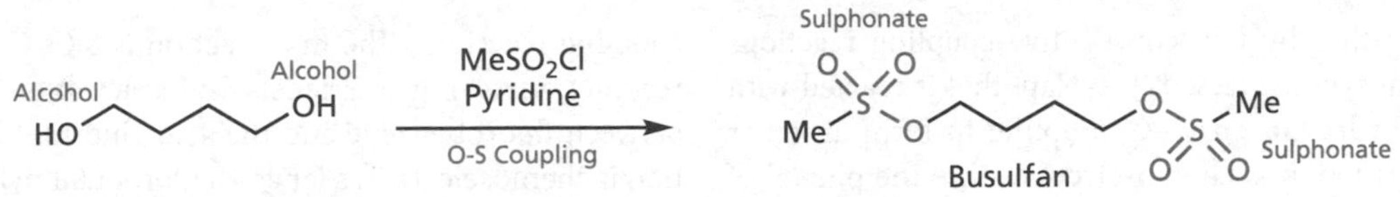

FIGURE 2.30 Synthesis of busulfan.

BOX 2.3 Synthesis of ibutilide

Ibutilide is a cardiovascular drug that is used to control irregular heart rhythms. Therefore, it is known as an anti-arrhythmic agent. The synthesis comprises three consecutive coupling reactions involving four building blocks (Fig. 1).

The first coupling reaction involves the reaction of aniline with methanesulphonyl chloride to form a sulphonamide functional group. The second stage is a C–C coupling reaction involving a Friedel–Crafts acylation with succinic anhydride. This is regioselective for the *para* position of the aromatic ring since the sulphonamide group is *ortho*, *para* directing. *Para* substitution is preferred over *ortho* substitution since the

latter is sterically hindered by the size of the sulphonamide group. Since a cyclic anhydride is used in this coupling reaction, the reaction reveals a carboxylic acid group which can now be used for the third coupling reaction. This is an N–C coupling between the carboxylic acid and the amine group of the final building block to form an amide, aided by the coupling agent dicyclohexylcarbodiimide (DCC). The ketone and amide groups that resulted from the second and third coupling reactions are not present in the final structure, but they can be easily reduced with lithium aluminium hydride to form the secondary alcohol and tertiary amine which are.

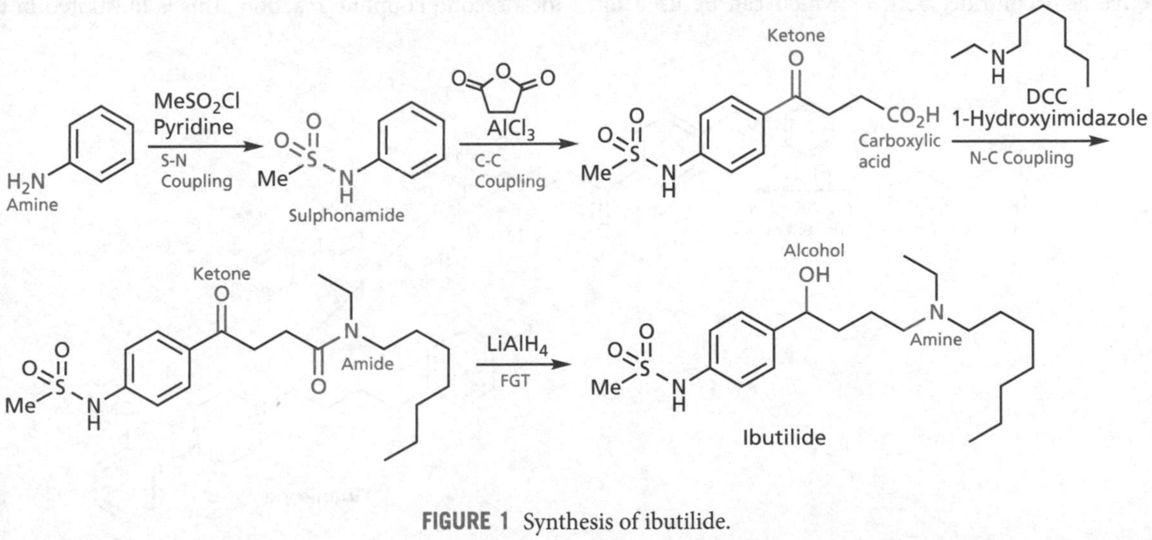

FIGURE 1 Synthesis of ibutilide.

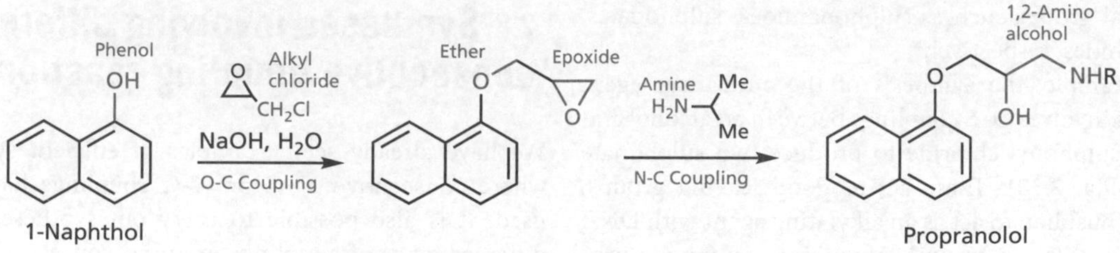

FIGURE 2.31 Synthesis of propranolol.

linked together by two consecutive coupling reactions of different types (Fig. 2.31). 1-Naphthol is reacted with epichlorohydrin in an O–C coupling to form an ether link. The reaction is chemoselective since the phenol of 1-naphthol reacts with the alkyl chloride rather than the epoxide. The resulting epoxide product is then treated with 2-aminopropane in an N–C coupling reaction. As well as coupling the third building block to form an amine, ring opening of the epoxide reveals the secondary alcohol group required in the final structure. Therefore, this reaction achieves two synthetic goals.

The key to the effective synthesis of propranolol lies in the choice of the epichlorohydrin building block, which contains two functional groups capable of undergoing coupling reactions, one of which is more reactive than the other. The alkyl chloride is more reactive than the epoxide, allowing the first reaction to be chemoselective and leaving the epoxide group available for the second coupling reaction.

A similar approach is illustrated in the synthesis of the antipsychotic agent **fluanisone** (Fig. 2.32). The key building block in this synthesis is 4-chlorobutanoyl chloride which contains an alkyl chloride group plus a more reactive acid chloride, both of which can be used for

coupling reactions. The first reaction is a C–C coupling reaction involving a Friedel–Crafts acylation reaction between fluorobenzene and the acid chloride. The reaction is chemoselective as far as 4-chlorobutanoyl chloride is concerned because the acid chloride is more reactive than the alkyl chloride. Therefore, the Friedel–Crafts acylation reaction is preferred over Friedel–Crafts alkylation. The reaction is also regioselective as far as the fluorobenzene is concerned, with the reaction occurring selectively at the *para* position of the aromatic ring. This is because the fluoro substituent directs electrophilic substitution to the *ortho* and *para* positions, rather than the *meta* position. Because of the directing effect of the fluorine substituent, a certain amount of *ortho*-disubstituted product is expected which has to be separated from the desired *para*-disubstituted product. An N–C coupling reaction can now be carried out by reacting the surviving alkyl chloride with an amine.

The preceding syntheses make use of a building block containing two different functional groups having different reactivities in order to carry out the consecutive coupling reactions. A different approach is to use a coupling reaction that creates the functional group required for a second coupling reaction. This is illustrated in the

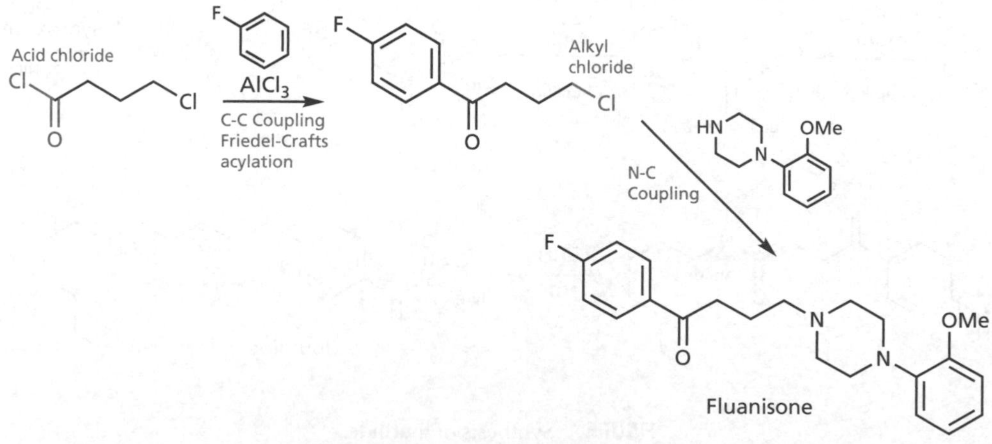

FIGURE 2.32 Synthesis of fluanisone.

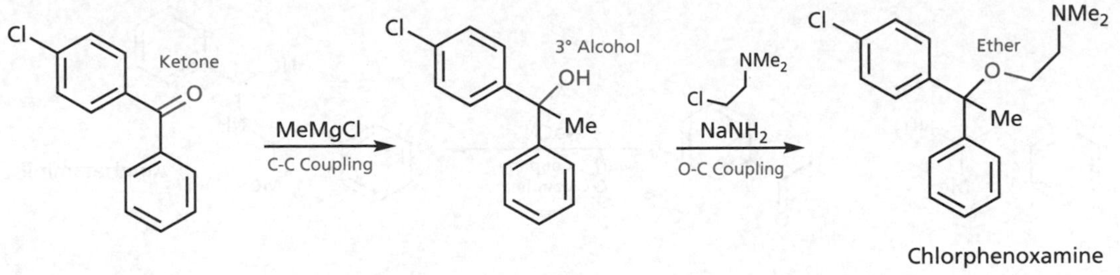

FIGURE 2.33 Synthesis of chlorphenoxamine.

synthesis of **chlorphenoxamine** which is an antihistamine used to treat itching. It also has anticholinergic activity and is used in the treatment of Parkinson's disease. The first stage of the synthesis is a C–C coupling involving a Grignard reaction (Fig. 2.33). A tertiary alcohol is formed as a result of the coupling, which can then be reacted with an alkyl chloride to form an ether in an O–C coupling.

An example of a synthesis involving three consecutive coupling reactions of different types can be seen in the synthesis of lapatinib (Case Study 2).

2.9 Syntheses involving two coupling reactions in one step

Reactions can be particularly useful if they result in two coupling reactions in one synthetic step. An interesting reaction that can couple three building blocks in the same reaction is the **Mannich reaction**. A C–C coupling reaction and an N–C coupling are both involved. This

is illustrated in the synthesis of the antipsychotic agent **molindone** from the reaction of a bicyclic starting material with morpholine and paraformaldehyde (Fig. 2.34).

Another example is the **Strecker reaction** where the reaction of a ketone with an amine in the presence of hydrogen cyanide or potassium cyanide results in a product that involves an N–C coupling with the amine plus a C–C coupling with the nitrile group (Fig. 2.35). The nitrile group can then be elaborated to introduce other functional groups or for further C–C couplings.

The Strecker reaction was used to synthesize **amphetaminil** (Fig. 2.36), which is a stimulant belonging to the amphetamine family and was formerly used for the treatment of obesity, ADHD, and narcolepsy. It is no longer recommended because of problems related to drug abuse.

The reaction has also been very useful for the synthesis of amino acids since the nitrile group can easily be hydrolysed to a carboxylic acid. The synthesis of **methyldopa** takes advantage of this approach (Fig. 2.37). Methyldopa is an adrenergic agonist that has been used to treat hypertension.

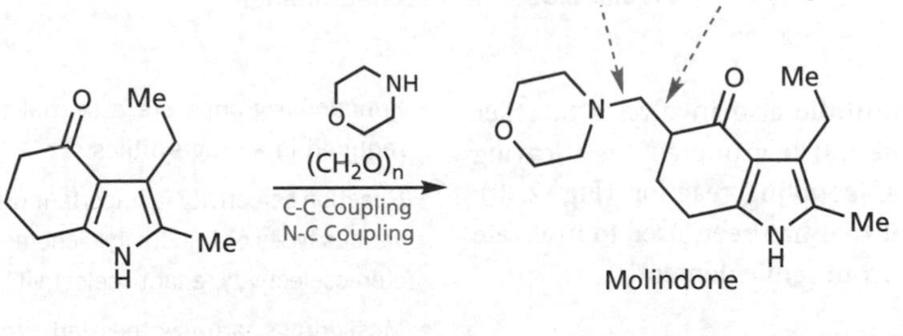

FIGURE 2.34 Synthesis of molindone.

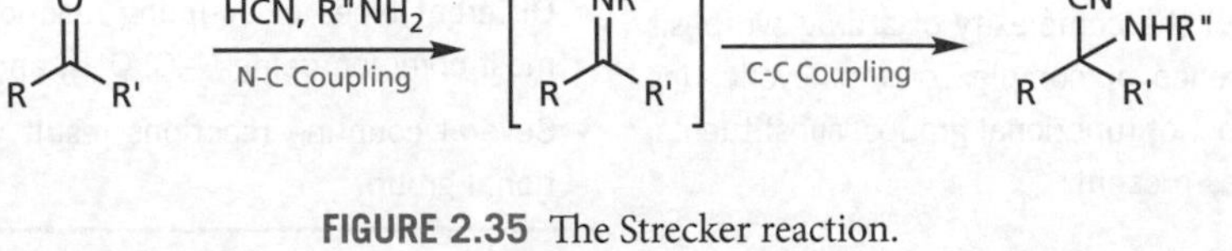

FIGURE 2.35 The Strecker reaction.

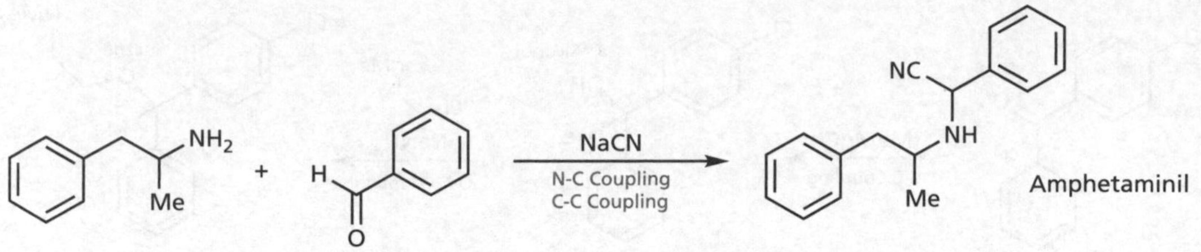

FIGURE 2.36 Synthesis of amphetaminil.

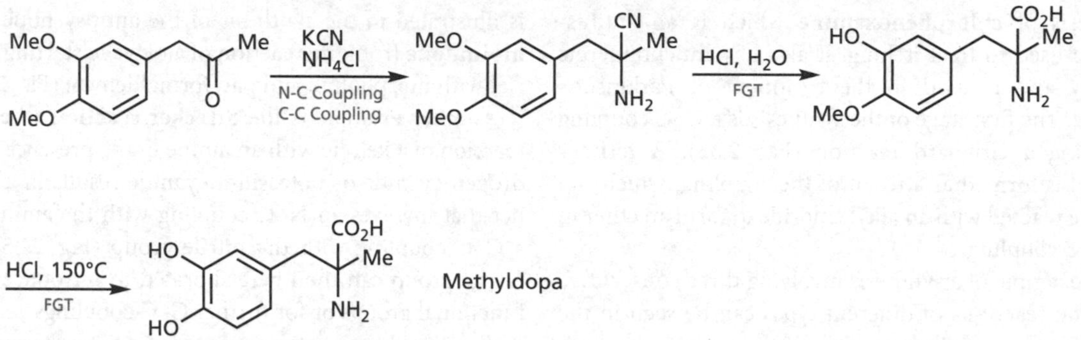

FIGURE 2.37 Synthesis of methyldopa.

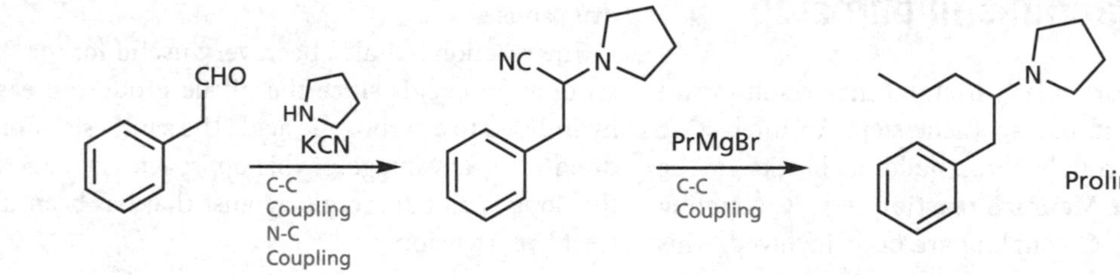

FIGURE 2.38 Synthesis of prolintane.

The synthesis of **prolintane** also involves a Strecker reaction. In this case, the nitrile group acts as a leaving group for a further C–C coupling reaction (Fig. 2.38). Prolintane is a stimulant that has been used to motivate elderly people suffering from senile dementia.

KEY POINTS

- Organic synthesis is important at various stages of the drug design and development programme.

- The features that affect the complexity of a drug synthesis are chirality, the presence of complex or multicyclic ring systems, and the number of functional groups, substituents, and asymmetric centres present.

- Functional groups are essential to carry out the reactions required in a drug synthesis.

- Reaction selectivity is important during a synthesis. There are various types of selectivity—chemoselectivity, regioselectivity, stereoselectivity, enantioselectivity, and diastereoselectivity.

- Most drugs are synthesized from simpler commercially available molecules that act as building blocks.

- Coupling reactions are used to link molecular building blocks together.

- Different types of coupling reaction are possible, with the most common being N–C, O–C, and C–C couplings.

- Several coupling reactions result in a characteristic functional group.

2.10 Functional group transformations

Up until now, we have considered drug syntheses that involve coupling reactions. These are highly efficient syntheses, but it is not always possible to carry out the synthesis of a drug purely by carrying out coupling reactions. In fact, it is usually the case that other reactions are required involving the modification of one functional group into another. We shall now consider examples of these.

2.10.1 Introducing a functional group in the final product

Ideally, all the functional groups required in the final product will be present in the building blocks used in the synthesis, or formed as a result of the coupling reactions. However, this is not always possible and it may be necessary to modify a functional group to introduce the functional group required in the final structure.

For example, the synthesis of the anticancer agent **uramustine** involves an N–C coupling reaction between 5-aminouracil and two equivalents of oxirane to give a diol (Fig. 2.39). A functional group transformation is now required to convert the diol to alkyl chlorides. The alkyl chlorides are essential to the mechanism of action of uramustine since the drug acts as an alkylating agent

of DNA in tumour cells. The drug is used in the treatment of lymphomas.

Similarly, the muscle relaxant **cyclobenzaprine** is synthesized using a coupling reaction that involves a Grignard reagent and formation of a tertiary alcohol (Fig. 2.40). This has to be dehydrated in order to obtain the required alkene group in the final structure. Cyclobenzaprine is used to treat muscle pain and injury.

Another example involves the synthesis of ibutilide where a ketone group is converted to an alcohol in the final stage (Box 2.3).

The preceding examples involve transforming a functional group that is formed as a result of a coupling reaction. Another possible scenario is where a functional group transformation involves a functional group that was present in one of the original building blocks. For example, the synthesis of the local anaesthetic **procaine** starts from 4-nitrobenzoyl chloride and involves two coupling reactions before the nitro group is reduced to an amino group (Fig. 2.41). We might ask why a starting material that already contains the amino substituent could not be used, or why the nitro group is not reduced to the amine at an earlier stage. The answer lies in the reactivity of the amino group. An amine is a good nucleophile and so it would undergo unwanted reactions with electrophilic functional groups during the synthesis, such as the acid chloride or the alkyl chloride used in the first stage of the synthesis. In contrast, a nitro group

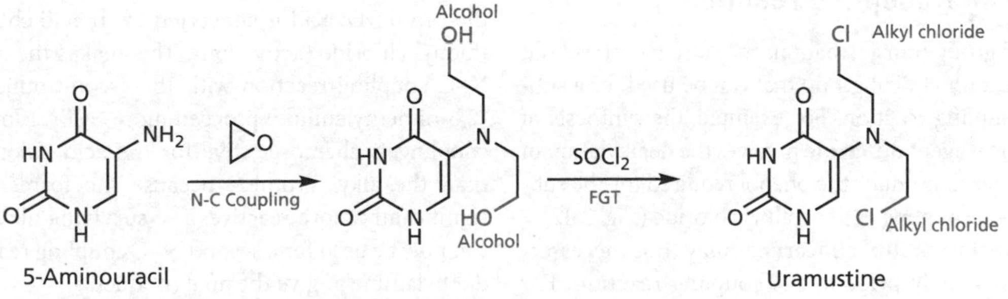

FIGURE 2.39 Synthesis of uramustine (uracil mustard).

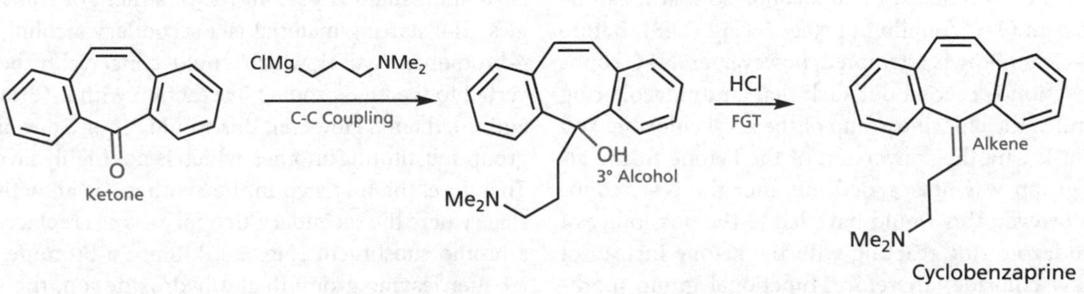

FIGURE 2.40 Synthesis of cyclobenzaprine.

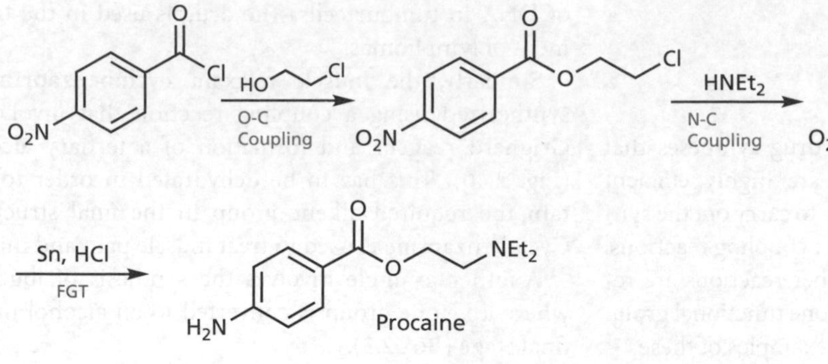

FIGURE 2.41 Synthesis of procaine.

is relatively unreactive to most reagents used in organic synthesis. For that reason, an aromatic nitro group is frequently used as a **latent group** for an aromatic amine. The conversion of the nitro group to the amine is carried out by reduction and is frequently the final stage of a synthesis. Another example of a nitro group being used in this manner can be found in the synthesis of mirabegron (see Fig. 2.52).

The functional group transformation required to introduce an important functional group does not necessarily have to be at the end of the synthetic sequence. For example, the synthesis of nadolol includes the transformation of an alkene to an important diol at an early part of the synthesis (Box 2.4).

2.10.2 Introducing a functional group for a further coupling reaction

Functional group transformations may also be carried out to produce a functional group that can be used for a subsequent coupling reaction. For example, the synthesis of the antifungal agent **dimazole** involves the dealkylation of an ethyl ether to produce the phenol required for the subsequent O–C coupling with an alkyl chloride (Fig. 2.42).

The transformation concerned may not necessarily be immediately prior to the coupling reaction. For example, the starting material used in the synthesis of the antifungal agent **fenticonazole** contains a ketone group which is reduced to an alcohol, so that it can be used for an O–C coupling in stage 3 (Fig. 2.43). Before the O–C coupling is attempted, however, an N–C coupling reaction is carried out such that an imidazole ring substitutes the chlorine group of the alkyl chloride. You might ask why the conversion of the ketone to the alcohol group was not carried out after the N–C coupling. However, this would have led to the possibility of the imidazole ring reacting with the ketone instead of the alkyl chloride. Therefore, functional group modifications designed to produce a functional group for a

subsequent coupling reaction may occur one or several steps before they are needed.

2.10.3 Activating a functional group

Some functional groups might be suitable for a coupling reaction, but the reaction will proceed more efficiently if they are converted to a more reactive functional group (activation). For example, a carboxylic acid can be used in N–C and O–C couplings to produce an amide and an ester link, respectively, but the reaction proceeds more easily if the carboxylic acid is converted to an acid chloride first.

This is illustrated in the synthesis of the local anaesthetic **etidocaine** (Fig. 2.44). Etidocaine has a long duration of action and is used in surgical operations and childbirth. The synthesis starts with 2-bromobutanoic acid, and the carboxylic acid is converted to an acid chloride with thionyl chloride (activation). This makes the subsequent N–C coupling reaction with the second building block (2,6-dimethylaniline) proceed more easily. Moreover, the coupling is chemoselective for the acid chloride rather than the alkyl bromide because the former group is significantly more reactive. The surviving alkyl bromide can now be used for a second N–C coupling reaction with diethylamine to give the final product.

A second example of activation can be seen in the synthesis of **bromazine** (bromodiphenhydramine) which is an antihistamine used in the treatment of some allergies. The starting material is a secondary alcohol called 4-bromobenzhydrol which could conceivably be converted to the final product by reaction with 2-(dimethylamino)ethanol. However, this would require one alcohol group substituting another, which is not highly favoured. Therefore, the first step in the synthesis is an activation stage where the secondary alcohol group is replaced with a bromo substituent (Fig. 2.45). Since a bromide ion is a better leaving group than a hydroxide ion, the subsequent substitution reaction proceeds far more easily.

BOX 2.4 Synthesis of nadolol

Nadolol is a non-selective beta-blocker that acts as an antagonist at β-adrenoceptors. It is used to treat high blood pressure and chest pain. The synthesis involves three building blocks which are combined together in the final two steps of the synthesis (Fig. 1). The first three steps of the synthesis are carried out in order to introduce the diol group that is needed in the final structure.

The starting material is a dihydronaphthol which lacks the diol functionality required in the final structure. Before introducing this, the phenol group is protected with an acetyl group (stage 1). The alkene is then converted into the diol (stage 2), followed by deprotection of the phenol group (stage 3). An O–C coupling reaction is now carried out which shows

chemoselectivity for both building blocks. As far as the epichlorohydrin is concerned, the reaction occurs at the alkyl chloride, leaving the epoxide untouched. This selectivity was seen in the synthesis of propranolol (Fig. 2.31). As far as the diol intermediate is concerned, the reaction occurs at the phenol group rather than the diol. This is because sodium methoxide was added to remove the slightly acidic proton of the phenol group. The phenoxide oxygen that is formed then proves more nucleophilic than the un-ionized diol.

The final stage is an N–C coupling where the third building block (*t*-butylamine) reacts with the epoxide. This reaction proceeds with regioselectivity, since the amine reacts at the less substituted carbon of the epoxide.

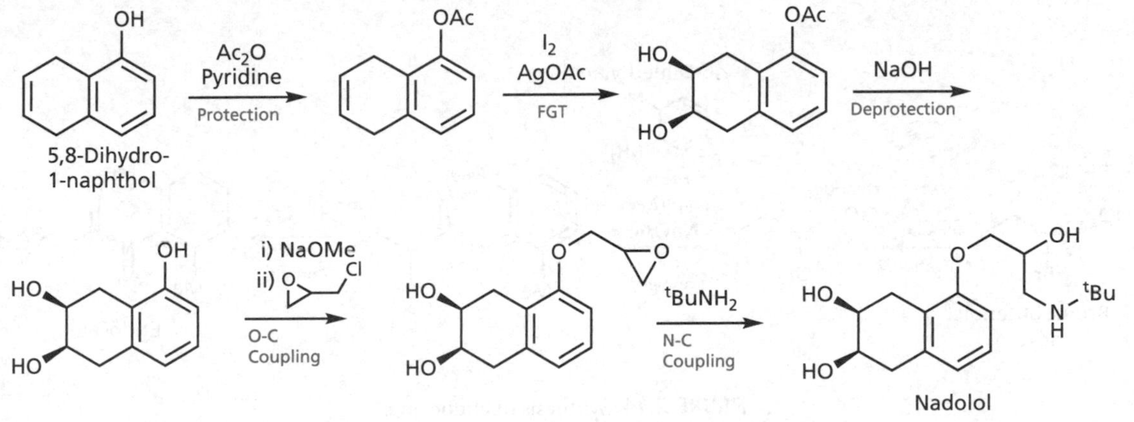

FIGURE 1 Synthesis of nadolol.

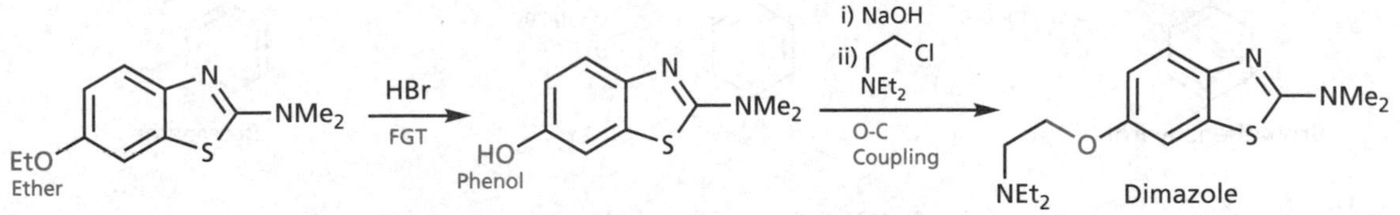

FIGURE 2.42 Synthesis of dimazole.

2.11 Functionalization and functional group removal

2.11.1 Functionalization

Some readily available starting materials may have the correct molecular skeleton for a particular drug target,

but lack a functional group at a key position. The functional group required might be needed in the final structure, or in order to carry out a coupling reaction. In such cases, it is necessary to introduce that functional group (**functionalization**) (see also Appendix 2). The functional group that is introduced does not have to be the one that is required in the final structure, as long as it is possible to transform it to the desired functional group

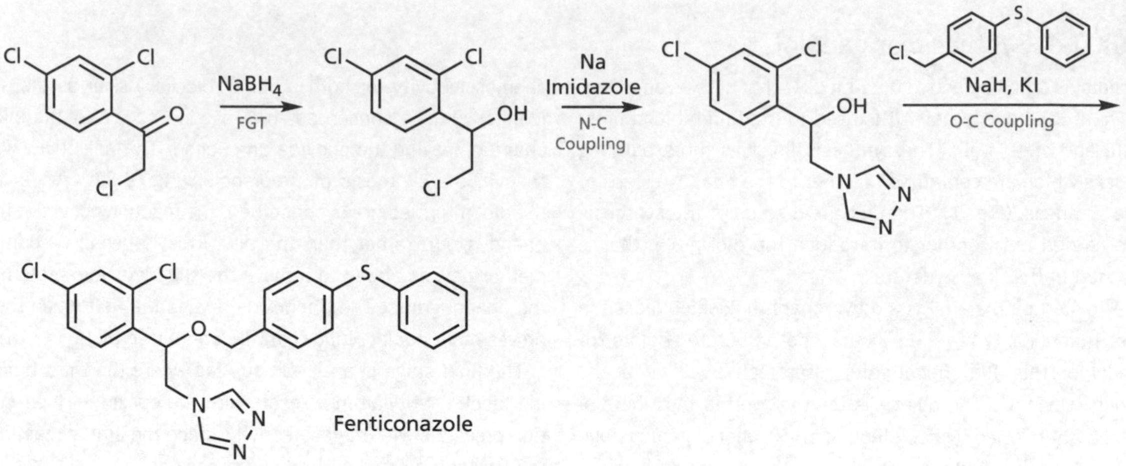

FIGURE 2.43 Synthesis of fenticonazole.

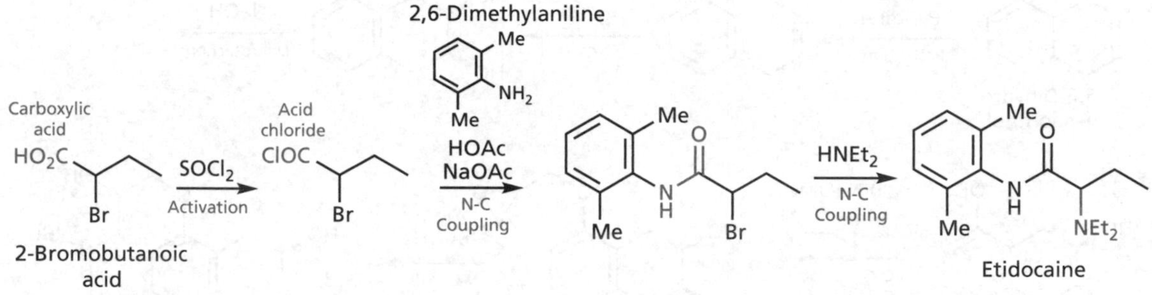

FIGURE 2.44 Synthesis of etidocaine.

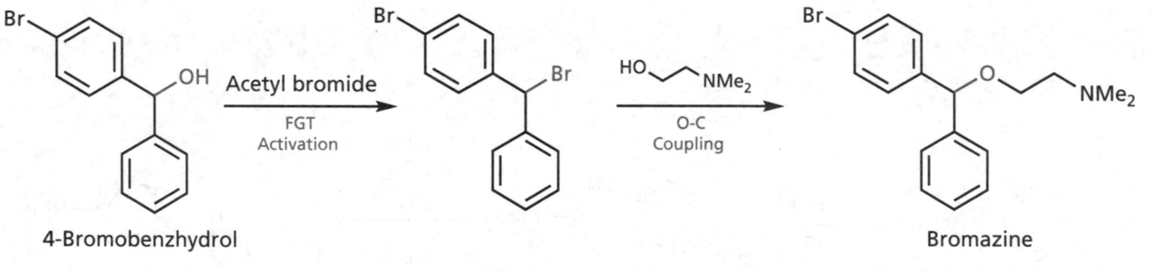

FIGURE 2.45 Synthesis of bromazine.

at a later stage. For example, **benzocaine** contains amine and ester functional groups at opposite ends of the ring (the *para* positions). Toluene is a common solvent that has the same carbon skeleton as benzocaine, but lacks the functional groups. Therefore, it is necessary to introduce functional groups at the *para* positions and then convert them to the desired amine and ester in the final product. This can be done using a four-step synthesis (Fig. 2.46). An electrophilic substitution reaction introduces a nitro group at the *para* position of toluene to give

p-nitrotoluene, which is then treated with hydrogen over a palladium charcoal catalyst to reduce the nitro group to the desired amine (FGT). The resulting *p*-tolylamine is then treated with potassium permanganate to oxidize the methyl group to a carboxylic acid (4-aminobenzoic acid), which is then esterified to the ethyl ester required in benzocaine. Thus, the synthetic route involves two functionalization reactions, a functional group transformation, and an O–C coupling. We could reasonably ask why it was not possible to introduce the NH_2 directly to the ring

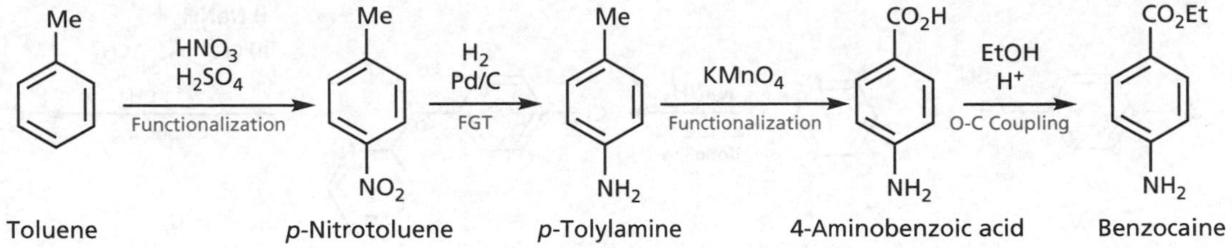

FIGURE 2.46 Synthesis of benzocaine.

to cut down the number of reactions involved. Quite simply, there is no reaction that can achieve that. Similarly, there is no reaction that can convert a methyl group directly to an ester.

We might also ask why the reaction steps were carried out in the way they were. For example, could the reaction be carried out by oxidizing the methyl group first, and then nitrating the ring (Fig. 2.47)?

In fact, carrying out the reactions in this order will result in an extremely poor yield because of the nitration step. The carboxylic acid group acts as a deactivating group for electrophilic substitution since the group withdraws electrons from the aromatic ring and decreases its nucleophilic strength. Therefore, the yield for the nitration will be lower compared with the equivalent step in Figure 2.42 where a methyl group is present for the nitration step. Alkyl groups act as activating groups for electrophilic substitution by 'pushing' electrons into the ring and increasing its nucleophilic strength.

Even more significant is the influence that these two substituents have on the position of electrophilic substitution. A methyl group will direct substitution to the *ortho* and *para* positions, whereas a carboxylic acid directs electrophilic substitution to the *meta* position (Fig. 2.48).

Therefore, even though the yield of *p*-nitrotoluene will be reduced as a result of competing *ortho* substitution, it will still be better than the yield obtained by nitrating benzoic acid, because the predominant product will be the *meta* isomer. To conclude, electrophilic substitution is regioselective for the *ortho* and *para* positions of an aromatic ring when a methyl group is present, but regioselective for the *meta* position when a carboxylic acid substituent is present.

2.11.2 Functional group removal

Instead of functionalizing a particular position within a molecular skeleton, it may be necessary to remove a functional group if it is not present in the final product.

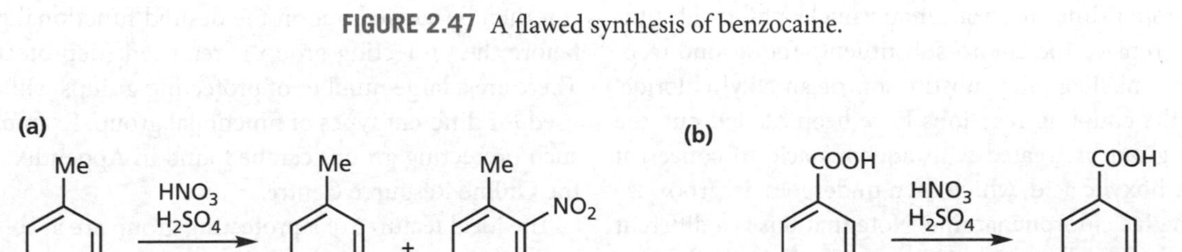

FIGURE 2.47 A flawed synthesis of benzocaine.

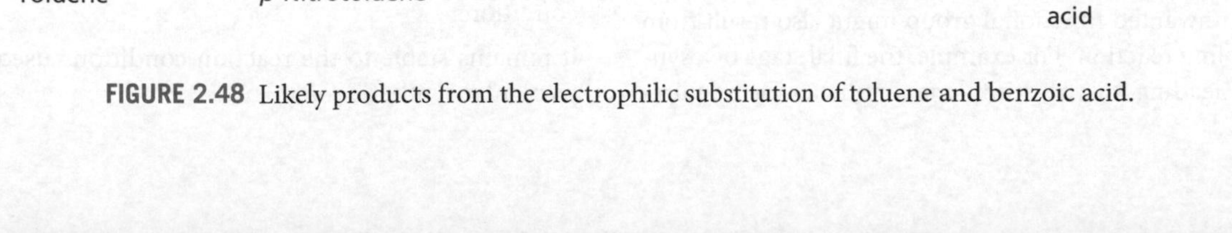

FIGURE 2.48 Likely products from the electrophilic substitution of toluene and benzoic acid.

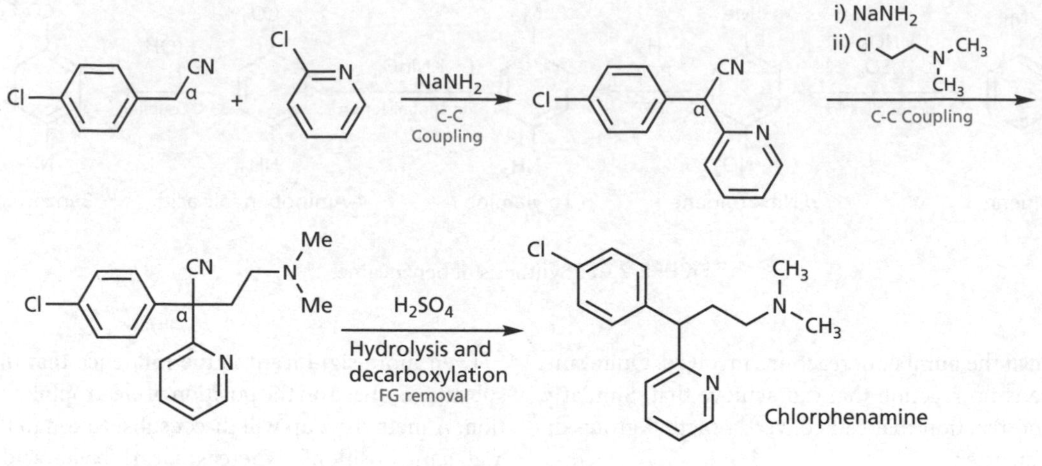

FIGURE 2.49 Synthesis of chlorphenamine.

Such a functional group might be present because of the building block that was used in the synthesis. For example, a nitrile was used as one of the building blocks in a synthesis leading to the antihistamine agent **chlorphenamine** (Fig. 2.49). The nitrile group is not present in the final structure and so the group has to be removed at the final stage of the reaction. But why was the nitrile group there in the first place? Could the synthesis not have been carried out using a starting material that lacked the nitrile group, and thus avoid the need for the final step? The answer is that the nitrile group is crucial to carrying out the C–C couplings. A nitrile group has a strong electron-withdrawing effect which is felt most acutely on the carbon atom to which it is attached (the α-carbon). As a result, the protons on this carbon are slightly acidic and can be removed by a strong base to form a carbanion. This is what happens in the two C–C coupling reactions. In each case, a strong base (sodium amide) is used to remove one of the protons from the α-carbon to form a carbanion, which is then reacted with a building block containing an electrophilic group. In the first coupling reaction, the carbanion reacts with 2-chloropyridine in an aromatic nucleophilic substitution to replace the chloro substituent. The second reaction is a nucleophilic substitution of an alkyl chloride. Once the coupling reactions have been carried out, the nitrile group is treated with aqueous acid to convert it to a carboxylic acid, which then undergoes decarboxylation to give chlorphenamine. Note that this is a different method of producing chlorphenamine from the synthesis shown in Figure 2.28. It is often the case that different synthetic routes can be used to produce a particular target compound.

An unwanted functional group might also result from a coupling reaction. For example, the final stage of a synthesis leading to the stimulant amphetamine requires the removal of an aromatic ketone group. This group is present as a result of an earlier C–C coupling reaction (Box 2.5). Examples of reactions that remove a functional group entirely are given in Appendix 3.

2.12 Protection and deprotection

In the previous sections, we have described coupling reactions which showed chemoselectivity between different functional groups. This is possible when one functional group is more reactive than the others. However, there are situations where it may be necessary to carry out a coupling reaction at a functional group which is less reactive than a competing group within the structure. This would result in mixtures of products, difficult purification procedures, and poor yields. In such cases, it may be necessary to 'disguise' or protect the competing group such that it will not react. This is achieved by carrying out a reaction that transforms the group concerned into a less reactive functional group. The coupling reaction can then be carried out on the desired functional group before the protecting group is removed (deprotection). There are a large number of protecting groups which are used for different types of functional group. Examples of such protecting groups can be found in Appendix 6 and the Online Resource Centre.

The ideal features of a protecting group are as follows.

• It can be easily added to the target functional group in high yield and under mild conditions.

• It can easily be removed in high yield and under mild conditions.

• It remains stable to the reaction conditions used in a particular synthetic route.

BOX 2.5 Synthesis of amphetamine

Amphetamine is a stimulant which has a relatively simple skeleton and can be synthesized from three building blocks by a three-stage process involving C–C coupling, N–C coupling, and removal of a functional group. The building blocks are a catechol ring, 2-chloroethanoyl chloride, and methylamine (Fig. 1).

The first stage is an example of a Fries rearrangement where the acid chloride initially reacts with a phenol group

to form an ester (Fig. 2). Rearrangement then occurs to create a new C–C bond and an aromatic ketone. The reaction is chemoselective in the sense that the alkyl chloride is unaffected by the reaction and is available for a subsequent N–C coupling reaction with methylamine. However, the ketone group resulting from the Fries reaction has to be removed in order to get the final product. This is achieved by a reduction using sodium amalgam.

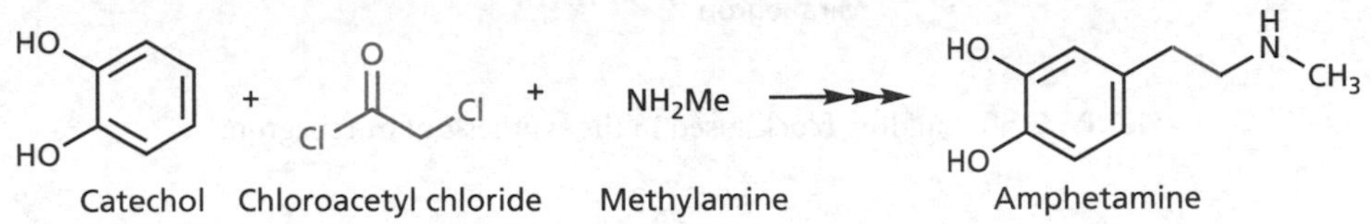

FIGURE 1 Building blocks for amphetamine.

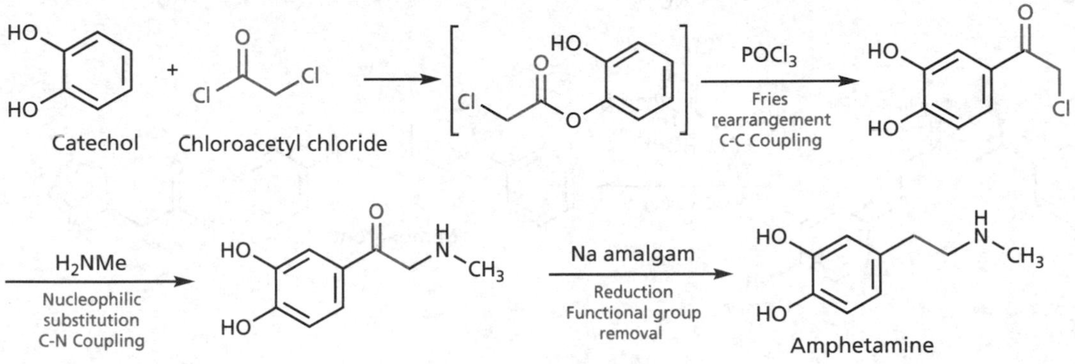

FIGURE 2 Synthesis of amphetamine.

- It can be removed under conditions that are different from those used to remove other types of protecting group—a property known as **orthogonality**.
- It is also useful to have a range of different protecting groups for the same type of functional group, such that they can be removed under different conditions. This allows selective deprotection of similar functional groups.

A protecting group was used effectively in the synthesis of **mirabegron**, which is an adrenergic agonist approved in 2012 for the treatment of overactive bladders. It works by acting as a muscle relaxant to increase the capacity of the bladder. Three building blocks were used in its synthesis—an epoxide, a primary amine, and a carboxylic acid (Fig. 2.50).

A possible synthesis which would link these building blocks is shown in Figure 2.51. However, there is a potential problem with the final coupling stage since the reaction could occur at the primary or the secondary amine. In this case, it is difficult to be sure what the relative reactivities of the primary and secondary amines would be. The primary amine is certainly less sterically hindered, but it is also an aromatic amine. Aromatic amines tend to be less reactive than aliphatic amines since the nitrogen's lone pair of electrons can interact with the pi system of the aromatic ring, making the nitrogen atom less nucleophilic.

It is not revealed whether the synthesis shown in Figure 2.51 was attempted, since the research team involved chose to protect the secondary amine to block any possibility of a competing reaction (Fig. 2.52).

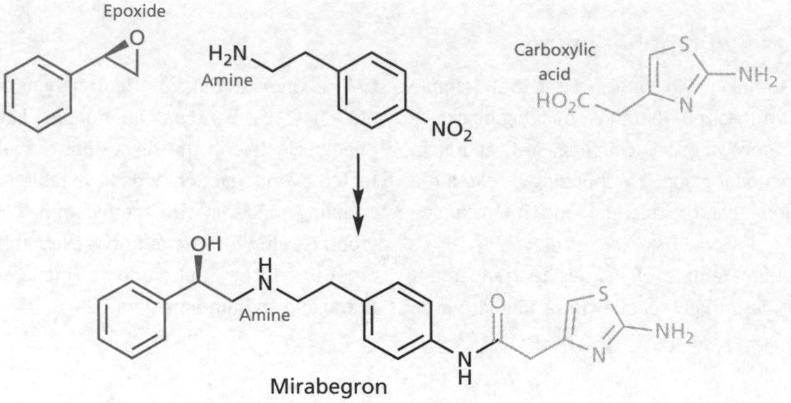

FIGURE 2.50 Building blocks used in the synthesis of mirabegron.

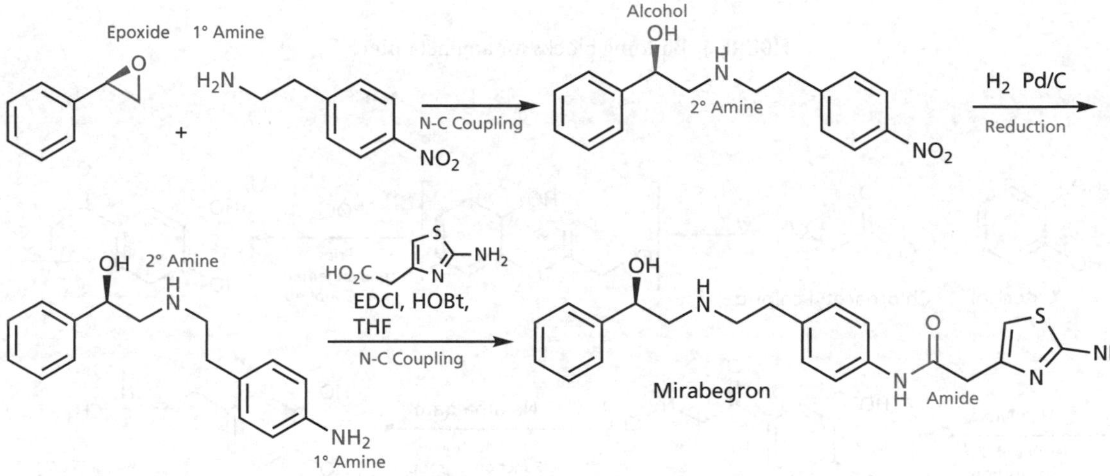

FIGURE 2.51 A proposed synthesis of mirabegron.

The first stage of the synthesis was an N–C coupling where the primary amine reacts with the epoxide to form an amino alcohol. The resulting secondary amine was then protected by treating it with di-*tert*-butyl dicarbonate to produce a urethane. The *third* stage involved reduction of the nitro group to a primary amine, which could now be used for the second coupling reaction. We have already seen that the nitro group is a useful latent group for an aromatic amine, since it is unreactive to most reagents but can easily be converted to the aromatic amine when the latter is needed. This is an alternative method to using a protecting group in order to 'disguise' an aromatic amine. The penultimate stage involved the formation of an amide by reacting the newly formed amine with the carboxylic acid of the third building block. The reagent EDCI (1-ethyl-3-(3-dimethylaminopropyl)carbodiimide) was used as a water-soluble coupling agent (see also Case Study 1). The final stage was

deprotection where hydrolysis of the urethane group restored the secondary amine that is present in mirabegron.

To conclude, the synthesis of mirabegron involves three building blocks which are linked together using two coupling reactions. The synthesis also involves the use of a latent group, a protecting group, and a coupling agent.

2.13 The decision to protect or not

When do you use a protecting group and which functional groups should be protected? This depends very much on the synthetic route involved and the reagents used at each stage. Therefore, protection may be necessary if a synthetic intermediate contains two or more functional groups that are susceptible to the same reagent.

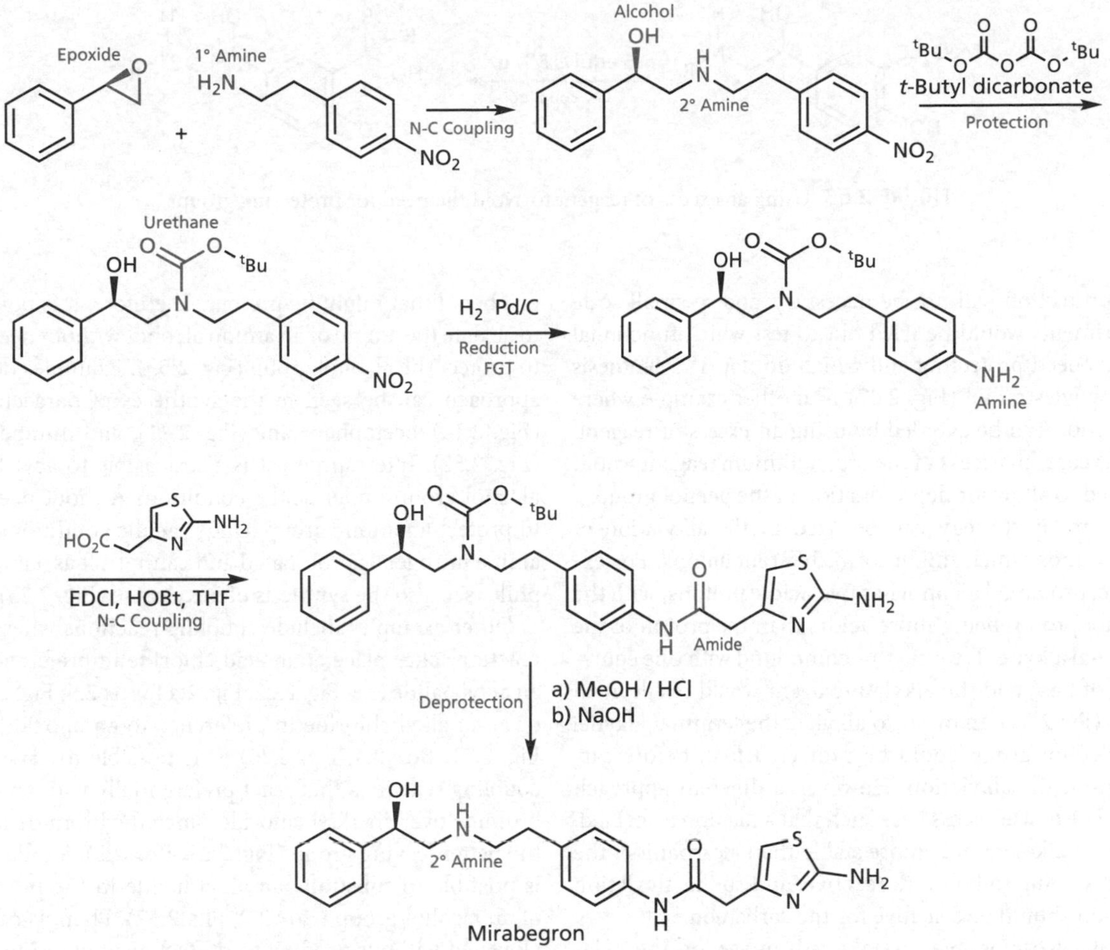

FIGURE 2.52 Synthesis of mirabegron: HOBt, 1-hydroxybenzotriazole.

If mildly basic or nucleophilic reagents are to be used, functional groups such as carboxylic acids and phenols may need to be protected, as these groups contain an acidic proton which will be removed by the reagent. With stronger bases and nucleophiles, it may be necessary to protect any functional group that has protons joined to a heteroatom, such as alcohols, amines, and thiols. Terminal alkynes also contain a proton which is likely to be removed by such reagents. Other functional groups capable of reacting with strong bases and nucleophiles are aldehydes, ketones, and esters.

If oxidizing agents are to be used, it may be necessary to protect groups that are easily prone to oxidation such as primary and secondary alcohols, aldehydes, and thiols. If electrophilic reagents such as alkyl halides or epoxides are being used, nucleophilic functional groups may need to be protected, particularly amines.

Having said all that, it has often been stated that the use of protecting groups is an admission of failure in not finding a synthetic route that avoids their need. Certainly,

their use adds extra synthetic steps which inevitably result in a decreased overall yield, and it would be preferable to avoid them if at all possible. Therefore, it is often worth considering whether protection is really required.

For example, is it really necessary to protect groups that can undergo acid–base reactions with a reagent? If the reagent is cheap and plentiful, it may be feasible to use an excess of the reagent to allow for the amount of reagent used up by the acid–base reaction. For example, the Grignard reaction carried out in Figure 2.53 involved the use of five equivalents of the Grignard reagent, three of which were used up in abstracting protons from the phenol, alcohol, and amine groups. The functional groups were restored following the aqueous work-up. The reaction is certainly wasteful on the Grignard reagent, but if the reagent is cheap and commercially available, it is preferable to protecting and then deprotecting three functional groups, especially if the starting material involves time and effort to prepare. Of course, it would be highly risky to commit an expensive starting material to a reaction in the hope

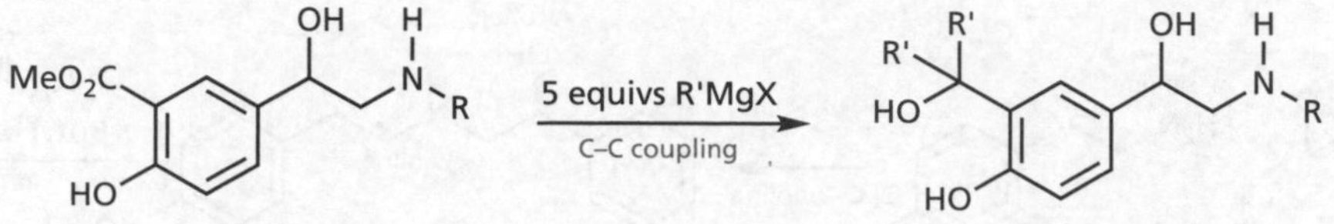

FIGURE 2.53 Using an excess of reagent to avoid the need for protecting groups.

that protection will not be necessary, and so small-scale experiments would be tried out to test which functional groups need protecting and which do not. The synthesis of ethynylestradiol (Fig. 2.27) is another example where protection can be avoided by using an excess of reagent. In this case, an excess of the organolithium reagent would be used to allow for deprotonation of the phenol group.

A similar strategy can be used in the alkylation of compounds which might form different anions. For example, propargyl alcohol has two acidic protons, with the alcohol proton being more acidic than the proton of the terminal alkyne. Treating this compound with one equivalent of base and the alkylating agent would result in an ether (Fig 2.54). In order to alkylate the terminal alkyne, the alcohol group could be protected first, before carrying out the alkylation. However, a different approach would be to use excess base such that a dianion is formed. Since an alkoxy ion is more stable than a carbanion, the latter is going to be more reactive, and so the alkylation reaction should be selective for the carbanion.

It is also possible to take advantage of the relative reactivities of different functional groups. For example, amines are more nucleophilic than alcohols or phenols, and so it is often possible to carry out a reaction on an amine without having to protect any alcohol or phenol that might be present. Therefore, it is possible to acylate the amine of an amino alcohol without needing to protect the alcohol group (Fig. 2.55). Examples of this approach can be seen in the syntheses of paracetamol (Fig. 2.14), acetophenazine (Fig. 2.21), and mirabegron (Fig. 2.52). Interestingly, it is also feasible to acylate an alcohol group under acidic conditions without needing to protect an amine group. Under acidic conditions, the amine nitrogen is protonated and cannot act as a nucleophile (see also the synthesis of hexylcaine (Fig. 2.25)).

Other examples include coupling reactions where the reaction takes place at an acid chloride in preference to an alkyl halide (see Fig. 2.22, Fig. 2.44, Box 2.5, Fig. 2.32), or at an alkyl chloride in preference to an epoxide (see Fig. 2.31, Box 2.4, Fig. 2.26). It is possible to carry out coupling reactions that react preferentially with an alkyl bromide over an alkyl chloride, since the bromide ion is the better leaving group (Fig. 2.21, Box 2.2). Similarly, it is possible to substitute an alkyl halide in the presence of an alcohol group (Box 2.2, Fig. 2.37). Phenols can be alkylated without needing to protect alcohol groups if a base is used to deprotonate the phenol group (Box 2.4).

It is also possible to carry out a reaction on an aliphatic amine in the presence of an aromatic amine since the latter tends to be less nucleophilic (Fig. 2.17). When it

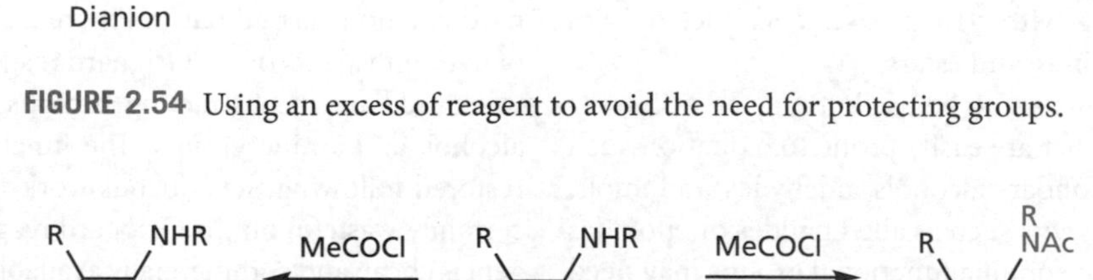

FIGURE 2.54 Using an excess of reagent to avoid the need for protecting groups.

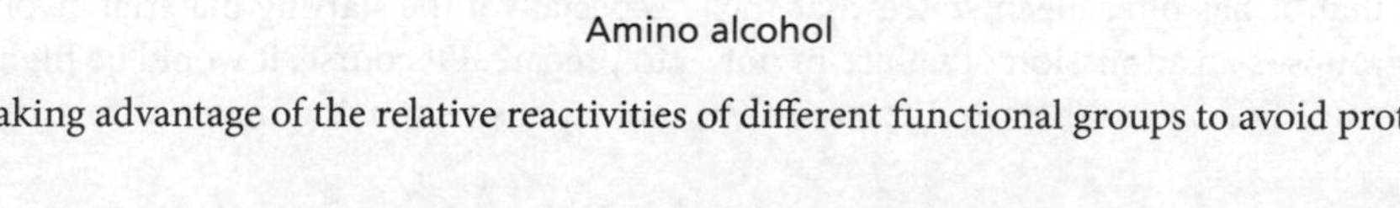

FIGURE 2.55 Taking advantage of the relative reactivities of different functional groups to avoid protecting groups.

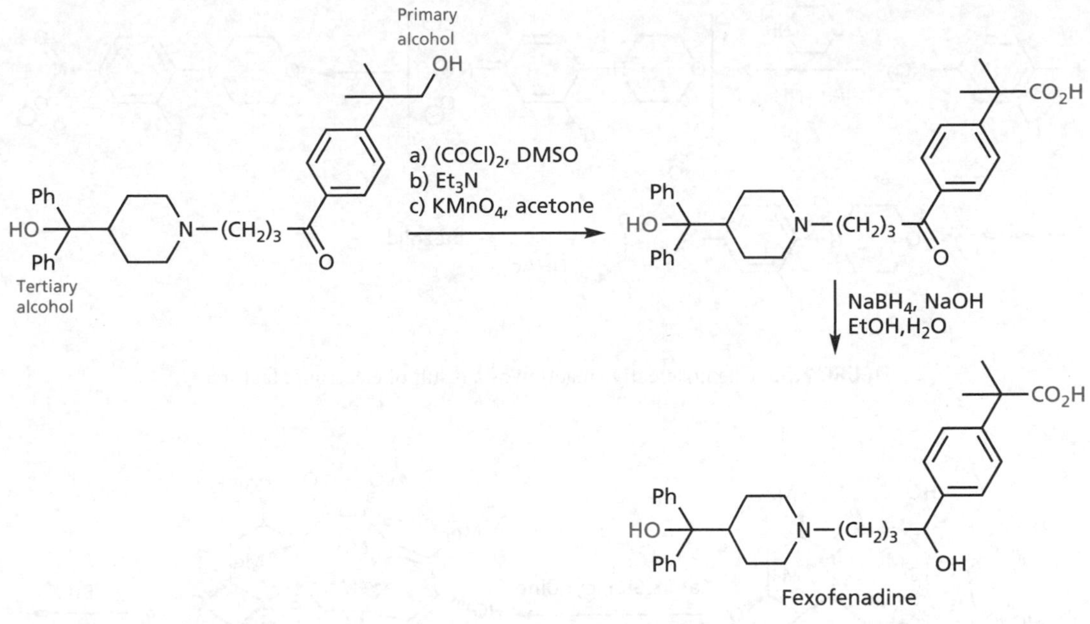

FIGURE 2.56 Synthesis of fexofenadine.

comes to the oxidation of alcohols, primary and secondary alcohols are susceptible to oxidation, whereas tertiary alcohols are not. Therefore, there is no need to protect a tertiary alcohol when an oxidation reaction is being carried out. An example of this is the oxidation of a primary alcohol during the synthesis of the antihistamine **fexofenadine** (Fig. 2.56). The oxidation involves a Swern oxidation of the primary alcohol to an aldehyde, followed by oxidation to a carboxylic acid with potassium permanganate. This proved more efficient than oxidizing directly to the carboxylic acid. There was no need to protect the tertiary alcohol during the process.

Chemoselectivity is often possible between similar functional groups if those groups have a different number of substituents attached. The more substituents that are present, the more sterically crowded the functional group will be, and this will lower its reactivity. For example, monosubstituted alkenes tend to be more reactive than di-, tri-, or tetrasubstituted alkenes. It may also be possible to carry out a reaction on a primary aliphatic amine in the presence of a secondary amine if the latter is sterically

hindered. A primary alcohol tends to react more readily than a secondary or tertiary alcohol. Therefore, it is possible to esterify a primary alcohol in the presence of a secondary alcohol by using one equivalent of an acylating agent under carefully controlled conditions (Fig. 2.57).

Electronic factors can influence the relative reactivities of similar functional groups. For example, a nucleophilic substitution of a difluoroaromatic ring was carried out as the first stage of a synthesis leading to the antibacterial agent **linezolid** (Fig. 2.58). One fluorine was substituted in preference to the other as a result of the intermediate being resonance stabilized by the electron-withdrawing nitro group. This was possible for the *para* substituent, but not for the *meta* substituent.

The reactivity of similar functional groups can be significantly affected by their position in a molecule, especially if that molecule has a complex structure. For example, **10-deacetylbaccatin III** (10-DAB) is a natural product that is used as the starting material in the synthesis of the important anticancer agent **paclitaxel** (Taxol) (Fig. 2.59). It contains four alcohol groups, three

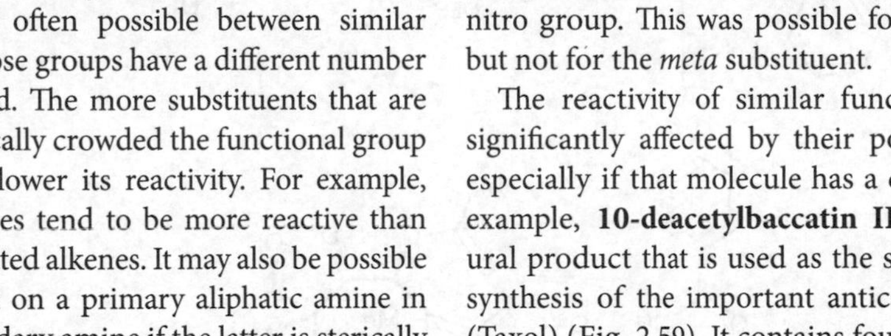

FIGURE 2.57 Chemoselective synthesis of an ester.

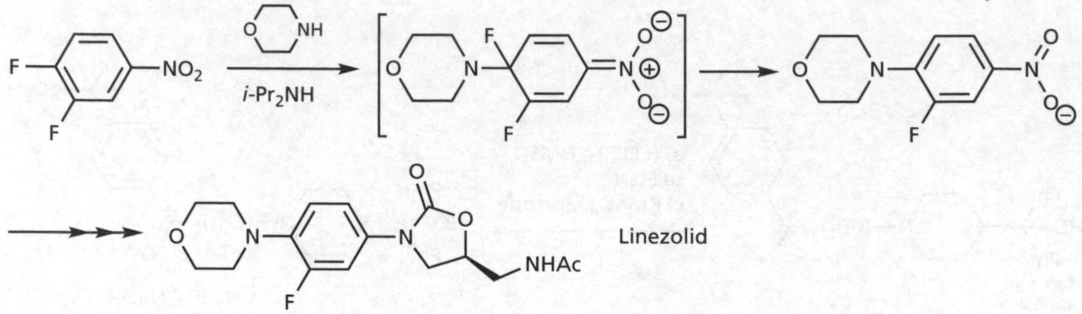

FIGURE 2.58 Chemoselective reaction as a result of electronic factors.

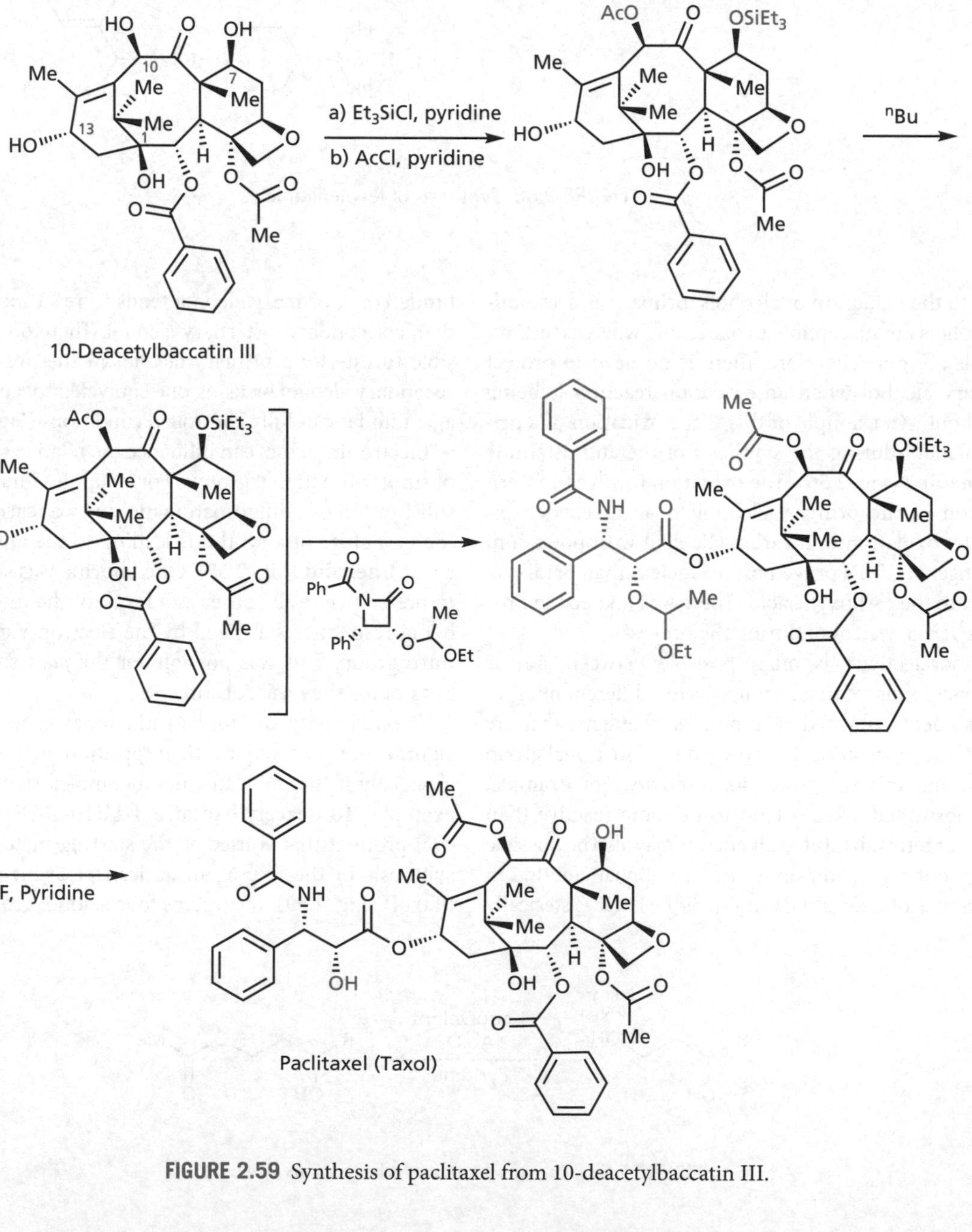

FIGURE 2.59 Synthesis of paclitaxel from 10-deacetylbaccatin III.

of which are secondary and one of which is tertiary. Investigations have shown that the secondary alcohol at position 7 is the most reactive, followed by the secondary alcohol at position 10, and then the secondary alcohol at position 13. The tertiary alcohol at position 1 is so unreactive that it can be left unprotected throughout the complete synthesis. The synthesis itself takes advantage of the relative reactivities of the four alcohol groups. The most reactive group at position 7 was first protected as a silyl ether. The alcohol at position 10 was then acetylated. Treatment with a strong base deprotonated the alcohol at position 13 which was then acylated with a β-lactam reagent. Finally, the silyl ether protecting group was removed with HF. Therefore, the complete synthesis required only one protecting group, despite the presence of four alcohols.

The synthesis of paclitaxel shows how a reactive secondary alcohol group had to be protected before less reactive alcohols were modified. This approach can also be applied to the selective esterification of a primary or a secondary alcohol when both groups are present. We have already seen that it is possible to esterify the primary alcohol selectively (Fig. 2.57). If we want to esterify the secondary alcohol instead, we could first protect the primary alcohol, carry out the esterification of the secondary alcohol, and then deprotect the primary alcohol. But that is not the only strategy we could use. In fact, it may be better to esterify both alcohols and then hydrolyse the ester of the primary alcohol selectively using a mild base (Fig. 2.60). This would involve fewer steps and might result in a better overall yield. However, the only way to find out the best approach is to try out both methods in the laboratory.

A useful method of synthesizing a protected aspartic acid for peptide synthesis involves the same kind of strategy (Fig. 2.61). The carboxylic acid in the side chain of aspartic acid needs to be protected to prevent it competing with the carboxylic acid in the 'head group' during peptide bond synthesis. The amine group is protected first; then, ideally, we would want to selectively protect the carboxylic acid in the side chain without protecting the carboxylic acid in the head group. However, it proves easier to protect both groups as benzyl esters and then selectively hydrolyse the ester in the head group.

Another example of this approach can be seen in the synthesis of a key intermediate required for the synthesis of **pemetrexed** (Fig. 2.62), an antifolate used in the treatment of lung cancer. The starting material contains a primary amino group which is protected as a pivalamide. An iodo substituent now has to be introduced at position 7 to allow a subsequent C–C coupling using a Sonogashira reaction (Case Study 2). However, when the iodination was carried out using one equivalent of N-iodosuccinimide (NIS), a mixture of mono-iodinated and di-iodinated products were obtained at positions 7 and 8. A better approach was to use 2.2 equivalents of NIS to give the di-iodinated product and then selectively remove the iodo-substituent at position 8 to give the desired intermediate.

A knowledge of the relative reactivities of functional groups can be helpful in determining whether it is possible to carry out reactions without the use of protecting groups. For example, the order of reactivity of functional groups to nucleophiles is

aldehydes > ketones > esters > amides > carboxylates .

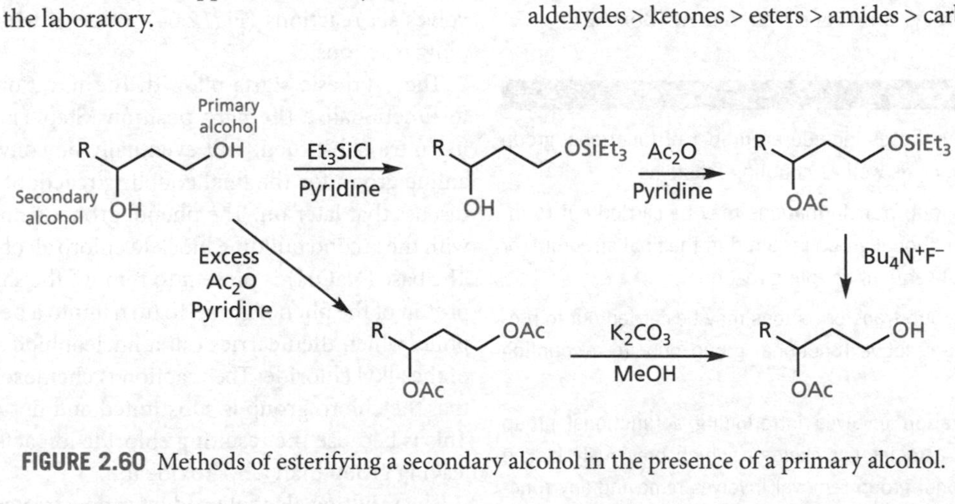

FIGURE 2.60 Methods of esterifying a secondary alcohol in the presence of a primary alcohol.

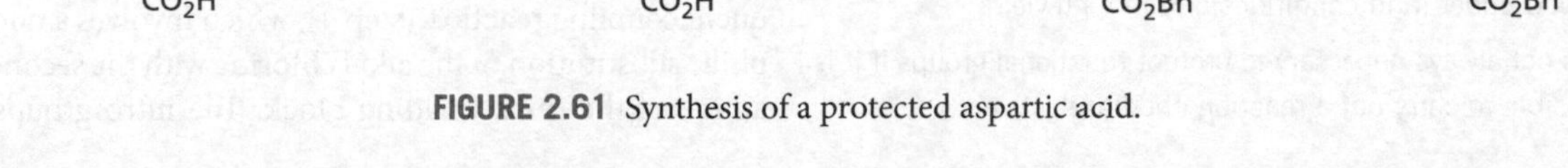

FIGURE 2.61 Synthesis of a protected aspartic acid.

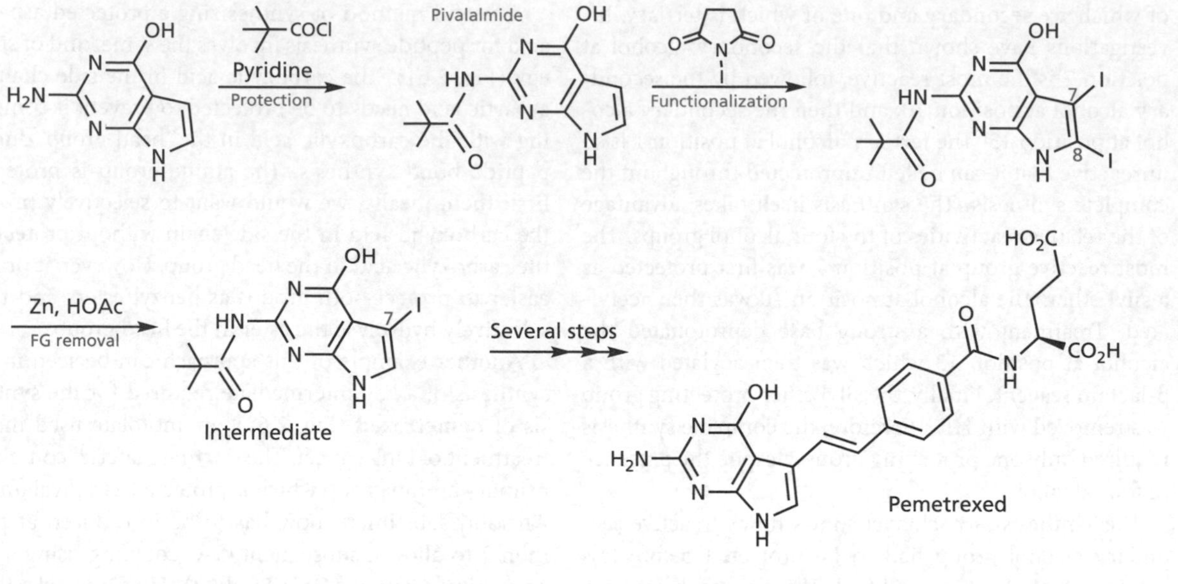

FIGURE 2.62 Synthesis of a key intermediate required in the synthesis of pemetrexed.

The order of reactivity of functional groups to hydrogenation is

acid chlorides > $ArNO_2$ > alkynes > aldehydes > alkenes > ketones > nitriles > ester > aromatic ring.

Finally, if a building block contains two identical groups, it may be possible to use it in large excess, such that the reaction occurs at one of the groups but not both. This is feasible if the reagent/building block is cheap and readily available. An example is the use of piperazine in large excess during the synthesis of perphenazine (Box 2.2).

KEY POINTS

- Most drug syntheses include a number of functional group transformations as well as coupling reactions.

- Functional group transformations may be carried out to introduce a functional group required in the final structure, or to allow a subsequent coupling reaction.

- Functional group transformations may be carried out to produce a more reactive functional group prior to a coupling reaction.

- Functionalization involves introducing a functional group into part of a molecular skeleton which previously lacked one. Functional group removal involves removing any functional groups that are not required in the final structure.

- Protecting groups are used to mask functional groups that do not undergo a reaction. They need to be added and removed under mild conditions and in high yield.

- It is not always necessary to protect functional groups if it is possible to carry out a reaction chemoselectively.

- Using an excess of one of the reactants in a reaction can sometimes avoid the need for protecting groups.

2.14 Case Study—Synthesis of dofetilide

Dofetilide is a cardiovascular drug that acts on the heart and is useful in the treatment of cardiac arrhythmia. Four building blocks are used (Fig. 2.63) and the synthesis involves six reactions (Fig. 2.64), three of which are coupling reactions.

The synthesis starts off with the nitration of phenol to functionalize the *para* position (Step 1). The resulting nitro substituent will eventually be converted to an amine group for the final coupling reaction, but we will discuss that later on. The phenol group is now alkylated with the second building block (a chloro alcohol) (step 2). The base (NaOH) is present to remove the slightly acidic proton of the phenol group to turn it into a better nucleophile, which then carries out a nucleophilic substitution of the alkyl chloride. The reaction is chemoselective such that the chloro group is substituted and not the alcohol. This is because the resulting chloride ion acts as a better leaving group than a hydroxide ion.

The resulting alcohol product is now treated with thionyl chloride to convert the alcohol group to an alkyl chloride (step 3). This is a functional group transformation designed to activate the functional group for the subsequent coupling reaction (step 4), which involves a nucleophilic substitution of the alkyl chloride with the secondary amine of the third building block. The nitro groups are

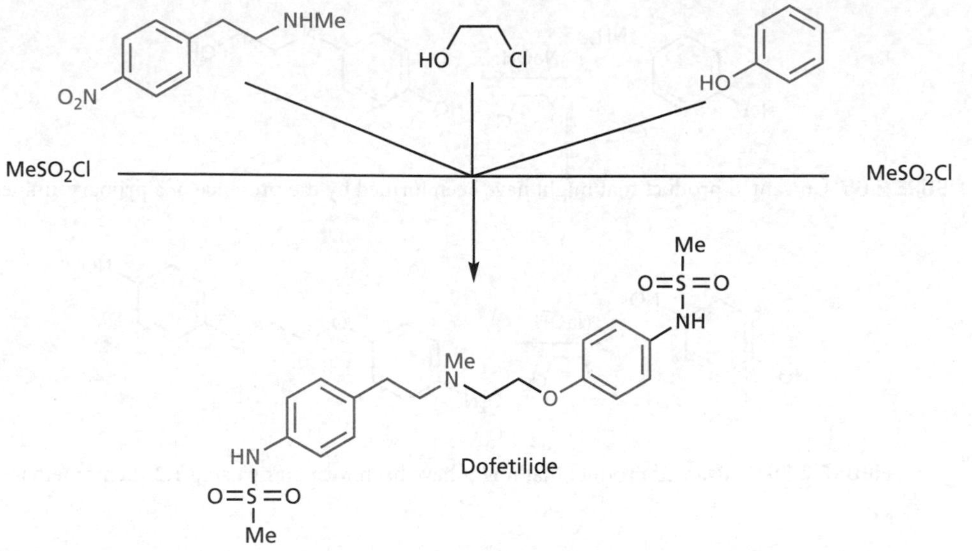

FIGURE 2.63 Molecular building blocks used for the synthesis of dofetilide.

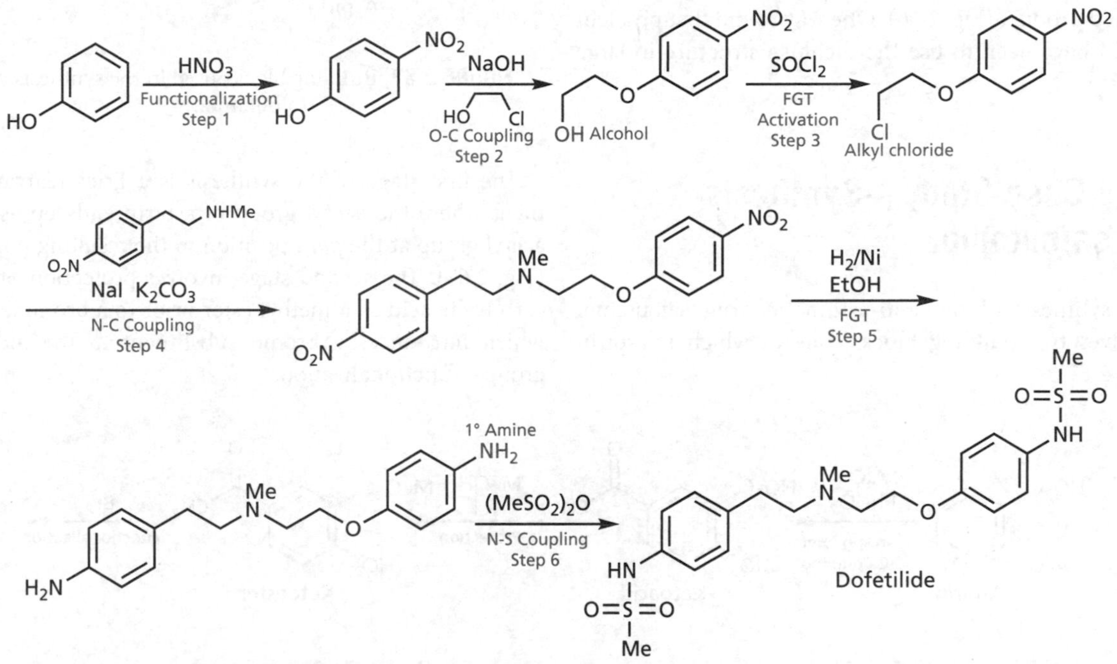

FIGURE 2.64 Synthesis of dofetilide.

now reduced to amines (step 5), which are treated with the final building blocks—two molecules of methanesulphonic anhydride—to give the two sulphonamide groups observed in the final structure (step 6).

Note that the reduction of the nitro groups was delayed until the penultimate step of the reaction sequence such that the amino groups were only revealed when they were needed. There was a good reason for that. A primary amine group is relatively reactive and could have resulted in unwanted reactions during the other stages

of the reaction sequence. For example, the amine could have reacted with the chloro alcohol in the first coupling reaction to give an unwanted product (Fig. 2.65). In contrast, a nitro group is relatively unreactive to most reagents and acts as a 'disguised' or 'latent' primary amine. The use of an aromatic nitro group as a latent group for a primary amine is a common strategy in drug synthesis.

Another point worth noting is the fact that a chloro alcohol was used as the second building block, rather than a dichloride. In theory, one could have attempted to use

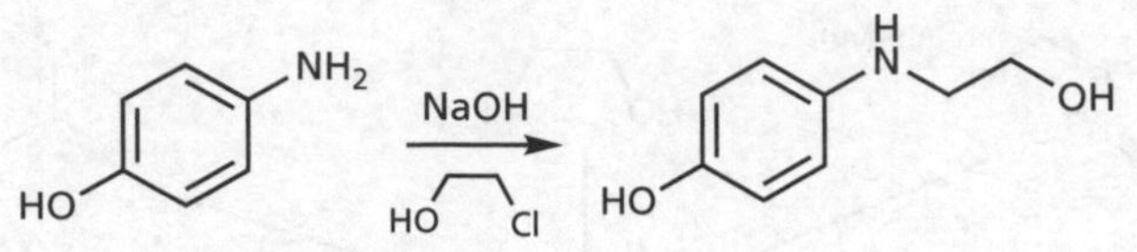

FIGURE 2.65 Unwanted product that might have been formed by the presence of a primary amine.

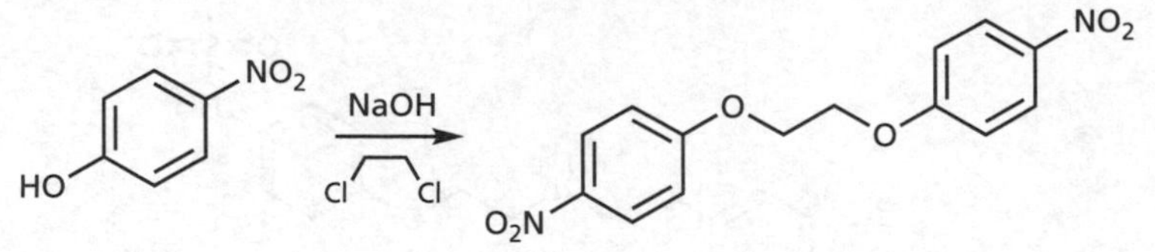

FIGURE 2.66 Unwanted product that might have been formed by using 1,2-dichloroethane.

a dichloride in order to miss out the later activation step involving thionyl chloride. However, this would have been difficult to control as the product obtained would have been susceptible to further reaction to give an unwanted product (Fig. 2.66). One way round that problem would have been to use the dichloro structure in large excess.

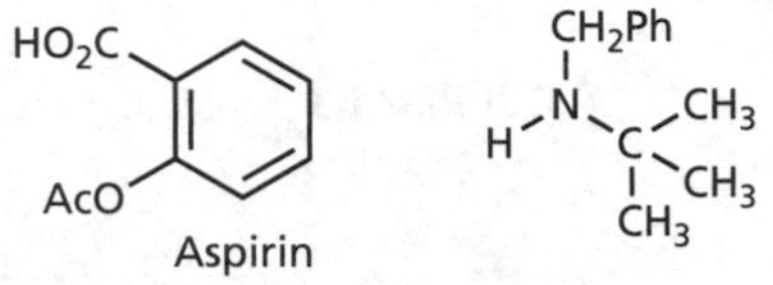

2.15 Case Study—Synthesis of salbutamol

The synthesis of the anti-asthmatic drug salbutamol involves two building blocks, one of which is aspirin (Fig. 2.67).

FIGURE 2.67 Building blocks used in the synthesis of salbutamol.

The first stage of the synthesis is a Fries rearrangement where the acetyl group of aspirin ends up as the acetyl group at the *para* position to the resulting phenol (Fig. 2.68). The second stage involves protection of the carboxylic acid as a methyl ester prior to a bromination which introduces a bromo substituent to the methyl group—functionalization.

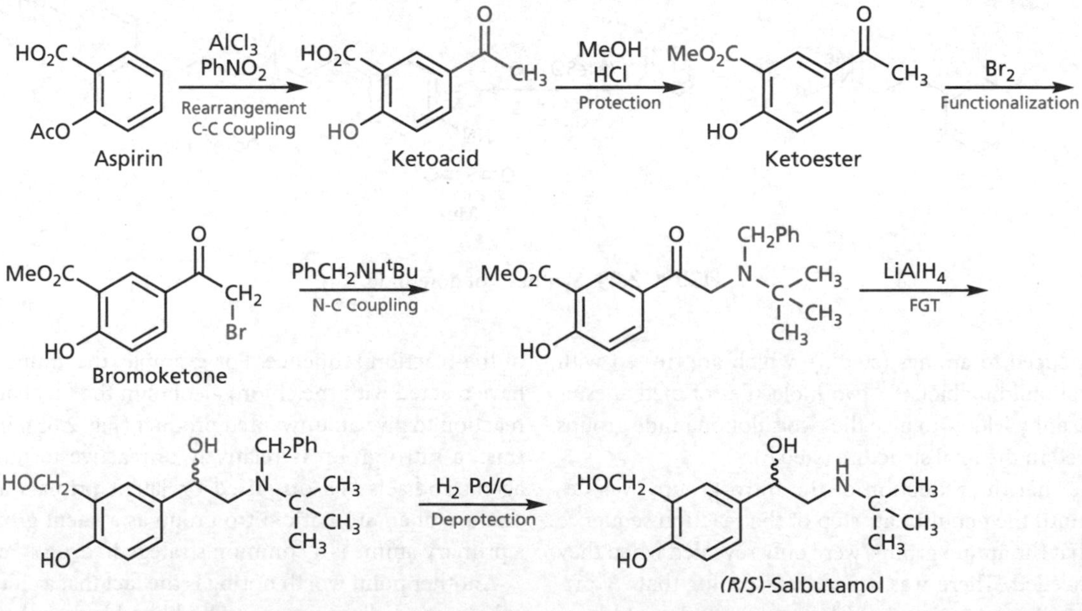

FIGURE 2.68 Synthesis of salbutamol.

Having introduced the bromo substituent, it is now possible to carry out a nucleophilic substitution reaction where the second building block (an amine) replaces the bromine. The amine that is used also contains a protecting group in the form of a benzyl moiety.

Treatment with lithium aluminium hydride now reduces the ketone group that resulted from the rearrangement to a secondary alcohol, and also reduces the ethyl ester to a primary alcohol. Finally, the benzyl protecting group is removed by hydrogenolysis.

QUESTIONS

1. Explain why the N–C coupling reaction used in the synthesis of the antimalarial agent **chloroquine** displaces one of the chlorine substituents in the bicyclic starting material rather than the other. What kind of selectivity is observed here?

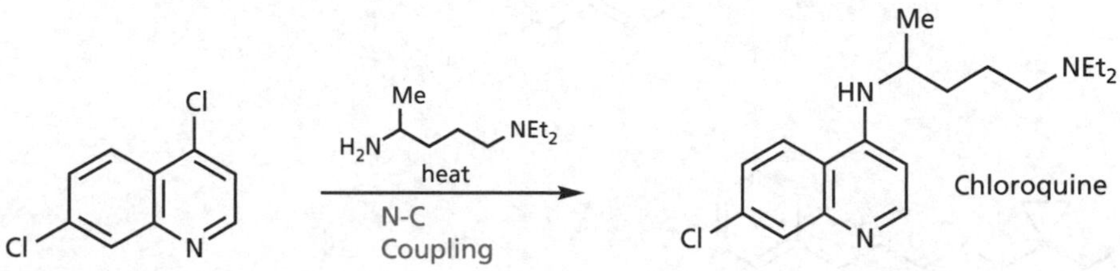

Chloroquine

2. Suggest how two consecutive coupling reactions could be used to synthesize the local anaesthetic **lidocaine**.

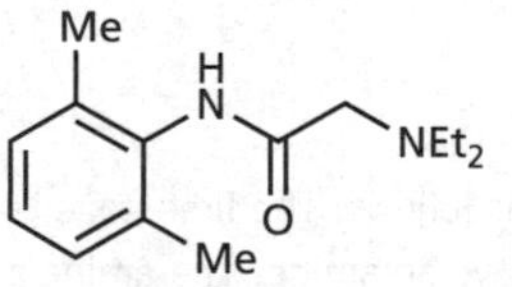

Lidocaine

3. Suggest how two consecutive coupling reactions could be used to synthesize **metoprolol**.

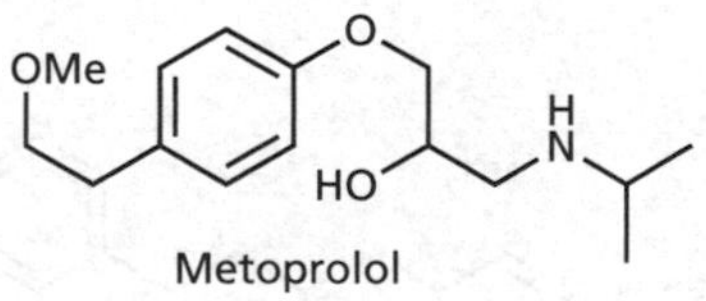

Metoprolol

4. The synthesis of the antibacterial agent **sulfadiazine** is shown. Why is a nitro substituent used in the starting material instead of an amine?

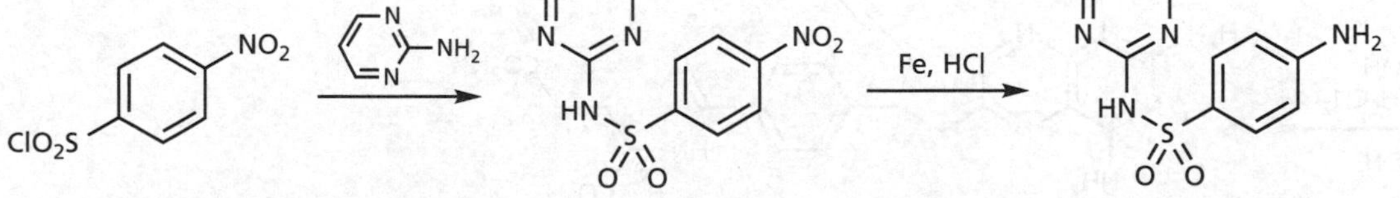

Sulfadiazine

5. In the synthesis of **procaine** shown in Figure 2.41, the starting material contains a nitro group which is converted to an amino group in the final stage. Discuss whether the synthesis would have been successful if the reduction of the nitro group had been carried out at the beginning of the synthesis instead of the end.

6. The two reactions shown in Figure 2.25 involve an alcohol reacting with an acid chloride, instead of an amine. However, an amine nitrogen is more nucleophilic than an alcohol oxygen. Why do these reactions produce esters rather than amides?

7. A synthesis of the anti-asthmatic agent **montelukast** involved an intermediate (I) containing two alcohol groups, and an intermediate (II) where one of the alcohol groups was protected. Suggest how intermediate I could be converted to intermediate II.

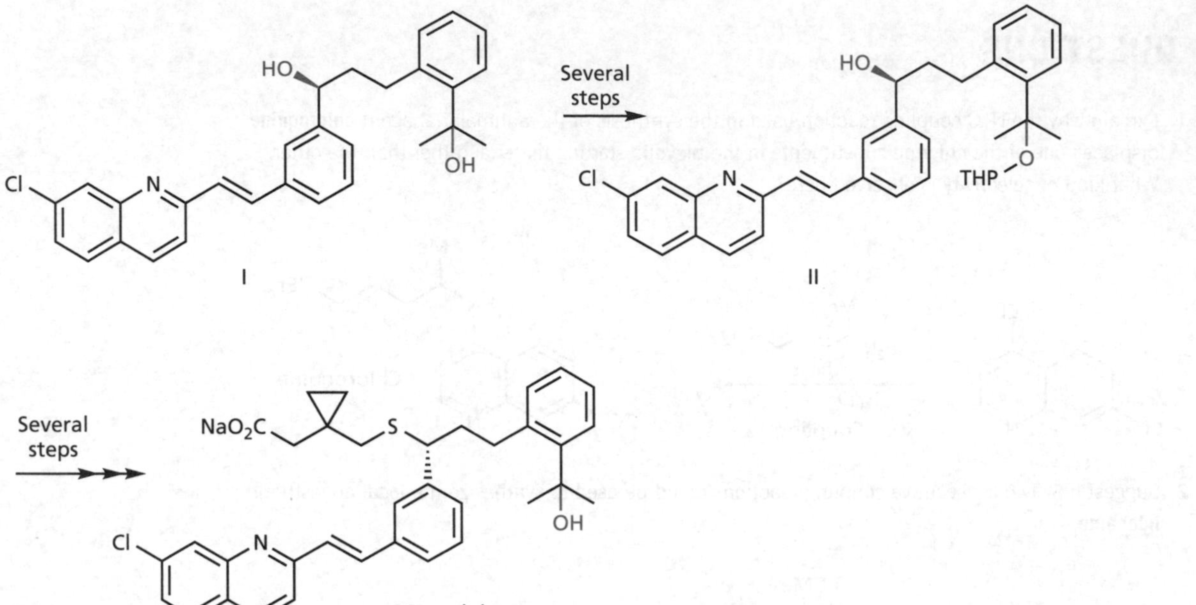

8. **Pralatrexate** is an anticancer agent which can be synthesized as follows. The first stage is a coupling reaction between an amine and a carboxylic acid to give an amide. The amine groups in the bicyclic starting material are not protected, so why do they not react with the carboxylic acid in a self-condensation reaction?

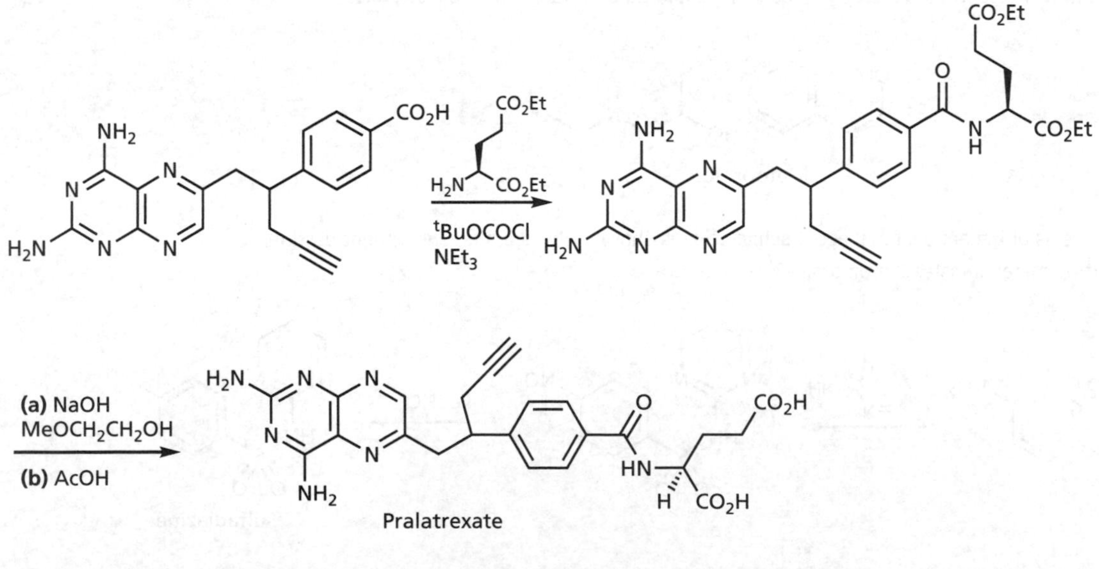

9. The following reaction was carried out as one of the early stages in a synthesis of fexofenadine. There is a high regioselectivity for the *para* position over the *ortho* and *meta* positions. Explain the reasons for this selectivity. Identify any chemoselectivity observed in the reaction.

FURTHER READING

Medicinal chemistry

Patrick, G.L. (2013) *An introduction to medicinal chemistry* (5th edn). Oxford University Press, Oxford (Chapter 20, 'Antiviral agents'; Chapter 21, 'Anticancer agents'; Chapter 22, 'Cholinergics, anticholinergics, and anticholinesterases'; Chapter 23, 'Drugs acting on the adrenergic nervous system'; Chapter 24, 'The opioid analgesics'; Chapter 25, 'Anti-ulcer agents').

Synthesis

Clayden, J., Greeves, N., and Warren, S. (2012) *Organic chemistry* (2nd edn). Oxford University Press, Oxford.

March, J. (1992) *Advanced organic chemistry: reactions, mechanisms and structure* (4th edn). Wiley, New York.

Roughey, S.D. and Jordan, A.M. (2011) 'The medicinal chemist's toolbox: an analysis of reactions used in the pursuit of drug candidates', *Journal of Medicinal Chemistry*, **54**, 3451–79.

Saunders, J. (2000) *Top drugs: top synthetic routes*. Oxford University Press, Oxford.

Specific compounds

Adams, H.J.F., et al. (1974) 'Acylxylidide local anaesthetics', US Patent 3,812,147 (etidocaine).

Bauer, D.J. and Sadler, P.W. (1964) 'Substituted isatin-β-thiosemicarbazones, their preparation and pharmaceutical preparations containing them', GB Patent 975,357 (methisazone).

Bieleford, H.A., et al. (1957) 'Basic ethers of substituted diphenylmethylcarbinols', US Patent 2,785,202 (chlorphenoxamine).

Cope, A.C. (1949) 'Benzoic acid esters of alicyclic (secondary) amino alcohols', US Patent 2,486,374 (hexylcaine).

Cross, P.E., et al. (1990) 'Anti-arrhythmic agents', US Patent 4,959,366 (dofetilide).

Cusic, J.W. (1956) 'N-(β-Acetoxyethyl)-N'-(chlorophenothiazinepropyl)piperazine', US Patent 2,766,235 (perphenazine).

Dehn, F.B. (1959) 'New tertiary amines and their salts and process for their preparation', GB Patent 807,835 (prolintane).

Einhorn, A. (1906) 'Alkamin esters of *para*-aminobenzoic acid', US Patent 812,554 (procaine).

Greenshields, J.N., et al. (1962) 'Pharmaceutical compositions', GB Patent 901,876 (etoglucid).

Hauck, F.P., et al. (1976) 'Tetrahydronaphthyloxyaminopropanols and salts thereof', US Patent 3,935,267 (nadolol).

Hester, J.B. (1992) 'Antiarrhythmic N-aminoalkylenealkyl and aryl sulfonamides', US Patent 5,155,268A (ibutilide).

Holton, R.A., et al. (1995) 'Semisynthesis of Taxol and Taxotere', in M. Suffness (ed.), *Taxol: science and applications*. CRC Press, Boca Raton, FL, pp 97–121 (paclitaxel).

Janssen, P.A.J. (1961) 'Heterocyclic derivatives of 1-phenyl-ω-(piperazine)alkanols', US Patent 2,979,508 (azaperone).

Janssen, P.A.J. (1961) '1-(Aroylalkyl)-4-arylpiperazines', US Patent 2,997,472 (fluanisone).

Jung, E., et al. (1971) '*p*-Acetamido-*N*-(2-diathylaminoathyl)-benzamid, seine salze mit sauren, verfahren zu seiner herstellung und arzneipraparate', DE Patent 2,062,978 (*N*-acetylprocainamide (acecainide)).

King, C.-H. and Kaminski, M.A. (1993) '4-Diphenyl piperidine derivatives and processes for their preparation', WO Patent 93/21156 (fexofenadine).

Klosa, J. (1961) 'A process for producing centrally acting stimulants of new aminonitriles', DE Patent 1,112,987 (amphetaminil).

Lunts, L.H.C. (1985) 'Salbutamol: a selective β₂-stimulant bronchodilator', in S.M. Roberts and B.J. Price (eds), *Medicinal chemistry: the role of organic chemistry in drug research*. Academic Press, London, pp 49–67 (salbutamol).

Lyttle, D.A. (1961) 'Derivatives of 5-amino uracil', US Patent 2,969,364 (uramustine).

Lyttle, D.A. and Petering, H.G. (1958) '5-Bis-(2-chloroethyl)-aminouracil, a new antitumor agent', *Journal of the American Chemical Society*, **80**, 6459–60 (uramustine).

Main, B.G. and Tucker, H. (1985) 'Beta blockers',in S.M. Roberts and B.J. Price (eds), *Medicinal chemistry: the role of organic chemistry in drug research*. Academic Press, London, pp 69–92 (propranolol).

Nardi, D., et al. (1980) 'Substituted dibenzyl ethers and pharmaceutical compositions containing said ethers for the treatment of infections', US Patent 4,221,803 (fenticonazole).

Pachter, I.J. and Schoen, K. (1970) 'Derivatives of 5 aminomethyl-4,5,6,7-tetrahydro-4-oxoindoles', US Patent 3,491,093 (molindone).

Pfister, K. and Stein, G.A. (1959) 'Alpha methyl phenylalanines', US Patent 2,868,818 (methyldopa).

Reasenberg, J. R. (1956) 'Beta (N-propylamino) beta, beta-dimethyl ethyl benzoate and its water soluble salts', US Patent 2,767,207 (meprylcaine).

Rieveschl, G. and Woods, G.P. (1950) '*p*-Dimethylamino-ethyl-*p*-halobenzydryl ethers and their salts', US Patent 2,527,963 (bromazine).

Rohrmann, E. (1950) 'Acid addition salts of 1-cyclopentyl-2-methyl-amino propane compounds and vasoconstrictor compositions thereof', US Patent 2,520,015 (cyclopentamine).

Ruschig H., et al. (1974) '3-Aminoacylamino thiophenes', US Patent 3,855,243 (articaine).

Sherlock, M.H. and Sperber, N. (1961) 'Piperazino derivatives and methods for their manufacture', US Patent 2,985,654 (acetophenazine).

Sperber, N., et al. (1951) 'Aryl-(2-pyridyl-amino alkanes and their production', US Patent 2,567,245 (chlorphenamine).

Stanley et al. (1973) '*p*-Acetamido-*N*-(2-diethylaminoethyl)-benzamide and its salts', GB Patent 1,319,980 (*N*-acetylprocainamide (acecainide).

Steiger, N., et al. (1951) '2-Tertiaryamino-6-(dialkylaminoalkoxy)-benzothiazoles and process for their manufacture', US Patent 2,578,757 (dimazole).

Takashima, M., et al. (1980) 'Benzamide derivatives', US Patent 4,210,660 (nemonapride).

Taylor, E.C., et al. (1992) 'A dideazatetrahydrofolate analogue lacking a chiral center at C-6, *N*-[4-[2-(2-amino-3,4-dihydro-4-oxo-7H-pyrrolo[2,3-*d*]pyrimidin-5-yl)ethyl] benzoyl]-L-glutamic acid is an inhibitor of thymidylate synthase', *Journal of Medicinal Chemistry*, **35**, 4450–4 (pemetrexid).

Thuresson, B. and Egner, B.P.H. (1957) 'Process of preparing amides of heterocyclic carboxylic acids', US Patent 2,799,679 (mepivacaine).

Timmis, G.M. (1959) 'Leukemia treatment', US Patent 2,917,432 (busulfan).

Vaughan, J.R., et al. (1949) 'Antihistamine agents; halogenated N,N-dimethyl-N-benzyl-N-(2-pyridyl_-ethylenediamines', *Journal of Organic Chemistry*, **14**, 228–34 (chloropyramine).

Villani, F.J., et al. (1962) 'Dialkylaminoalkyl derivatives of 10,11-dihydro-5H-dibenzo[a,d]cycloheptane and related compounds', *Journal of Medicinal and Pharmaceutical Chemistry*, **5**, 373 (cyclobenzaprine).

Whitaker, W.D. (1961) 'Novel dibenzocycloheptaenes and salts thereof and a process for the manufacture of same', GB Patent 858,187 (cyclobenzaprine).

3 Retrosynthesis

3.1 Introduction

One of the skills required in drug synthesis is to design effective synthetic routes to a target structure. It may be that the structure has been obtained from the natural world, in which case it may never have been synthesized before. Alternatively, the structure may already have been synthesized, but the synthetic route is not suitable for the preparation of a range of analogues, or is inefficient in terms of yield and cost. Whatever the reason, the chemist is required to come up with a synthetic route that is practical and useful for the task in hand. **Retrosynthesis** is a strategy by which such routes can be designed. 'Retro' means 'backwards', and so 'retrosynthesis' should indicate a 'backward synthesis'. In fact, this is misleading and it is better to view retrosynthesis as meaning 'backward synthetic planning' as there are no actual synthetic procedures involved. Retrosynthesis is a mind exercise where the research chemist plans a synthetic route by considering the structure of the product (structure A) and identifying what that product could be synthesized from (structure B). Having identified structure B, the process is repeated to find what structure B could be made from, and so on until it is possible to identify simple commercially available starting materials (Fig 3.1). Note that the arrows used in a retrosynthetic plan are different from the normal arrows used in a synthetic scheme. This is to avoid any confusion between a retrosynthetic plan and an actual synthesis.

3.2 Disconnections of C–C bonds

The purpose of retrosynthesis is to identify a feasible synthetic route to a complex target molecule, starting from simple starting materials. Therefore, the retrosynthesis must inevitably include one or more disconnection stages where the target structure is split into two simpler molecules; for example, the disconnection of structure B into structures C and D (Fig. 3.1). It must be emphasized again that a disconnection in retrosynthetic terms is not an actual chemical reaction which splits the molecule. In reality, you are trying to identify two structures C and D which can be linked to form structure B. This can be quite hard to get your head around until you are used to the concept, and so we are going to use an example to illustrate this. Suppose that our target structure is **acetophenone** (Fig. 3.2). In order to identify simple starting materials we have carry out a disconnection which splits the carbon skeleton, so which bonds should we disconnect? We can discount any possibility of disconnecting bonds within the aromatic ring as it is not easy to synthesize aromatic rings from simple starting materials. This leaves us with only two disconnection options—disconnecting the bond between the carbonyl group and the aromatic ring, or disconnecting the bond between the carbonyl group and the methyl group. We are going to look at each of these in turn.

First of all, we disconnect the C–C bond between the carbonyl group and the aromatic ring (Fig. 3.2). This is indicated by the wiggly line. The two fragments that

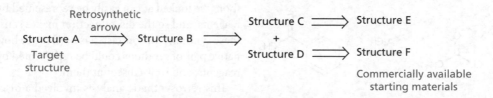

FIGURE 3.1 Retrosynthetic planning.

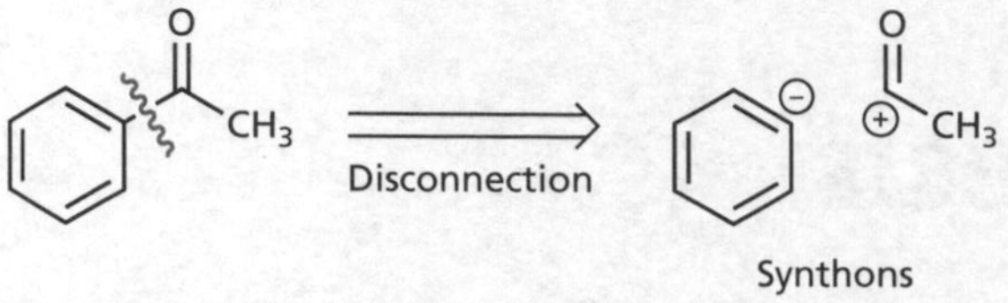

FIGURE 3.2 Disconnection to form a pair of synthons.

result are called **synthons**. One of these is given a positive charge and the other is given a negative charge. This is because most reactions involve two reagents where one acts as a nucleophile and the other acts as an electrophile. However, it is important to appreciate that the synthons we have identified are not real species or reagents, which is why they are described as synthons. Therefore, we now need to identify reagents which will react as if these two synthons are really present in a reaction.

To do that, we add a hydrogen (or a metal atom) to the nucleophilic synthon and a good leaving group (halogen or OH) to the electrophilic synthon, and then assess what we get. In this case, adding a hydrogen to the nucleophilic synthon gives us benzene, while addition of a chlorine to the electrophilic synthon gives us acetyl chloride (Fig. 3.3).

Are these suitable reagents which would react together to give the desired product acetophenone? Indeed they are. Acetyl chloride contains an electrophilic acid chloride group, whereas benzene is nucleophilic in nature. Therefore, they should react together. We can now identify the forward reaction as being a Friedel–Crafts acylation (Fig. 3.4) and so this disconnection is a valid one for a retrosynthetic scheme.

Now, you may have already noticed that this disconnection could give an alternative pair of synthons where the charges are swapped about (Fig. 3.5). This can also be analysed to see if there are corresponding reagents.

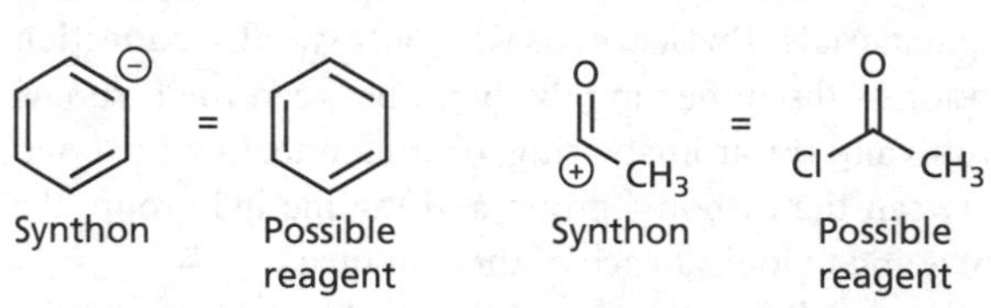

FIGURE 3.3 Possible reagents that would represent the synthons.

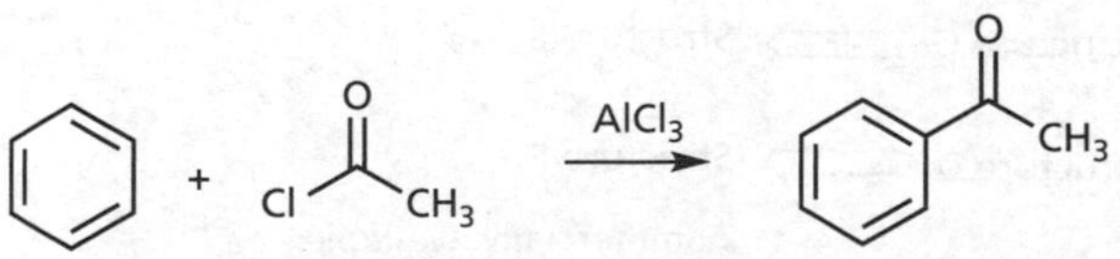

FIGURE 3.4 The Friedel–Crafts acylation of benzene to give acetophenone.

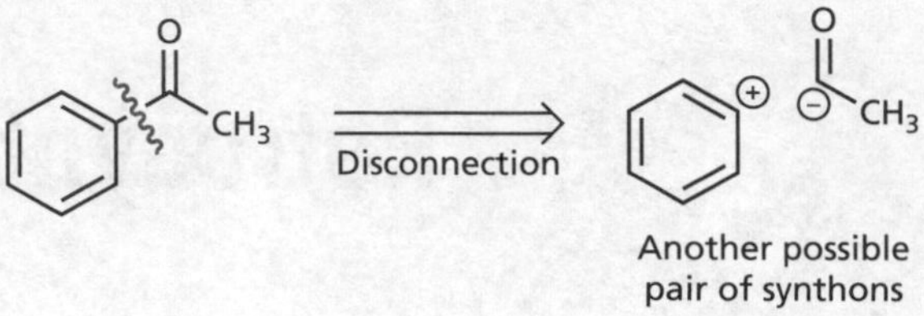

FIGURE 3.5 Alternative synthons for the disconnection in Figure 3.2.

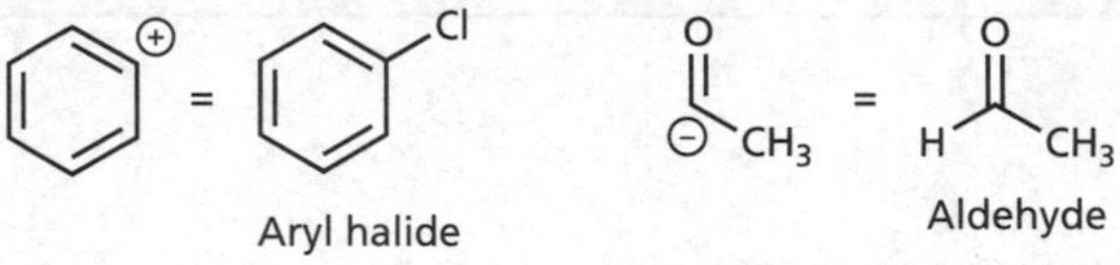

FIGURE 3.6 Possible reagents corresponding to synthons.

Once again, we add a hydrogen to the nucleophilic synthon and a leaving group to the electrophilic synthon (Fig. 3.6). This gives us an aryl halide and an aldehyde.

Are these reagents likely to represent the desired synthons? If the aryl halide was to act as an electrophile, it would have to undergo nucleophilic substitution, but aromatic rings do not readily undergo this reaction unless there are other electron-withdrawing substituents present in the ring. Therefore, the aryl halide is unlikely to act as a suitable electrophile.

What about the aldehyde? If the aldehyde was to act as the nucleophilic synthon shown, it would have to lose the aldehyde proton. In other words, the aldehyde proton would have to be slightly acidic. However, the acidic hydrogen is actually on the methyl group, and so acetaldehyde is more likely to act as a nucleophile at the methyl carbon than at the aldehyde group. We should not find that surprising. The carbonyl group is polarized such that the oxygen is slightly negative and the carbon is slightly positive. This makes the carbonyl carbon electrophilic rather than nucleophilic. Therefore, it is difficult to see how we could reverse that natural polarity. In conclusion, neither of these reagents is likely to give us the reaction we want and so we can discount this approach.

This example illustrates the importance of analysing how feasible the forward reaction is for each retrosynthetic step as we consider it in turn. The first pair of synthons we looked at can easily be represented by common reagents, and so the disconnection makes sense if we use these synthons. In contrast, it is hard to see how the alternative pair of synthons could be represented by common reagents, and so we discount them.

This retrosynthetic analysis involved a disconnection between the aromatic ring and the carbonyl group, but what about the other possible disconnection between

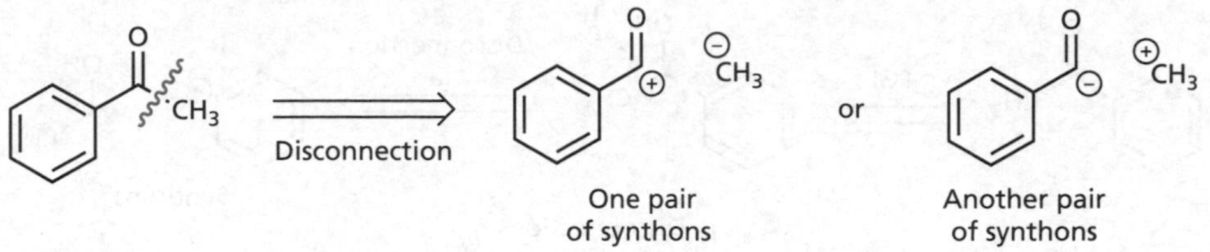

FIGURE 3.7 Synthons resulting from the disconnection between the carbonyl and methyl groups of acetophenone.

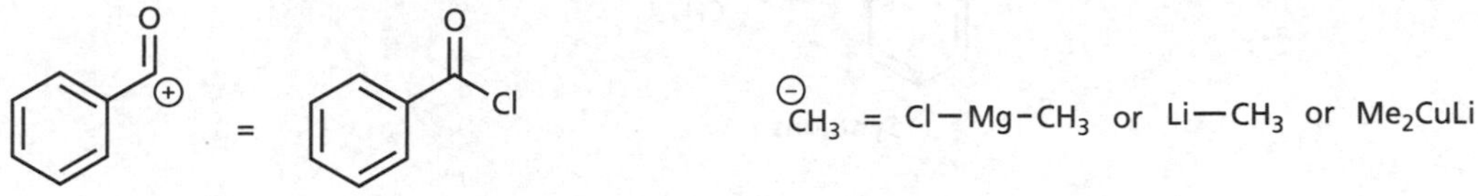

FIGURE 3.8 Reagents representing synthons.

the carbonyl group and the methyl group? We can repeat the retrosynthetic analysis to produce two pairs of synthons (Fig. 3.7).

Let us look at the first pair of synthons (Fig. 3.8). If we add a chlorine to the electrophilic synthon we get an acid chloride, which will certainly act as an electrophile and undergo nucleophilic substitution. What about the nucleophilic synthon? If we add hydrogen to it we get methane (CH_4), which is certainly not going to act as a nucleophile. At first sight, this seems like a dead end, but what if a metal was linked to the methyl group instead of a hydrogen? A metal would polarize the bond such that the methyl carbon was slightly negative and act as a nucleophile. Therefore, an organometallic reagent looks as if it would be suitable. We can now consider several possible organometallic reagents, such as Grignard, organolithium, or organocuprate reagents. But which of these would be best to use?

To answer that, we have to look at the reaction we want to carry out. The desired reaction is between an organometallic reagent and an acid chloride to give a ketone. A knowledge of the reactions undergone by organometallic reagents with acid chlorides tells us that Grignard reagents and organolithium reagents react twice with acid chlorides to give tertiary alcohols. In contrast, an organocuprate reagent reacts only once with the acid chloride to give the desired ketone (Fig. 3.9).

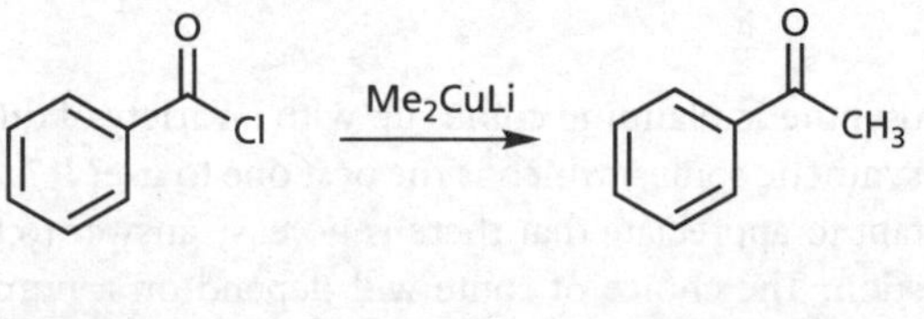

Let us now look at the second possible pair of synthons to see whether they are reasonable (Fig. 3.7). The electrophilic synthon is perfectly reasonable and could be represented by a simple alkyl halide (CH_3I). However, the other synthon involves a carbonyl group with a negative charge on the carbon atom. We have already seen that this goes against the natural polarity of the carbonyl bond, and so we would discount this approach in favour of the first pair of synthons.

3.3 Functional group interconversions

One of the strengths of retrosynthesis is the way it can help a synthetic chemist come up with different synthetic approaches. In our example using acetophenone, we have now identified two possible syntheses using different reagents, but are there any more? Our product has a ketone group, so could acetophenone be obtained from another structure by a functional group transformation? It is well known that secondary alcohols are easily oxidized to ketones, so a legitimate retrosynthetic functional group interconversion (FGI) is to change the ketone group to a secondary alcohol (Fig. 3.10).

Having identified the alcohol, we can now study possible disconnections of that structure. Reagents corresponding to the four synthons identified are shown in Figure 3.11. Note that the addition of a leaving group to both electrophilic synthons produces the equivalent of an aldehyde. Both nucleophilic synthons could be represented by Grignard reagents.

Therefore, another two possible routes to acetophenone could involve a Grignard reaction with an aldehyde, followed by oxidation of the product to the ketone (Fig. 3.12).

FIGURE 3.9 Conversion of an acid chloride to a ketone.

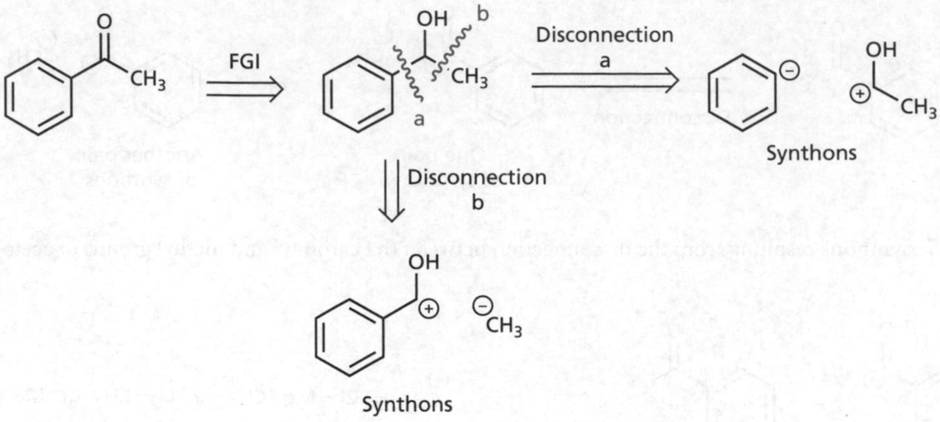

FIGURE 3.10 Retrosynthesis via an alcohol.

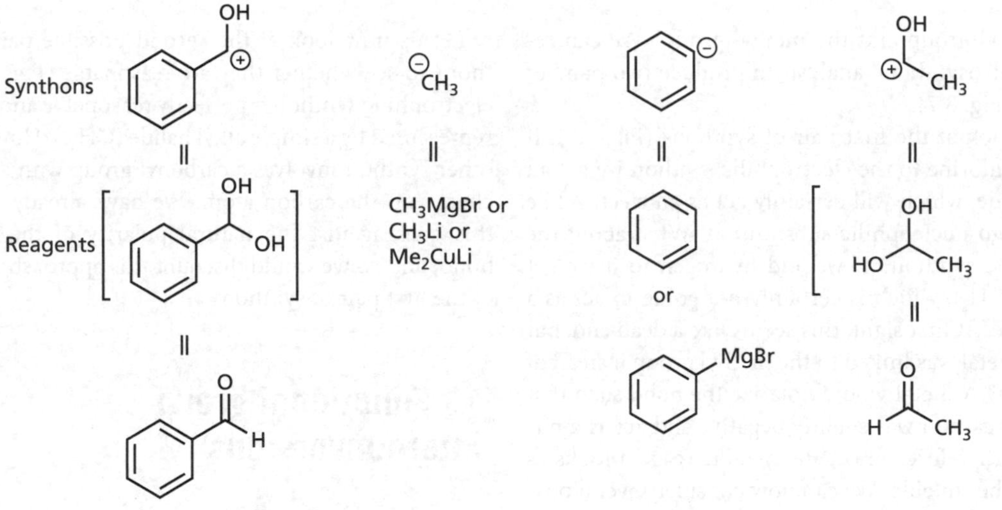

FIGURE 3.11 Reagents corresponding to possible synthons.

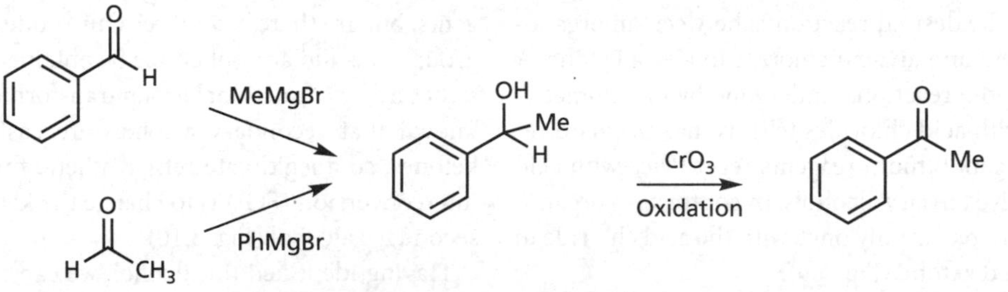

FIGURE 3.12 Two possible synthetic routes to acetophenone.

It should now be apparent how retrosynthesis can be used as a tool to identify different synthetic routes to a target structure. In this very simple example, we have identified four different routes. With more complex structures, the number of possible routes increases enormously. So if retrosynthetic planning comes up with a variety of different synthetic routes, which is the best one to use? It is important to appreciate that there is no easy answer to that question. The choice of route will depend on a number of factors, such as the number of steps in each synthetic route, the cost and availability of reagents, and the practicality and safety of the reactions involved. Such considerations may well rule out certain routes, but it is usually the case that the best route is only discovered by actually trying out the different syntheses in the laboratory.

Therefore, retrosynthetic planning does not provide us with a single correct solution to the synthesis of a drug structure. The appeal of the method is that it allows us to identify several possible routes. These can then be tried out in practice to see what advantages one route has over another.

In our discussions, we have considered the retrosynthesis of a simple aromatic ketone. The antipsychotic agent haloperidol also contains an aromatic ketone and its retrosynthesis is discussed in section 3.12.

3.4 Umpolung

In section 3.2, we described how it is preferable to choose synthons which have charges that match the natural polarity of the group involved. For example, the natural charge for a synthon representing the aromatic ring should be negative, as most aromatic rings are nucleophilic in nature. The synthon representing a carbonyl group should have a positive charge on the carbonyl carbon as this is inherently electrophilic. However, there are occasions where it might be desirable to use a reagent

that represents an 'unnatural' synthon. Such reagents do exist. For example, a dithiane structure (Fig. 3.13) can be synthesized from an aldehyde and then treated with a base to give a negatively charged species, where the charge is stabilized by the neighbouring sulphur atoms. Alkylation followed by hydrolysis then gives a ketone. Therefore, converting the aldehyde to the dithiane structure reverses the polarity of the carbonyl carbon, allowing reactions to be carried out where that carbon behaves as a nucleophile rather than as an electrophile. This reversal of polarity is called **umpolung**.

Applying the principles of umpolung provides us with yet another possible synthesis of acetophenone (Fig. 3.14).

An alternative approach is to consider the disconnection shown in Figure 3.5 which involves umpolung in both synthons. The dithiane of acetaldehyde is a feasible reagent for the carbonyl synthon, but is there a reagent that could correspond to an electrophilic aromatic ring? In fact, a diazonium salt acts in this manner and so one might be tempted to consider the reaction of a dithiane with the diazonium salt (Fig. 3.15). In order to judge whether this is a feasible reaction, it would be necessary to search the literature to see whether diazonium salts react with carbanions. A cursory look at the literature reveals that a reaction is possible, but that a coupling reaction takes place rather than the desired substitution reaction (Fig. 3.16). Therefore, it does not seem likely that this approach would work and it would be better to concentrate on one of the previous routes.

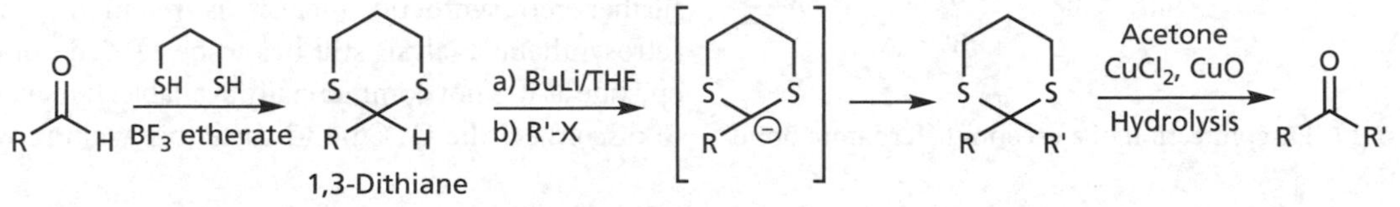

FIGURE 3.13 An example of umpolung.

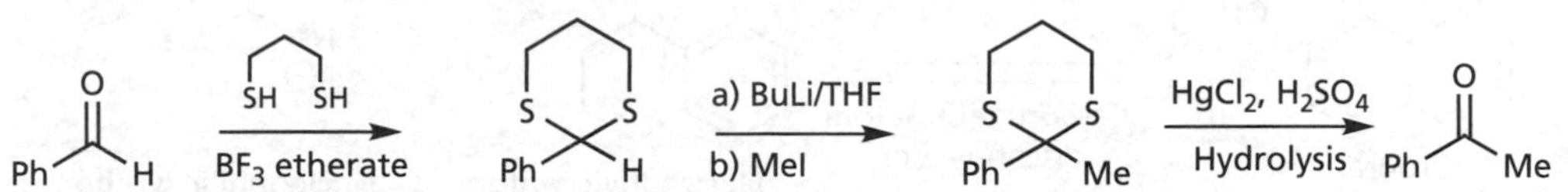

FIGURE 3.14 Synthesis of acetophenone via a dithiane.

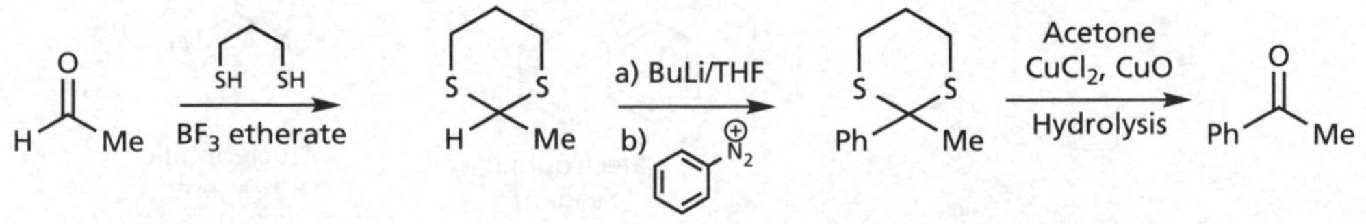

FIGURE 3.15 Proposed synthesis of acetophenone via the alternative dithiane.

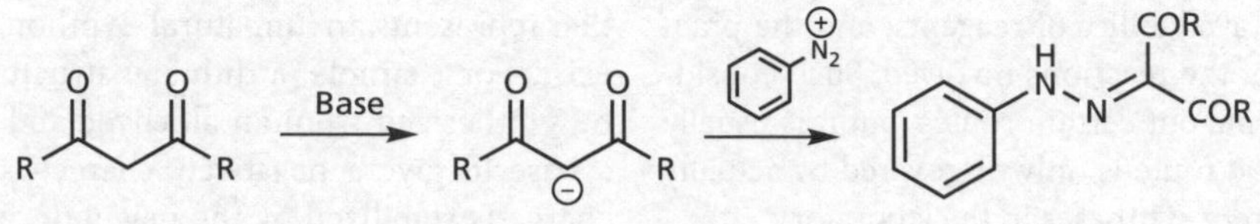

FIGURE 3.16 Known reaction of a carbanion with a diazonium salt.

3.5 Disconnections of carbon–heteroatom bonds

In the previous sections, we disconnected C–C bonds to study the retrosynthesis of acetophenone. However, retrosynthesis is not limited to C–C disconnections. Indeed, a particularly good disconnection is between a carbon atom and a heteroatom (X), as the latter will provide a nucleophilic synthon that corresponds to a naturally nucleophilic reagent (Fig. 3.17).

Therefore, any retrosynthetic analysis of a structure containing heteroatoms should look closely at the possibility of such C–X disconnections. For example, consider the first-generation β-blocker **propranolol** (Fig. 3.18). By choosing to cleave a carbon–heteroatom

bond, we are guaranteed a good nucleophilic reagent in the form of an amine which is inherently nucleophilic. Of course, we have to make sure that the electrophilic synthon also corresponds to a reasonable reagent. Normally, we could identify a suitable reagent by adding a leaving group to the electrophilic centre. In this case, we would end up with the leaving group at the α-position to the alcohol, which is likely to undergo a cyclization to an epoxide (Fig. 3.19). That might seem to be a problem, until we appreciate that the epoxide itself could act as the desired electrophilic reagent. Therefore, reacting the epoxide with the amine should provide propranolol, and has the advantage that the alcohol functional group is produced as a consequence of the addition reaction.

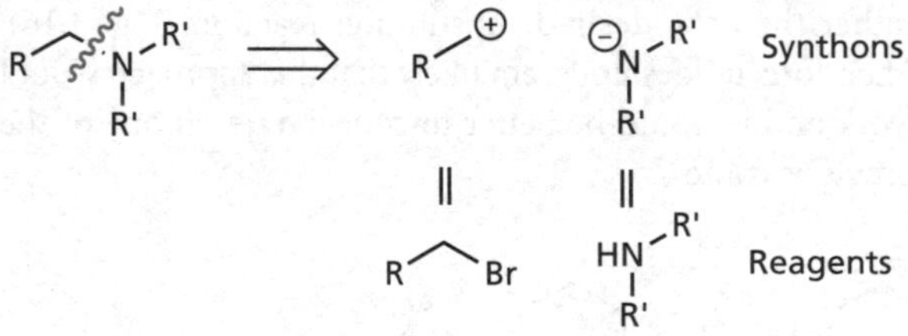

Therefore, the synthetic step corresponding to the retrosynthetic disconnection of the C–X bond is treatment of an epoxide with an amine (Fig. 3.20). We would expect the amine to react at the less substituted position of the epoxide to give the desired product—an example of regioselectivity.

The amine is a readily available reagent and so no further retrosynthetic analysis is required. However, retrosynthetic analysis still has to be carried out on the epoxide as it is not commercially available. It makes sense to disconnect the O–C bond since this again provides a

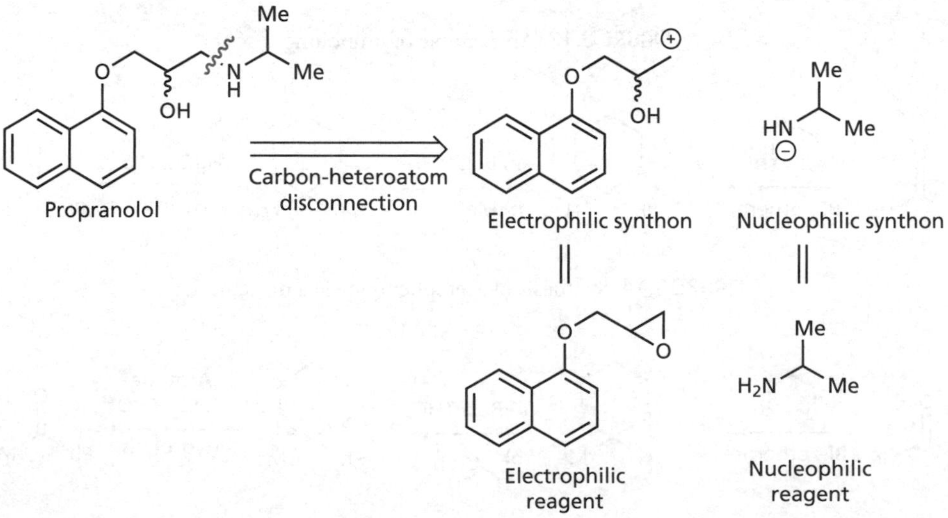

FIGURE 3.17 Disconnection of a carbon–heteroatom bond.

FIGURE 3.18 Retrosynthetic disconnection of a C–X bond in propranolol.

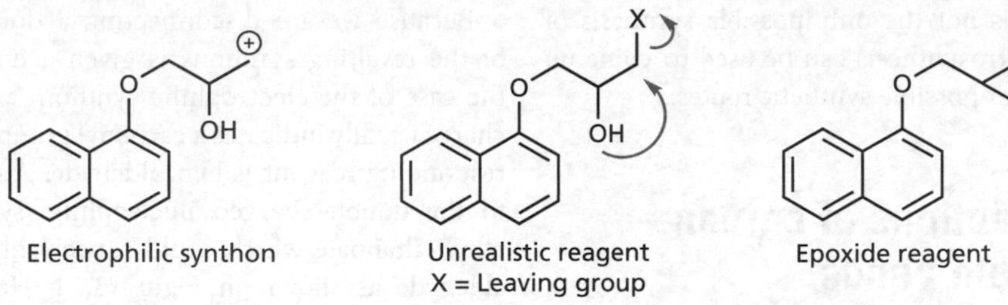

FIGURE 3.19 Comparison of reagents corresponding to the electrophilic synthon.

Electrophilic synthon

Unrealistic reagent
X = Leaving group

Epoxide reagent

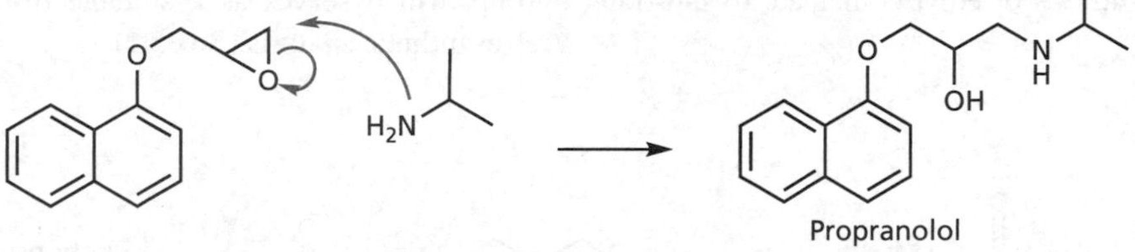

FIGURE 3.20 Synthesis of propranolol.

natural nucleophile where an oxygen atom serves as a nucleophilic centre (Fig. 3.21). α-Naphthol would be the equivalent of the nucleophilic synthon, while an α-chloro epoxide (or epichlorohydrin) would be the equivalent of the electrophilic synthon. Both these reagents are commercially available.

We now have a feasible synthesis of propranolol using three commercially available reagents (Fig. 3.22).

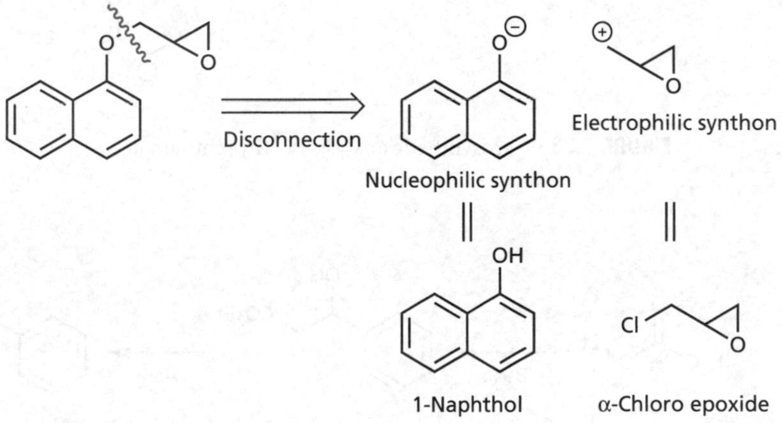

Disconnection

Nucleophilic synthon

Electrophilic synthon

1-Naphthol

α-Chloro epoxide

FIGURE 3.21 Carbon–heteroatom disconnection for the epoxide intermediate.

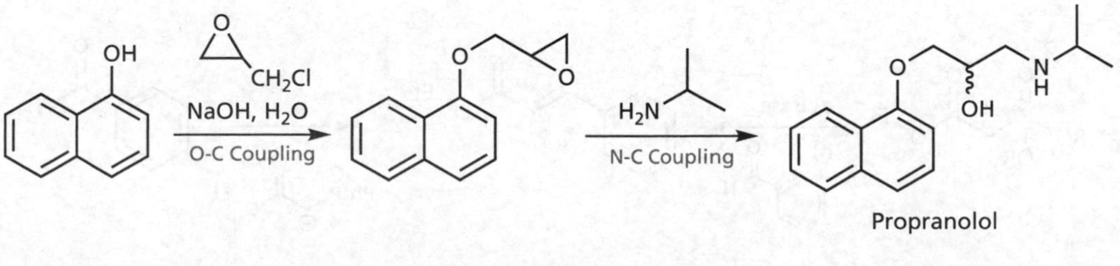

O-C Coupling

N-C Coupling

Propranolol

FIGURE 3.22 Synthesis of propranolol.

Of course, this is not the only possible synthesis of propranolol and retrosynthesis can be used to come up with a range of other possible synthetic routes.

3.6 **Disconnections of carbon–carbon double bonds**

Disconnections are not limited to single bonds and it is perfectly feasible to disconnect a carbon–carbon double bond as part of a retrosynthetic analysis. We shall consider the retrosynthesis of ethyl cinnamate to illustrate this (Fig. 3.23).

Because we are disconnecting a double bond, each of the resulting synthons is given a double charge. In the case of the electrophilic synthon, a double positive charge usually indicates a carbonyl group, and so the corresponding reagent is benzaldehyde. Adding hydrogens to the double-charged nucleophilic synthon gives us ethyl ethanoate which could conceivably react with the aldehyde as shown in Figure 3.24. However, a much better approach would be to carry out a Wittig or a Horner–Wadsworth–Emmons reaction using phosphorus reagents (Figs 3.23 and 3.25).

The anticancer agent **bexarotene** contains an alkene group, which serves as a suitable disconnection in a retrosynthetic analysis (Box 3.1).

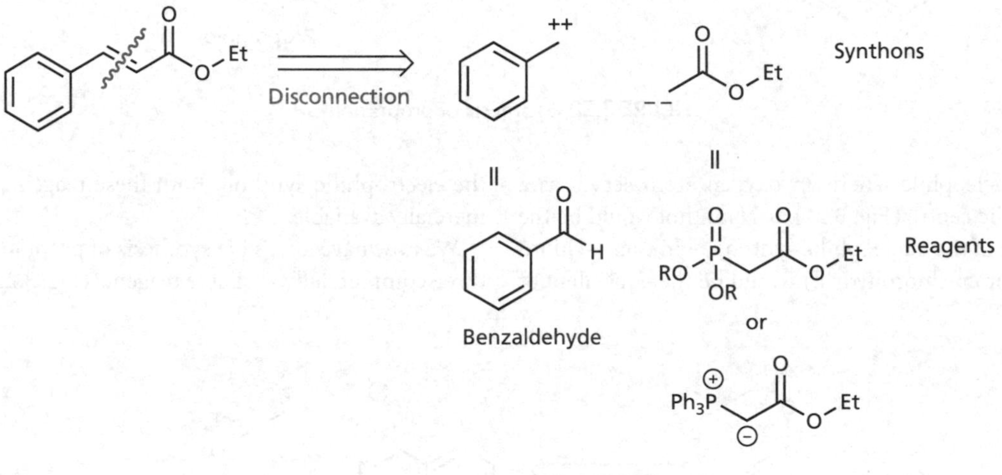

FIGURE 3.23 Retrosynthetic analysis of ethyl cinnamate.

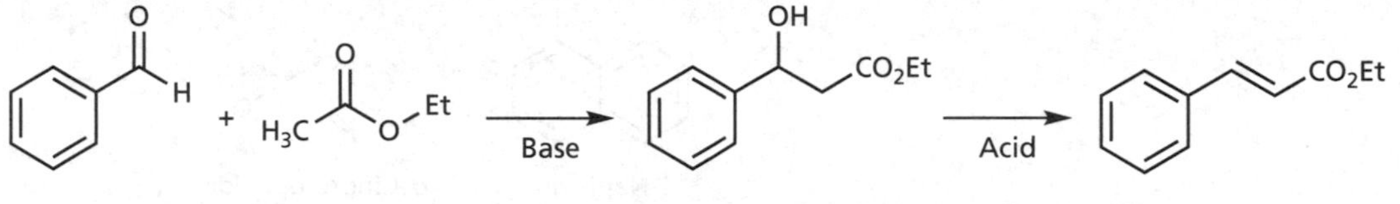

FIGURE 3.24 Possible approach to the synthesis of ethyl cinnamate.

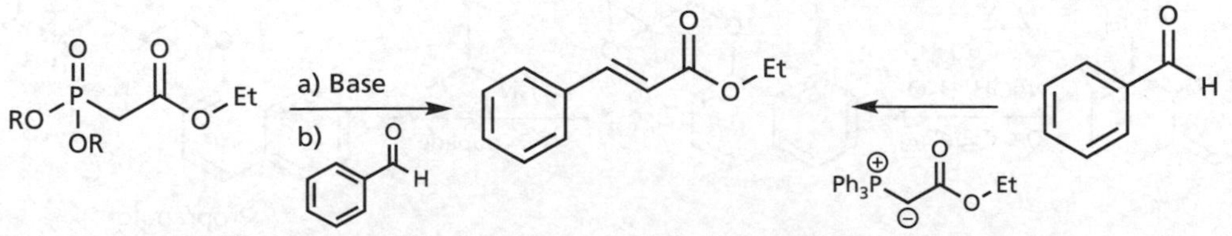

FIGURE 3.25 Synthesis of ethyl cinnamate using phosphorus reagents.

BOX 3.1 Retrosynthetic analysis of bexarotene

Bexarotene (Fig. 1) is an anticancer agent which includes a terminal alkene and a carboxylic acid.

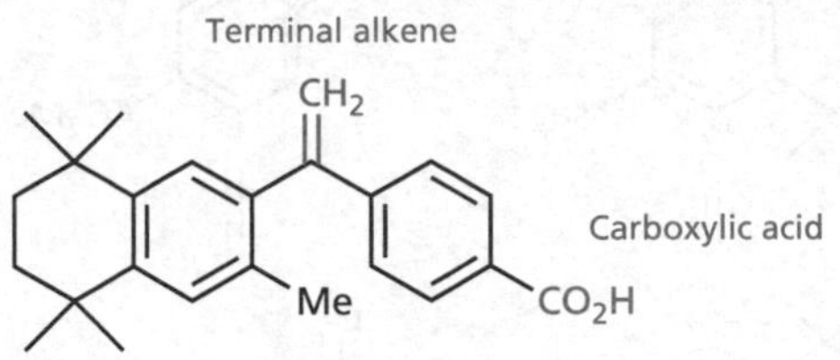

Terminal alkene

Carboxylic acid

FIGURE 1 Bexarotene.

Since a carboxylic acid has an acidic proton which could react with any basic or nucleophilic reagents used in the synthesis, it makes sense to devise a synthesis where a protecting group is used, with deprotection occurring at the final stage. Therefore, the first stage of

our retrosynthetic analysis would involve a functional group transformation (Fig. 2). It is then a case of identifying what kind of functional group could easily be converted back to the carboxylic acid. A simple methyl ester should be sufficient.

Disconnection of the terminal alkene group then results in two synthons with double charges (Fig. 3). These correspond to a ketone and a Wittig reagent. Therefore, the reverse reaction would be a Wittig reaction.

The Wittig reagent is commercially available but the ketone is not, and so the ketone is further analysed (Fig. 4). A C–C disconnection in the heart of the structure results in synthons that correspond to a known bicyclic tetrahydronaphthalene structure, and an acid chloride that is commercially available.

The corresponding synthesis involves a Friedel–Crafts acylation followed by a Wittig reaction, with the methyl ester being converted to the carboxylic acid at the final stage (Fig. 5).

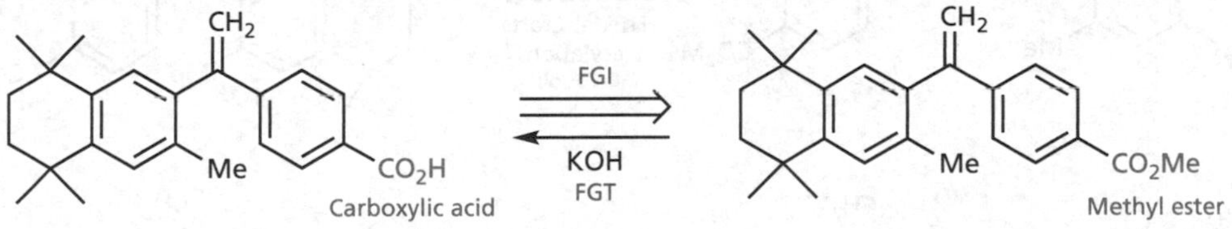

FIGURE 2 Identification of a methyl ester as the penultimate intermediate in the synthesis.

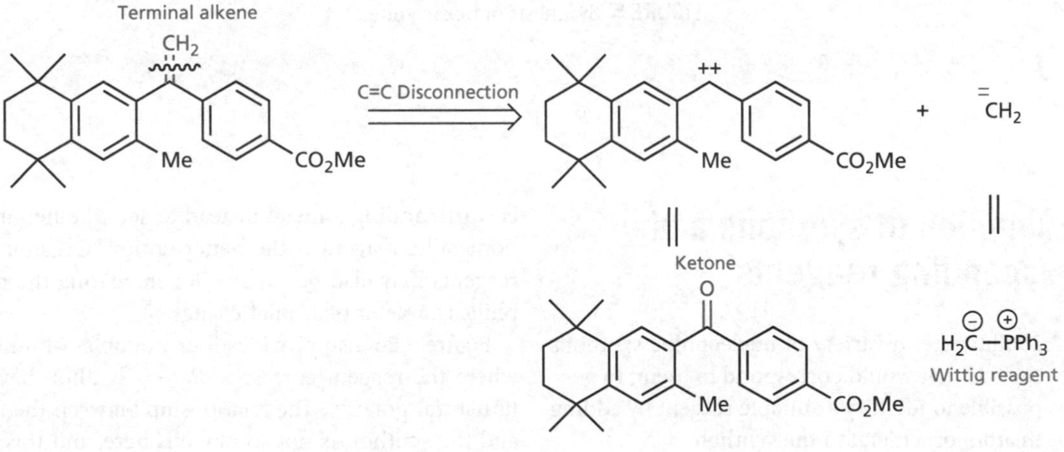

FIGURE 3 C=C Disconnection of a terminal alkene.

BOX 3.1 Retrosynthetic analysis of bexarotene *(Continued)*

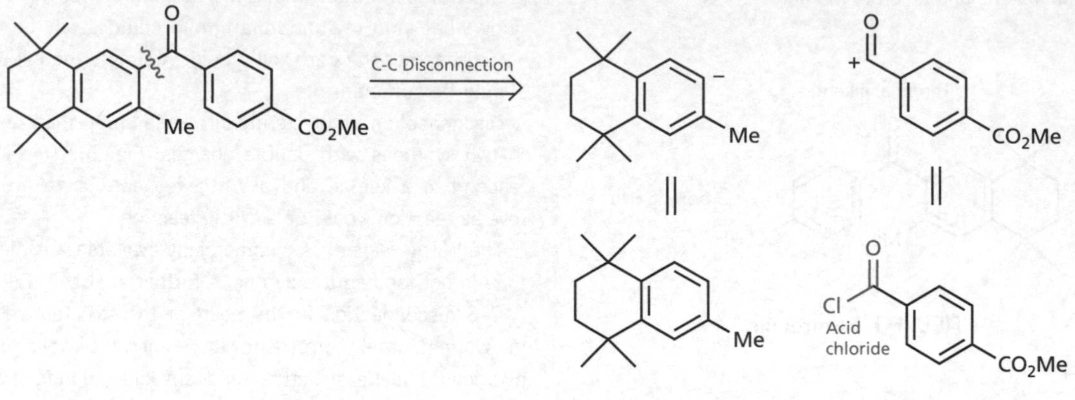

FIGURE 4 C–C Disconnection of an aromatic ketone.

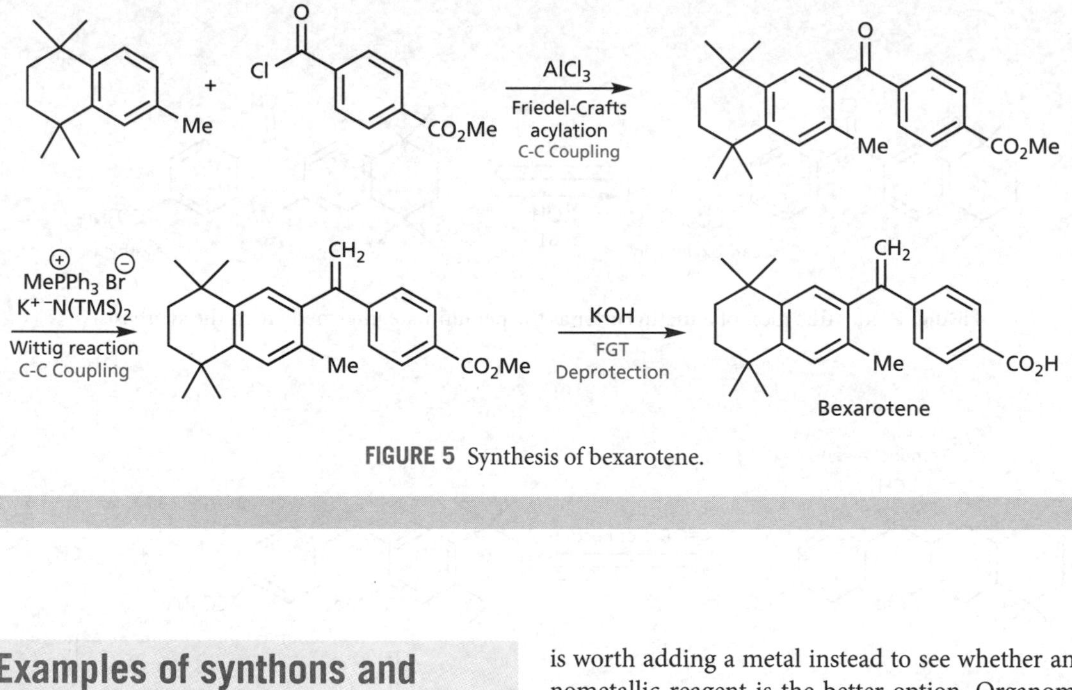

FIGURE 5 Synthesis of bexarotene.

3.7 Examples of synthons and corresponding reagents

Figure 3.26 illustrates a variety of nucleophilic synthons and the reagents that would correspond to them. In general, it is possible to identify a suitable reagent by adding a hydrogen atom or a metal to the synthon.

The addition of a hydrogen atom may result in a molecule that is known to be nucleophilic, such as an aromatic ring, an amine or a phenol.

If the addition of a hydrogen atom to a nucleophilic synthon does not provide a nucleophilic reagent, then it is worth adding a metal instead to see whether an organometallic reagent is the better option. Organometallic reagents may also be suitable for increasing the nucleophilic character of aromatic rings.

Figure 3.26 also provides four examples of umpolung where the reagent corresponds to a synthon having an unnatural polarity. The relationship between the reagent and the synthon is not so obvious here, and this means that two or more reactions may be involved before the relationship becomes evident. We have already discussed 1,3-dithianes as reagents which represent a nucleophilic acyl group. A reaction is required to synthesize the dithiane from the parent aldehyde before an alkylation takes

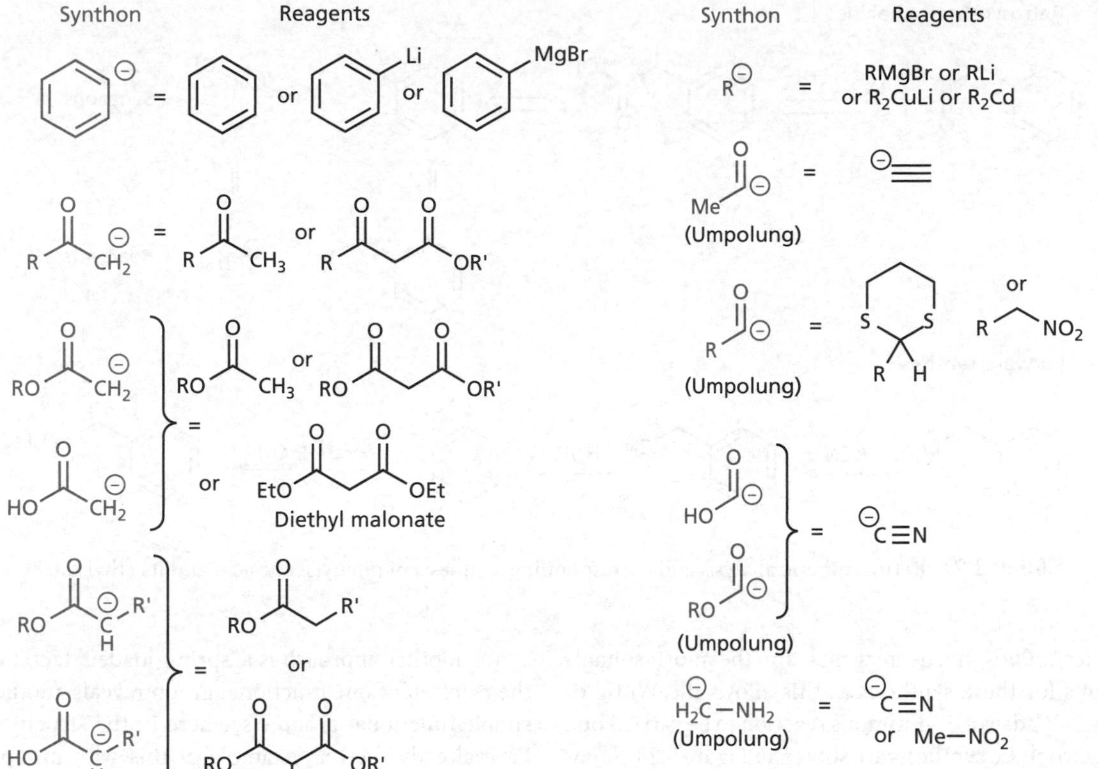

Synthon	Reagents	Synthon	Reagents

FIGURE 3.26 Nucleophilic synthons and corresponding reagents (synthons are shown in blue).

place; then the dithiane has to be hydrolysed to restore the original carbonyl group.

Another example of umpolung is the cyanide group. This is commonly used as a reagent for nucleophilic synthons representing a carboxylic acid or ester ($^-CO_2H$ and $^-CO_2R$). Again, the relationship between the reagent and the synthon may not appear obvious, but it is easy to hydrolyse a nitrile group to a carboxylic

acid and then convert the acid to an ester (Fig. 3.27). The nitrile group can also be used to represent the synthon $^-CH_2NH_2$, since it can be reduced to a primary amine using $LiAlH_4$. In all these cases, the original carbon chain will be extended by one carbon unit; this is known as a homologation.

Synthons with a double negative charge will result from disconnection of carbon–carbon double bonds

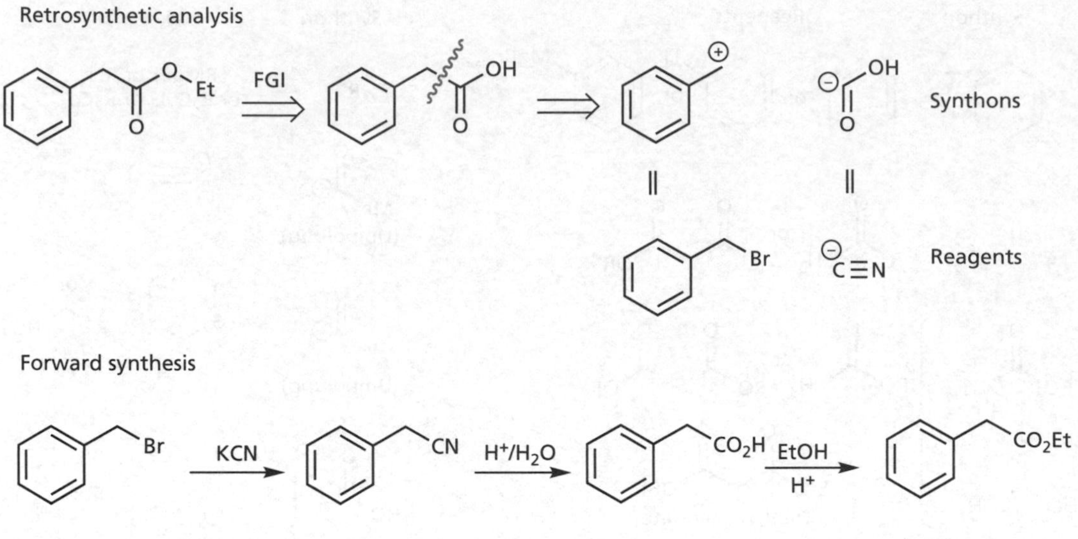

Retrosynthetic analysis

Forward synthesis

FIGURE 3.27 Retrosynthetic analysis and corresponding synthesis of phenylacetic acid and its ethyl ester.

(alkenes). Phosphorus reagents are the most suitable reagents for these synthons as this allows the Wittig or Horner–Wadsworth–Emmons reaction to be carried out.

Electrophilic synthons are shown in Figure 3.28. Suitable reagents are often identified by adding a good leaving group to the synthon. Good leaving groups are typically halogens, mesylates, and tosylates, but acyl, alkoxy, and hydroxy groups are also relevant groups to consider.

Several synthons in Figure 3.28 are doubly charged. This normally indicates that a reagent should have a carbonyl group, which corresponds to the addition of two OH groups to the synthon (Fig. 3.29).

3.8 Protecting groups and latent groups

Whenever a molecule contains two or more functional groups, it is likely that protecting groups may have to be considered as part of the retrosynthetic process. This will usually become apparent when the feasibility of the forward reaction is related to a particular retrosynthetic step. If it is likely that competing reactions involving different functional groups can occur in the forward reaction, then additional retrosynthetic steps will have to be considered to introduce the relevant protecting group.

A different approach is to consider the introduction of a functional group which is relatively unreactive, but which could be converted to the desired functional group when required. For example, an aromatic nitro group is a latent group for an amine, and a nitrile group serves as a latent group for carboxylic acids, esters, and amides.

Yet another approach is a 'spring-loaded' tactic where the reaction of one functional group reveals another desirable functional group elsewhere in the structure. We have already seen an example of this with an epoxide (Fig. 3.20). The epoxide serves as the electrophilic group required for addition of an amine, and as a consequence of the reaction an alcohol group is revealed in the structure. This avoids the need to protect the alcohol as it does not appear until the reaction takes place. Other cyclic reagents can be used in this manner. For example, consider the retrosynthesis of the keto acid in Figure 3.30. An obvious disconnection is the one between the carbonyl group and the aromatic ring, as this would result in reagents that could be linked together by a Friedel–Crafts acylation—the benzene ring and an acid chloride.

However, we have a problem. The acid chloride reagent also includes a carboxylic acid group. Since acid chlorides are synthesized from carboxylic acids, we would have to generate the desired acid chloride from a diacid, such that only one of the carboxylic acids is converted to the acid chloride. This is not easy and it is likely that both carboxylic acid groups will be converted to acid chlorides (Fig. 3.31). We could try to react this reagent with benzene, but it is highly likely that the reagent will react with two molecules of benzene to give us the wrong product.

So why don't we protect one of the carboxylic acids as an ester, convert the remaining carboxylic acid to the acid chloride, carry out the acylation, and then hydrolyse the ester back to the carboxylic acid (Fig. 3.32)?

We are certainly more likely to get the correct product, but again we are faced with the problem of trying to distinguish between two identical functional groups, this time in an attempt to esterify one of the groups but not

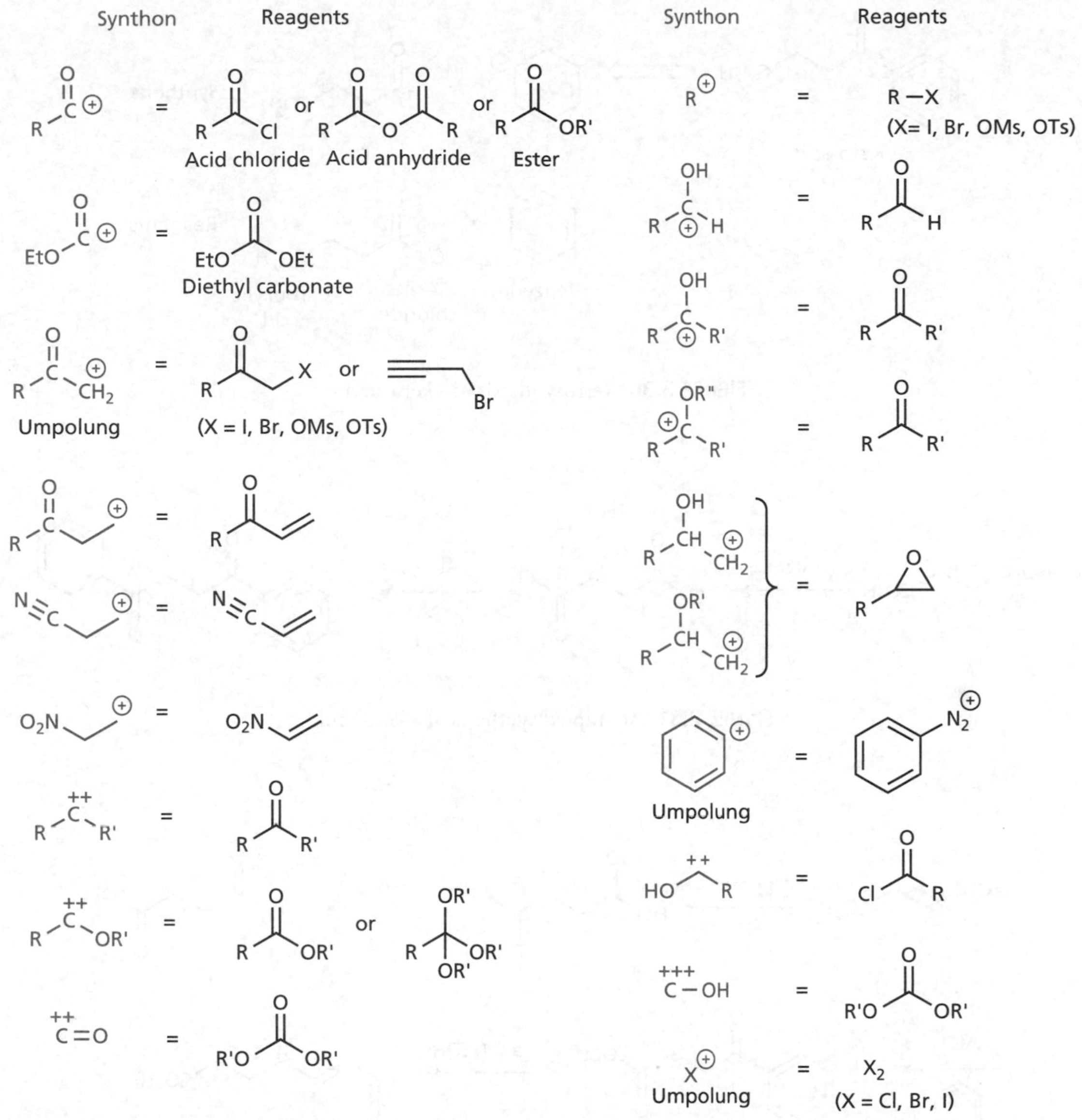

Synthon	Reagents		Synthon	Reagents

FIGURE 3.28 Electrophilic synthons and corresponding reagents (synthons are shown in blue).

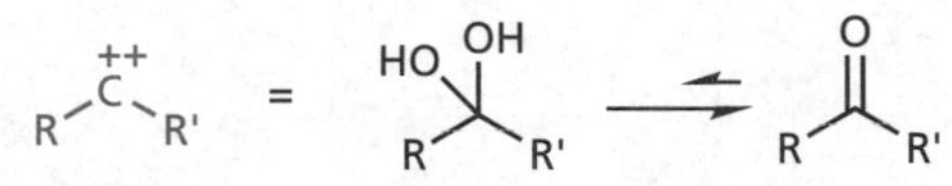

FIGURE 3.29 Reagents for doubly charged synthons.

the other. Again, it will be difficult to control the reaction to get the mono-ester.

Another approach would be to use a different starting material which contains the carboxylic acid we want plus a different functional group that could be converted to the other carboxylic acid once the Friedel–Crafts reaction has been carried out. We have already seen that a

cyanide group can act as a latent group for a carboxylic acid, so one possible synthesis would be to start with 3-cyanopropanoic acid (Fig. 3.33).

This certainly looks feasible, but a better method involves a 'spring-loaded' tactic where a cyclic anhydride is used for the Friedel–Crafts reaction (Fig. 3.34). Not only is the aromatic ketone introduced by the acylation reaction, but a carboxylic acid is automatically released as a result of the reaction. Therefore, the target compound is obtained in a single reaction. This is clearly more efficient than any of the previously proposed routes. Moreover, the succinic anhydride costs only 28 pence per gram. Compare that with the cost of 3-cyanopropanoic acid (Fig. 3.33), which is £790 per gram! Succinic anhydride has been used in a

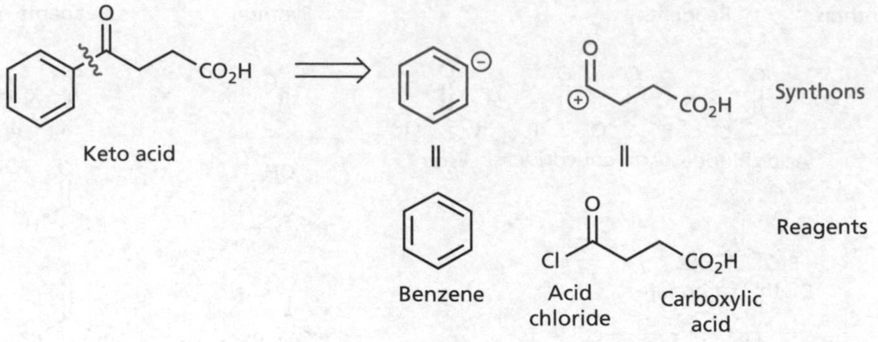

FIGURE 3.30 Retrosynthesis of a keto acid.

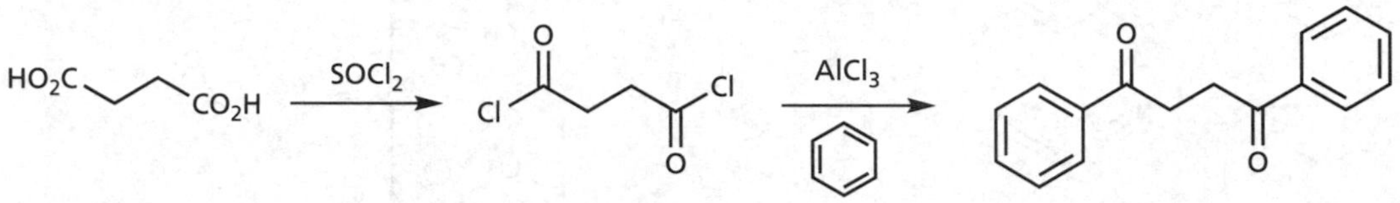

FIGURE 3.31 Attempted synthesis of a keto acid.

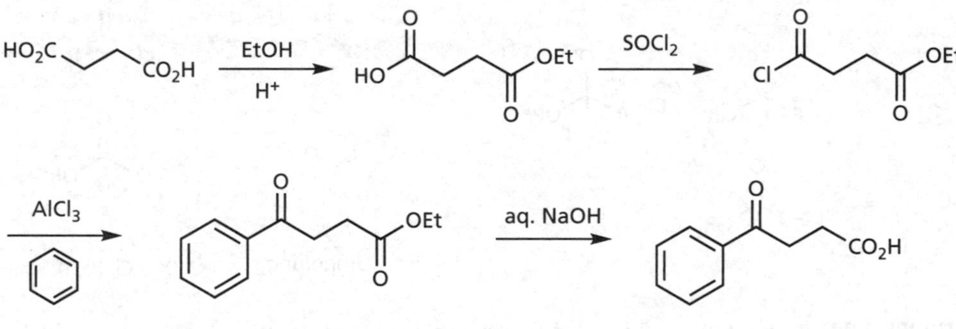

FIGURE 3.32 Attempted synthesis of a keto acid.

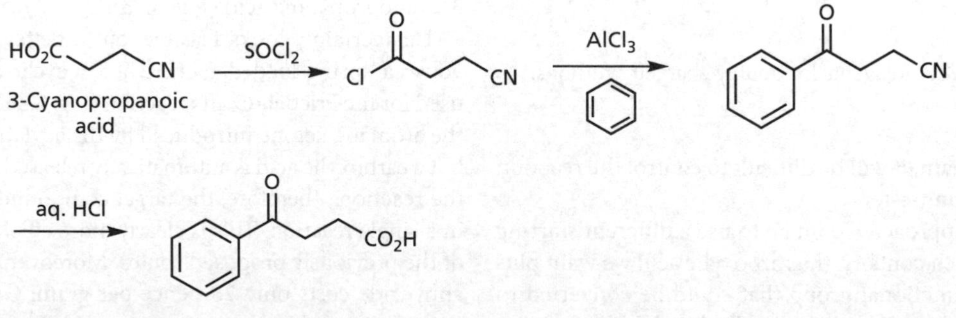

FIGURE 3.33 Possible synthesis of a keto acid from 3-cyanopropanoic acid.

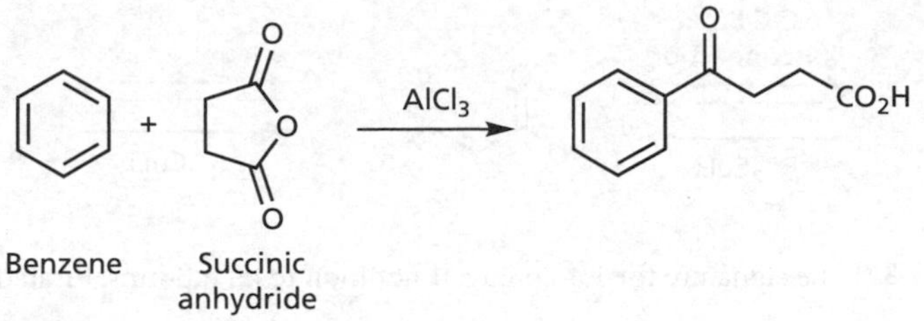

FIGURE 3.34 Synthesis of a keto acid from benzene and succinic anhydride.

Friedel–Crafts reaction for this very purpose during the synthesis of the cardiovascular drug **ibutilide** (Box 2.3).

3.9 Molecular signatures

The larger or more complex the target structure, the greater the number of possible disconnections, resulting in a mammoth number of possible retrosynthetic schemes. This may be an advantage in allowing us to explore different possible solutions, but it can also be a disadvantage as the number of possible options becomes overwhelming. Therefore, it is helpful to identify key features (molecular 'signatures') within a structure that indicate which disconnections are the most sensible to make.

The main features to look out for are functional groups. Functional groups are key to the chemical reactions involved in C–C and C–X bond formation, and there is usually a functional group present as a result of those reactions. In other words, the functional group acts as a signature for a particular reaction. Therefore, it makes sense to identify

the functional groups present in the target structure and to assess what kind of reactions could be carried out to produce them. For example, a secondary alcohol group in the target molecule could result from the reduction of a ketone (a functional group transformation) or from a Grignard reaction (a C–C bond formation) (Fig. 3.35).

Molecular signatures for particular reactions may include more than just a functional group. For example, a tertiary alcohol with two identical alkyl groups can be viewed as a 'signature' for a Grignard reaction carried out on an ester group (Fig. 3.36a). A double bond within a six-membered ring (a cyclohexene structure) is a recognizable signature for the Diels–Alder reaction (Fig. 3.36b).

The position of a substituent relative to a functional group may also act as a signature. For example, an alkyl substituent located at the β-position to a ketone could be the signature for conjugate addition to an α,β-unsaturated ketone (Fig. 3.37).

If two or more functional groups are present, their relative positions may provide a signature for a particular reaction. We have already seen that an amine and an alcohol on adjacent carbons indicate the reaction of an

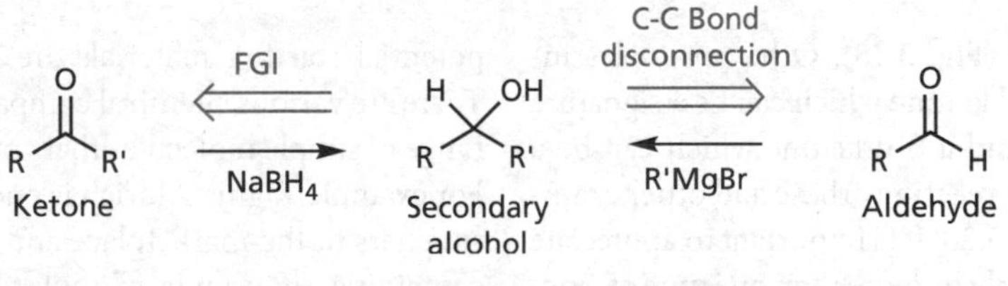

FIGURE 3.35 Reactions that are likely to result in a secondary alcohol.

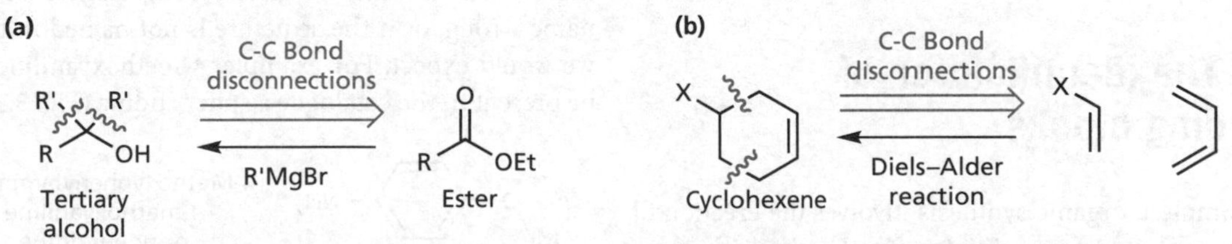

FIGURE 3.36 The signatures for (a) a Grignard reaction carried out on an ester and (b) the Diels–Alder reaction.

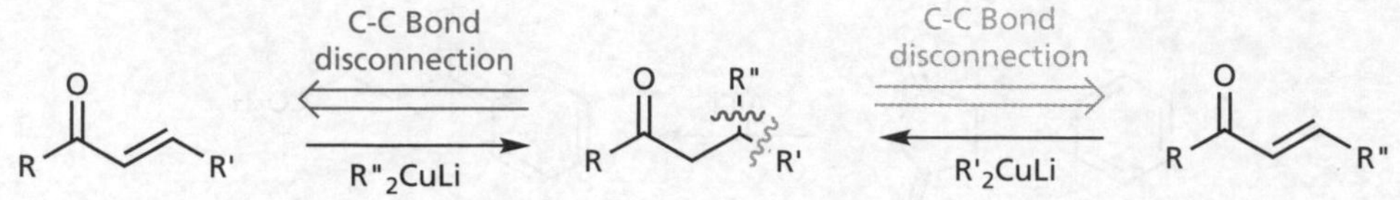

FIGURE 3.37 The signature for 1,4-conjugate addition to an α,β–unsaturated ketone.

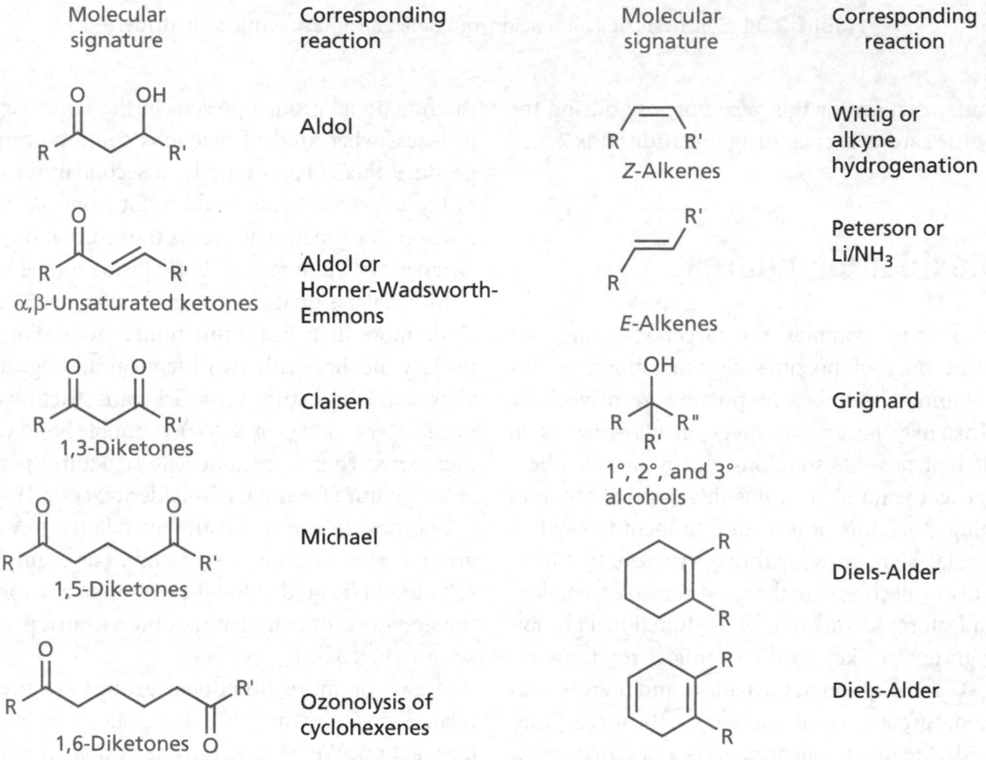

FIGURE 3.38 Various molecular 'signatures' and some of the reactions that can be used to form them (see also Appendices 4 and 5).

epoxide with an amine (Fig. 3.18). Other examples include an α,β-unsaturated ketone which can be a signature for the aldol reaction, and a β-diketone which can be a signature for the Claisen reaction. These and other examples are shown in Figure 3.38. It is important to appreciate that a 'signature' may apply to more than one type of reaction. For example, there are other ways of synthesizing an α,β-unsaturated ketone than just the aldol reaction.

3.10 The identification of building blocks

At its simplest, organic synthesis involves the creation of a target molecule from simpler structures which serve as the building blocks for the synthesis. Therefore, when planning a synthesis, it is crucial to identify whether

potential starting materials are commercially available. There are various chemical companies that provide a large range of simple molecules that could be used in synthesis. For example, Sigma Aldrich is one of the biggest chemical suppliers in the marketplace and has an online catalogue containing thousands of molecules. The catalogue can be searched for specific compounds by entering the name of the compound we want. However, there is a risk that we will not find the desired compound if we get the name wrong, or if the structure is not named in the way we would expect. For example, 4-methoxyaniline might be present in the catalogue as phenetidine (Fig. 3.39).

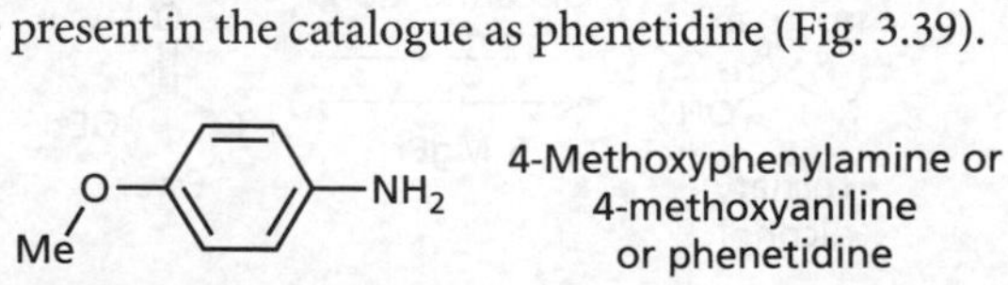

4-Methoxyphenylamine or
4-methoxyaniline
or phenetidine

FIGURE 3.39 Different ways of naming a compound.

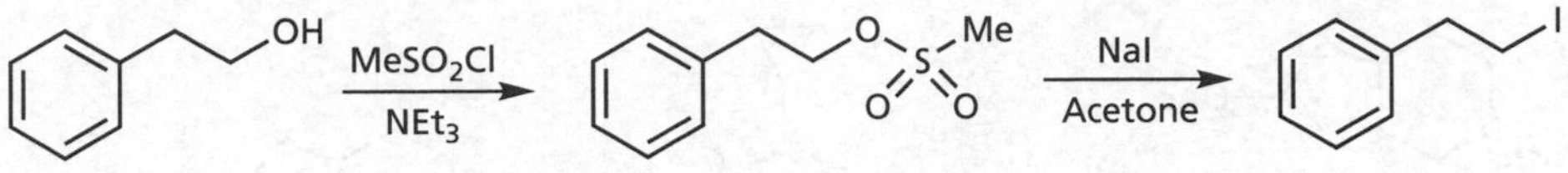

FIGURE 3.40 Synthesis of phenethyl iodide from phenethyl alcohol.

If there is uncertainty about how the structure is named, we can search the catalogue by the molecular formula. For example, searching for the formula C_7H_9NO would show up all the structures having that formula, including phenetidine.

Another method of searching the catalogue is to carry out a structure search. The search can be controlled such that it searches for the specific compound wanted. Alternatively, it can be carried out to find all structures that are similar to the desired compound. This can be useful as it may throw up structures that might be useful alternative starting materials to the one we originally intended. For example, we might have been looking for phenethyl iodide as a starting material. This may not be commercially available, but the search would indicate phenethyl bromide and phenethyl alcohol which could both be easily converted to phenethyl iodide (Fig. 3.40).

3.11 Useful strategies in retrosynthesis

There are a number of guidelines which can be considered when carrying out a retrosynthetic analysis.

- Identify suitable disconnections by recognizing molecular signatures in the target structure (section 3.9).
- Identify key disconnections that are likely to result in feasible synthons and achievable reactions such as disconnections involving:
  - a bond between an aromatic ring and a substituent
  - a bond between other forms of ring and their substituents
  - a C–X bond where X is a heteroatom (N, O, or S), including RCOX
  - bonds within rings that would be formed by a favoured cyclization.
- Identify favoured disconnections that are in the centre of the structure, rather than those at the periphery. It is more efficient to develop a synthesis that creates two similarly sized compounds which can be linked together towards the end of the synthesis. This is known as a **convergent synthesis**.

- Favour disconnections that result in symmetrical synthons and reagents.
- Favour disconnections at branch points (substituents) in the target structure. These are more likely to result in simple synthons and reagents.
- At each stage of the retrosynthetic analysis, ensure that reagents for the synthons or structures produced are available, and that the reaction involving these reagents is feasible.
- Modify the retrosynthesis scheme in the light of the preceding findings and introduce protection strategies, latent functional groups, or 'spring-loaded' functional groups as required.
- Identify a retrosynthetic scheme that leads to simple commercially available starting materials with the minimum number of steps, such that the actual synthesis is as short as possible.

We can see some of these principles at work in the retrosynthetic analysis of the platelet inhibitor **prasugrel** (Fig. 3.41). There are several bond disconnections that we could consider in this molecule, but there is one disconnection which fits in particularly well with several of the key points stated above. The disconnection shown in Figure 3.41 has the following advantages.

- It is between a carbon atom and a heteroatom, and so the synthon with the heteroatom is a natural nucleophile.
- The disconnection is in the heart of the molecule and so two synthons of roughly equivalent size are produced, which allows a convergent synthesis.
- The disconnection is between a branch point and a ring, resulting in synthons with relatively simple structures.
- There are feasible structures that could represent these synthons—an α-halo ketone and a tetrahydrothienopyridine ring system.
- The reaction required to link these structures also looks feasible. The secondary amine group of the tetrahydrothienopyridine ring should be capable of substituting the halogen of the α-halo ketone in a nucleophilic substitution reaction.

We can now subject the α-halo ketone to a retrosynthetic analysis (Fig. 3.42).

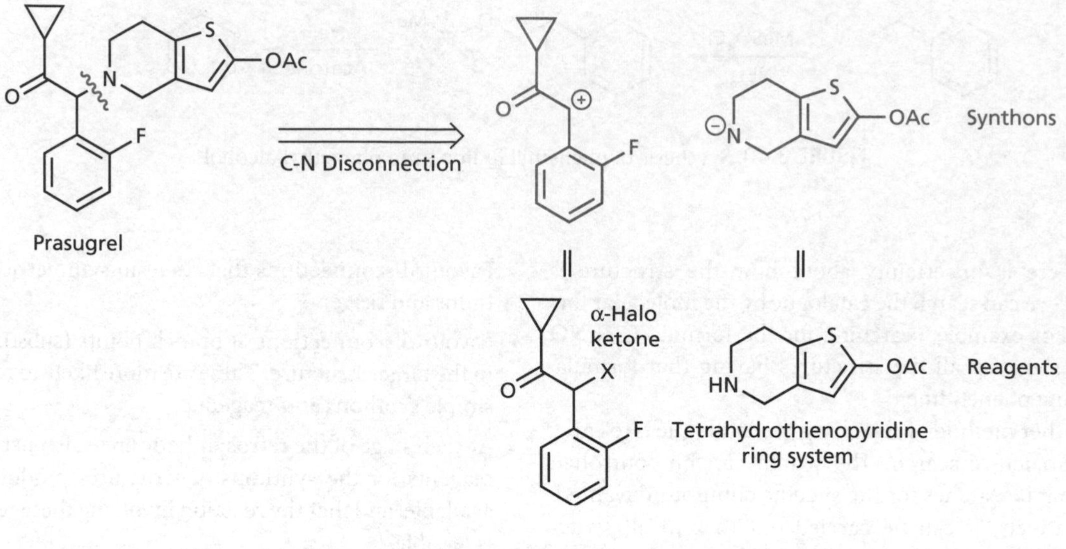

FIGURE 3.41 Retrosynthetic analysis of prasugrel.

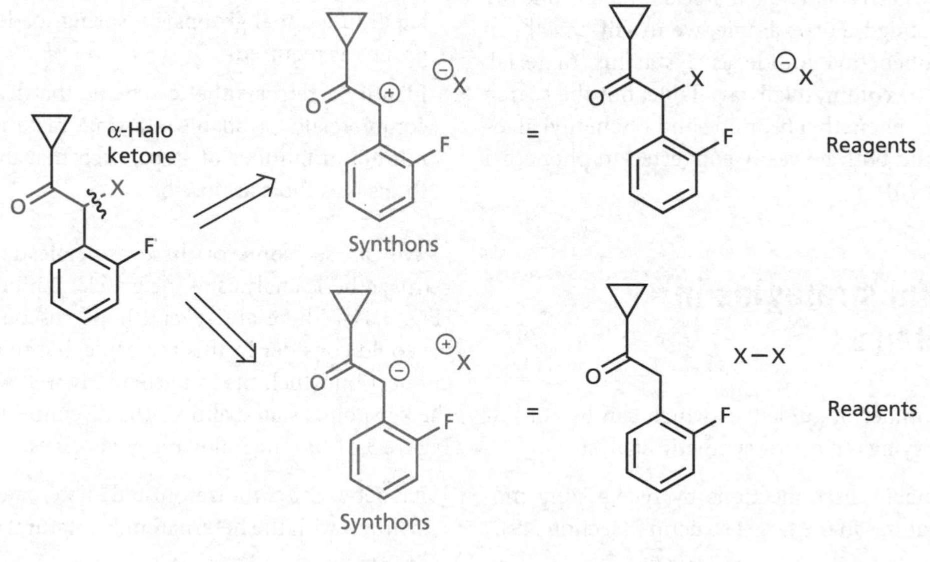

FIGURE 3.42 Retrosynthetic analysis of the α-halo ketone.

The most obvious disconnection would be the C–X bond, with the halogen corresponding to a nucleophilic halide ion. However, the other synthon would be another halo ketone, so that takes us back to square one.

Therefore, we should consider the other possible pair of synthons from this disconnection. This gives us a ketone which would act as a nucleophile at the α-carbon. That fits in well with the known chemistry of ketones which can lose an α-proton in the presence of a base to form a carbanion at that position. The positively charged halogen looks a bit of a problem at first sight. However, it is not really a problem at all, as it is well known that bromine, chlorine, and iodine are diatomic molecules

that can act as electrophiles when they react with nucleophiles (e.g. alkenes) (Fig. 3.43).

A retrosynthetic analysis can now be carried out on the ketone and results in quite a number of retrosynthetic schemes (Fig. 3.44). Deciding which is the best scheme would involve identifying whether the reagents required for the corresponding synthesis are commercially

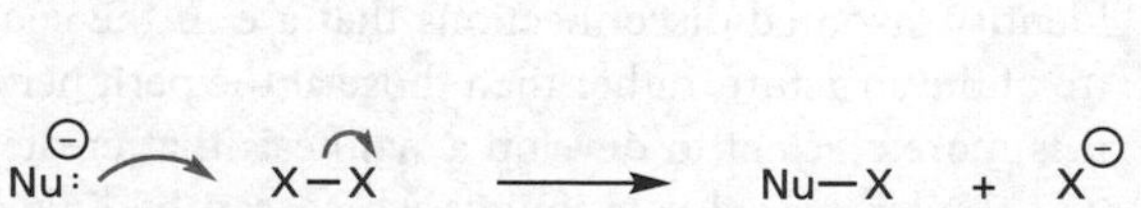

FIGURE 3.43 Reaction of halogens with nucleophiles.

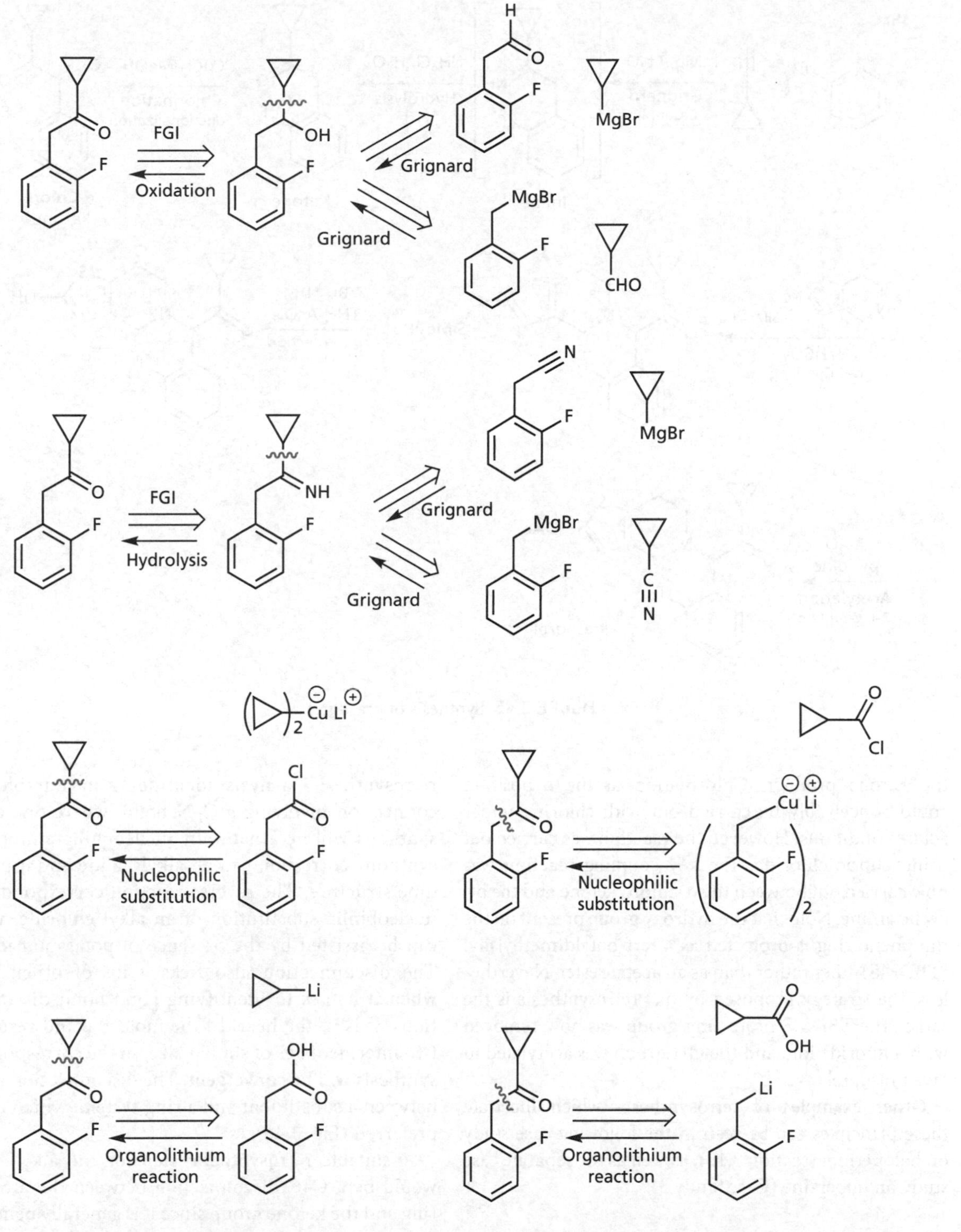

FIGURE 3.44 Examples of different retrosynthetic analyses carried out on the ketone.

available and whether the reactions involved are feasible. A search of the literature is always worth carrying out to investigate whether similar reactions have been carried out already, but it is often the case that the best method is only discovered by trying the schemes out in practice.

The actual synthetic route used to produce prasugrel started with a Grignard reaction between a benzyl bromide and a nitrile (Fig. 3.45). This corresponds to one of the retrosyntheses described in Figure 3.44. The imine product from the reaction was hydrolysed in situ to give

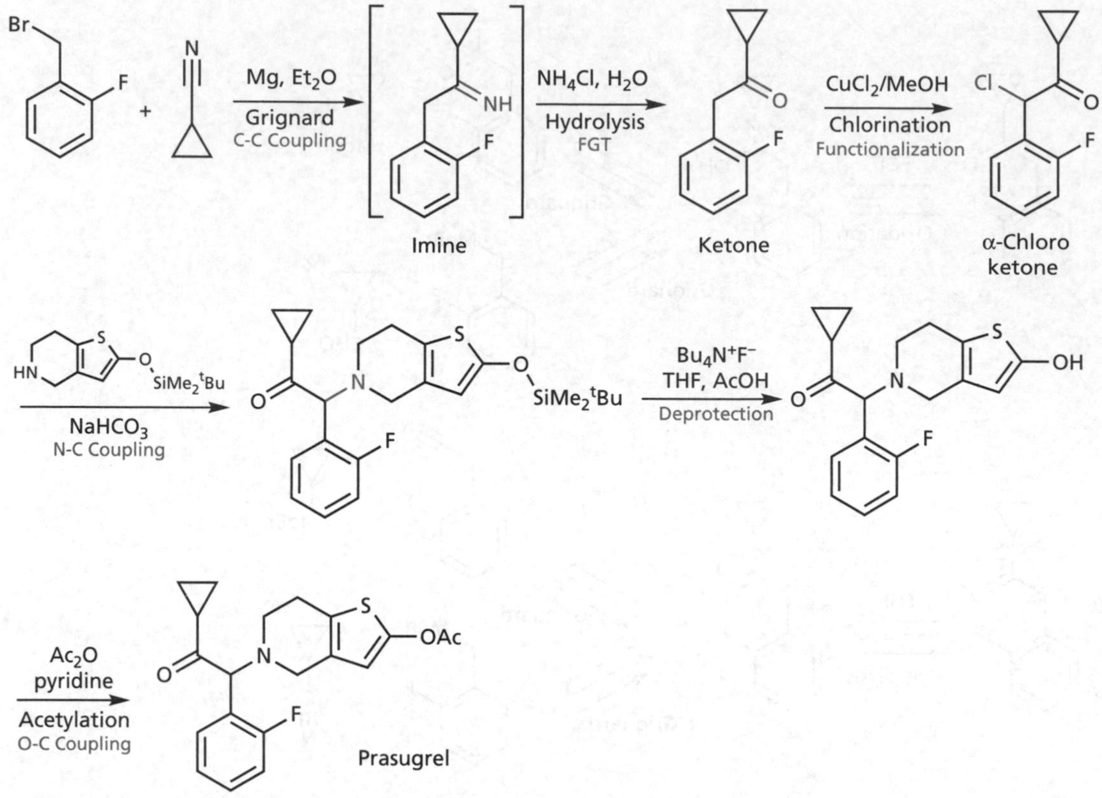

FIGURE 3.45 Synthesis of prasugrel.

the ketone product. Chlorination at the α-position could conceivably be carried out with chlorine under acidic conditions. However, the reaction was carried out using cupric chloride. The key coupling reaction was now carried out between the α-chloro ketone and the bicyclic amine. Note that the hydroxy group present on the thiophene ring is protected as a tert-butyldimethylsilyl (TBDMS) ether rather than as an acetate ester. Nevertheless, the strategy proposed by our retrosynthesis is the same. The TBDMS protecting group was now removed with a fluoride ion, and the OH group was acetylated to give prasugrel.

Other examples of retrosyntheses which illustrate these principles can be seen in the following case study on haloperidol (section 3.12), as well as the separate case study on huperzine (Case Study 3).

3.12 Case Study—Retrosynthetic analysis of haloperidol

Haloperidol, which was first synthesized in the late 1950s, is an early example of an antipsychotic agent and is used in the treatment of schizophrenia. A retrosynthesis analysis identifies a favourable disconnection involving a C–N bond, where one of the synthons will be a naturally nucleophilic amine. The synthons correspond to an alkyl halide and a piperidine structure. The reverse reaction corresponds to a nucleophilic substitution of an alkyl chloride, which can be assisted by the presence of potassium iodide. The disconnection also ticks a lot of other boxes when it comes to identifying good bond disconnections. It is in the heart of the molecule and results in two intermediates of similar size, so the corresponding synthesis will be convergent. The disconnection is also between a substituent and a ring system, which is also preferred (Fig. 3.46).

A suitable retrosynthetic step for the alkyl halide would be a C–C disconnection between the aromatic ring and the ketone group since it is generally beneficial to have a disconnection between an aromatic ring and a substituent (Fig. 3.47). The aromatic ring then acts as a natural nucleophile. The reagents corresponding to the synthons are fluorobenzene and 4-chlorobutyryl chloride, which could be linked together by a Friedel–Crafts acylation (Fig. 3.48). Since an acid chloride is more reactive than an alkyl chloride, it should be possible to carry out this reaction without any competing

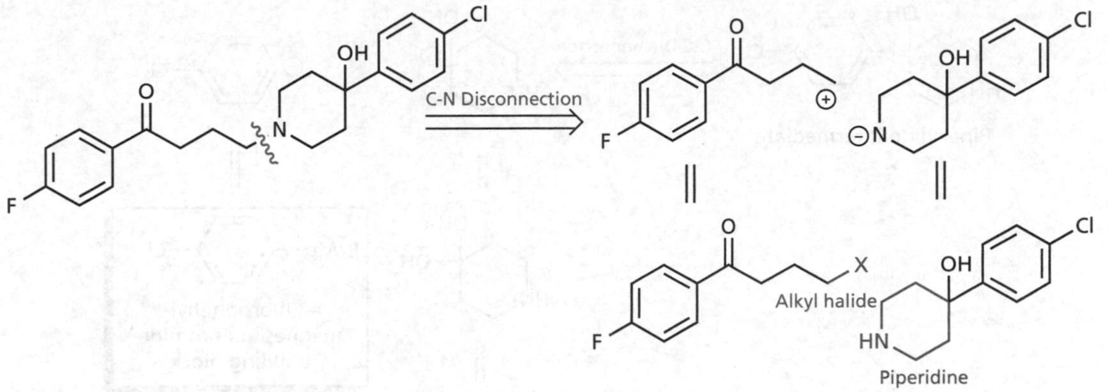

FIGURE 3.46 Retrosynthesis of haloperidol—stage 1.

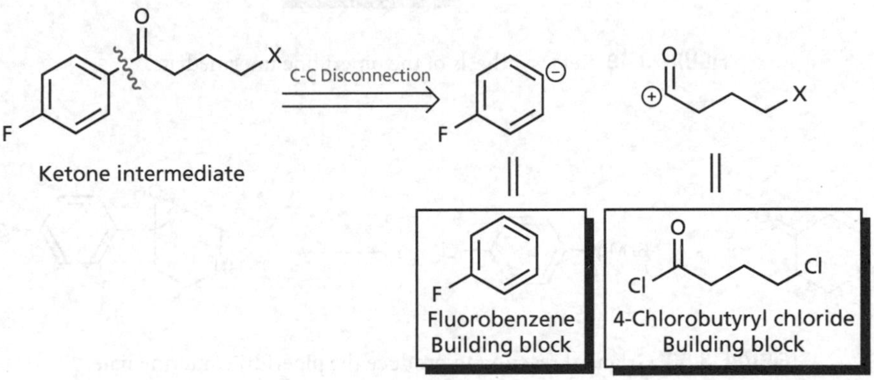

FIGURE 3.47 C–C Disconnection of the alkyl chloride intermediate.

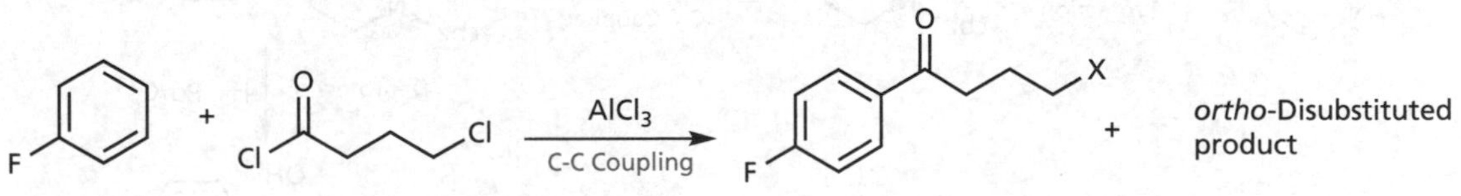

FIGURE 3.48 Corresponding C–C coupling.

Friedel–Crafts alkylation. In other words, the reaction should be chemoselective. It should also be regioselective since a fluoro substituent on the aromatic ring will direct substitution to the *ortho* and *para* positions and not the *meta* position. If any *ortho*-disubstituted product is formed, a separation of the *ortho*- and *para*-disubstituted products will be necessary after the

reaction. However, this is acceptable since it will be the first stage in the synthesis and the reagents are cheap and readily available.

A good retrosynthetic step for the piperidine structure would be a C–C disconnection between the piperidine and aromatic rings (Fig. 3.49). The disconnection is between two rings and in the heart of the molecule. The

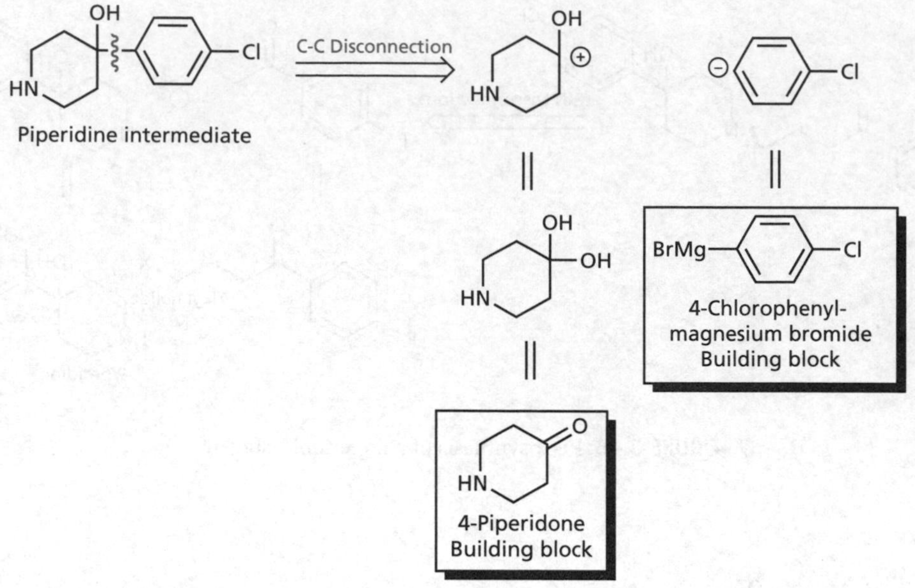

FIGURE 3.49 Retrosynthesis of the piperidine intermediate.

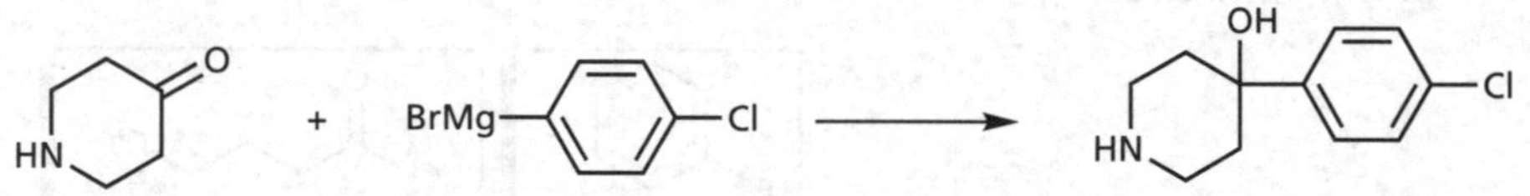

FIGURE 3.50 Grignard reaction to produce the piperidine intermediate.

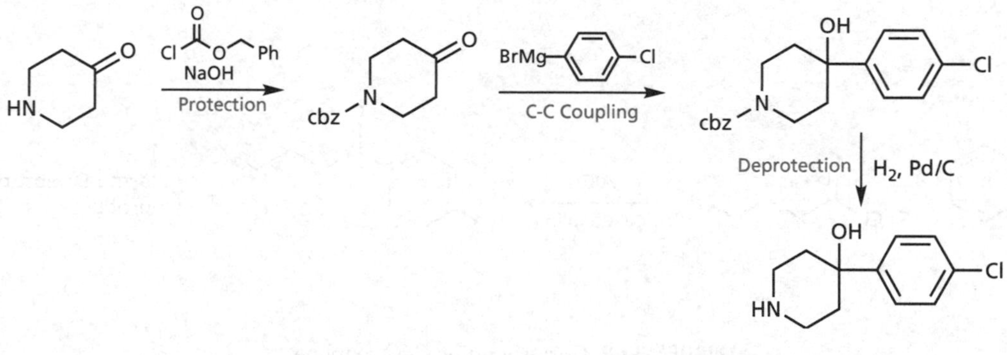

FIGURE 3.51 Inclusion of a protecting group strategy.

synthons correspond to 4-piperidone and 4-chlorophenyl magnesium bromide, which are both simple molecules.

A Grignard reaction should provide the desired product (Fig. 3.50), but it would be necessary to use two equivalents of the Grignard reagent since one equivalent would be used up reacting with the NH proton in an acid–base reaction.

To avoid this, it may be worth protecting the amine group of 4-piperidone prior to the Grignard reaction, and then removing the protecting group prior to the N–C coupling reaction (Fig. 3.51).

Therefore, the actual synthesis would be convergent and involve four building blocks (Figs. 3.52 and 3.53).

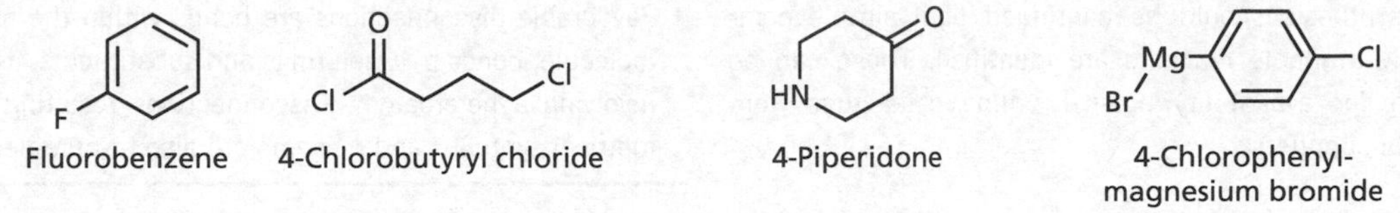

FIGURE 3.52 Building blocks required for the synthesis of haloperidol.

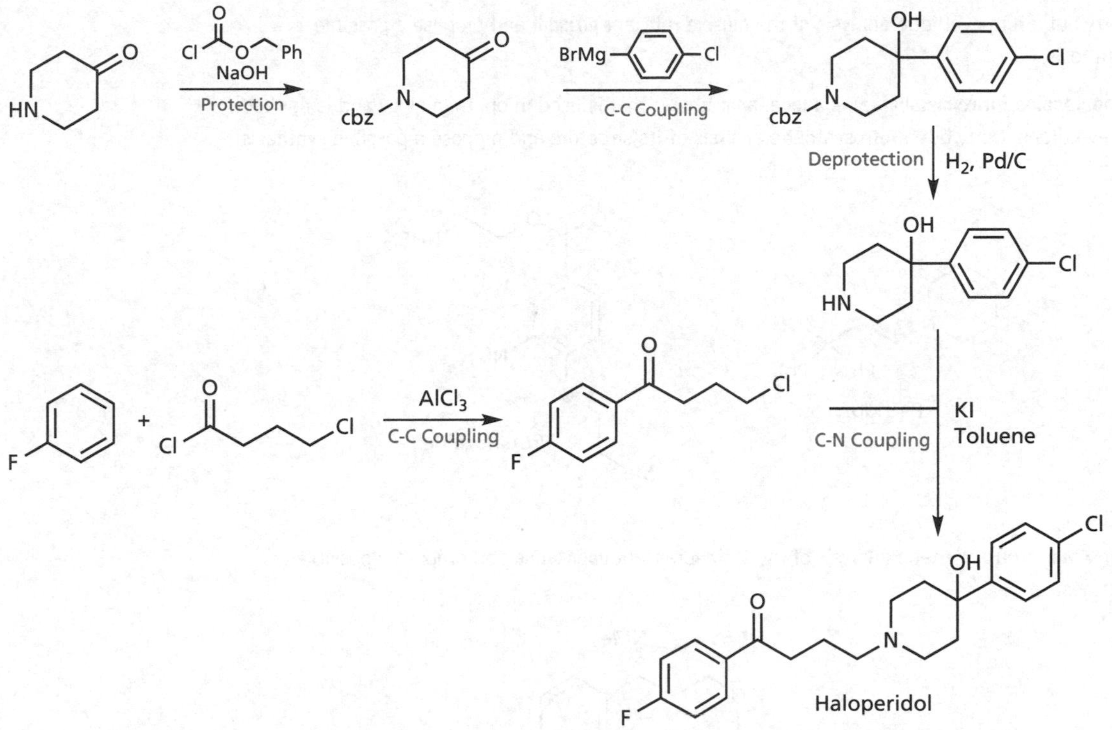

FIGURE 3.53 Synthesis of haloperidol.

KEY POINTS

- Retrosynthesis is used in the design of synthetic routes. The target structure is analysed for feasible disconnections in order to identify what simpler building blocks could be used for its synthesis.

- Each step in a retrosynthesis must be validated by identifying a feasible reverse reaction.

- The fragments resulting from a disconnection are called synthons. They are not real species or reagents, but can be used to identify reagents that correspond to them.

- Reagents corresponding to a positively charged synthon can usually be identified by adding a proton or metal to the synthon.

- Reagents corresponding to a negatively charged synthon can usually be identified by adding a hydroxyl group or a halogen.

- Each disconnection offers the possibility of two sets of synthons.

- The preferred synthons are those where the charge is compatible with the electronic nature of the structure.

- Favourable disconnections include C–X bonds where X is a heteroatom.

- Functional group interconversions are included in retrosynthetic schemes.

- Umpolung refers to synthons which have charges that do not match the natural polarity of the group involved.

- Considering the reverse reaction for a disconnection may indicate the need for protecting groups or latent groups. The retrosynthesis can be modified accordingly.

- Molecular signatures are patterns of functional groups and substituents that are characteristic of a particular reaction. They help to identify suitable disconnections for a target molecule.

• A retrosynthesis should be continued until simple commercially available reagents are identified. These can be checked for availability against online catalogues from chemical suppliers.

• Favourable disconnections are bonds within the heart of a molecule, bonds between rings and substituents, and bonds involving a heteroatom. Disconnections resulting in symmetrical synthons and reagents are also favourable.

QUESTIONS

1. Carry out a retrosynthetic analysis of the muscle relaxant **pirindol** and propose a possible synthesis.

2. **Proparacaine (proxymetacaine)** is a local anaesthetic that is used in ophthamology and is applied in eye drops. Carry out a retrosynthetic analysis of its structure and propose a possible synthesis.

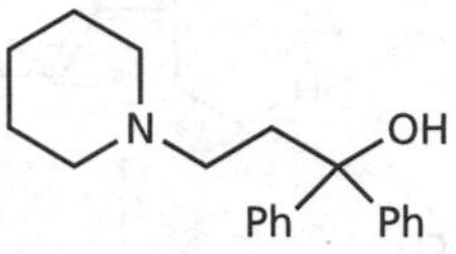

Pirindol

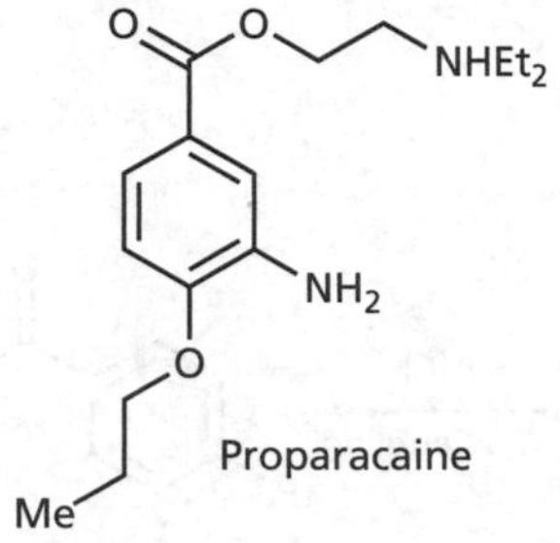

Proparacaine

3. Carry out a retrosynthetic analysis of the antidepressant **dapoxetine** and propose a possible synthesis.

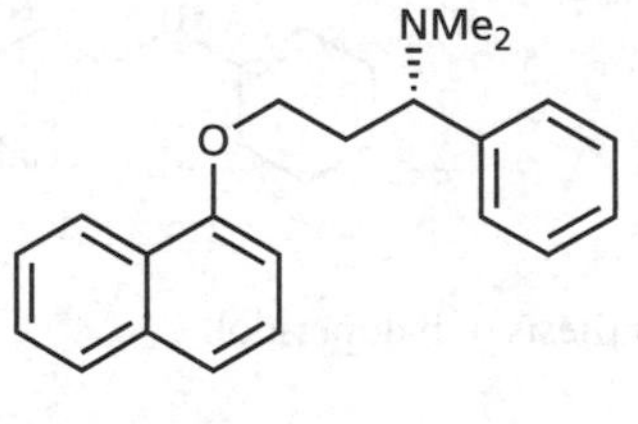

Dapoxetine

FURTHER READING

General reading

Clayden, J., Greeves, N., and Warren, S. (2012) *Organic chemistry* (2nd edn). Oxford University Press, Oxford, Chapter 28.

Warren, S. and Wyatt, P. (2008) *Organic synthesis: the disconnection approach* (2nd edn). Wiley, Chichester.

Warren, S. and Wyatt, P. (2008) *Workbook for Organic synthesis: the disconnection approach* (2nd edn). Wiley, Chichester.

Specific compounds

Dawson, M.L., et al. (1995) 'Bridged bicyclic aromatic compounds and their use in modulating gene expression of retinoid receptors', US Patent 5,466,861 (bexarotene).

Janssen, P.A.J. (1962) 'Pyrrolidine and piperidine derivatives', GB Patent 895,309 (haloperidol).

Janssen, P.A.J. (1969) '1-Aroylalkyl derivatives of arylhydroxypyrrolidines and arylhydroxy-piperidines', US Patent 3,438991 (haloperidol).

Liu, K.-C., et al. (2011) 'Synthetic approaches to the 2009 new drugs', *Bioorganic and Medicinal Chemistry*, **19**, 1136–54 (prasugrel).

Main, B.G. and Tucker, H. (1985) 'Beta blockers', in S.M. Roberts and B.J. Price (eds), *Medicinal chemistry: the role of organic chemistry in drug research*. Academic Press, London, pp 69–92 (propranolol).

4 Cyclic systems in drug synthesis

4.1 Introduction

Cyclic systems play a vital role in drug structures and can be defined as **carbocyclic** or **heterocyclic**. The former ring systems are made up totally of carbon atoms, while the latter contain at least one heteroatom—usually N, O, or S. Rings can be of various sizes. They can also be monocyclic or multicyclic, where two or more rings are fused together. This means that there is an infinite variety of possible ring systems in chemistry. Having said

that, a relatively small number of different ring systems have been used in drug structures to date. Examples can be seen in the clinically important drugs shown in Figure 4.1.

There are pharmacodynamic and pharmacokinetic reasons why ring systems are so important in drugs. When it comes to binding interactions with a binding site, it is important that all the binding groups in a drug are correctly positioned. Rings are ideal scaffolds in order to achieve this. The rigidity of the ring

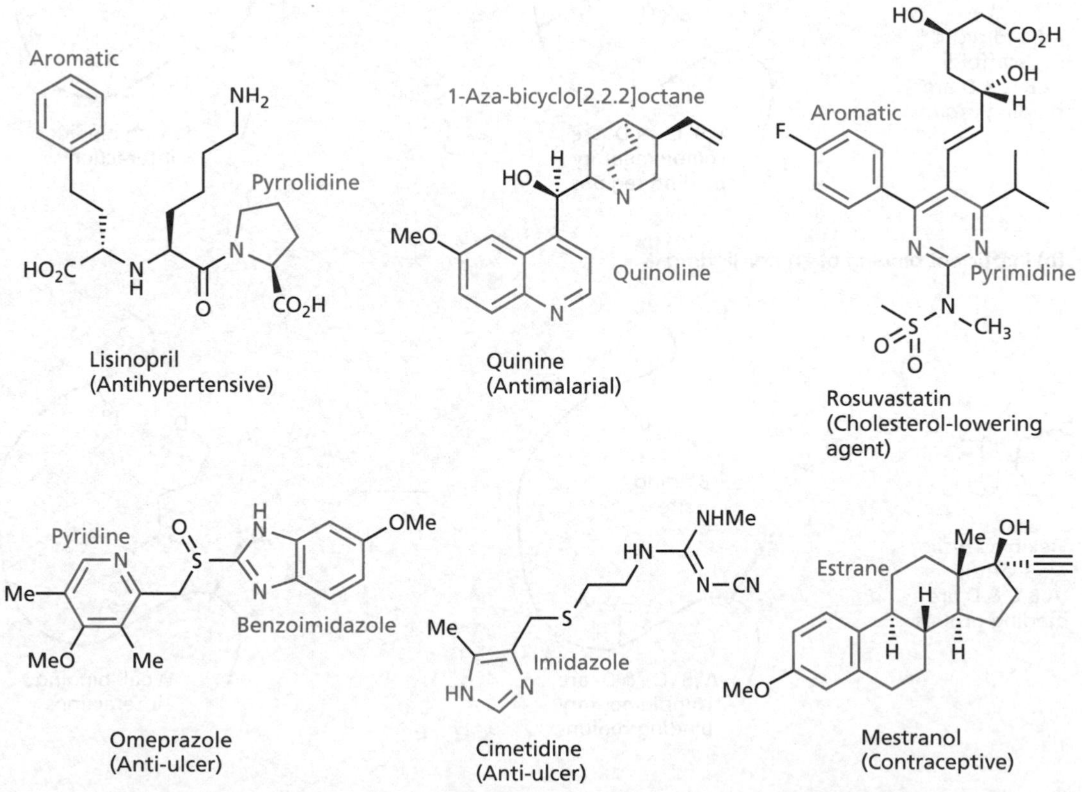

FIGURE 4.1 Examples of clinically important drugs with ring systems highlighted.

system allows binding groups to be positioned on the scaffold in a particular orientation with respect to each other. If this corresponds to the required pharmacophore, then the groups will interact simultaneously with complementary binding regions within the target binding site (Fig. 4.2). In other words, the active conformation for a particular drug is locked into place within the cyclic system.

With a flexible acyclic system containing several rotatable bonds, there are a large number of possible conformations, and only one of these will be the active conformation. Therefore, there is much less chance of the drug being in the active conformation when it enters the binding site. This means that the drug will have to drift in and out of the binding site several times before it adopts the active conformation, resulting in much weaker activity.

There is also an entropic factor at play here. When a flexible acyclic drug eventually binds to a target binding site, it becomes locked into its active conformation during the period that it is bound. This involves a significant entropy penalty as the structure can no longer adopt the many other conformations that are normally available to it. In contrast, there is no entropy penalty to be paid when a rigid cyclic structure binds. Since the overall binding energy is determined by the beneficial interactions gained on binding, minus any detrimental entropic effects, a rigid structure having the same binding interactions as a flexible structure invariably binds more effectively and has greater activity.

The rings themselves can also play an important role in the binding process. For example, a carbocyclic structure is hydrophobic in nature and is likely to form favourable van der Waals interactions with a hydrophobic binding

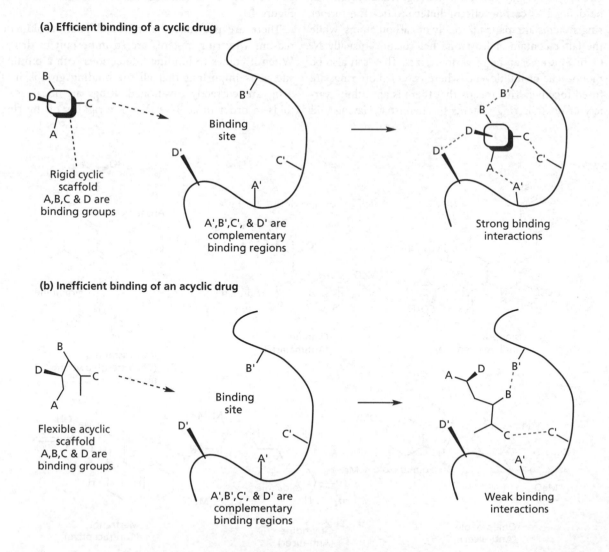

FIGURE 4.2 (a) Efficient binding of a drug containing a rigid cyclic scaffold. (b) Inefficient binding of a drug containing a flexible acyclic scaffold.

site. Heterocycles can also do this, but heterocycles have the added advantage that the heteroatom(s) present may form important hydrogen bonding interactions with the binding site. Nitrogen-containing heterocycles are particularly prevalent in drugs. In some of these agents the nitrogen is protonated and charged, and this allows the possibility of a strong ionic interaction as well as hydrogen bonding. The group is also beneficial for the drug's aqueous solubility. It is no coincidence that the number of heterocyclic rings in drugs is much greater than the number of purely carbocyclic rings.

In addition to their pharmacodynamic advantages, ring structures often have better pharmacokinetic properties than comparable acyclic compounds. Since the skeleton of a ring structure is mostly hydrophobic, it helps to offset the polarity of the polar functional groups that need to be present as binding groups. This allows the drug to have the correct balance of hydrophobicity and hydrophilicity required if it is to cross cell membranes and also be soluble in aqueous environments such as the gastrointestinal tract and the blood supply. Moreover, it has been found that rigid drugs with cyclic scaffolds are often absorbed from the gastrointestinal tract more efficiently than flexible drugs. Too many rotatable bonds can be detrimental to oral absorption (section 1.5.3).

Estradiol is a natural steroid hormone that contains a rigid carbocyclic ring system, and is commonly used alongside other drugs in various contraceptive treatments. The tetracyclic ring structure acts as a rigid scaffold which is roughly planar in shape, and contains phenol and alcohol substituents at opposite ends of the structure. Both of these functional groups are crucial to activity since they form hydrogen bonds with specific amino acids within the binding site of the estradiol receptor (Fig. 4.3a). Because of the rigidity of the tetracyclic

scaffold, both binding groups are ideally positioned such that they bind simultaneously to the respective amino acids. In addition, the region of the binding site that is occupied by the tetracyclic ring system is hydrophobic in nature, and so there are additional van der Waals binding interactions between the tetracyclic ring system and the binding site. Finally, there is very little entropic penalty involved in the binding process as the steroid structure is already locked into the active conformation. The tetracyclic ring system is also crucial in a pharmacokinetic sense since estradiol has to cross hydrophobic cell membranes in order to reach its target receptor within the cell. The phenol and alcohol groups are polar in nature, which is good for aqueous solubility but should hinder the crossing of hydrophobic barriers. However, the hydrophobic ring system counteracts this polarity, allowing the molecule to pass through the cell membrane.

It is important that medicinal chemists appreciate nature's design of endogenous compounds if they are to design analogues that act as either agonists or antagonists. For example, we could easily design an acyclic system that contains a phenol group and an alcohol group which could be held in the correct relative positions for binding (Fig. 4.3b). However, this active conformation would only be one of a large number of possible conformations, and so it is highly unlikely that the compound will be in the active conformation when it enters the binding site. There would also be a massive entropy penalty involved in binding, Moreover, there are fewer carbon atoms present in the structure, which means that the molecule will be less hydrophobic. This would be detrimental to the van der Waals interactions with the target binding site, as well as to the ability of the structure to cross the cell membrane. All in all, the acyclic structure is likely to have very weak or negligible activity.

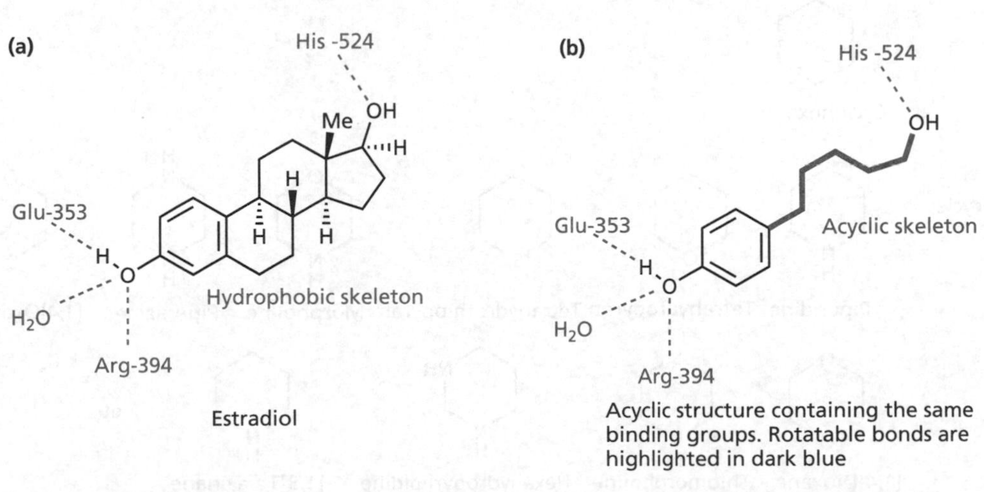

(a)

Glu-353

H_2O

Arg-394

His -524

Me OH

H

H

H

Hydrophobic skeleton

Estradiol

(b)

Glu-353

H_2O

Arg-394

His -524

OH

Acyclic skeleton

Acyclic structure containing the same binding groups. Rotatable bonds are highlighted in dark blue

FIGURE 4.3 Binding interactions for estradiol compared with an acyclic analogue.

4.2 Carbocycles versus heterocycles

There are several endogenous compounds and drugs which contain purely carbocyclic scaffolds, particularly in the field of steroids. Examples include **mestranol** (Fig. 4.1) and **estradiol** (Fig. 4.3). Many drugs also contain aromatic substituents—for example, **lisinopril** and **rosuvastatin** (Fig. 4.1). However, these are vastly outnumbered by the number of drugs containing heterocyclic ring systems of various types. For example, five of the six drugs shown in Figure 4.1 contain heterocyclic ring systems. There are a number of reasons for the prevalence of heterocycles in drugs.

First of all, the variety of heterocyclic ring systems that are possible for any particular type of ring is much larger than for carbocycles. If we consider fully saturated six-membered rings, there is only one possible carbocyclic structure and that is cyclohexane. In contrast, a large number of heterocyclic structures are possible if the number, type, and positions of heteroatom present are varied (Fig. 4.4). Therefore, heterocycles provide far more scope for varying scaffolds and substituents. This can be extremely useful when attempting to optimize the pharmacodynamic and pharmacokinetic properties of drugs. It is also extremely valuable in allowing pharmaceutical companies to design drugs that have been termed as **'me too' drugs** (Fig. 4.5). If the initial drug has a number of functional groups and substituents attached to a recognizable heterocyclic scaffold, a competing company could design a drug containing the same substituents and functional groups attached to a different heterocyclic

scaffold to get round patent restrictions. Such drugs might also have improved properties (**'me better' drugs**).

Secondly, heterocyclic ring systems tend to be more easily synthesized than carbocyclic ring systems. The presence of one or more heteroatoms is a 'handle' that can allow cyclization reactions to occur. In contrast, carbocyclic ring systems lack that functionality. Although it is certainly possible to synthesize carbocyclic ring systems, it is usually preferable to identify a readily available structure that already contains the desired carbocyclic ring system and then modify it to obtain the target structure.

Despite the vast number of heterocyclic ring systems that are possible in chemistry, a much smaller variety of heterocyclic rings are found in most drugs and drug candidates. These include saturated heterocyclic rings such as piperazine, piperidine, pyrrolidine, and morpholine, as well as unsaturated heteroaromatic rings such as pyridine, pyrimidine, pyrazole, pyrrole, and imidazole. Several fused ring systems are commonly found, such as indole, benzimidazole, benzoxazole, quinoline, and quinazoline (Fig. 4.6).

In the past, there has been a particular interest in using heteroaromatic ring systems to create chemical libraries which can be used to identify active compounds ('hits'). Such ring systems have the advantage that they are planar, and so the presence of substituents around the ring does not result in asymmetric centres and chiral molecules. This means that it is easier to synthesize such structures without having to carry out an asymmetric synthesis, or separate enantiomers or diastereoisomers. However, there is a downside. Since the substituents all 'sprout' from a planar core, there is a limited amount of three-dimensional space occupied by the molecule,

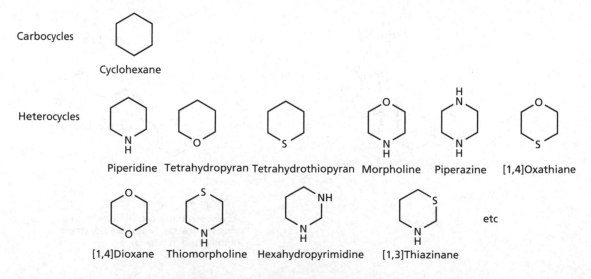

FIGURE 4.4 Saturated six-membered ring systems.

Indole
Fluvastatin
Pyridine
Cerivastatin
Pyrimidine
Rosuvastatin
Pyrrole
Atorvastatin
Quinoline
Pitavastatin

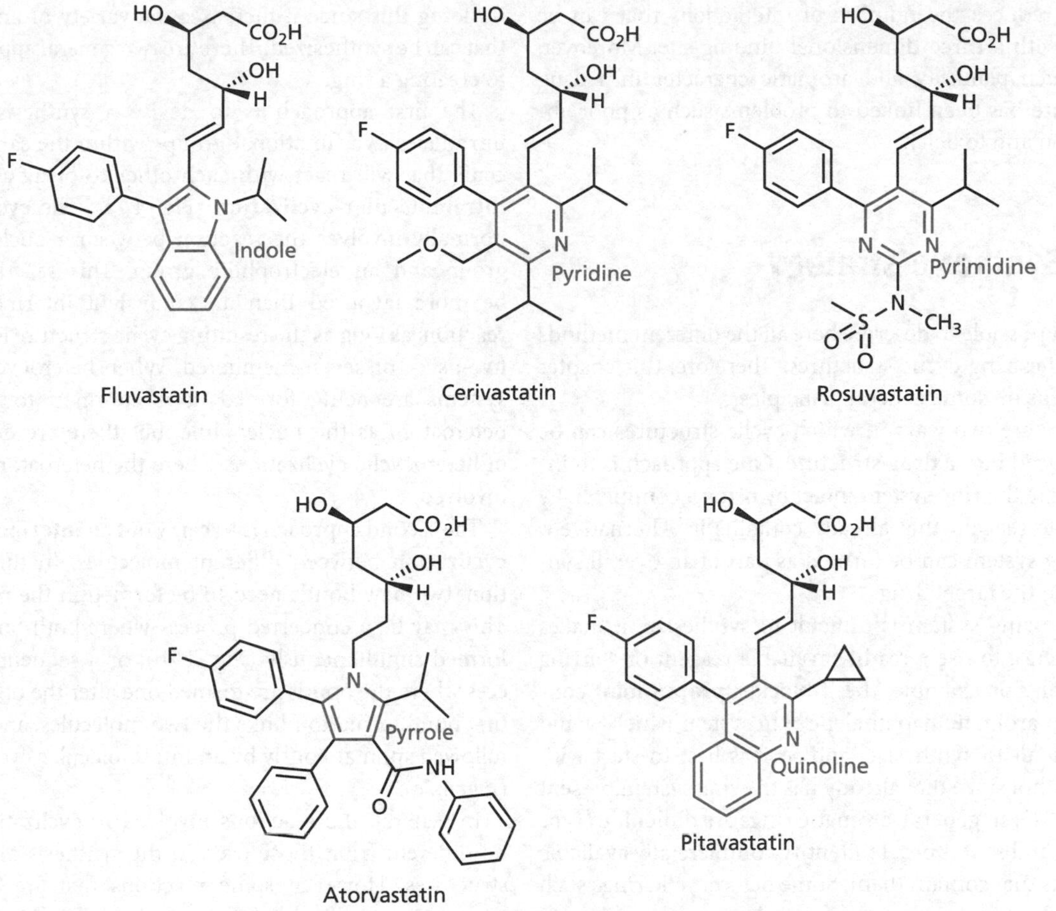

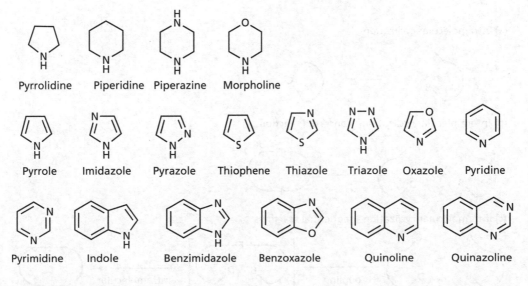

FIGURE 4.5 Various statins with different heterocyclic scaffolds.

Pyrrolidine
Piperidine
Piperazine
Morpholine
Pyrrole
Imidazole
Pyrazole
Thiophene
Thiazole
Triazole
Oxazole
Pyridine
Pyrimidine
Indole
Benzimidazole
Benzoxazole
Quinoline
Quinazoline

FIGURE 4.6 Examples of heterocyclic ring systems that are commonly present in drug structures.

which reduces the number of interactions that can be made with a three-dimensional binding site. Moreover, too much planarity and aromatic character in a drug structure has been linked to problems such as poor absorption and toxicity.

4.3 Synthetic strategy

It is not possible to describe here all the different methods of synthesizing cyclic structures. Therefore, this chapter will focus on some general principles.

There are two ways in which cyclic structures can be introduced into a drug structure. One approach is to incorporate the ring system intact by using a commercially available reagent that already contains it. Alternatively, the ring system can be formed as part of an overall synthesis to the target drug.

If the ring system is difficult to synthesize, it makes more sense to use a readily available reagent or starting material. For example, the β-blocker **propranolol** contains an aromatic naphthalene ring system which would be difficult to synthesize, and so it is best to start with 1-naphthol since that already has the ring system present (Fig. 2.31). In general, aromatic rings are difficult to synthesize and so it is best to identify commercially available reagents that contain them. Some heterocyclic rings such as pyridine and morpholine are also incorporated into drugs by using readily available reagents.

On the other hand, many heterocyclic ring systems are not commercially available and so it is necessary to construct them from scratch. There is actually an advantage in doing this since it increases the variety of analogues that can be synthesized. There are two general approaches to creating a ring.

The first approach is to devise a synthesis which introduces two functional groups within the same molecule that will react with each other to bring about an **intramolecular cyclization** (Fig. 4.7a). The cyclization normally involves the reaction between a nucleophilic group and an electrophilic group. This reaction will be more favoured than an equivalent intermolecular reaction, as long as the resulting cyclic structure is three-, five-, six-, or seven-membered. When heterocyclic ring systems are being formed, it is common to use the heteroatom as the nucleophile, but there are examples of heterocyclic cyclizations where the heteroatom is not involved.

The second approach is to carry out an **intermolecular cyclization** between different molecules. In this situation, two new bonds need to be formed in the reaction. This may be a concerted process where both bonds are formed simultaneously (Fig. 4.7b), or a sequential process where the bonds are formed one after the other. The first bond formation links the two molecules, and this is followed spontaneously by an intramolecular cyclization (Fig. 4.7c).

In general, the reactions involved in cyclizations are no different from those used in the synthesis of acyclic structures. However, some reactions that are difficult to carry out between different molecules can proceed far more easily in intramolecular cyclization reactions. This is because there is a far greater chance of reaction between the two functional groups concerned if they are within the same molecule.

(a) Intramolecular cyclization

(b) Intermolecular cyclization—concerted reaction

(c) Intermolecular cyclization—sequential reactions

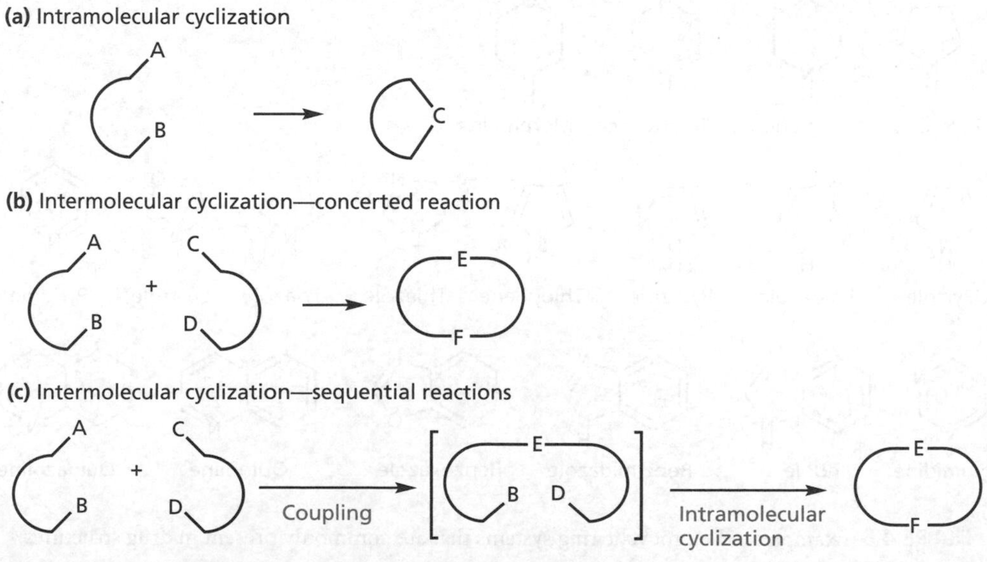

FIGURE 4.7 General approaches to carrying out a cyclization.

4.4 Examples of syntheses using preformed ring systems

There are several examples of drugs containing multicyclic ring systems as scaffolds. Examples include the opioids, tetracyclines, and steroids. Building up a multicyclic ring system from simple building blocks is challenging, and is likely to involve a large number of different reaction steps and low overall yields. Therefore, it makes sense to make use of naturally occurring compounds that

contain the desired molecular scaffold, and then modify these to obtain the desired products. Such an approach is called a semi-synthetic procedure. Examples of this approach can be seen in sections 9.3, 12.4, and 13.3.

At the opposite end of the spectrum, there are many drugs which contain a simple monocyclic ring as part of a substituent. It is often best to use a commercially available reagent which contains that ring rather than to synthesize it from scratch (Box 4.1). For example, a phenyl ring is extremely prevalent in drugs, but is not easily synthesized. However, there are a large number of commercially available chemicals that contain this feature and can be used as

---

BOX 4.1 Opioids containing a substituent involving a three or four-membered ring

Three- and four-membered rings are strained systems, and so one might be tempted to think that they are unlikely to be present in drugs. In fact, several important drugs contain these ring systems, including various opioid structures (Fig. 1). The cyclopropane ring is more prevalent in drugs than the cyclobutane ring, and is used as a bio-isostere for an alkene group. In other words, it is often possible to replace an alkene group in a drug with a cyclopropane ring without losing activity.

There are several methods of synthesizing 3- and 4-membered rings. For example, it is possible to synthesize cyclopropane rings from alkenes, α,β-unsaturated

ketones or allylic alcohols using di-iodomethane with a zinc catalyst—a reaction known as the **Simmons-Smith reaction**. However, it is much simpler to use a commercially available building block that already contains the required ring.

The synthesis of opioids such as the ones shown in figure 1 involves starting from a naturally occurring opioid and modifying the functional groups and substituents that are present. For example, the substituent on the amine nitrogen can be introduced by demethylating the *N*-methyl group which is normally present and replacing it with the desired substituent (Fig. 2).

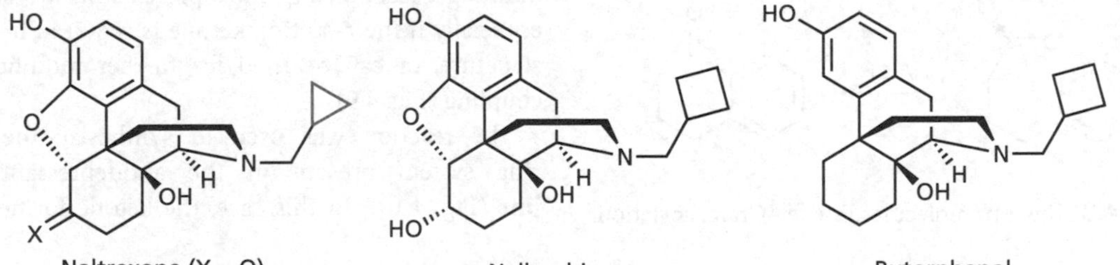

Naltrexone (X = O)
Nalmefene (X = CH$_2$)

Nalbuphine

Butorphanol

FIGURE 1 Opioids containing three- and four-membered rings.

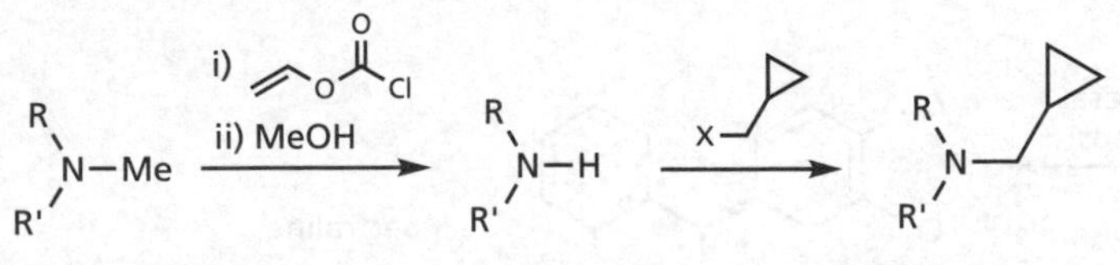

FIGURE 2 Modifying an *N*-methyl substituent to a different *N*-substituent (X = halogen).

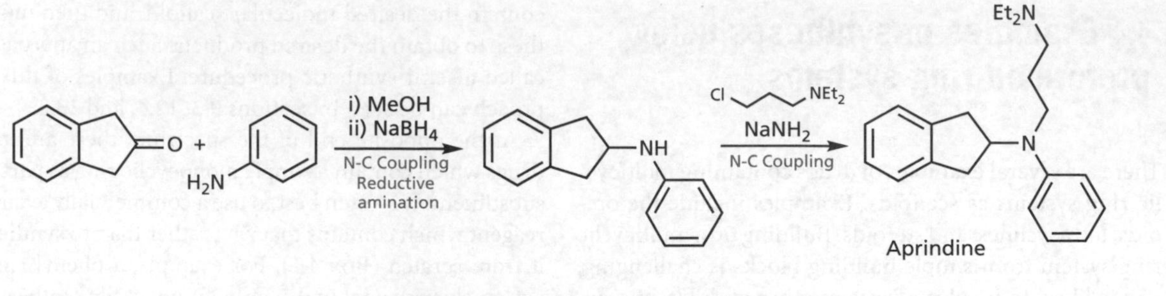

FIGURE 4.8 Synthesis of aprindine.

reagents or building blocks. One example is the synthesis of **aprindine**, which is used in cardiovascular medicine as an anti-arrhythmic. The structure contains a bicyclic ring system as well as an aromatic ring. Both rings can be incorporated into the structure by choosing readily available reagents that already contain them (Fig. 4.8).

• Carbocyclic structures are rings made up of carbon atoms. Heterocyclic structures are rings that include one or more heteroatoms.

• Heterocyclic systems act as good scaffolds for drugs because of their rigidity and relative ease of synthesis.

• Far more heterocyclic ring systems than carbocyclic systems are possible.

• Ring systems can be incorporated into drugs from starting materials or reagents that already include them.

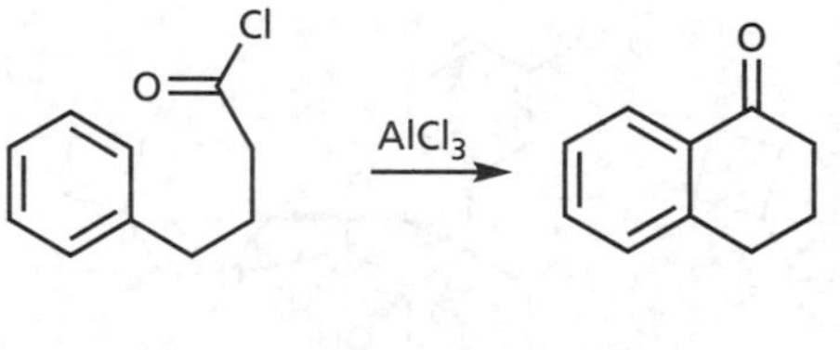

FIGURE 4.9 The intramolecular Friedel–Crafts acylation.

• Cyclic systems can be created by intramolecular or intermolecular cyclizations.

4.5 Examples of syntheses involving intramolecular cyclizations

In this section, we will consider a number of reactions that are commonly used in intramolecular cyclizations. They are by no means the only reactions, but they do give a flavour of how these cyclizations are achieved.

4.5.1 The Friedel–Crafts reaction

The Friedel–Crafts acylation is an excellent method of building a second ring onto a pre-existing aromatic ring, especially if the resulting ketone is required in the final structure, or can be used for further modification or coupling (Fig. 4.9).

The reaction was used to synthesize the bicyclic ring system present in the antidepressant **sertraline** (Fig. 4.10). In this case, the ketone formed by the

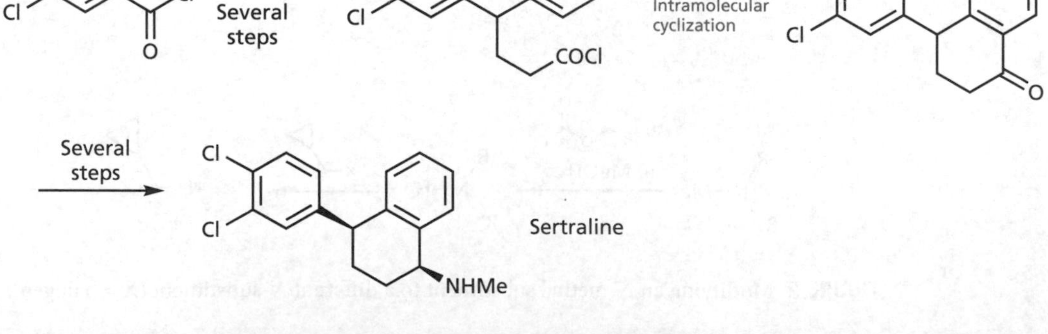

FIGURE 4.10 Synthesis of the bicyclic ring system in sertraline.

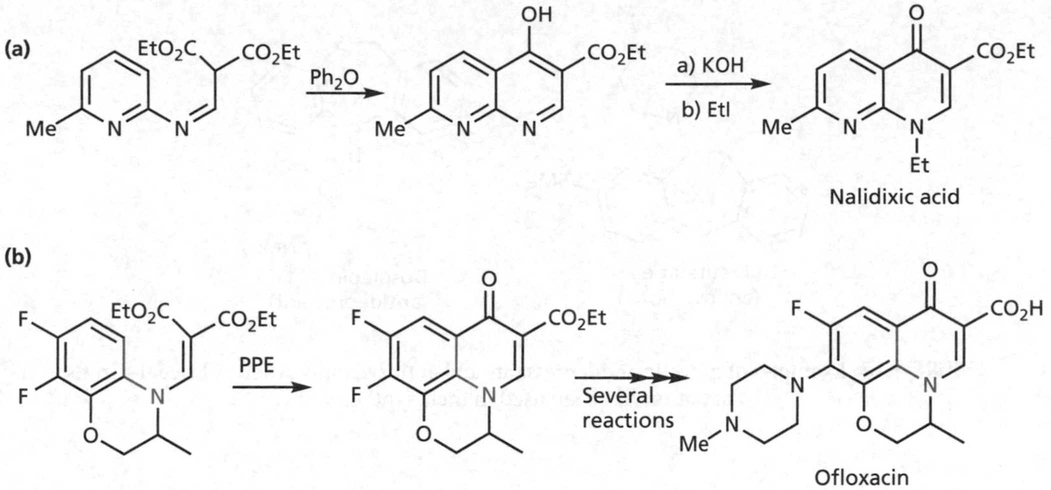

FIGURE 4.11 Synthesis of the quinolone ring system in the synthesis of nalidixic acid and ofloxacin.

acylation was useful since it allowed the incorporation of a methylamine substituent at that position.

Friedel–Crafts acylations have also been useful in creating a variety of heterocyclic systems, such as those present in the antibacterial agents **nalidixic acid** and **ofloxacin** (Fig. 4.11; see also Chapter 14). The intramolecular reaction can be carried out with an ester group by heating it in diphenyl ether or polyphosphoric ethyl ester (PPE).

Another example is the synthesis of the tricyclic ring system required in the structure of the antihistamine **loratadine** (Fig. 4.12). In this case, a seven-membered ring was formed as a result of the cyclization. The resulting ketone group allowed the coupling of the final piperidine building block through a Grignard reaction.

Friedel–Crafts acylation has also proved useful in synthesizing a range of tricyclic systems present in various antidepressants and antipsychotics (Fig. 4.13).

Instead of the Friedel–Crafts acylation, a Friedel–Crafts alkylation can be used when there is no need to have a ketone group present in the final product. For example, the bicyclic system present in **tazarotene** was synthesized from an alkene by a Friedel–Crafts alkylation (Fig. 4.14). Tazarotene binds to retinoic acid receptors and is used for the treatment of acne.

Similarly, the synthesis of the antidepressant **nomifensine** (now discontinued) involved a cyclization as the

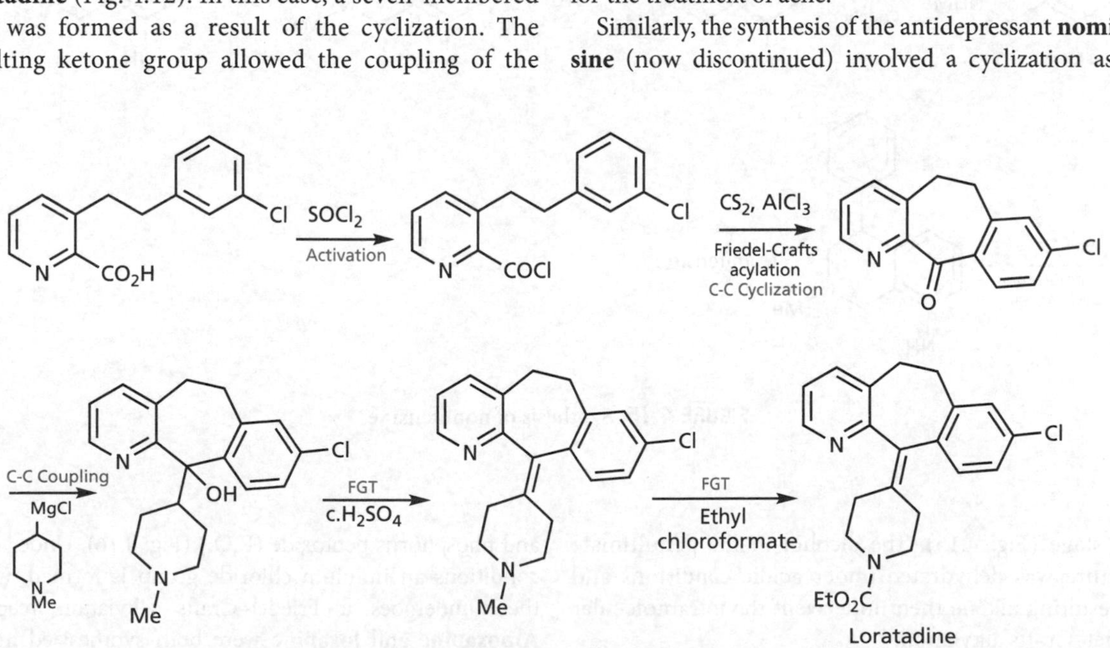

FIGURE 4.12 Synthesis of the tricyclic ring system in loratadine.

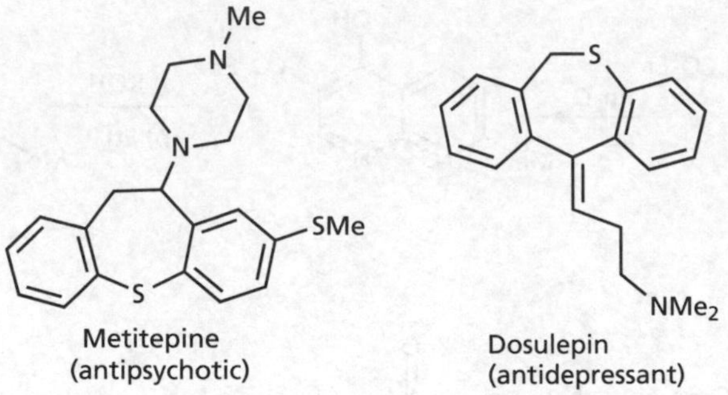

FIGURE 4.13 Examples of tricyclic antidepressants and antipsychotics where a Friedel–Crafts acylation has been used in their synthesis.

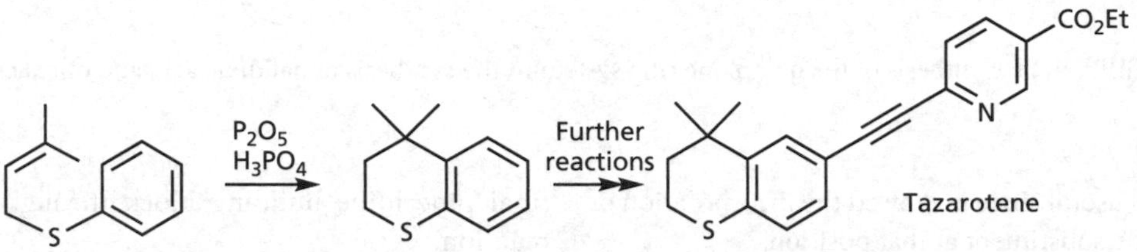

FIGURE 4.14 Synthesis of the bicyclic system present in tazarotene.

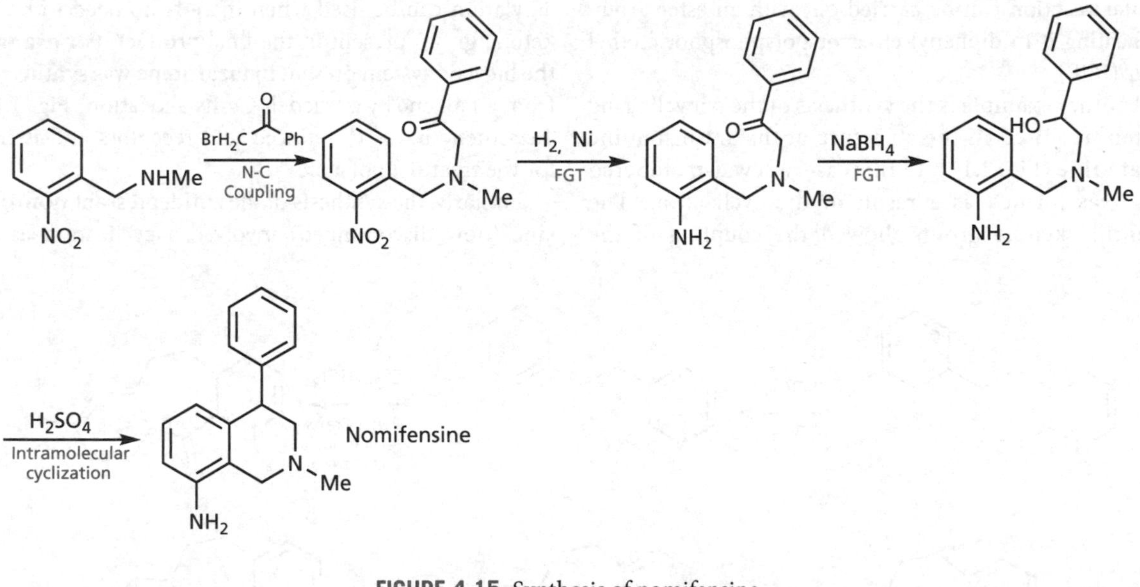

FIGURE 4.15 Synthesis of nomifensine.

final stage (Fig. 4.15). The alcohol in the penultimate structure was dehydrated under acidic conditions and the resulting alkene then underwent the intramolecular Friedel–Crafts alkylation.

It has even been possible to carry out cyclization reactions with amides in the presence of phosphoryl chloride (POCl$_3$)

and phosphorus pentoxide (P$_2$O$_5$) (Fig. 4.16). Under these conditions an iminium chloride group is formed, which then undergoes a Friedel–Crafts alkylation reaction. **Amoxapine** and **loxapine** were both synthesized in this fashion. Amoxapine has been used as an antidepressant, while loxapine has been used as an antipsychotic agent.

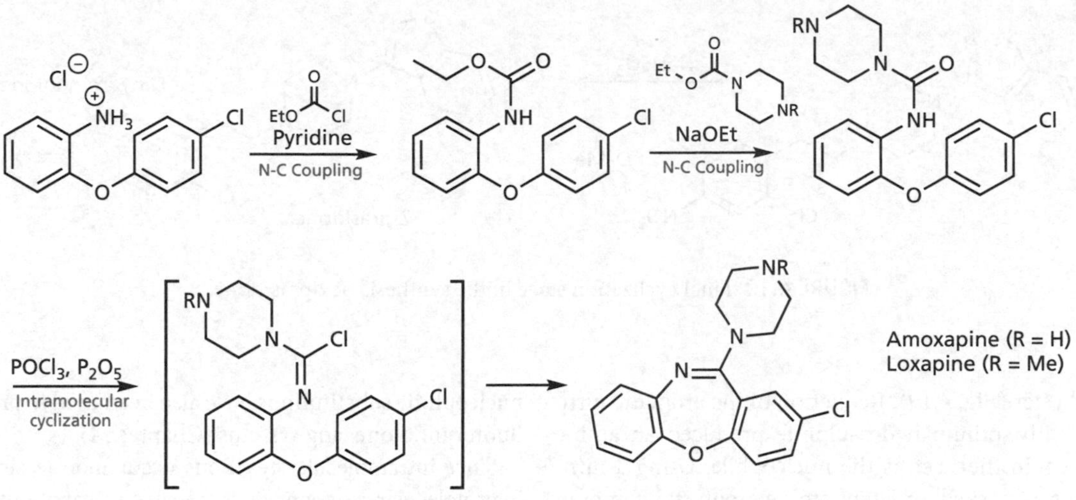

FIGURE 4.16 Synthesis of amoxapine and loxapine.

4.5.2 Nucleophilic substitution

Cyclization reactions have been carried out which involve the nucleophilic substitution of an alkyl halide, alkyl mesylate, alkyl tosylate, aryl halide, or carboxylic acid derivative. In the case of heterocyclic ring systems, the heteroatom is normally involved as the nucleophile (Fig. 4.17). However, there are some cases where a carbanion has been used as the nucleophilic group.

An example of this approach can be seen in a synthesis of the antidepressant **paroxetine** (Fig. 4.18), where a Boc-protected amine was treated with a base to remove the proton from the nitrogen. The resulting negatively charged nitrogen then displaced a mesylate group to form a piperidine ring.

In the final step of a synthesis leading to the antipsychotic agent **ziprasidone**, the dihydroindolone ring system was created by a nucleophilic substitution of a

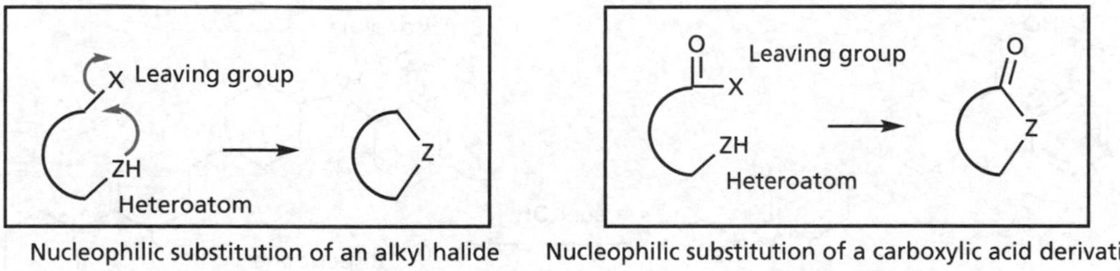

Nucleophilic substitution of an alkyl halide   Nucleophilic substitution of a carboxylic acid derivative

FIGURE 4.17 Intramolecular cyclizations involving nucleophilic substitutions.

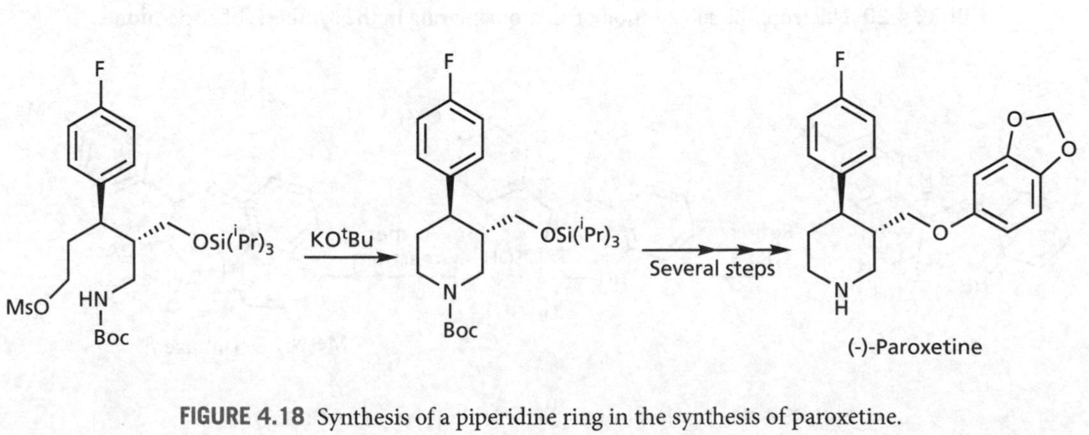

FIGURE 4.18 Synthesis of a piperidine ring in the synthesis of paroxetine.

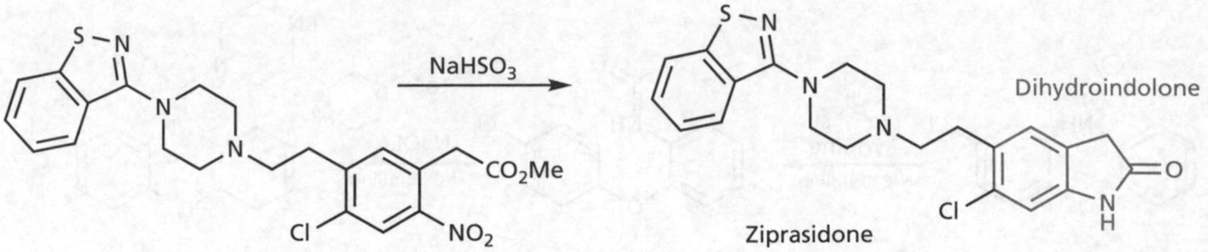

FIGURE 4.19 Final cyclization stage in the synthesis of ziprasidone.

methyl ester (Fig. 4.19). Reduction of the aromatic nitro group with sodium hydrosulphite produced an amine which could then act as the nucleophile. Using a nitro group as a 'masked' or latent amine group is a common strategy in synthetic design (see also sections 2.10.1, 2.14, and 3.8). The nitro group is relatively unreactive and can be introduced early on in a synthesis without having to worry about it reacting with reagents used in the other stages of the synthesis. The molecule can then be set up for cyclization, such that a spontaneous cyclization occurs as soon as the nitro group has been reduced.

Nucleophilic substitutions of an aryl halide are feasible if other electron-withdrawing substituents are present. The nucleophilic substitution of a fluoride ion from an aromatic ring was the final step in a synthesis of the antipsychotic **risperidone** (Fig. 4.20). This produced a benzisoxazole ring system. Aromatic nucleophilic substitutions have also been used to produce fluoroquinolone ring systems (Chapter 14).

Since intramolecular reactions occur more easily than intermolecular reactions, it is possible to carry out some reactions that would be very difficult to carry out when making acyclic molecules. For example, the reaction of a primary amine with a carboxylic acid to form an amide usually requires a coupling agent for an intermolecular reaction, but can proceed relatively easily in an intramolecular reaction. The amide formation that created the bicyclic system present in the cardiovascular drug **diltiazem** was achieved without the need for a coupling reagent (Fig. 4.21).

Although a heteroatom is normally used as the nucleophile to create heterocyclic rings, there are examples of successful cyclizations where carbanion chemistry has been used. For example, the quinolone ring system that is present in the antibacterial fluoroquinolone

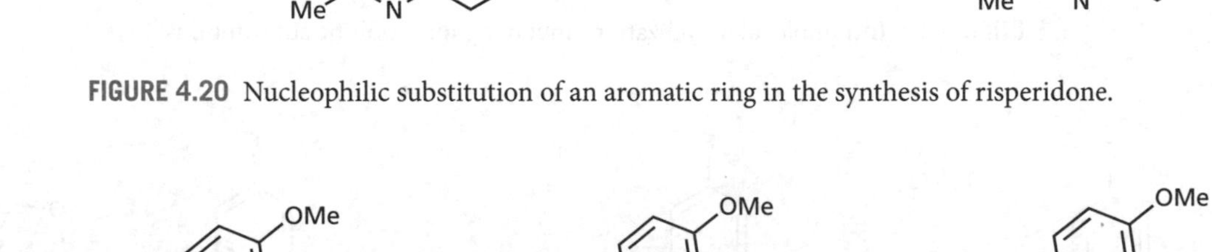

FIGURE 4.20 Nucleophilic substitution of an aromatic ring in the synthesis of risperidone.

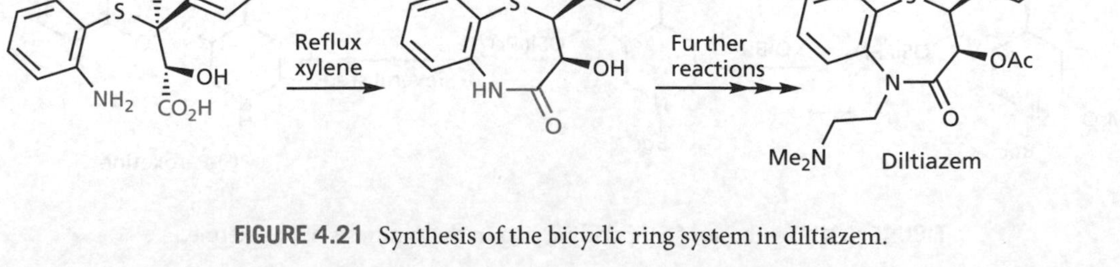

FIGURE 4.21 Synthesis of the bicyclic ring system in diltiazem.

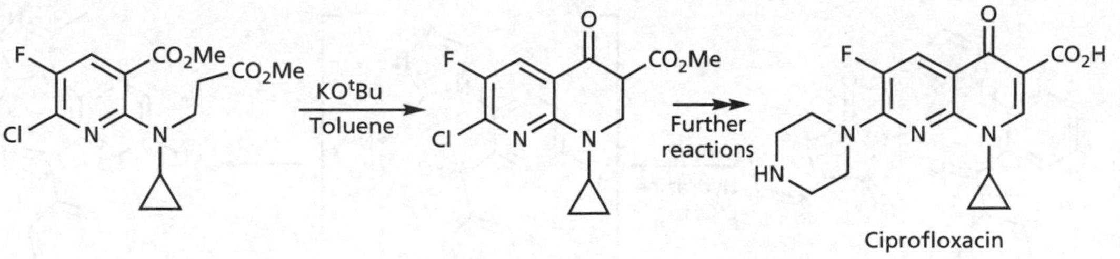

FIGURE 4.22 Synthesis of the quinolone ring system of ciprofloxacin.

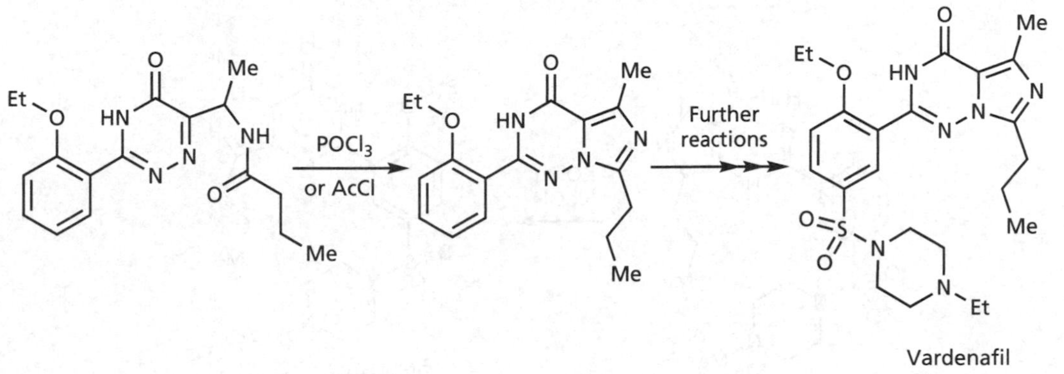

FIGURE 4.23 Synthesis of vardenafil.

ciprofloxacin has been synthesized using carbanion chemistry (Fig. 4.22). The reaction is similar to the **Dieckmann condensation**, where a carbanion reacts with an ester group.

Amides can be activated with $POCl_3$ to form a reactive chloroimine structure (compare Fig. 4.16). For example, the bicyclic ring system present in the second-generation anti-impotence drug **vardenafil** was created by activating an amide group with $POCl_3$, resulting in an intramolecular cyclization (Fig. 4.23).

4.5.3 Nucleophilic addition and elimination

Imine formation involves the reaction between a primary amine and a ketone, and is another reaction that has frequently been used in intramolecular cyclizations (Fig. 4.24). Normally, the reaction is initiated by deprotecting the amine group, or by converting a functional group such as a nitro, nitrile, or amide group to the amine. Once the amine is formed, imine formation occurs spontaneously (compare Fig. 4.19). Depending on the structure of the compound formed, the double bond might shift to a more favourable position—for example, conjugation with any unsaturated group that might be present.

The tricyclic ring system of the anxiolytic **alprazolam** was synthesized by an imine reaction. The primary amine group required for this reaction was protected as a phthalimide. Treatment with hydrazine deprotected the primary amine group, allowing a spontaneous cyclization to take place (Fig. 4.25).

As mentioned earlier, an intramolecular reaction can occur which would be difficult or almost impossible in an intermolecular reaction. We have already seen this in figure 4.21. Another example is in a synthesis of the anti-impotence drug **sildenafil**, which involved a final cyclization reaction where the nitrogen of a primary amide group underwent a nucleophilic addition and elimination reaction with the carbonyl group of a secondary amide (Fig. 4.26).

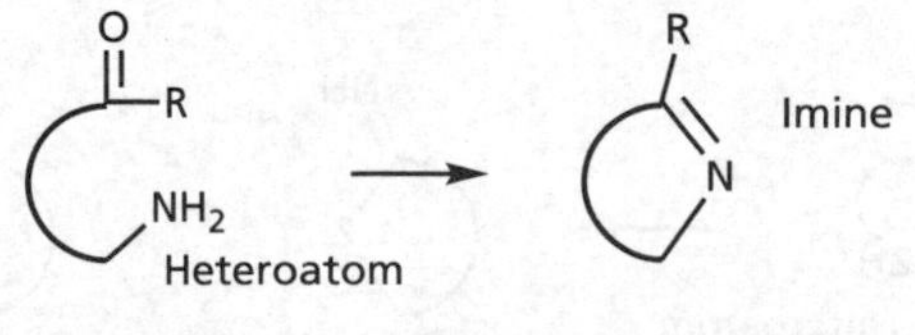

FIGURE 4.24 Imine formation in intramolecular cyclizations.

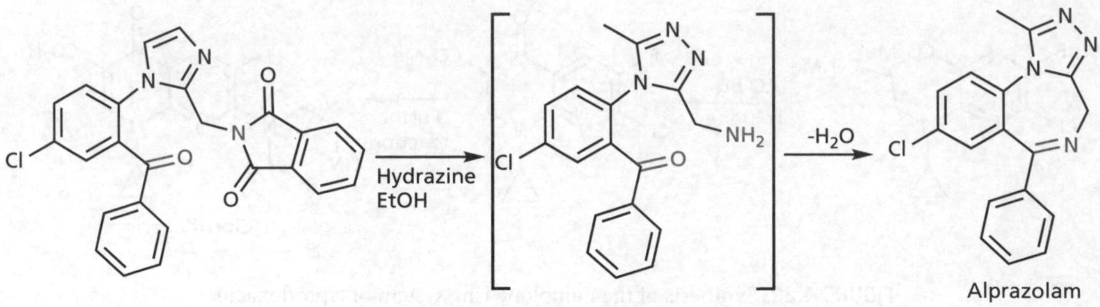

FIGURE 4.25 Synthesis of alprazolam.

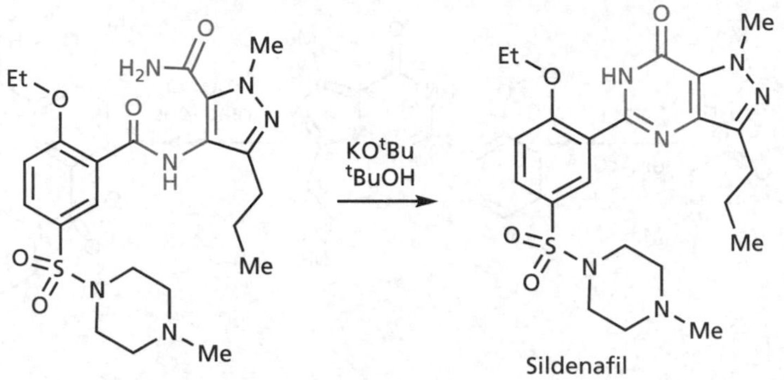

FIGURE 4.26 Synthesis of sildenafil.

4.5.4 Nucleophilic addition

A number of cyclizations have been carried out by a nucleophilic addition of the heteroatom onto a nitrile group (Fig. 4.27).

For example, the penultimate step in a synthesis of the antipsychotic agent **olanzapine** involved reduction of an aromatic nitro substituent to a primary amine (Fig. 4.28). This added spontaneously to the nitrile group to invoke ring closure. The resulting primary amino group was then substituted with a piperazine ring to give the final structure.

4.5.5 Palladium-catalysed couplings

Palladium-catalysed couplings (Case Study 2) have proved useful in several cyclization reactions. For

example, the **Heck reaction** has been used to synthesize the indole rings present in the anti-migraine agents **almotriptan** and **eletriptan** (Figs 4.29a and 4.29b). Intramolecular cyclizations involving trisubstituted and tetrasubstituted alkenes have proved possible, and the reaction has been used to form cyclic systems containing quaternary carbon centres (Fig. 4.29c).

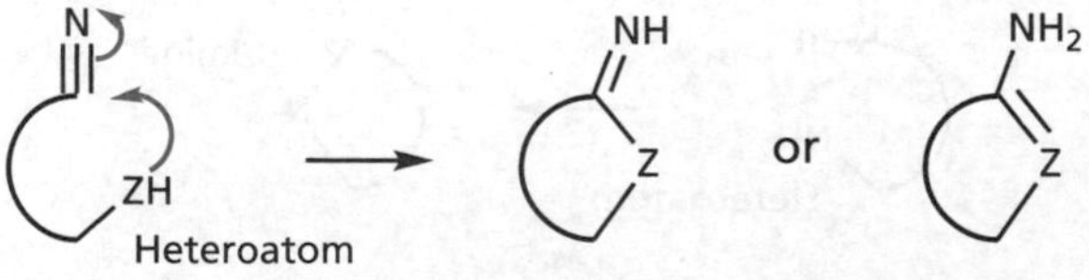

4.6 Examples of syntheses involving concerted intermolecular cyclizations

4.6.1 The Diels–Alder reaction

A well-known example of a concerted cyclization reaction is the Diels–Alder reaction. The reaction proceeds best when an electron-rich diene reacts with an electron-deficient dienophile. In truth, the reaction has not been widely used in drug synthesis. Nevertheless, the reaction has been crucial in the synthesis of a number of very important drugs, especially in the opioid field. In particular, the Diels–Alder reaction was instrumental in

FIGURE 4.27 Ring closures involving nitrile groups.

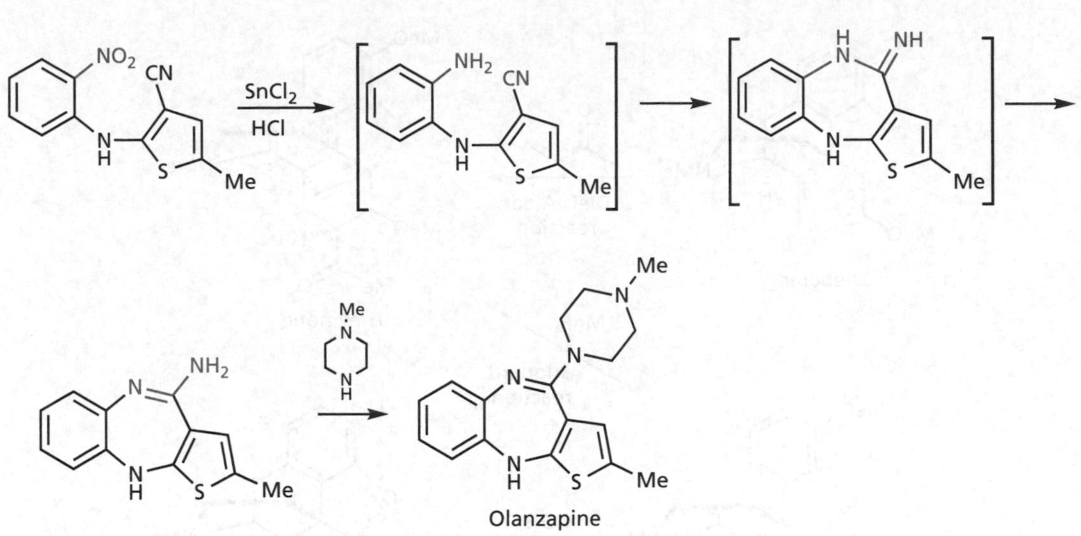

FIGURE 4.28 Synthesis of olanzapine.

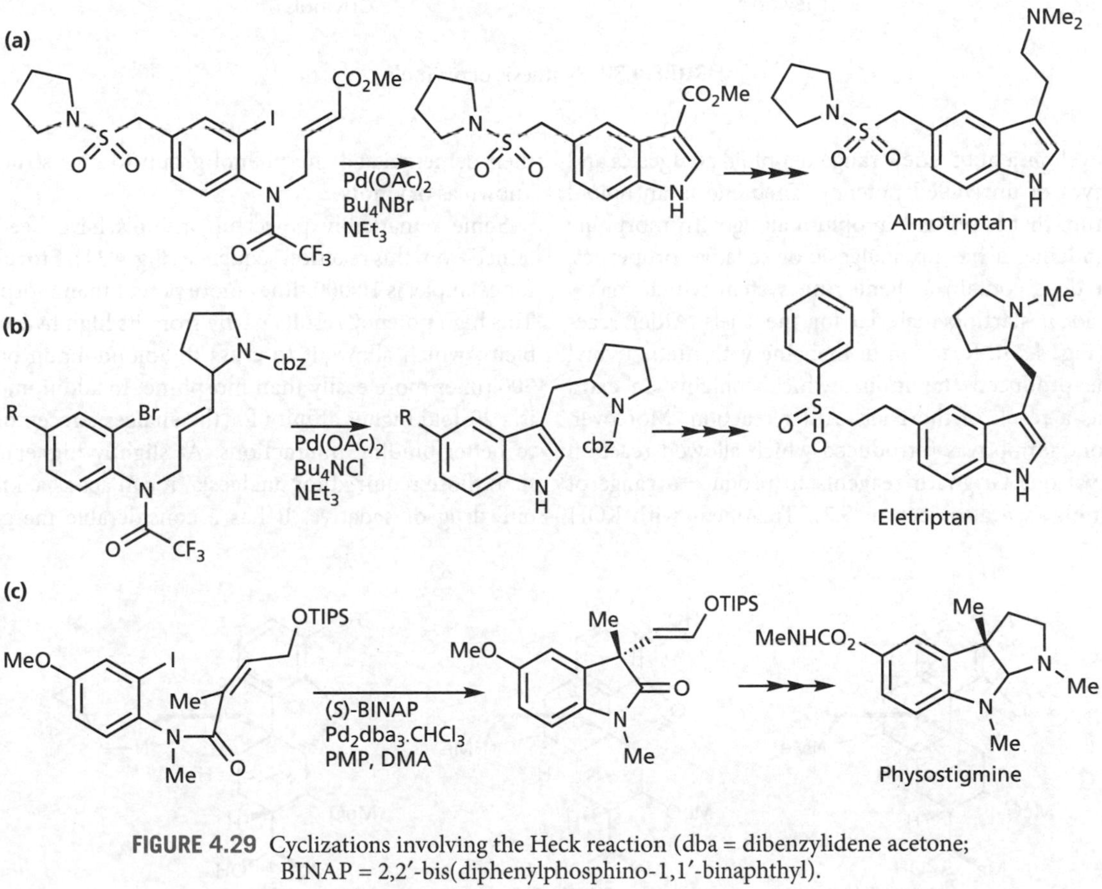

FIGURE 4.29 Cyclizations involving the Heck reaction (dba = dibenzylidene acetone; BINAP = 2,2′-bis(diphenylphosphino-1,1′-binaphthyl).

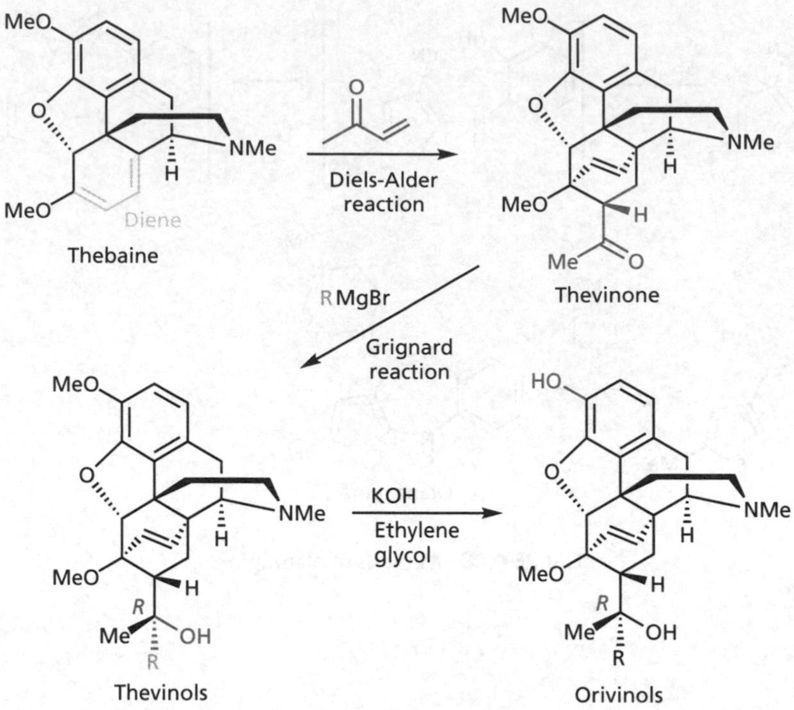

FIGURE 4.30 Synthesis of orvinols.

the development of a new range of opioid analgesics and sedatives of unrivalled potency. **Thebaine** is an opioid structure that is present in opium along with morphine and codeine. It has no analgesic or sedative properties, but it does contain a diene ring system which makes it an ideal starting material for the Diels–Alder reaction (Fig. 4.30). Reaction of thebaine with methyl vinyl ketone produced **thevinone**, which contains an extra ring as a result of the Diels–Alder reaction. Moreover, a ketone group was introduced which allowed reaction with various Grignard reagents to produce a range of **thevinols** (see also section 5.7). Treatment with KOH

then demethylated the phenol group to give structures known as **orvinols**.

Some remarkably powerful orvinols have been obtained from this reaction sequence (Fig. 4.31). **Etorphine**, for example, is 10 000 times more potent than morphine. This high potency results partly from its high hydrophobicity, which allows it to cross the blood–brain barrier 300 times more easily than morphine. In addition, there is a 20-fold higher affinity for the analgesic receptor due to better binding interactions. At slightly higher doses than those required for analgesia, it can act as a 'knock-out' drug or sedative. It has a considerable margin of

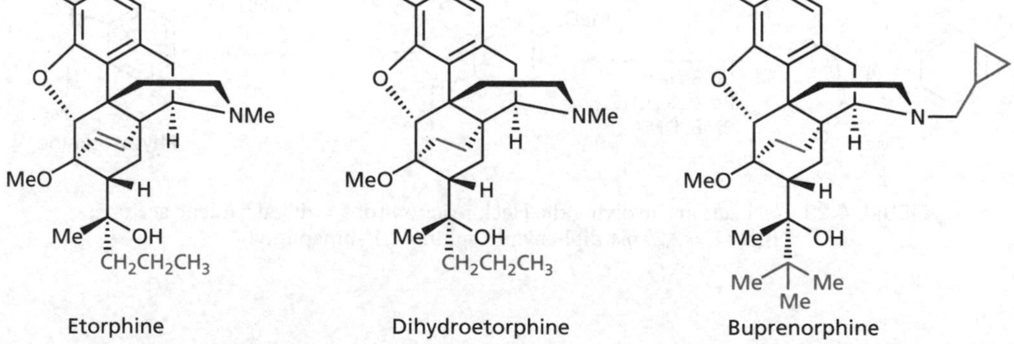

FIGURE 4.31 Examples of orvinols used in human and veterinary medicine.

safety and is used to immobilize large animals such as elephants, rhinoceros, and hippopotami. Since the compound is so active, only very small doses are required. These can be dissolved in small volumes (1 ml) so that they can be placed in darts which can be fired into the hide of the animal. Reducing the double bond of etorphine increases activity over tenfold, and the resulting structure (**dihydroetorphine**) is one of the most potent analgesics ever discovered. Dihydroetorphine is used in China as an analgesic and as a treatment for opioid addiction. **Buprenorphine** has been used in UK hospitals as an analgesic to treat patients recovering from surgery, as well as those suffering from cancer. It has also been used as an alternative to methadone for weaning addicts off heroin.

A Diels–Alder reaction was also used to create the bicyclic ring system present in **sertraline** (Fig. 4.32). In this case, benzyne was generated from a dihalo aromatic ring with *n*-butyl lithium, then underwent the Diels–Alder reaction with furan. A reductive ring opening was then carried out to give a bicyclic ring system which was further modified to give the final product. The secondary alcohol in the bicyclic ring system served to functionalize the ring system such that the methylamine substituent could be introduced

at that position. This represents an alternative synthesis of sertraline from the one described in Figure 4.10.

4.6.2 The hetero Diels–Alder reaction

The hetero Diels–Alder reaction refers to a Diels–Alder reaction where a heteroatom is part of the diene or the dienophile. For example, an aldehyde can act as a dienophile with electron-rich dienes (Fig. 4.33a). However, if the heteroatom is in the diene, it may be necessary to use an electron-rich dienophile (Fig. 4.33b). This is a reverse of the normal situation, where usually an electron-rich diene reacts with an electron-deficient dienophile.

Cycloadditions are not limited to six-membered rings. [3 + 2]-Cycloadditions can be carried out to synthesize five-membered rings using a three-atom moiety that contains both a positive and negative charge. The three-atom dipolar moiety acts as the dienophile, and the reaction is also called a 1,3-dipolar cycloaddition because of the two charges in the three-atom unit.

An example of a [3 + 2]-cycloaddition involved the creation of the central pyrrole scaffold of **atorvastatin lactone** (a cholesterol-lowering agent) (Fig. 4.34). It is

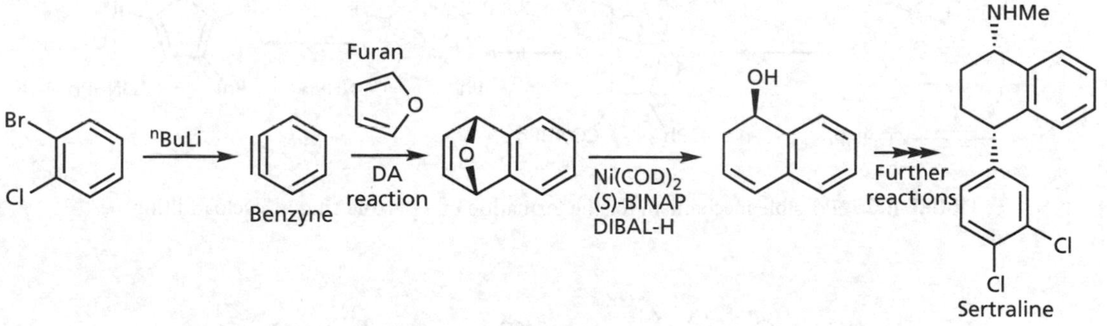

FIGURE 4.32 A Diels–Alder (DA) reaction used in a synthesis of sertraline.

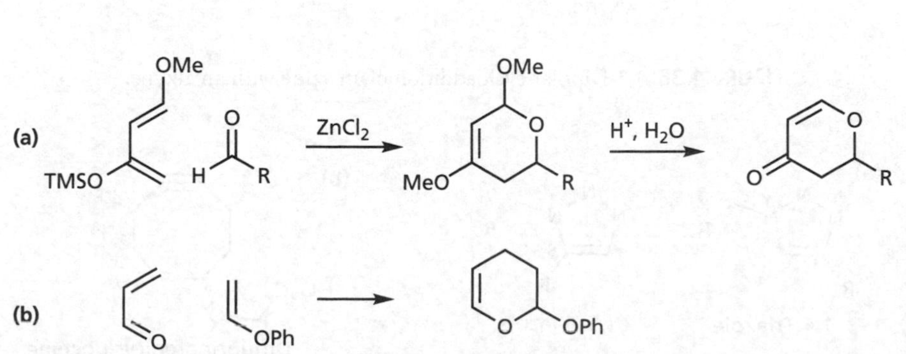

FIGURE 4.33 Examples of heterocyclic systems formed by the hetero Diels–Alder reaction.

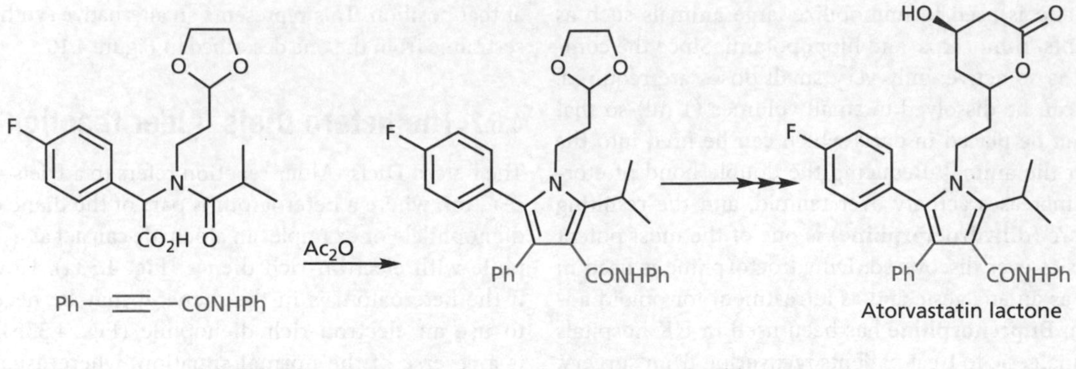

FIGURE 4.34 Formation of a pyrrole ring by a cycloaddition reaction.

thought that the cycloaddition involves an oxazolone intermediate with the required dipolar nature (Fig. 4.35).

A particularly popular example of 1,3-dipolar cycloadditions has been the reaction of azides with alkynes to produce triazole rings (Fig. 4.36). The reaction is thermodynamically favoured and occurs under mild conditions to produce impressive yields of high purity product. Because the reaction occurs so easily, it is seen as a classic example of a category of reactions which have been termed as '**click chemistry**'. In other words, combining

the correct reagents results in the units 'clicking' together with virtually no effort. Click chemistry has been seen as a useful approach for developing lead compounds for future drug discovery (see section 7.9).

A copper salt is needed as a catalyst to allow mild reaction conditions and product selectivity. In the absence of catalyst, the reaction requires heating which produces mixtures of two regioisomers—the 1,4 and 1,5 isomers (Fig. 4.37a). In the presence of the catalyst, only the 1,4-triazole regioisomer is obtained. However,

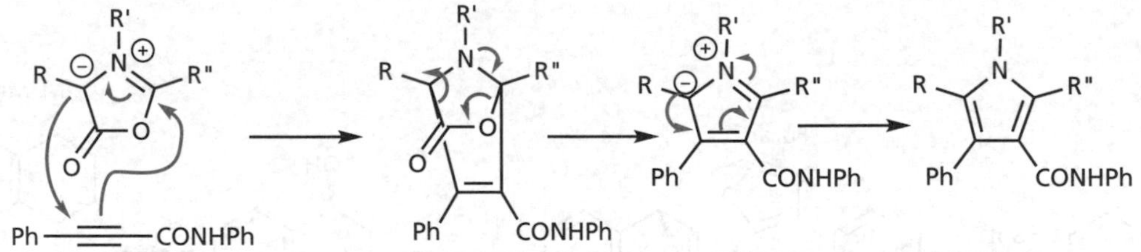

FIGURE 4.35 Possible mechanism for the formation of a pyrrole ring by cycloaddition.

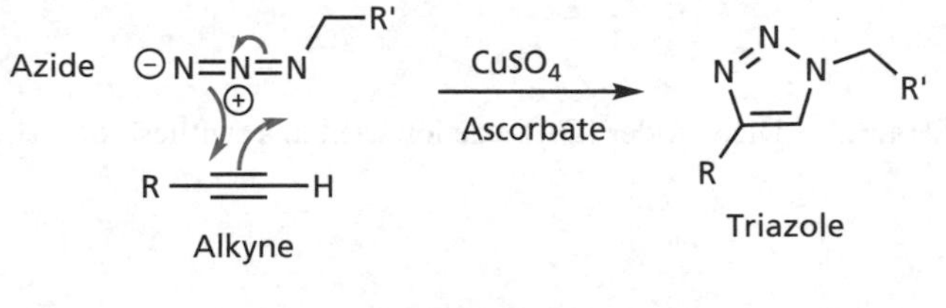

FIGURE 4.36 1,3-Dipolar cycloaddition of an azide with an alkyne.

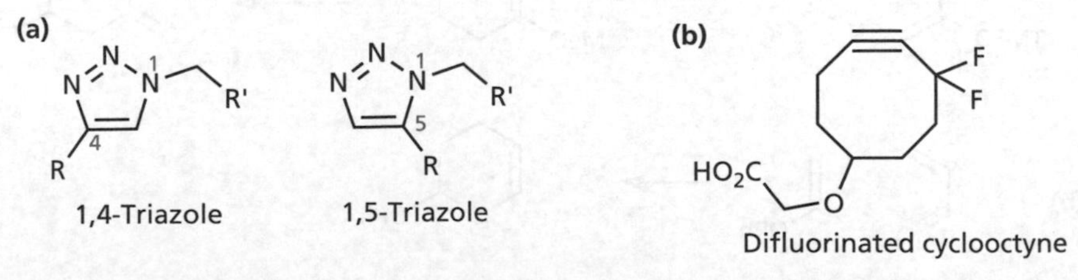

FIGURE 4.37 (a) Regioisomers formed in the absence of copper (II) catalyst. (b) An example of an activated alkyne.

increasing the reactivity of the reagents allows the reaction to occur in the absence of catalyst. This is useful when carrying out studies on living systems, since the copper catalyst is toxic to cells. One way of increasing the reactivity of the alkyne group is to incorporate it into a strained ring system. Another method is to add electron-withdrawing groups. For example, a difluorinated cyclooctyne (Fig. 4.37b) is sufficiently reactive to undergo the cycloaddition without the need for a copper catalyst.

- Intramolecular reactions occur more readily than intermolecular reactions.
- Common reactions used in intramolecular reactions include the Friedel–Crafts acylation, nucleophilic substitution of alkyl halides, aryl halides and carboxylic acid derivatives, imine formation, and nucleophilic addition to ketones and nitriles.
- Nitrogen is frequently used as a nucleophile in cyclization reactions leading to nitrogen-containing heterocycles.
- Concerted reactions such as the Diels–Alder and hetero Diels–Alder reactions have been used to synthesize the cyclic systems in some drugs.
- 1,3-Dipolar cyclizations have been used to create triazole rings and represent a category of reactions known as 'click chemistry'.

4.7 Examples of syntheses involving intermolecular coupling and cyclization reactions

In this section, we will consider cyclization reactions between two separate molecules. Two sequential reactions are involved—one which links the molecules together, and then a cyclization which occurs spontaneously once the molecules have been linked. The two reactions involved may be similar or different in nature. A number of cheap and commercially available reagents are frequently used in these reactions, and the following sections have been defined by the number of atoms incorporated into the new ring by these reagents.

4.7.1 Cyclization reactions which insert a one-atom unit into the resulting heterocycle

In this type of reaction, we start with a molecule that has a chain of n atoms containing reactive functional groups at each end of the chain. The other reagent contains the atom

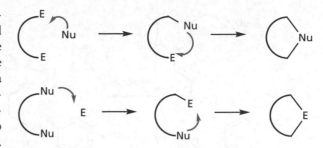

FIGURE 4.38 General mechanisms by which a cyclization can occur with incorporation of a one-atom unit.

which is going to react with these functional groups to produce the cyclic structure. There are two possible approaches. The one-atom unit can be chosen to act twice as a nucleophile, in which case the other molecule has to contain two electrophilic groups. Alternatively, the one-atom unit acts twice as an electrophile, in which case the other molecule contains two nucleophilic groups (Fig. 4.38).

4.7.1.1 Dual-acting nucleophiles

Primary amines are commonly used in intermolecular reactions such that the amine nitrogen acts as a dual-acting nucleophile and is incorporated into the ring. Typical reactions include nucleophilic addition/ elimination with ketones and nucleophilic substitution with alkyl halides, mesylates, tosylates, and carboxylic acid derivatives.

Nucleophilic addition and elimination of a primary amine with a diketone is a good method of synthesizing a pyrrole ring, as seen in the synthesis of **atorvastatin lactone** (Fig. 4.39). The reaction is driven by the formation of the stable heteroaromatic system and offers an alternative method of synthesizing atorvastatin lactone to the procedure described in Figure 4.34.

If nucleophilic substitution reactions are carried out, a saturated or partially saturated heterocyclic ring is obtained. For example, the saturated piperidine ring present in **gacyclidine** is formed by reacting a primary amine with a dibromide (Fig. 4.40). Gacyclidine is being studied for the treatment of tinnitus.

The nucleophilic reactions do not need to be identical in nature. For example, the creation of a piperidinone ring was achieved during a synthesis of (-)-**paroxetine** (Fig. 4.41). Both reactions involved the amine acting as the nucleophile in nucleophilic substitutions, one with the mesylate group and one with the ester. This represents a different approach to the synthesis of the piperidine ring in paroxetine to the method described in Figure 4.18.

It is also possible to use ammonia or hydroxylamine in these kinds of cyclization. For example, hydroxylamine was used as a dual-acting nucleophile to create the

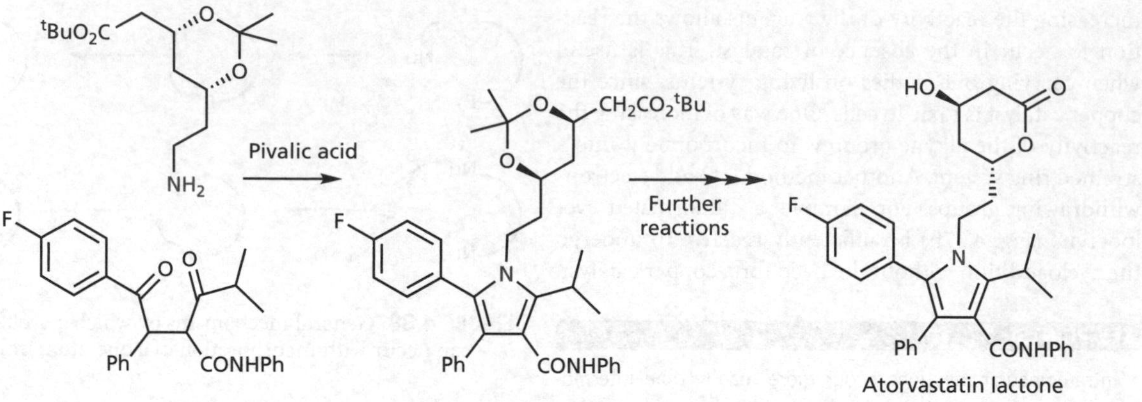

FIGURE 4.39 Formation of the pyrrole ring in atorvastatin lactone.

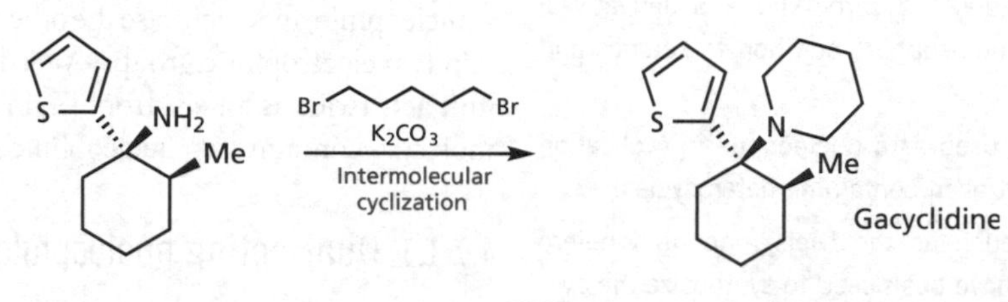

FIGURE 4.40 Formation of a piperidine ring in gacyclidine.

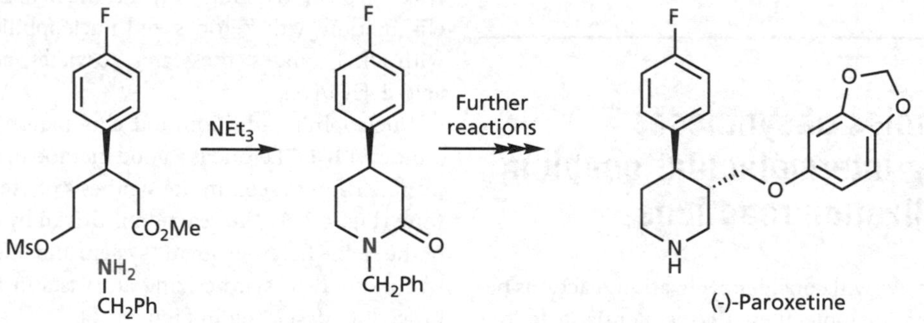

FIGURE 4.41 Formation of the piperidinone ring in the synthesis of (-)-paroxetine.

pyrimidinol ring present in the hair restorer **minoxidil**. The hydroxylamine undergoes a nucleophilic addition with two nitrile groups (Fig. 4.42).

4.7.1.2 Dual-acting electrophilic reagents

A number of reagents can act as dual-acting electrophiles such that a single atom is incorporated into a new ring. Formaldehyde and aldehydes are particularly good reagents for this as they can react twice as an electrophile at the same carbon centre.

For example, formaldehyde was used in order to fuse a piperidine ring to a pre-existing thiophene ring

in the final stage of a synthesis leading to the anti-thrombotic agent **clopidogrel** (Fig. 4.43). In this case, a Mannich base was initially formed which underwent a Friedel–Crafts type reaction with the thiophene ring. This latter step is also known as a **Pictet–Spengler** reaction.

Aldehydes can be used similarly. For example, an aromatic aldehyde was reacted with an indole structure to create a tricyclic ring system as part of a synthesis leading to **tadalafil** (Fig. 4.44). This drug is used to treat erectile dysfunction.

Reagents such as phosgene or diethyl carbonate have been used as dual-acting electrophiles and

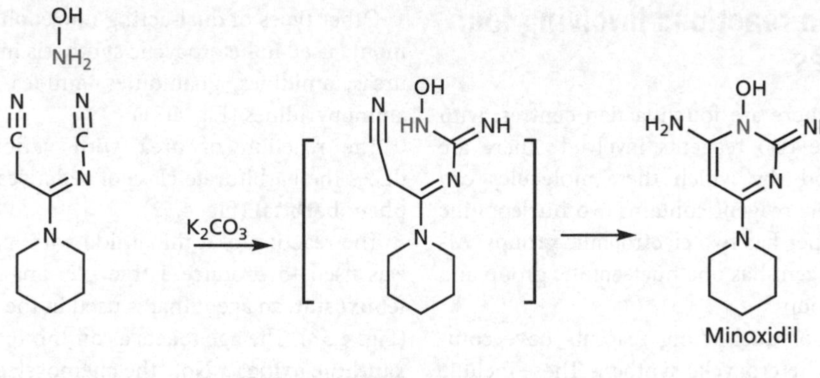

FIGURE 4.42 Formation of the bicyclic ring present in minoxidil.

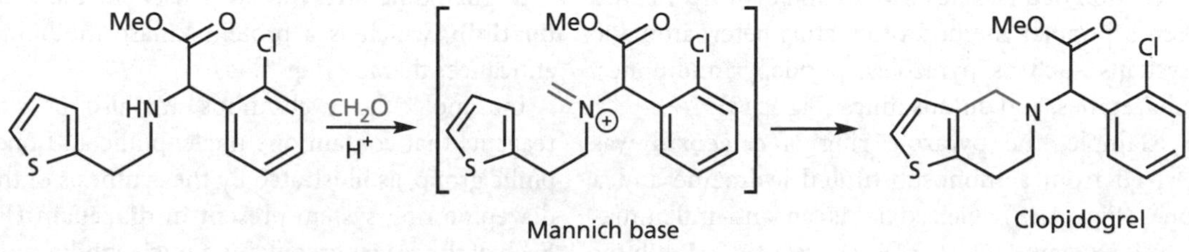

FIGURE 4.43 Final cyclization reaction in a synthesis of clopidogrel.

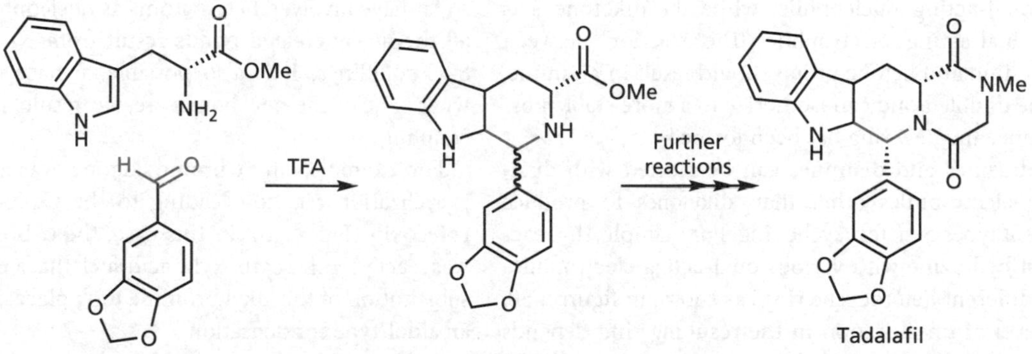

FIGURE 4.44 Creation of a tricyclic ring system as part of a synthesis leading to tadalafil.

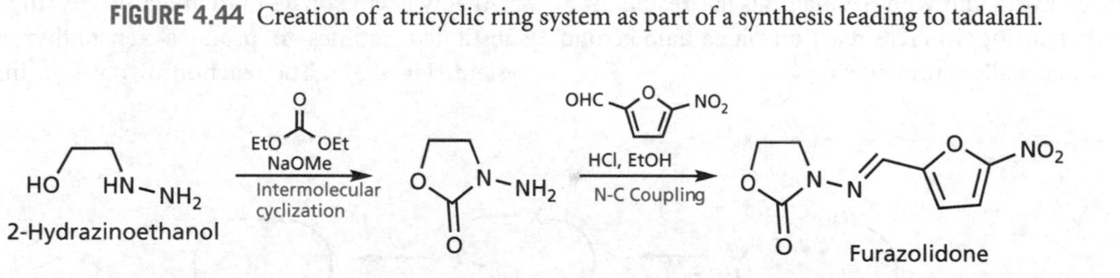

FIGURE 4.45 Synthesis of furazolidone.

introduce a carbonyl unit into the newly formed ring. For example, the oxazolidone ring system present in the antibacterial agent **furazolidone** was synthesized by reacting 2-hydrazinoethanol with diethyl carbonate (Fig. 4.45).

Other reagents that have proved useful as dual-acting electrophiles in heterocyclic synthesis include potassium ethyl xanthate, carbon monoxide, cyanogen bromide, acid chlorides, and acid anhydrides (see Online Resource Centre).

4.7.2 **Cyclization reactions involving four reaction centres**

In these reactions there are four reaction centres, with two on each of the two reagents involved. There are two general methods by which these molecules can react (Fig. 4.46). One reagent contains two nucleophilic groups while the other has two electrophilic groups. Alternatively, each reagent has one nucleophilic group and one electrophilic group.

A large number of dual-acting reagents have commonly been used in heterocyclic synthesis These include hydrazines, ureas, thioamides, diketones, diesters, keto esters, and halo ketones (Fig. 4.47).

The reaction of diketones with diamines or hydrazines has been a popular method of creating heteroaromatic ring systems such as pyrazoles, pyridopyrimidinones, benzodiazepines, and quinoxalines (Fig. 4.48).

For example, the pyrazole ring of **celecoxib** was synthesized from a monosubstituted hydrazine and a diketone (Fig. 4.49). Celecoxib is an anti-inflammatory cyclooxygenase-2 (COX-2) selective inhibitor which blocks the synthesis of **prostaglandin H2** from **arachidonic acid**.

In these reactions, the diamine or hydrazine acts as a dual-acting nucleophile, while the diketone acts as a dual-acting electrophile. The reaction between each amino and ketone group should result in an imine, but the double bond can isomerize to a more stable position once the new ring has been formed.

Hydrazines and diamines can be reacted with dual-acting electrophiles other than diketones to produce different types of heterocyclic ring. For example, the reaction of hydrazine with various dual-acting electrophiles gives different heterocyclic rings as shown in figure 4.50. The level of unsaturation in the resulting ring depends on the types of reaction taking place. For example, the reaction of hydrazine with a dihalo alkane results in a fully saturated ring, whereas reaction with a halo ketone results in a partially saturated ring.

Other types of dual-acting nucleophiles that are commonly used in heterocyclic synthesis include ureas, thioureas, amidines, guanidines, amides, thioamides, and aminopyridines (Fig. 4.51).

The reaction of urea with various diesters produces the barbiturate class of sedatives, one of which is **phenobarbital** (Fig. 4.52).

The reaction of a thioamide with a bromo keto ester was used to produce a thiazole ring in a synthesis of **febuxostat**, an agent that is used in the treatment of gout (Fig. 4.53). The agent acts as an inhibitor of the enzyme **xanthine oxidase**. Note the chemoselectivity of the reaction where the ketone and alkyl halide are more reactive than the ester.

A guanidine structure was used in the synthesis of **imatinib**, which is a protein kinase inhibitor used in anticancer therapy (Fig. 4.54).

Intermolecular cyclizations are also possible using reagents that contain one nucleophilic and one electrophilic group, as illustrated by the synthesis of the benzodiazepine ring system present in **diazepam** (Fig. 4.55). Each of the reagents contains a nucleophilic amine group and an electrophilic group (ketone or carboxylic acid derivative).

All the intermolecular cyclization reactions described so far have involved heteroatoms as nucleophiles, so that all the newly created bonds result from N–C, O–C, or S–C couplings. It is also possible to have cyclizations where one of the new bonds created results from a C–C coupling.

For example, an α-bromo ketone was involved in a cyclization reaction leading to the COX-2 inhibitor **rofecoxib** (Fig. 4.56). In this case, the α-bromo ketone was reacted with a carboxylic acid such that a nucleophilic substitution of the alkyl bromide took place, followed by an aldol-type condensation.

Another example involves the reaction of diethyl ethoxymethylenemalonate (a dual-acting electrophile) with substituted anilines to produce a naphthyridinol ring system (Fig. 4.57). The reaction involves an initial N–C

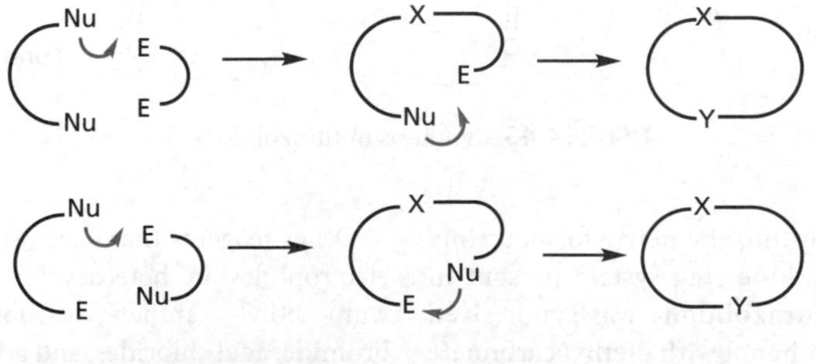

FIGURE 4.46 Methods of cyclization that involve four reaction centres.

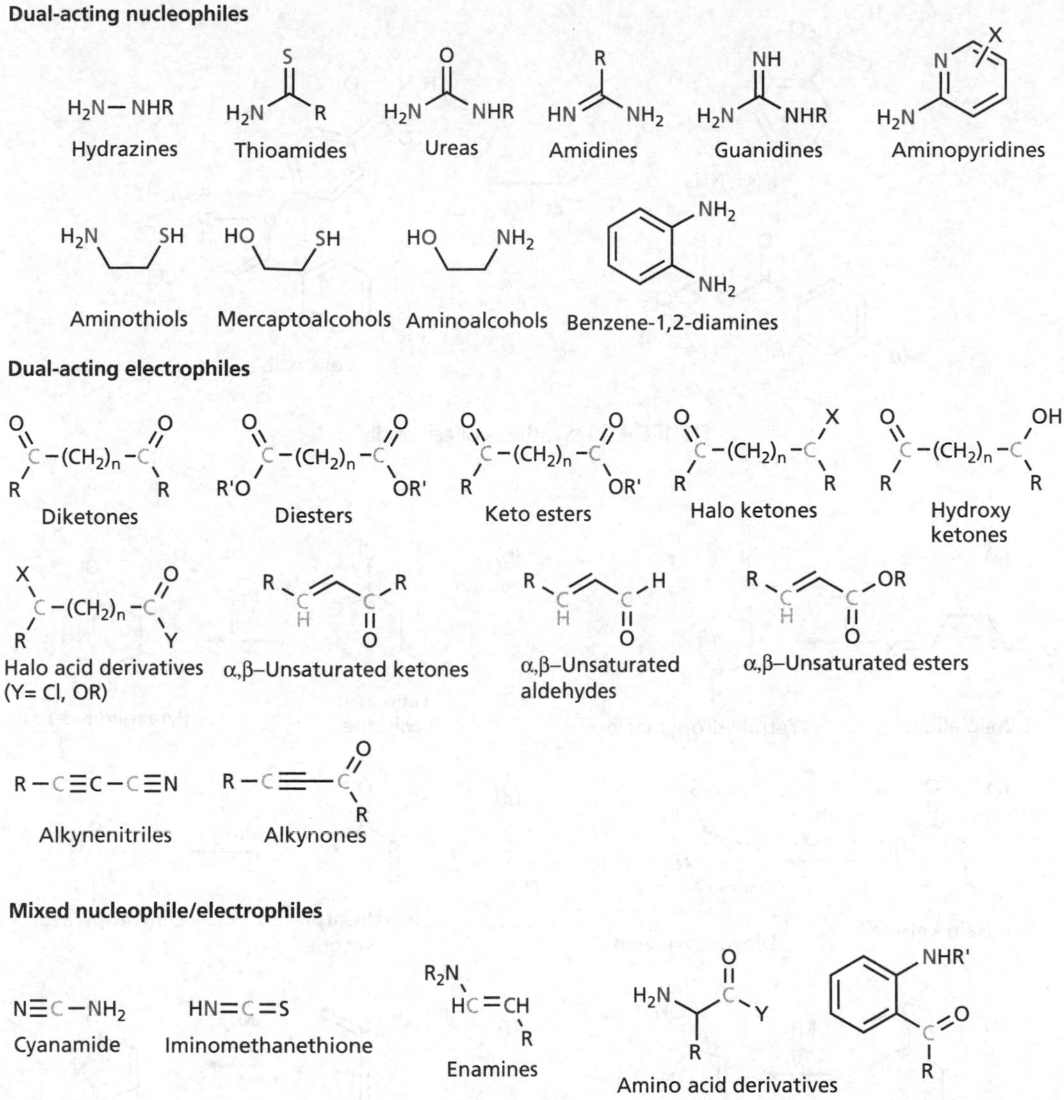

Dual-acting nucleophiles

H$_2$N—NHR
Hydrazines

H$_2$N—C(=S)—R
Thioamides

H$_2$N—C(=O)—NHR
Ureas

Amidines

Guanidines

Aminopyridines

Aminothiols   Mercaptoalcohols   Aminoalcohols   Benzene-1,2-diamines

Dual-acting electrophiles

Diketones

Diesters

Keto esters

Halo ketones

Hydroxy ketones

Halo acid derivatives (Y= Cl, OR)

α,β–Unsaturated ketones

α,β–Unsaturated aldehydes

α,β–Unsaturated esters

R—C≡C—C≡N
Alkynenitriles

Alkynones

Mixed nucleophile/electrophiles

N≡C—NH$_2$
Cyanamide

HN=C=S
Iminomethanethione

Enamines

Amino acid derivatives

FIGURE 4.47 Examples of dual-acting reagents that are commonly used in synthesizing heterocyclic ring systems.

(a)

Pyrazoles

(b)

Pyridopyrimidinones

(c)

Benzodiazepine

(d)

Quinoxalines

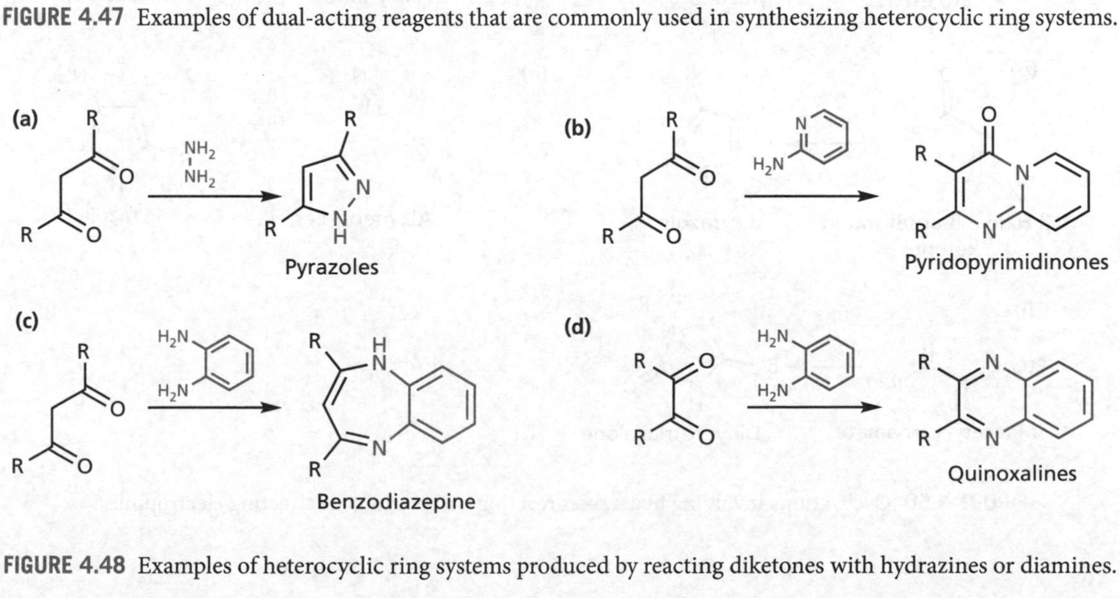

FIGURE 4.48 Examples of heterocyclic ring systems produced by reacting diketones with hydrazines or diamines.

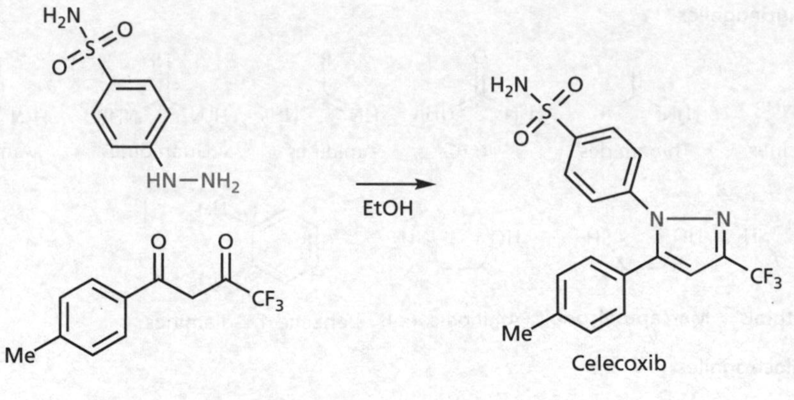

FIGURE 4.49 Synthesis of celecoxib.

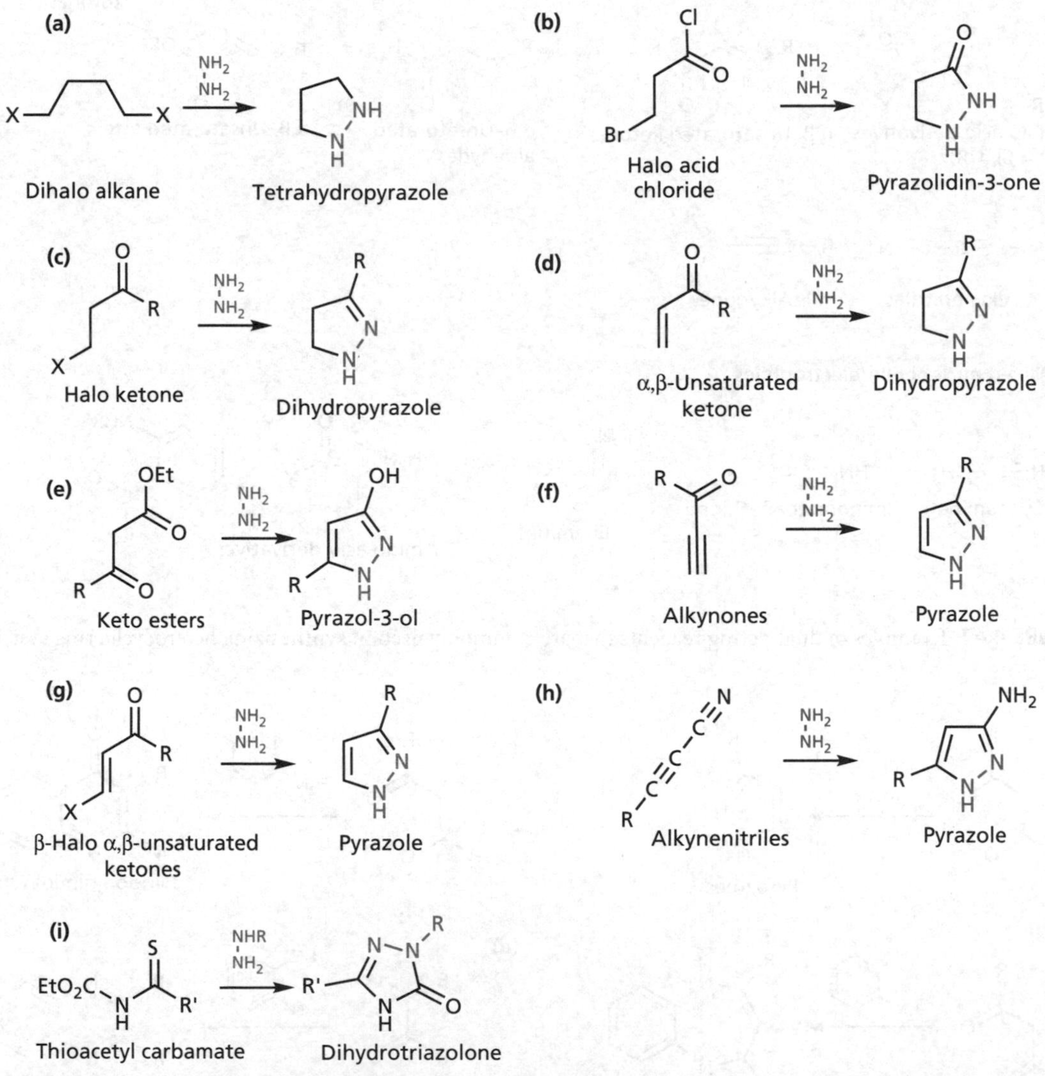

FIGURE 4.50 Cyclizations involving hydrazines reacting with various dual-acting electrophiles.

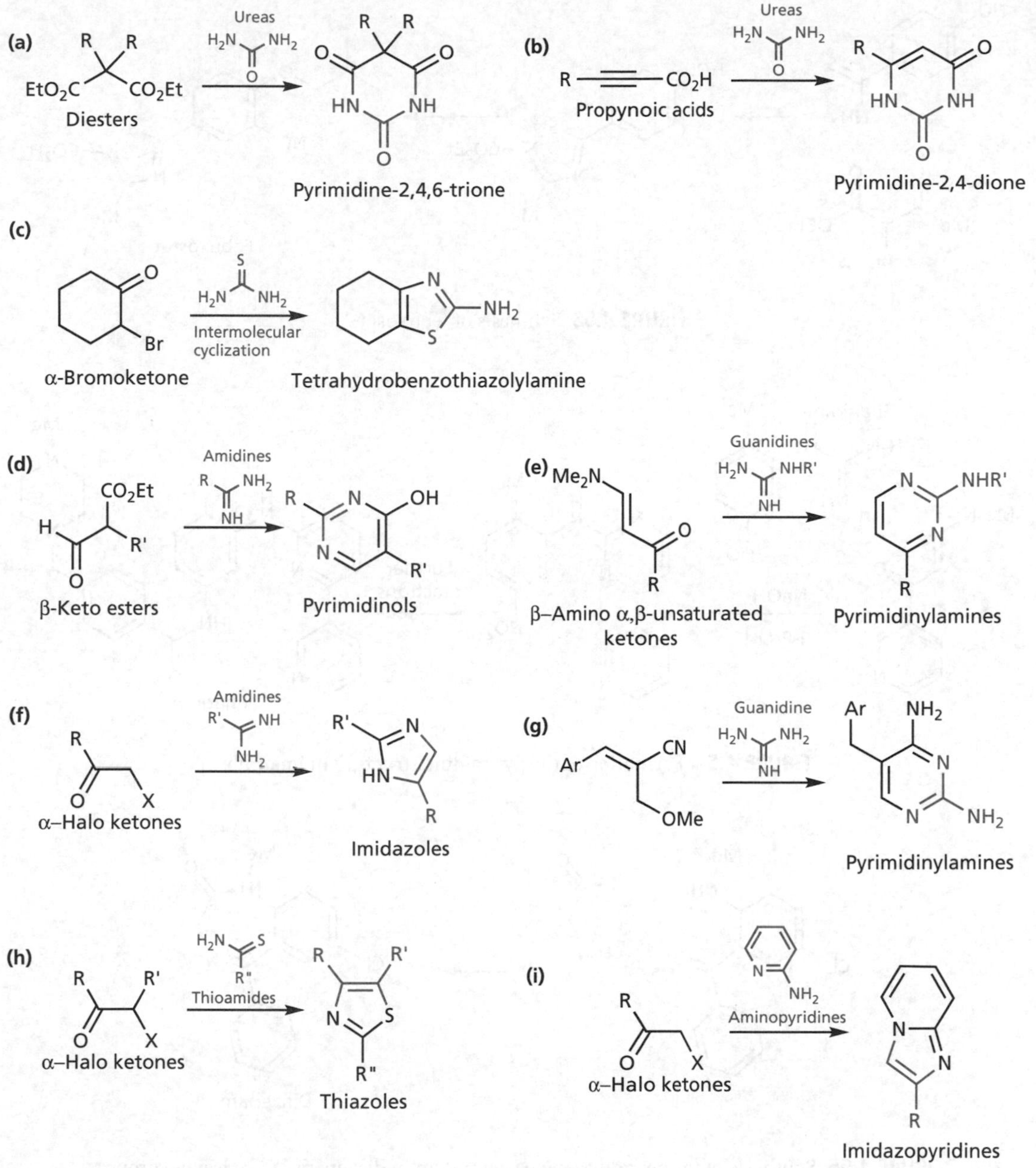

FIGURE 4.51 Examples of heterocyclic ring systems obtained from other dual-acting nucleophiles (indicated in blue).

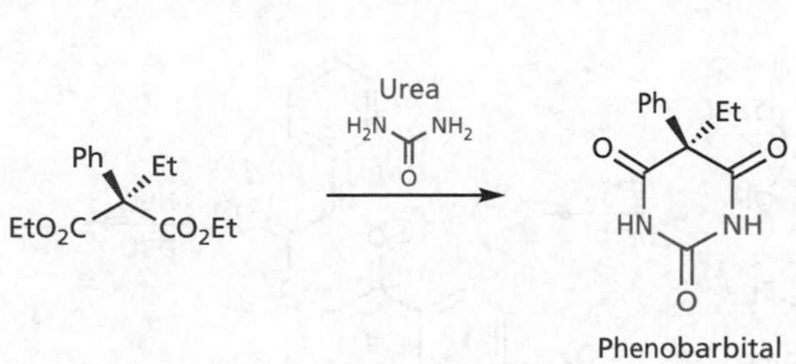

FIGURE 4.52 Synthesis of phenobarbital.

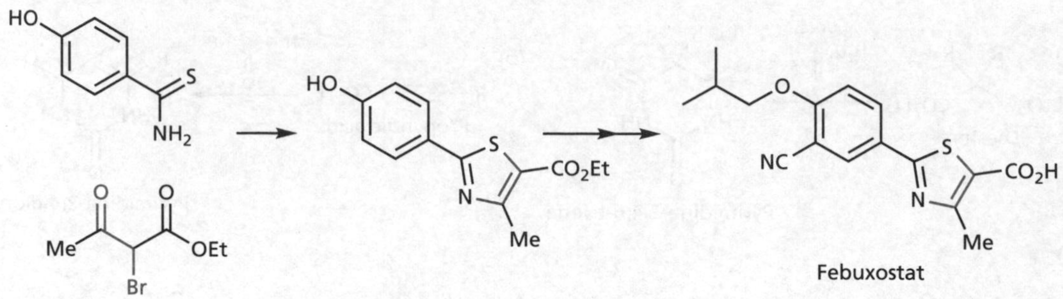

FIGURE 4.53 Synthesis of febuxostat.

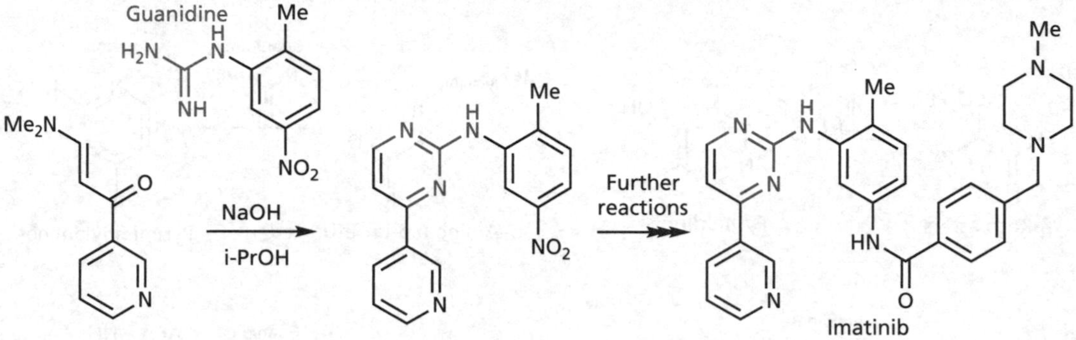

FIGURE 4.54 Formation of the pyrimidine structure in imatinib.

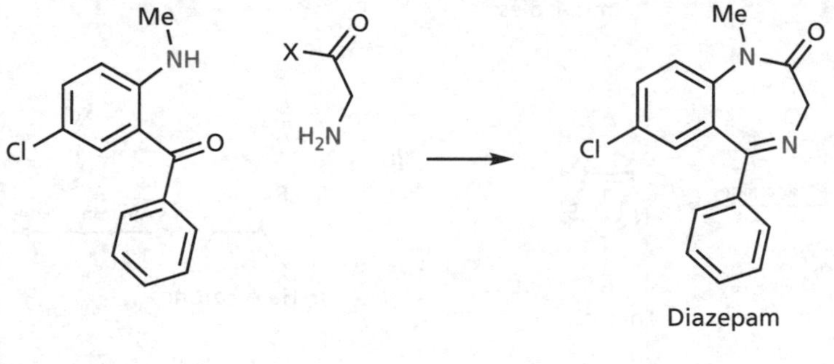

FIGURE 4.55 Synthesis of the benzodiazepine ring system in diazepam (X = a leaving group).

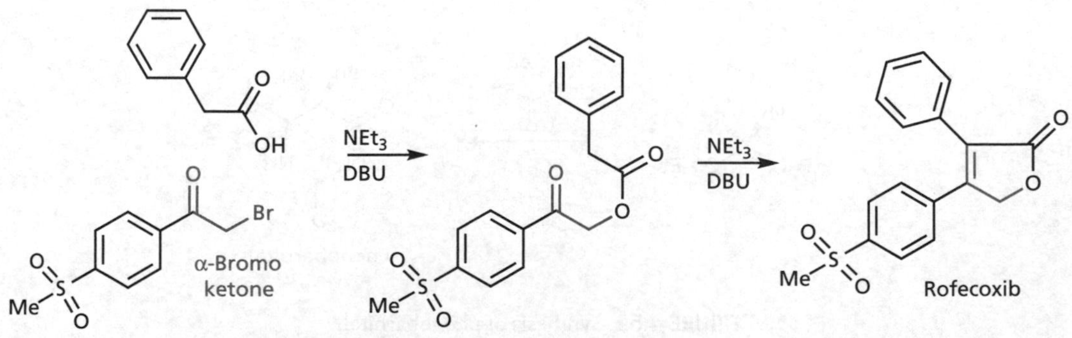

FIGURE 4.56 Synthesis of the five-membered lactone ring in rofecoxib.

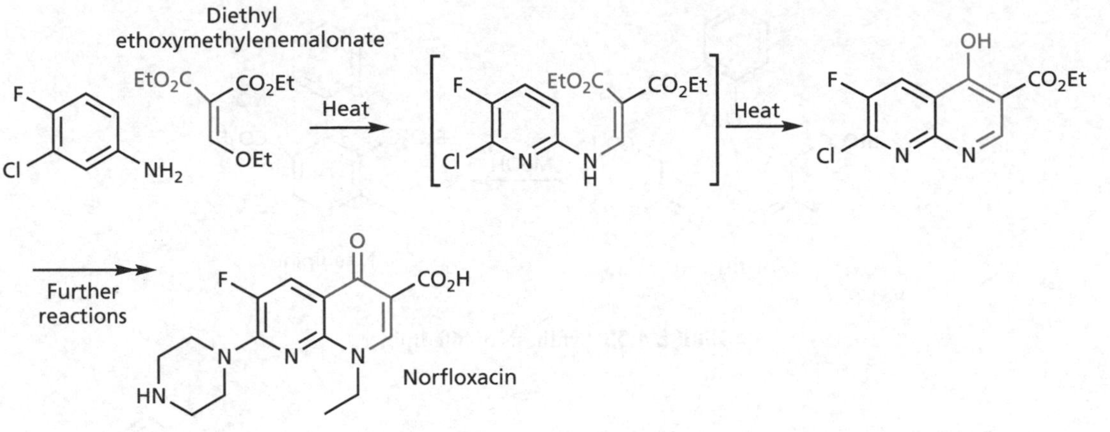

FIGURE 4.57 Synthesis of a naphthyridinol ring system during a synthesis leading to norfloxacin.

coupling, followed by an electrophilic substitution on the aromatic ring (C–C coupling). The naphthyridinol ring system can then be modified to form the naphthyridinone ring present in the fluoroquinolone antibacterial agents, an example of which is **norfloxacin** (see also Chapter 14).

KEY POINTS

- Intermolecular cyclization reactions involve an initial coupling/linking of two separate molecules, followed by spontaneous intramolecular cyclization.

- Each reagent used in an intermolecular cyclization reaction must be capable of reacting twice.

- Primary amines are commonly used as dual-acting nucleophiles in the formation of nitrogen-containing heterocycles. The other reagent must be a dual-acting electrophile, such as a diketone, dihalo alkane, or halo ester. The amine nitrogen atom is incorporated into the newly created ring.

- Aldehydes are commonly used as dual-acting electrophiles such that the carbon atom of the aldehyde group is incorporated into the new ring. The other reagent must be a dual-acting nucleophile.

- A carbonyl unit can be incorporated into a ring by using a dual-acting electrophile such as phosgene or diethyl carbonate.

- A large number of dual-acting nucleophiles have been used in intermolecular cyclizations, including diamines, hydrazines, ureas, thioureas, amidines, guanidines, amides, and thioamides.

- Dual-acting electrophiles that are commonly used in intermolecular cyclizations include diketones, diesters, halo ketones, keto esters, and α,β-unsaturated carbonyl compounds.

- Intermolecular cyclizations are possible with reagents that contain one nucleophilic and one electrophilic group.

4.8 Synthesis of dihydropyridines

The **Hantzsch condensation** is a particularly good method for synthesizing a class of pharmaceutical agents containing a dihydropyridine ring. These agents have proved to be useful in cardiovascular medicine as a treatment of hypertension, and work by blocking calcium-ion channels. Calcium-ion channels are present in the cell membranes of vascular smooth muscle and cardiac muscle. When the ion channels in vascular smooth muscles are open, calcium ions flood into the cells, resulting in smooth muscle contraction and constriction of the blood vessels, which produces an increase in blood pressure and hypertension. The opening of calcium-ion channels in heart muscle results in an increase in the rate and force of contraction. The dihydropyridine calcium-channel blockers show selectivity for the ion channels in vascular smooth muscle and therefore are used mainly for the treatment of hypertension.

The Hantzsch condensation can be carried out by reacting three reagents together—an aldehyde, two equivalents of a β-keto ester, and ammonia. For example, the calcium-channel blocker **nifedipine** was synthesized in this manner (Fig. 4.58).

Although it is possible to synthesize nifedipine in one step, as shown in Figure 4.58, better yields are obtained when the synthesis is carried out in two stages (Fig. 4.59).

A problem associated with the Hanstzsch condensation, as shown in Figures 4.58 and 4.59, is that the resulting structure is symmetrical with identical substituents on each half of the ring. Therefore, the synthesis has to be modified in order to obtain asymmetrical dihydropyridines. For example, **amlodipine** was synthesized as shown in Figure 4.60.

By synthesizing a large number of dihydropyridine analogues, it has been found that the activity of the

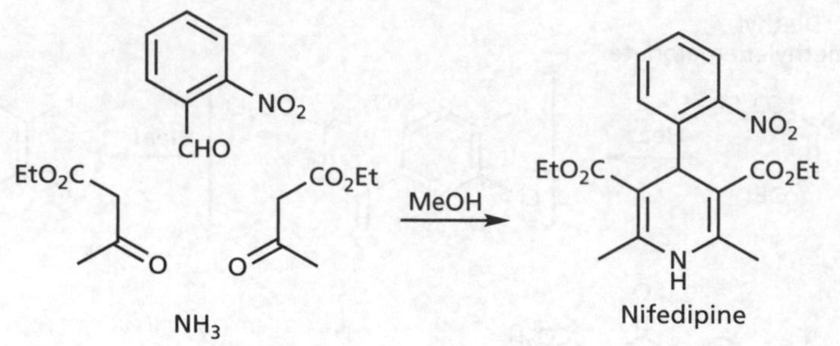

FIGURE 4.58 Synthesis of nifedipine.

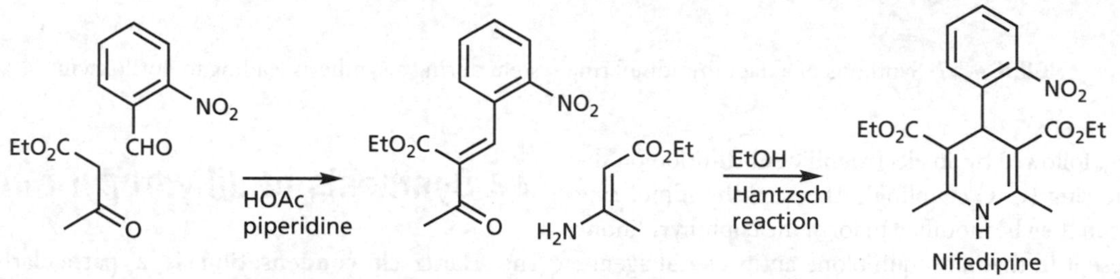

FIGURE 4.59 Synthesis of nifedipine in two stages.

dihydropyridines is dependent on the following factors (Fig. 4.61).

- The nitrogen in the dihydropyridine ring must be unsubstituted.
- Small hydrophobic alkyl groups are preferred at positions 2 and 6.
- Ester groups are preferred at positions 3 and 5.
- There must be an aryl substituent at position 4.
- The aryl ring normally contains a substituent at the *ortho* or *meta* position. A substituent at the *para* position is bad for activity.

The dihydropyridine ring is essential for activity. If a pyridine ring is present, activity is lost. It has been shown that the preferred conformation of the dihydropyridine ring is a flattened boat structure, with the aryl substituent in a pseudo-axial position. The aromatic ring itself is perpendicular to the dihydropyridine ring such that one of the *ortho* protons hangs over the face of the dihydropyridine ring, while any substituent on the aromatic ring points away from the dihydropyridine ring. This is thought to correspond to the active conformation (Fig. 4.62). In contrast, a pyridine ring is planar. This has a significant effect on the orientation of the substituents around the ring system, particularly the orientation of the aromatic substituent.

The nature of the substituent on the aromatic ring has a role in the activity of the dihydropyridines. Larger substituents are generally better for activity and appear to have a greater flattening effect on the dihydropyridine ring.

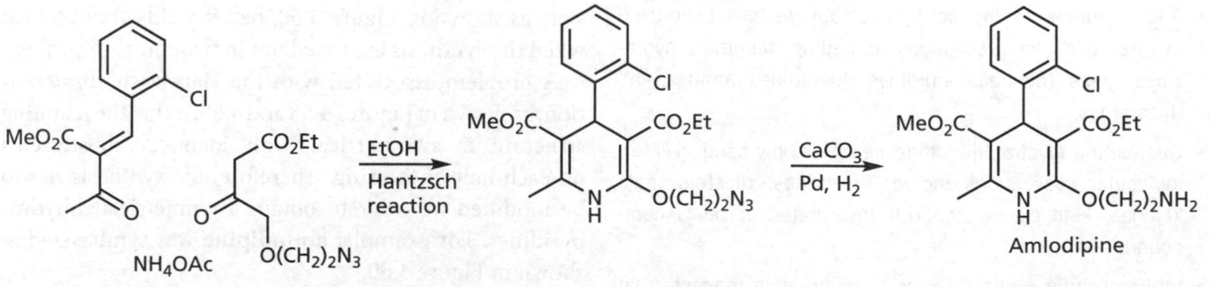

FIGURE 4.60 Synthesis of amlodipine.

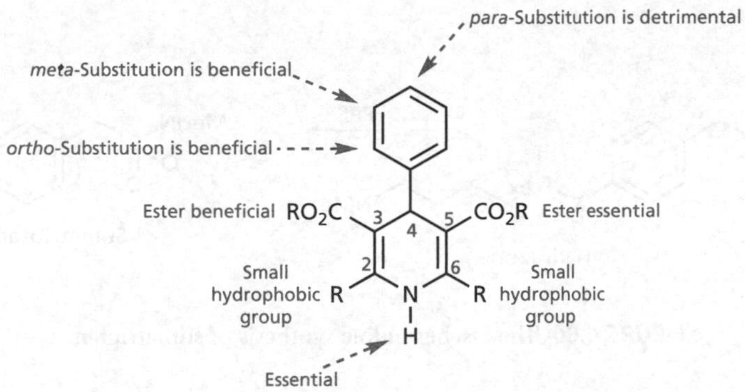

FIGURE 4.61 SAR results for dihydropyridines as calcium-ion channel blockers.

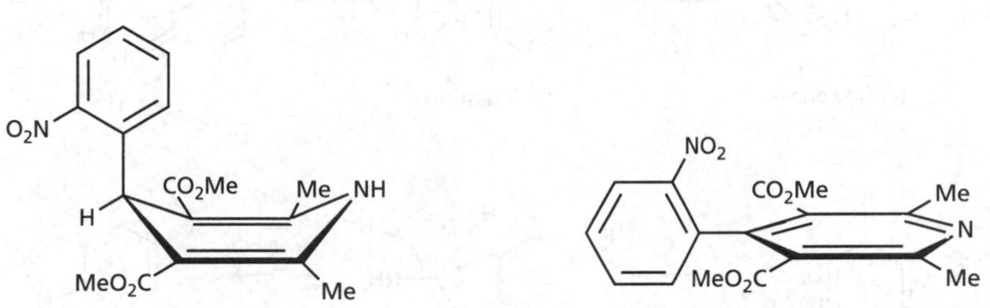

FIGURE 4.62 Preferred conformation of nifedipine compared with a comparable pyridine structure.

Asymmetrical dihydropyridines contain an asymmetric centre at position 4, and it is found that one enantiomer is more active than the other. In such structures, a distinction can be made between the two halves of the dihydropyridine ring (Fig. 4.63), which can be labelled port and starboard. It has been demonstrated that the ester group on the port side is not essential, although it plays a role in determining whether the agent acts as an agonist or an antagonist. Modelling studies carried out on a homology model of the L-type calcium-ion channel have also provided evidence that there are three groups which

act as important hydrogen bond donors (HBDs) or hydrogen bond acceptors (HBAs) with the binding site.

Finally, dihydropyridines are being considered as scaffolds to create libraries of structures that may prove effective in other fields of medicine, such as Alzheimer's disease. The dihydropyridine ring has been defined as a **privileged scaffold**, as a wide range of different activities are observed by varying the types of substituent that are present around the ring.

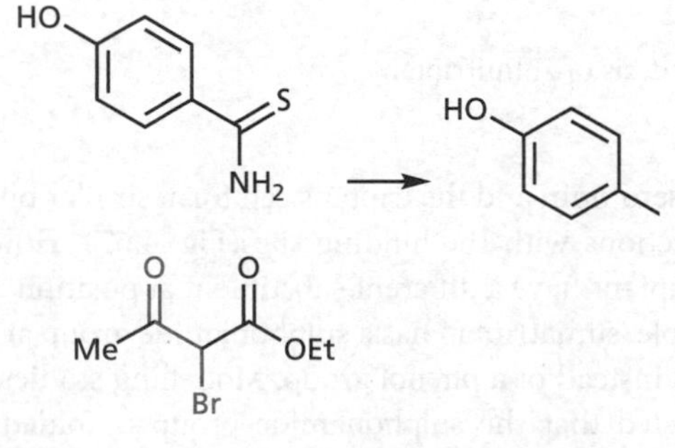

4.9 The Fischer indole synthesis

The Fischer indole synthesis is a popular method of synthesizing indole rings from a hydrazone, and was used in the final step of a synthesis leading to the anti-migraine agent **sumatriptan** (Fig. 4.64).

The mechanism for the reaction is thought to involve the isomerization of the hydrazone to form an enamine followed by a [3,3]-sigmatropic rearrangement and ring closure with loss of ammonia to form the indole ring (Fig. 4.65). The reaction is carried out in the presence of polyphosphate ester (PPE), polyphosphoric acid (PPA), or a Lewis acid.

The indole ring for another anti-migraine agent **zolmitriptan** was also synthesized by means of a Fischer indole cyclization (Fig. 4.66). In this case, the hydrazone

FIGURE 4.63 Proposed binding groups required for activity in nifedipine.

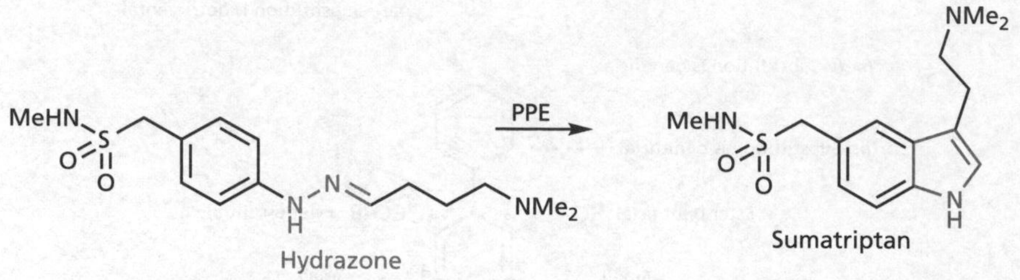

FIGURE 4.64 The Fischer indole synthesis of sumatriptan.

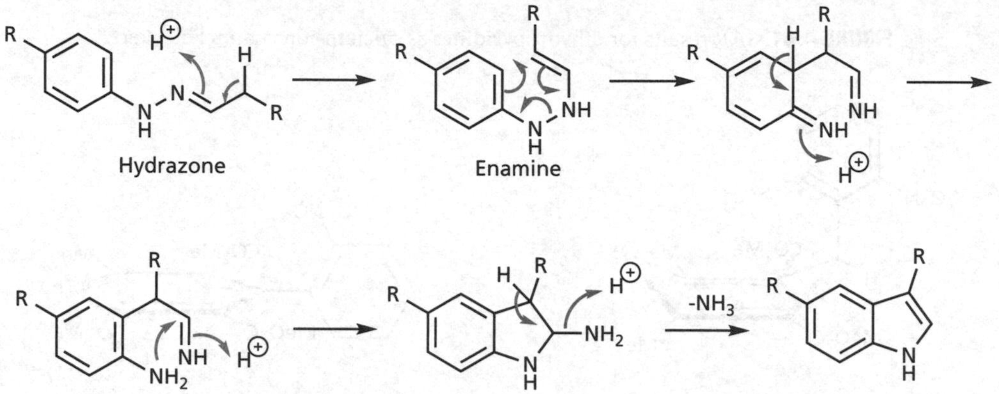

FIGURE 4.65 Mechanism of the Fischer indole synthesis.

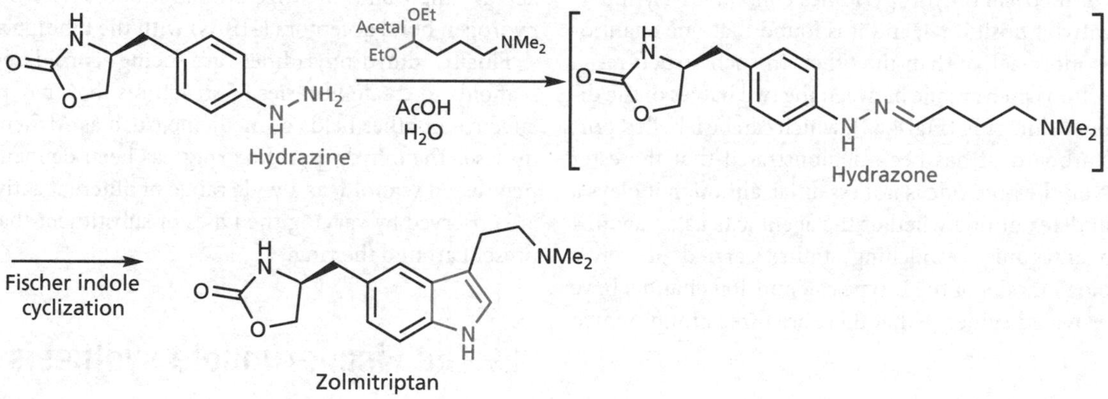

FIGURE 4.66 The Fischer indole synthesis of zolmitriptan.

was synthesized in situ by reacting a hydrazine with an acetal. The acetal hydrolysed under the reaction conditions to produce an aldehyde which then reacted with the hydrazine to form the hydrazone. Once formed, the hydrazone underwent the cyclization reaction.

The triptans act as selective agonists of serotonin receptors in the brain, resulting in vasoconstriction of dilated blood vessels. In addition, the release of inflammatory neuropeptides is inhibited. The structure of the triptans shares the same indole scaffold as serotonin itself, as well as a protonated amine substituent at position 3. Therefore,

both serotonin and the triptans can form similar binding interactions with the binding site (Fig. 4.67). However, the triptans have a different substituent at position 5. For example, sumatriptan has a sulphonamide group at position 5 instead of a phenol group. Modelling studies have suggested that the sulphonamide group is folded over the indole ring when sumatriptan is bound to the $5HT_{1D}$ receptor, and is held there by hydrogen bonding interactions with serine and threonine residues. This causes the indole ring to sink deeper into the binding pocket compared with serotonin, such that different aromatic

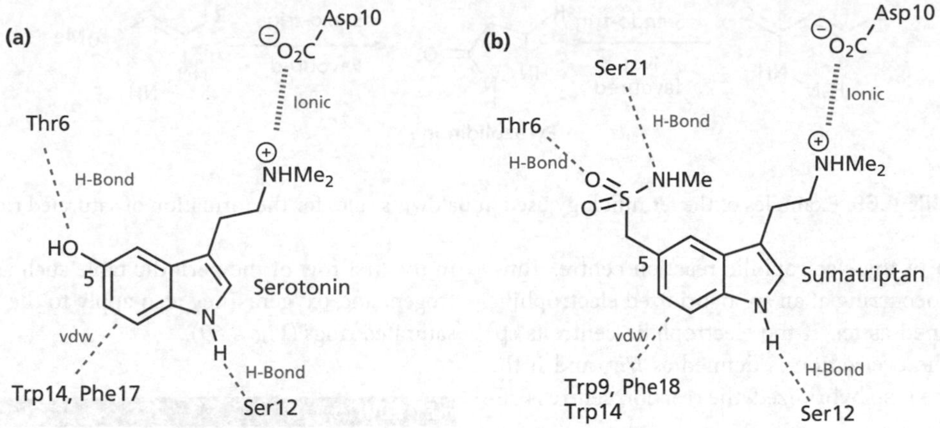

FIGURE 4.67 Model binding studies of serotonin and sumatriptan with the $5HT_{1D}$ receptor.

residues are involved in forming hydrophobic interactions with the indole ring.

4.10 Enzyme-catalysed cyclizations

Super-enzymes such as the polyketide synthases and non-ribosomal peptide synthases are multiprotein complexes responsible for the biosynthesis of many multicyclic secondary metabolites in microbial cells. These super-enzymes contain an array of active sites, one of which is a thioesterase catalytic site responsible for the release and cyclization of the biosynthetic products from these super-enzymes. Genetic engineering has been used to produce a protein which contains this thioesterase catalytic site and catalyses the formation of macrocyclic lactones and lactams (section 9.6.8). The substrates used so far have been quite limited, but it may be possible in the future to use genetically engineered enzymes in a general synthesis of macrocyclic structures.

4.11 Baldwin's rules

Baldwin's rules are a guide to predicting which types of cyclization reaction are most likely to occur. To begin with, each cyclization can be defined as being *endo* or *exo*. An *exo*-type reaction is where the bond being broken does not end up as part of the ring system, whereas an *endo*-type reaction is where the bond does end up as such (Fig. 4.68). For example, in Figure 4.68a, a bond is broken between the carbonyl group and the methoxy group with the electrons of that bond ending up on the methoxide leaving group. Therefore, the electrons are not part of the resulting ring. In figure 4.68b, the bond being broken is the pi bond of an alkene. The electrons in that bond end up forming a C–H bond which is part of the ring.

Each *exo* or *endo* cyclization is also defined by the size of the ring that is formed. Therefore, the cyclizations in Figure 4.68 are defined as 5-*exo* and 5-*endo*. In addition, each cyclization can be defined as tetrahedral (*tet*), trigonal (*trig*), or digonal (*dig*), depending on the

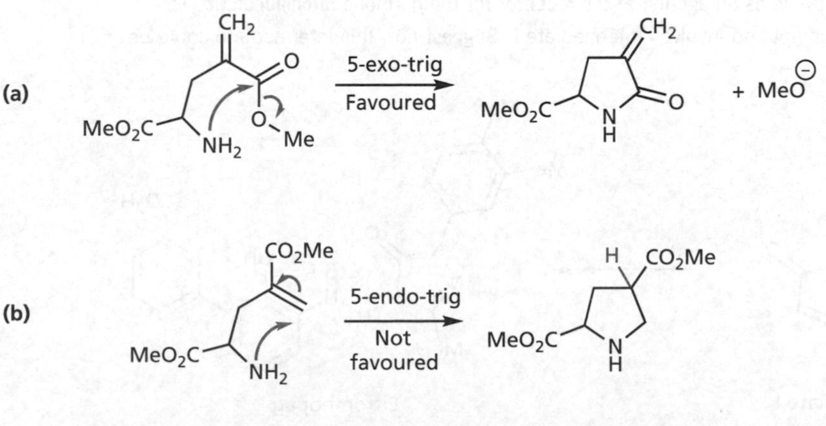

FIGURE 4.68 Examples of terminology used in Baldwin's rules.

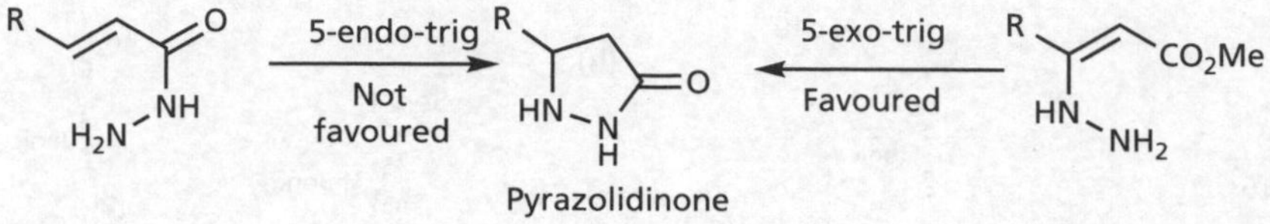

FIGURE 4.69 Examples of the terminology used in Baldwin's rules for the formation of saturated rings.

hybridization of the electrophilic reaction centre. Thus, a cyclization occurring at an sp³ hybridized electrophilic centre is defined as tet. If the electrophilic centre is sp² hybridized, the cyclization is defined as *trig*, and if the reaction centre is sp hybridized, the reaction centre is defined as *dig*.

The success of a cyclization depends on the stereochemistry of the transition state, and Baldwin's studies of these transition states led to the following conclusions.

- For tetrahedral systems (*tet*) with sp³ electrophilic centres:
  – 3- to 7-*exo* reactions are favoured but 5- and 6-*endo* are not.
- For trigonal systems (*trig*) with sp² electrophilic centres:
  – 3- to 7-*exo* are favoured, as are 6- and 7-*endo*, but 3- to 5-*endo* are not.
- For digonal systems (*dig*) with sp electrophilic centres:
  – 5- to 7-*exo* are favoured, as are 3- to 7-*endo*, but 3- and 4-*exo* are not.

These rules apply to cyclizations leading to both carbocycles and heterocycles and hold true for nucleophiles in the first row of the periodic table such as carbon, nitrogen, and oxygen. They also apply to the formation of saturated rings (Fig. 4.69).

KEY POINTS

- The Hantzsch condensation is effective in synthesizing dihydropyridine rings.
- Asymmetrical dihydropyridines can be synthesized by modifying the Hantzsch condensation procedure.
- The dihydropyridine ring has been described as a privileged scaffold since different structures interact with different targets depending on the number, nature, and position of the substituents present.
- The Fischer indole synthesis has proved an effective method of creating indole rings from aromatic hydrazones.
- Drugs containing indole rings have been designed to interact with serotonin receptors.
- Enzyme-catalysed cyclizations are being investigated as a method of synthesizing macrocycles.
- Baldwin's rules are used to determine whether particular intramolecular cyclizations are favoured or not.

QUESTIONS

1. In the synthesis of **sertraline** shown in Figure 4.10, a Friedel–Crafts acylation took place on one aromatic ring but not the other. Why?

2. **Eltrombopag** is a drug that was approved in 2008 for the treatment of patients having low platelet counts. It acts as an agonist at the receptor for the hormone **thrombopoietin**. The synthesis of the compound involves intermediate I. Suggest how this intermediate could be synthesized.

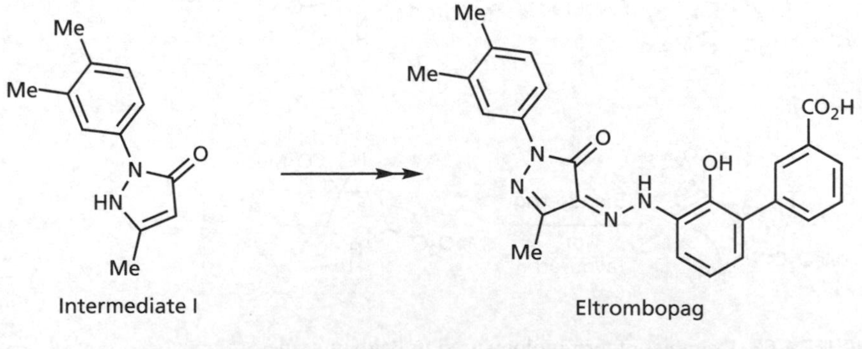

Intermediate I   Eltrombopag

3. The following synthesis of a benzofuran (II) was involved in the synthesis of **dronedarone**, which is used in the treatment of cardiac arrythmias. Propose a mechanism by which the benzofuran intermediate is formed.

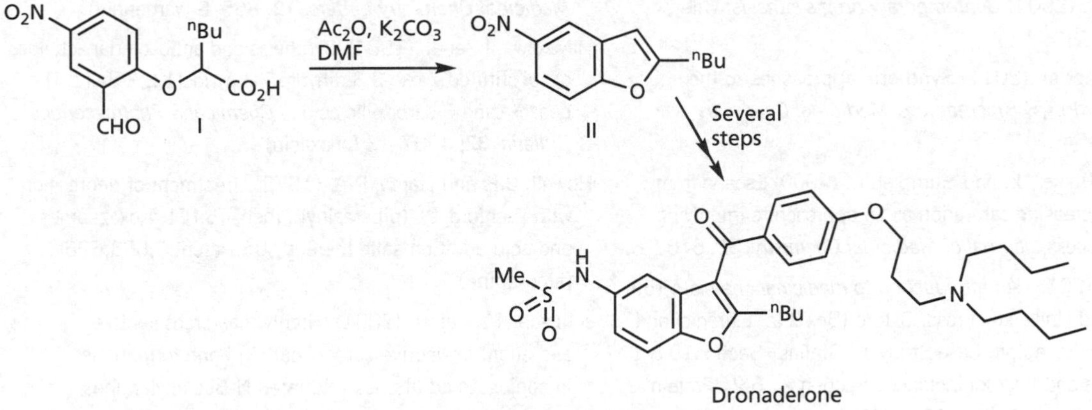

Dronaderone

4. Structure III was prepared as part of a synthesis of **asenapine**, which has been approved in the USA as an atypical antipsychotic agent. Structure III was converted in two stages to the tetracyclic structure IV. Suggest how this might have been carried out.

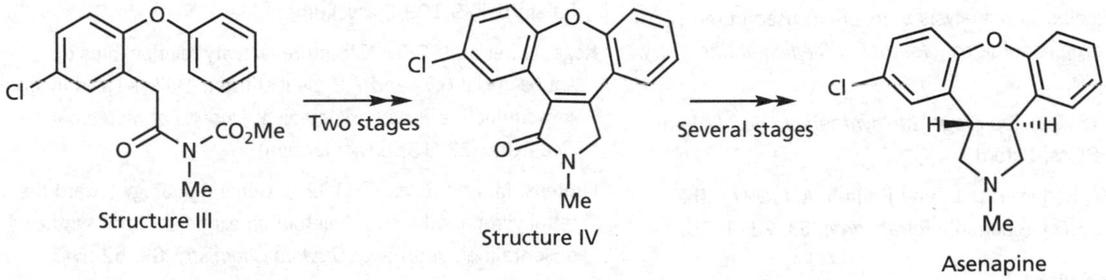

Structure III   Structure IV   Asenapine

FURTHER READING

General reading

Baskin, J.M., et al. (2007) 'Copper-free click chemistry for dynamic *in vivo* imaging', *Proceedings of the National Academy of Sciences of the USA*, **104**, 16 793–7.

Birch, H. (2010) 'Biology meets click chemistry', *Chemistry World*, October, pp 36–39.

Clayden, J., Greeves, N., and Warren, S. (2012) *Organic chemistry* (2nd edn). Oxford University Press, Oxford.

Crisp, G.T. (1998) 'Variations on a theme: recent developments on the mechanism of the Heck reaction and their implications for synthesis', *Chemical Society Reviews*, **27**, 427–36.

DeForest, C.A., Polizzotti, B.D., and Anseth, K.S. (2009) 'Sequential click reactions for synthesising and patterning three-dimensional cell microenvironments', *Nature Materials*, **8**, 659–64.

Dounay, A.B. and Overman, L.E. (2003) 'The asymmetric intramolecular Heck reaction in natural product total synthesis', *Chemical Reviews*, **103**, 2945–63.

Ertl, P., et al. (2006) 'Quest for the rings: *in silico* exploration of ring universe to identify novel bioactive heteroaromatic scaffolds', *Journal of Medicinal Chemistry*, **49**, 4568–73.

Evans, M.J., et al. (2005) Target discovery in small-molecule cell-based screens by in situ proteome reactivity profiling, *Nature Biotechnology*, **23**, 1303–7.

Kalesh, K.A., Liu, K., and Yao, S.Q. (2009) 'Rapid synthesis of Abelson tyrosine kinase inhibitors using click chemistry', *Organic & Biomolecular Chemistry*, **7**, 5129–36.

Kilbourn, M.R. and Shao, X. (2009) 'Fluorine-18 radiopharmaceuticals', in I. Ojima (ed.), *Fluorine in medicinal chemistry and chemical biology*. Blackwell, Winchester, Chapter 14.

Li, J.-J., et al. (2004) *Contemporary drug synthesis*. Wiley, New York.

Liu, K.K.-C., et al. (2011) Synthetic approaches to the 2009 new drugs, *Bioorganic & Medicinal Chemistry*, **19**, 1136–54.

Lovering, F., Bikker, J., and Humblet, C. (2009) Escape from flatland: increasing saturation as an approach to improving clinical success, *Journal of Medicinal Chemistry*, **52**, 6752–6.

Patrick, G.L. (2013) *An introduction to medicinal chemistry* (5th edn). Oxford University Press, Oxford (Box 8.2, 'Estradiol and the estrogen receptor'; Case Study 1, 'Statins'; Section 19.8.1, 'Quinolones and fluoroquinolones'; Section 21.6.2, 'Protein kinase inhibitors'; Section 22.15, 'Anticholinesterases'; Chapter 23, 'Drugs acting on the adrenergic nervous system'; Chapter 24, The opioid analgesics).

Pitt, W.R., et al. (2009) 'Heteroaromatic rings of the future', *Journal of Medicinal Chemistry*, **52**, 2952–63.

Roughey, S.D. and Jordan, A.M. (2011) 'The medicinal chemist's toolbox: an analysis of reactions used in the pursuit of drug candidates', *Journal of Medicinal Chemistry*, **54**, 3451–79.

Saunders, J. (2000) *Top drugs, top synthetic routes*, Oxford University Press, Oxford.

Shibasaki, M., Boden, C.D.J., and Kojima, A. (1997) 'The asymmetric Heck reaction', *Tetrahedron*, **53**, 7371–95.

Specific compounds

Bosch, J., et al. (2001) 'Synthesis of 5-(sulfamoylmethyl) indoles', *Tetrahedron*, **57**, 1041–8 (almotriptan).

Bossart, F. and Vater, W. (1969) '4-Aryl-1,4-dihydropyridines', US Patent 3,485,847 (nifedipine).

Calligaro, D.O., et al. (1997) 'The synthesis and biological activity of some known and putative metabolites of the atypical antipsychotic agent olanzapine (LY170053)', *Bioorganic & Medicinal Chemistry Letters*, **7**, 25–30 (olanzapine).

Chandraratna, R.A.S. (1992) 'Disubstituted acetylenes bearing heteroaromatic and heterobicyclic groups having retinoid like activity', US Patent 5,089,509 (tazarotene).

Dale, D.J., et al. (2000) 'The chemical development of the commercial route to sildenafil: a case history', *Organic Process Research & Development*, **4**, 17–22 (sildenafil).

Daugan, A. C-M. (1999) 'Tetracyclic derivatives: process of preparation and use', US Patent 5,859,006 (tadalafil).

Ducharme, Y., et al. (1995) 'Phenyl heterocycles as COX-2 inhibitors', WO Patent 95/18799 (rofecoxib).

Gever, G. (1956) 'New N-(5-nitro-2-furfurylidene)-3-amino-2-oxazolidines', US Patent 2,742,462 (furazolidone).

Haning, H., et al. (2002) 'Imidazo[5,1-f][1,2,4]triazin-4(3H)-ones, a new class of potent PDE5 inhibitors', *Bioorganic & Medicinal Chemistry Letters*, **12**, 865–8 (vardenafil).

Hayakawa, I., et al. (1984) 'Synthesis and antibacterial activities of substituted 7-oxo-2,3-dihydro-7H-pyrido[1,2,3-de][1,4]benzoxazine-6-carboxylic acids', *Chemical & Pharmaceutical Bulletin*, **32**, 4907–13 (ofloxacin).

Howell, C.F. and Hardy, R.A. (1972) 'Treatment of depression with 2-chloro-11-(piperazinyl)dibenz-[b,f][1,4]oxazepines and acid addition salts thereof', US Patent 3,663,696 (amoxapine).

Johnson, T.A., et al. (2001) 'Highly diastereoselective and enantioselective carbon–carbon bond formations in conjugate additions of lithiated N-Boc allylamines to nitroalkenes: enantioselective synthesis of 3,4- and 3,4,5-substituted piperidines including (–)-paroxetine', *Journal of the American Chemical Society*, **123**, 1004–5 (paroxetine).

Kamenka, J.M., et al. (1993) 'Pharmaceutical compositions for neuroprotection containing arylcyclohexylamines', US Patent 5,179,109 (gacyclidine).

Koga, H., et al. (1980) 'Structure–activity relationships of antibacterial 6,7- and 7,8-disubstituted 1-alkyl-1,4-dihydro-4-oxoquinoline-3-carboxylic acids', *Journal of Medicinal Chemistry*, **23**, 1358 (norfloxacin).

Lautens, M. and Rovis, T. (1997) 'General strategy toward the tetrahydronaphthalene skeleton: an expedient total synthesis of sertraline', *Journal of Organic Chemistry*, **62**, 5246–7 (sertraline).

Lesher, G.Y., et al. (1962) '1,8-Naphthyridine derivatives', *Journal of Medicinal Chemistry*, **5**, 1063 (nalidixic acid).

Macor, J.E. and Wythes, M.J. (1996) 'Indole derivatives', US Patent 5,545,644 (eletriptan).

Marquillas Olondriz, F., et al. (1994) 'Procedure for obtaining 3-[2-[4-(6-fluoro-1,2-benzisoxazol-3-yl)piperidin]ethyl]-2-methyl-6,7,8,9-tetrahydro-4H-pyrido[1,2-a]pyrimidin-4-one', ES Patent 2,050,069 (risperidone).

McCall, J. M., et al. (1975) 'A new approach to triaminopyrimidine N-oxides', *Journal of Organic Chemistry*, **40**, 3304–6 (minoxidil).

Oxford, A.W. (1991) 'Indole derivative', US Patent 5,037,845 (sumatriptan).

Penning, T.D., et al. (1997) 'Synthesis and biological evaluation of the 1,5-diarylpyrazole class of cyclooxygenase-2 inhibitors: identification of 4-[5-(4-methylphenyl)-3-(trifluoromethyl)-1H-pyrazol-1-yl]benzenesulfonamide (SC-58635, celecoxib)', *Journal of Medicinal Chemistry*, **40**, 1347–65 (celecoxib).

Robertson, A.D., et al. (1995) 'Indolyl compounds for treating migraine', US Patent 5,466,699 (zolmitriptan).

Schwartz, A., et al. (1992) 'Enantioselective synthesis of calcium channel blockers of the diltiazem group', *Journal of Organic Chemistry*, **57**, 851–6 (diltiazem).

Urban, F.J. (1994) 'Processes and intermediates for the preparation of 5-[2-(4-benzoisothiazol-3-yl)piperazin-1-yl)ethyl]-6-chloro-1,3-dihydro-indol-2-one', US Patent 5,359,068 (ziprasidone).

Vanhoof, P.M. and Clarebout, P.M. (1975) 'Derivatives of 2-aminoindanes', US Patent 3,923,813 (aprindane).

Welch, W., et al. (1985) 'Antidepressant derivatives of *cis*-4-phenyl-1,2,3,4-tetrahydro-1-naphthalenamine', US Patent 4,536,518 (sertraline).

Zimmerman, J., et al. (1997) 'Potent and selective inhibitors of the ABL-kinase: phenylaminopyrimidine (PAP) derivatives', *Bioorganic Medicinal Chemistry Letters*, **7**, 187 (imatinib).

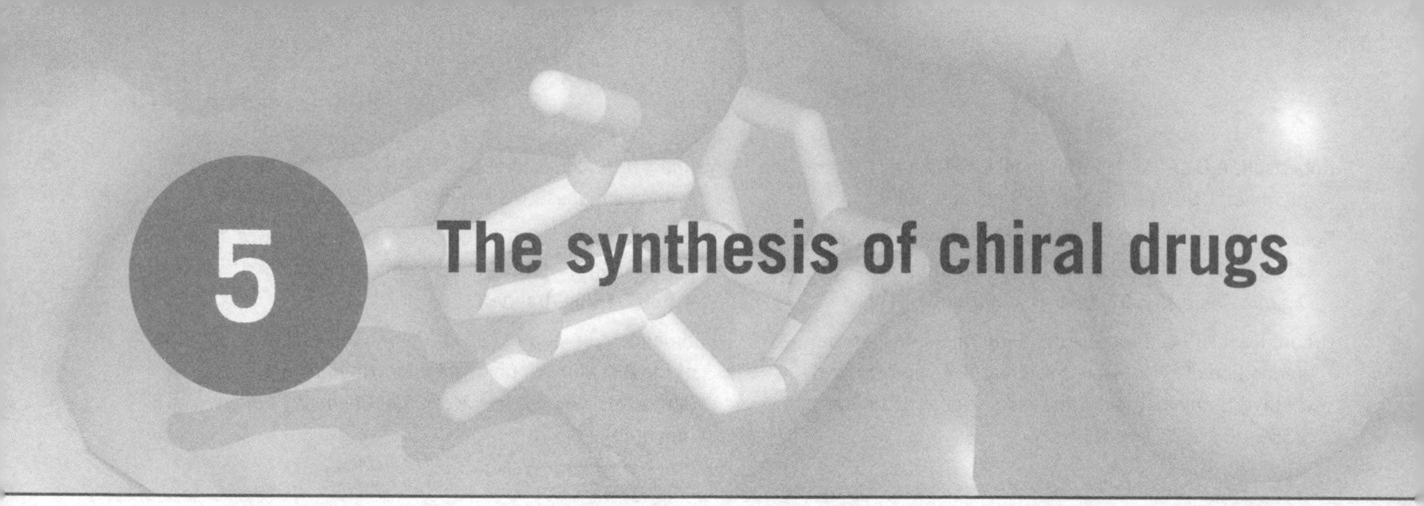

5 The synthesis of chiral drugs

5.1 Introduction

Chiral drugs are asymmetric compounds that can exist as two non-superimposable mirror images known as **enantiomers**. This property is related to a lack of symmetry (**asymmetry**) in the molecule, and is known as **chirality**. Thus, asymmetric molecules are defined as **chiral**, whereas symmetrical molecules are defined as **achiral**. To be precise, a molecule must have no more than one axis of symmetry to be chiral. In most cases, it is possible to identify whether a molecule is chiral or not by identifying an asymmetric carbon centre, which is generally a carbon atom having four different substituents. For example, the asymmetric centres in **propranolol** and **salbutamol** are marked with a star in Figure 5.1. There are situations where molecules can be chiral without having an asymmetric centre and where molecules can have asymmetric centres but not be chiral (*meso* structures), but in most situations chiral molecules can be identified by the presence of an asymmetric centre.

Enantiomers have identical chemical and physical properties, with two exceptions. First, if plane polarized light is shone through a solution of one enantiomer, it is rotated clockwise or anticlockwise. If it is shone through a solution of the opposite enantiomer, the light will be rotated to the same extent, but in the opposite direction. Hence, enantiomers are often defined as being **optical isomers**. This is a form of **configurational isomerism**, where the orientation of substituents attached to an asymmetric centre is defined by the nomenclature *R* or *S* (Box 5.1). It is important to appreciate that the *R* and *S* nomenclature does not define the direction in which the chiral compound rotates plane polarized light. That can only be found out by experimentation, and is defined as levorotatory (l) and dextrorotatory (d), corresponding to anticlockwise and clockwise rotation respectively. Alternatively, the symbols (–) and (+) are used.

The second exception is the ability of enantiomers to differ in their interactions with other chiral molecules. This has two important consequences. Because the molecules of life are chiral, enantiomers of drugs often interact differently with protein targets and this results in different biological and pharmacological properties for each enantiomer. It also means that enantiomers may react differently with chiral reagents, allowing the possibility of asymmetric syntheses where one enantiomer is synthesized in preference to the other.

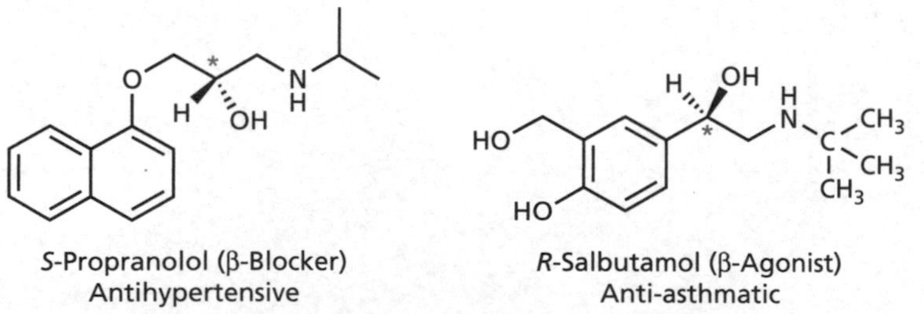

S-Propranolol (β-Blocker)
Antihypertensive

R-Salbutamol (β-Agonist)
Anti-asthmatic

FIGURE 5.1 The active enantiomers of propranolol and salbutamol.

BOX 5.1 Nomenclature for defining absolute configuration

The configuration at an asymmetric centre is defined as being R or S according to a set of rules known as the **Cahn–Ingold–Prelog sequence rules**. To illustrate the use of these rules, we will determine the configuration of the asymmetric centre in the hormone adrenaline (Figure 1a). The four different substituents attached to the asymmetric centre are highlighted in Figure 1b.

First, we have to prioritize the four different groups attached to the asymmetric centre. The first step in this process is to identify the atoms directly linked to the asymmetric centre and assign their atomic numbers (stage 1, Fig. 2). Next, the atoms are prioritized with respect to their atomic numbers (stage 2, Fig. 2). Oxygen has the highest atomic number and so it takes priority 1. Hydrogen has the smallest atomic number and so it takes priority 4. However, we have two identical carbon atoms, and so we cannot distinguish between them at this stage. We now have to identify the atom of highest priority attached to each of the carbon atoms (stage 3, Fig. 2). Therefore, the atoms linked to the carbon atoms have to be identified and assigned atomic numbers. One of the carbons is an aromatic carbon and so there are another two carbons linked to it. However, the rules state that any atom linked by a double bond can be counted twice. Therefore, a nominal three carbons are attached. The other carbon has a nitrogen and two hydrogens attached. In the final stage (stage 4, Fig. 2), we compare the atoms of highest priority in

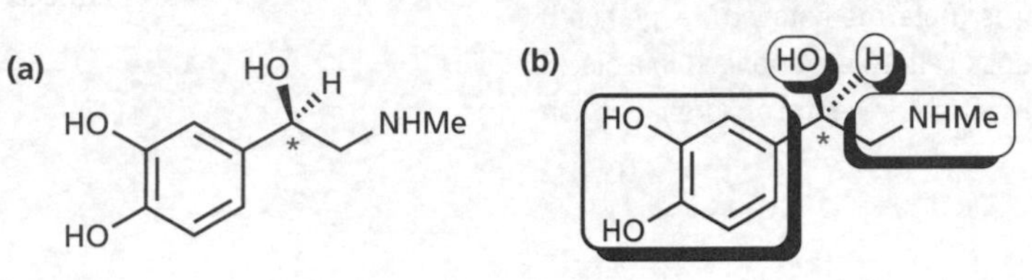

FIGURE 1 Adrenaline.

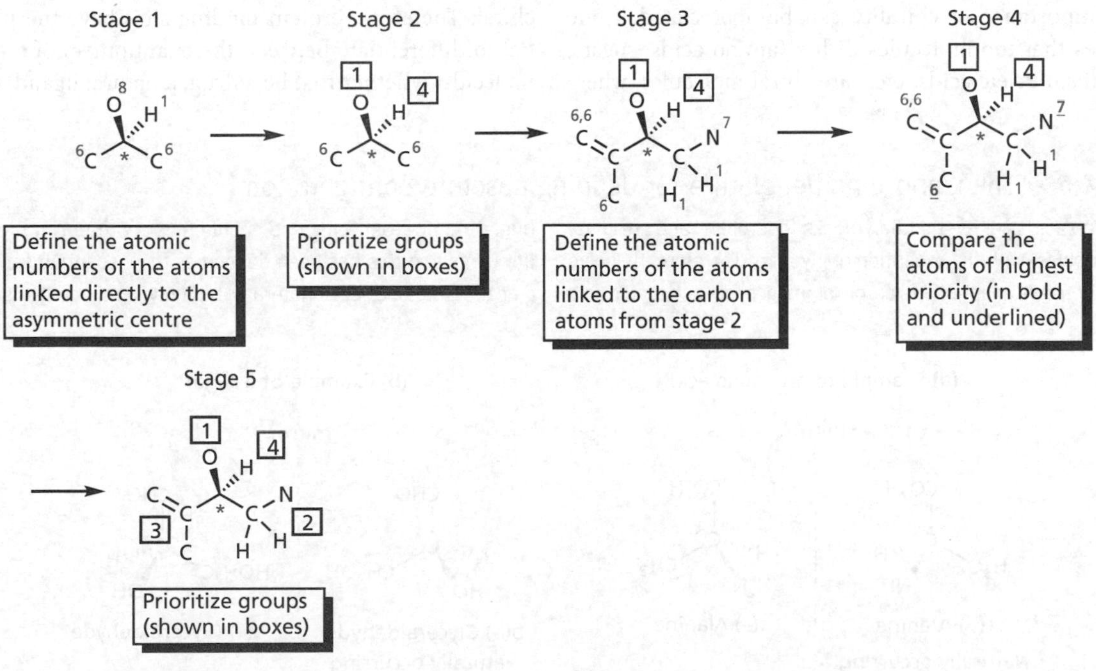

FIGURE 2 Prioritizing the groups attached to the asymmetric centre of adrenaline.

(Continued)

BOX 5.1 Nomenclature for defining absolute configuration *(Continued)*

each of the groups. In this case, we are comparing nitrogen with carbon. Nitrogen has the higher atomic number and so the group containing nitrogen takes priority. Note that it is not a case of 'adding up' the atomic numbers of the atoms concerned in each group. Therefore, priorities are *not* based on how large a particular group is. For example, the aromatic ring in adrenaline is larger than the CH_2NHMe group, but takes a lower priority as a result of the prioritization process.

Having identified the priorities of the groups, the molecule is orientated such that the group of lowest priority is pointing backwards. If the remaining three groups are arranged clockwise from highest to lowest priority, the centre is defined as *R*. If they are arranged anticlockwise, the centre is defined as *S*. In our example, the group of lowest priority is already pointing backwards (note the hatched wedge bond) (Fig. 3). The priority groups (1–2–3) are arranged in a clockwise fashion and so the asymmetric centre of adrenaline can be assigned as *R*.

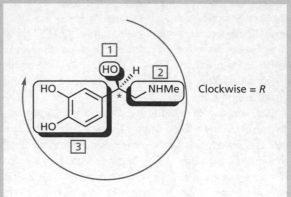

FIGURE 3 Assigning the asymmetric centre of adrenaline as *R*.

5.2 **Relevance of chirality to the pharmaceutical industry**

The importance of chirality can be appreciated if one realizes that the molecules of life (amino acids, sugars, proteins, nucleic acids, etc.) are chiral molecules, where only one of the two possible enantiomers is naturally occurring (Box 5.2).

As the macromolecules of life are made up of chiral building blocks, drug targets in the body must also be chiral. Therefore, protein binding sites have the potential to differentiate between the enantiomers of a chiral molecule, whether that be a drug, a natural ligand, or an

BOX 5.2 The ʟ and ᴅ nomenclature for defining absolute configuration

Amino acids are the building blocks for proteins and are chiral molecules (with the exception of glycine). Therefore, there are two possible enantiomers or mirror images for each amino acid

(Fig. 1a). However, only one enantiomer is present in higher life forms and this has been defined as the ʟ-enantiomer. The ᴅ-enantiomer does exist naturally, but only in microorganisms.

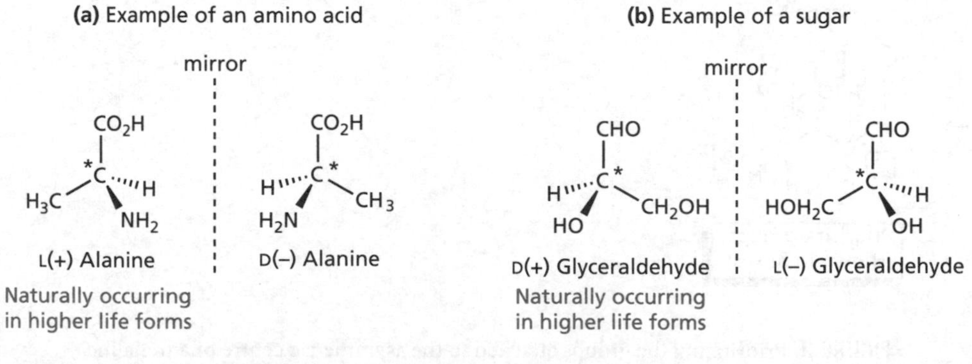

(a) Example of an amino acid

mirror

ʟ(+) Alanine
Naturally occurring
in higher life forms

ᴅ(−) Alanine

(b) Example of a sugar

mirror

ᴅ(+) Glyceraldehyde
Naturally occurring
in higher life forms

ʟ(−) Glyceraldehyde

FIGURE 1 The enantiomers of alanine and glyceraldehyde.

BOX 5.2 The L and D nomenclature for defining absolute configuration *(Continued)*

Similarly, the sugar building blocks involved in the structure of DNA, RNA, and carbohydrate polymers are chiral molecules capable of existing as two enantiomers. Here, too, only one of the two enantiomers exists in higher life forms—the D-enantiomer (Fig. 1b).

The L and D nomenclature is an old method of defining the absolute configuration of an asymmetric centre, and is still used commonly for amino acids and carbohydrates. To define whether the centre is L or D, a Fischer diagram of the structure is drawn such that the carbon chain is drawn vertically with the main functional group at the top (Fig. 2). The direction of the other functional group at the asymmetric centre then determines whether the centre is L or D. If the group is pointing to the left, it the L-enantiomer. If it is pointing to the right, it is the D-enantiomer. Most L-amino acids have the S-configuration, the exception being cysteine.

The symbols (d) and (l) indicate whether the enantiomer rotates plane polarized light clockwise (+) or anticlockwise (–). They are not related to the symbols D and L which represent absolute configuration.

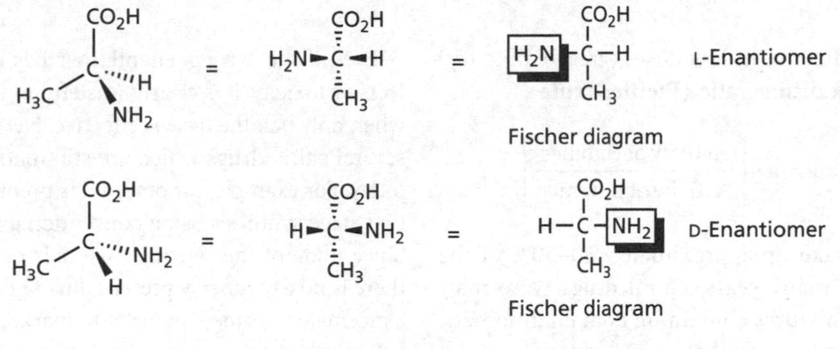

FIGURE 2 Defining the asymmetric centre of alanine as L or D.

enzyme substrate. For example, the enzyme **lactate dehydrogenase** (LDH) can catalyse the oxidation of *R*-lactic acid, but not *S*-lactic acid (Fig. 5.2).

This is a consequence of how effectively the active site recognizes and binds the two enantiomers. If we assume that there are at least three binding interactions between the active site and lactic acid, then only one of the two enantiomers can bind such that all three binding interactions take place simultaneously (Fig. 5.3). This means that the *R*-enantiomer is bound strongly and undergoes the enzyme-catalysed reaction, whereas the *S*-enantiomer is bound weakly and departs from the active site before a reaction can occur.

The activity of most drugs depends on their ability to bind to protein binding sites, and so the same principle holds true for the activity of chiral drugs. One enantiomer may bind strongly, resulting in good activity, whereas the other enantiomer binds more weakly and results in weak or no activity. Moreover, the 'wrong enantiomer' may have the ability to bind to a different protein that is not recognized by the active enantiomer, resulting in side effects or toxicity. In such cases, the more active enantiomer is called the **eutomer**, whereas the less active enantiomer is called the **distomer**. The **eudismic ratio** is a measure of how the eutomer and distomer differ in their activities. It has often been observed that chiral

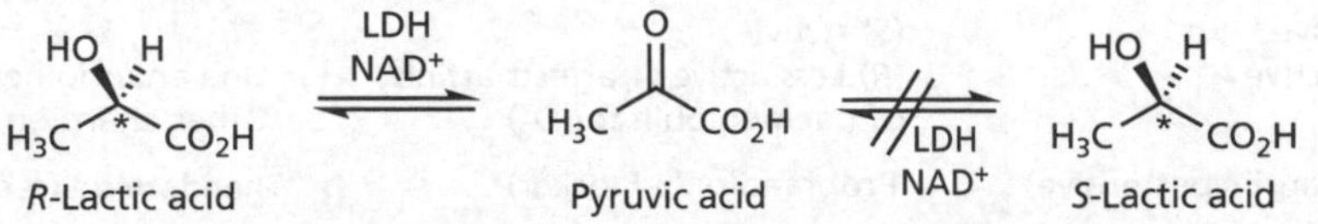

FIGURE 5.2 Enzyme-catalysed oxidation of *R*-lactic acid, but not *S*-lactic acid. LDH is the enzyme lactate dehydrogenase. NAD$^+$ is the cofactor nicotinamide adenine dinucleotide.

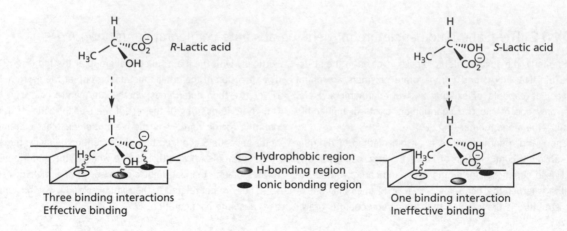

FIGURE 5.3 Different binding interactions allowing differentiation between two enantiomers.

drugs which are effective in low doses (greater potency) have the greatest eudismic ratio (**Pfeiffer's rule**).

$$\text{Eudismic ratio} = \log\left[\frac{\text{Activity of eutomer}}{\text{Activity of distomer}}\right]$$

Chiral drugs make up approximately 30–50% of the drug market. For many years, chiral drugs were marketed as racemic mixtures containing both enantiomers. However, it is often the case that one of the enantiomers produces the desired pharmaceutical effect, while the other enantiomer is less active or has no effect at all. For example, the S-enantiomer of α-**methyldopa** acts as an antihypertensive, whereas the R-enantiomer is inactive (Fig. 5.4).

With some drugs, the 'wrong' enantiomer may have a different pharmaceutical action and may even be toxic. The S-enantiomer of **propranolol** is a β-blocker used in cardiovascular medicine. However, the R-enantiomer is less active and causes side effects by affecting the affinity of haemoglobin for oxygen. Finally, one enantiomer of **thalidomide** acts as a sedative, whereas the other is teratogenic.

Even if the 'wrong' enantiomer has no serious side effects or toxicity, it is clearly wasteful to use a racemic drug when only half the dose is effective. Nevertheless, there are several chiral drugs which are still marketed as racemates today. For example, propranolol is prepared by an efficient two-stage synthesis using conventional reagents (Fig. 5.5). Since none of the starting materials or reagents is chiral, there is no asymmetry present during the synthesis and so a racemate is formed. In order to market propranolol as the active enantiomer, a more complex synthesis or separation process would be involved which would inevitably increase the cost of production. Therefore, the company has to weigh up the increased cost of production versus the clinical benefits of the pure enantiomer, and decide whether to produce the racemate or the pure enantiomer. In the case of propranolol, the clinical advantages of marketing the active enantiomer are outweighed by the cost factors.

For other racemic drugs, there may be several therapeutic and economic advantages in producing and marketing the more active enantiomer. For example, **omeprazole** (Fig. 5.6) was the first proton pump inhibitor to reach the market in 1988 and was marketed as **Losec** for the treatment of ulcers. In 1996 it became the biggest selling

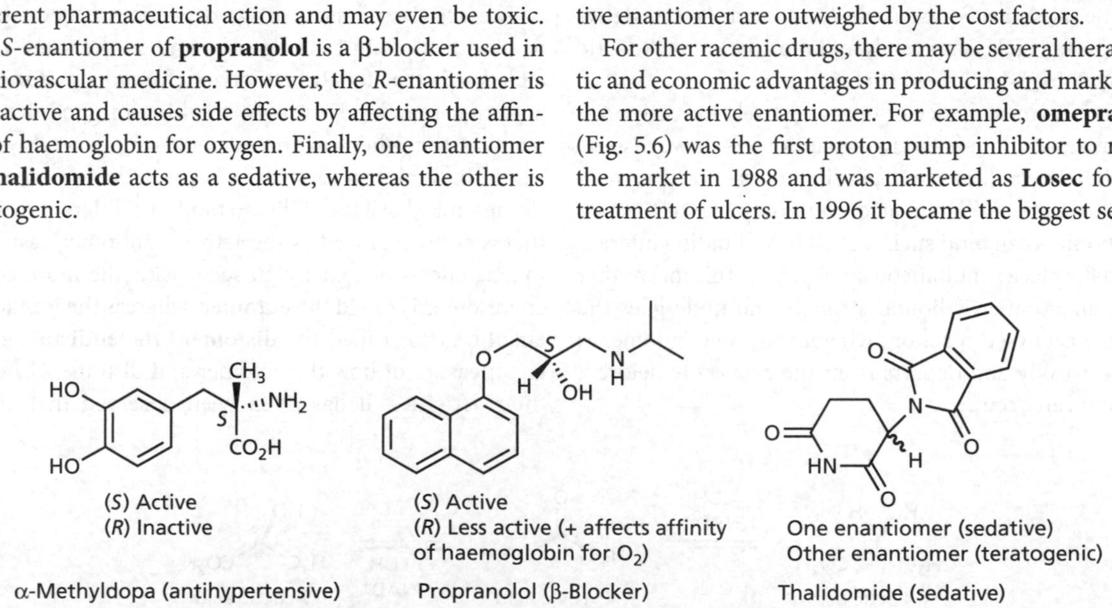

(S) Active
(R) Inactive

α-Methyldopa (antihypertensive)

(S) Active
(R) Less active (+ affects affinity of haemoglobin for O$_2$)

Propranolol (β-Blocker)

One enantiomer (sedative)
Other enantiomer (teratogenic)

Thalidomide (sedative)

FIGURE 5.4 Comparison of enantiomers and their activities.

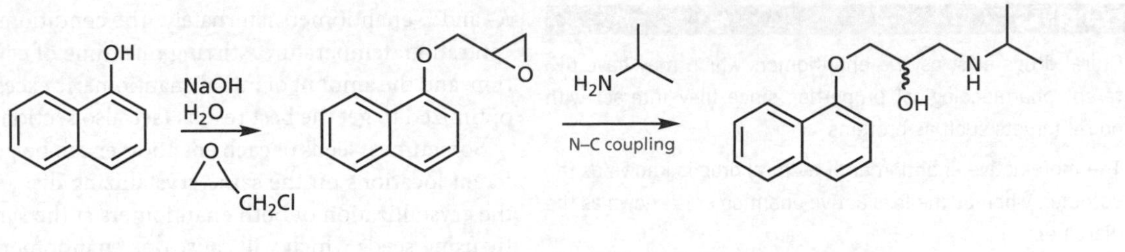

FIGURE 5.5 Synthesis of propranolol (the squiggly bond indicates that both enantiomers are present).

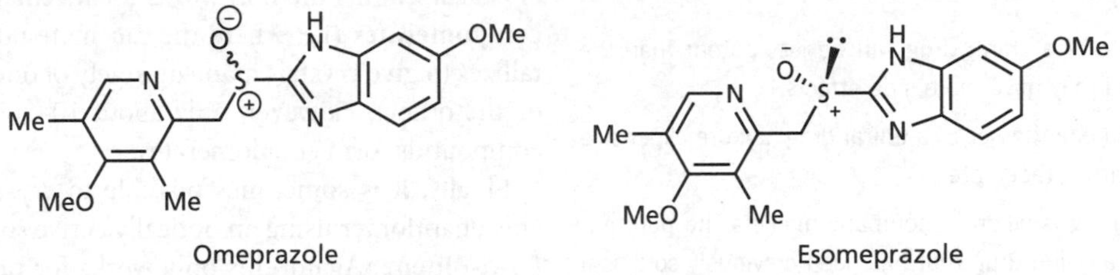

FIGURE 5.6 Omeprazole and esomeprazole.

pharmaceutical ever, and in 2000 it earned its makers (Astra) worldwide sales of $6.2 billion (£3.6 billion). The patents on omeprazole ran out in Europe in 1999 and in the USA in 2001, and so Astra launched the *S*-enantiomer of omeprazole—**esomeprazole (Nexium)**—on the basis that it had therapeutic advantages over omeprazole.

At first sight, it may not be evident that omeprazole has an asymmetric centre. In fact, the sulphur atom is an asymmetric centre as it is tetrahedral and contains a lone pair of electrons. Unlike the nitrogen atom of an amine, sulphur atoms do not undergo pyramidal inversion, and so it is possible to isolate both enantiomers of omeprazole. The *S*-enantiomer was found to be superior to the *R*-enantiomer in terms of its pharmacokinetic profile, and was launched in Europe and the USA in 2000 and 2001 respectively. The story of esomeprazole is an example of **chiral switching**, where a racemic drug is replaced on the market with a single enantiomer.

Chiral switching is only possible for drugs that were brought to the market some years ago. Nowadays, pharmaceutical companies are required to synthesize the pure enantiomers of any novel chiral drug in order to study the properties of both enantiomers, and then decide whether to sell the drug as the pure enantiomer or as a racemate. It is worth noting that the pure enantiomer of a chiral drug is not always therapeutically superior to the racemate. For example, the more active *R*-enantiomer of **fluoxetine** (Prozac) (Fig. 5.7) was found to be less beneficial than the racemic mixture because of an unexpected increase in side effects.

There are three approaches to obtaining single enantiomers. The first is to synthesize a racemic mixture of

the drug using conventional methods and then separate the enantiomers by a process known as **resolution**. However, this procedure is wasteful since half the final product is discarded. For that reason, it is better to carry out a resolution at the very start of a synthetic sequence, rather than at the end. It is then a case of ensuring that the chiral centre is not epimerized or racemized by any of the subsequent reactions in the synthesis. In some cases, the 'wrong' enantiomer can be converted to the desired enantiomer after resolution has taken place.

A second approach is to start from a pure enantiomer which already contains the required chiral centres, and to convert that compound through a series of reactions to the final product. Again, it is important that all the reactions concerned maintain the integrity of the chiral centres present.

Finally, one can carry out an asymmetric synthesis which includes one or more asymmetric reactions—reactions which produce a new chiral centre and produce one enantiomer or diastereoisomer in preference to another. We shall consider all these approaches in the following sections.

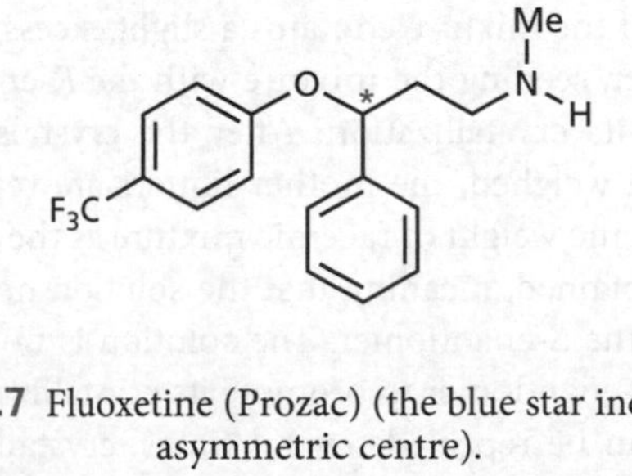

FIGURE 5.7 Fluoxetine (Prozac) (the blue star indicates the asymmetric centre).

- Chiral drugs exist as two enantiomers which may have different pharmacological properties since they interact with chiral targets such as proteins.

- The more active enantiomer of a chiral drug is known as the eutomer, whereas the less active enantiomer is known as the distomer.

- Most chiral drugs contain one or more asymmetric centres.

- The absolute configuration of an asymmetric centre is defined as *R* or *S*.

- The racemate of a chiral drug will be less potent than the eutomer, and may have more side effects.

- Producing an enantiomer of a chiral drug is more expensive than producing a racemate.

- Chiral switching is where a company markets the pure enantiomer of a chiral drug that has been previously sold as a racemate.

- Pure enantiomers can be obtained by the resolution of a racemic mixture or by an asymmetric synthesis.

- An asymmetric synthesis can be achieved by starting with an optically pure starting material that already contains the required asymmetric centres, or by including an asymmetric reaction as part of the synthetic sequence.

- An asymmetric reaction produces a new asymmetric centre with selectivity for one enantiomer over the other.

5.3 Asymmetric synthesis—resolution of racemates

There are a number of ways in which a racemic mixture can be resolved into its two constituent enantiomers.

5.3.1 Preferential crystallization

Preferential crystallization involves the crystallization of one enantiomer in preference to the other. This can be done by adding a small amount of pure enantiomer to the racemate, such that there is a slight excess of one enantiomer over the other, and then seeding a supersaturated solution of the mixture with that enantiomer. For example, if the mixture contains a slight excess of the *R*-enantiomer, seeding the mixture with the *R*-enantiomer promotes its crystallization. After the crystals are collected and weighed, the mother liquors are replenished with the same weight of racemic mixture as the weight of crystals obtained, meaning that the solution now has an excess of the *S*-enantiomer. The solution is now seeded with the *S*-enantiomer to promote its crystallization. The process can be repeated several times, crystallizing the

R- and *S*-enantiomers alternately. The conditions of concentration, temperature, stirring rate, time of crystallization, and the amount of initial enantiomeric excess can be optimized to get the best results (see also section 5.9.1).

Sometimes, seeds of each enantiomer can be put in different locations on the same crystallizing dish, allowing the crystallization of both enantiomers at the same time. By using seeds which will cause one enantiomer to form smaller crystals than the other, the final crystals can be sifted to separate the two enantiomers based on their crystal size. This method works for racemates that form **conglomerates** (i.e. where the racemate normally crystallizes to give crystals made up solely of one enantiomer or the other). However, only about 10–20% of organic compounds form conglomerates.

Finally, it is sometimes possible to crystallize a specific enantiomer using an optically active solvent such as (-)-α-**pinene**. Again, this only works for racemates that form conglomerates.

5.3.2 Chromatography

Chromatography is another method that can be used to separate enantiomers, as long as there is a chiral compound linked to the silica support (Fig. 5.8). For example, L-amino acids can be covalently linked to the silica, or the silica can be impregnated with the single enantiomer of a chiral organic acid such as tartaric, malic, or camphorsulphonic acid. The racemate is then passed down the chiral column and each enantiomer forms a reversible complex with the chiral compound present on the stationary phase. This complex is **diastereomeric** rather than enantiomeric since there is more than one asymmetric centre present, and so each diastereomeric complex will have different chemical and physical properties. This means that the complexation/decomplexation rate of each enantiomer will differ as the racemate passes down the column, resulting in one enantiomer travelling faster than the other.

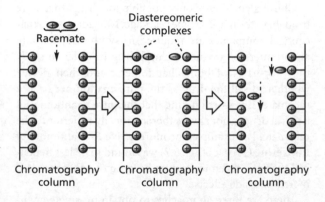

FIGURE 5.8 Resolution by chromatography.

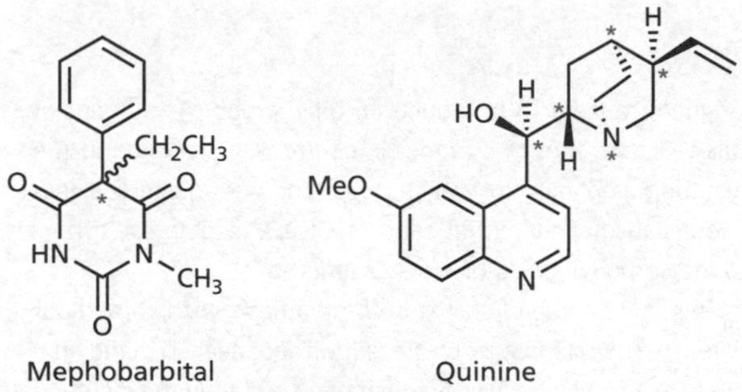

Mephobarbital   Quinine

FIGURE 5.9 Mephobarbital and quinine.

Another method is to use a naturally occurring chiral polysaccharide such as cellulose as the stationary phase. For example, a racemic mixture of **mephobarbital** (Fig. 5.9) was resolved by passing it through a cellulose column (see also section 5.9.2).

Chiral chromatography involving preparative HPLC has been used on a production scale to obtain pure enantiomers of important drugs such as **sertraline**, **escitalopram**, **levetiracetam**, **radafaxine**, and **pagoclone** in batches varying from several tons to several hundred tons. The chromatographic process can be made more efficient by carrying out a continuous chromatographic process where the racemate is circulated through a number of separate columns, with the pure enantiomer being collected as it elutes from each column. The racemate is not collected and continues to recycle through the columns.

5.3.3 Formation of diastereomeric derivatives

The classical method of resolving racemates is to form **diastereomeric derivatives** of each enantiomer. Since diastereomers have different properties, they can be separated by conventional methods such as crystallization or distillation. After separation, the derivative can be hydrolysed or neutralized to recover the pure enantiomer.

For example, a chiral drug containing a carboxylic acid functional group can be treated with the single enantiomer of an optically active amine such as quinine (Figs 5.9 and 5.10). A diastereomeric salt is formed which can be separated by crystallization. Once the diastereomers have been separated, the pure enantiomers are recovered by treating the salt with dilute hydrochloric acid to remove the basic quinine.

A chiral drug containing an amine group can be treated similarly with the single enantiomer of a chiral carboxylic acid such as tartaric acid or malic acid. The diastereomeric salts are separated by crystallization and then treated with base to remove the acid and recover each enantiomer of the amine (see Box 5.3).

If the chiral compound lacks a carboxylic acid or amine group, it may be possible to form a diastereomeric derivative by temporarily linking a chiral compound to another functional group. For example, if the racemic drug contains an alcohol functional group, it can be treated with a single enantiomer of **camphorsulphonyl chloride** to give a camphorsulphonate derivative (Fig. 5.11) as a pair

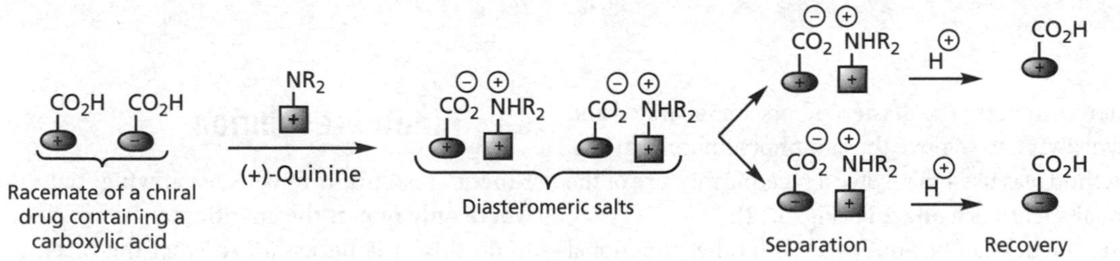

FIGURE 5.10 Resolution of a chiral carboxylic acid.

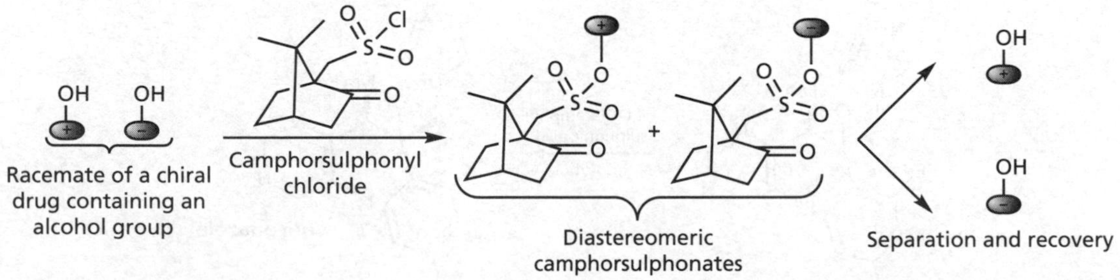

FIGURE 5.11 Resolution of a chiral alcohol via camphorsulphonyl derivatives.

BOX 5.3 Synthesis of tapentadol

Tapentadol was approved in 2008 as an analgesic that acts as an agonist at opioid receptors. It has similar efficacy to opioids but without their side effects. It is also an inhibitor of noradrenaline reuptake by transport proteins at nerve synapses.

The first step in the synthesis involves **Eschenmoser's salt** and acetyl chloride to produce a chiral amine as a racemic mixture. The racemate was crystallized along with L-(-)-dibenzoyl tartaric acid monohydrate (DBTA) in ethanol to give the desired S-enantiomer as the tartrate salt. The free base was generated by reaction with aq. NaOH and then treated with a

Grignard reagent to introduce an ethyl group. As a result of this reaction, a new asymmetric centre was produced at the resulting alcohol. However, the reaction was diastereoselective because of the chiral centre that is already present next to the carbonyl group (see also section 5.7).

The final couple of stages in the synthesis were carried out to remove the alcohol group since it was not needed in the final product. The alcohol was first esterified and then the resulting ester was removed under reducing conditions. The reaction conditions also resulted in deprotection of the phenol group.

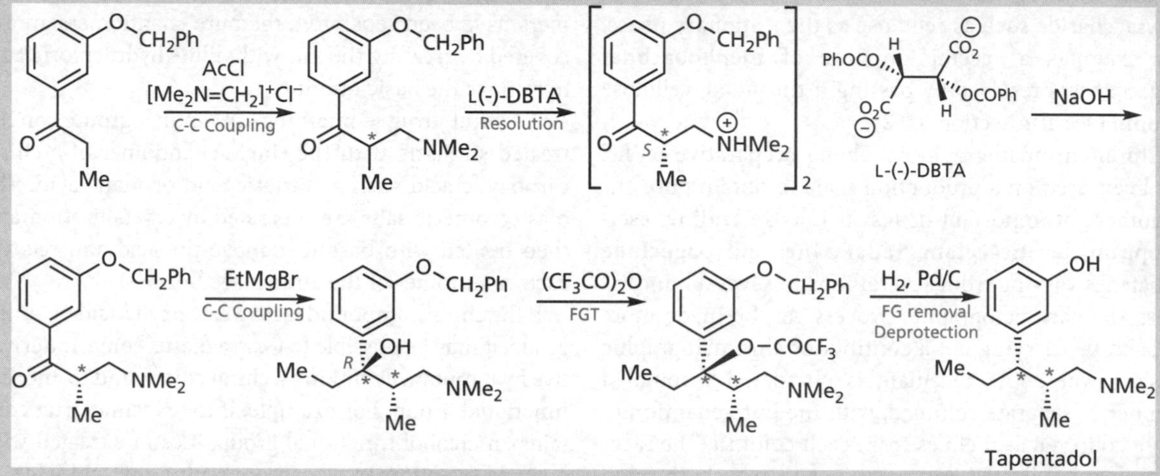

FIGURE 1 Synthesis of tapentadol (the asterisks mark chiral centres).

of diastereoisomers. The diastereomers are separated and then hydrolysed to remove the camphorsulphonyl group. This method was used to resolve a racemic mixture of the antifungal agent **voriconazole** (Fig. 5.12).

Similar tactics can be employed with other functional groups. For example, a diastereoisomeric hydrazone derivative could be formed if a ketone is present in the chiral drug.

5.3.4 **Kinetic resolution**

Kinetic resolution involves carrying out a reaction where only one of the enantiomers is affected. In order to do this, it is necessary to react the racemic mixture with an enantiopure reagent. The method relies on the formation of **diastereomeric transition states** which will have different stabilities for each enantiomer,

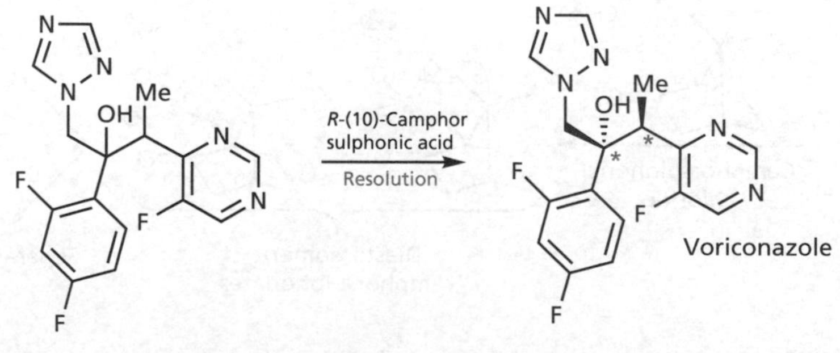

FIGURE 5.12 Resolution of voriconazole.

resulting in different activation energies for the reaction. If there is a large difference in transition state stability and activation energy, the reaction will favour one enantiomer over the other.

Enzymes have been particularly useful in resolving racemates. For example, chiral esters can be resolved using **lipase enzymes** which catalyse the hydrolysis of one enantiomeric ester rather than the other. It is then a simple procedure to separate the product carboxylic acid from the unreacted ester. Similarly, **proteases** and **peptidases** can be used to separate the enantiomers of chiral amides or peptides. In some cases, proteases can be used to hydrolyse chiral esters. For example, a racemic mixture of N-acyl-m-tyrosine methyl ester was resolved by treating it with a bacterial protease (alcalase) which catalysed ester hydrolysis of the S-enantiomer but not the R-enantiomer (Fig. 5.13). The two structures could then be easily separated since the resulting N-acyl-(S)-m-tyrosine was soluble in bicarbonate solution, whereas the remaining fully protected (R) enantiomer was not. Hydrolysis of the resulting N-acylated amino acid then generated L-m-tyrosine in 42% yield over the two steps. The D-enantiomer could also be obtained by deprotection of the remaining (R)-N-acyl-m-tyrosine methyl ester.

Resolution is an efficient process if both enantiomers of the resolved structure are useful. For example, both enantiomers of m-tyrosine are useful in the synthesis of **peptidomimetic drugs**—drugs which mimic peptide lead compounds in their ability to bind to a particular drug target, but are more resistant to metabolism.

However, the resolution process is wasteful if no use can be found for one of the enantiomers.

This waste can be minimized if the 'unwanted' enantiomer can be racemized and the resolution process repeated. For example, D-**phenylglycine** is used in the synthesis of the antibacterial agent **ampicillin**, but is synthesized as a racemate. In order to separate the enantiomers, the racemate can be converted to a primary amide (D,L-phenylglycinamide), which is then treated with an aminopeptidase enzyme to hydrolyse the L-enantiomer but not the D-enantiomer (Fig. 5.14). The D-enantiomeric amide can then be separated from L-phenylglycine and hydrolysed to the desired D-phenylglycine.

The unwanted L-phenylglycine is then racemized with sulphuric acid and converted back to racemic D,L-phenylglycinamide with ammonia. The resolution–separation–racemization cycle can be repeated until virtually all the original racemate has been converted to the desired enantiomer.

An alternative approach to this resolution–racemization cycle is to carry out a reaction which inverts the asymmetric centre of the 'wrong' enantiomer once the resolution has been carried out. For example, the enantiomeric alcohol (II) was required as an intermediate for a drug synthesis (Fig. 5.15). The racemic ester (I) was resolved by treating it with a microorganism called *Athrobacter*, where the microbial enzymes present catalysed the hydrolysis of the 'unwanted' enantiomer. The products were then separated and the desired enantiomer (II) isolated following chemical hydrolysis of the ester group. The 'unwanted'

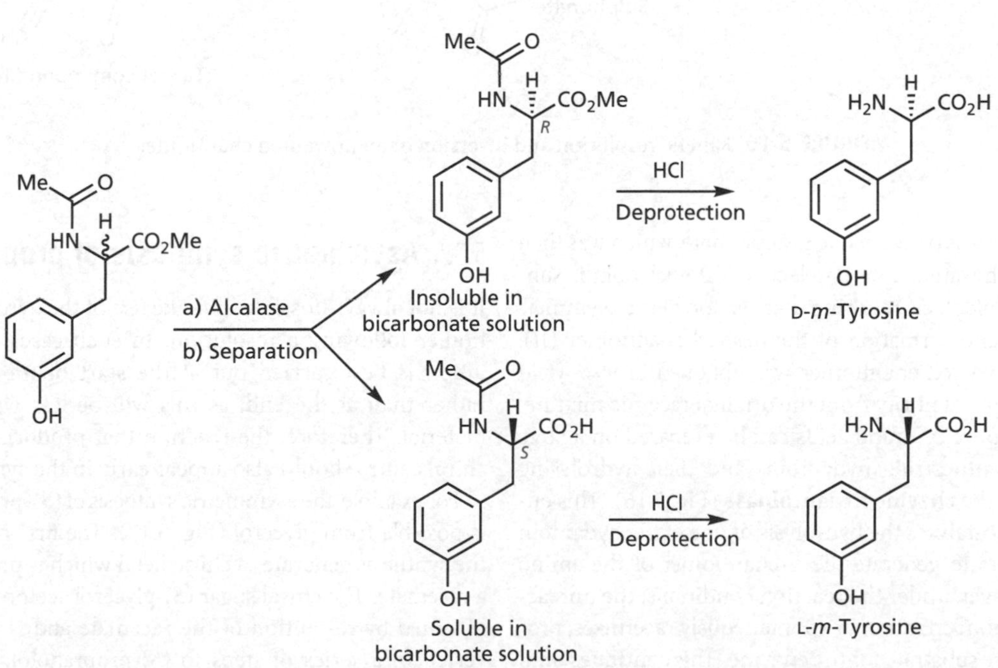

FIGURE 5.13 Resolution of an unnatural amino acid.

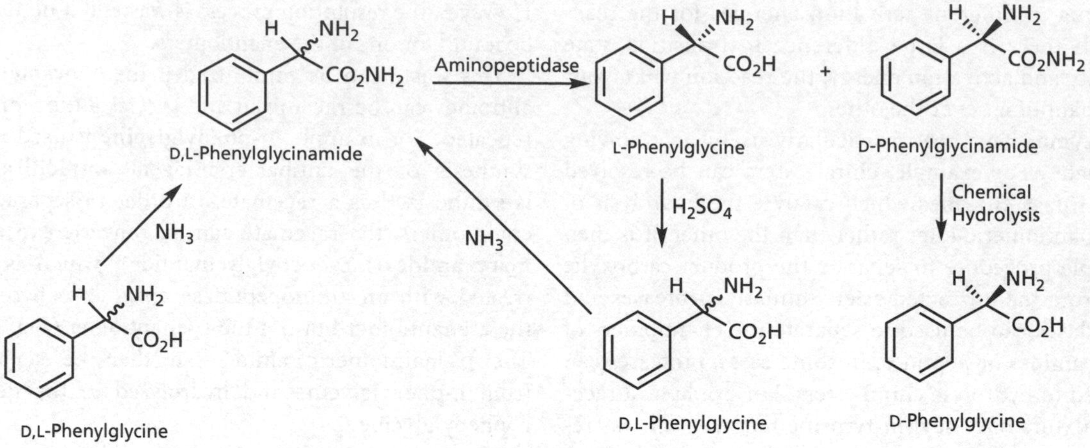

FIGURE 5.14 Kinetic resolution of D,L-phenylglycine.

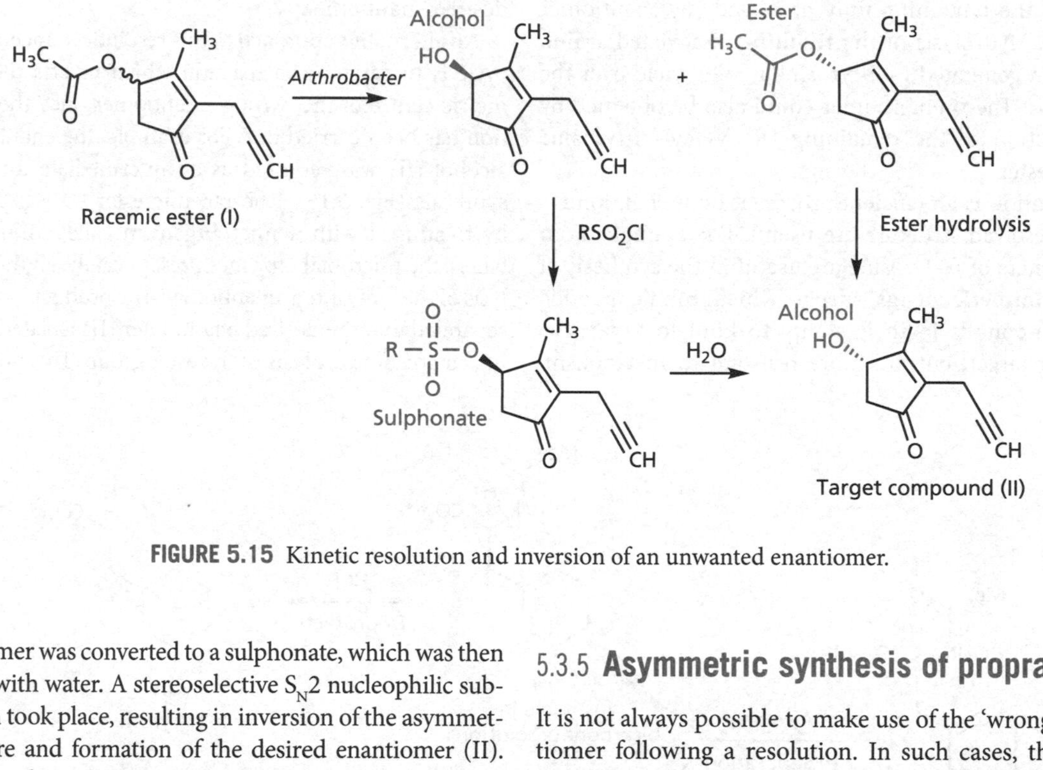

FIGURE 5.15 Kinetic resolution and inversion of an unwanted enantiomer.

enantiomer was converted to a sulphonate, which was then treated with water. A stereoselective S_N2 nucleophilic substitution took place, resulting in inversion of the asymmetric centre and formation of the desired enantiomer (II). Thus, the desired enantiomer was obtained in 99% yield and 99% optical purity from the original racemic mixture.

Enantiopure D-amino acids can be prepared on a large scale by synthesizing hydantoins and then hydrolysing them with the enzyme **hydantoinase** (Fig. 5.16). This enzyme only catalyses the hydrolysis of one of the hydantoin enantiomers to generate the D-enantiomer of the amino acid. However, under the reaction conditions, the unreactive hydantoin enantiomer spontaneously racemizes, providing more substrate for the enzyme. This continues until the entire sample is resolved to the D-enantiomer.

5.3.5 Asymmetric synthesis of propranolol

It is not always possible to make use of the 'wrong' enantiomer following a resolution. In such cases, the resolution is best carried out at the start of the synthesis rather than at the end, as this will be less wasteful on material. Therefore, the reaction that produces the new chiral centre should also appear early in the synthesis.

For example, the asymmetric synthesis of (**S**)-**propranolol** is possible from glycerol (Fig. 5.17). The first reaction in the synthesis generates a chiral ketal which is produced as a racemate. The chiral sugar (S)-glycerol acetonide can be obtained by resolution of the racemate and is then converted in a series of steps to (S)-propranolol. However, the overall route involves seven stages, compared with

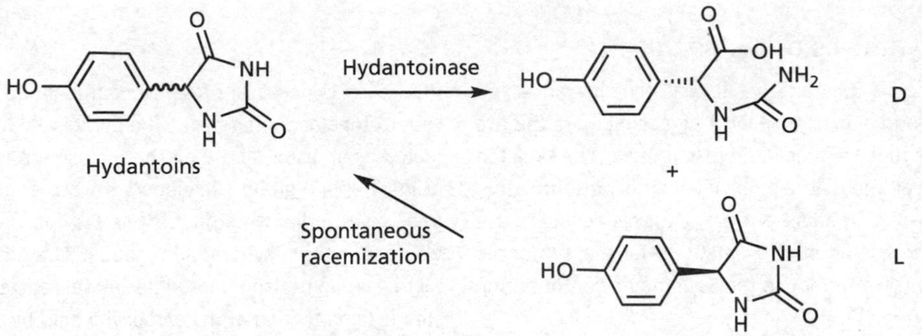

FIGURE 5.16 Synthesis of D-amino acids.

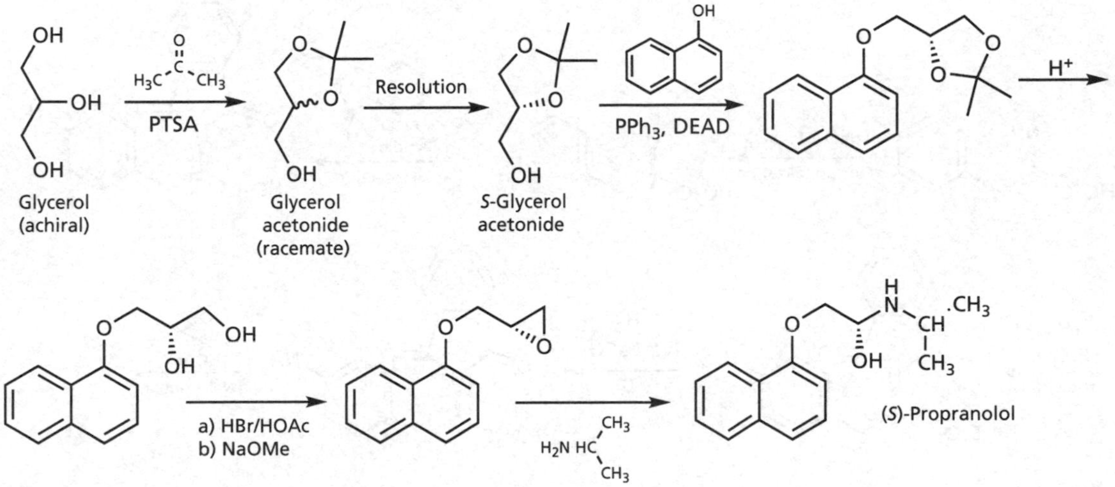

FIGURE 5.17 Asymmetric synthesis of *S*-propranolol from a chiral sugar.

the two-stage racemic synthesis mentioned in section 5.2 (Fig. 5.5).

KEY POINTS

- Racemates can be resolved by preferential crystallization, or by chromatography using a chiral stationary phase.

- A chiral drug that contains an amine can be resolved by forming a salt with the pure enantiomer of a chiral carboxylic acid and then carrying out a crystallization.

- A chiral drug that contains a carboxylic acid can be resolved by forming a salt with the pure enantiomer of a chiral amine and then carrying out a crystallization.

- Chiral drugs containing other types of functional groups can be resolved by linking the enantiomer of a chiral molecule to a particular functional group and then separating the resulting diastereoisomers by crystallization or chromatography.

- Kinetic resolution involves carrying out a reaction where one enantiomer reacts more readily than the other. The resulting product of the reaction can then be separated from the unreacted enantiomer. Enzymes are commonly used in kinetic resolution.

5.4 Asymmetric syntheses from a chiral starting material

In section 5.3 we discussed how an enantiomer can be obtained by resolving a racemate produced by conventional reactions. Another approach, which avoids the need for resolution, is to start the synthesis with a pure enantiomer that already contains the asymmetric centres required in the final product. It is then a case

BOX 5.4 Synthesis of melphalan

Melphalan is an anticancer agent that contains an asymmetric centre and acts as an alkylating agent due to the two alkyl chlorides that are present in its structure. The skeleton of the natural amino acid L-phenylalanine is present within the structure of melphalan and so it makes sense to use L-phenylalanine as the starting material for the synthesis, rather than trying to introduce the asymmetric centre during the synthesis (Fig. 1).

The aromatic ring of L-phenylalanine is first nitrated in order to functionalize the *para* position, and then the primary amino group is protected with a phthalimide group.

Protection is necessary to prevent this group from reacting with oxirane at a later stage. The carboxylic acid is now protected as an ester, and then the nitro group is reduced to a primary amino group. Reaction with two equivalents of oxirane gives a diol which is treated with phosphoryl chloride to produce the alkyl chloride groups. The phthalimide protecting group is also removed during this stage, and the final step is deprotection of the carboxylic acid by acid-catalysed hydrolysis of the ethyl ester. The integrity of the asymmetric centre originally present in L-phenylalanine remains intact throughout the synthesis.

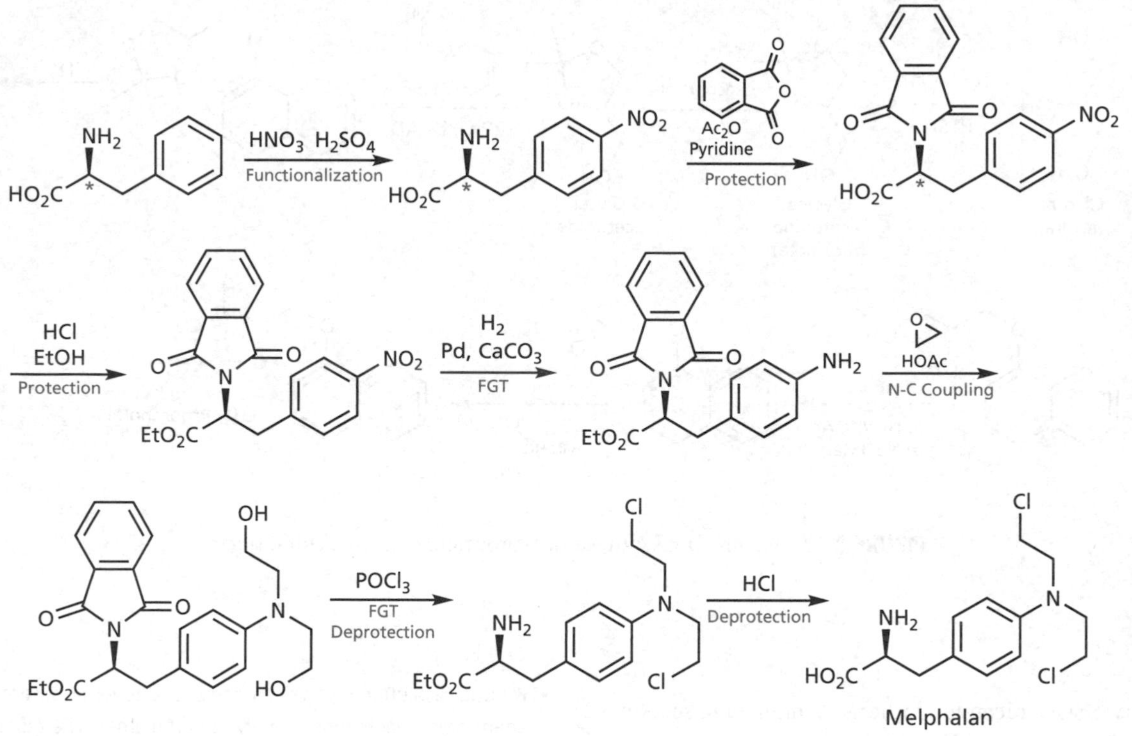

FIGURE 1 Synthesis of melphalan.

of carrying out the various reactions in the synthetic route whilst preserving the integrity of the asymmetric centres. In other words, none of the reactions involved in the synthesis should result in racemization or epimerization.

Chiral natural products typically exist as single enantiomers and so these compounds have proved popular starting materials for asymmetric syntheses, especially if they are cheap and readily available. For example, several asymmetric syntheses have been carried out

starting with a readily available amino acid (see Box 5.4) or carbohydrate.

The natural product used will depend on the number and relative positions of the asymmetric centres present in the target product. For example, the antiviral agent **oseltamivir** (Tamiflu) (Fig. 5.18) contains three chiral centres that are adjacent to each other. Therefore, a suitable natural product for the synthesis of oseltamivir should also contain three adjacent asymmetric centres. **Shikimic acid** (Fig. 5.18) fits the bill nicely. It is a

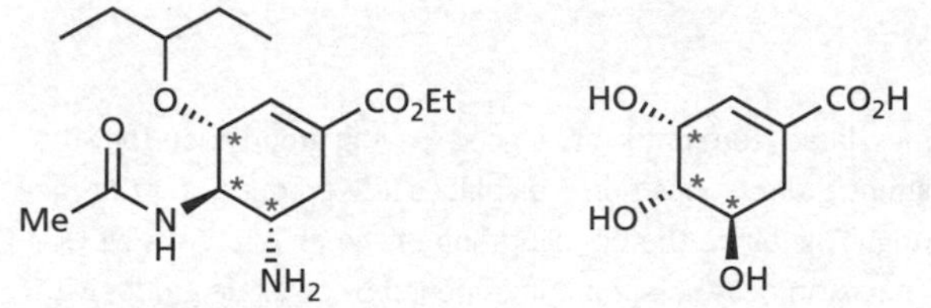

FIGURE 5.18 Comparison of oseltamivir and shikimic acid.

naturally occurring structure which acts as an important intermediate in the biosynthesis of a large number of aromatic natural products, such as the amino acid tryptophan. Not only does shikimic acid contain the necessary three asymmetric centres, but it contains them within a similar cyclohexene ring system to oseltamivir, which includes a carboxylic acid group that can easily be converted to the ester in the target structure. The synthesis now comes down to modifying the functional groups in shikimic acid to those in oseltamivir, as well as inverting two of the asymmetric centres (Box 5.5).

Synthesizing a chiral compound from a chiral natural product has several advantages, the key one being the presence of important asymmetric centres. However, there are disadvantages as well. A large number of synthetic stages are often required in order to convert the functional groups in the starting material to those in the target structure, as well as achieving the correct configurations at the asymmetric centres (see Box 5.5). In addition, a complex series of protection and deprotection

BOX 5.5 Synthesis of oseltamivir

A number of syntheses of oseltamivir have been developed from shikimic acid. The one shown in Figure 2 was developed in China and involves 12 stages. The synthesis can be viewed as consisting of five phases that are designed to synthesize four key intermediates (Fig. 1).

The first phase is simply esterifying shikimic acid to its ethyl ester. This not only protects the carboxylic acid group, but also introduces the ester group required in the final product.

In the second phase, a series of steps are carried out in order to introduce the 3-pentyloxy substituent required in the final structure. First, the *cis*-diol system is protected as a ketal and the remaining *trans*-alcohol is mesylated.

The rationale for this second step will become evident later on. A ketal exchange reaction is then carried out and the ketal is treated with $TiCl_4$ and Et_3SiH to introduce the required ether at position 3 of the ring system.

The third phase of the synthesis involves a series of steps to introduce a primary amine group at position 5 of the ring system. Treatment with a mild base induces a ring closure where the alcohol group displaces the mesylate group to form an epoxide. We can now see why the mesylate group was introduced, as this is a much better leaving group than a hydroxyl group. The epoxide is then ring-opened by an azide ion with regioselectivity for position 5. The regioselectivity results from the large ether group which helps to

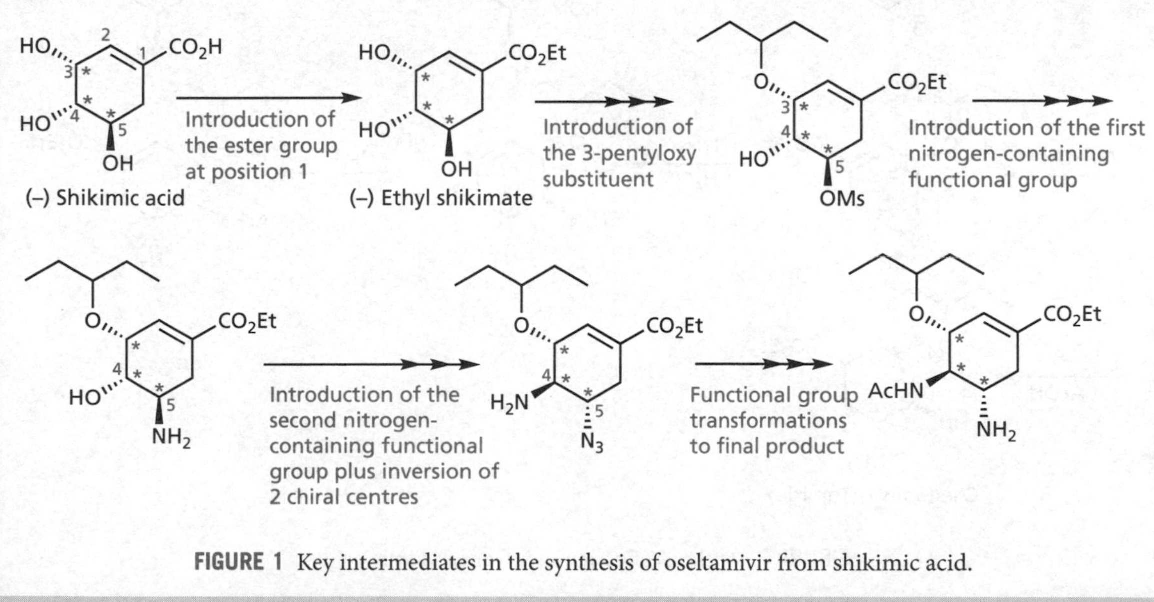

FIGURE 1 Key intermediates in the synthesis of oseltamivir from shikimic acid.

(Continued)

BOX 5.5 Synthesis of oseltamivir *(Continued)*

shield the epoxide from attack at position 4. The reaction is also stereoselective, with the azide group adding from the opposite side of the cyclohexene ring system from the epoxide. Reduction with triphenylphosphine (PPh₃) and triethylamine (NEt₃) then provides the primary amine group. At first sight, this looks as if it will correspond to the 5-amino group in the final product, but it will actually end up as the amide group at position 4 following the next series of reactions.

In phase 4, the second nitrogen-containing group is introduced in such a way that the asymmetric centres at positions 4 and 5 are inverted to the configurations required in the final product. The alcohol group at position 4 is

mesylated, turning it into a good leaving group such that the amino group at position 5 displaces it to produce an aziridine ring. This alters the configuration of the chiral centre at C4. The aziridine ring is then ring-opened by an azide group, with the azide reacting at C5 in another regioselective and stereoselective reaction. The nitrogen group that was originally introduced at position 5 now ends up at position 4, with the azide group ending up at position 5. The chiral centre at C5 has now been inverted.

Phase 5 is now a case of converting the amine at position 4 to an amide by acetylation, and the azide at position 5 to a primary amine by reduction.

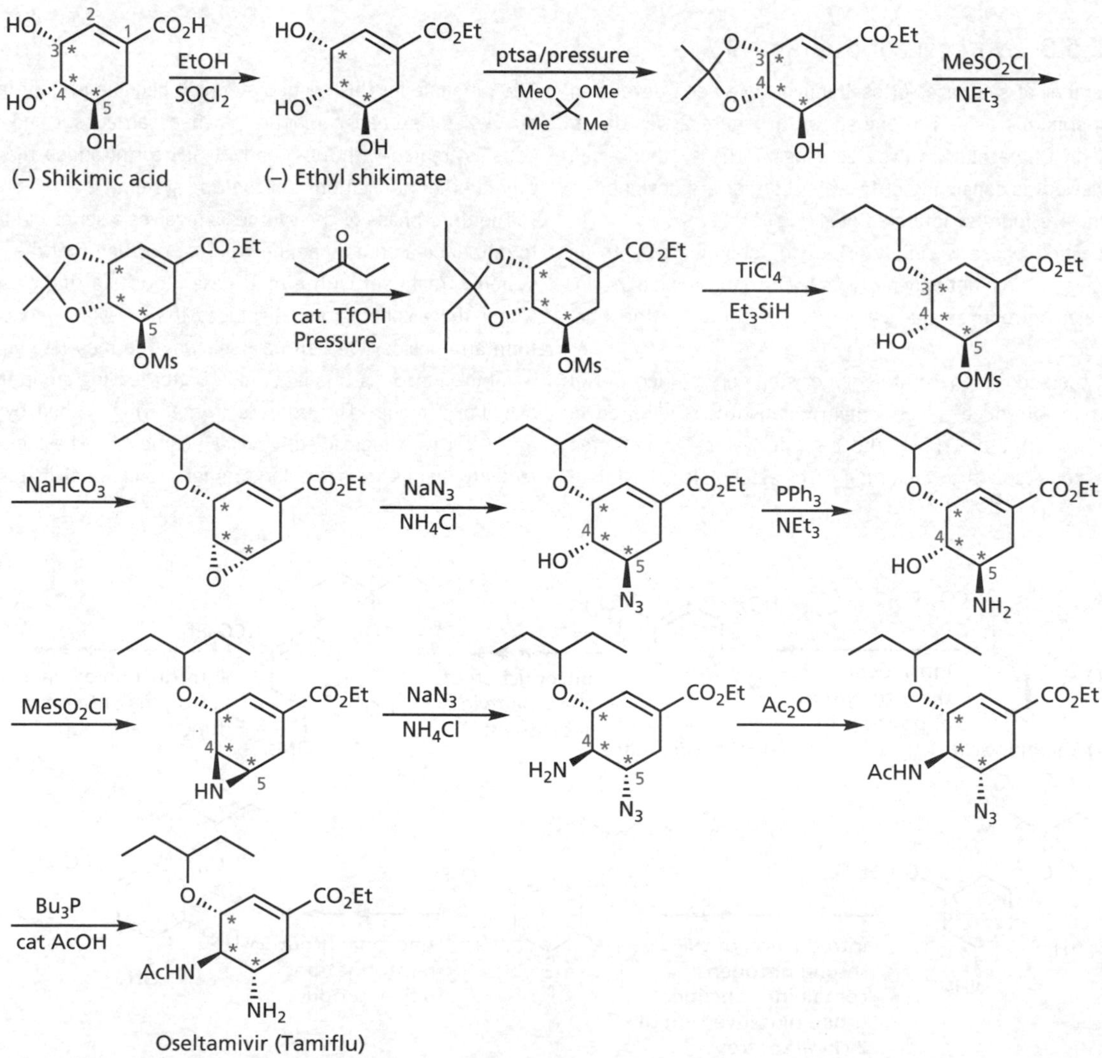

FIGURE 2 Synthesis of oseltamivir from shikimic acid.

BOX 5.6 Asymmetric synthesis of dapoxetine

Dapoxetine is a selective serotonin reuptake inhibitor (SSRI) that was originally developed as an antidepressant, but has now been marketed in several countries as a treatment for premature ejaculation. (*R*)-3-Chloro-2-phenyl-1-propanol was used as a chiral starting material for the asymmetric synthesis and the reactions were carried out without racemization (Fig. 1). The final stage involved an S_N2 nucleophilic substitution of the mesyl group with dimethylamine, which resulted in inversion of the asymmetric centre to provide the desired absolute configuration.

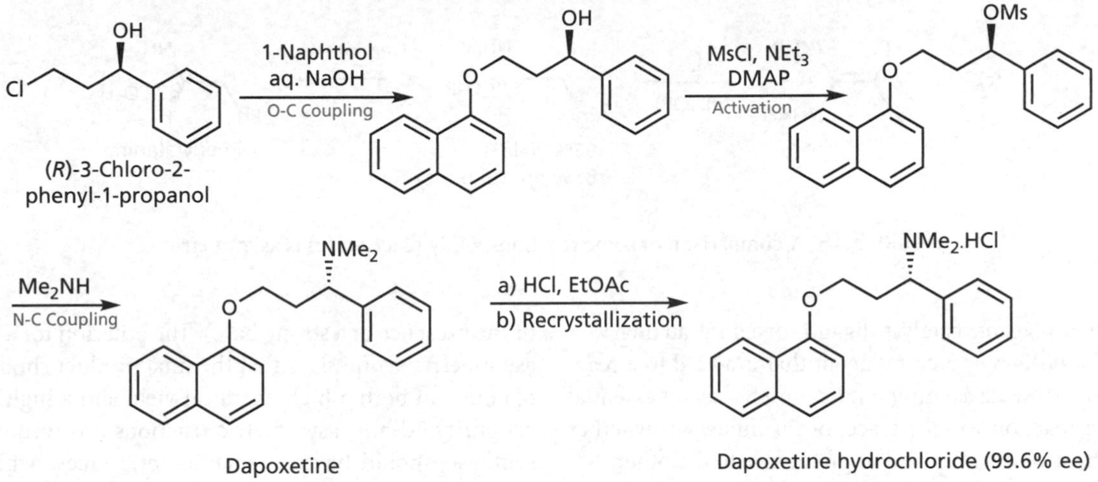

FIGURE 1 Asymmetric synthesis of dapoxetine.

steps are often needed in order to carry out these transformations. Other problems could include the availability and cost of the starting material. A natural product has to be extracted from its natural source, and if that source is not readily available, harvesting might be slow and expensive. The shikimic acid used in the synthesis of oseltamivir is obtained from a Chinese spice called **star anise**. The yields are reasonable, but would not be sufficient to provide the amount of shikimic acid required should a worldwide pandemic of flu require vastly increased stocks of oseltamivir. Consequently, work is being carried out to produce shikimic acid in genetically modified cells of the bacterium *Escherichia coli*.

Another problem associated with the use of natural products as chiral starting materials can be the difficulty in identifying whether they are suitable or not. For example, carbohydrates have been popular starting materials for asymmetric syntheses because of the number of chiral centres they contain. However, the structural relationships between a carbohydrate and the target compound are rarely obvious and extra synthetic steps are inevitably required.

Another example of an asymmetric synthesis that starts from a chiral starting material is the synthesis of the selective serotonin reuptake inhibitor dapoxetine (Box 5.6).

5.5 Asymmetric syntheses involving an asymmetric reaction—general principles and terminology

5.5.1 Introduction

In sections 5.3 and 5.4 we looked at methods of asymmetric synthesis which either involve the resolution of a racemate or the use of a natural product that already contains the chiral centres needed in the final product. In this and following sections, we will look at asymmetric syntheses that include one or more **asymmetric reactions**—reactions that introduce one or more new asymmetric centres with a preference for one enantiomer/diastereoisomer over another. Over the last couple of decades, synthetic procedures and reagents that allow a large variety of asymmetric reactions to be carried out have been successfully developed. For this to occur, some aspect of asymmetry must already be present in the reaction. In other words, there must be a chiral molecule present as a single enantiomer or diastereoisomer. The asymmetric molecule could be any of the following—starting

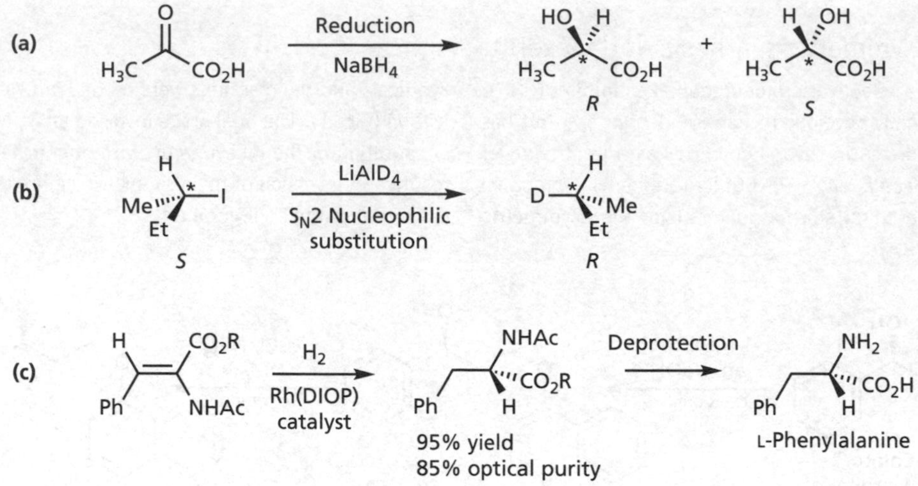

FIGURE 5.19 A comparison of some reactions. Only reaction (c) is asymmetric.

material, reagent, catalyst, ligand, or chiral auxiliary. A **chiral auxiliary** is a chiral agent that is added to a reaction to introduce asymmetry. Its presence is not essential for the reaction to take place, but it influences whether one enantiomer is formed in preference to the other.

In order to fully understand what is and what is not an asymmetric reaction, we will look at the reactions in Figure 5.19. The borohydride reduction of a ketone containing two different substituents (Fig. 5.19a) will result in an alcohol with a new asymmetric centre, but both possible enantiomers are formed in equal quantities (a racemate) and so this is not an asymmetric reaction.

The second reaction (Fig. 5.19b) is a nucleophilic substitution of the S-enantiomer of a chiral iodide. In this reaction, asymmetry is present in the starting material and a single enantiomer is formed, but this is a consequence of the reaction mechanism and no new asymmetric centre is formed. Therefore, this is not an asymmetric reaction.

The third example (Fig. 5.19c) shows a synthesis of the amino acid phenylalanine. The first reaction in this synthesis is an asymmetric reaction as a new asymmetric centre has been generated with a preference for the S-enantiomer (shown) over the R-enantiomer (not shown). A measure of this preference is given by the **optical purity** (also known as the **enantiomeric excess**). An optical purity of 85% indicates that 85% of the product is the S-enantiomer, with the remaining 15% being a racemate of both enantiomers. Therefore, the ratio of enantiomers present is 92.5% S to 7.5% R.

Asymmetric syntheses which involve an asymmetric reaction are usually longer than conventional syntheses and involve more expensive reagents. They are also more demanding, since it is necessary to avoid conditions which are likely to cause racemization (e.g. strong heat

or the presence of a strong base). The criterion for a good asymmetric synthesis is that the final product should be obtained in both a high chemical yield and a high optical purity. Ideally, asymmetric reactions involved in the synthesis should have an enantiomeric excess which is greater than 98%. In addition, any expensive chiral reagents used in the reaction must be present in catalytic amounts, or else easily recoverable so that they can be used again.

5.5.2 **Definition of the *re* and *si* faces of a planar prochiral molecule**

An asymmetric reaction carried out on an achiral starting material produces a new chiral centre and is selective for one enantiomer over the other. In such cases, the achiral starting material can be described as being **prochiral**; in other words, it has the potential to become chiral. An example of such a reaction is the enzyme-catalysed reduction of pyruvic acid to give lactic acid (Fig. 5.20). The enzyme involved is **lactate dehydrogenase** (LDH) and the biological reducing agent is a structure called nicotinamide adenine dinucleotide (NADH). The enzyme is chiral and is responsible for the asymmetric control in the reaction such that only the R-enantiomer of lactic acid is formed, unlike the sodium borohydride reduction

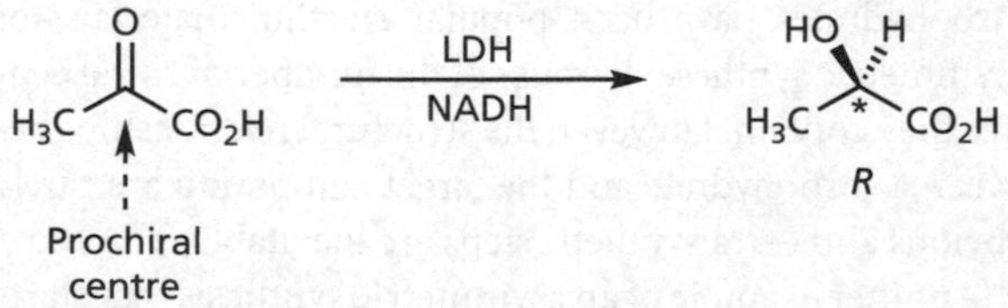

FIGURE 5.20 Example of an asymmetric reaction.

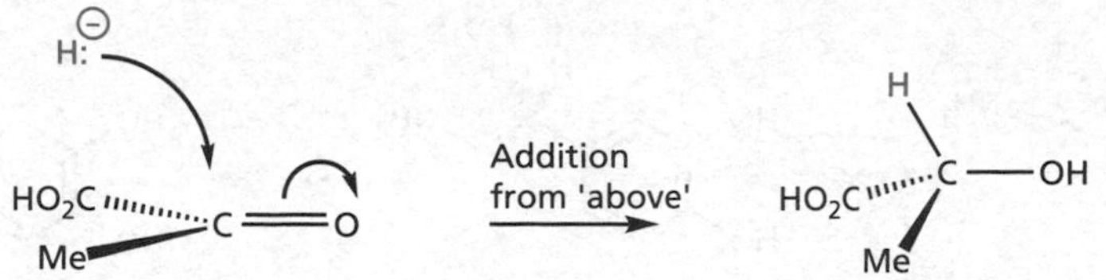

FIGURE 5.21 Preferred direction of hydride approach in the enzyme-catalysed reduction of pyruvic acid.

(Fig. 5.19a). Pyruvic acid is the prochiral starting material, and the carbonyl carbon atom is both the reaction centre and the prochiral centre since this is the carbon that becomes chiral in lactic acid.

In order to form the *R*-enantiomer, the incoming hydrogen has to approach one face of the carbonyl group and not the other. Figure 5.21 shows the planar nature of the carbonyl group in pyruvic acid with the hydride nucleophile approaching from above in order to generate the *R*-enantiomer. Clearly, the enzyme does not permit approach from below since none of the *S*-enantiomer is formed. In other words, the enzyme is able to differentiate between the two faces of the planar carbonyl group. Such faces are defined as being **enantiotopic** if reaction at one of the faces produces one enantiomer, and reaction at the other face produces the opposite enantiomer.

The enantiotopic faces of a prochiral ketone can be defined as *re* (Latin *rectus*) or *si* (Latin *sinister*) by applying the same priority rules that are used to assign *R* or *S* asymmetric centres (Box 5.1). For example, Figure 5.22 shows the two faces of pyruvic acid with their assignment as *re* or *si*. Having defined the enantiotopic faces, it is now possible to state that the hydride nucleophile in the enzyme-catalysed reduction of pyruvic acid approaches the *re* face of the ketone to produce the *R*-enantiomer of lactic acid.

In some reactions, it is possible to identify **enantiotopic groups** where reaction at one group results in one enantiomer, while reaction at the other group results in the opposite enantiomer. For example, the enzyme-catalysed hydrolysis of the achiral diester shown in Figure 5.23 results in the *S*-enantiomer of the product as the enzyme targets hydrolysis of one of the methyl ester groups

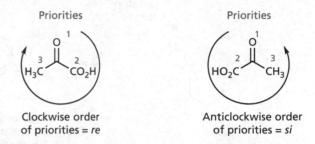

Clockwise order of priorities = *re*

Anticlockwise order of priorities = *si*

FIGURE 5.22 Defining the enantiotopic faces of pyruvic acid as *re* and *si*.

over the other. Therefore, we can define the ester groups as enantiotopic groups. In this example, reaction at the ester shown leads to an *S*-enantiomer, and so we can define this ester as the Pro-*S* enantiotopic group. This means that the other ester can be defined as the Pro-*R* enantiotopic group. Note that in this reaction the prochiral centre is not the same as the reaction centre (the ester group).

Enantiotopic groups can be assigned as Pro-*R* or Pro-*S* without defining a particular reaction. For example, the benzylic protons present in the structure shown in Figure 5.24 can be defined as Pro-*R* or Pro-*S* by identifying what enantiomer would be formed if each proton was replaced with deuterium. Deuterium is a heavier isotope of hydrogen and takes priority over hydrogen in the priority rules.

5.5.3 Diastereoselective reactions

Prochiral centres are not restricted to achiral molecules. They may also be present on molecules which are already chiral. For example, the ketone shown in Figure 5.25 has an asymmetric centre next to the carbonyl carbon.

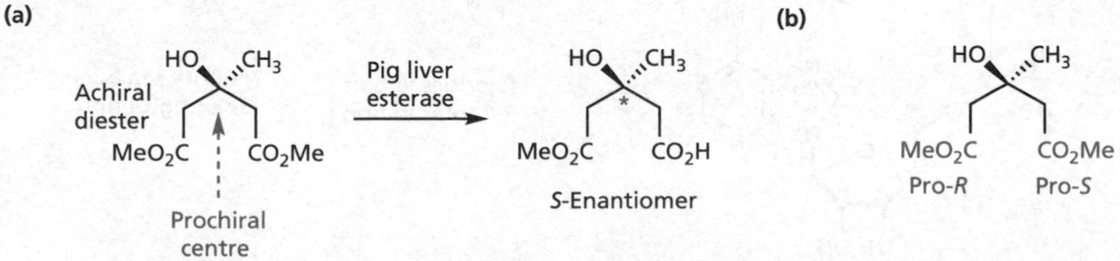

(a)

Achiral diester

Pig liver esterase

S-Enantiomer

Prochiral centre

(b)

Pro-*R*   Pro-*S*

FIGURE 5.23 Asymmetric hydrolysis of a diester.

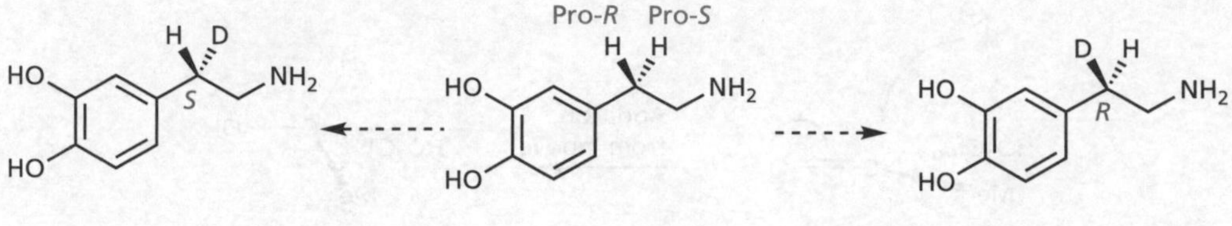

FIGURE 5.24 Assigning benzylic hydrogens as Pro-*R* and Pro-*S*.

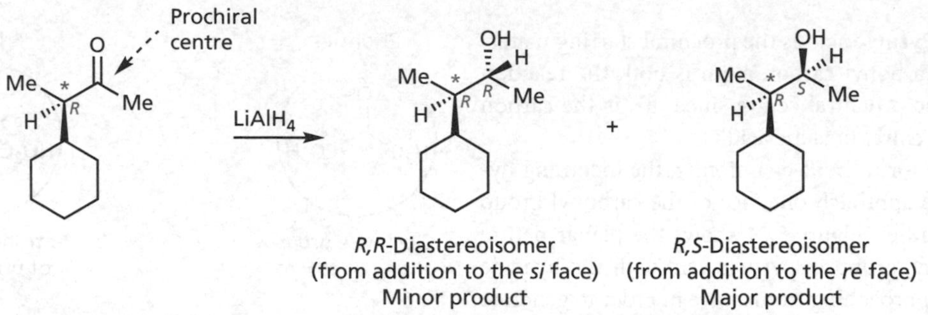

R,R-Diastereoisomer
(from addition to the *si* face)
Minor product

R,S-Diastereoisomer
(from addition to the *re* face)
Major product

FIGURE 5.25 Asymmetric reduction of a chiral ketone.

Reaction of LiAlH$_4$ with the ketone results in a new chiral centre and the generation of two possible products depending on which face of the ketone is attacked by the incoming nucleophile. These cannot be mirror images as an asymmetric centre is already present, and so the products are diastereoisomers having different chemical and physical properties. Since diastereoisomers are formed in the reaction, the faces of the ketone are described as being **diastereotopic** rather than enantiotopic. However, they can be defined as *re* or *si* in the same way. In this reaction, nucleophilic attack at the *si* face results in the *R,R*-diastereoisomer, whereas attack at the *re* face generates the *R,S*-diastereoisomer.

By the same token, two prochiral groups in a chiral molecule are defined as **diastereotopic** rather than enantiotopic since reaction at each group will produce diastereoisomers rather than enantiomers. For example, the reducing agent NADH involved in the enzyme-catalysed reduction of pyruvic acid has diastereotopic protons, defined as Pro-*R* and Pro-*S* (Fig. 5.26). The definition can be made by imagining what enantiomer would be formed if one of the enantiotopic protons was replaced with a deuterium isotope. The groups must be diastereotopic rather than enantiotopic as other chiral centres are present in the two ribose sugar molecules.

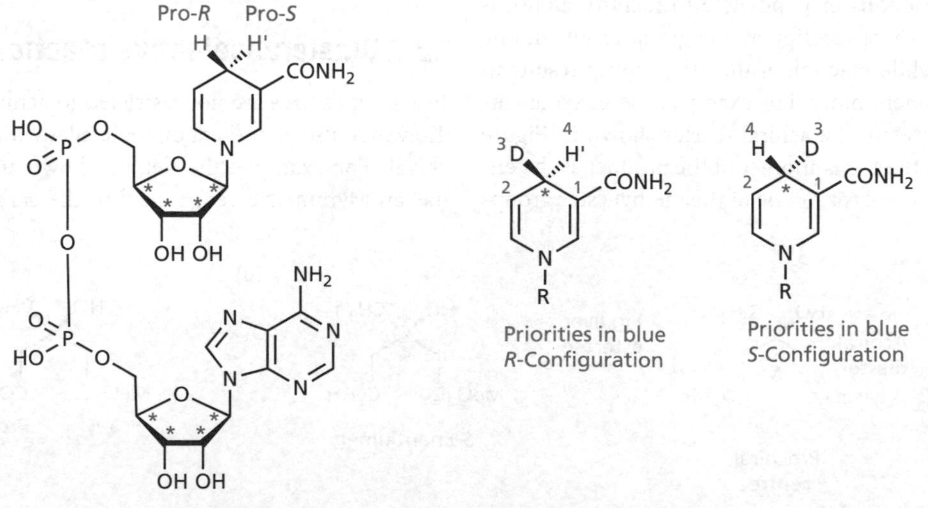

Priorities in blue
R-Configuration

Priorities in blue
S-Configuration

FIGURE 5.26 Nicotinamide adenine dinucleotide (NADH).

Only one of the diastereotopic protons in NADH is involved in the reduction mechanism catalysed by LDH (Fig. 5.20), and is transferred from the NADH cofactor to the substrate as nature's equivalent of a hydride ion. It is possible to identify which of the diastereotopic protons is involved by labelling the NADH with a deuterium isotope. This demonstrates that it is the pro-*R* hydrogen which is transferred exclusively to pyruvic acid (Fig. 5.27).

The observed selectivity is a consequence of the way in which the substrate and the NADH cofactor are bound in the enzyme's active site. This results in the *re* face of the carbonyl group being oriented towards one face of the nicotinamide ring of the cofactor. As a result, the pro-*R* hydrogen of the nicotinamide ring is transferred as it is closer to the *re* face of the substrate (Fig. 5.28).

<hr>

KEY POINTS

- The asymmetric synthesis of a chiral drug can be achieved by using a starting material that already contains the asymmetric centres required in the final structure. The reactions in the synthesis must not cause epimerization or racemization.

- Many natural products are chiral and exist naturally as a single enantiomer. They serve as useful starting materials for an asymmetric synthesis.

- In order to carry out an asymmetric reaction, an asymmetric compound must be present. This could be a starting material, enzyme, reagent, ligand, or chiral auxiliary.

- A chiral auxiliary is an asymmetric molecule that is introduced into a reaction to introduce asymmetry. Chiral auxiliaries often act as metal ligands to create a chiral metal complex that acts as a catalyst.

- The optical purity of a compound defines how much of one enantiomer is present rather than the other. This is also known as the enantiomeric excess.

- Prochiral molecules are not asymmetric, but have the potential to become chiral depending on the reaction carried out.

- The two planes of a planar prochiral molecule are known as enantiotopic faces and are defined as *re* or *si*.

- Two identical groups in an achiral molecule are defined as enantiotopic if a reaction at one of the groups results in the enantiomer of a chiral molecule. The groups can be defined as Pro-*R* or Pro-*S*.

- Enantioselective reactions are asymmetric reactions which result in one enantiomer being formed in preference to the other. Diastereoselective reactions are reactions that produce one diastereoisomer in preference to another.

- Diastereoselective reactions involve starting materials that are already chiral, and result in the creation of one or more asymmetric centres.

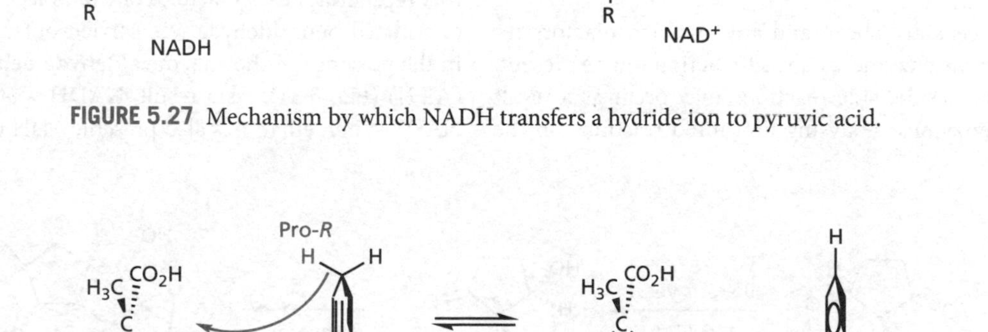

FIGURE 5.27 Mechanism by which NADH transfers a hydride ion to pyruvic acid.

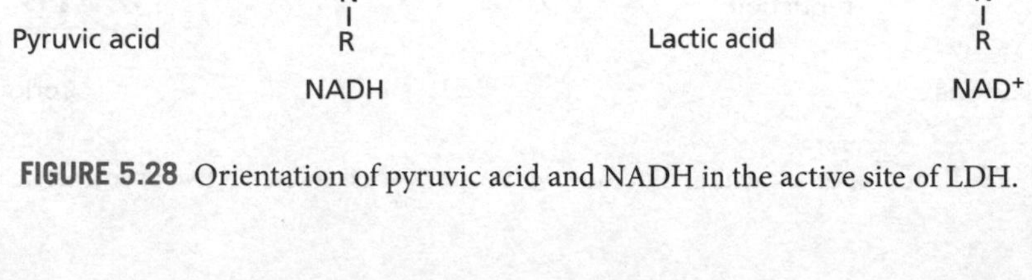

FIGURE 5.28 Orientation of pyruvic acid and NADH in the active site of LDH.

5.6 Asymmetric reactions using enzymes

5.6.1 Natural enzymes

The asymmetric reactions that are fundamental to life are catalysed by enzymes which are themselves chiral and exist as a single enantiomer. Enzymes have also been used effectively to carry out asymmetric reactions in the laboratory, particularly the hydrolysis of esters and the reduction of ketones. The advantages of an enzyme-catalysed reaction include high levels of chemoselectivity, regioselectivity, stereoselectivity, and enantioselectivity (see section 2.3.1). Moreover, the catalytic properties of an enzyme allow reactions to be carried out under mild conditions of temperature and pH. However, there are drawbacks. Not all enzymes are suitable for laboratory-based synthesis. For example, some enzymes have a high substrate specificity, which means that they will only accept a limited number of structures as substrates. The cost of enzymes can also be a disadvantage, and there may be problems in scaling up enzyme-catalysed reactions. Another problem might be solubility. Enzymes exist naturally in aqueous environments and the intended substrates and products are likely to be insoluble under such conditions. Cosolvents, where a solvent mixture that can dissolve the enzyme, substrate, and product is used, can be one way round this problem. However, one has to be careful that the enzyme is not denatured as a result. Another way round many of these problems is to use enzymes that are attached to a solid support.

There are two general ways in which enzymes can be used in organic synthesis. One method is to use whole cells which contain the enzyme of interest, for example yeast cells. Alternatively, purified enzymes can be used. There are advantages and disadvantages to both approaches.

Whole cells are cheap and any enzyme cofactors required for the enzyme-catalysed reaction will be present. On the debit side, side reactions may occur as a result of other enzymes catalysing unwanted reactions on the substrate. Moreover, there are practical difficulties as large glassware is normally used, and the removal of cells and cell debris can be messy and difficult.

An example of the use of whole cells is the enantioselective reduction of an achiral diketone as part of a synthesis leading to the anticancer agent **coriolin** (Fig. 5.29). An enzyme present in the yeast cell reacts selectively at one of the ketone groups to produce a new asymmetric centre with the S-configuration. This results in loss of the symmetry of the ring, which means that the carbon atom next to the new chiral centre also becomes an asymmetric centre, but with the R-configuration. Therefore, two asymmetric centres have been formed simultaneously. Despite the fact that two asymmetric centres have been produced, the ketone groups of the achiral diketone can be described as enantiotopic because reaction at the other ketone would have created the opposite enantiomer.

Baker's yeast is frequently used to reduce ketone groups, and it is possible to predict which enantiomer is more likely to be formed by using **Prelog's rule**. This is based on the observation that the configuration of the new asymmetric centre is determined by the relative sizes (not the priorities) of the two groups attached to the carbonyl carbon (Fig. 5.30).

Isolated enzymes have several advantages over whole cells. The reactions are easier to carry out and involve simpler apparatus. Work-up and purification procedures are easier, and there is less chance of side reactions and impurities. However, isolated enzymes can be expensive. It may also be necessary to add expensive cofactors to the reaction.

One way round the problem of expensive cofactors is to design experiments such that the cofactor is recycled during the reaction. This means that the cofactor can be used in catalytic quantities. Recycling can be achieved by coupling the desired reaction with a second reaction that regenerates the cofactor. For example, the reduction of tritiated benzaldehyde was carried out using NADH in the presence of the enzyme **aldehyde dehydrogenase** (ADH) (Fig. 5.31). As a result, NADH was oxidized to NAD$^+$. When ethanol is also present, ADH catalyses the

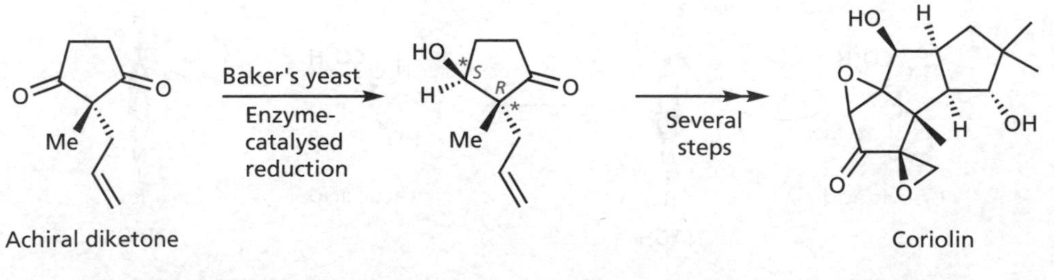

FIGURE 5.29 Reaction of an achiral diketone with baker's yeast to create two asymmetric centres.

FIGURE 5.30 Prelog's rule (S = small group; L = large group).

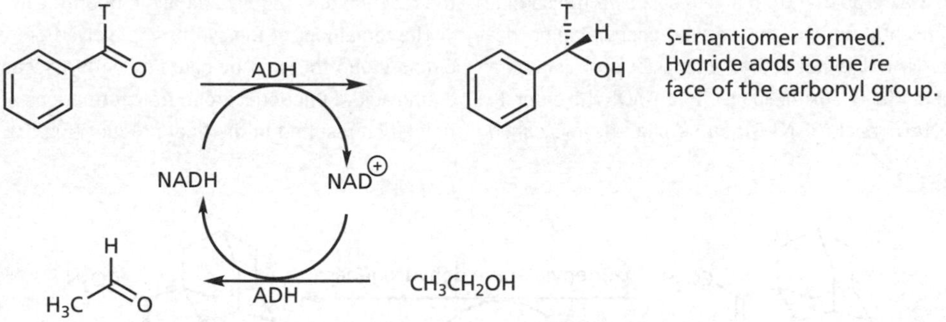

S-Enantiomer formed. Hydride adds to the *re* face of the carbonyl group.

FIGURE 5.31 Coupling two enzyme-catalysed reactions.

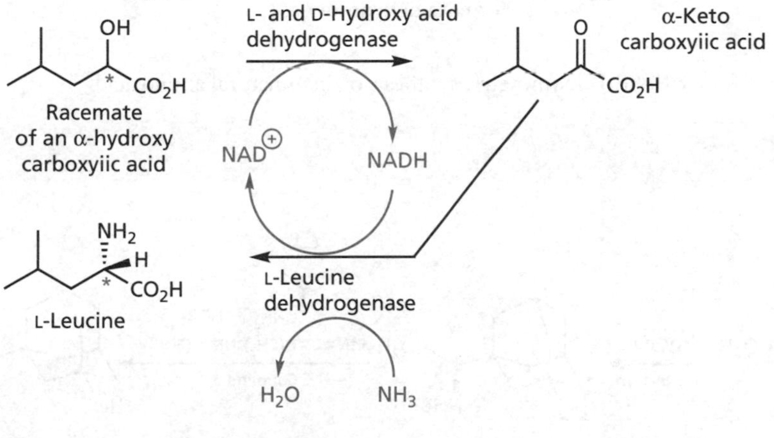

FIGURE 5.32 One pot synthesis of L-leucine from an α-hydroxy carboxylic acid.

oxidation of ethanol to ethanal, thus reducing NAD⁺ back to NADH. This means that catalytic quantities of the cofactor can be used.

A second example is a one-pot synthesis of the amino acid L-leucine from a racemic mixture of an α-hydroxy carboxylic acid (Fig. 5.32). Three enzymes are present in the reaction along with the cofactor NADH and ammonia. The first stage of the reaction involves oxidation of the α-hydroxy acid to an α-keto acid. As soon as this is formed, it acts as the substrate for L-leucine dehydrogenase which catalyses a reductive amination to L-leucine. Once again, the cofactor can be used in catalytic quantities. NAD⁺ is reduced by the first enzyme-catalysed reaction, and the resulting NADH is oxidized

back to NAD⁺ by the second enzyme-catalysed reaction (see also Box 5.7).

5.6.2 Genetically modified enzymes

Genetic engineering and mutation studies have led to the design of genetically modified enzymes that can accept substrates which the native enzyme would not normally accept. For example, **transaminase enzymes** catalyse the conversion of a ketone to an amine and would be useful in the synthesis of **sitagliptin**—a drug that is prescribed for the treatment of diabetes. The conventional chemical approach to carrying out this transformation involves converting the ketone to an enamine and then carrying

BOX 5.7 Synthesis of saxagliptin

Saxagliptin inhibits an enzyme called dipeptidyl pepti-dase-4 (DPP-4), and was approved in 2009 as an oral hypo-glycaemic for the treatment of diabetes. A key stage in the synthesis involves the formation of the S-enantiomer of an unnatural amino acid containing an adamantine ring system (Fig. 1). This was prepared from a keto acid by a transamin-ation reaction catalysed by the enzyme phenylalanine de-hydrogenase. NADH is needed as a cofactor and acts as a reducing agent, being oxidized itself to NAD+. In order to recycle the NAD+ back to NADH, a second enzyme called

formate dehydrogenase was included in the reaction mix-ture along with ammonium formate. Formate dehydrogenase catalyses the oxidation of formate to carbon dioxide, using NAD+ as a cofactor. Thus, NAD+ is reduced back to NADH. Since the cofactor NAD+ is recycled as soon as it is formed, the reaction only requires catalytic quantities.

The remainder of the synthesis involved protection of the amino group followed by coupling with a second unnatural amino acid. Functional group transformations and deprotec-tions then resulted in the final product (Fig. 2).

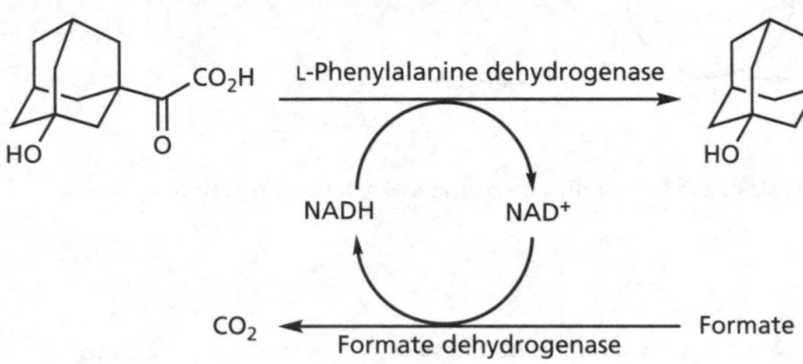

FIGURE 1 Asymmetric synthesis of an unnatural amino acid.

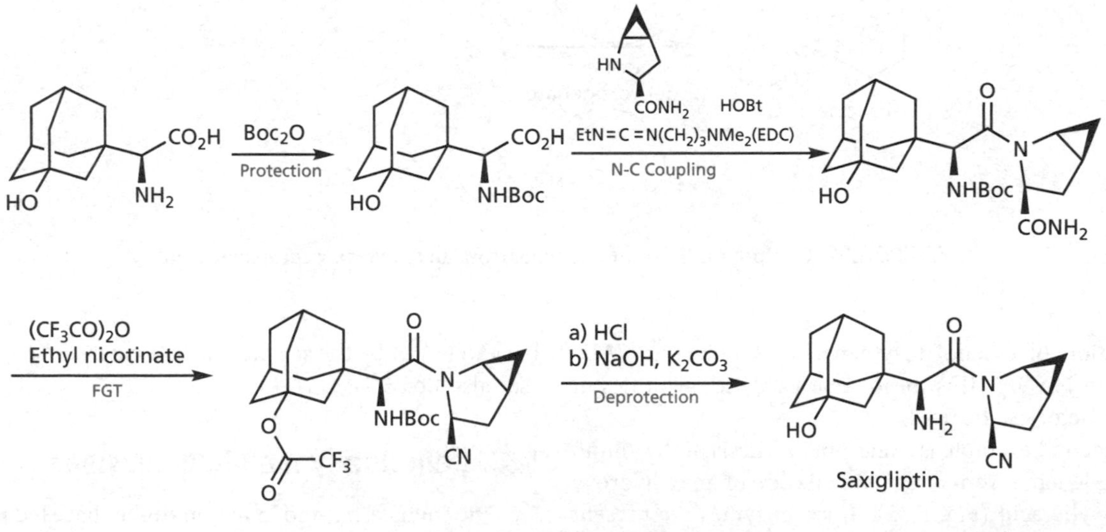

FIGURE 2 Asymmetric synthesis of saxagliptin.

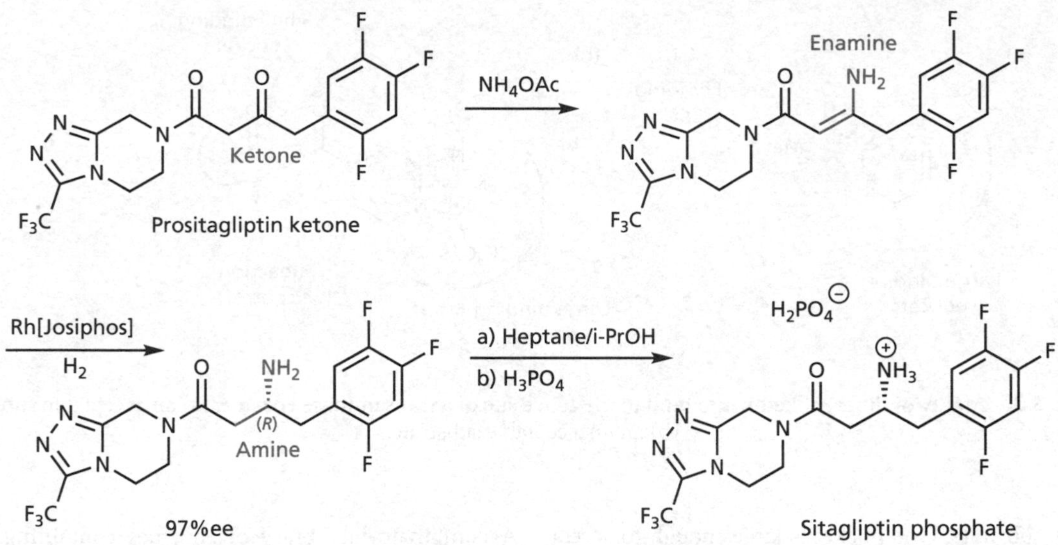

FIGURE 5.33 Chemical synthesis of sitagliptin phosphate including a rhodium-catalysed hydrogenation (see also sections 5.8.1 and 5.8.2).

out an asymmetric hydrogenation with a rhodium cata-lyst (Fig. 5.33) (sections 5.8.1 and 5.8.2). However, the reaction has to be done at high pressure, the catalyst is expensive, and there are significant purification prob-lems. A certain amount of the opposite enantiomer is produced which has to be removed, and it is also import-ant to remove all traces of the potentially toxic heavy metal rhodium from the final product. As a result, the final yields have not been spectacular.

In theory, a transaminase enzyme could catalyse this kind of reaction and avoid many of the problems

described above (Fig. 5.34). However, in practice, nat-urally occurring transaminase enzymes fail to catalyse the reaction particularly well and the resulting yields are poor.

This is not too surprising since the range of substrate structures that transaminase enzymes can accept is limited. Indeed, substrates are normally restricted to methyl ketones because steric interactions prevent binding of larger substrates. Molecular modelling studies of a model transaminase active site have shown that the active site contains two binding pockets for the two substituents of

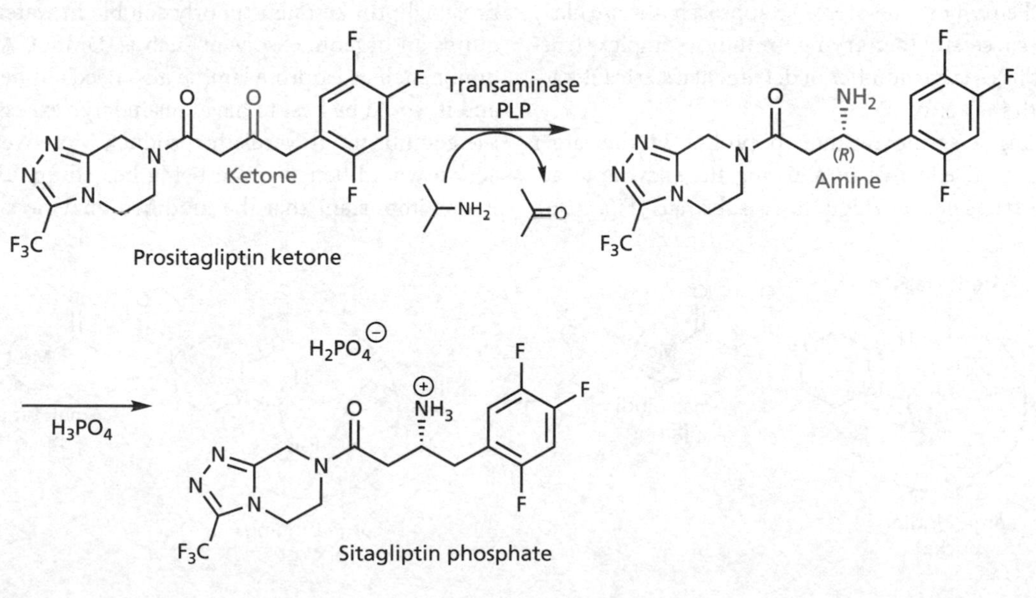

FIGURE 5.34 Synthesis of sitagliptin phosphate using a transaminase enzyme (PLP = pyridoxal phosphate).

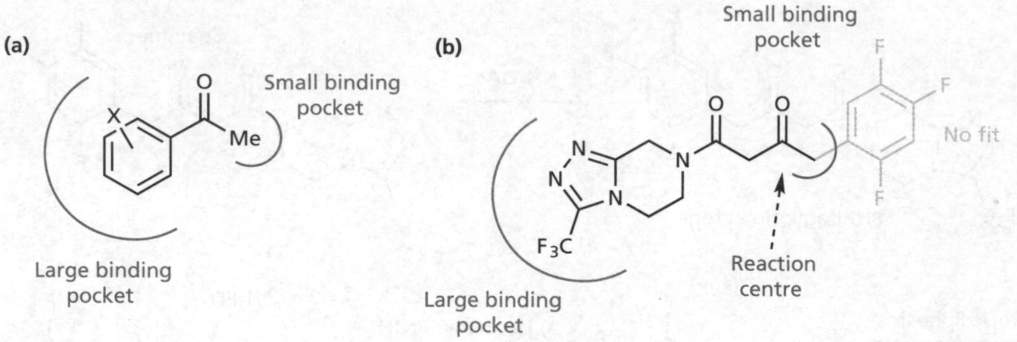

FIGURE 5.35 Ability of different ketones to bind to the active site of a transaminase enzyme: (a) an acceptable substrate; (b) an unacceptable substrate.

a ketone substrate. One pocket is large enough to accept a wide range of groups, but the other is so small that anything larger than a methyl group will not fit (Fig. 5.35). Therefore, the trifluorinated aromatic ring of prositagliptin ketone is much too large. Indeed, modelling studies have also revealed that the bicyclic ring system on the other side of the carbonyl group is too big for its binding pocket.

Therefore, it was decided to modify the amino acids lining the active site to increase the size of the binding pockets so that they could accept larger substrates. The first stage was to increase the size of the larger binding pocket, and so a 'cut-down' truncated analogue of prositagliptin ketone was synthesized to test whether the strategy worked (Fig. 5.36). This represents an approach called **substrate walking** where a simpler substrate than the target substrate is used to find a modified enzyme that will accept it. Once that has been achieved, more complex substrates can be studied on further genetically modified enzymes. This stepwise approach is more likely to prove successful than trying to study a complex structure that has a large number of detrimental steric interactions with the active site.

Mutating a serine residue to proline in the larger pocket proved effective in allowing the enzyme to accept the truncated analogue as a substrate (Fig. 5.36).

A combinatorial library of enzymes containing different mutations within the small binding site was then produced. From this library, a number of modified enzymes were identified which could accept prositagliptin ketone as a substrate. The enzymes concerned had the previously described mutation in the large pocket, as well as three mutations in the small pocket where phenylalanine, valine, and alanine residues were replaced with smaller amino acids such as glycine (Fig. 5.37).

The most active of these enzymes was then studied in a second phase of evolutionary mutations, which led to the identification of a new variant having six additional mutations and a 75-fold increase in activity.

Encouraging though these results were, there were many other factors to consider before a modified enzyme could be considered as a realistic approach to transamination on a production scale. For example, the natural enzyme exists in an aqueous environment, but prositagliptin ketone is poorly soluble in water and requires an organic cosolvent such as DMSO. An amine donor such as isopropylamine also needs to be present, and it would be best to have this in large excess to push the equilibrium towards the product. Moreover, the reaction would benefit from being heated up. Therefore, it was important that the modified enzyme should be

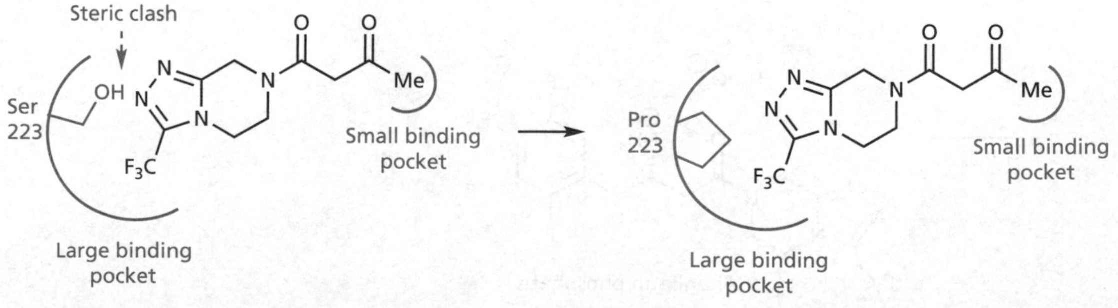

FIGURE 5.36 Beneficial mutation of Ser-223 to Pro-223 in the binding of a truncated analogue of prositagliptin ketone.

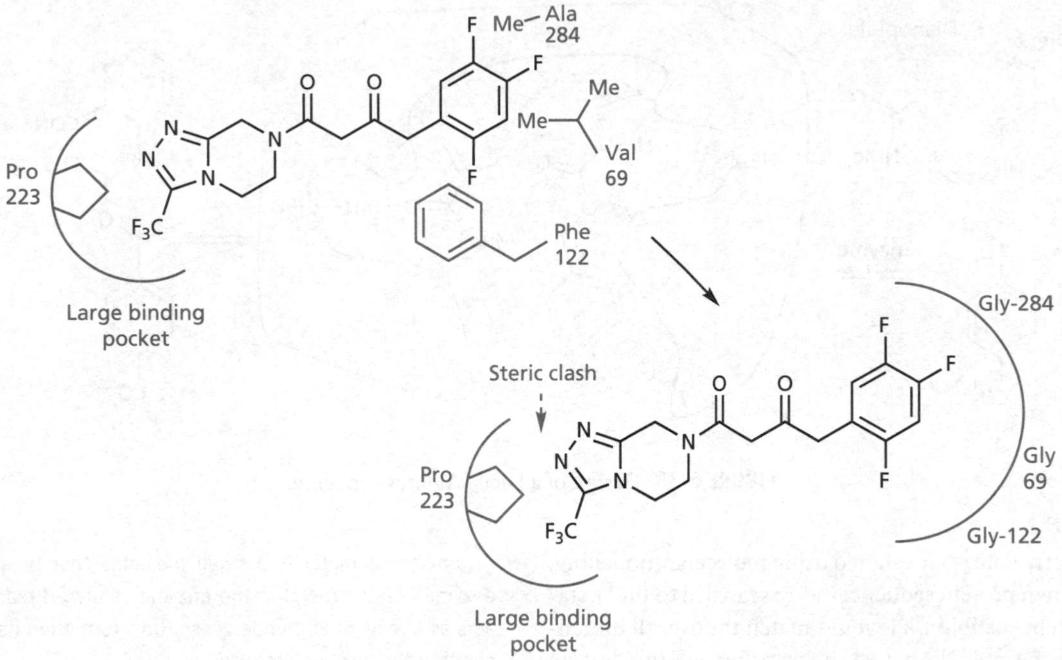

FIGURE 5.37 Beneficial mutations allowing binding of prositagliptin ketone.

stable in those experimental conditions, whilst still catalysing the transamination with a high enantiomeric excess.

To tackle these issues, further libraries of mutated enzymes were generated to see which would perform best under different experimental conditions. Eleven evolutionary rounds of modifications were carried out and the reaction conditions were altered gradually such that they became increasingly more demanding—a process that can be described as **directed evolution**. This involved gradual increases in substrate and isopropanol concentration, DMSO levels, pH, and temperature. The enzymes also had to be capable of being expressed in the bacterium *E.coli* which would be the cellular host.

The final enzyme from these studies contained 27 mutations in total, including five mutations in the small pocket and five mutations in the large pocket. The mutations in the binding pockets allow the desired substrate to bind to the active site. The remaining mutations play a role in stabilizing the enzyme under different experimental conditions. The enzyme exists as a protein dimer and it is believed that these latter mutations stabilize the enzyme by strengthening the interactions between the protein subunits.

The final enzyme was found to catalyse the transamination of prositagliptin ketone 27,000 times more effectively than the original enzyme, with an enantiomeric excess of over 99.95%. There was also a 10–13% increase in overall yield compared with the previous chemical process, and so the overall process proved more economical.

By avoiding the hydrogenation step, there is no need for specialized high pressure hydrogenation equipment, and there is a 19% reduction in total waste, with no heavy metals being used in the process. Finally, the modified enzyme has a higher substrate range than the natural enzyme, and so it may be possible to use it to convert other prochiral ketones to chiral amines.

Molecular modelling and genetic engineering studies have been used to design enzymes that can catalyse reactions that are not catalysed by any natural enzyme. For example, there is no naturally occurring enzyme that catalyses the Diels–Alder reaction. The ideal enzyme to catalyse this reaction would have an active site that binds both the diene and dienophile in the correct orientation for the concerted reaction (Fig. 5.38). Furthermore, amino acid residues would interact with the diene and the dienophile to lower the energy of the transition state, and thus increase the reaction rate. This can be achieved by incorporating an amino acid residue that acts as a hydrogen bond acceptor (HBA) with the diene, as this would raise the energy of the highest occupied molecular orbital (HOMO) and stabilize the positive charge developing in the transition state. An amino acid residue capable of acting as a hydrogen bond donor (HBD) with the dienophile would also be beneficial, because this would lower the energy of the lowest unoccupied molecular orbital (LUMO) and stabilize a developing negative charge in the transition state. The overall effect of both interactions would be to narrow the gap between the HOMO and LUMO energies, thus increasing the rate of reaction.

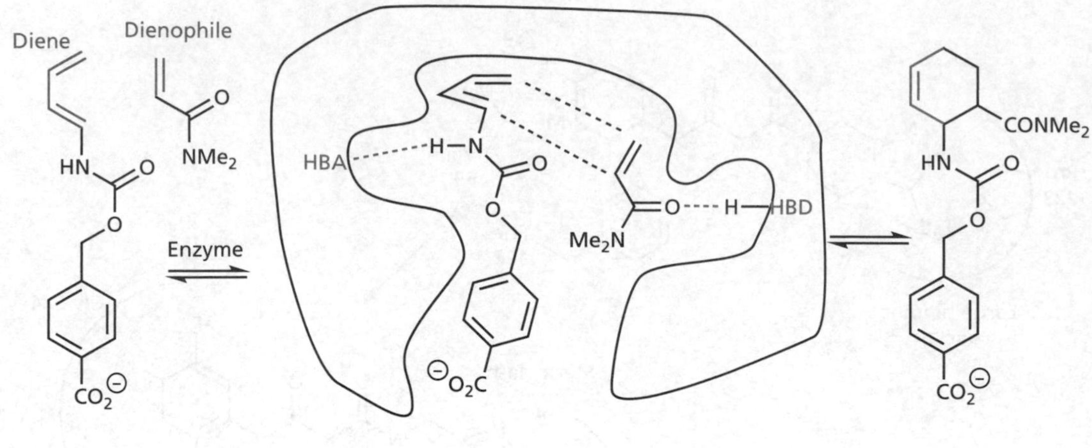

FIGURE 5.38 Design of a Diels–Alderase enzyme.

An active site was designed using molecular modelling, and known protein sequences were searched to find a stable protein scaffold that would match the overall dimensions of the site. The active site was then modified using molecular modelling software such that the model would be capable of accommodating the desired substrates and would contain amino acid residues capable of acting as the required HBD and HBA.

Genes were synthesized that would code for these proteins and expressed in *E.coli*. Fifty proteins were generated as possible candidates, and two of these were found to have catalytic activity for the Diels–Alder reaction shown in Figure 5.38.

One of these was a **diisopropylfluorophosphatase** from *Loligo vulgaris* which contained 13 mutations including a glutamine and a tyrosine residue. The glutamine carbonyl group acted as the HBA, while the tyrosine hydroxyl group acted as the HBD. The other enzyme was derived from **ketosteroid isomerase** and contained 14 mutations.

Further mutations were then carried out on both enzymes to improve binding interactions and to produce the 3R,4S *endo* structure as the only product. In contrast, the uncatalysed reaction resulted in four isomers, with the desired structure representing only 47% of the total product.

Currently, the modified enzymes are very specific in terms of the substrates they accept, but it is thought that different mutations could be carried out to allow reactions to be catalysed on alternative substrates. With further research, it may well be possible to produce a *de novo* designed enzyme that could be used to catalyse reactions on the production scale.

KEY POINTS

• Enzymes act as catalysts and contain an active site that binds the substrates for an enzyme-catalysed reaction.

• An enzyme cofactor is a small molecule that binds to an enzyme and is crucial to the enzyme-catalysed reaction. It acts as a reagent and undergoes a transformation itself as a result of the enzyme-catalysed reaction.

• An asymmetric reaction involving an enzyme can be carried out with the isolated enzyme or with whole cells that contain the enzyme. There are advantages and disadvantages to both approaches.

• It is necessary to add any cofactors required if an isolated enzyme is being used in an asymmetric reaction. The cofactor can be recycled in situ by coupling the desired reaction with a second reaction.

• Genetically modified enzymes have been designed to accept substrates that the natural enzyme will not accept.

• Genetically modified enzymes have been designed to catalyse reactions that do not occur in nature.

5.7 Asymmetric reactions with an asymmetric starting material

If the starting material is already chiral, any reaction which produces a new chiral centre will result in two possible diastereoisomers rather than enantiomers. The fact that diastereoisomers have different chemical and physical properties may influence the selectivity of the reaction to favour one of these diastereoisomers. In addition, the asymmetry already present in the starting material may influence the outcome of the reaction to favour one diastereoisomer over the other.

The Grignard reaction shown in Figure 5.39 was carried out as part of a synthesis leading to an opioid sedative called **etorphine**, which is used in veterinary medicine (section 4.6.1). The starting material is a chiral thevinone with five

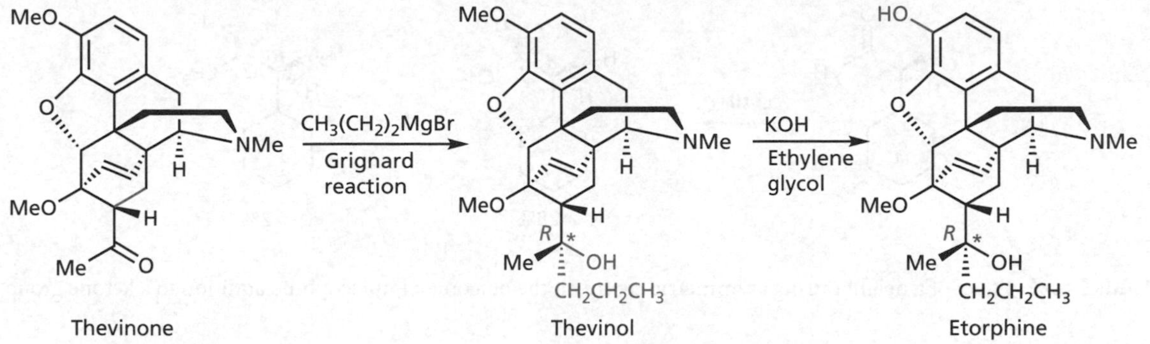

MeO

O

NMe

H

MeO

H

Me

O

Thevinone

→ CH₃(CH₂)₂MgBr

Grignard reaction

MeO

O

NMe

H

MeO

H

R

Me   OH

CH₂CH₂CH₃

Thevinol

→ KOH

Ethylene glycol

HO

O

NMe

H

MeO

H

R

Me   OH

CH₂CH₂CH₃

Etorphine

FIGURE 5.39 Synthesis of etorphine. The star indicates the new asymmetric centre.

asymmetric centres, and the new asymmetric centre resulting from this reaction has exclusively the *R*-configuration. This can be explained by the asymmetric magnesium complex which is formed during the Grignard reaction prior to the alkyl group being transferred to the least hindered face of the carbonyl group (Figs 5.40 and 5.41).

The preceding example involves a reasonably complex chiral starting material derived from a natural product—an example of a semi-synthetic synthesis (section 9.3).

The presence of chirality in a simpler molecule does not necessarily mean that an asymmetric reaction will occur elsewhere in the structure. In general, asymmetry is most likely when the new chiral centre is created close to the one which is already present. For example, the presence of a chiral centre next to a ketone group

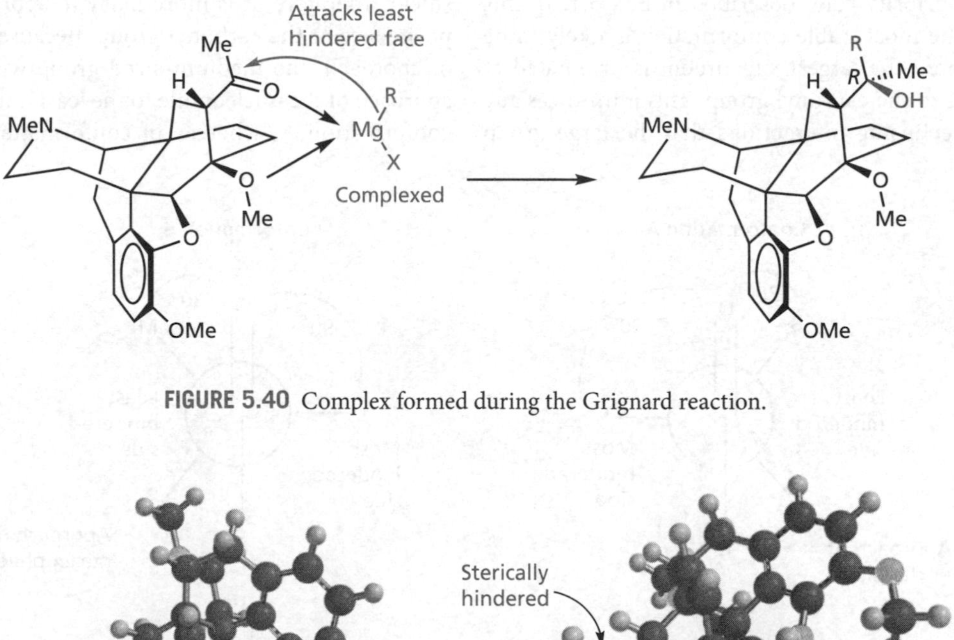

Attacks least hindered face

H

Me

O

MeN

O

Me

O

OMe

R

Mg

X

Complexed

→

H

R

Me

OH

MeN

O

Me

O

OMe

FIGURE 5.40 Complex formed during the Grignard reaction.

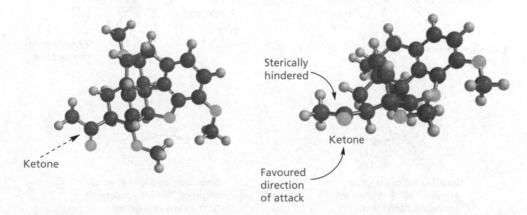

FIGURE 5.41 3D models of the thevinone from two perspectives. The second one is looking into the plane of the methyl ketone group.

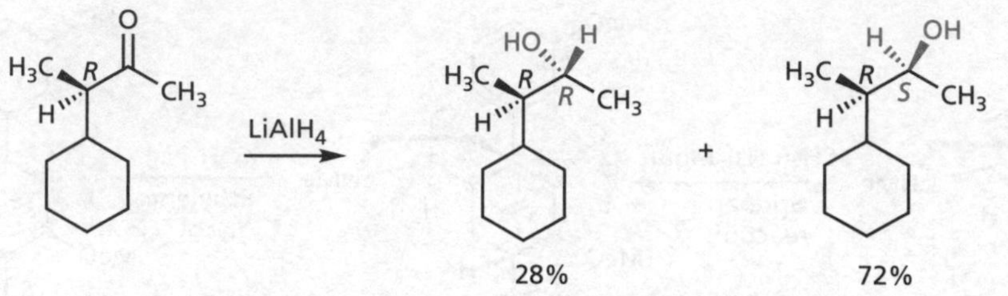

FIGURE 5.42 Effect of a neighbouring asymmetric centre on the outcome of nucleophilic addition to a ketone group.

has an influence over the addition of nucleophiles to the carbonyl group. The favoured diastereoisomer is the R,S-diastereoisomer rather than the R,R-diastereoisomer when the ketone shown in Figure 5.42 is reduced with lithium aluminium hydride.

The selectivity observed can be explained if the reaction is more likely to occur when the ketone is in a stable conformation. This makes sense as the molecule will spend more of its time in a stable conformation than in unstable conformations. In order to identify likely stable conformations, we define the three groups attached to the chiral centre as small (S), medium (M), and large (L). This is based on their actual size and not on the priority rules described in Box 5.1. In this situation, the most stable conformation is likely to be the one where the largest substituent is orientated at right angles to the carbonyl group. This minimizes any steric and eclipsing interactions that the large group

might have with the carbonyl group (Fig. 5.43). There are two possibilities. One of these has the small-sized group at a dihedral angle of 30° from the carbonyl group (conformation A). The other conformation has the medium-sized group at a dihedral angle of 30° from the carbonyl group (conformation B). We now consider how the nucleophile will approach each of these conformations. The angle of approach for nucleophilic addition to a carbonyl group is 107° from the plane of the carbonyl group.

In both conformations A and B, the nucleophile is hindered from approaching one face of the carbonyl group because of the large substituent acting as a steric shield. Therefore, it is more likely to approach the opposite face of the carbonyl group. Because of the angle of approach, the medium-sized group will hinder the approach of the nucleophile to the least hindered side of conformation A, whereas in conformation B it is the

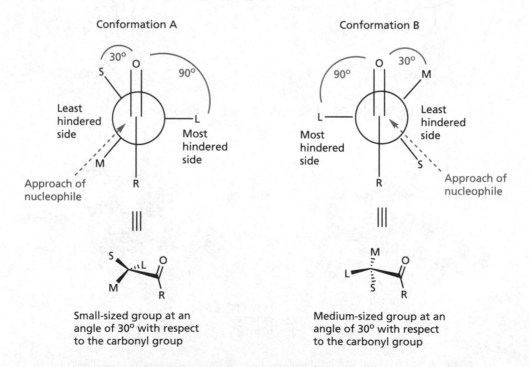

FIGURE 5.43 Comparison of two stable conformations.

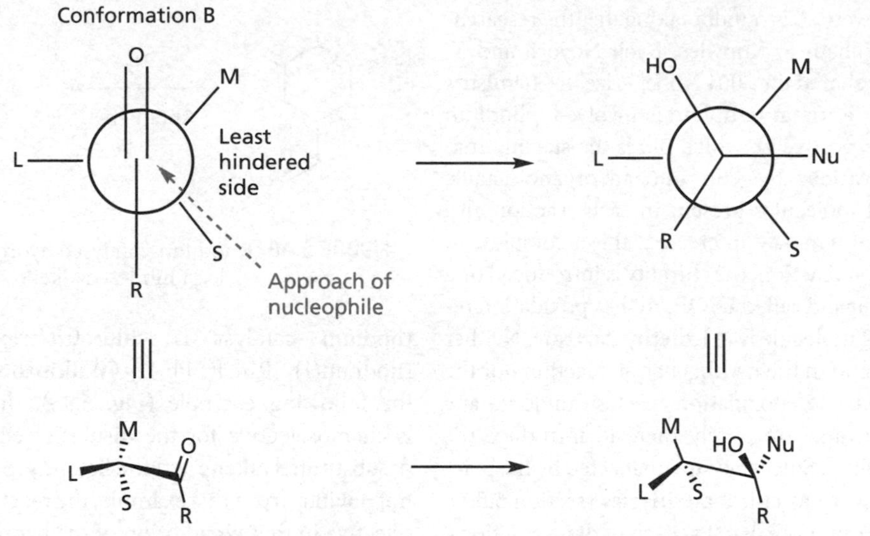

Conformation B

FIGURE 5.44 The Felkin–Anh model for explaining diastereoselectivity.

small-sized group that hinders approach. If there is a significant difference in size between the small- and medium-sized groups, we would expect the majority of the reaction to occur when the molecule is in conformation B (Fig. 5.44).

This is known as the **Felkin–Anh model** and it holds for a range of reagents involved in nucleophilic addition such as LiAlH$_4$, organolithium reagents (RLi), Grignard reagents (RMgX), KCN, and lithium acetylide. It is important that the chiral centre already present is next to the reaction centre and, hence, the developing chiral centre.

5.8 Asymmetric reactions using chiral reagents

5.8.1 Introduction

In the last couple of decades, there have been huge advances in the development of chiral reagents that allow many well-known organic reactions to be carried out under asymmetric conditions. Two early examples are the **rhodium-catalysed hydrogenation** of alkenes and the **Sharpless epoxidation** of allylic alcohols (Fig. 5.45).

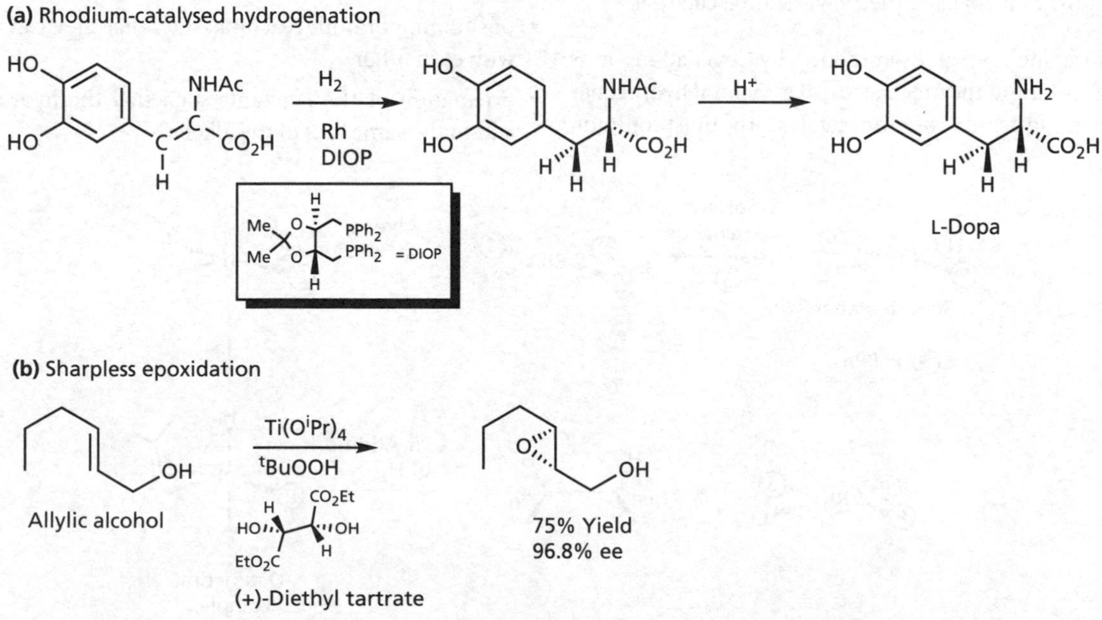

(a) Rhodium-catalysed hydrogenation

(b) Sharpless epoxidation

FIGURE 5.45 Early asymmetric syntheses using chiral reagents.

These reactions were so groundbreaking that the researchers involved (William S. Knowles, Ryoji Noyori, and K. Barry Sharpless) shared the 2001 Nobel Prize in Chemistry.

In both cases, a **metal template** is involved (rhodium and titanium, respectively), which binds the starting material and the various reagents into an organometallic complex. Chiral molecules present in each reaction also bind to the metal template to create a chiral complex. In the hydrogenation reaction, the chirality is introduced by a bidentate chiral ligand called **DIOP**. In the epoxidation reaction, the chiral molecule is (+)-**diethyl tartrate**. Neither the bidentate ligand in the hydrogenation reaction nor the diethyl tartrate in the epoxidation reaction undergo any form of reaction and are purely there to introduce the asymmetry required. Since they are unaffected by the reaction, they are known as **chiral auxiliaries** (section 5.5.1). We will now look in more detail at each of these reactions.

5.8.2 Asymmetric hydrogenations with rhodium catalysts

Catalytic hydrogenations can be either heterogeneous or homogeneous. In heterogeneous catalysis, the catalyst (e.g. Pd/C) is insoluble and provides a surface to which the reactants can bind and then react. In homogeneous catalysis, the catalyst (e.g. RhCl(PPh$_3$)$_3$) is in solution and reacts with the reactants on a molecular level. There are several advantages of homogeneous catalysis over heterogeneous catalysis:

- better selectivity for functional groups
- less double-bond migration
- improved stereoselectivity
- the possibility of asymmetry with chiral catalysts.

Before looking at asymmetric hydrogenation, it is worth looking at the mechanism of an achiral hydrogenation using an achiral rhodium catalyst. The most common

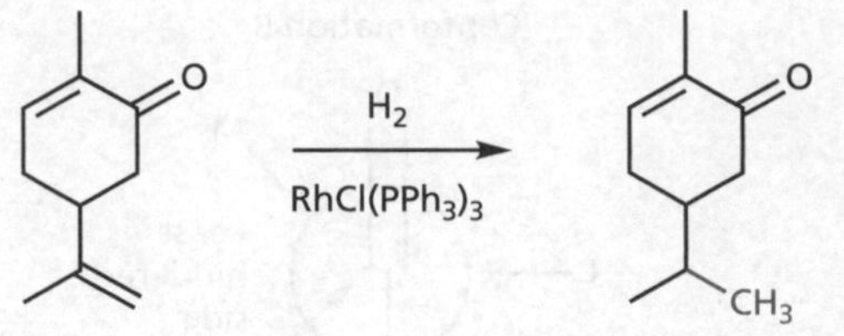

FIGURE 5.46 Rhodium-catalysed hydrogenation of the least hindered alkene.

rhodium catalyst is chlorotris(triphenylphosphine) rhodium(I) RhCl(PPh$_3$)$_3$ (**Wilkinson's catalyst**). In the following example (Fig. 5.46), the hydrogenation is chemoselective for the disubstituted alkene over the trisubstituted alkene and the ketone group. Although it is not obvious from this example, the reaction is also stereoselective in that *cis* addition of the hydrogen atoms takes place to the double bond. However, the reaction is not asymmetric since the hydrogen atoms can add equally well to either face of the double bond.

The mechanism (Fig. 5.47) involves substitution of one of the rhodium ligands by a solvent molecule, followed by substitution of the solvent molecule with an alkene. Hydrogen then bonds to rhodium and the bond between the two hydrogens is split, resulting in the formation of an octahedral complex. Thus, the catalyst has carried out one of its crucial roles in catalysing the splitting of the hydrogen molecule. Moreover, the hydrogen atoms are now held close to the alkene starting material and can be transferred to the same face of the alkene. Thus, the rhodium catalyst serves various catalytic roles:

- binding of the reagents
- splitting of the hydrogen–hydrogen bond
- positioning of both reactants such that they can react with each other
- orientation of the reactants such that the hydrogens add to the same face of the alkene.

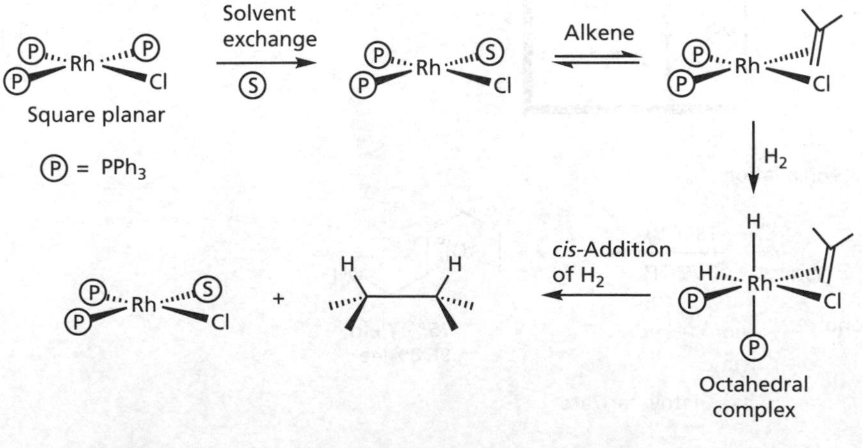

FIGURE 5.47 Proposed mechanism for the rhodium-catalysed hydrogenation of an alkene under achiral conditions.

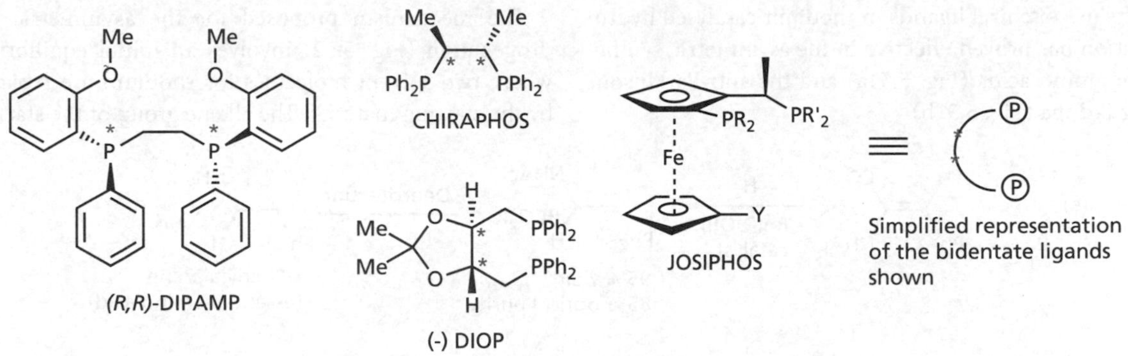

Me   Me

Ph₂P ——— PPh₂

CHIRAPHOS

(R,R)-DIPAMP

Me
   O   PPh₂
Me
   O   PPh₂
      H

(−) DIOP

JOSIPHOS

Simplified representation
of the bidentate ligands
shown

FIGURE 5.48 Chiral ligands for the rhodium atom (see also section 5.9.2).

The only thing that is missing for an asymmetric reaction is some asymmetry to ensure that hydrogen transfer takes place at only one of the two possible faces of the alkene.

The asymmetric rhodium-catalysed hydrogenation is made possible by using chiral ligands for the rhodium atom. There are several chiral ligands which can serve that purpose, such as **DIOP, DIPAMP, CHIRAPHOS,** and **JOSIPHOS** (Fig. 5.48). All of these chiral ligands contain two asymmetric centres and are **bidentate**; in other words each ligand can undergo two interactions with the central rhodium atom.

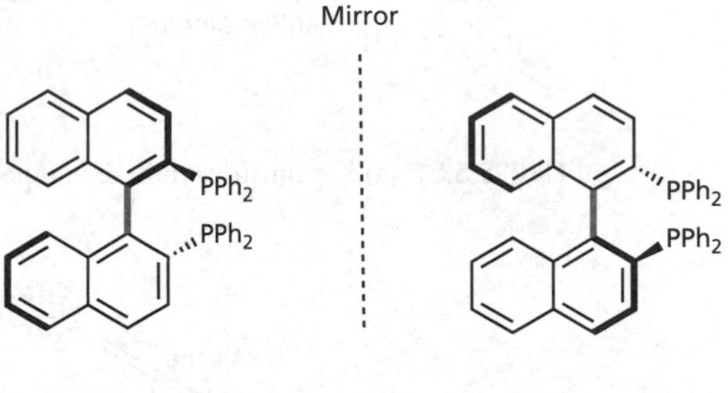

Another commonly used bidentate chiral ligand is **BINAP** (Fig. 5.49). This is an interesting example of a chiral structure that lacks an asymmetric centre. Chirality occurs because there is restricted rotation around the single bond connecting the two naphthalene moieties. This is due to the high steric strain that would occur if the two aromatic systems were to become planar, since the two diphenylphosphine groups would have to occupy the same region of space. Therefore, the molecule is locked

Mirror

PPh₂
PPh₂ | PPh₂
         | PPh₂

FIGURE 5.49 The two non-superimposable mirror images of BINAP—a chiral ligand without asymmetric centres. The bond with restricted rotation is shown in bold blue.

into a conformation where the two naphthalene rings are at right angles (orthogonal) to each other. Restricted rotation prevents the molecule from adopting the mirror-image conformation, and since the mirror images are not superimposable, the molecule is chiral and can exist as two isolable enantiomers (Fig. 5.50).

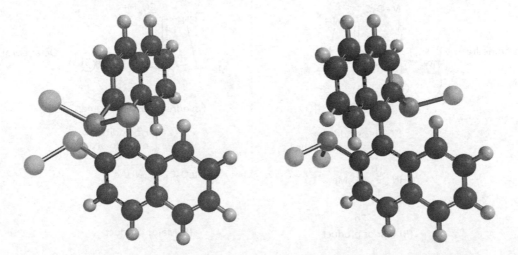

FIGURE 5.50 3D models of the two enantiomers of BINAP (the phenyl groups of the two diphenylphosphine substituents have been shown as light blue spheres for reasons of clarity).

The use of chiral ligands in rhodium-catalysed hydrogenation has proved effective in the asymmetric synthesis of amino acids (Fig. 5.51a) and the anti-Parkinson's drug L-dopa (Fig. 5.51b).

The mechanism proposed for the asymmetric hydrogenation (Fig. 5.52) involves an initial equilibrium where two solvent molecules for rhodium are replaced by the starting material. The alkene group of the starting

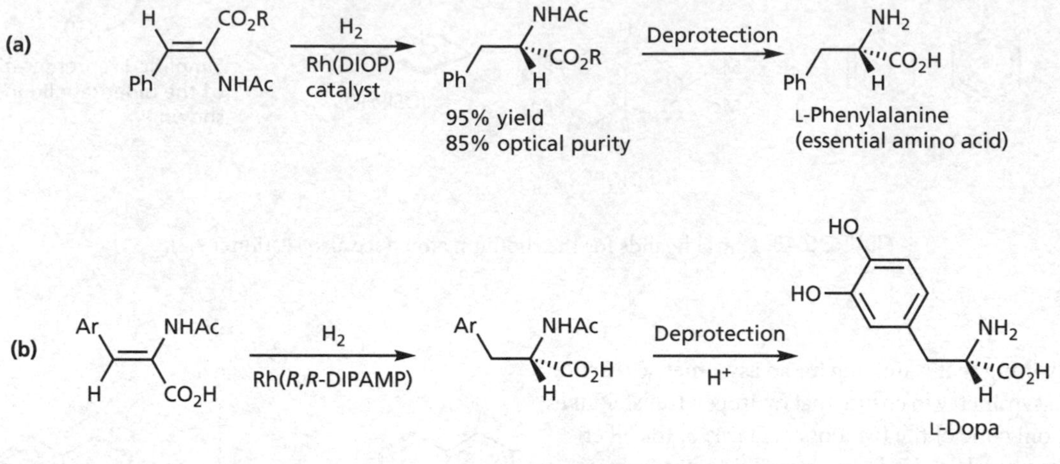

FIGURE 5.51 (a) Asymmetric synthesis of L-phenylalanine. (b) Asymmetric synthesis of L-dopa.

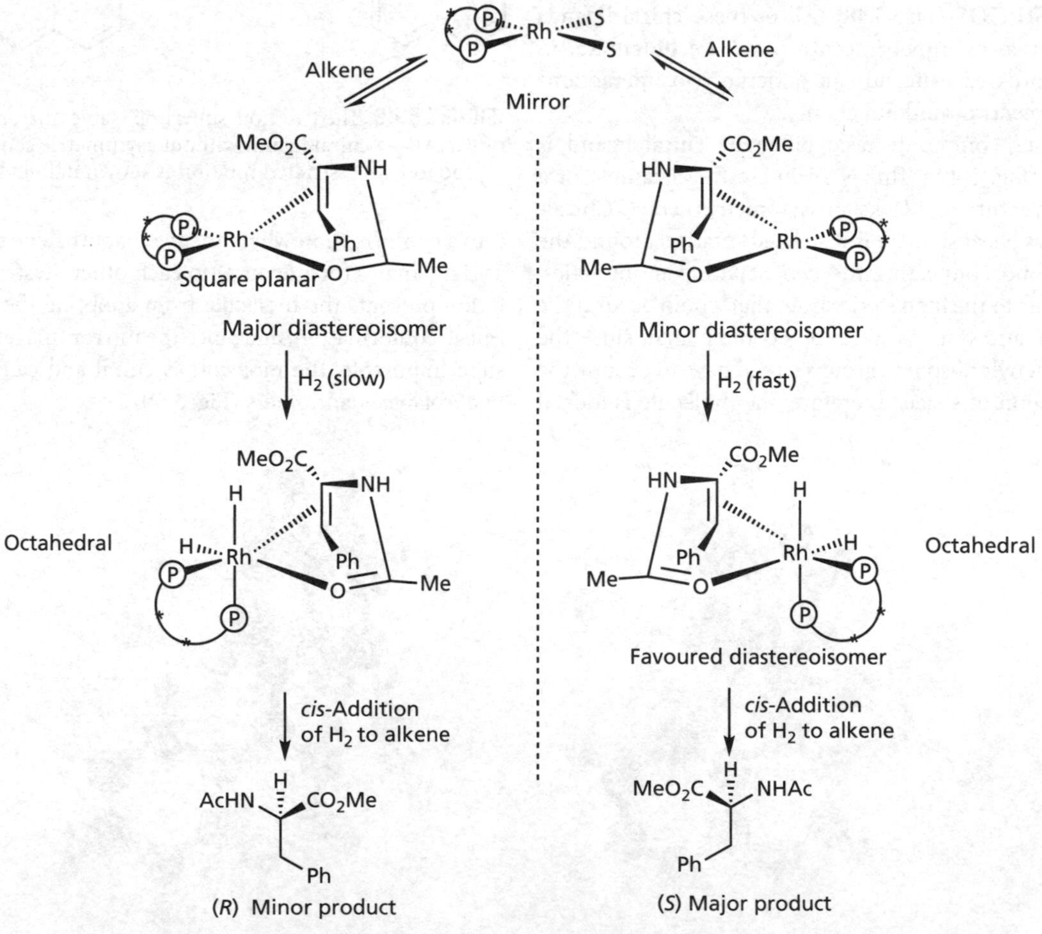

FIGURE 5.52 Mechanism of the asymmetric hydrogenation.

material acts as one ligand and the amide carbonyl group acts as another. This results in two possible square-planar complexes which are asymmetric in their overall shape and can be viewed as non-superimposable mirror images. In fact, they are not truly mirror images since the chiral centres in the chiral bidentate ligand are the same configuration in both possible complexes. Therefore, these complexes are actually diastereoisomers as a result of the additional asymmetry of the square-planar complex. Since they are diastereoisomers, they do not have identical chemical and physical properties. As a result, one of the diastereoisomers is more stable than the other and is formed preferentially. The more stable diastereoisomer is labelled as the major diastereoisomer in Figure 5.52, but it is important to appreciate that an equilibrium exists between the two square-planar complexes and the initial rhodium catalyst.

The next stage in the mechanism is the binding and splitting of the hydrogen molecule to form the octahedral complex. The rate at which this occurs differs for each of the two possible square-planar complexes, and the faster reaction takes place with the minor diastereoisomer. Since an equilibrium is still taking place between the square-planar complexes, the reaction of the minor diastereoisomer upsets the equilibrium and the major diastereoisomer is funnelled through the original square-planar complex to form more of the minor diastereoisomer. This continues to occur such that most of the reaction takes place through the less stable square-planar diastereoisomer. As a result, the major product is the *S*-enantiomer rather than the *R*-enantiomer.

The success of the reaction relies crucially on the fact that the square-planar complexes are in equilibrium with each other and are diastereoisomers with different properties. It is important to appreciate that it is the formation of the square-planar complex that introduces the extra asymmetry required for these two square-planar complexes to be diastereoisomers. Figure 5.53 is an attempt to illustrate this extra asymmetry.

In the preceding mechanism, both the alkene and the amide protecting group of the substrate act as ligands

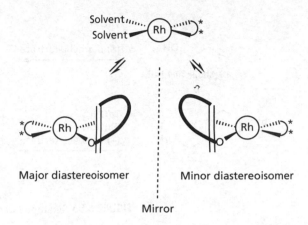

FIGURE 5.53 The additional asymmetry introduced by the binding of the starting material.

to the rhodium metal. This is important in creating the additional asymmetry on which this reaction depends, but the two-point attachment limits the type of molecule which can be successfully hydrogenated in an asymmetric manner. Further research has developed other types of chiral ligand and metals which allow a greater range of molecules to be successfully hydrogenated in an asymmetric manner. For example, a ruthenium catalyst with a BINAP chiral ligand has proved effective (Fig. 5.54).

5.8.3 The Sharpless epoxidation

The Sharpless epoxidation of allylic alcohols has proved useful in generating chiral centres with high enantiomeric selectivity (Fig. 5.55).

The catalyst in the reaction is titanium, which is introduced as titanium tetra-isopropoxide. The epoxidizing agent is *tert*-butyl hydroperoxide, while (+)-diethyl L-tartrate (DET) introduces asymmetry into the reaction. DET acts as a bidentate ligand and displaces two of the isopropoxide ligands from titanium while the allylic alcohol and *tert*-butylhydroperoxide displace the remaining two isopropoxide groups (Fig. 5.56). The order in

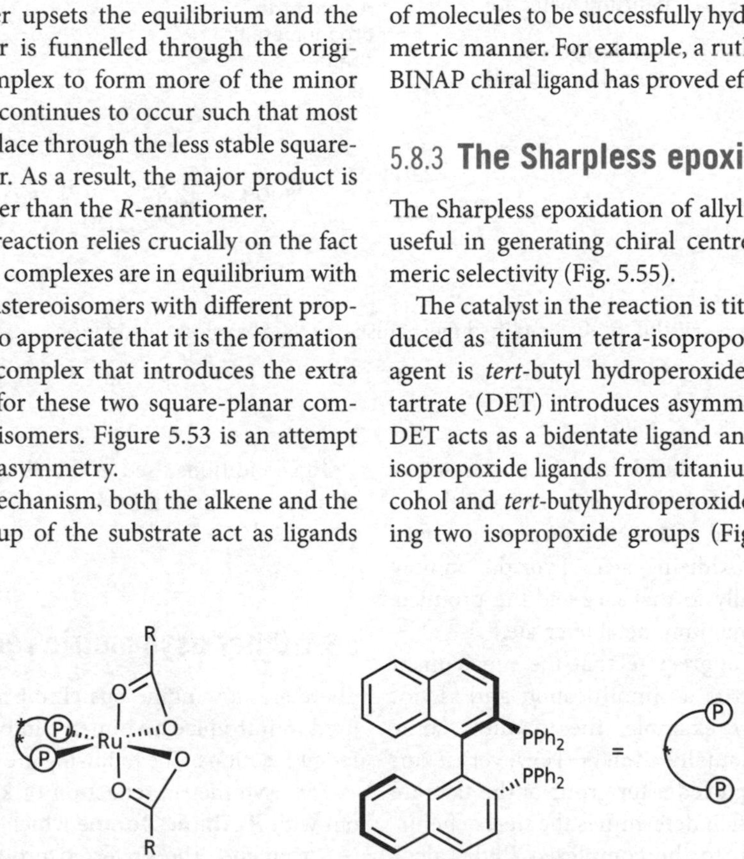

Ruthenium catalyst

BINAP (chiral ligand)

FIGURE 5.54 Ruthenium catalyst with a chiral bidentate ligand.

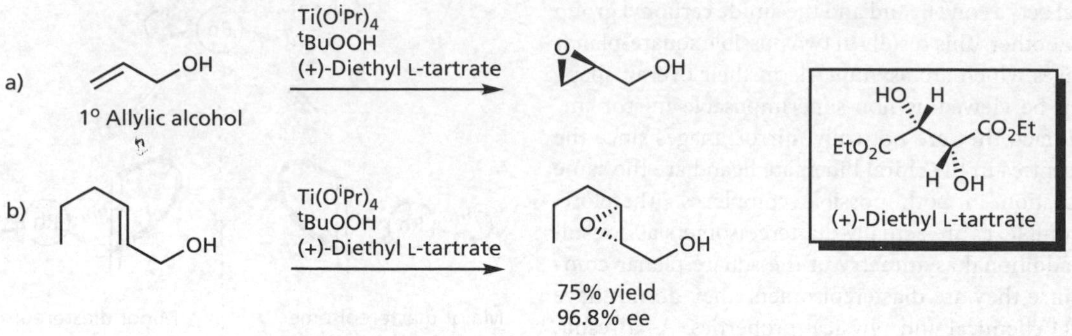

FIGURE 5.55 Examples of the Sharpless epoxidation.

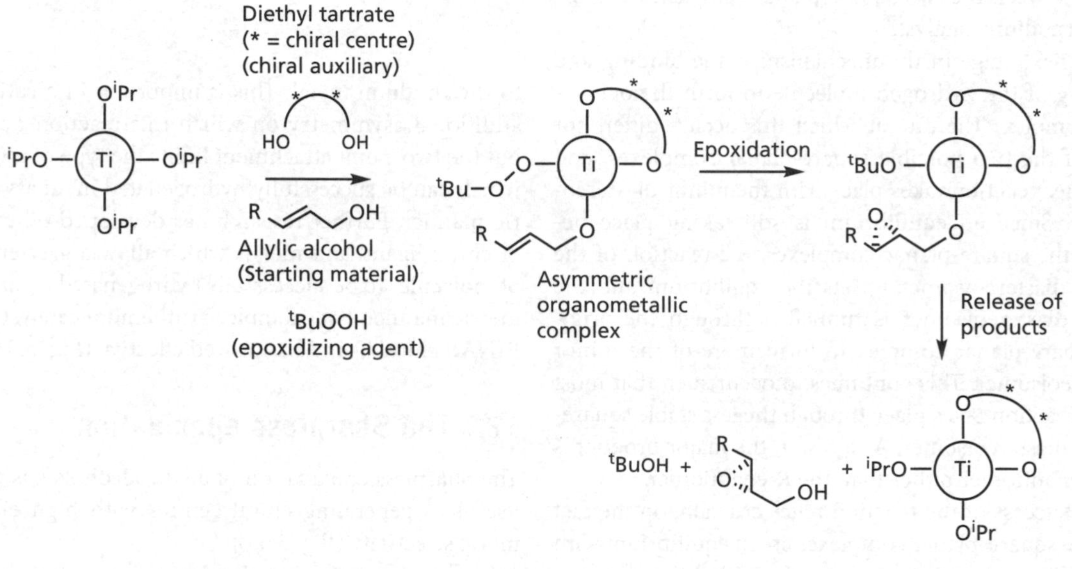

FIGURE 5.56 Simplified mechanism for the Sharpless epoxidation.

which these substitutions take place is not important. A chiral organometallic complex is now formed and the allylic alcohol is positioned such that it presents one face of the alkene to the epoxidizing agent. Epoxidation now takes place preferentially to that face and the products then depart from the titanium metal template.

It is important to appreciate that the mechanism shown in Figure 5.56 is a simplification and is not fully understood. For example, the titanium complexes involved are actually dimers. However, it appears that the uncomplexed ester group of the tartrate ligand is the group which determines the stereochemical delivery of oxygen to the complexed allylic alcohol. Simple alkenes lacking the allylic group are also epoxidized by these reaction conditions, but with no enantioselectivity.

The conditions used in the Sharpless epoxidation can also be used to oxidize sulphides to chiral sulphoxides (Box 5.8).

5.8.4 Other asymmetric reducing agents

There are now numerous chiral reagents which can be used to introduce asymmetry into a wide variety of different reactions. The following are a few examples.

The asymmetric reduction of ketones can be carried out with **R-Alpine-Borane** which can be prepared from (+)-α-pinene. The preferred product is determined by the relative sizes of the substituents on the ketone as shown in Figure 5.57. This reaction is particularly enantioselective for alkynyl ketones.

BOX 5.8 Synthesis of omeprazole and esomeprazole

Omeprazole is a proton pump inhibitor which is synthesized by linking the two halves of the molecule through a nucleophilic substitution reaction (Fig. 1). The benzimidazole half of the molecule has a thiol substituent which is treated with sodium hydroxide to give a thiolate. On reaction with the chloromethylpyridine, the thiolate group displaces the chloride ion to link up the two halves of the molecule. Subsequent oxidation of the sulphur atom with *meta*-chloroperbenzoic acid gives omeprazole.

As described in section 5.2, omeprazole is actually a chiral molecule since the sulphur atom is an asymmetric centre. The synthesis involving mcpba shows no enantioselectivity and so omeprazole is a racemate of the two possible enantiomers. However, the route can be modified to carry out the asymmetric synthesis of esomeprazole (the *S*-enantiomer of omeprazole) by employing asymmetric conditions for the final sulphoxidation step. Early attempts to carry out this reaction involved the Sharpless reagent formed from Ti(O-^{i}Pr)$_4$, the oxidizing agent cumene hydroperoxide PhC(Me)$_2$OOH, and the chiral auxiliary (*S*,*S*)-diethyl tartrate. Although sulphoxidation took place, it required almost stoichiometric quantities of the titanium reagent and

there was little enantioselectivity. The reaction conditions were modified in three ways to improve the selectivity to over 94% enantiomeric excess, with less titanium reagent (4–30 mol%).

- Formation of the titanium complex was carried out in the presence of the sulphide starting material.

- The solution of the titanium complex was equilibrated at an elevated temperature for a prolonged time period.

- The oxidation was carried out in the presence of an amine such as *N*,*N*-diisopropylethylamine (DIPEA). The role of the amine is not fully understood, but it may participate in the titanium complex.

The enantiomeric excess can be enhanced further by preparing a metal salt of the crude product and carrying out a crystallization which boosts the enantiomeric excess to more than 99.5%.

A similar process was used to synthesize **dexlansoprazole**—the *R*-enantiomer of the proton pump inhibitor lansoprazole (Fig. 2). In order to obtain the *R*-enantiomer, (*R*)-(+)-DET was used as the chiral auxiliary instead of (*S*)-(−)-DET. Dexlansoprazole has also been marketed as a proton pump inhibitor.

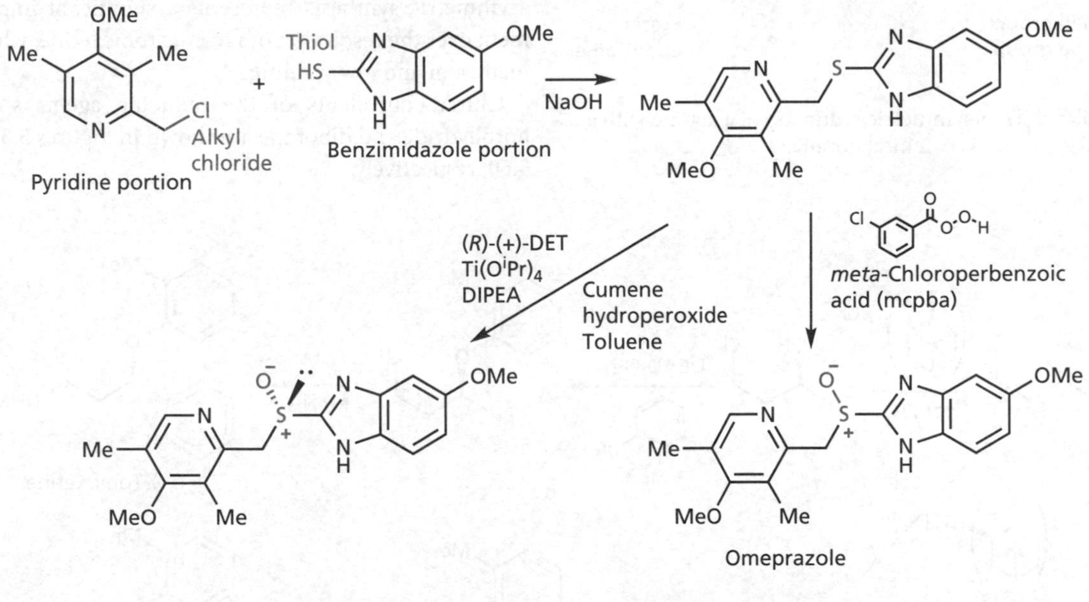

FIGURE 1 Synthesis of omeprazole and esomeprazole.

(Continued)

BOX 5.8 Synthesis of omeprazole and esomeprazole *(Continued)*

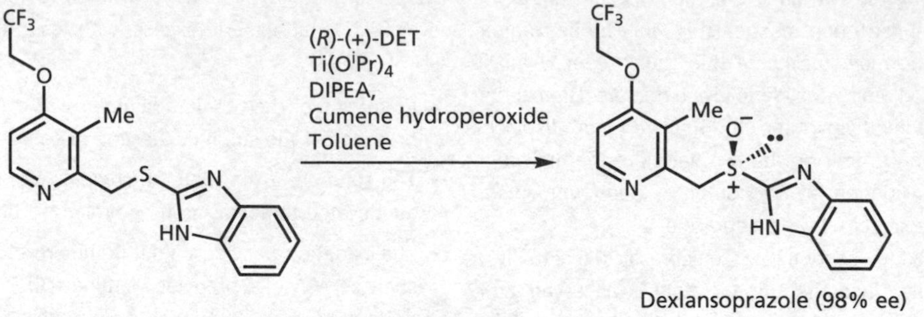

FIGURE 2 Asymmetric synthesis of dexlansoprazole.

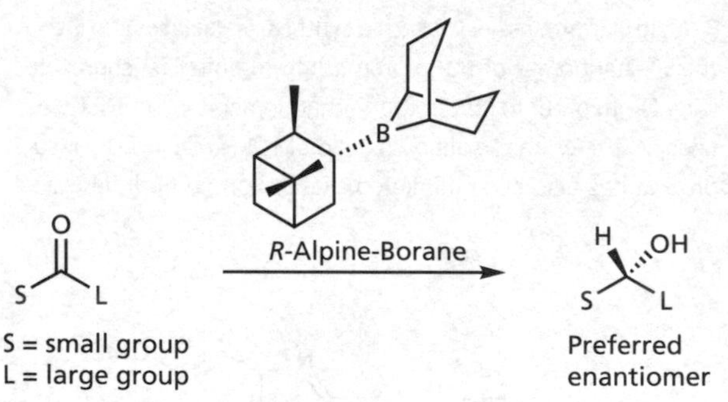

FIGURE 5.57 Asymmetric reduction of a ketone with a chiral borane.

The asymmetric reduction of aryl ketones can be carried out with good enantioselectivity using either enantiomer of **diisopinocampheylchloroborane** (^{d}Ipc$_2$BCl or ^{l}Ipc$_2$BCl). For example, the first stage of a synthesis leading to both enantiomers of the antidepressant **tomoxetine** involved an asymmetric reduction of 3-chloropropiophenone with ^{d}Ipc$_2$BCl or ^{l}Ipc$_2$BCl (Fig. 5.58). The asymmetric synthesis represents a significant improvement over the resolution of racemic tomexetine, which is inefficient and low yielding.

Chiral equivalents of the reducing agents sodium borohydride and diborane are shown in Figures 5.59 and 5.60, respectively.

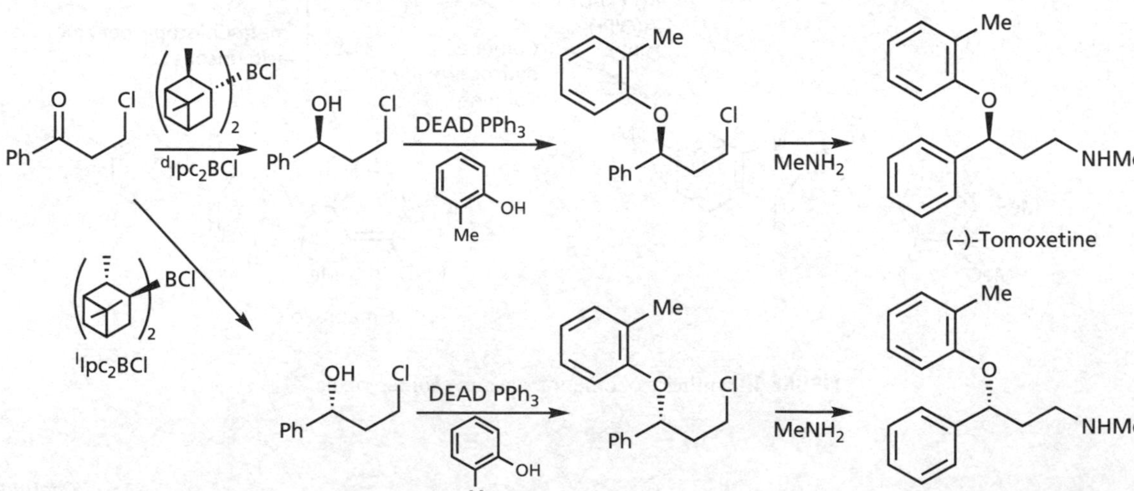

FIGURE 5.58 Reduction of 3-chloropropiophenone as a key stage in the synthesis of enantiomers of tomoxetine (DEAD = diethyl azodicarboxylate).

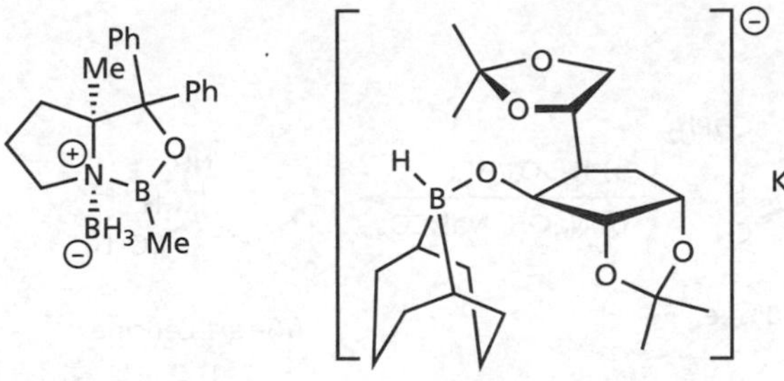

FIGURE 5.60 Chiral equivalents of diborane.

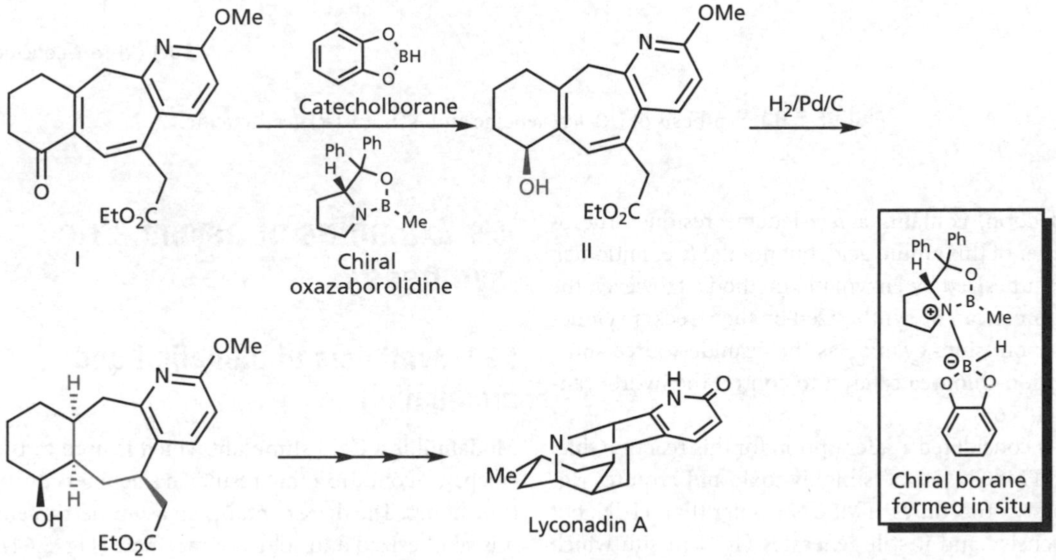

FIGURE 5.59 Chiral equivalents of sodium borohydride.

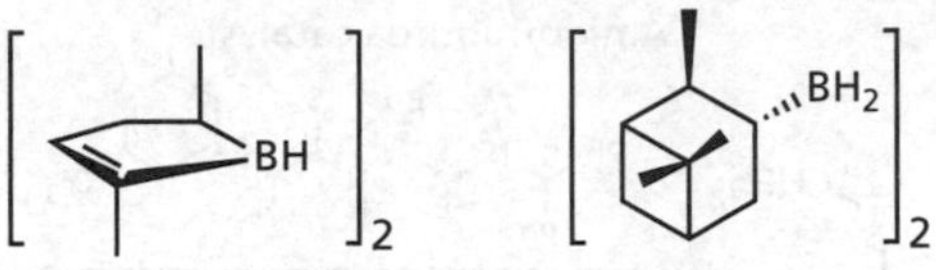

FIGURE 5.61 Use of the Corey–Bikashi–Shibata reaction.

An example of the use of a chiral borane can be seen in the synthesis of the natural product **lyconadin A**, which is of interest in the treatment of Alzheimer's disease (Fig. 5.61). The ketone group in the tricyclic structure (I) is reduced enantioselectively with a catecholborane in the presence of a chiral oxazaborolidine in 85% yield. The mechanism involves the formation in situ of the chiral borane structure shown in the box (Fig. 5.61) and is known as the **Corey–Bikashi–Shibata reaction**. The presence of the chiral alcohol in structure II is important in controlling the diastereoselectivity of the hydrogenation reaction that followed.

5.8.5 The asymmetric Strecker synthesis

The asymmetric rhodium-catalysed hydrogenation described in section 5.8.2 is an effective enantioselective method of synthesizing amino acids. However, this method is not possible if the target amino acid has an aryl or quaternary alkyl substituent at the α-position (Fig. 5.62).

An alternative approach is to carry out an asymmetric Strecker synthesis in the presence of a catalyst. This allows the synthesis of unnatural amino acids such as *tert*-leucine, which contains a *tert*-butyl side chain and has been an important constituent in a number of clinically important drugs. For example, the protease inhibitor **atazanavir**, which is used in the treatment of

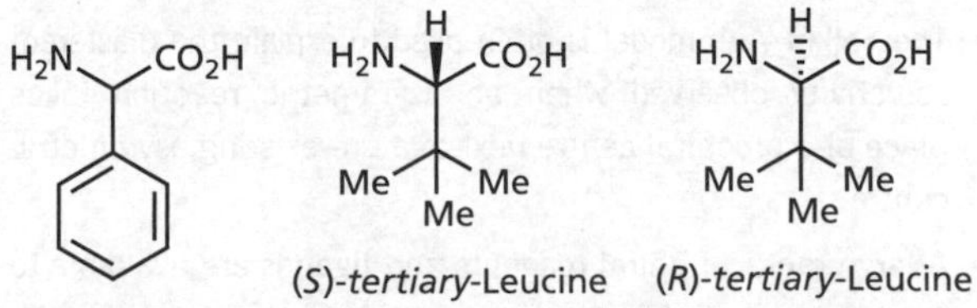

(*S*)-*tertiary*-Leucine (*R*)-*tertiary*-Leucine

FIGURE 5.62 Examples of amino acids that cannot be synthesized by rhodium-catalysed hydrogenation.

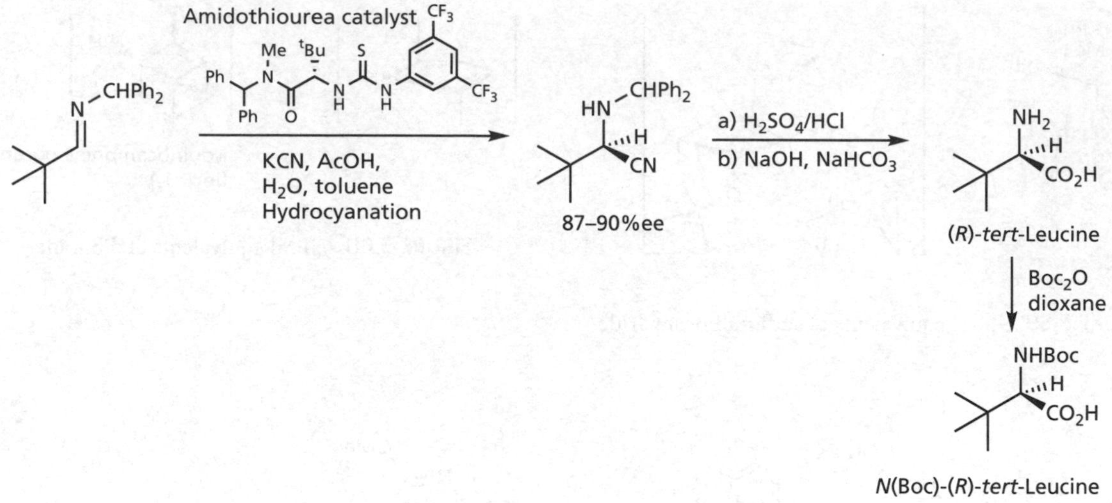

FIGURE 5.63 Synthesis of (R)-*tert*-leucine and N(Boc)-(R)-*tert*-leucine.

HIV infection, contains a *tert*-leucine residue. The S-enantiomer of this amino acid, but not the R-enantiomer, can be synthesized by enzymatic methods. However, the R-enantiomer can be synthesized by the Strecker synthesis using potassium cyanide as the cyanide source and a chiral amido-thiourea catalyst to control the hydrocyanation (Fig. 5.63).

KCN is considered a safer option for this reaction than HCN or TMSCN. HCN is highly toxic and requires rigorous safety procedures. TMSCN is safer than HCN, but it is expensive and it still generates HCN in situ which limits the extent to which this reaction can be scaled up. Unfortunately, KCN is not particularly soluble in organic solvents and the catalyst is not particularly soluble in aqueous solution. Therefore, a biphasic system is used for the reaction. The mechanism involves a complex which is formed between the catalyst, the cyanide ion, and the protonated iminium ion.

KEY POINTS

- An asymmetric reaction may be possible using a chiral starting material if the chiral centre already present is close enough to the reaction centre to affect the stereochemical outcome. Alternatively, the overall bulk and shape of a chiral molecule may affect the stereochemical outcome.

- The Felkin–Anh model is often used to explain the diastereoselectivity observed when an asymmetric reaction takes place at a prochiral centre next to a pre-existing asymmetric centre.

- A large range of chiral reagents and ligands are available to carry out asymmetric reactions on achiral starting materials.

- Many asymmetric reactions involve the formation of a chiral metal complex, which acts as the catalyst for the reaction. Chirality is introduced in the form of chiral bidentate ligands.

5.9 Examples of asymmetric syntheses

5.9.1 Synthesis of modafinil and armodafinil

Modafinil is a CNS stimulant which is used to treat narcolepsy, a condition that results in sleepiness during daytime hours. The drug contains an asymmetric centre, but it is synthesized and sold as a racemate (Fig. 5.64).

Benzhydrol is the starting material for the synthesis and the first stage is a functional group transformation of a secondary alcohol to a secondary alkyl bromide. The bromide ion is a better leaving group than the hydroxide ion, and so the next stage is set up for an S_N2 nucleophilic substitution reaction with thiourea. This produces an S-alkylisothiouronium salt which is then hydrolysed with aqueous base to give the thiol. The thiol is alkylated with 2-chloroethanoic acid, and the resulting thioether is oxidized to the sulphoxide with hydrogen peroxide. The carboxylic acid is now converted to a more reactive acid chloride and then treated with ammonia to provide the primary amide in the final structure.

Armodafinil (Nuvigil) is the R-enantiomer of modafinil, and was approved in 2007 for the treatment of excessive sleepiness. It has a slower metabolism and excretion than the racemate, resulting in a longer duration of action. Initially, the compound was obtained by resolving modafinil through a continuous chiral separation method known as **VARICOL** using a chiral polysaccharide stationary phase (section 5.3.2). However, this was costly and so a crystallization method was developed using modafinic acid, which is one of the synthetic intermediates in Figure 5.64. Modafinic acid

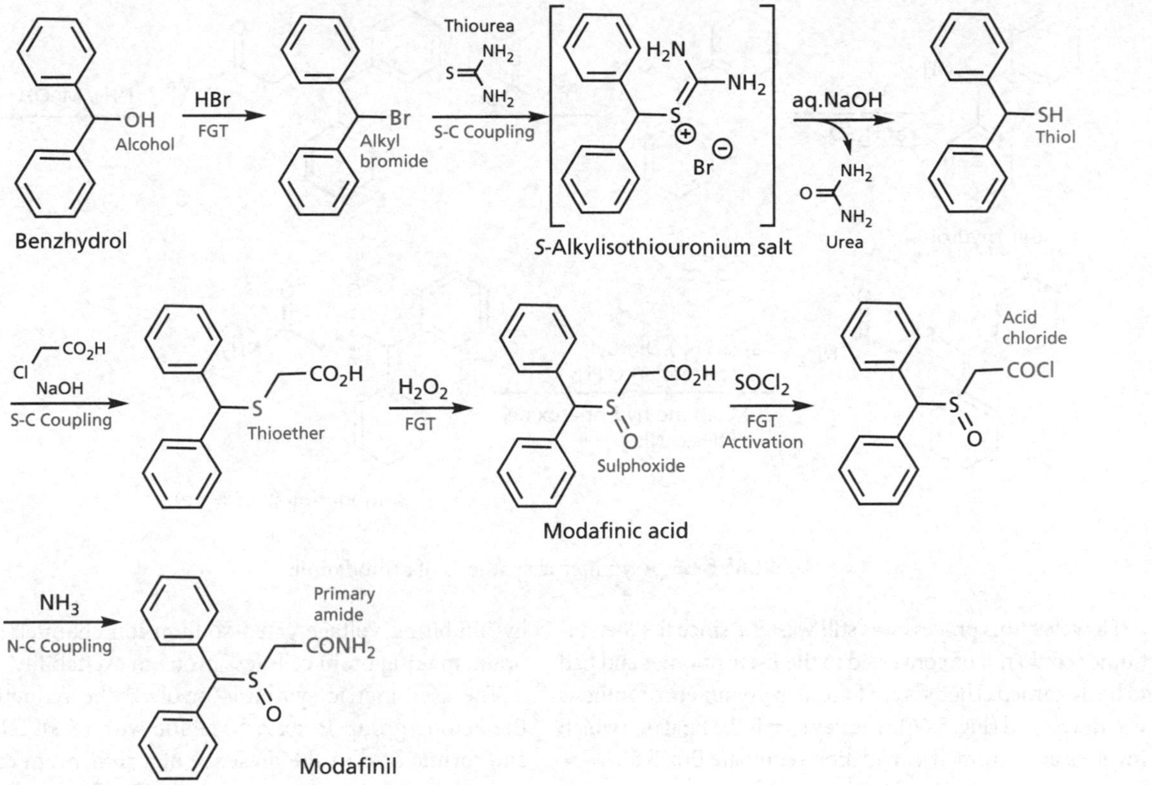

FIGURE 5.64 Synthesis of modafinil.

was found to exist as a eutectic mixture or conglomerate which could be seeded with the *R*-enantiomer to provide the pure enantiomer. Additional racemate was then added to the mother liquor which was now rich in the *S*-enantiomer. Seeding with the *S*-enantiomer then produced a crop of the pure *S*-enantiomer. The process was repeated for up to 35 cycles to gain alternate crops of the *R*- and *S*-enantiomers (see also section 5.3.1). The *R*-enantiomer was then converted to armodafinal (Fig. 5.65).

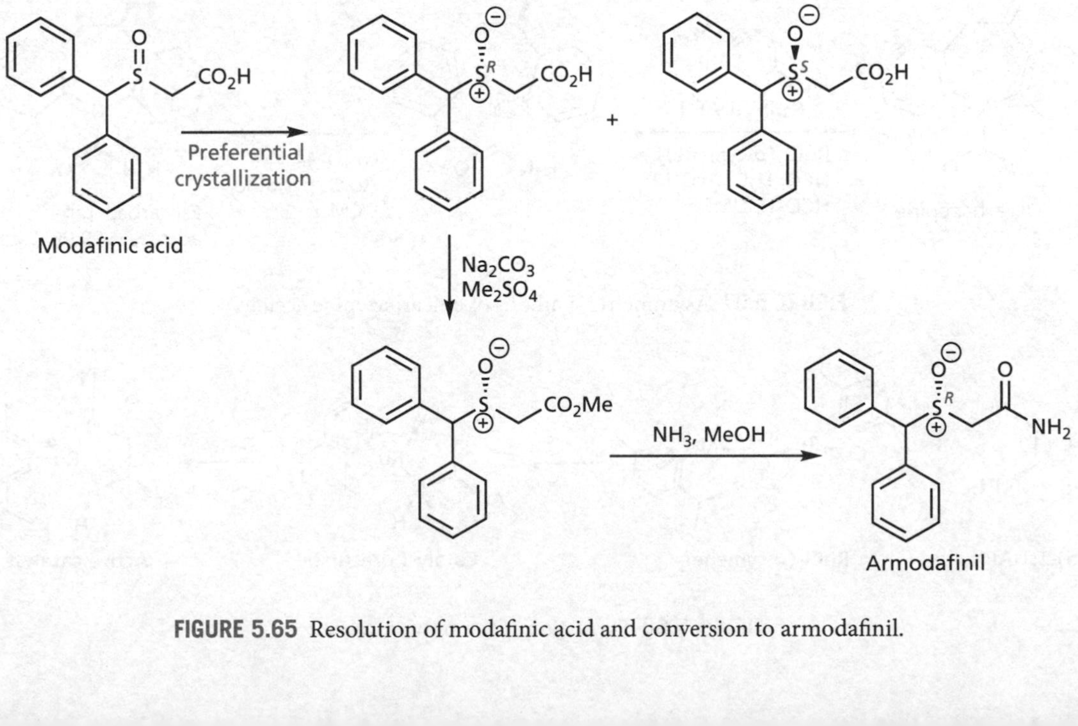

FIGURE 5.65 Resolution of modafinic acid and conversion to armodafinil.

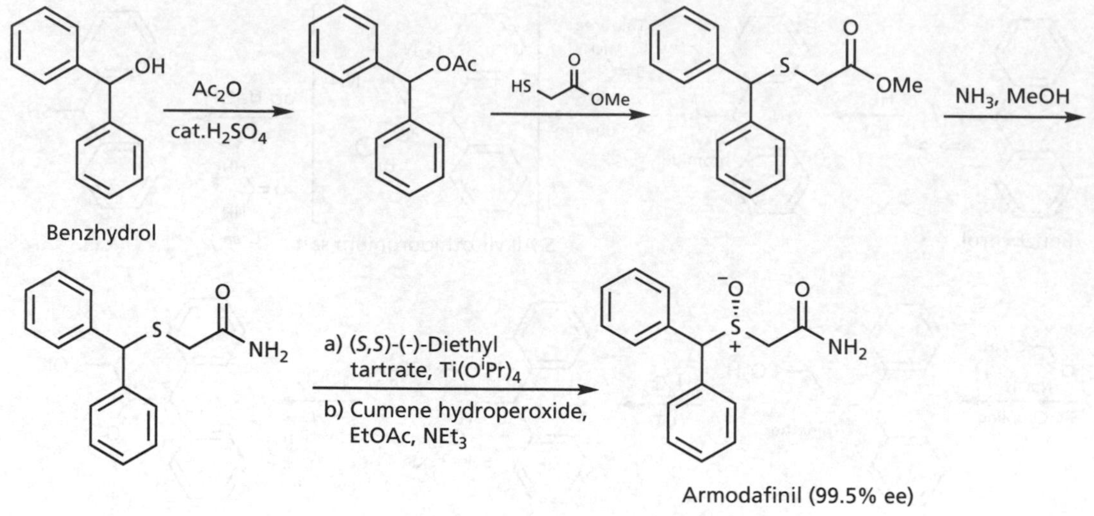

FIGURE 5.66 Asymmetric synthesis of armodafinil.

However, this process was still wasteful since the S-enantiomer could not be converted to the R-enantiomer and had to be discarded. Therefore, a four-step asymmetric synthesis was developed (Fig. 5.66). The key step is the final one which involves an asymmetric oxidation (compare Box 5.8).

5.9.2 Asymmetric synthesis of eslicarbazepine acetate

Eslicarbazepine acetate is a prodrug of **eslicarbazepine**—a metabolite of the anti-epileptic drug **oxcarbazepine**—and is used in the treatment of seizures. It acts

by inhibiting voltage-gated sodium ion channels in the brain, making brain cells less prone to excitability.

The asymmetric synthesis involved the reduction of the ketone group in oxcarbazepine with (S,S)-TsDAEN and formic acid in the presence of a ruthenium catalyst and a mixed solvent system (Fig. 5.67). The resulting alcohol was then esterified to give eslicarbazepine acetate.

The N-tosylated diamine (S,S-TsDAEN) is the chiral auxiliary in the asymmetric reduction and forms a complex with the ruthenium metal to form a catalyst precursor, where the diamine acts as a bidentate ligand (Fig. 5.68). The active catalyst is then formed under the

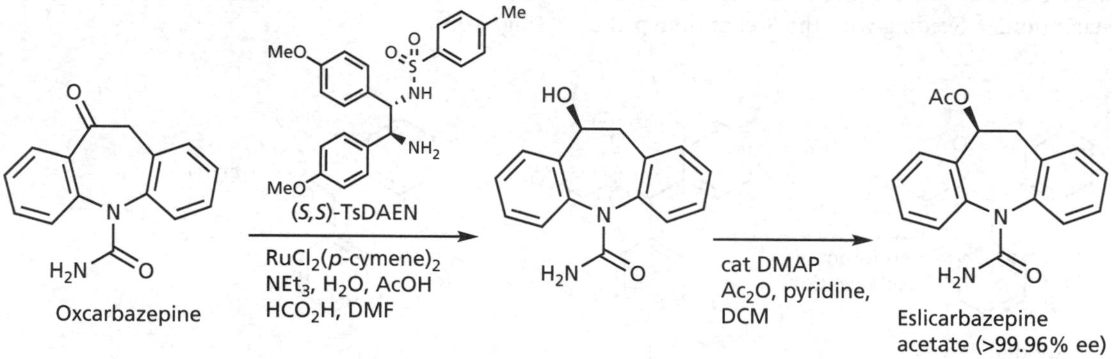

FIGURE 5.67 Asymmetric synthesis of eslicarbazepine acetate.

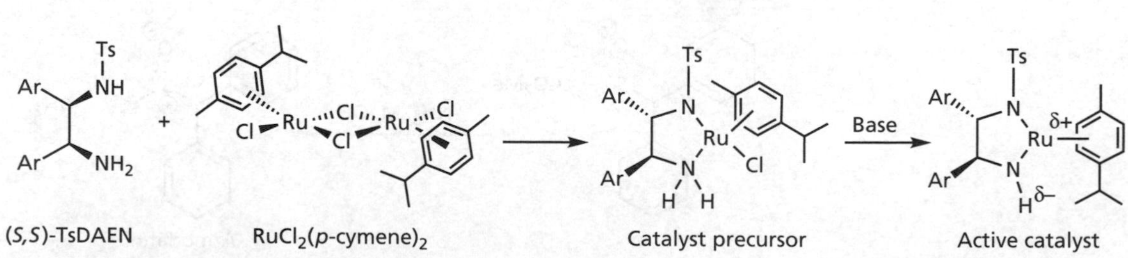

FIGURE 5.68 Formation of the active catalyst.

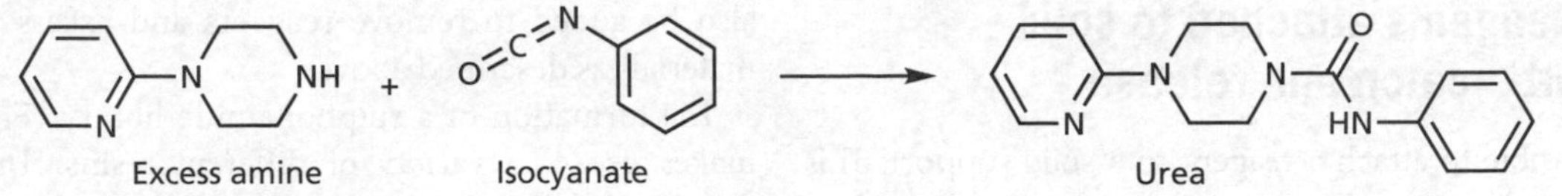

FIGURE 6.14 Reaction of an isocyanate with excess amine to produce a urea.

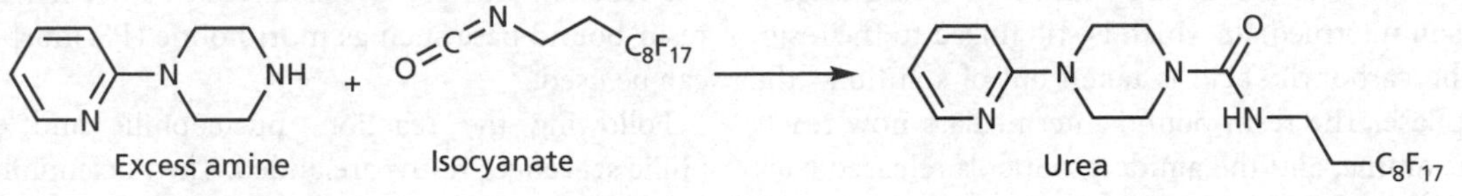

FIGURE 6.15 Removal of excess amine by reaction with a fluorinated isocyanate followed by F-SPE.

linked to alkyl chains containing a large number of fluoro substituents. The highly fluorinated silica has a high affinity for fluorinated molecules and can be used to separate fluorinated compounds from non-fluorinated compounds. For example, consider the reaction shown in Figure 6.14 where an isocyanate is treated with an amine to give a urea product. The amine is used in excess in order to drive the reaction to completion, but the amine left over has to be removed. In order to do this, a fluorinated isocyanate is added which reacts with the excess amine to produce a fluorinated urea (Fig. 6.15). The crude solution is passed through an F-SPE column which acts as a scavenger resin to retain the highly fluorinated urea and allow the desired unfluorinated urea product to pass through.

Sometimes an aqueous work-up cannot be avoided. For example, an aqueous work-up is required following a Grignard reaction, which means that the aqueous and organic phase have to be separated. Fortunately, there are efficient methods of carrying out such separations in parallel.

One such method is to use a **lollipop phase separator**. A pin is inserted into a mixture of the two phases, and the mixture is rapidly cooled in a dry ice–acetone bath at –78°C. The aqueous phase freezes onto the pin to form a 'lollipop'. The pin and its lollipop can then be removed from the reaction vial, leaving the organic phase behind. Up to 96 such separations can be performed in parallel with specially designed units.

Another method is to use **phase separation columns**, which can be used to separate a dense chlorinated organic layer from an aqueous phase. The lower organic layer passes through a hydrophobic frit by gravity, whereas the upper aqueous layer is retained on the frit. It is important not to apply pressure, otherwise the aqueous phase may be forced through the frit as well.

6.4.3 The use of resins in solution phase organic synthesis (SPOS)

By carrying out a parallel synthesis in solution, it is easy to monitor the reaction by ^{1}H NMR spectroscopy, thin-layer chromatography, or HPLC. Work-up procedures can be greatly simplified by the use of a variety of resins. Since resins are solid-supported, there is little interaction between different types, allowing a variety of resins to be used in the same reaction. Thus, it is common to have a reaction cocktail which includes nucleophilic and electrophilic resins or acidic and basic resins without any problems arising.

Reactions are carried out such that one of the reagents—usually the cheaper and more readily available—is used in excess in order to drive the reaction to completion (Fig. 6.16). The crude reaction mixture will comprise the product AB and excess starting material A. The crude mixture is treated with a solid-supported scavenger resin that is capable of reacting with the excess reagent (A). As a result, excess reagent becomes attached to the resin and can be removed by filtering the resin. Removal of the solvent then leaves the pure product (AB).

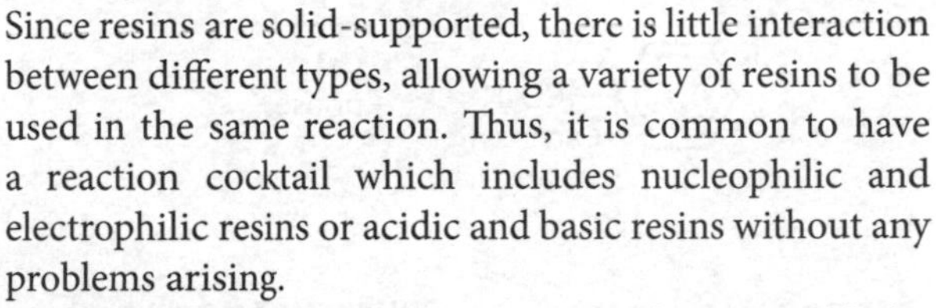

FIGURE 6.16 The use of scavenger resins in solution phase organic synthesis.

6.4.4 **Reagents attached to solid support—catch and release**

It is possible to attach a reagent to a solid support. This has the advantage that the reagent or its by-product can be easily removed at the end of the reaction. For example, the coupling agent used for amide synthesis can be attached to a resin instead of being present in solution (Fig. 6.17). The reaction involves a carboxylic acid starting material reacting with the coupling reagent to form an intermediate which is still linked to the resin. Thus, the carboxylic acid is taken out of solution—the 'catch' phase. The resin-bound intermediate now reacts with the amine, and the amide product is released back into solution. The urea by-product which is formed remains bound to the resin and is easily removed when the resin is finally filtered. Acidic and basic resins can

also be added to remove reagents and excess starting materials as described above.

The formation of a sulphonamide library (Fig. 6.18) makes use of a variety of different resins. The reaction involves an amine being treated with an excess of a sulphonyl chloride. A basic catalyst is required for the reaction, and triethylamine is normally used in a conventional synthesis. However, this is quite a smelly volatile compound and would have to be removed once the reaction was complete. Instead of triethylamine, a resin-bound base such as morpholine (PS-morpholine) can be used.

Following the reaction, nucleophilic and electrophilic scavenger resins are added. The nucleophilic resin PS-trisamine reacts with excess sulphonyl chloride to remove it from solution, while the electrophilic resin PS-isocyanate removes unreacted amine (Fig. 6.19)

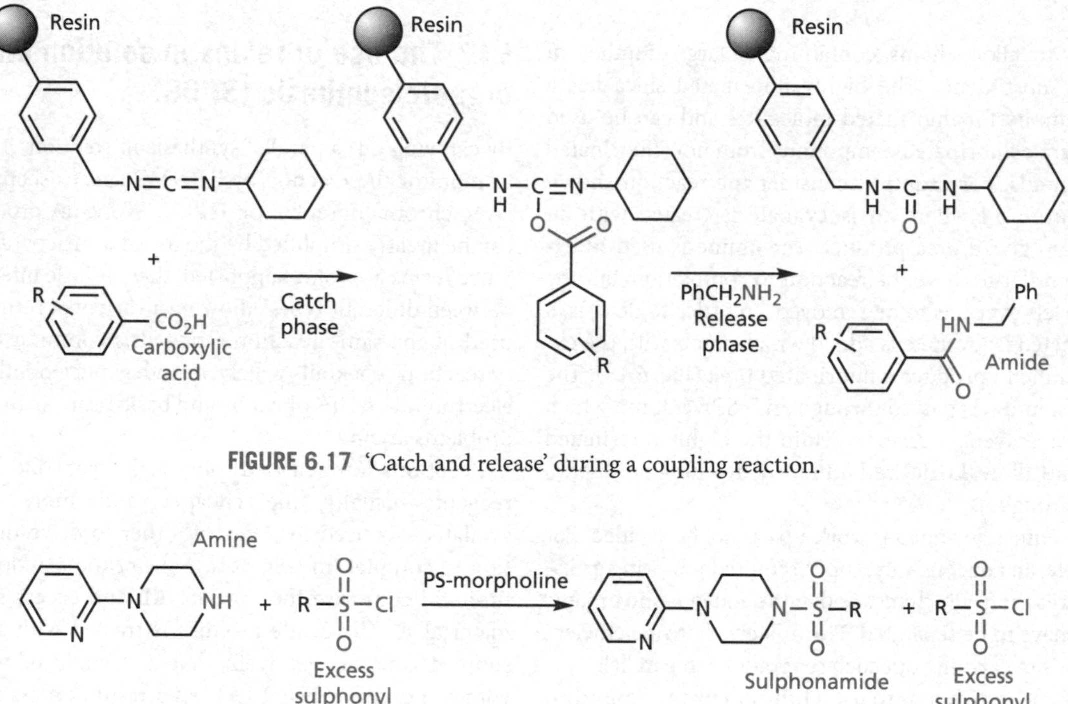

FIGURE 6.17 'Catch and release' during a coupling reaction.

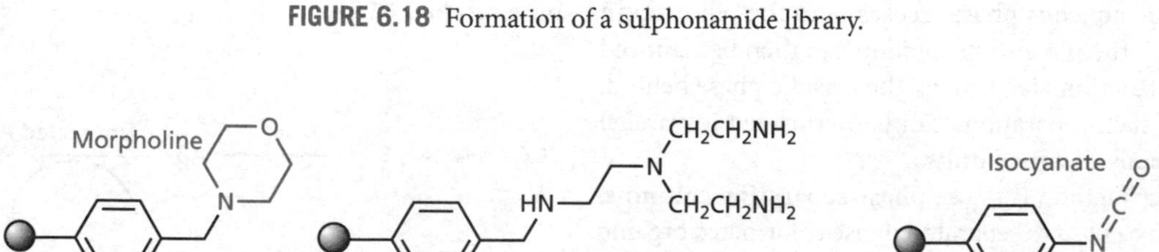

FIGURE 6.18 Formation of a sulphonamide library.

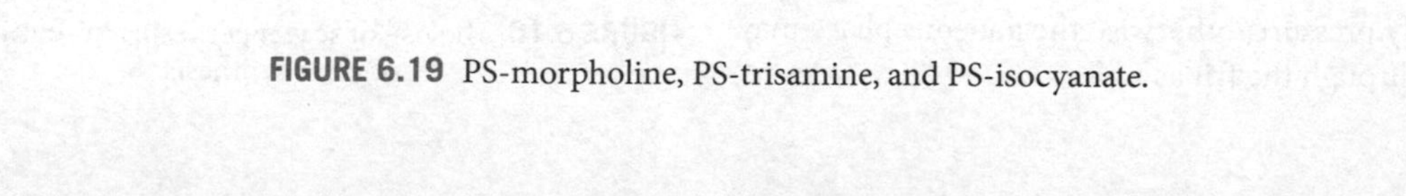

FIGURE 6.19 PS-morpholine, PS-trisamine, and PS-isocyanate.

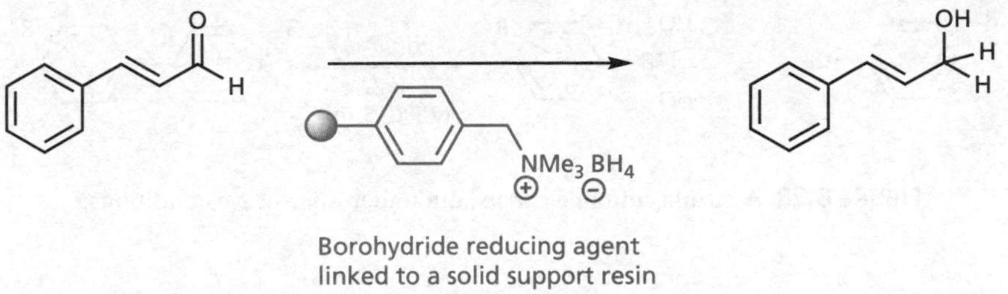

FIGURE 6.20 Reduction of an aldehyde with a solid-supported borohydride reagent.

Filtration to remove the resins leaves the pure sulphon-amide in solution.

Solid-supported reagents can be used in a variety of very common synthetic reactions. For example, a solid-supported borohydride can be used to reduce carbonyl groups (Fig. 6.20). In some reactions, it is also possible to reduce the toxicity and odour of reagents and their by-products. For example, the normal Swern oxidation involves the formation of dimethyl sulphide as a by-product—a compound which has a pungent cabbage odour! This is avoided by using a solid-supported reagent instead (Fig. 6.21).

6.4.5 Microwave technology

Drug discovery is a very expensive process and **microwave-assisted organic synthesis** (MAOS) is proving to be a very useful tool for accelerating syntheses and making the process more efficient. There are many examples of thermal reactions that take several hours to complete using heaters or oil baths, but which are carried out in minutes using microwave conditions. There is a much greater efficiency of energy transfer using microwave technology, which accounts for the faster reaction times. Moreover, reaction times are shortened by the additional pressure obtained when using sealed reaction vessels. In addition, yields can sometimes be dramatically improved with less decomposition and side reactions. Specially de-signed microwave units are now commonly employed in library syntheses (Fig. 6.13). Examples of reactions that have been carried out using microwave technology include the formation of amides from acids and amines without the need for coupling agents (Fig. 6.22), metal-catalysed Suzuki couplings which can be performed even on usually unreactive aryl chlorides (Fig. 6.23), and metal-mediated reductions and aminations (Fig. 6.24). The reduction shown in Figure 6.24a took 24 hours using conventional heating, and only 15 minutes using microwave heating.

6.4.6 Microfluidics in parallel synthesis

The science of **microfluidics** involves the manipulation of tiny volumes of liquids in a confined space. Companies are devising microreactors that can be used to carry out parallel syntheses on microchips (Fig. 6.25) using a continuous flow of reactants in microfluidic channels.

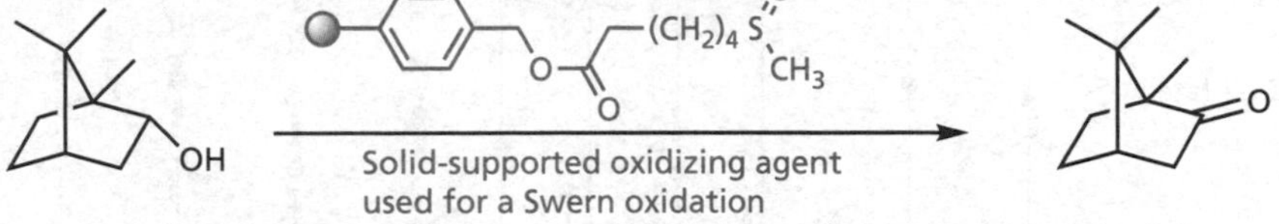

FIGURE 6.21 Swern oxidation using a solid-supported reagent.

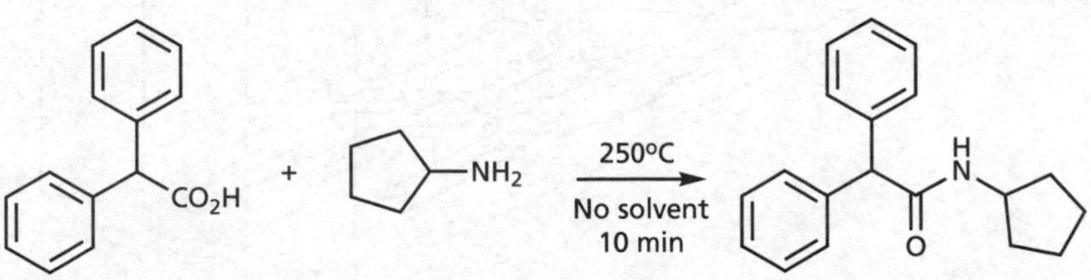

FIGURE 6.22 Amide formation using microwave technology.

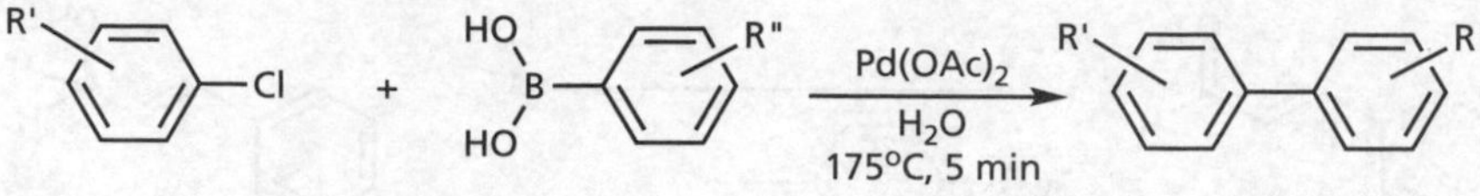

FIGURE 6.23 A Suzuki coupling carried out under microwave conditions.

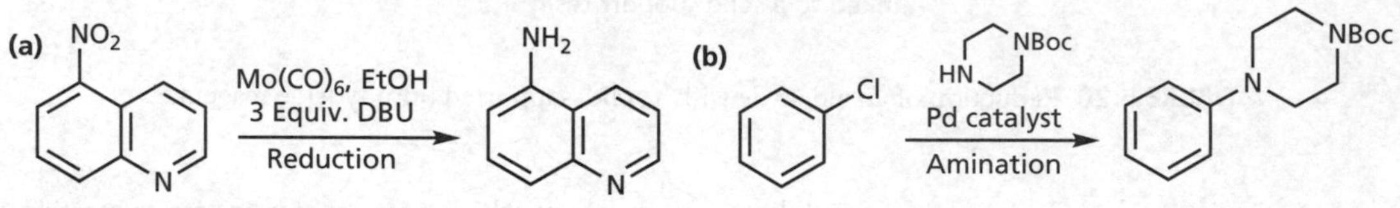

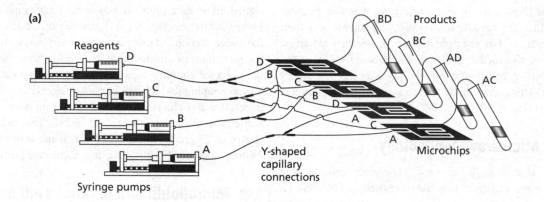

FIGURE 6.24 Microwave-assisted transition-metal-mediated reactions: (a) reductions and (b) aminations.

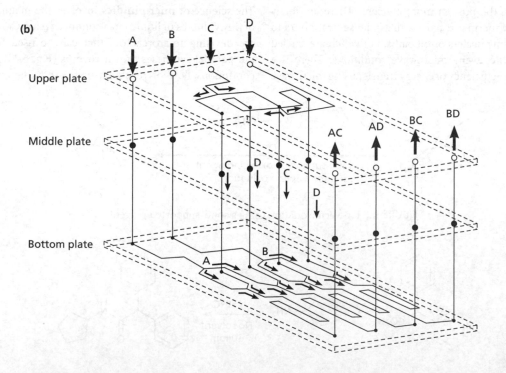

FIGURE 6.25 Parallel synthesis on a microchip. Parallel synthesis of four products using (a) four separate 2D microchips and (b) a 3D microchip. (Reprinted by permission of Macmillan Publishers Ltd, *Nature Reviews Drug Discovery*, **5**, 210–18 (2006).

The channels are designed such that various reactants are mixed and reacted as they flow through the microchip. Several reactions have already been successfully carried out on microscale, and it is found that many reaction times are shortened from hours to minutes. Some reactions take place in higher yield and with less side products. It is also possible to control the temperature of each reaction extremely accurately. Another advantage of microreactors is the potential to handle a vast number of parallel reactions on microchips. The channels through each chip can be fabricated to allow all possible mixing combinations of the various reactants, either on separate microchips or on a 3D microchip. The example in Fig 6.25 is a simple illustration of how a microreactor system could be set up to create a mini-library from the reaction of A or B with C or D.

A disadvantage with this system is that separate microchannels are required for each reaction, which inevitably requires increasingly large numbers of microchannels as one increases the number of reactions. Therefore, further research has looked at developing a system where reactions can be carried out one after another through the same microchannel. Microvalves can be used to control which solutions enter the microchannel at any one time, and micropumps allow controlled volumes of solutions to be pushed through in an automated fashion. The microchannels have also been designed to allow efficient mixing of the reagents as they pass through.

For example, a microreactor has been designed that can carry out 1024 reactions one after the other (Fig. 6.26). A huge advantage of carrying out these reactions on a microfluidic device is the fact that much smaller quantities of reagents are required for these studies compared with conventional parallel synthesis. Other advantages include automation, faster results, and more sensitive methods of identifying hits.

As an illustration of this efficiency, a series of 1,3-dipolar cycloadditions were carried out between different azides and alkynes to produce a library of triazoles (Fig. 6.27). The time required for each reaction was 17 seconds using a reaction volume of about 400 nl containing 20 nl (120 pmol) of both the acetylene and azide. A miniature reverse phase clean-up stage was included to remove polar reagents (DMSO and PBS salts) and the product was analysed offline by electrospray ionization mass spectrometry to identify hits. The time taken for each analysis was 15 seconds per reaction.

The chip consisted of the following components (Fig. 6.28).

a) A pair of multiplexers

The multiplexers controlled which reagent was used in each reaction. There was one multiplexer for the azoles and one multiplexer for the alkynes. Each multiplexer had a series of 16 microchannels, allowing the possibil-

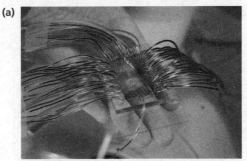
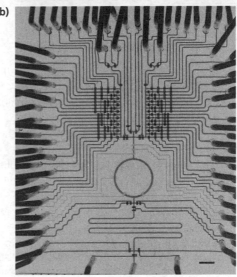

FIGURE 6.26 (a) Lab on a chip; (b) microchannel circuitry of the chip.

Part (a) from *Chemistry World*, August 2009 issue, p 29. Part (b) reproduced from Wang et al (2009) 'An integrated microfluidic device for large-scale in situ click chemistry screening' Lab on a Chip Vol. 9, Iss. 16. 2281-2285. Both RSC materials.

FIGURE 6.27 1,3-Dipolar cycloaddition.

ity of using 16 different azides and 16 different alkynes. Microvalves determined which microchannel was open at any particular time and controlled which reagent was added for each reaction.

b) A rotary mixer

For each reaction, the azide and alkyne were added from the multiplexers to a circular 150 nl microchannel and then circulated round the circuit in order to get efficient mixing of the reagents.

c) A serpentine channel

After the reagents were mixed in the rotary mixer, they were fed into a 250 nl serpentine microchannel—so

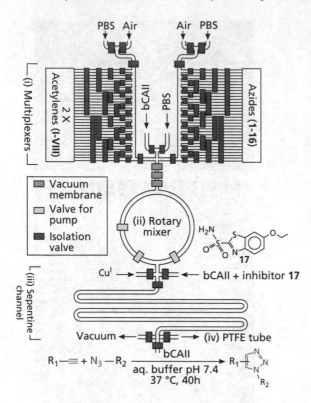

FIGURE 6.28 Components in the microchip.
Reproduced from Wang et al. 'Integrated Microfluidics for
Parallel Screening of an In Situ Click Chemistry Library'
Angewandte Chemie International Edition. Vol. 45, Iss. 32 © 2006
WILEY-VCH Verlag GmbH & Co. KGaA, Weinheim

called because it winds back and forward towards an
exit leading to a PTFE tube. Additional solvent was
added to the rotary mixer to ensure that all the contents
were washed out into the serpentine channel to make a
total volume of 400 nl. Further mixing took place as the
solution moved through the serpentine channel.

d) A PTFE tube

Once the solution had passed through the serpentine
channel, it was pushed into a detachable PTFE tube as a
400 nl 'slug' of the reaction mixture.

After each reaction, air and buffer solutions were
passed though the rotary mixer and serpentine channel to prevent cross-contamination between the different reactions. The air also served to separate each 'slug'
in the PTFE tube. Once eight reaction mixture slugs had
been collected in the detachable PTFE tube, the tube was
manually removed and replaced with a new PTFE tube to
collect the next eight reaction mixtures.

In this way, it was possible to carry out 1024 reactions in about 290 minutes and store them in 128 separate PTFE tubes. The tubes were incubated for 40 hours
at 37°C to complete the reactions, and then the slugs
were expelled with water into 200 µl centrifuge tubes.

A miniaturized solid phase extraction procedure was
carried out to separate the products from DMSO and
PBS salts, and then the products were analysed by mass
spectrometry.

KEY POINTS

- In parallel synthesis, a reaction or series of reactions is carried out in a series of wells to produce a range of analogues.
  Each reaction well contains a single product.

- Parallel synthesis can be carried out on solid phase or in
  solution.

- Parallel synthesis allows the synthesis of a large number
  of easily identifiable analogues which can be quickly and
  easily tested to speed up the optimization process.

- Solid phase extraction is often used in parallel synthesis for
  work-up. It involves the use of columns to remove impurities
  and excess reagents.

- An aqueous phase can be separated from an organic phase
  using phase separation columns or by freezing the aqueous
  phase onto a solid surface.

- Catch and release strategies involve reagents which are
  linked to a solid support. Reactants are taken out of solution when they react with the reagent, and are then released
  when a subsequent reaction takes place.

- Solid-supported reagents are easily removed at the end of
  a reaction. The potential toxicity of the reagent or its by-
  product is reduced when attached to a solid support.

- Microwave technology can prove advantageous over conventional heating.

- Microreactors have been designed that allow reactions to be
  carried out speedily on a very small scale.

6.5 **Combinatorial synthesis**

6.5.1 **Introduction**

In combinatorial synthesis, mixtures of compounds are
deliberately produced in each reaction vessel, allowing
chemists to produce thousands or even millions of novel
structures in the time that they would take to synthesize a few dozen by conventional means. This method
of synthesis goes against the grain of conventional organic synthesis, where chemists set out to produce a
single identifiable structure which can be purified and
characterized. The structures in each reaction vessel of
a combinatorial synthesis are not separated and purified,
but are tested for biological activity as a whole. If there is
no activity, then there is no need to continue studies on
that mixture and it is stored. If activity *is* observed, one
or more components in the mixture are active, although
false positives can sometimes be an issue. Overall, there

is an economy of effort because a negative result for a mixture of 100 compounds saves the effort of synthesizing, purifying, and identifying each component of that mixture. On the other hand, identifying the active component of an active mixture is not straightforward.

In a sense, combinatorial synthesis can be looked upon as the synthetic equivalent of nature's chemical pool. Through evolution, nature has produced a huge number and variety of chemical structures, some of which are biologically active. Traditional medicinal chemistry dips into that pool to pick out the **active principles** and develop them. Combinatorial synthesis produces pools of purely synthetic structures that we can dip into for active compounds. The diversity of structures from the natural pool is far greater than that likely to be achieved by combinatorial synthesis, but isolating, purifying, and identifying new agents from natural sources is a relatively slow process and there is no guarantee that a lead compound will be discovered against a specific drug target. The advantage of combinatorial chemistry is that it produces new compounds faster than those derived from natural sources, and can produce a diversity not found in the traditional banks of synthetic compounds held by pharmaceutical companies.

A few words of caution are necessary here with regard to negative assays. There is always the possibility that a combinatorial mixture does not contain all the structures expected. This can happen if some of the starting materials or intermediates in the synthesis do not react as expected. A negative assay would then lead to the conclusion that these compounds are inactive, even though they are not present. This could mean that an active compound is missed. Assays might also be affected adversely if the individual components of a mixture interact with each other or have conflicting activities.

6.5.2 The mix and split method in combinatorial synthesis

A combinatorial synthesis is designed to produce a mixture of products in each reaction vessel, starting with a wide range of starting materials and reagents. This does not mean that all possible starting materials are thrown together in the one reaction flask. If this was done, a black tarry mess would result. Instead, molecular structures are synthesized on solid supports such as beads. Each individual bead may contain a large number of molecules, but all the molecules on that bead are identical—'the one-bead–one-compound concept'. Different beads have different structures attached and can be mixed together in a single vial such that the molecules attached to the beads undergo the same reaction. In this way, each vial contains a mixture of structures, but each structure is physically distinct from the others since it is attached to a different bead.

Planning has to go into designing a combinatorial synthesis to minimize the effort involved and to maximize the number of different structures obtained. The strategy of **mix and split** is a crucial part of this. As an example, suppose that we wish to synthesize all the possible dipeptides of five different amino acids. Using orthodox chemistry, we would synthesize these one at a time. There are 25 possible dipeptides, and so we would have to carry out 25 separate syntheses(Fig. 6.29).

By using a mix and split strategy, the same products can be obtained with far less effort (Fig. 6.30). First, the beads are split between five reaction vials. The first amino acid is attached to the beads, using a different amino acid for each vial. The beads from all five flasks are collected, mixed together, and then split back into the five vials. This means that each vial now has the same mixture. The second amino acid is now coupled, with a different amino acid being used for each vial. Each vial now contains five different dipeptides, with no single vial containing the same dipeptide as another. Each of the five mixtures can now be tested for activity. If the results are positive, the emphasis is on identifying which of the dipeptides is active. If there is no activity present, the mixture can be ignored.

In studies such as these, one can generate large numbers of mixtures, many of which are inactive. However, these mixtures are not discarded. Although they may not contain a lead compound on this particular occasion, they may provide the necessary lead compound for a different target in medicinal chemistry. Therefore, all the mixtures (both active and inactive) resulting from a combinatorial synthesis are stored as **compound libraries.**

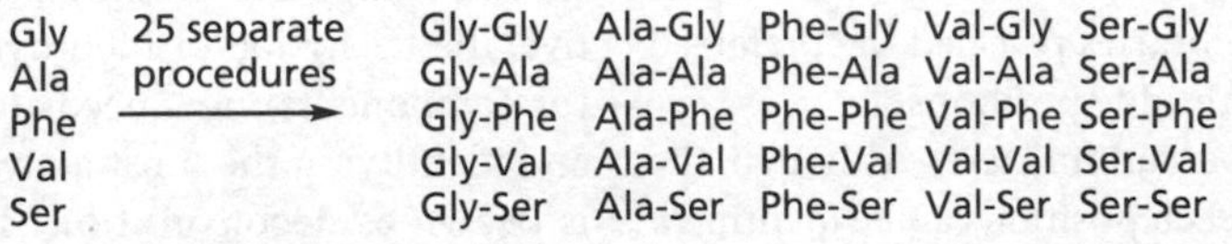

Gly	25 separate	Gly-Gly	Ala-Gly	Phe-Gly	Val-Gly	Ser-Gly
Ala	procedures	Gly-Ala	Ala-Ala	Phe-Ala	Val-Ala	Ser-Ala
Phe	→	Gly-Phe	Ala-Phe	Phe-Phe	Val-Phe	Ser-Phe
Val		Gly-Val	Ala-Val	Phe-Val	Val-Val	Ser-Val
Ser		Gly-Ser	Ala-Ser	Phe-Ser	Val-Ser	Ser-Ser

FIGURE 6.29 The possible dipeptides that can be synthesized from five different amino acids. Each procedure involves protection, coupling, and deprotection stages.

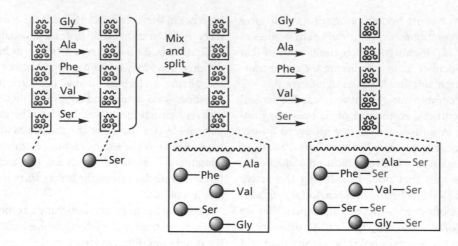

FIGURE 6.30 Synthesis of five different dipeptides using the mix and split strategy. Note that the addition of each amino acid involves protection, coupling, and deprotection steps.

The example above produced 25 compounds in five mixtures. However, combinatorial synthesis can be used to produce several thousand compounds.

In this example, amino acids were linked together, but any monomer unit or combination of chemical structures could be linked together using the same strategy, and so the process is not limited to peptide synthesis.

6.5.3 Mix and split in the production of positional scanning libraries

A variation of the mix and split method allows the creation of **positional scanning libraries**. In this method, the same library compounds are prepared several times and in each library a different residue in the sequence is held constant. For example, a series of hexapeptide libraries totalling 34 million compounds was produced, with each library consisting of six sets of mixtures and each mixture containing 1 889 568 peptides. In each of the mixtures one of the amino acid positions was held constant.

The first library consisted of six mixtures where the first amino acid was constant *within* each individual mixture but was different *between* mixtures. The most active mixture was then identified, and since the amino acid at position 1 was constant in that mixture it could also be identified. The second library consisted of a series of six mixtures where the second amino acid was constant within each mixture but different between the mixtures. Thus, it was known which amino acid was present at position 2 in the most active of these mixtures. Testing all the mixtures in all the libraries revealed the preferred residues at each of the six residue positions.

However, one has to be careful here. Although the most active amino acid at each position can be identified, it does not mean that the most active structure is the one

linking each of these amino acids. After all, the activity being measured is the combined effect of several different hexapeptide structures in the active mixtures. Therefore, it is quite possible that the most active hexapeptide is in a mixture which has a lower overall activity than another mixture, since the latter might contain a large number of compounds with moderate activity.

One way around this is to identify the top three to four amino acids at each position rather than just the most active one. Once these have been identified, all the possible variations of these amino acids could be synthesized. For example, several libraries of a hexapeptide were prepared in a search for structures which would bind to the mu-opiate receptor. The most active amino acids at each position were found to be Tyr (Y), Gly (G), Phe (F), Phe (F), Leu (L), and Arg (R). Linking these together gave a hexapeptide (YGFFLR) which was only weakly active. However, the results showed that mixtures containing Gly or Phe at position 3, Phe, Tyr, Met (M), or Leu at position 5, and Phe, Tyr, or Arg at position 6 were also active.

Therefore, all 24 hexapeptides having the sequence Y,G,G/F,F,F/M/L,F/Y/R were made, with the most active being YGGFMY. It is interesting to note that the endogenous analgesic Met-enkephalin is the pentapeptide YGGFM.

6.5.4 Isolating and identifying the active component in a mixture

Once a compound mixture proves to be biologically active, the tricky job of identifying the active component (or components) now needs to be carried out. Isolating and identifying the most active compound in a mixture is known as **deconvolution**. There are several methods of doing this.

6.5.4.1 Micromanipulation

Each bead in a mixture contains only one type of structural product. Therefore, the individual beads can be separated, and the product can be cleaved from the bead and then tested. This procedure can be aided by colorimetric analysis where products are tested for activity when they are still bound to the beads. The active beads are distinguished by a colour reaction and can then be 'picked out' by micromanipulation.

6.5.4.2 Recursive deconvolution

Micromanipulation is tedious and has serious drawbacks when handling large quantities of beads. A method known as recursive deconvolution can be useful in cutting down the amount of work involved and can be illustrated by considering a library of tripeptides produced from three different amino acids (Gly, Val, and Ala) by the 'mix and split' approach as described in the following stages.

Stage 1

Each amino acid is linked to a solid support (Fig. 6.31). (To clarify the diagrams, each sphere represents a resin bead with suitable linker unit.)

Stage 2

The beads are mixed together and separated into three equal mixtures (Fig. 6.32).

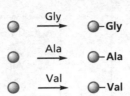

FIGURE 6.31 Linking the first amino acid to the solid support. Note that protecting groups have not been specified.

Stage 3

Each mixture is reacted with a different amino acid (Fig. 6.33)

All possible nine dipeptides have now been synthesized in three separate mixtures. Samples of each portion are retained at this stage for **recursive deconvolution**.

Stage 4

All the beads are collected and mixed together, and then split into three equal mixtures. Each mixture will now contain all nine possible dipeptides (Fig. 6.34).

Stage 5

Each mixture is reacted with one of the three amino acids (Fig. 6.35). All 27 possible tripeptides have now been synthesized in three mixtures.

Having prepared three mixtures containing the 27 tripeptides, let us assume that one of the three final

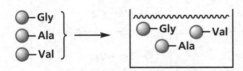
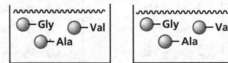

FIGURE 6.32 The mix and split stage.

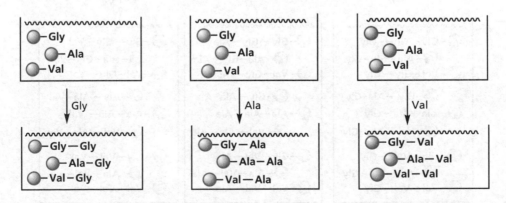

FIGURE 6.33 Coupling of the next amino acid. Note that the addition of each amino acid involves protection, coupling, and deprotection steps.

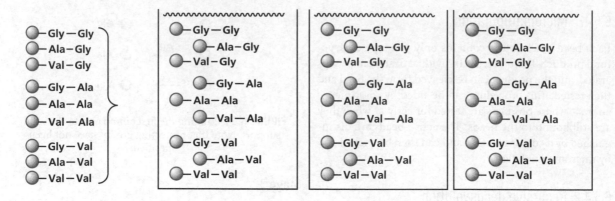

FIGURE 6.34 The second mix and split stage.

mixtures shows activity. How do we find out which of the nine possible tripeptides in that mixture is the active component? We could synthesize all nine possible tripeptides separately and test each one. However, there is no need to do this if we have retained samples of the mixtures produced during earlier mix and split phases.

Let us assume that the third tripeptide mixture from stage 5 shows activity (Fig. 6.35). This means that the

active tripeptide has valine at the *N*-terminus. The next stage is to take the three dipeptide mixtures which were retained from stage 3 and link valine to each mixture (Fig. 6.36). This gives us the nine tripeptides we need in three separate mixtures. Note that the second and third amino acids are now the same in each mixture.

We now test these mixtures for activity. If one of these mixtures is active, then the three component tripeptides can be individually synthesized and tested.

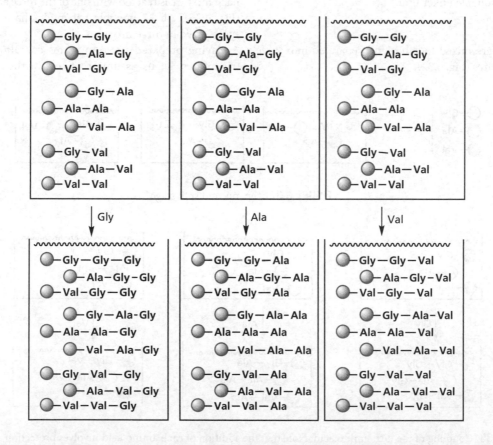

FIGURE 6.35 Synthesis of 27 tripeptides. Note that the addition of each amino acid involves protection, coupling, and deprotection steps.

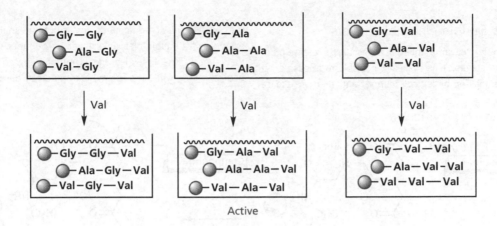

FIGURE 6.36 Recursive deconvolution.

In this example, we looked at tripeptides involving three different amino acids, but typical combinatorial syntheses involve a larger variety of monomeric units, and the larger the number of monomers, the larger the number of compounds produced. For example, a series of 34 million hexapeptides was synthesized by this method using 18 amino acids and 324 mixtures.

Appropriate use of the mix and split method, and the retention of intermediate mixtures for recursive deconvolution, is crucial in economizing the effort involved in identifying active components.

6.5.4.3 Sequential release

Linkers have been devised which allow release of a certain percentage of the product from a bead (Box 6.1). The process can be repeated, releasing the product sequentially rather than all at once. Therefore, a mixture of beads can be treated to release some of the bound product for testing. If the mixture is active, the same beads are split into smaller mixtures and further product is released and tested (Fig. 6.37). This process can be repeated several times until the active bead is identified (Fig. 6.38).

KEY POINTS

- Most combinatorial syntheses are carried out using automated or semi-automated synthesizers.

- The mix and split method allows the efficient synthesis of large numbers of compounds with a minimum number of operations.

- The compounds synthesized in a combinatorial synthesis are stored as combinatorial libraries.

- Positional scanning libraries involve the synthesis of sets of libraries where a component is kept constant in each subset of each library. This allows the identification of components which are good for activity.

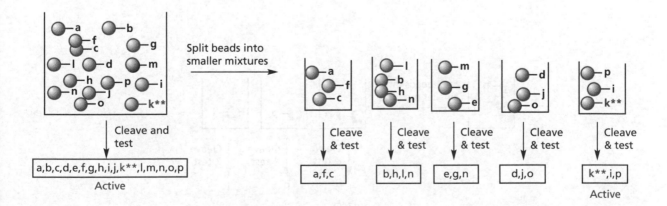

FIGURE 6.37 Sequential release of a product from a bead.

BOX 6.1 Double-cleavable linkers

With a double-cleavable linker, it is possible to release products in two stages. This makes it possible to release a set of compounds for biological testing, and then to release the second set of compounds for structural analysis. An example of a double-cleavable linker for peptides is shown in Figure 1. The first cleavage is initiated by addition of neutral buffer, while the second is initiated by base.

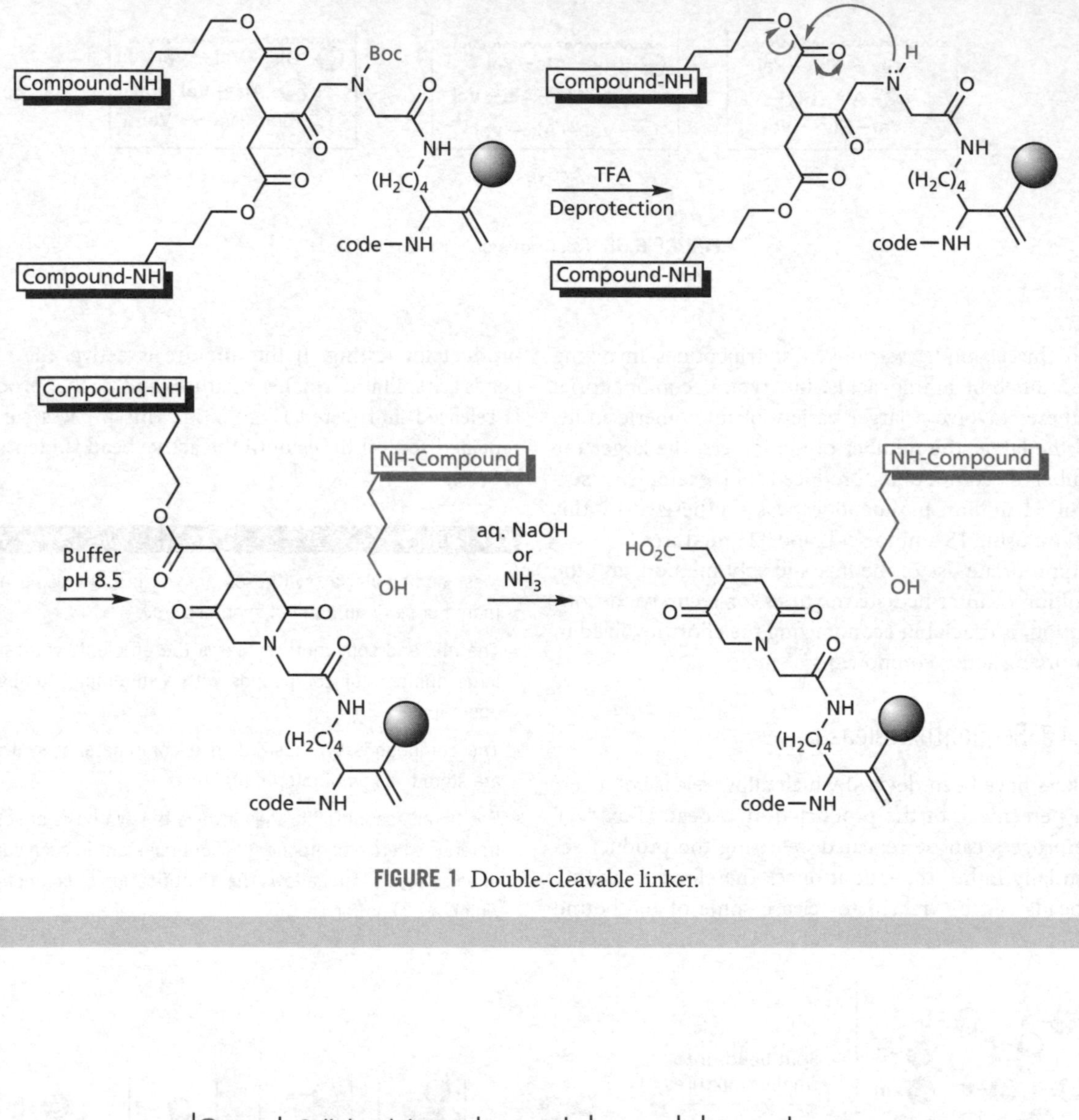

FIGURE 1 Double-cleavable linker.

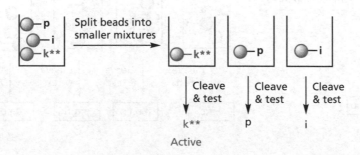

FIGURE 6.38 Identification of active bead.

- Deconvolution is the process by which the active component in a combinatorial synthetic can be identified.

- Recursive deconvolution involves the use of product mixtures obtained at different stages of a mix and split combinatorial synthesis in order to narrow down the likely structure of the active component in a final mixture.

- Linkers are available which permit the sequential release of a product bound to a particular resin.

6.5.5 Structure determination of the active compound(s)

The direct structural determination of components in a compound mixture is no easy task, but advances have been made in obtaining interpretable mass, NMR, Raman, infrared, and ultraviolet spectra on products attached to a single resin bead. Peptides can be sequenced while still attached to the bead. Each 100 μm bead contains about 100 pmole of peptide, which is enough for microsequencing. With non-peptides, the structural determination of an active compound can be achieved by deconvolution methods. Alternatively, **tagging** procedures can be used during the synthesis.

6.5.5.1 Tagging

In this process, two molecules are built up on the same bead. One of these is the intended structure; the other is a molecular tag (usually a peptide or oligonucleotide) which will act as a code for each step of the synthesis. For this to work, the bead must have a multiple linker capable of linking both the target structure and the molecular tag. A starting material is added to one part of the linker, and an encoding amino acid or nucleotide to another part. After each subsequent stage of the combinatorial synthesis, an amino acid or nucleotide is added to the growing tag to indicate what reagent was used. One example of a multiple linker is called the **Safety CAtch Acid-Labile Linker (SCAL)** (Fig. 6.39), which includes lysine and tryptophan. Both these amino acids have a free amino group.

The target structure is constructed on the amino group of the tryptophan moiety, and after each stage of the synthesis a tagging amino acid is built onto the amino group of the lysine moiety. Figure 6.40 illustrates the procedure for a synthesis involving three reagents, so that by the end of the process there is a tripeptide tag where each amino acid defines the identity of the variable groups R, R′, and R″ in the target structure.

The non-peptide target structure can be cleaved by reducing the two sulphoxide groups in the SCAL and then treating with acid. Under these conditions, the tripeptide tagging sequence remains attached to the bead. This can now be sequenced on the bead to identify which reagents were used in the reaction sequence and thus identify the structure of the final compound.

The same strategy can be used with an oligonucleotide as the tagging molecule. The oligonucleotide can be amplified by replication and the code read by DNA sequencing.

There are drawbacks to tagging processes. They are time consuming and require elaborate instrumentation. Building the coding structure itself also adds extra restraints on the protection strategies that can be employed and may impose limitations on the reactions that are used. In the case of oligonucleotides, their inherent instability may prove a problem. Another problem with tagging is the possibility of an unexpected reaction taking place, resulting in a different structure from that expected. Nevertheless, the tagging procedure is still valid since it identifies the reagents and the reaction conditions, and when these are repeated on a larger scale any unusual reactions would be discovered.

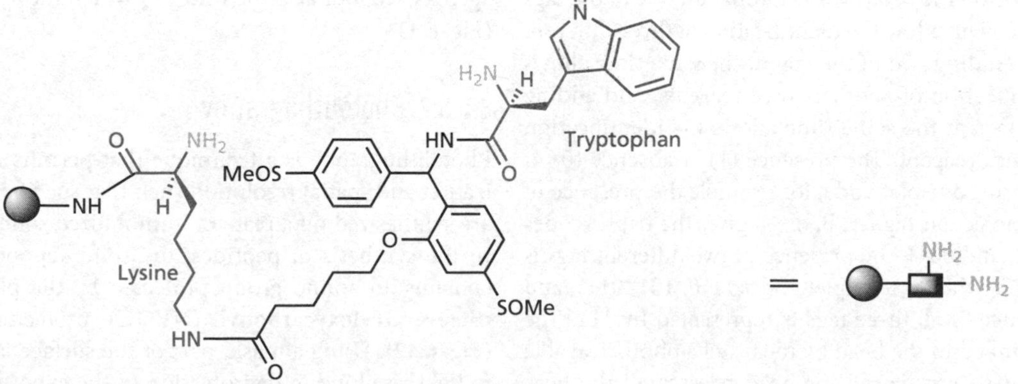

FIGURE 6.39 SCAL (Safety CAtch Acid-Labile Linker).

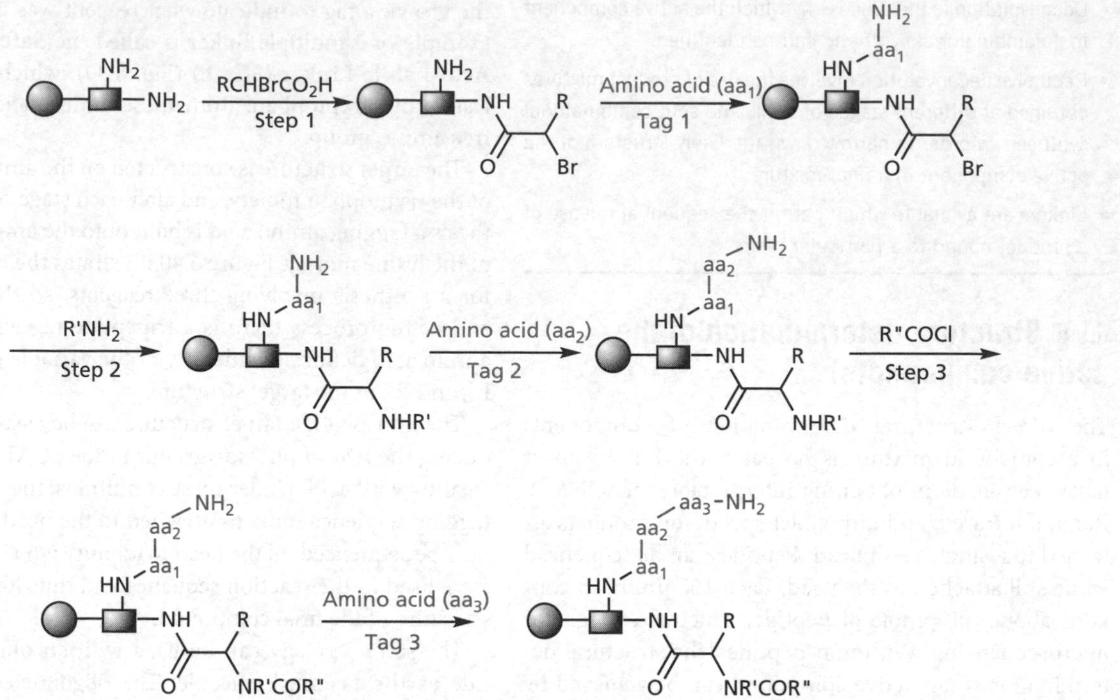

FIGURE 6.40 Tagging a bead to identify the structure being synthesized. Note that the reaction sequence has been simplified here to illustrate the principle of tagging. Amino acids (aa) are *N*-protected when coupled and the protecting group is removed before the next coupling. Coupling agents are also present. An orthogonal protection strategy is also required to distinguish between the amino groups of the SCAL.

These tagging methods require the use of a specific molecular tag to represent each reagent used in the synthesis. Moreover, the resultant molecular tag has to be sequenced at the end of the synthesis. A more efficient method of tagging and identifying the final product is to use some form of encryption or '**bar code**'. For example, it is possible to identify which of seven possible reagents has been used in the first stage of a synthesis with the use of only three molecular labels (A–C). This is achieved by adding different combinations of the three tags to set up a triplet code on the bead. Thus, adding just one of the tags (A, B, or C) will allow the identification of three different reagents. Adding two of the tags at the same time allows the identification of another three reagents, and adding all three tags at the same time allows the identification of a seventh reagent. The presence (1) or absence (0) of the tag forms a triplet code; for example the presence of a single molecular tag (A, B, or C) gives the triplet codes (100, 010, and 001), the presence of two different tags is indicated by the three triplet codes (110, 101, 011), and the presence of all three tags is represented by 111. The tags are linked to the bead by means of a photocleavable bond, and so irradiating the bead releases all the tags. These can then be passed through a gas chromatograph and identified by their retention time.

Three different molecular tags could now be used to represent seven reagents in the second stage of the synthetic sequence, and so on. All the tags used to represent the second reagent would have longer retention times than the tags used to represent the first reagent. Similarly, all subsequent tags would have longer retention times. Once the synthesis is complete, all the tags are released simultaneously and passed through the gas chromatograph as before. The 'bar code' is then read from the chromatograph in one go, identifying not only the reagents used, but also the order in which they were used (Fig. 6.41).

6.5.5.2 Photolithography

Photolithography is a technique that permits miniaturization and spatial resolution such that specific products are synthesized on a plate of immobilized solid support. In the synthesis of peptides, the solid support surface contains an amino group protected by the photolabile **nitroveratryloxycarbonyl (NVOC)** protecting group (Fig. 6.42). Using a mask, part of the surface is exposed to light resulting in deprotection of the exposed region. The plate is then treated with a protected amino acid and the coupling reaction takes place only on the region of

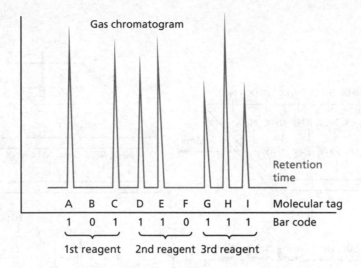

FIGURE 6.41 Identification of reagents and their order of use by bar coding.

the plate which has been deprotected. The plate is then washed to remove excess amino acid. The process can be repeated on a different region using a different mask, and so different peptide chains can be built on different parts of the plate; the sequences are known from the record of masks used.

Incubation of the plate with a protein receptor can then be carried out to detect active compounds which bind to the binding site of the receptor. A convenient method to assess such interactions is to incubate the plate with a fluorescently tagged receptor. Only those regions of the plate which contain active compounds will bind to the receptor and fluoresce. The fluorescence intensity can be measured using fluorescence microscopy and is a measure of the affinity of the compound for the receptor. Alternatively, testing can be carried out such that active compounds are detected by radioactivity or chemiluminescence.

The photodeprotection described above can be achieved in high resolution. At a 20 μm resolution, plates can be prepared with 250 000 separate compounds per square centimetre.

6.5.6 Dynamic combinatorial synthesis

Dynamic combinatorial synthesis is an exciting development which has been used as an alternative to the classic mix and split combinatorial syntheses in the search for new lead compounds. The aim of dynamic combinatorial synthesis is to synthesize all the different compounds in one flask at the same time, screen them in situ as they are being formed, and thus identify the most active compound in a much faster time period (Box 6.2). How can this be achieved? Several important principles are followed.

- The best way of screening the compounds is to have the desired target present in the reaction flask along with the building blocks. This means that any active compounds can bind to the target as soon as they are formed. The trick is then to identify which of the products are binding.

- The reactions involved should be reversible. If this is the case, a huge variety of products are constantly being formed in the flask and then breaking back down into their constituent building blocks. The advantage of this may not seem obvious, but it allows the possibility of 'amplification' where the active compound is present to a greater extent than the other possible products. By having the target present, active compounds become bound and are effectively removed from the equilibrium mixture. The equilibrium is now disturbed such that more of the active product is formed. Thus, the target serves not only to screen for active compounds, but to amplify them as well.

- In order to identify the active compounds, it is necessary to 'freeze' the equilibrium reaction such that it no longer takes place. This can be done by carrying out a further reaction which converts all the equilibrium products into stable compounds that cannot revert to the starting materials.

A simple example of dynamic combinatorial synthesis involved the reversible formation of imines from aldehydes and primary amines (Fig. 6.43). A total of three aldehydes and four amines were used in the study (Fig. 6.44), allowing the possibility of 12 different imines in the equilibrium mixture.

The building blocks were mixed together with the target enzyme **carbonic anhydrase** and allowed to interact. After a suitable period of time, sodium

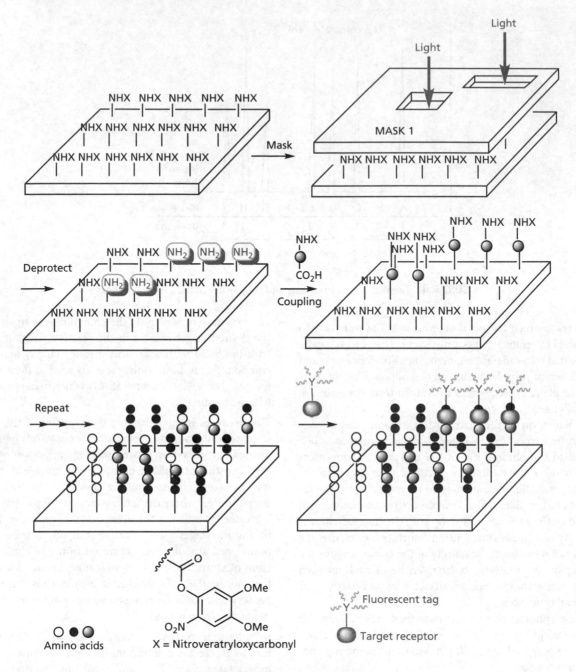

FIGURE 6.42 Photolithography.

cyanoborohydride was added to reduce all the imines present to secondary amines so that they could be identified (Fig. 6.45). The mixture was separated by reverse-phase HPLC, allowing each product to be quantified and identified. These results were compared with those obtained when the experiment was carried out in the absence of carbonic anhydrase, making it possible to identify which products had been amplified. In this experiment, the sulphonamide shown in Figure 6.45 was significantly amplified,

which demonstrated that the corresponding imine was an active compound.

The preceding example illustrates a simple case involving one reaction and two sets of building blocks, but it is feasible to have more complex situations. For example, a molecule with two or more functional groups could be present to act as a scaffold on to which various substituents could be added from the building blocks available (Fig 6.46). The use of a central scaffold has another benefit because it helps in the amplification process. If

BOX 6.2 Dynamic combinatorial synthesis of vancomycin dimers

Vancomycin is an antibiotic that works because it masks the building blocks required for bacterial cell wall synthesis. Binding takes place specifically between the antibiotic and a peptide sequence (L-Lys-D-Ala-D-Ala) which is present in the building block. It is also known that this binding promotes dimerization of the vancomycin–target complex, which suggests that covalently linked vancomycin dimers might be more effective antibacterial agents than vancomycin itself. A dynamic combinatorial synthesis was carried out to synthesize a variety of different vancomycin dimers covalently linked by bridges of different lengths. The vancomycin monomers used had been modified such that they contained long-chain alkyl substituents with double bonds at the end. Reaction between the double bonds in the presence of a catalyst then led to bridge formation through a reaction known as olefin metathesis (Fig. 1).

The tripeptide target was present to accelerate the rate of bridge formation and to promote formation of vancomycin dimers having the ideal bridge length. As shown in Figure 2, the vancomycin monomers bind the tripeptide, which encourages the self-assembly of non-covalently linked dimers. Once formed, those dimers having the correct length of substituent are more likely to react together to form the covalent bridge (Fig. 2).

Having established the optimum length of bridge, another experiment was carried out on eight vancomycin monomers which had the correct length of 'tether' but varied slightly in their structure. The mixture of 36 possible products was analysed by mass spectrometry to indicate the relative proportion of each dimer formed. Eleven of the 36 compounds were then synthesized separately and it was found that their antibacterial activity matched their level of amplification, i.e. the compounds present in greater quantities had greater activity.

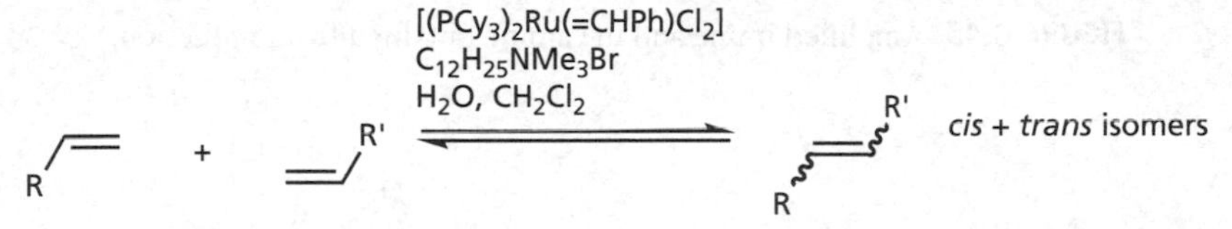

FIGURE 1 Olefin metathesis.

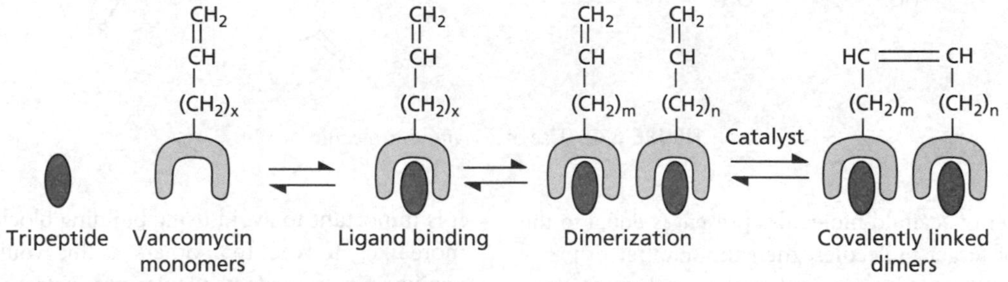

FIGURE 2 Formation of covalently linked dimers.

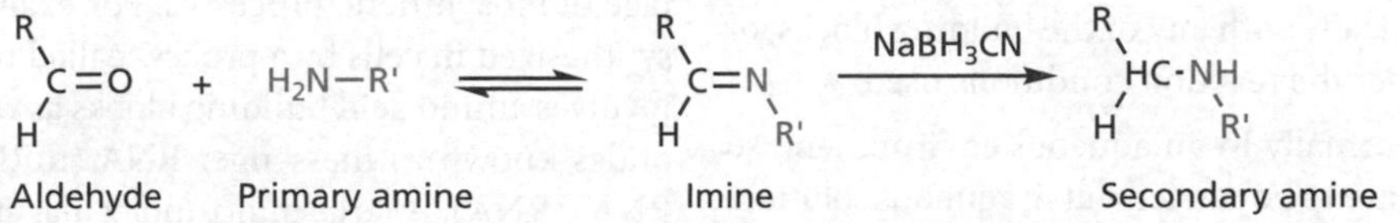

FIGURE 6.43 Example of dynamic combinatorial synthesis.

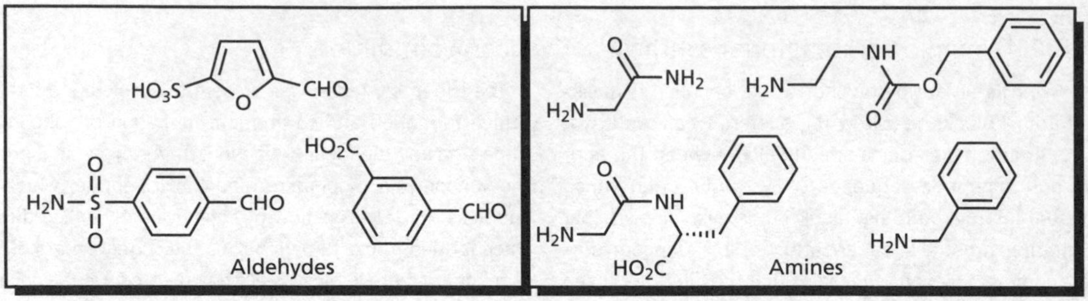

FIGURE 6.44 Aldehyde and amine building blocks used in the dynamic combinatorial synthesis of imines.

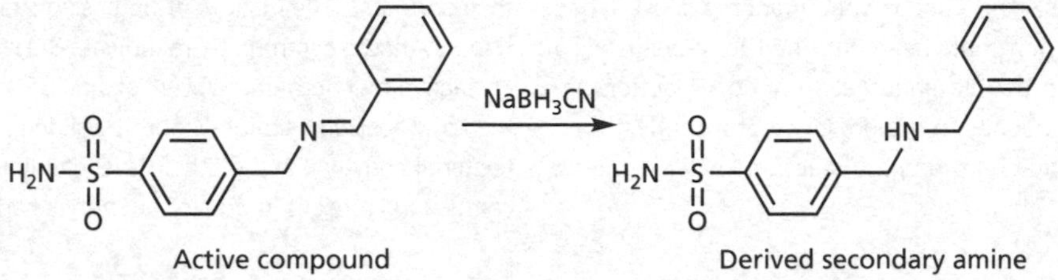

FIGURE 6.45 Amplified imine and the amine obtained from reduction.

FIGURE 6.46 Use of a scaffold molecule.

the number of scaffold molecules present is equal to the number of target molecules, then the number of products formed cannot be greater than the number of targets available. If any of these products binds to the target, the effect on the equilibrium will be greater than if there were more products than targets available.

There are certain limitations to dynamic combinatorial chemistry.

• Conditions must be chosen such that the target does not react chemically with any of the building blocks, or is unstable under the reaction conditions used.

• The target is normally in an aqueous environment, so the reactions have to be carried out in aqueous solution.

• The reactions themselves have to undergo fast equilibration rates to allow the possibility of amplification.

• It is important to avoid using building blocks that are more likely to react than others, as this would bias the equilibrium towards particular products and confuse the identification of the amplified product.

6.5.7 Compound libraries using biological processes

Chemical libraries have been prepared by taking advantage of biosynthetic processes. For example, proteins are synthesized in cells by a process called translation, which involves amino acid building blocks as well as oligonucleotides known as messenger RNA (mRNA) and transfer RNA (tRNA). A large multi-molecular structure made up of protein and ribosomal RNA (the ribosome) acts as the protein 'factory' and binds mRNA, which serves as the

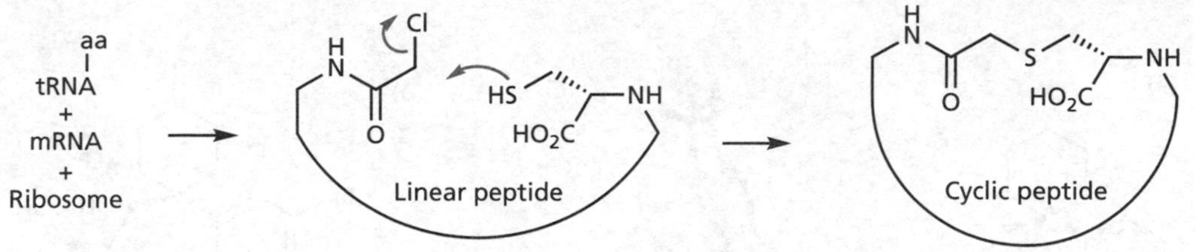

FIGURE 6.47 Random peptide integrated discovery.

genetic code for a new protein. Once the ribosome has bound mRNA, the mRNA determines which tRNA molecules are bound to the ribosome and in which order. The amino acids are linked to the tRNA molecules, and so when a specific tRNA becomes bound to the ribosome–mRNA complex, a specific amino acid is brought to the protein factory and linked to the growing protein chain.

It has proved possible to 'hijack' the molecules involved in this process and to reproduce translation in test tubes by adding all the required components in a non-cellular environment. For example, a vast number of novel cyclic peptides have been synthesized by providing bacterial ribosomes and tRNAs with random mRNA sequences (Fig. 6.47). Moreover, it has been possible to carry out genetic engineering to attach unnatural amino acids to some of the tRNA molecules such that the cyclic peptides produced contain both natural and unnatural amino acids. This adds to the structural diversity of the compounds produced and also helps to make them more resistant to metabolism. The process was also designed to produce short chains containing 8–11 amino acids with a chlorinated D-amino acid at one end and a cysteine residue at the other end. Once the linear peptide is released from the ribosome, it is set up for cyclization. The process has been tagged as **R**andom **p**eptide **i**ntegrated **d**iscovery (**Rapid**).

6.5.8 **Diversity-orientated synthesis**

6.5.8.1 Introduction

One of the problems associated with the early chemical libraries produced by combinatorial synthesis was the lack of structural diversity in the compounds represented. It is now recognized that a chemical library containing a large number of very similar structures has a low chance of providing hits for a wide range of target binding sites. Therefore, it is important that combinatorial syntheses are devised that create as much structural diversity as possible. It has also been recognized that many of the scaffolds used in early chemical libraries are simple heteroaromatic ring systems which result in molecules being generally two-dimensional in nature. As such, they are unlikely to occupy a 3D binding site as effectively as a structure that is more 'globular' in shape. Therefore, research has been carried out into devising chemical libraries that contain structurally diverse molecules with saturated multicyclic ring systems as scaffolds. This is known as diversity-orientated synthesis (see also section 7.7).

6.5.8.2 Synthesis of complex 'globular' scaffolds

Galantamine is a natural product that acts as an inhibitor of the acetylcholinesterase enzyme and has been used in the treatment of Alzheimer's disease. It contains a rigid tetracyclic scaffold made up of one aromatic ring and three saturated rings (Fig. 6.48). The cyclic system is far less planar than most conventional scaffolds as one of the rings is orientated at right angles to the other three. Therefore, it has a rigid but more 3D shape, allowing the scaffold and its substituents to occupy a greater volume within the binding site than a more planar structure.

The full synthesis of galantamine is not straightforward. However, a chemical library based on a similar tetracyclic structure has been devised (Fig. 6.49). The functional groups and the stereochemistry of the ring system are different, but the aim of the library is not to

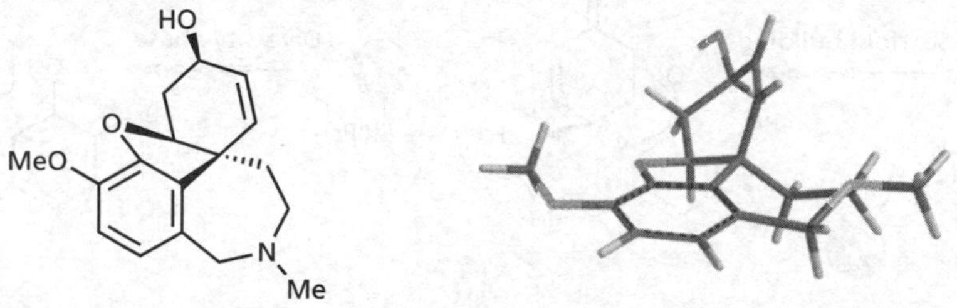

FIGURE 6.48 Galantamine.

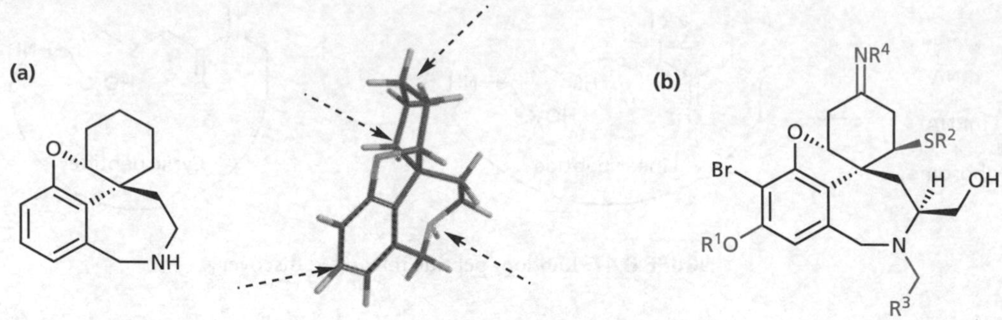

FIGURE 6.49 (a) Scaffold of the target chemical library (arrows indicate where the variation in substituents will occur). (b) Scaffold plus substituents.

synthesize galantamine or galantamine analogues, but to create complex structures by an efficient process to see whether the products could act as hits and potential lead compounds for any type of target, not just the acetylcholinesterase enzyme. The target compound was also designed to have substituents at four different positions to introduce structural diversity.

A solid phase synthesis was devised whereby the first phase of reactions created the tetracyclic ring (the scaffold building phase), leaving the last four reactions for the introduction of the various substituents (the diversity phase) (Fig. 6.50). This was an efficient approach as it meant that mix and split procedures would only occur during the final four reactions and the scaffold building process could be carried out on a single batch of resin-bound structure.

As previously stated, the first phase of the synthesis was to create the scaffold, but in such a way that functionality was introduced for the subsequent diversity phase (Fig. 6.51).

The synthesis began with the attachment of a chiral alcohol (I) to the solid support (this alcohol had previously been prepared from L-tyrosine). Treatment with a palladium catalyst then removed the protecting groups from the phenol and the amine to give the primary amine (II). The amine was then reacted with the aromatic aldehyde

(III) in the presence of trimethyl orthoformate to form an imine, which was immediately reduced with sodium cyanoborohydride to give the amine (IV).

This was treated with allyl chloroformate to protect the amine as a urethane (V). An intramolecular cyclization reaction was then carried out in the presence of the oxidizing agent iodobenzene diacetate ($PhI(OAc)_2$) to give a spiro structure (VI). The allyl protecting groups on the amine and the two phenol groups were removed in the presence of $Pd(PPh_3)_4$ as a catalyst to give the phenol (VII). This was now set up for a spontaneous cyclization, where one of the phenol groups carried out a stereoselective Michael-type addition on the neighbouring α,β-unsaturated ketone to give the tetracyclic scaffold (VIII).

The diversity phase was now carried out to introduce various substituents at four different positions of the scaffold (Fig. 6.52).

The batch of resin-bound structure (VIII) was split into equal portions and each portion (bar one) was treated with a different alcohol in a **Mitsunobu coupling** reaction to give the ethers (IX).

The second diversity reaction was to mix the batches and resplit them into equal portions, with each portion (bar one) being reacted with a different thiol. This involved another Michael addition with the remaining α,β-unsaturated ketone to provide the thioethers (X).

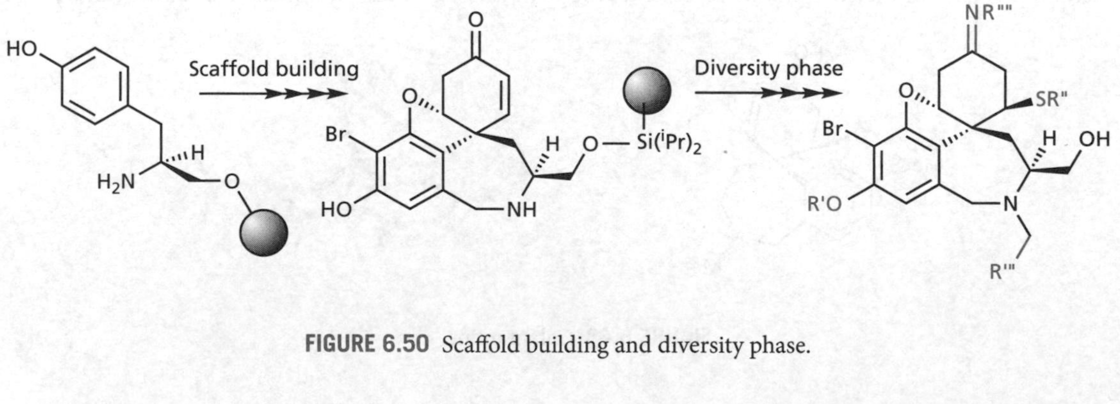

FIGURE 6.50 Scaffold building and diversity phase.

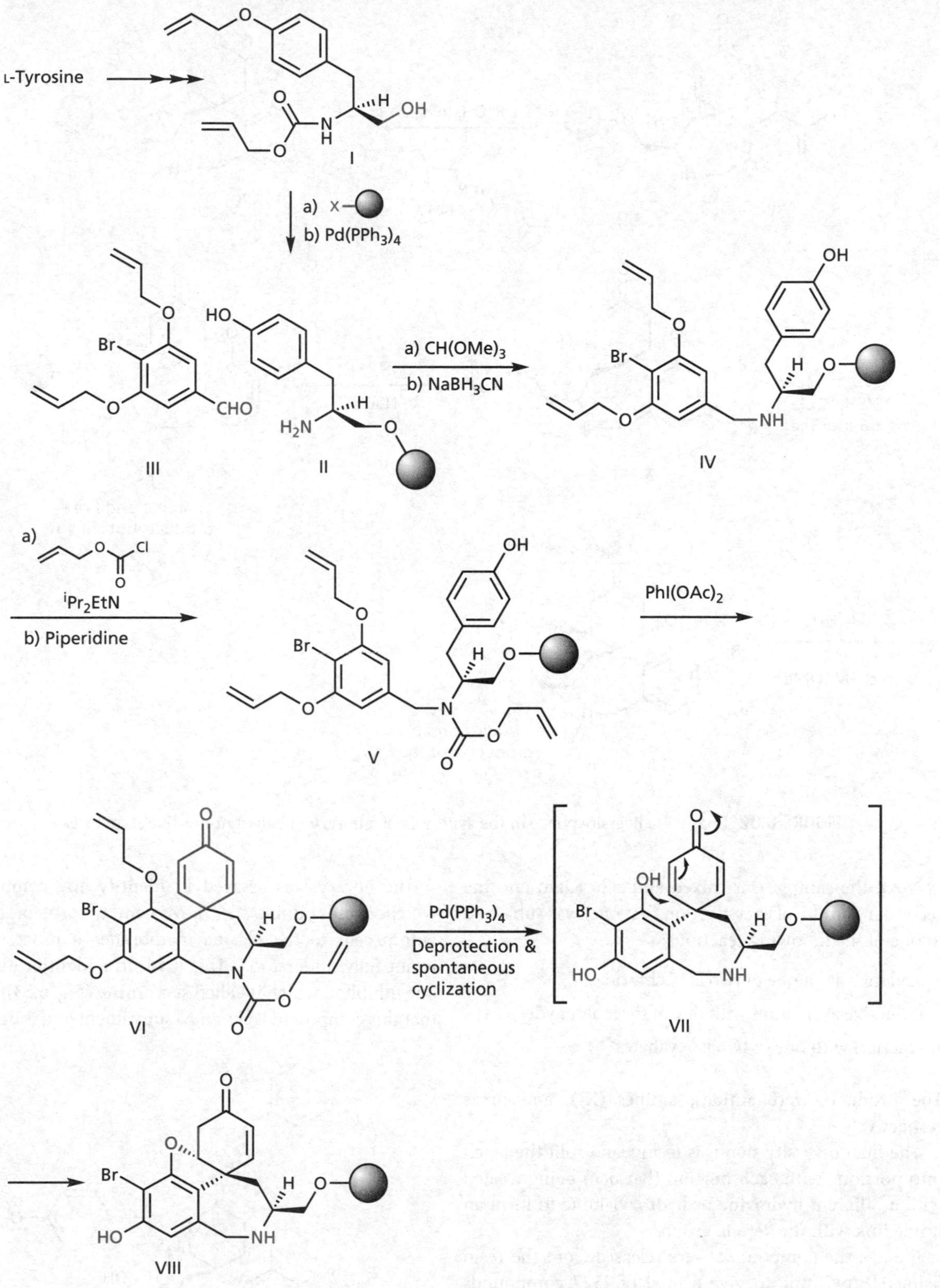

FIGURE 6.51 Phase 1—scaffold building.

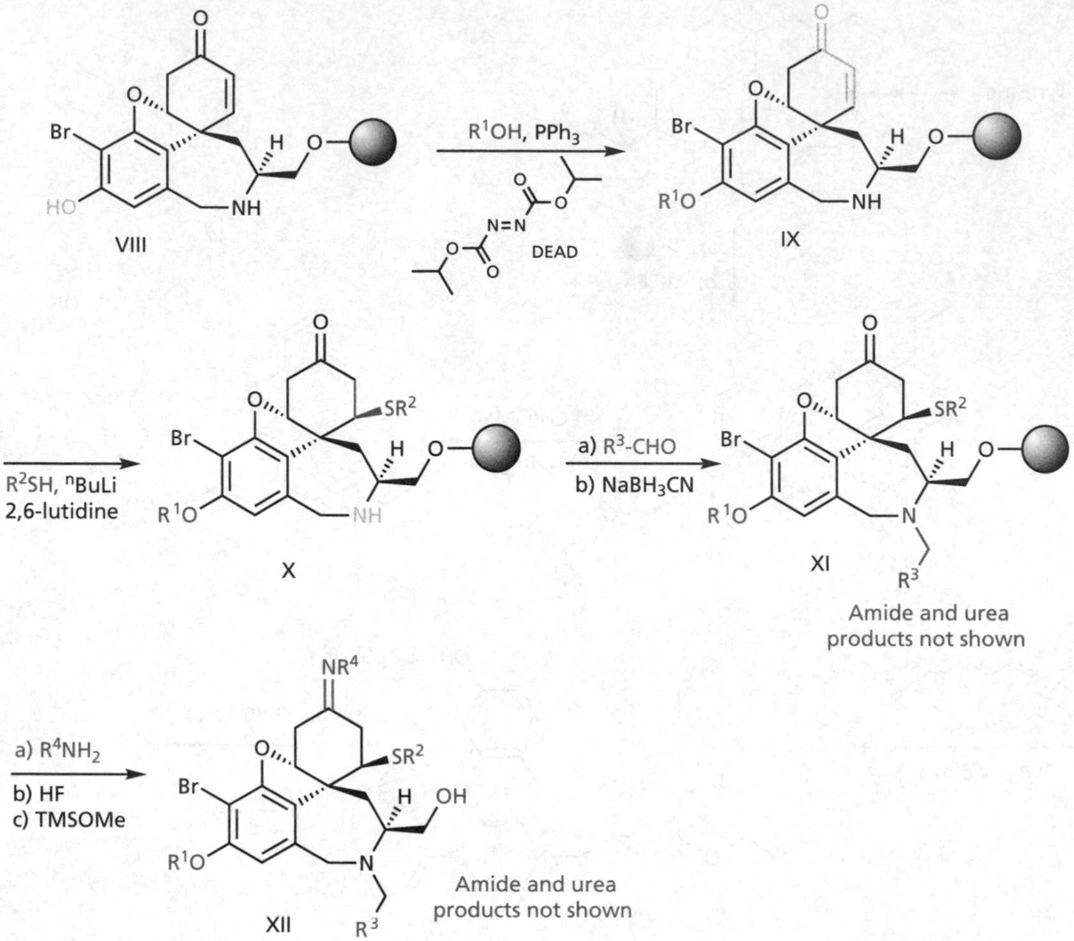

FIGURE 6.52 Phase 2—diversity phase in the synthesis of a library of galantamine-like structures.

Next, the sample was mixed and split again and the secondary amine of each portion (bar one) was subjected to one of the following reactions;

- acylation with one of two acid chlorides;
- reductive amination with one of three aldehydes;
- reaction with one of two isocyanates.

The products were amides, amines (XI), and ureas respectively.

The final diversity step was to mix and split the batch into portions with each portion (bar one) being treated with a different hydrazine or hydroxylamine to form an imine link with the ketone group.

Finally, the compounds were released from the resin using HF/pyridine to give a total of 2527 compounds identified by mass spectrometry.

Note that at each of the mix and split phases, one portion was set aside and not treated with a reagent, such that products would be obtained that lacked a substituent at each of the four variable positions.

The library was assayed to identify any compound which could inhibit the movement of proteins from within cells to the plasma membrane—a process that is not fully understood. This led to the identification of one inhibitor, which is called **secramine** (Fig. 6.53). Note that this compound lacks an N-substituent and is derived

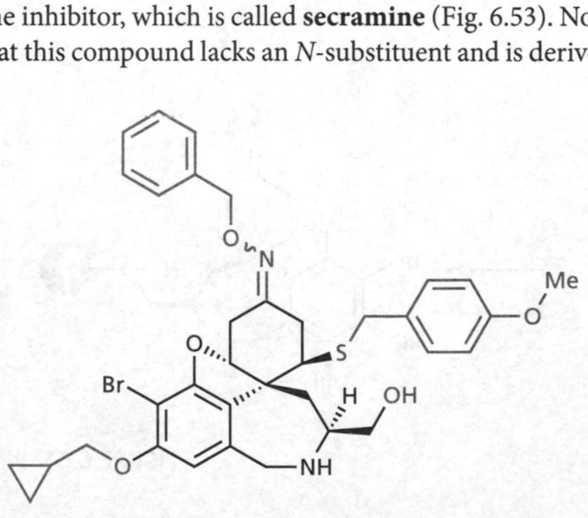

FIGURE 6.53 Secramine.

from the portion that was not reacted in the third diversity stage of the synthesis.

6.5.8.3 Synthesis of a 'globular' scaffold using tandem reactions

A good method of developing a multicyclic system with the minimum number of synthetic steps is to involve a **tandem reaction** where the product from one reaction is set up for a second reaction. For example, a multicomponent **Ugi coupling** reaction was carried out, such that the product formed would be set up for a spontaneous **Diels–Alder reaction** (Fig. 6.54). The Ugi reaction is carried out with a mixture of an amine, an aldehyde or

ketone, an isocyanide, and a carboxylic acid. The amine and the carbonyl group react first to give an imine (II). The isocyanide and the carboxylic acid then react with the imine, possibly through an intermediate such as III to give structure IV which undergoes an intramolecular nucleophilic substitution to give a bis amide (V). By choosing furfural and the unsaturated carboxylic acid as components, the final product from the Ugi reaction contains a diene and a dienophile which are positioned sufficiently close to each other that the spontaneous Diels–Alder reaction occurs to give the multicyclic structure (VI) in 67% yield as a single stereoisomer.

N-Allylation was then carried out on both secondary amides, followed by an olefin metathesis in the presence

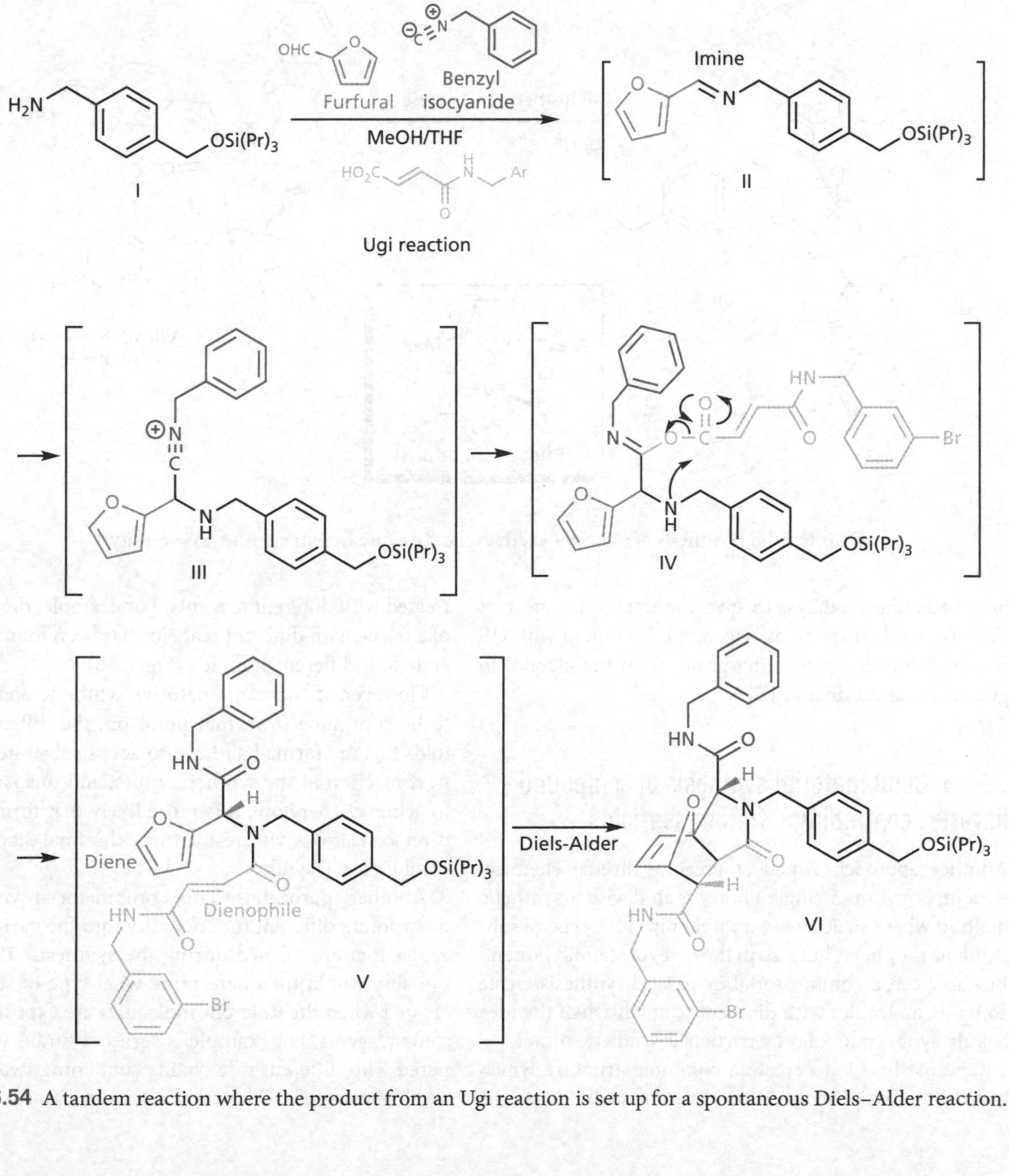

FIGURE 6.54 A tandem reaction where the product from an Ugi reaction is set up for a spontaneous Diels–Alder reaction.

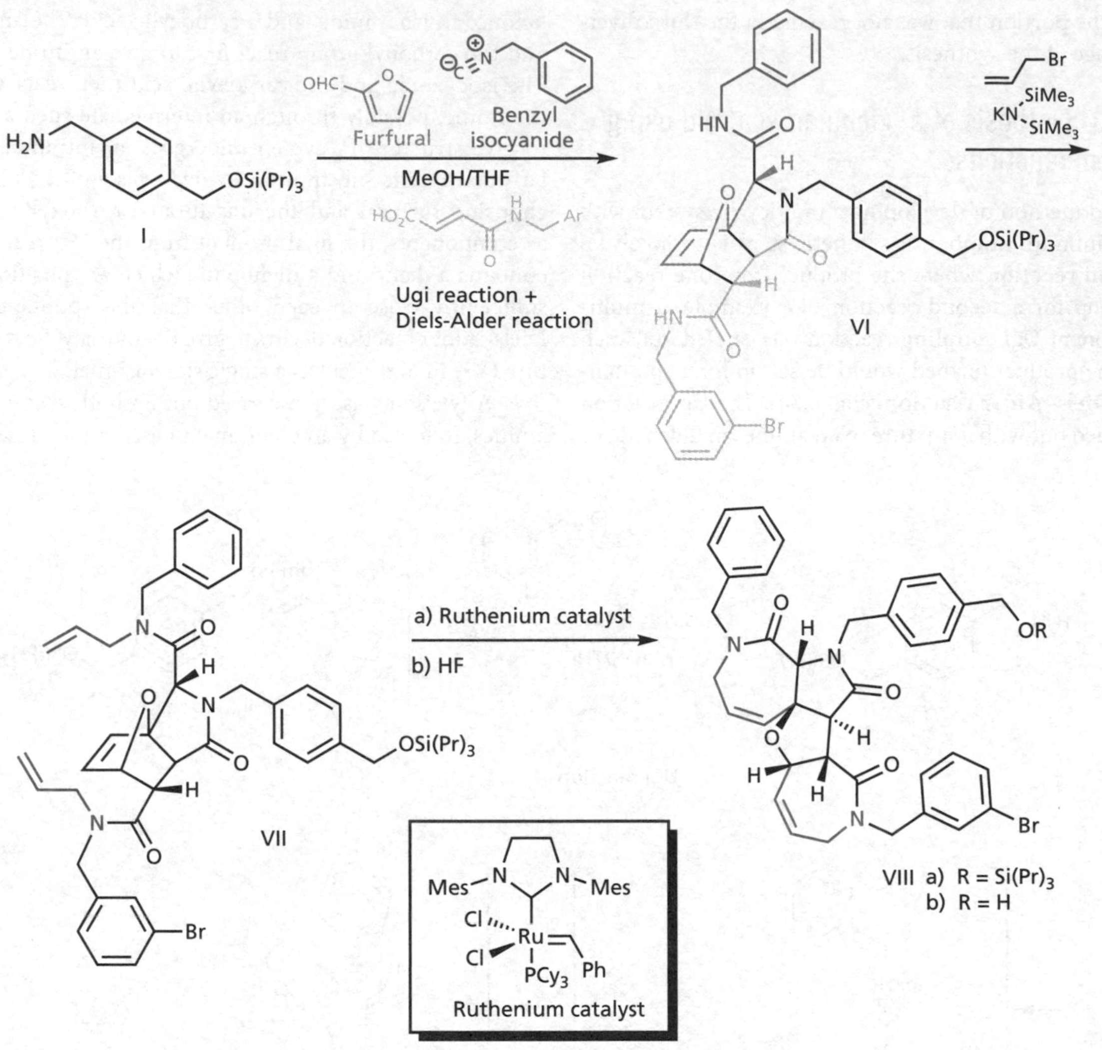

FIGURE 6.55 Synthesis of a complex tetracyclic structure in four steps (Mes = mesityl).

of a ruthenium catalyst to give the tetracyclic product (VIIIa) in only three steps (Fig. 6.55). Treatment with HF removed the silyl protecting group from the alcohol to give the final product (VIIIb).

6.5.8.4 Combinatorial synthesis of compound libraries containing different scaffolds.

Another approach aimed at creating diverse chemical structures within a single library is to devise a synthetic method where structures vary not only in the type of substituents they have, but also in the types of scaffold present. But how can a combinatorial or parallel synthesis create complex molecules with different scaffolds when the idea has always been to follow a particular synthetic route?

One method is to create a common structure which has the potential to generate different scaffolds if it is

treated with different reagents. For example, the reaction of a triene with different reagents has been found to generate four different scaffolds (Fig. 6.56).

However, if the combinatorial synthetic approach is to be continued from that point on, the different scaffolds that are formed all need to act as substrates for the next reaction in the synthetic route, and this is not easy to achieve. Therefore, it is more likely that formation of diverse scaffolds will result from the final stage in the combinatorial synthesis.

Another approach is to use combinatorial synthesis to incorporate different functionality into the various molecules that are formed during the synthesis. This functionality would then determine what type of scaffold is created when the different molecules are exposed to the same reagents. For example, a series of furans were prepared with different side chains containing two, one, or no alcohol groups. Treatment with N-bromosuccinimide

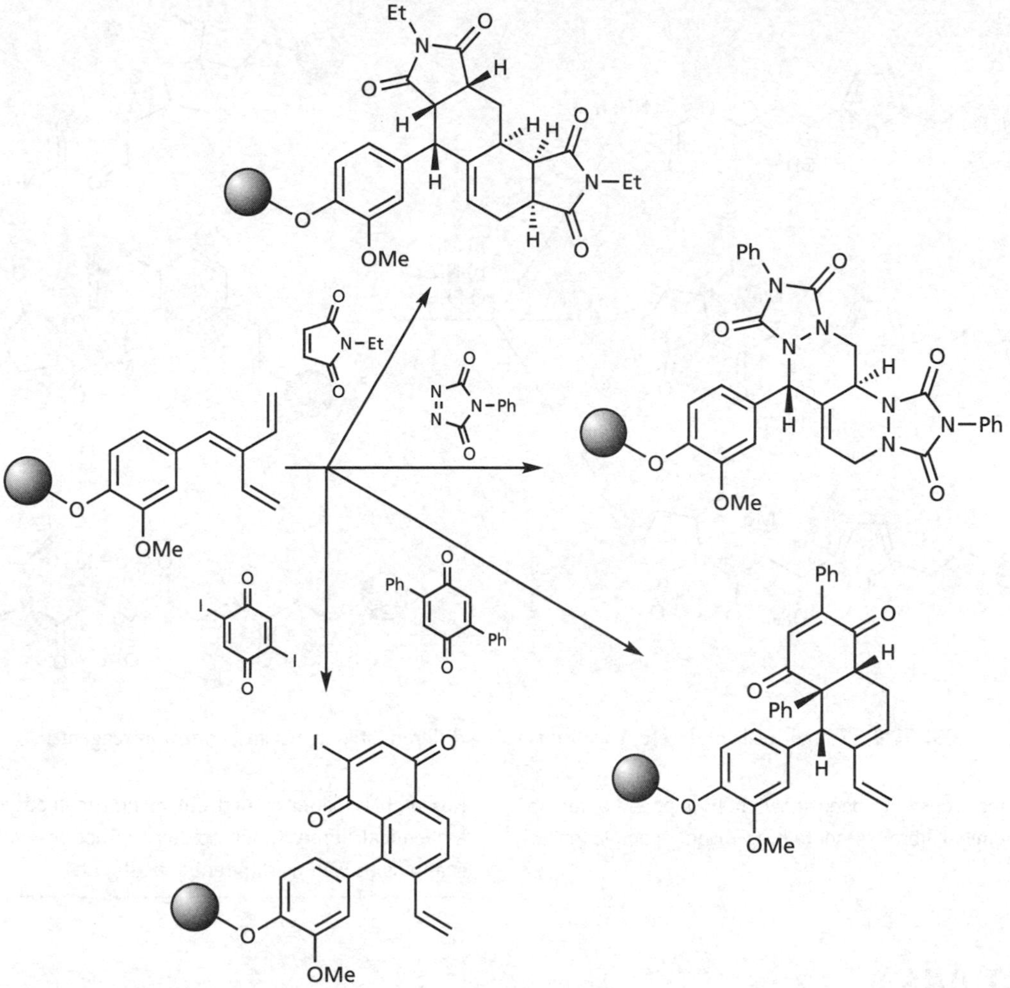

FIGURE 6.56 Synthesis of different cyclic structures from the same triene.

followed by pyridinium 4-toluenesulphonate resulted in the furan ring opening up to form a *cis*-enedione intermediate. Depending on the presence or otherwise of the side-chain alcohols, different scaffolds were formed as a result of recyclization, dehydration, and/or isomerization (Fig. 6.57).

Further research is still in progress to see whether such approaches are feasible for producing a library of compounds having diverse substituents and scaffolds.

KEY POINTS

• Tagging involves the construction of a tagging molecule on the same solid support as the target molecule. Tagging molecules are normally peptides or oligonucleotides. After each stage of the target synthesis, the peptide or oligonucleotide is extended and the amino acid or nucleotide used defines the reactant or reagent used in that stage.

• Photolithography is a technique involving a solid support surface containing functional groups protected by photolabile groups. Masks are used to reveal defined areas of the plate to light, thus removing the protecting groups and allowing a reactant to be linked to the solid support. A record of the masks used determines what reactions have been carried out at different regions of the plate.

• Combinatorial synthesis has been used for the synthesis of peptides, peptoids, and heterocyclic structures. Most organic reactions are feasible.

• Dynamic combinatorial chemistry involves the equilibrium formation of a mixture of compounds in the presence of a target. Binding of a product with the target amplifies that product in the equilibrium mixture.

• Diversity-orientated synthesis aims to produce compounds with as wide a diversity as possible in order to fully explore the conformational space around a molecule when it interacts with a target binding site.

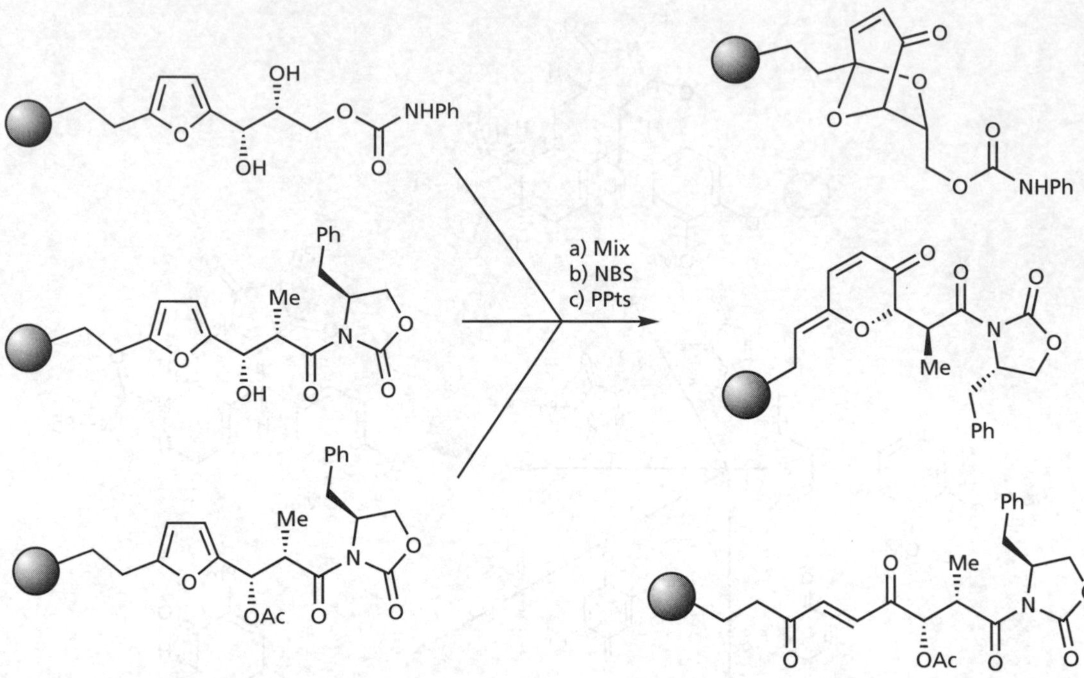

FIGURE 6.57 Synthesis of different scaffolds using different structures and common reagents.

- Tandem reactions have been shown to have potential for creating chemical libraries containing diverse complex structures.

- Research is being carried out to find methods of creating a chemical library which contains structures with different scaffolds as well as different substituents.

QUESTIONS

1. Identify three stages of the drug discovery, design, and development process where combinatorial chemistry or parallel synthesis is of importance.

2. A pharmaceutical laboratory wishes to synthesize all the possible dipeptides containing the amino acids tyrosine, lysine, phenylalanine, and leucine. Identify the number of possible dipeptides and explain how the laboratory would carry this out using combinatorial techniques.

3. What particular precautions have to be taken with the amino acids tyrosine and lysine in the synthesis in Question 2?

4. Identify the advantages and disadvantages of the structures in the figure as scaffolds.

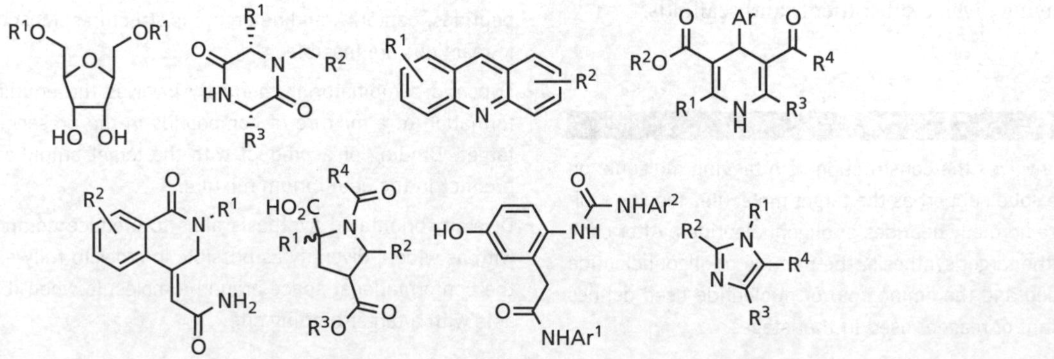

5. You wish to carry out the combinatorial synthesis shown in Figure 6.40 using bar coding techniques rather than the conventional tagging scheme shown in the figure. You have nine molecules suitable for tagging purposes (A–I), seven bromo acids (B1–B7), seven amines (A1–A7), and seven acid chlorides (C1–C7). Construct a suitable coding system for the synthesis.

6. Based on your coding scheme from Question 5, what product is present on the bead if the released tags resulted in the gas chromatograph shown in Figure 6.41?

FURTHER READING

Beck-Sickinger, A. and Weber, P. (2002) *Combinatorial strategies in biology and chemistry.* Wiley, New York.

Bhalay, G., Dunstan, A., and Glen, A. (2000) 'Supported reagents: opportunities and limitations', *Synlett*, **12**, 1846–59.

Booth, R.J. and Hodges, J.C. (1997) 'Solid-supported quenching reagents for parallel purification', *Journal of the American Chemical Society*, **119**, 4482–6.

Braeckmans, K., et al. (2002) 'Encoding microcarriers: present and future technologies', *Nature Reviews Drug Discovery*, **1**, 447–56.

Broadwith, P. (2011) 'Cells hijacked to turn out cyclic peptides', *Chemistry World*, September, p 14.

Burke, M.D. and Schreiber, S.L. (2004) 'A planning strategy for diversity-oriented synthesis', *Angewandte Chemie, International Edition*, **43**, 46–58.

DeWitt, S.H., et al. (1993) ' "Diversomers": an approach to nonpeptide, nonoligomeric chemical diversity', *Proceedings of the National Academy of Sciences of the USA*, **90**, 6909–13.

Dittrich, P.S., and Manz, A. (2006) 'Lab-on-a-chip: microfluidics in drug discovery', *Nature Reviews Drug Discovery*, **5**, 210–18.

Dobson, C.M. (2004) 'Chemical space and biology', *Nature*, **432**, 824–8.

Dolle, R.E. (2003) 'Comprehensive survey of combinatorial library synthesis: 2002', *Journal of Combinatorial Chemistry*, **5**, 693–753.

Geysen, H.M., et al. (2003) 'Combinatorial compound libraries for drug discovery: an ongoing challenge', *Nature Reviews Drug Discovery*, **2**, 222–30.

Guillier, F., Orain, D., and Bradley, M. (2000) 'Linkers and cleavage strategies in solid-phase organic synthesis and combinatorial chemistry', *Chemical Reviews*, **100**, 2091–158.

Houlton, S. (2002) 'Sweet synthesis', *Chemistry in Britain*, April, pp 46–9.

Jung, G. (ed.) (1996) *Combinatorial peptide and nonpeptide libraries.* VCH, Weinheim.

Kappe, C.O. (2004) 'Controlled microwave heating in modern organic synthesis', *Angewandte Chemie, International Edition*, **43**, 6250–84.

Le, G.T., et al. (2003) 'Molecular diversity through sugar scaffolds', *Drug Discovery Today*, **8**, 701–9.

Lee, D., Sello, J.K., and Schreiber, S.L. (2000) 'Pairwise use of complexity-generating reactions in diversity-orientated organic synthesis', *Organic Letters*, **2**, 709–12.

Ley, S.V. and Baxendale, I.R. (2002) 'New tools and concepts for modern organic synthesis', *Nature Reviews Drug Discovery*, **1**, 573–86.

Mavandadi, F. and Pilotti, A. (2006) 'The impact of microwave-assisted organic synthesis in drug discovery', *Drug Discovery Today*, **11**, 165–74.

Nicolaou, K.C., et al. (2000) 'Target-accelerated combinatorial synthesis and discovery of highly potent antibiotics effective against vancomycin-resistant bacteria', *Angewandte Chemie, International Edition*, **39**, 3823–8.

Pelish, H.E., et al. (2001) 'Use of biomimetic diversity-orientated synthesis to discover galantamine-like molecules with biological properties beyond those of the natural product', *Journal of the American Chemical Society*, **123**, 6740–1.

Ramstrom, O. and Leh, J.-M. (2002) 'Drug discovery by dynamic combinatorial libraries', *Nature Reviews Drug Discovery*, **1**, 26–36.

Reader, J.C. (2004) 'Automation in medicinal chemistry', *Current Topics in Medicinal Chemistry*, **4**, 671–86.

Terret, N.K. (1998) *Combinatorial chemistry.* Oxford University Press, Oxford.

Wang, Y., et al. (2009) 'An integrated microfluidic device for large-scale in situ click chemistry screening', *Lab on a Chip*, **9**, 2281–5.

CASE STUDY 1
Peptide synthesis

CS1.1 Introduction

Peptide synthesis has been an important area of organic synthesis for many years. Many of the body's neurotransmitters and hormones are peptides or proteins, and the ability to carry out peptide synthesis has allowed the medicinal chemist to prepare these structures, as well as their analogues. This provided an understanding of structure–activity relationships and led to useful drugs. The same holds true for peptides and proteins that have been obtained from the natural world. Several of these have potent pharmacological activity and have served as important lead compounds for the design of structures intended to mimic or block their action at target receptors. Many peptides and proteins also serve as substrates for enzyme-catalysed reactions, and peptide synthesis has been vital in synthesizing analogues that bind to the active site of the enzyme but do not undergo the enzyme-catalysed reaction. Thus, these analogues act as enzyme inhibitors.

Unfortunately, very few of these peptide analogues ever reached the clinic since they tended to have poor pharmacokinetic properties, such as metabolic susceptibility and poor oral absorption. Normally, a further process of design and development was required to synthesize compounds where the peptide nature of these first-generation agents was disguised by using amino acids with unnatural side chains, or replacing the peptide links which are susceptible to metabolic hydrolysis with more stable functional groups. The resulting structures are known as **peptidomimetics**—agents that mimic a peptide structure in its ability to bind to a target protein, but lack the adverse pharmacokinetics associated with peptide structures. The ability to synthesize peptides efficiently is important in the early stages of this process and, although the aim is often to develop a non-peptide drug,

there are several examples where peptide-like drugs have proved clinically useful.

CS1.2 Amino acids—the building blocks for peptide synthesis

Amino acids are the building blocks used for the biosynthesis and synthesis of peptides and proteins. They all contain an amine and a carboxylic acid functional group, which are normally linked to the same carbon atom— the α-carbon. Therefore they are called α-amino acids (Fig. CS1.1). The amine, α-carbon, and carboxylic acid are consistent for all the α-amino acids, and this arrangement is called the **head group**. A side chain (R) is also attached to the α-carbon which means that the α-carbon is an asymmetric centre, except in the case of glycine where R = H. This means that two enantiomers are possible, but only one of these (the L-amino acid) is present in higher life forms. The D-enantiomer is often referred to as the 'unnatural' enantiomer, but this is not wholly true as D-amino acids are present in microorganisms.

There are 20 common amino acids found in humans. These are listed in Table CS1.1, with the three-letter and one-letter codes often used to represent them. The structures of the amino acids are shown in Appendix 7; they differ in the nature of the side chain (R) attached to the head group.

CS1.3 Peptide structure

Peptides are the condensation products obtained when amino acids are linked together. The condensation involves the amino group of one amino acid reacting with

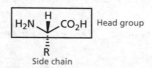

FIGURE CS1.1 General structure of an α-amino acid.

TABLE CS 1.1 The 20 common amino acids found in humans.

Synthesized in the human body			Essential to the diet		
Amino acid	Codes 3-letter	1-letter	Amino acid	Codes 3-letter	1-letter
Alanine	Ala	A	Histidine	His	H
Arginine	Arg	R	Isoleucine	Ile	I
Asparagine	Asn	N	Leucine	Leu	L
Aspartic acid	Asp	D	Lysine	Lys	K
Cysteine	Cys	C	Methionine	Met	M
Glutamic acid	Glu	E	Phenylalanine	Phe	F
Glutamine	Gln	Q	Threonine	Thr	T
Glycine	Gly	G	Tryptophan	Trp	W
Proline	Pro	P	Valine	Val	V
Serine	Ser	S			
Tyrosine	Tyr	Y			

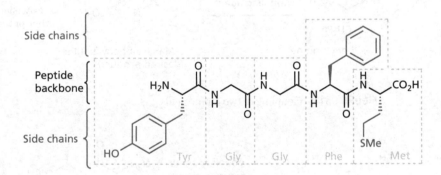

FIGURE CS1.2 Structure of Met-enkephalin. The shorthand notation for this peptide is H-Tyr-Gly-Gly-Phe-Met-OH or YGGFM.

the carboxylic acid group of another amino acid, resulting in the formation of an amide or peptide bond. The primary structure of a peptide defines the order in which the amino acids are linked. It consists of a regularly repeating peptide backbone with side chains attached to the various α-carbons. The structure of **Met-enkephalin** (an endogenous pain killer) is shown in Figure CS1.2. By convention, the peptide is represented such that the terminal amine group is positioned to the left of the structure, and the terminal carboxylic acid is positioned to the right. The naming of the structure follows the same convention by identifying the amino acids from the left to right as drawn.

CS1.4 Peptide synthesis—factors to consider

In theory, it should be possible to synthesize peptides by reacting one amino acid directly with another. However, this is not possible in practice. For a start, the reaction of an amine with a carboxylic acid is more likely to result in the formation of a salt (Fig. CS1.3). In order to get peptide formation, strong heating would be required. However, this is likely to have detrimental effects on structures that are sensitive to heat, and is also likely to epimerize asymmetric centres.

Fortunately, it is possible to get round this problem by using **coupling agents** which activate the carboxylic acid and allow the reaction to take place under milder conditions. Nevertheless, we are still faced with difficulties. Since each amino acid has an amino group and a carboxylic acid group, there is no control over which group reacts. For example, if we wanted to synthesize L-alanyl-L-phenylalanine directly from L-alanine and L-phenylalanine with the aid of a coupling agent, we would obtain a variety of different products (Fig. CS1.4).

The desired product involves a coupling reaction between the amino group of phenylalanine and the carboxylic acid group of alanine, but there is no reason why the coupling reaction could not equally take place between the amino group of alanine and the carboxylic acid of phenylalanine to couple the amino acids 'the wrong way round' (Fig. CS1.5).

Moreover, there is nothing to stop coupling between two alanine molecules or two phenylalanine molecules to give the dipeptides L-alanyl-L-alanine or L-phenylalanyl-L-phenylalanine (Fig. CS1.6). Indeed, the products that are formed from any of these reactions could react further to form tripeptides, tetrapeptides, etc.

Therefore a protection strategy is required which will mask the functional groups that we do not want to react. Once the desired coupling reaction has been completed,

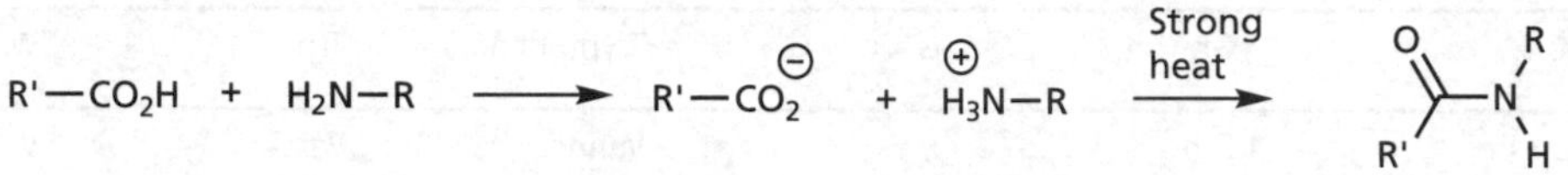

FIGURE CS1.3 Reaction of an amine with a carboxylic acid.

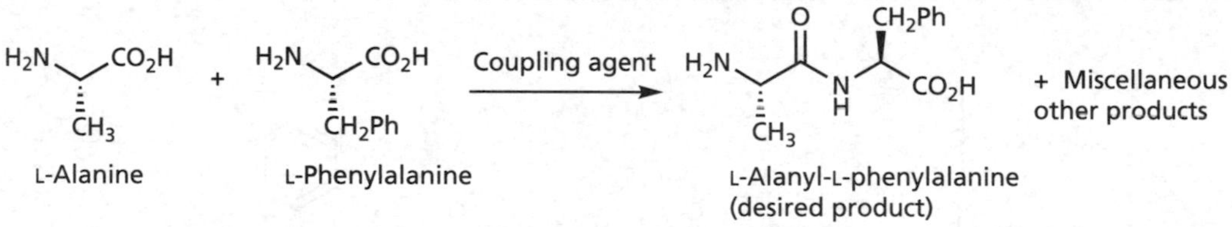

FIGURE CS1.4 Coupling of two amino acids.

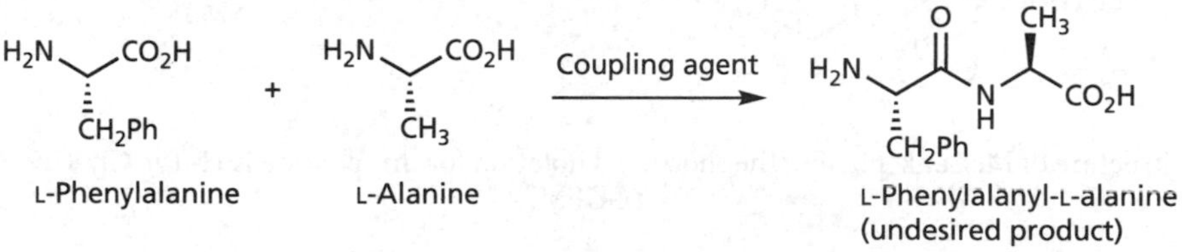

FIGURE CS1.5 'Incorrect' coupling.

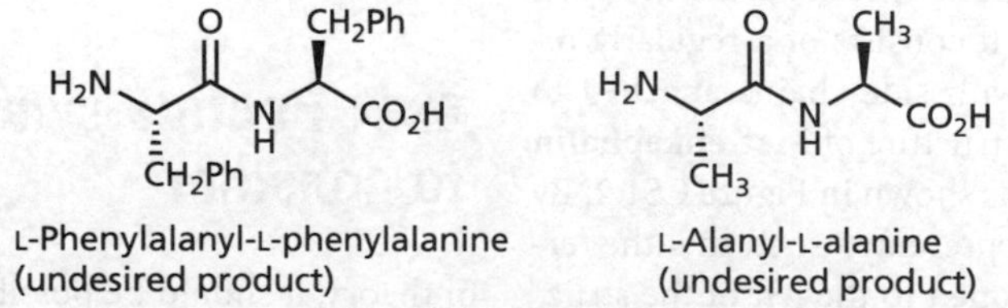

FIGURE CS1.6 Other dipeptide products.

FIGURE CS1.7 Synthesis of L-alanyl-L-phenylalanine (X, Y = protecting groups).

the protecting groups can be removed to give the desired dipeptide (Fig. CS1.7).

Suppose now that we want to make a longer peptide than a dipeptide. We can apply the same principles to link the first two protected amino acids to create a protected dipeptide. However, this time we do not want to remove both protecting groups. Instead, we want to remove one of the protecting groups, such that we can add the next protected amino acid to the deprotected terminus. This implies that there are two approaches to peptide synthesis—building the peptide up one amino acid at a time starting from the C-terminus, or building the peptide up one amino acid at a time starting from the N-terminus.

If we wish to build the peptide up from the C-terminus (the end containing the carboxylic acid group) to the N-terminus (the end containing the amino group), we start by coupling the two protected amino acids that would be

found at the C-terminus. The amino group in the dipeptide is then deprotected before coupling the next N-protected amino acid (Fig. CS1.8). Once the protected tripeptide has been prepared, the amino group can be deprotected to allow the addition of a fourth N-protected amino acid. The process can then be repeated until the full peptide chain has been synthesized, whereupon all the protecting groups are removed to provide the desired peptide or protein. Note that the group protecting the carboxylic acid of the C-terminal amino acid remains attached throughout the whole synthesis and is only removed at the final stage.

The other strategy involves building the peptide chain from the N-terminus to the C-terminus, in which case the group protecting the amine of the N-terminal amino acid is retained throughout the synthesis and it is the carboxylic acid protecting group that is removed before each coupling stage (Fig. CS1.9).

FIGURE CS1.8 Building a polypeptide from the C-terminus to the N-terminus (X, Y = protecting groups).

FIGURE CS1.9 Building a polypeptide from the N-terminus to the C-terminus.

Clearly, if either of these strategies are to work effectively, it is necessary to deprotect the amine or the carboxylic acid group selectively without removing the other protecting group.

CS1.5 Coupling agents used in peptide synthesis

One of the most popular coupling agents used in peptide synthesis is **dicyclohexylcarbodiimide (DCC)**. This structure reacts with a carboxylic acid to form an activated intermediate which then reacts with the amine (Fig. CS1.10). However, there can be problems with side reactions and epimerization. These problems can be eased to a large extent by adding a further reagent called **1-hydroxybenzotriazole (HOBt)**. This reagent reacts

with the activated intermediate to form a second intermediate, which then reacts with the amine with less risk of epimerization.

Another problem with the use of DCC is the formation of **dicyclohexylurea (DCU)**, which is poorly soluble and difficult to remove from the peptide product. Therefore a large number of alternative coupling agents have been developed, some of which are shown in Figure CS1.11 along with the ureas that are formed from them. For example, **diisopropylcarbodiimide (DIC)** is a popular choice as a coupling agent for solid phase peptide synthesis since the resulting urea (DIU) is soluble in dichloromethane and is easily washed from the solid phase. Another example is **1-ethyl-3-(3-dimethylaminopropyl)carbodiimide (EDC or EDCI)**, which is a water-soluble coupling agent.

Different types of peptide coupling agent from the carbodiimides are phosphorus reagents such as

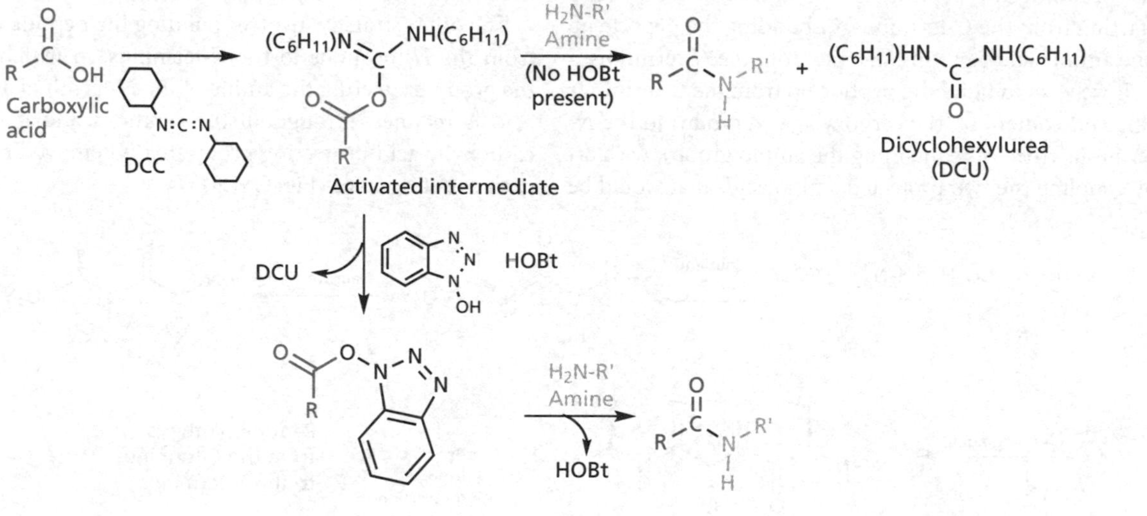

FIGURE CS1.10 Coupling of an amine with a carboxylic acid in the presence of dicyclohexylcarbodiimide (DCC).

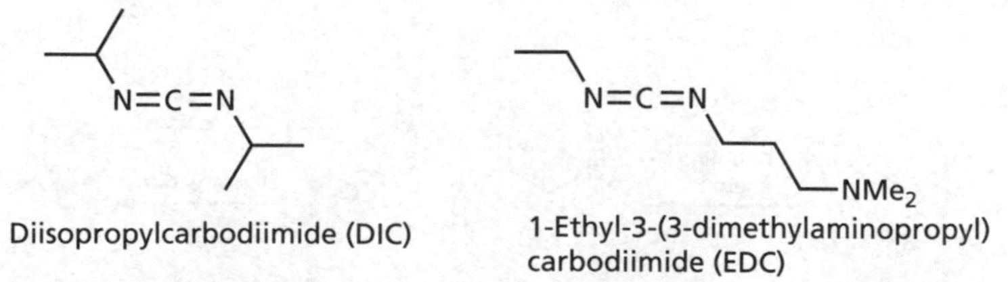

FIGURE CS1.11 Structures of coupling agents and the ureas formed.

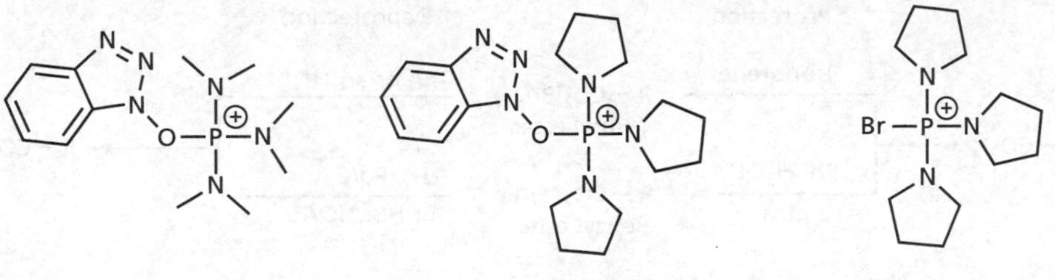

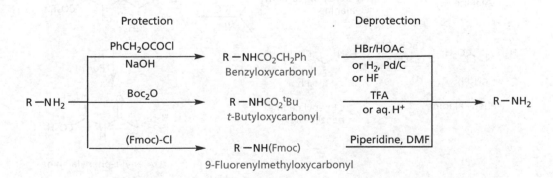

BOP

PyBOP

PyBrOP

FIGURE CS1.12 Structures of phosphorus coupling agents.

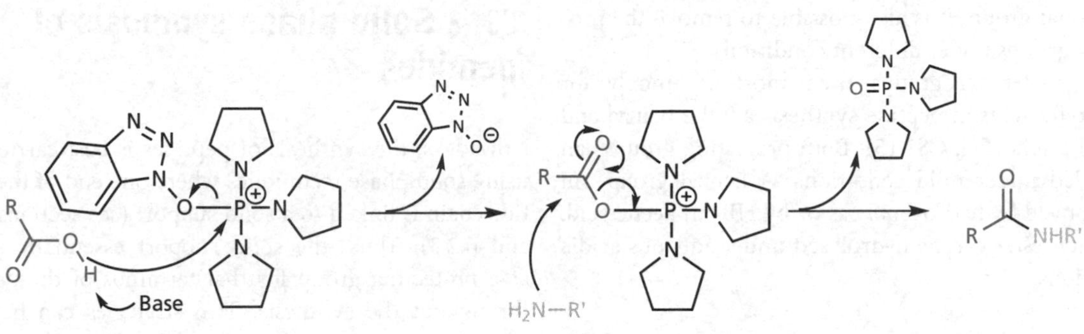

FIGURE CS1.13 The coupling mechanism using PyBOP.

(benzotriazol-1-yloxy)tris(dimethylamino)phosphonium hexafluorophosphate (BOP), benzotriazol-1-yloxytripyrrolidinophosphonium hexafluorophosphate (PyBOP), and bromotripyrrolidinophosphonium hexafluorophosphate (PyBrOP) (Fig. CS1.12). These reagents have been commonly used in solid phase peptide synthesis, but the use of BOP is now discouraged as the by-product from the reaction is hexamethylphosphoramide (HMPA), which is carcinogenic.

The mechanism of the coupling reaction is shown in Figure CS1.13 and takes advantage of the very strong phosphorus oxygen double bond that is formed in the phosphorus by-product.

CS1.6 Protecting groups in peptide synthesis

A range of protecting groups are available for both the amino group and the carboxylic acid group. As far as the amino group is concerned, three popular protecting groups are the benzyloxycarbonyl group (Z or Cbz), the *t*-butyloxycarbonyl group (Boc), and the fluorenylmethyloxycarbonyl group (Fmoc) (Fig. CS1.14). The addition of these groups results in the amine being incorporated into a relatively unreactive urethane

FIGURE CS1.14 Common protecting groups for amines.

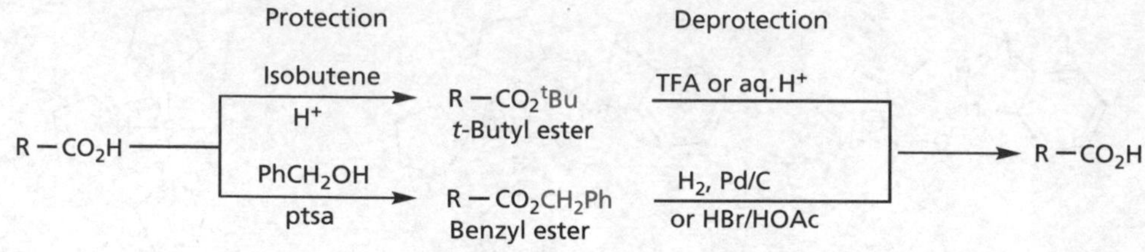

FIGURE CS1.15 Esters as protecting groups for carboxylic acids.

functional group. It is also possible to remove the protecting groups under different conditions.

The protecting groups used most commonly for carboxylic acids in peptide synthesis are the benzyl and *t*-butyl esters (Fig. CS1.15). Both protecting groups can be added under mild conditions. A benzyl group can be removed by hydrogenolysis or by HBr in acetic acid. A *t*-butyl ester can be hydrolysed under aqueous acidic conditions.

CS1.7 **Synthesis of L-alanyl-L-phenylalanine**

We are now in a position to design a synthesis of the dipeptide L-alanyl-L-phenylalanine described earlier. Protecting the amino group of L-alanine with a benzyloxycarbonyl group (Z) and the carboxylic acid group of L-phenylalanine with a benzyl ester (Bz) would be ideal, since both protecting groups can be removed simultaneously once the coupling stage has been carried out (Fig. CS1.16).

CS1.8 **Solid phase synthesis of peptides**

Nowadays, the synthesis of peptides is best carried out using solid phase techniques where one end of the peptide chain is linked to a solid support (see sections 6.2.2 and 6.2.3). Thus, the solid support essentially serves as a protecting group for that terminus of the peptide throughout the synthesis. Two strategies can be used where the structure is built from the *C*-terminus to the *N*-terminus. The type of protecting group strategy used depends on the strategy used.

The Boc/benzyl protection strategy

The *N*-terminus of each amino acid to be used in the synthesis is protected by a *t*-butyloxycarbonyl (Boc) group. The carboxylic acid of the first amino acid in the sequence is linked to the solid support where it will remain tethered throughout the synthesis (Fig. CS1.17). The *N*-Boc protecting group is now removed with TFA and the next *N*-Boc amino acid is coupled. After each coupling

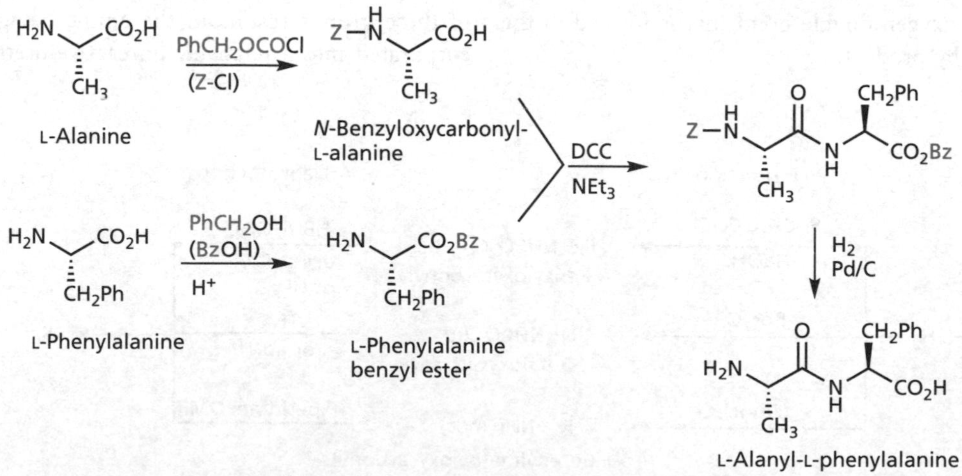

FIGURE CS1.16 Synthesis of L-alanyl-L-phenylalanine.

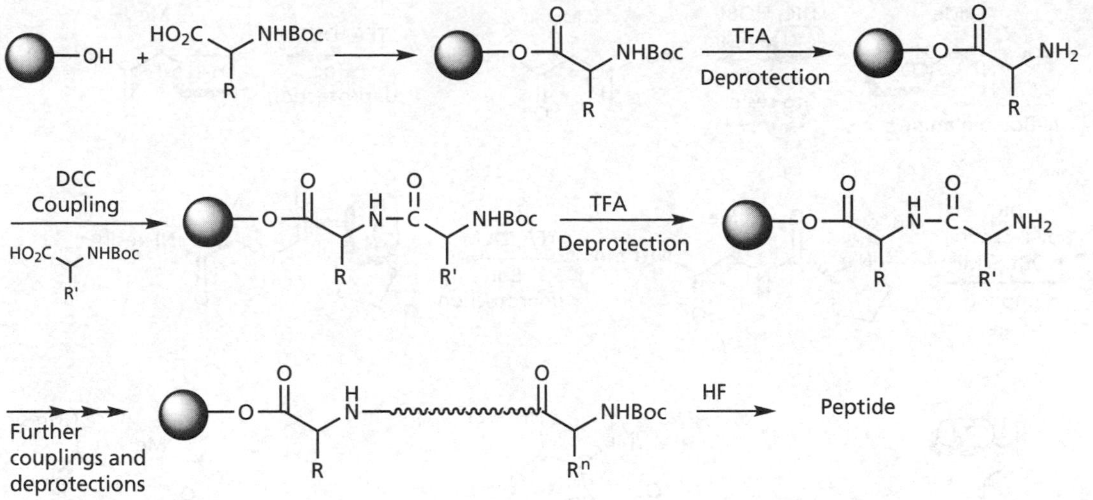

FIGURE CS1.17 Strategy involved in peptide synthesis on a solid support.

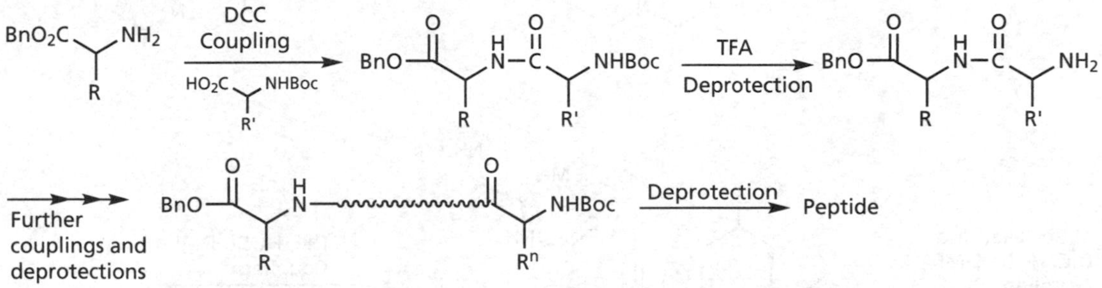

FIGURE CS1.18 Strategy involved in solution-phase peptide synthesis.

stage, the Boc protecting group is removed to free up the amine group for the next coupling reaction.

Functional groups on the amino acid side chains may also have to be protected during the synthesis, which means that any protecting groups on the side chain will have to be stable to TFA. Benzyl protecting groups fit the bill. They are stable to TFA but susceptible to hydrofluoric acid, which also detaches the peptide from the solid phase.

If the peptide is being synthesized in solution, rather than on a solid support, the carboxylic acid of the *C*-terminal amino acid is protected as a benzyl ester and is coupled to an *N*-Boc protected amino acid. The same process of deprotection and coupling now takes place, but this time the *C*-terminal amino acid has its carboxylic acid protected by a benzyl group throughout the synthesis (Fig. CS1.18).

The Fmoc/*t*-Bu protection strategy

The alternative Fmoc/*t*-Bu strategy can also be used when synthesizing a peptide from the *C*-terminus to the *N*-terminus. The 9-fluorenylmethyloxycarbonyl (Fmoc)

group is used to protect the terminal amine and can be removed using a mild base such as piperidine. Functional groups on amino acid side chains can be protected with a *t*-butyl group which can be removed by TFA. Since totally different reaction conditions are involved in removing the Fmoc group and the *t*-Bu group—one basic and one acidic—the protecting group strategy is defined as being orthogonal.

CS1.9 Synthesis of degarelix acetate

Degarelix acetate is a decapeptide structure that was approved in Europe in 2009 for the treatment of prostate cancer, and includes a mixture of natural and unnatural amino acids. The structure was synthesized in 25 stages on a solid support, building from the *C*-terminus to the *N*-terminus, which meant that the *C*-terminus of the growing peptide chain remained linked to the solid phase resin throughout

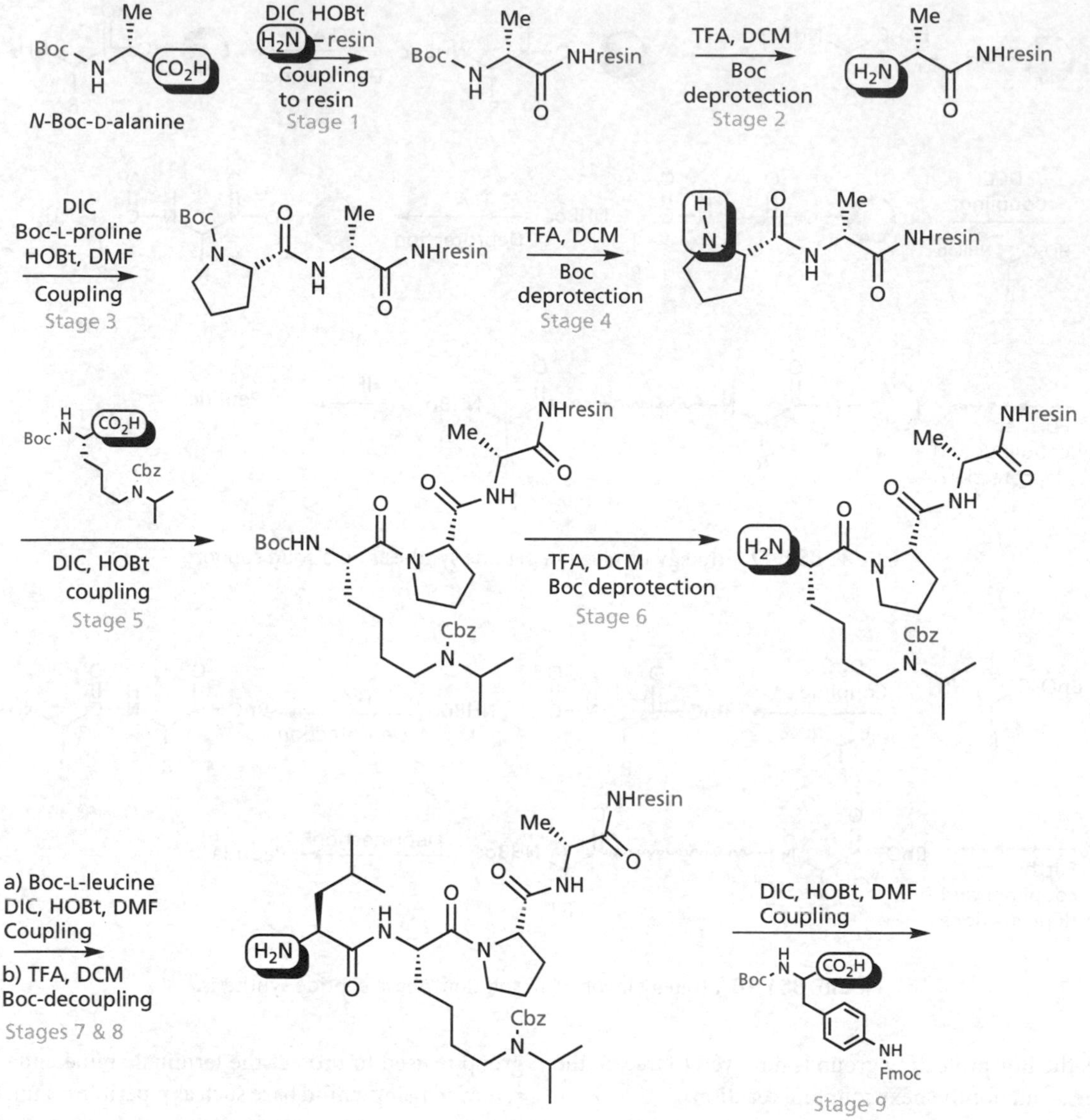

FIGURE CS1.19 Synthesis of degarelix acetate. Protecting groups are coloured blue. Reaction centres are boxed.

the process. The *t*-butyloxycarbonyl group (Boc) was used as the amine protecting group and was removed with TFA prior to each coupling reaction (Fig. CS1.19).

The real challenge in this synthesis involved choosing the correct protecting groups for the functional groups on the various side chains. This involved the use of one Cbz group, two Fmoc groups, one *t*-butyl group, and one benzyl protecting group.

The first four stages of the synthesis are straightforward as the amino acids concerned have no functional groups on their side chains. Therefore the first challenge comes at stage 5 where the third amino acid has a side chain with an amino group. It is important that this group is protected to prevent it competing in the peptide

couplings, and so a benzyloxycarbonyl group was used as this is stable to TFA and remains in place throughout the rest of the synthesis.

A more interesting challenge comes at stage 9, where the next amino acid contains an aromatic amine. This time an Fmoc protecting group has been used to protect the aromatic amine. At first sight this might seem an odd choice as an Fmoc group can be deprotected under milder conditions than either the Boc or Cbz groups. However, that was exactly what was required, as it was necessary to remove the Fmoc group without affecting the Boc group, such that the aromatic amine could be modified to a urea group (stages 10 and 11). The resulting urea was synthesized such that it was protected by a tertiary butyl group,

FIGURE CS1.19 (Continued)

which would be sufficiently stable to remain in place for the rest of the synthesis.

At stage 12, another amino acid was introduced with an aromatic amine protected by Fmoc. Again, this allowed the aromatic amine to be selectively deprotected and modified—this time to form an amide link to a piperidinone ring (stages 13 and 14).

The next complication comes at stage 16 when a serine residue containing an alcohol group in the side chain

was coupled. A benzyl group was used to protect the alcohol as it was stable to TFA and would remain in place throughout the remaining stages of the synthesis.

At stage 23, the *N*-terminus was deprotected in the normal manner, but this time it was acetylated instead of being coupled to another amino acid (stage 24). The final stage involved treatment with HF to remove the decapeptide from the resin, whilst deprotecting the Cbz, benzyl, and *t*-butyl protecting groups.

FURTHER READING

Jiang, G., et. al. (2001) 'GnRH antagonist: a new generation of long lasting analogues incorporating *p*-ureido-phenylalanines at positions 5 and 6', *Journal of Medicinal Chemistry*, **44**, 453–67 (degarelix acetate).

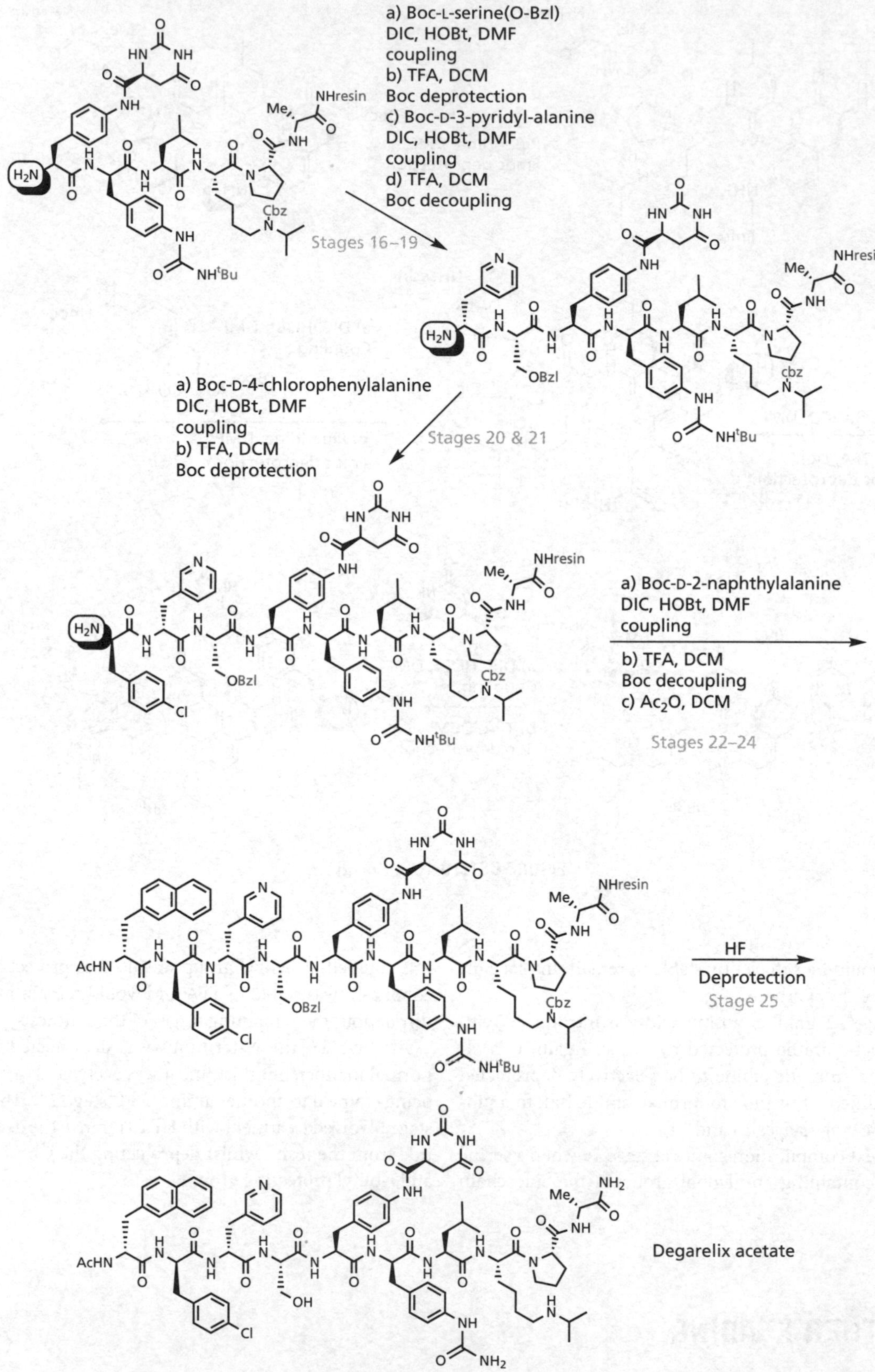

a) Boc-L-serine(O-Bzl)
DIC, HOBt, DMF
coupling
b) TFA, DCM
Boc deprotection
c) Boc-D-3-pyridyl-alanine
DIC, HOBt, DMF
coupling
d) TFA, DCM
Boc decoupling

Stages 16–19

a) Boc-D-4-chlorophenylalanine
DIC, HOBt, DMF
coupling
b) TFA, DCM
Boc deprotection

Stages 20 & 21

a) Boc-D-2-naphthylalanine
DIC, HOBt, DMF
coupling
b) TFA, DCM
Boc decoupling
c) Ac₂O, DCM

Stages 22–24

HF
Deprotection
Stage 25

Degarelix acetate

FIGURE CS1.19 (Continued)

■ CASE STUDY 2
Palladium-catalysed reactions in drug synthesis

CS2.1 Introduction

Palladium-catalysed reactions have been a relatively recent innovation in drug synthesis and have proved extremely useful in linking together molecules containing unsaturated carbon centres, and achieving reactions that would be difficult to do any other way. For example, they have proved successful in linking an aromatic ring with another aromatic or heteroaromatic ring to produce a biaryl structure, and have also been used to couple a terminal alkyne to an aromatic ring. In this case study, we shall consider some examples of such reactions used in drug syntheses.

CS2.2 Types of palladium-catalysed C–C coupling reactions

There are a number of palladium-catalysed reactions that can carry out C–C coupling reactions between different unsaturated carbon centres. They have been named after the chemists who developed them and include the Suzuki–Miyaura, Kosugi–Migita–Stille, Karasch–Kumada–Corriu–Tamao, Negeshi, Hiyama, and Sonagashera reactions. The most commonly used of these reactions is the Suzuki–Miyaura reaction, which is more commonly referred to as the Suzuki reaction.

CS2.3 The Suzuki–Miyaura reaction

The Suzuki–Miyaura reaction is more commonly called the Suzuki reaction and is an effective method of synthesizing biaryl structures, where a C–C bond is created between two aromatic or heteroaromatic rings (Fig. CS2.1). The reaction proceeds with good yield and selectivity under milder conditions than those used by alternative coupling methods involving Grignard reagents or tin compounds. The reaction involves the coupling of an aryl halide with an aryl boronate (or aryl boronic acid) in the presence of a palladium catalyst. Because of the popularity of this reaction, a large variety of aryl halides and aryl boronic acids are commercially available. The latter have proved to be relatively safe reagents, and the boron-containing by-products are easy to handle and remove. Moreover, the reaction can be carried out in the presence of a range of other functional groups.

Aryl iodides and bromides are normally used, but it is also possible to use aryl triflates. An aryl iodide is more reactive than an aryl bromide, and so it is possible to be selective in the coupling if more than one of these halogens is present. Aryl chlorides react too slowly to be effective, unless special reaction conditions are used such as nickel catalysts.

The solvent is usually ether or THF, but it is also possible to carry out the reaction with water, which is beneficial on large scale. The base is normally sodium carbonate, but other bases have been used.

The most frequently used catalyst is $Pd(PPh_3)_4$. The palladium essentially captures the aryl rings from both reagents, with displacement of the halide and boronate ions. The rings are then coupled and the palladium catalyst is released to catalyse another coupling reaction. The palladium catalyst can be used in very small quantities, which is particularly important for large-scale synthesis. Cross-coupling between aromatic and heteroaromatic rings to produce products such as aryl-substituted pyridines, pyrroles and indoles can also be carried out.

$$Ar\!-\!X \quad + \quad \begin{array}{c} Ar'\!-\!B(OH)_2 \\ \text{or} \\ Ar'\!-\!B(OR)_2 \end{array} \quad \xrightarrow[\text{C-C Coupling}]{\substack{Pd(PPh_3)_4 \\ Na_2CO_3}} \quad Ar\!-\!Ar'$$

Aryl halide   Aryl boronic acid
(X = I, Br)   or boronate

FIGURE CS2.1 Synthesis of biaryls using the Suzuki–Miyaura reaction (Ar = aromatic or heteroaromatic ring).

FIGURE CS2.2 Stages involved in the Suzuki reaction.

Unlike aryl chlorides, heteroaryl chlorides are sufficiently reactive to undergo reaction under the normal reaction conditions.

The mechanism of the reaction involves three stages (Fig. CS2.2). An initial oxidative addition takes place between the palladium catalyst and the aryl halide where the palladium is inserted between the aryl ring and the halogen atom. As a result, the oxidation state of the palladium atom changes from Pd(0) to Pd(II). The resulting palladium complex now reacts with the aryl boronate in a transmetallation reaction where the aryl group of the boronate is transferred from the boron atom to the palladium atom with substitution of the halogen. The palladium has now been inserted between the two aryl groups. Finally, a reductive elimination takes place where the palladium is eliminated and restored to its original oxidation state, while the two aryl groups are linked together in the C–C coupling.

The reaction has proved successful in a number of drug syntheses. For example, the synthesis of the antibacterial agent garenoxacin includes a Suzuki coupling towards the end of a multistage synthesis (section 14.5.3).

Another example can be seen in the synthesis of **solabegron** (Fig. CS2.3), which is currently undergoing clinical trials as a treatment for incontinence resulting from an overactive bladder. Solabegron works as a selective agonist of β_3 adrenoceptors, resulting in the relaxation of smooth muscle in the bladder. The compound has also been found to have analgesic properties and is being studied as a possible treatment for irritable bowel syndrome.

The synthesis starts with a Suzuki coupling reaction between an aryl bromide and an arylboronic acid to give a biaryl product (Fig. CS2.3, stage 1). A nitro substituent is present as a latent amino group and is now reduced under hydrogen to reveal the amine (stage 2) which is alkylated with an alkyl bromide (stage 3). The side chain

FIGURE CS2.3 Synthesis of solabegron.

FIGURE CS2.4 Possible competing reaction in the absence of the protecting group.

includes another amine group which is protected by a *t*-butyloxycarbonyl group. Without the protecting group, there could be a competing reaction where two molecules of the alkyl bromide react with each other (Fig. CS2.4). The protecting group is now removed under acid conditions (stage 4). The final stage involves an N–C coupling between the amine and an epoxide to alkylate the amine. This also reveals a secondary alcohol group which is needed in the final structure.

Overall, the synthesis uses four building blocks which are linked together in three coupling reactions, the last of which illustrates an example of a 'spring-loaded' reaction (section 3.8). The reaction also illustrates the use of the nitro group as a latent group for an aromatic amine and the use of a protecting group for the aliphatic amine.

CS2.4 The Kosugi–Migita–Stille reaction

The Kosugi–Migita–Stille reaction is normally called the Stille reaction and is another method of synthesizing biaryl compounds, which can be obtained in high yield and with good selectivity (Fig. CS2.5). The reaction takes place between an arylstannane and an aryl halide (or triflate) in the presence of a palladium catalyst. The addition of copper salts is often beneficial and it is thought that these may result in a more reactive organocopper reagent being formed in situ. When the organostannane is being linked to an aryl triflate, lithium chloride is normally added. Like the Suzuki reaction, the Stille reaction can be carried out in the presence of water. However, a serious disadvantage with the Stille reaction is the fact that the organotin reagents and by-products are toxic, making the reaction unsuitable for large-scale syntheses. Moreover the organostannanes are expensive.

The Stille reaction has been used in a synthesis of **lapatinib**, which is an anticancer agent that acts as a

kinase enzyme inhibitor. Kinase enzymes are responsible for catalysing the phosphorylation of amino acid residues in protein substrates. These phosphorylated proteins trigger a series of signal transduction processes which ultimately result in cell growth and division. By inhibiting kinase enzymes in tumour cells, it is possible to inhibit tumour growth.

The synthesis of lapatinib involves four building blocks which are linked together in three consecutive coupling reactions (Fig. CS2.6).

The first reaction involves an N–C coupling where a primary amine group substitutes a chlorine group from a bicyclic heteroaromatic starting material. The reaction is chemoselective since the iodine substituent is unaffected. This chemoselectivity is possible because of the nitrogen atoms in the right-hand ring which can stabilize the negative charge resulting from the initial addition of the amine (Fig. CS2.7). This stabilization is not possible if the amine added to the carbon bearing the iodine substituent.

The second reaction is the Stille C–C coupling reaction which involves the aryl iodide and an organostannane in the presence of a palladium catalyst.

The final coupling reaction is an example of an N–C coupling reaction between an amine and an aldehyde, carried out under reducing conditions. This results in an amine being formed rather than an imine. The reducing agent used here is sodium triacetoxyborohydride ($NaBH(OAc)_3$).

CS2.5 The Negeshi reaction

The Negeshi reaction (Fig. CS2.8) involves the reaction of an alkyl or aryl zinc reagent with an alkyl or aryl halide (or triflate) in the presence of a palladium or nickel catalyst. The reaction has good yield and selectivity, and the organozinc reagent can be generated in situ. However, the reaction cannot be carried out in water, and there are safety issues when it is carried out on large scale.

The Negeshi reaction was used in a synthesis of **elvitegravir**, which is an antiviral drug that was approved in 2012 for the treatment of HIV-infected patients. The drug acts as an enzyme inhibitor and inhibits a viral enzyme called **integrase**. This enzyme is responsible for incorporating viral DNA into host DNA as part of the

FIGURE CS2.5 Synthesis of biaryls by the Stille reaction (Ar = aromatic or heteroaromatic ring).

FIGURE CS2.6 The final three coupling stages in the synthesis of lapatinib.

FIGURE CS2.7 Mechanism for the nucleophilic substitution reaction of stage 1.

life-cycle of the human immunodeficiency virus. Inhibiting this enzyme interrupts the life-cycle of the virus and slows down the development of AIDS.

The synthesis involves nine stages (Fig. CS2.9). The first stage is a functionalization reaction where an iodine substituent is added to 2,4-difluorobenzoic acid. This iodine substituent will eventually direct the Negeshi coupling reaction to that position of the aromatic ring. The two fluorine atoms also serve a synthetic purpose as neither

of them will be present in the final structure. One will act as a leaving group in a later cyclization reaction, while the other will act as a leaving group to allow the introduction of a methoxy group at the final stage of the synthesis.

The second stage is an activation reaction which converts the carboxylic acid to a more reactive acid chloride, such that a C–C coupling reaction can take place in stage 3. This is then followed by an N–C coupling reaction where a primary amine group in the next building

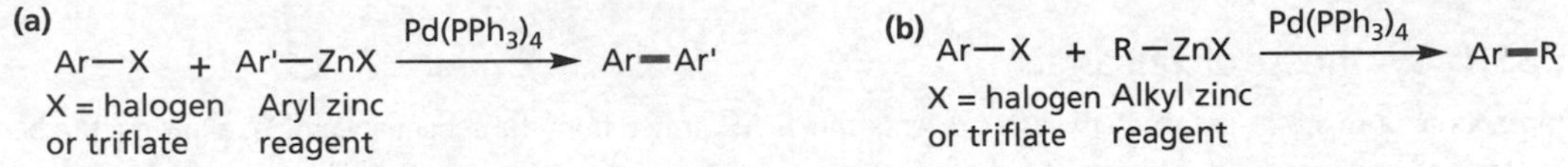

FIGURE CS2.8 The Negeshi reaction.

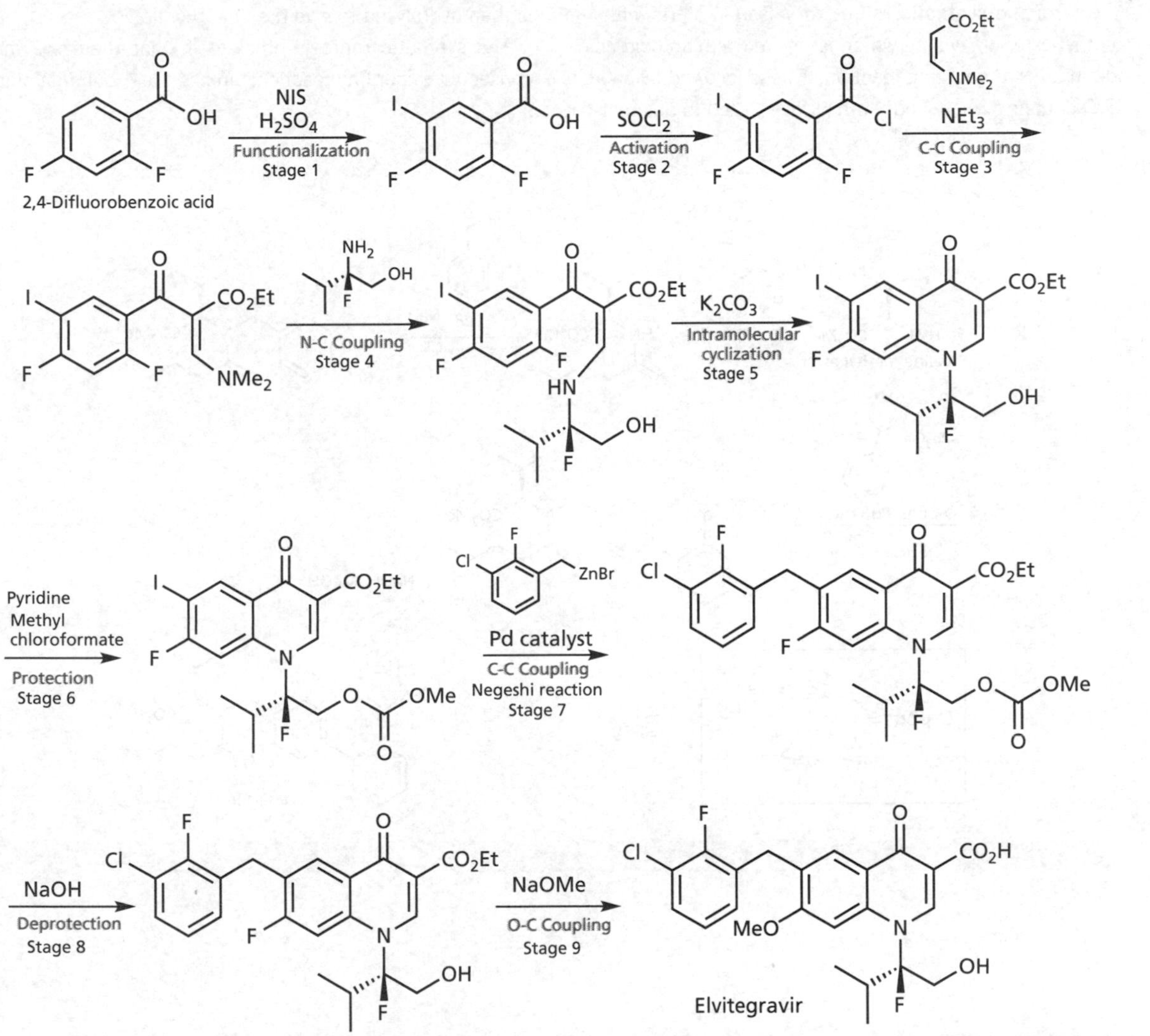

FIGURE CS2.9 Synthesis of elvitegravir.

block displaces the dimethylamino group (stage 4). The building block bearing the primary amine contains an unprotected alcohol, but there is no need to protect this as the primary amine is more nucleophilic, and the reaction goes with good chemoselectivity.

An intramolecular cyclization is now carried out (stage 5) which involves a nucleophilic substitution by the amine on the aromatic ring to displace a fluorine substituent. The alcohol group is now protected in stage 6, prior to the Negeshi coupling reaction between the aryl

iodide and an organozinc reagent (stage 7). The alcohol group is deprotected at stage 8, and then the fluoro group in the bicyclic ring is substituted with a methoxy group.

Overall, the synthesis involves five building blocks which are linked together in four coupling reactions and a cyclization. The remaining four reactions involve functionalization of the initial starting material, an activation, and the addition and removal of a protecting group.

The Negeshi reaction has also been used in the synthesis of valsartan (Box CS2.1).

BOX CS2.1 Synthesis of valsartan.

Valsartan acts as a receptor antagonist which prevents the peptide hormone angiotensin II interacting with its receptor to promote vasoconstriction of the blood vessels. As a result, valsartan is used as an antihypertensive agent to lower blood pressure.

One of the synthetic routes to valsartan starts with the methyl ester of L-valine which is reacted with an acid chloride in an N–C coupling reaction. This is followed by another N–C coupling where sodium hydride is used to abstract the

NH proton from the secondary amide, allowing it to be alkylated with 4-bromobenzyl bromide.

The Negeshi coupling reaction is now carried out in the presence of a catalytic amount of Q-phos and palladium acetate to produce the biaryl ring system. The methyl ester is then hydrolysed to give the final product.

The synthetic route is efficient in that there are three consecutive coupling reactions linking the 4 building blocks involved.

CS2.6 The Hiyama reaction

The Hiyama reaction involves a coupling reaction between organosilanes and aryl halides in the presence of a palladium catalyst (Fig. CS2.10). The aryl halide used is normally an aryl iodide or an aryl bromide. The reaction

proceeds in good yield and selectivity, and the organosilanes have low toxicity. However, special reactors are required if a fluoride ion is being used as an activating group.

CS2.7 The Kharasch-Kumada-Corriu-Tamao reaction

Biaryls can be synthesized by reacting an aryl Grignard reagent with an aryl halide in the presence of a palladium or a nickel catalyst. However, the reaction suffers the disadvantage that it is not possible to have nitrile, ester, or cyano

FIGURE CS2.10 The Hiyama reaction.

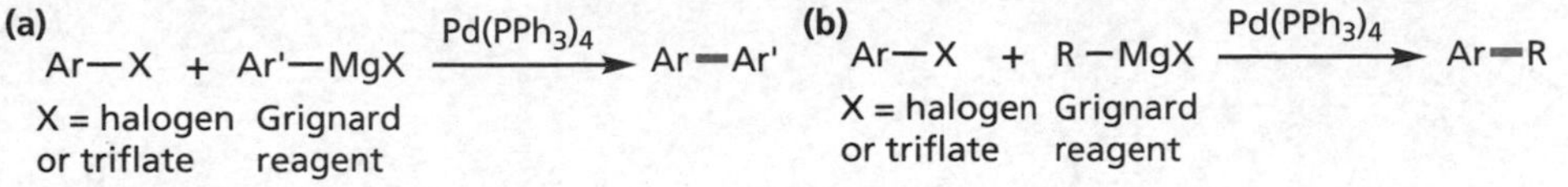

(a)

Ar—X + Ar'—MgX $\xrightarrow{Pd(PPh_3)_4}$ Ar—Ar'

X = halogen   Grignard
or triflate   reagent

(b)

Ar—X + R—MgX $\xrightarrow{Pd(PPh_3)_4}$ Ar—R

X = halogen   Grignard
or triflate   reagent

FIGURE CS2.11 The Karasch–Kumada–Corriu–Tamao reaction.

groups present in the starting materials, as these would react with the Grignard reagent. Nevertheless, the method is convenient for the synthesis of relatively simple biaryl structures (Fig. CS2.11). The reaction can also be used to couple alkyl Grignard reagents with aryl halides.

The reaction was used to create a biaryl ring system in an early synthesis of the antiviral agent **atazanavir**, which is used as a protease inhibitor in the treatment of AIDS. The aldehyde group of 4-bromobenzaldehyde was first protected as an acetal and then reacted with magnesium to create a Grignard reagent (Fig. CS2.12). This was reacted with 2-bromopyridine in the presence of a nickel catalyst and a catalytic amount of DIBAL to provide the desired biaryl ring system. The DIBAL is present to reduce the nickel catalyst to the correct oxidation level. A further five reactions were required to complete the synthesis.

Although the yield from the Kumada coupling was high, later studies showed that it was more efficient to generate the biaryl ring system using the Suzuki reaction.

CS2.8 **The Sonogashira reaction**

A recent review of reactions used in drug discovery over the last few years has shown that the Sonogashira reaction is second only to the Suzuki reaction as the most common method of creating C–C bonds. The reaction involves a palladium-catalysed coupling of a terminal alkyne to an aryl halide (Fig. CS2.13). Like the Suzuki reaction, the starting materials are generally stable and free from unacceptable toxicity. However, care has to be taken to remove all traces of palladium from the product to avoid toxic effects in bioassays. There are also safety issues related to the use of alkynyl derivatives on the large scale.

Aryl iodides react more effectively than aryl bromides, and both of these react more effectively than aryl chlorides. Not surprisingly, aryl iodides and bromides are normally preferred for the reaction. The palladium catalysts are usually $PdCl_2(PPh_3)_2$ or $Pd(PPh_3)_4$.

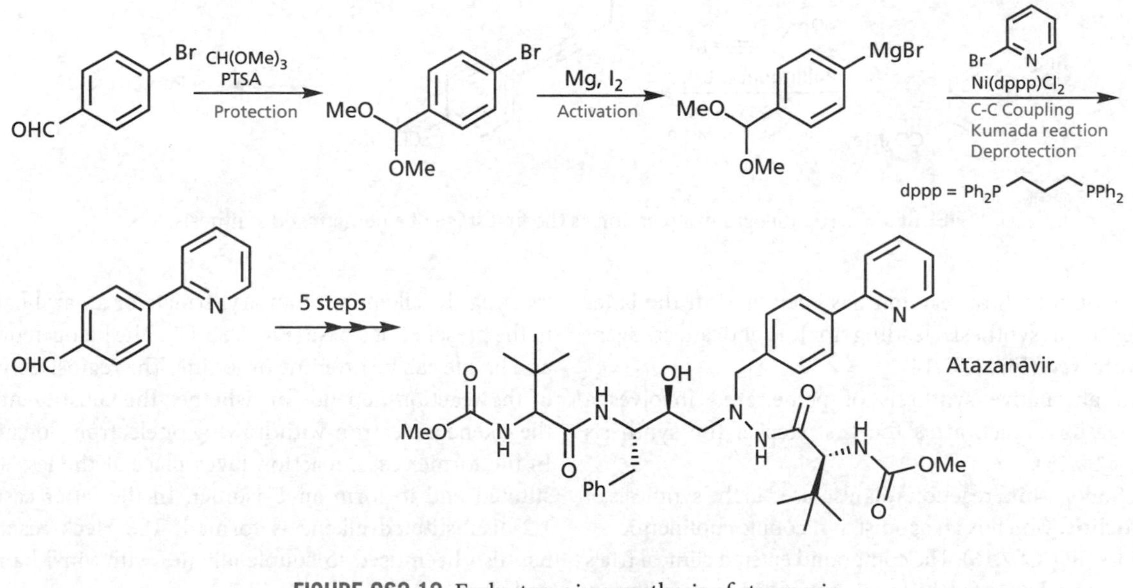

FIGURE CS2.12 Early stages in a synthesis of atazanavir.

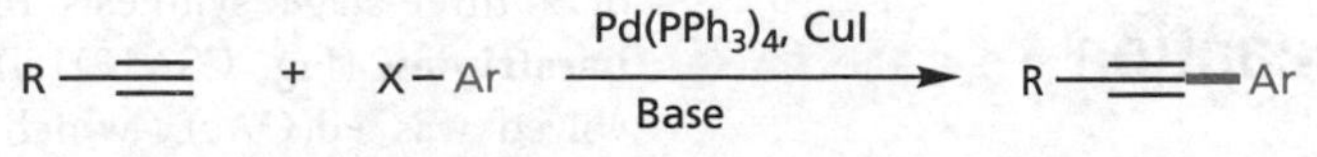

R$\equiv$ + X—Ar $\xrightarrow[\text{Base}]{Pd(PPh_3)_4,\ CuI}$ R$\equiv$Ar

FIGURE CS2.13 The Sonogashira reaction.

FIGURE CS2.14 Final stages in a synthesis of pemetrexed.

FIGURE CS2.15 Sonogashira reaction as the first stage of a pemetrexed synthesis.

The Sonogashira reaction has been used in the latter stages of a synthesis leading to the anticancer agent **pemetrexed** (Fig. CS2.14).

An alternative synthesis of pemetrexed involves a Sonogashira reaction as the first step of the synthesis (Fig. CS2.15).

A Sonogashira reaction was also used in the synthesis of **altinicline**, which is an agonist at nicotinic cholinergic receptors (Fig. CS2.16). The compound entered clinical trials as a potential drug for the treatment of Parkinson's disease.

CS2.9 The Heck reaction

The Heck reaction is a palladium-catalysed reaction where an alkene is coupled to an aromatic ring. It involves reacting the alkene with an aryl triflate or an aryl halide in the presence of a base (Fig. CS2.17). The halogen in the aryl halide can be bromine or iodine. The regioselectivity of the reaction depends on whether the substituent on the alkene is electron-withdrawing or electron-donating. In the former case, reaction takes place at the less substituted end to form an *E*-isomer. In the latter case, a 1,2-disubstituted alkene is formed. The Heck reaction has also been used to couple alkenes with vinyl halides and alkyl halides.

The Heck reaction was used as the second stage of a three-stage synthesis of the antimigraine agent **naratriptan** (Fig. CS2.18). The palladium catalyst involved was Pd(OAc)$_2$, which underwent reduction in situ to form a palladium(0) complex with two tri(o-tolyl) phosphine ligands.

FIGURE CS2.16 Synthesis of altinicline.

FIGURE CS2.17 The Heck reaction (X = Br, I, or triflate).

FIGURE CS2.18 Three-stage synthesis of naratriptan.

FURTHER READING

General reading

Chemistry World (2010) 'Palladium-catalysed couplings', November, pp 40–3.

Corbet, J.-P. and Mignani, G. (2006) 'Selected patented cross-coupling reaction technologies', *Chemical Reviews*, **106**, 2651–710.

Crisp, G.T. (1998) 'Variations on a theme—recent developments on the mechanism of the Heck reaction and their implications for synthesis', *Chemical Society Reviews*, **27**, 427–36.

Dounay, A.B. and Overman, L.E. (2003) 'The asymmetric intramolecular Heck reaction in natural product total synthesis', *Chemical Reviews*, **103**, 2945–63.

Fan, X. et al. (2008) 'An efficient and practical synthesis of the HIV protease inhibitor atazanavir via a highly diastereoselective reduction approach', *Organic Process Research and Development*, **12**, 69–75.

Kotha, S. (2008) 'Recent applications of the Suzuki-Miyaura cross-coupling reaction in organic synthesis', *Tetrahedron*, **58**, 9633–95.

Miyaura, N. and Suzuki, A. (1995) 'Palladium-catalysed cross-coupling reactions of organoboron compounds', *Chemical Reviews*, **95**, 2457–83.

Negishi, E. and Anastasia, L (2003) 'Palladium-catalysed alkynylation', *Chemical Reviews*, **103**, 1979–2017.

Shibasaki, M. (1997) 'The asymmetric Heck reaction', *Tetrahedron*, **53**, 7371–95.

Stanforth, S.P. (1998) 'Catalytic cross-coupling reactions in biaryl synthesis, *Tetrahedron*, **54**, 263–303.

Wagner, F.F. and Comins, D.L. (2006) 'Expedient five-step synthesis of SIB-1508Y from natural nicotine', *Journal of Organic Chemistry*, **71**, 8673–5.

Specific compounds

Carter, M.C., et al. 'Bicyclic heteroaromatic compounds as protein tyrosine kinase inhibitors', US Patent 6,727,256 (lapatinib).

Donaldson, K.H., et al. 'Therapeutic biaryl derivatives', US Patent 6,251,925 (solabegron).

Harbeson, S.L. (2009) '4-Oxoquinoline derivatives', US Patent 143,427 A1 (elvitegravir).

Oxford, A.W., et al. 'Indole derivatives', US Patent 4,997,841 (naratriptan).

Taylor, E.C. (1992) 'A dideazatetrahydrofolate analogue lacking a chiral center at C-6, *N*-[4-[2-(2-amino-3,4-dihydro-4-oxo-7H-pyrrole[2,3-d]pyrimidin-5-yl)ethyl]benzoyl]-L-glutamic acid, is an inhibitor of thymidylate synthase', *Journal of Medicinal Chemistry*, **35**, 4450–4 (pemetrexid).

CASE STUDY 3
Synthesis of (–)-huprzine A

(–)-Huperzine A (Fig. CS3.1) is a natural product that is obtained from a Chinese herb called *Huperzia serrata*. It is of interest because it inhibits an enzyme called acetylcholinesterase which normally hydrolyses the neurotransmitter acetylcholine and terminates its action (Fig. CS3.2). Inhibiting acetylcholinesterase increases acetylcholine levels in the brain, which is of benefit in alleviating the symptoms of Alzheimer's disease. Huperzine A has also been studied as a possible method of counteracting the nerve agents that normally act as irreversible inhibitors of the acetylcholinesterase enzyme.

The compound has been tested clinically and is effective with a good therapeutic ratio. However, it is not easy to obtain significant quantities of the compound for further testing as the yield obtained from harvesting the plant is only 0.011%. To add to this difficulty, the plant has to reach maturity before it can be harvested, and this takes 20 years!

In many cases, it is possible to synthesize a natural product from a related natural product—often a biosynthetic precursor (section 9.3). This is known as a semi-synthetic approach. Unfortunately, semi-synthetic methods of producing huperzine A have not been developed as it has not been possible to isolate biosynthetic precursors in sufficiently high yield.

Therefore, there would be a huge benefit in designing a full synthesis of huperzine A, as it would allow the structure to be synthesized in greater quantities. Moreover, such a synthesis would allow analogues to be made, which might have improved properties. Various synthetic routes have already been devised, but they are not feasible for large-scale synthesis. For example, the most efficient route involved 16 steps with an overall yield of only 2.8%. However, a new synthetic route has recently been devised which involves fewer stages and produces vastly increased yields of 35–45%. The design of this synthesis was based on a retrosynthetic analysis, where a key stage involved disconnection of bonds a and b of the structure.

First, it was important to identify the functional groups that are present. These consist of a primary amino group, an endocyclic trisubstituted alkene, an exocyclic trisubstituted alkene and a pyridinone ring (Fig. CS3.1).

A primary amino group is quite a reactive nucleophilic group which will react with most electrophiles, and so it makes sense to protect or 'mask' this group such that it is not revealed until the end of the synthesis. In addition, it

FIGURE CS3.1 (–)-Huperzine A.

FIGURE CS3.2 Hydrolysis of acetylcholine catalysed by acetylcholinesterase.

FIGURE CS3.3 First two stages of the retrosynthesis.

was decided that the endocyclic alkene group should be protected until the final stages. Therefore, the first stage in the retrosynthesis involves a couple of functional group interconversions to give a structure where both the amino and endocyclic alkene groups are protected or present as 'latent' groups. This is indicated by X and Y in Figure CS3.3.

Having disguised these groups, the exocyclic alkene group is now ripe for disconnection as it should be possible to create this bond by means of a Wittig reaction. This corresponds to the second stage of the retrosynthesis.

The research team then chose to disconnect bonds a and b to generate two monocyclic synthons (Fig. CS3.4). Note that there are distinct advantages to these disconnections. Both bonds are next to branch points or ring junctions, which means that significantly simpler looking

monocyclic structures are generated as synthons. The disconnections are also within the heart of the structure, and so the two synthons produced are roughly equivalent in size. This means that the corresponding synthesis will be convergent and more efficient.

The nucleophilic synthon corresponds to a cyclohexanone ring. This looks feasible as the required nucleophilic centres are on the α-carbons next to the carbonyl group, matching their natural polarity. Treatment of the ketone with a base should remove protons from these positions to produce carbanions. The electrophilic synthon corresponds to a pyridinone reagent with halogen substituents.

Further retrosynthetic analysis led to the conclusion that a chiral cyclohexenone and a dibromo pyridine structure could act as simple starting materials (Fig. CS3.5). The asymmetric centre in the cyclohexanone structure is

FIGURE CS3.4 Key bond disconnections.

FIGURE CS3.5 Starting materials for the synthesis.

FIGURE CS3.6 Synthesis of the chiral cyclohexenone starting material from (+)-pulegone.

important as it was expected to control the stereochemistry of various stages in the synthesis such that a single enantiomer of huperzine would be obtained.

As the chiral cyclohexenone was not commercially available, it was synthesized from a chiral compound called (+)-**pulegone** (Fig. CS3.6). (+)-Pulegone is a naturally occurring plant monoterpene and is commercially available.

The rest of the synthesis was then carried out as follows (Fig. CS3.7).

The first stage of the reaction scheme is a busy looking one! It links the two starting materials, generates two new chiral centres with stereocontrol, and introduces the protecting group for the endocyclic alkene in the final product. The reaction involves a conjugate addition of lithium dimethylphenylsilylcuprate to the α,β-unsaturated ketone group of the cyclohexenone. This gives an enolate intermediate which is then alkylated with the pyridine reagent. As a result of this reaction, two new chiral centres are formed, with the chiral centre in the cyclohexenone starting material influencing

the diastereoselectivity such that only the desired diastereoisomer is obtained. The dimethylphenylsilyl group that is introduced now acts as the protecting group for the alkene group which will eventually be formed at that position.

Treatment of the product with a strong non-nucleophilic base ((TMS)$_2$NLi) under kinetic control forms the carbanion at the least substituted α-carbon, and this is treated with para-toluenesulphonyl cyanide (TsCN) to introduce a cyanide group. The cyanide group is added as it will eventually be converted to the primary amine group in the final product. Therefore it can be viewed as a latent amino group.

A palladium-catalysed intramolecular heteroarylation is then carried out to create the ring skeleton of huperzine. This is followed by the Wittig reaction, which is carried out using a lithium ylide. Selectivity for the desired E-isomer is observed under kinetic conditions to form the complete carbon skeleton of huperzine.

It is now a case of carrying out functional group transformations to reveal the necessary functional groups.

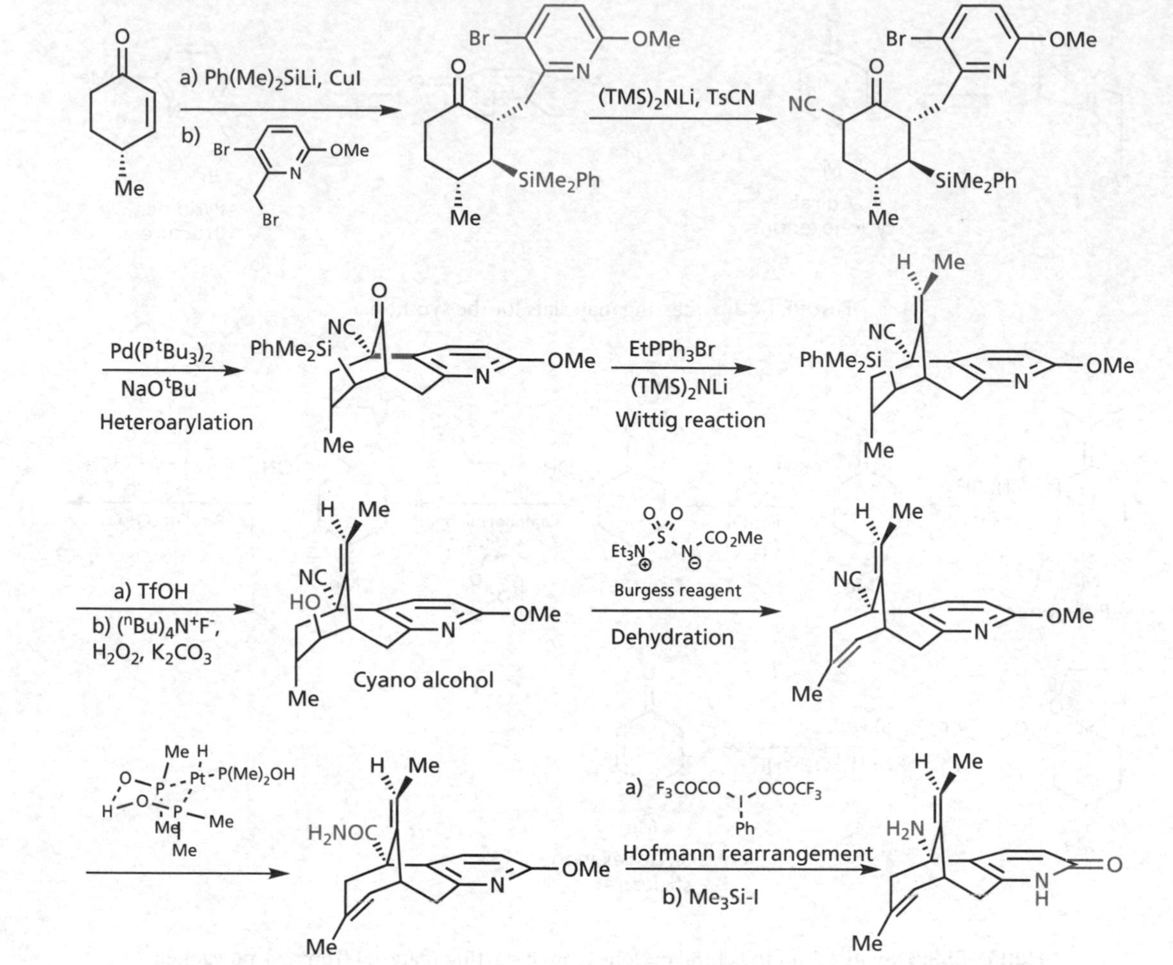

FIGURE CS3.7 Synthesis of huperzine A.

Treatment with trifluoromethanesulphonic acid followed by oxidative desilylation with a fluoride salt gives a cyano alcohol. Dehydration of the alcohol is carried out by heating the compound with **Burgess reagent**, and then the dehydrated product is treated with a platinum catalyst to convert the nitrile to an amide.

A **Hofmann rearrangement** is then carried out in the presence of bis(trifluoroacetoxy)iodobenzene. Deprotection with trimethysilyl iodide completes the synthesis. Batches of up to 3.6 g have been prepared in 35–45% overall yield which is a 16-fold improvement on previous methods.

To conclude, the full synthesis of a natural product often poses difficult challenges because of the presence of complex ring systems, several functional groups, and chiral centres. Early synthetic routes used to produce such a compound may involve a large number of steps and result in low overall yields, making them impractical for the production of analogues or significant quantities of the compound itself. However, there are usually several approaches that one can try in order to prepare complex compounds. Intelligent use of retrosynthetic strategies can lead to novel synthetic approaches that result in more efficient syntheses involving fewer steps and higher overall yields.

FURTHER READING

Tun, M.K.M., et al. (2011) 'A robust and scalable synthesis of the potent neuroprotective agent (−)-huperzine A', *Chemical Science*, **2**, 2251–3.

PART B

Applications of drug synthesis in the drug development process

Part B contains five chapters which describe how organic synthesis plays a crucial role at different stages of the drug discovery and development process.

Chapter 7 is related to the search for novel structures which might act as 'hits' for specific drug targets such as enzymes or receptors. From such hits, it may be possible to identify a 'lead compound' which can act as a basis for further study. The synthetic priority here is to prepare compounds that are structurally diverse.

Once a promising lead compound has been identified, it is important to prepare a large number of analogues in order to identify structure–activity relationships, and to try to find structures having improved properties. Chapter 8 describes how the synthetic routes used will determine what kinds of analogues can be made and the positions in a structure that are most amenable to change.

Several important drugs originate from the natural world. Chapter 9 focuses on the preparation of such natural prod-

ucts and describes different methods of preparing analogues. These methods include synthesis, biosynthesis, and genetic engineering.

Chapter 10 looks at chemical and process development. Once a promising compound has been identified, it goes forward to preclinical and clinical trials. At the same time, development and process chemists are tasked with finding an efficient cost-effective synthetic route that will allow the drug to be manufactured on large scale. This chapter describes the various factors that have to be considered in achieving that goal.

Chapter 11 covers the synthesis of isotopically labelled drugs, involving heavy isotopes and radioisotopes. Labelled drugs are particularly important in drug metabolism studies carried out as part of preclinical and clinical trials. However, labelled drugs are also important in biosynthetic studies, drug testing, and diagnostics. Some labelled drugs also have a therapeutic benefit.

7 Synthesis of lead compounds

7.1 Introduction

A lead compound is a molecule which shows some activity or property of interest, and which has the potential to be developed into a useful drug. Many lead compounds show some kind of useful pharmacological activity such as enzyme inhibition or receptor activation. However, the presence of pharmacological activity is not necessarily essential. Just the ability to bind to a molecular target that is implicated in a disease process—such as a protein or a nucleic acid—can often be enough to identify the compound as a useful lead compound.

Some lead compounds that *do* have pharmacological activity may be sufficiently active to be used in medicine—for example, several of the natural products isolated from the natural world (Chapter 9)—but most lead compounds would make poor medicines because of various problems such as lack of activity, the presence of side effects, toxicity issues, chemical instability, poor absorption, metabolic susceptibility, rapid excretion, etc. However, the existence of such problems does not rule out the structure being adopted as a lead compound. The value in a lead compound is the fact that it 'recognizes' or binds to the molecular target. Once such a lead compound has been identified, the medicinal chemist can use its structure as the basis for designing analogues which bind more effectively to the target and have better pharmacokinetic properties—the aim being to introduce or increase the desired activity, whilst removing or diminishing any adverse effects.

Lead compounds have always been crucial in drug design and discovery. Without them, it would not be possible to begin the long process of developing a clinically useful drug. In the past, lead compounds were often found by chance, either from the natural world or from the stocks of synthetic compounds that have been prepared by pharmaceutical companies over the years.

As a result, the discovery of novel lead compounds was relatively slow and inefficient. With the genomic and proteomic revolution, there was a massive increase in the number of proteins discovered in humans and other organisms. In order to establish whether these proteins could serve as useful drug targets, there was an urgent need to find compounds that would interact with them and show useful activity. This placed a huge responsibility on synthetic chemists to come up with approaches that would increase the rate at which novel structures were synthesized. The development of combinatorial and parallel synthesis (Chapter 6) proved crucial in achieving that goal, and there was a dramatic acceleration in the number of novel compounds that were synthesized. Chemical libraries have been created that contain hundreds, thousands, or even millions of structures, and the compounds in these libraries can be tested to find 'hits' which can be assessed see whether any of them would be suitable lead compounds for a novel drug design project. There have been many successes, but the 'strike rate' from these libraries can often be disappointing, with a relatively low number of lead compounds being identified from the thousands or millions of compounds synthesized. As a result, there has been an increasing awareness that the quality of structures present within a chemical library is just as important as the quantity. In other words, the structures present should be as diverse as possible in order to increase the chances of finding good lead compounds. The term **chemical space** is often used to indicate structural diversity and refers to the vast number of possible structures that could theoretically be synthesized. It has been estimated that there are 10^{60} such possible structures, and so there is plenty of scope for future novel structures. We will now investigate why it is important to synthesize structurally diverse compounds, and identify methods of how to achieve that diversity.

7.2 **Characteristics of a lead compound**

What kind of characteristics are we looking for in a lead compound? The first thing to appreciate is that a synthetic lead compound is a small molecule, and is often significantly smaller than the final drug developed from it. Like the final drug, the lead compound has to bind to a much larger macromolecular target (section 1.7), such as a protein or nucleic acid, and so it will interact with only a small portion of that target. There are important lead compounds/drugs which bind to DNA and RNA; for example, the antibiotic tetracycline binds to bacterial RNA (Case Study 2). However, the majority of drug targets are proteins, such as receptors, enzymes, and transport proteins. Therefore, we will now focus on these.

The 'portion' of the protein target that we are interested in is the binding site, which is normally a cleft or hollow on the surface of the protein. This is where an endogenous ligand will bind. In the case of an enzyme, the endogenous ligand is called a substrate and undergoes an enzyme-catalysed reaction. In the case of a receptor, the endogenous ligand is a chemical messenger which acts as a 'messenger' and switches the receptor 'on'. A lead compound that is capable of binding to those binding sites could be developed into a drug that acts as an enzyme inhibitor, receptor agonist, or receptor antagonist, depending on the target involved.

As the binding site is a hollow or cleft, the lead compound has to have the correct shape and size to fit it. However, merely fitting into a binding site is not sufficient. After all, water can fit into binding sites, but is not going to be a sensible lead compound for the design of any drug. The lead compound also has to 'stick' to the binding site and be held there for a reasonable amount of time. In other words, there must be some kind of 'recognition' between the binding site and the lead compound. This recognition takes the form of intermolecular binding interactions between the lead compound and the binding site, such as hydrogen bonds, ionic bonds, and van der Waals interactions. For that to happen, both the binding site and the lead compound have to have the relevant functional groups and substituents to form these interactions. Moreover, they must be in the correct positions within both molecules so that several interactions can be made at the same time. One or two interactions will be far too weak to bind the ligand to the binding site for any significant time.

As far as the binding site of a protein is concerned, the functional groups and substituents involved in binding a ligand are provided by the peptide backbone and the amino acid side chains lining the surface of the binding site. The peptide links in the peptide backbone have the potential to form hydrogen bonding interactions where the carbonyl oxygen acts as a hydrogen bond acceptor and the NH proton acts as a hydrogen bond donor (Fig. 7.1). The side chains of amino acids such as serine, threonine, lysine, aspartic acid, glutamic acid, glutamine, asparagine, and tyrosine contain functional groups that are also capable of forming hydrogen bonds (Fig. 7.1). Amino acids such as lysine, glutamate, and aspartate have side chains capable of forming ionic or electrostatic bonds as well as hydrogen bonds (Fig. 7.2), while amino acids such as alanine, valine, leucine, isoleucine, tryptophan, and phenylalanine have side chains capable of forming van der Waals interactions (Fig. 7.3).

Those side chains and functional groups which 'stick out' into the binding site are accessible to any visiting ligand. Therefore, there will be particular regions in the binding site where hydrogen bonding groups are

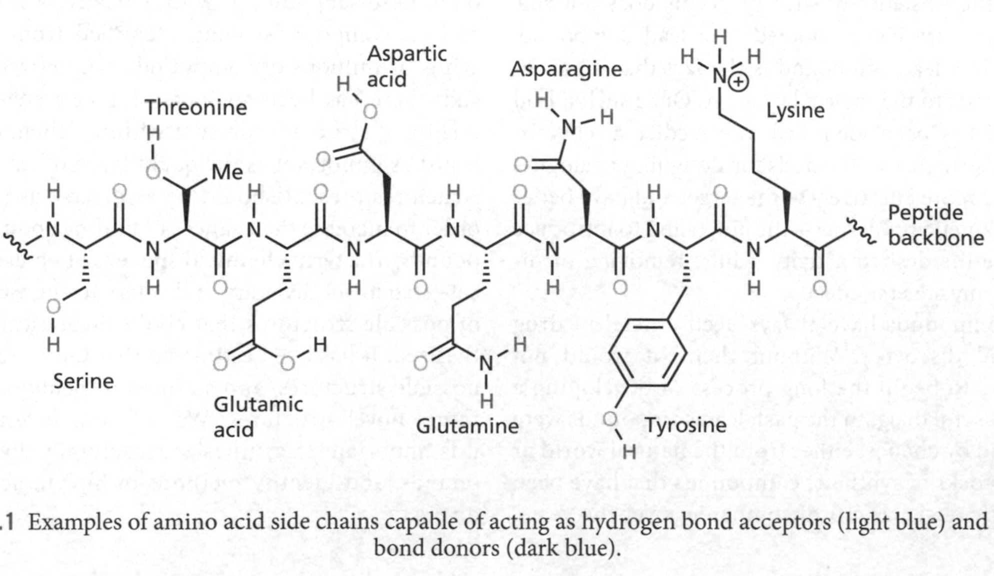

FIGURE 7.1 Examples of amino acid side chains capable of acting as hydrogen bond acceptors (light blue) and hydrogen bond donors (dark blue).

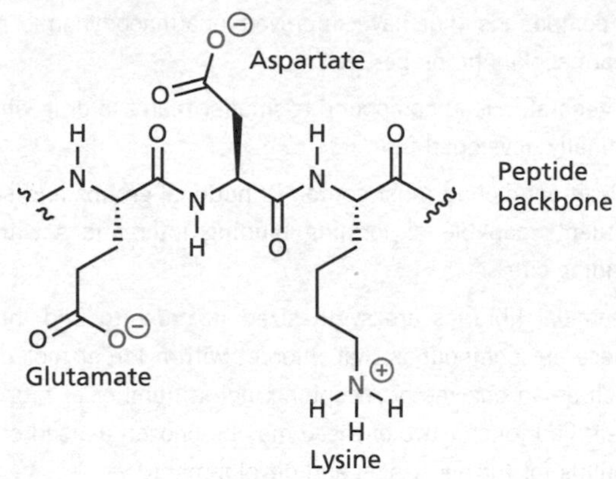

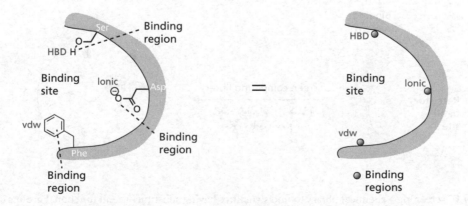

FIGURE 7.2 Examples of amino acid side chains capable of forming ionic or electrostatic interactions.

available, other regions where hydrophobic side chains are present, and yet other regions where an ionic group may be revealed. These represent the **binding regions** within the binding site (Fig. 7.4), and are the regions which can form interactions with endogenous ligands.

The initial aim in drug design is to find a lead compound that bears the necessary groups and substituents to interact with some of these binding regions.

So how do we improve our chances of finding such a compound? What kind of structures should be synthesized in a chemical library? Put at its simplest, a lead compound should have two features.

First, it should have a molecular skeleton which defines the size and shape of the molecule. This is known as the **core** or the **scaffold** (Fig. 7.5).

Secondly, it should have substituents and functional groups capable of forming different intermolecular interactions. These are the potential **binding groups** that, it is hoped, will interact with the binding regions in the binding site.

If we had a detailed knowledge of the binding site structure, we could conceivably design our lead compound such that it would have the correct binding groups in the correct positions around its scaffold. However, a detailed knowledge of the binding site is often unavailable, especially for novel targets, and so the process of

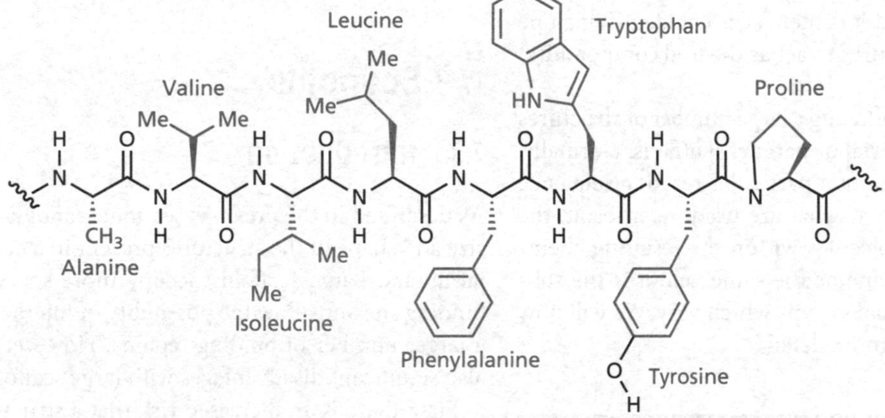

FIGURE 7.3 Examples of amino acid side chains capable of forming van der Waals interactions.

FIGURE 7.4 Representation of a binding site and binding regions.

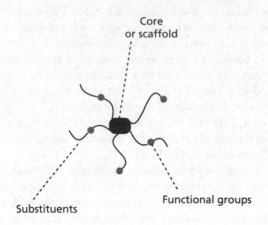

FIGURE 7.5 The components of a lead compound.

finding the lead compound is a bit of a hit and miss affair. To increase the chances of success, it is necessary to create as many compounds as possible, varying the nature and position of the substituents and functional groups around the scaffold in the hope that some of the structures will strike lucky (Fig. 7.6). Those structures are described as 'hits', and it is then a case of identifying one or two of these structures to act as the lead compound(s) for further study.

The best way of producing a large number of structures is through combinatorial or parallel synthesis. Normally, this involves carrying out a particular reaction sequence and varying the reagents that are used. As a result, the scaffold of all the molecules within the resulting chemical library usually remains the same, and it is the substituents and functional groups which vary. We will now consider scaffolds in more detail.

KEY POINTS

- A lead compound shows some activity or property of interest, and is the starting point for the design and development

of compounds that have improved pharmacodynamic and pharmacokinetic properties.

- In general, a lead compound is smaller than the drug which is finally developed from it.

- A lead compound must contain functional groups and substituents capable of forming binding interactions with a binding site.

- Chemical libraries are synthesized in order to find 'hits'. These are compounds that interact with a target molecule such as an enzyme or receptor. Once a number of hits are identified, one or two of these may be chosen as lead compounds for further design and development.

- A chemical library usually includes structures that share a common molecular skeleton or scaffold, with different substituents and functional groups around the scaffold.

- The diversity of the compounds within a library is crucial in increasing the chances of finding compounds that will interact with a particular target.

- Varying the number, nature, and positions of substituents and functional groups around a scaffold increases the diversity of the compounds present.

7.3 Scaffolds

7.3.1 Introduction

As described in the previous section, scaffolds define the size and shape of the structures present in a combinatorial library. Large scaffolds occupy more space within a binding site and allow the possibility of interactions with a larger number of binding regions. However, there are also significant disadvantages with large scaffolds.

First, there is an increased risk that a structure with a large scaffold will experience steric clashes that prevent it entering the target binding site. Even if the structure

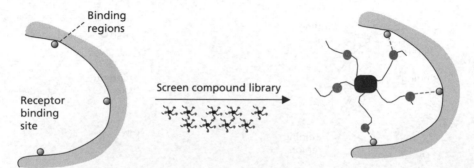

FIGURE 7.6 Screening a chemical library to find structures having substituents and functional groups capable of interacting with binding regions. Blue dots represent groups in each structure that can form different types of intermolecular interaction.

does enter the binding site, steric clashes may prevent it binding efficiently.

Secondly, large lead compounds usually lead to even larger drugs. That may be an issue when it comes to oral absorption. In general, drug designers prefer to design drugs that have a molecular weight of about 500 amu or less, because experience has shown that larger drugs tend to be poorly absorbed when taken orally. Since the process of developing a drug from a lead compound usually involves an increase in molecular weight, most researchers prefer lead compounds to have a molecular weight of about 300 or less.

Thirdly, the larger the structure, the more complex the synthesis is likely be.

Therefore, small scaffolds are generally favoured over large ones when preparing compound libraries aimed at finding 'hits' for a particular target. This means that the lead compound chosen from those hits may only interact with a small part of the binding site, and further drug development will allow the 'growth' of the structure to fill up more of the space available and increase the number of binding interactions.

Scaffolds serve as the core onto which different substituents and functional groups are placed, and the positioning and nature of these substituents will be determined by the synthetic route used and the choice of reagents available. The scaffold itself may also be involved in binding interactions. Those portions of the scaffold that are hydrocarbon in nature will be capable of forming van der Waals interactions with hydrophobic regions in the binding site. Other portions of the scaffold may contain functional groups or heteroatoms as an integral part of the structure, and these too can serve as potential binding groups.

There are an infinite number of potential scaffolds that could be used in the preparation of a compound library, but in general we can define them as being linear or cyclic in nature.

7.3.2 Linear scaffolds

Linear structures are generally much easier to synthesize than cyclic structures, especially if the linear structure is built up using a series of building blocks. Indeed, most of the early chemical libraries produced by combinatorial

chemistry consisted of thousands or millions of peptide structures built up using the naturally occurring L-amino acids. There are certainly several advantages in creating peptide libraries. Peptide synthesis is straightforward and there are an infinite number of possible peptide structures that can be synthesized by varying the length of the peptide chain, the type of amino acid building blocks used, and the order in which they are linked. The number and variety of potential peptides can be increased even further by including D-amino acids and synthetic 'unnatural' amino acids in the choice of building blocks. The side chains of both natural and unnatural amino acids provide a wide variety of functional groups capable of different types of interaction, and the peptide backbone itself provides a regular sequence of hydrogen bonding groups as we saw in Figures 7.1–7.3. Another potential advantage is the fact that many of the substrates used by the body's enzymes are peptides or proteins, and so there is a reasonable chance that a peptide-like structure may be capable of binding to the active site of the enzyme and provide a potential lead compound for the design of an enzyme inhibitor.

Unfortunately, there are also disadvantages. Peptides tend to be poorly absorbed from the gut, and are susceptible to both digestive and metabolic enzymes. This means that a lot of research has to be carried out in order to modify a peptide lead compound into a structure that avoids those problems. Such a structure is called a **peptidomimetic**—it mimics the peptide lead compound in its ability to bind to a target protein, but has much better pharmacokinetic properties.

There have been notable successes in developing peptidomimetics—for example the development of the protease inhibitor **saquinavir** used in the treatment of HIV (Fig. 7.7). However, most of these successes were based on identifying the substrate for the enzyme and using that as the lead compound. The use of peptide libraries to find lead compounds for other types of target has not been particularly effective or efficient.

Another approach has been to use polyamines as a source of lead compounds. Polyamines contain protonated nitrogens and have the ability to form ionic interactions with the carboxylate groups of aspartate or glutamate

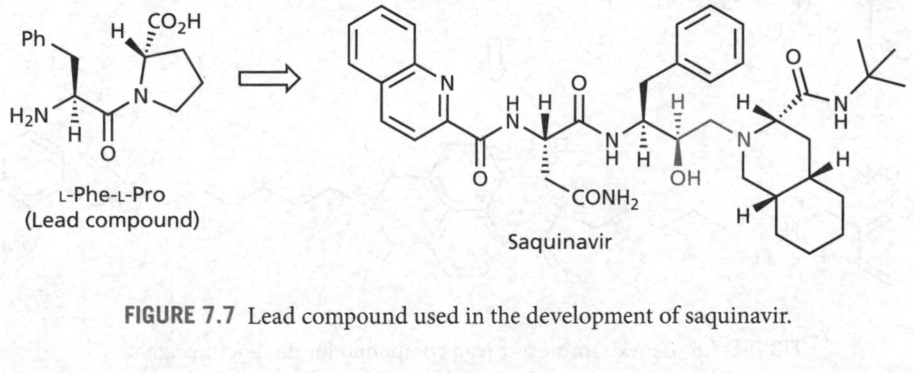

FIGURE 7.7 Lead compound used in the development of saquinavir.

side chains, as well as pi-cation interactions with aromatic residues in virtually any protein target. Moreover, the flexible linear structure allows the polyamine to adopt a huge number of different conformations, making it more likely that one of those conformations will allow the molecule to form interactions within a particular target binding site.

Indeed, it is possible for a single polyamine to interact with the binding sites of different protein targets by adopting different conformations. Such a compound is defined as a **promiscuous ligand** and can be a useful lead compound for the development of multi-targeted drugs. For example, a polyamine called **benextramine** was identified as a lead compound capable of interacting with the α-adrenoceptor, the muscarinic (M2) receptor, and the acetylcholinesterase enzyme. Further development led to a couple of dual-acting structures that showed selectivity for the last two targets, acting as antagonists at the muscarinic receptor and inhibitors of the acetylcholinesterase enzyme (Fig. 7.8). Both actions served to increase the levels of acetylcholine, which has potential in the treatment of Alzheimer's disease.

There are clear advantages in linear compounds being able to adopt different conformations, as there is a better chance of each individual structure being able to adopt a conformation that will be 'recognized' by a target binding site. However, there are also distinct disadvantages.

The more flexible the molecule, the more conformations can be formed. However, only one of these may represent the active conformation. If all the conformations have similar stability, relatively few molecules are in the active conformation compared with those that are in inactive conformations. That means that the majority of compounds will float into the binding site and then float back out again without binding. If a molecule can form a hundred different conformations, and only one of these is the active conformation, there is only a 1% chance of the molecule being in the correct conformation when it enters the binding site. Inevitably, this will result in low binding affinity—perhaps too low to be detected.

Moreover, when the molecule eventually does bind, there is a high entropic penalty to be paid since the structure is constrained to the active conformation, and is unable to move between the other possible conformations. This represents a large decrease in disorder which is detrimental to the overall binding process. A large entropic penalty counteracts the stabilization energy obtained from the binding interactions, leading to a poor overall binding affinity.

Another disadvantage is that the number of different conformations available to a linear structure increases the chances of interactions with different targets. This is useful when trying to find a multi-targeted drug, as described earlier, but the problem then comes of developing the drug such that it only interacts with the desired targets and not those that will result in side effects and toxicity.

Finally, linear flexible molecules with a large number of rotatable bonds have been found to be poorly absorbed from the gut, making it unlikely that such compounds will be orally active.

Because of these problems, a large amount of work often has to be carried out to modify a linear lead compound before it can become a feasible drug. One of the main strategies to achieve this is to incorporate rings to rigidify the structure and decrease the number of possible conformations.

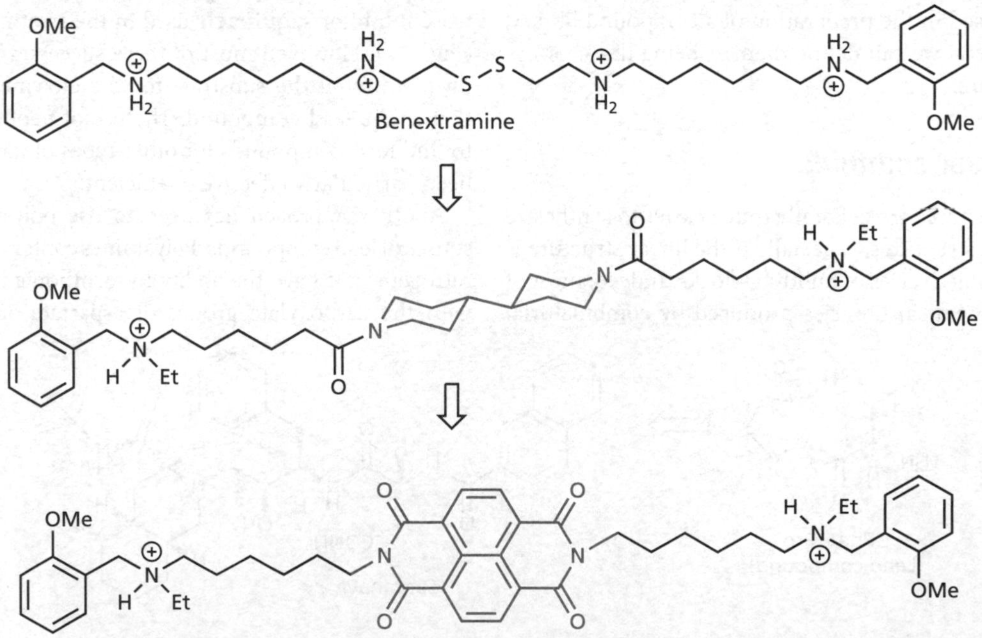

FIGURE 7.8 Benextramine as a lead compound for dual-acting agents.

If the introduction of cyclic systems into drugs is beneficial to binding affinity, activity, selectivity, and absorption, then why not start from a cyclic system in the first place? We shall now consider cyclic scaffolds.

7.3.3 Cyclic scaffolds

The advantages of cyclic scaffolds can be seen by the number of compound libraries which contain them. The cyclic system serves as a relatively rigid core or scaffold onto which different substituents and functional groups can be placed, and avoids many of the problems previously identified with linear flexible structures. There are far fewer rotatable bonds and less possible conformations. If a particular structure can adopt the active conformation, there is more chance that the structure will be in that conformation when it enters the binding site. Therefore, binding should be quicker. There is also less of an entropy penalty involved and so binding should be stronger. Finally, drugs eventually derived from the lead compound should be absorbed from the gut more effectively because of the small number of rotatable bonds present.

The downside of this greater rigidity is that there is less chance of any specific cyclic compound in the library being able to find a suitable conformation that will allow it to bind to a target binding site. However, that is not a major problem if a sufficiently large number of compounds are synthesized such that different substituents and binding groups are displayed around the scaffold of each component structure.

Having identified the merits of a cyclic scaffold, what kind of cyclic system should be used? In general, the preference is for a heterocyclic ring system rather than an carbocyclic ring system. There are several reasons for this. First, many of the popular heterocyclic ring systems used in chemical libraries contain one or more nitrogen atoms. This has the advantage that the nitrogen atom (or any hydrogen attached to that nitrogen) serves as a possible hydrogen bonding group capable of interacting with the binding site. Secondly, the heteroatoms within a cyclic system help to offset the hydrophobic nature of the ring. It is important that drugs should have a good balance of hydrophobic and hydrophilic properties if they are to be effective drugs with good pharmacokinetic properties (section 1.5.3).

Scaffolds containing one or more heteroaromatic rings are particularly popular scaffolds. This is because structures containing such scaffolds are usually easier to synthesize than those containing saturated ring scaffolds. This is a consequence of the many palladium-catalysed coupling reactions that have been developed over the last 20 years or so, such as the Suzuki, Heck, Stille, and Sonogashira reactions (Case Study 2). Another synthetic advantage is that heteroaromatic ring systems are planar in nature, and so placing substituents at different positions of the ring will not create chiral centres (Fig. 7.9). With saturated ring

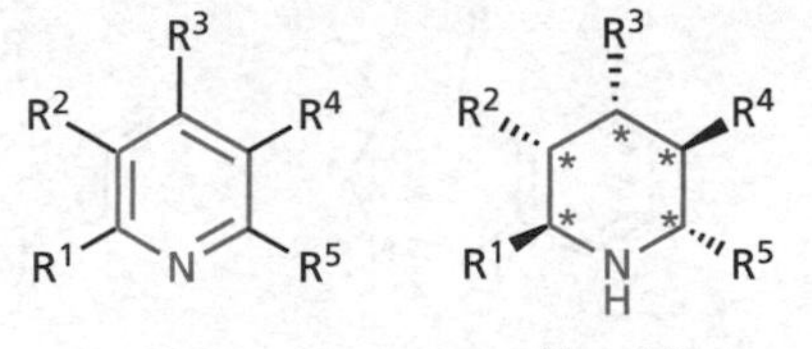

FIGURE 7.9 Unsaturated versus saturated heterocyclic rings.

systems, the introduction of a substituent inevitably creates a chiral centre, which also introduces the problem of controlling stereochemistry during the synthesis. For example, there are no chiral centres present in a pyridine ring containing five substituents. In contrast, a piperidine ring having five substituents would have five chiral centres, resulting in 32 possible stereoisomers.

Rigid heteroaromatic scaffolds are also more likely to show selectivity between different targets. They are capable of forming pi–pi interactions with other aromatic or heteroaromatic rings in the binding site, or pi–cation interactions with any positively charged aminium ions that might be present in the side chains of amino acids such as lysine and arginine.

However, it is not all plain sailing with heteroaromatic scaffolds. The planar hydrophobic nature of these rings can lead to drugs that have poor aqueous solubility. Also, the orientation of substituents from a rigid cyclic system is generally in the plane of the ring, limiting the scope for developing analogues capable of forming additional interactions with the binding site (section 8.4.1). Finally, many of the easily prepared heteroaromatic ring systems have already been synthesized and patented.

Some cyclic scaffolds are particularly common in medicinal chemistry (e.g. benzodiazepine, hydantoin, tetrahydroisoquinoline, benzenesulphonamide, biaryl, pyrimidine, and diphenylmethane ring systems) (Fig. 7.10). Drugs containing these scaffolds are associated with a diverse range of activities, and such scaffolds are termed as '**privileged**' scaffolds.

7.3.4 Positioning of substituents on scaffolds

A compound library is typically generated by using a specific reaction sequence and then varying the reagents used. However, the very fact that the same reaction sequence is being used means that many of the compounds produced are likely to be similar to each other. Indeed, in most chemical libraries the core scaffold is identical. And yet it is important that the molecules in these libraries are structurally diverse to increase the chances of finding hits. Therefore, the structural variety must come from varying the numbers, types, and positions of substituents attached to the scaffold

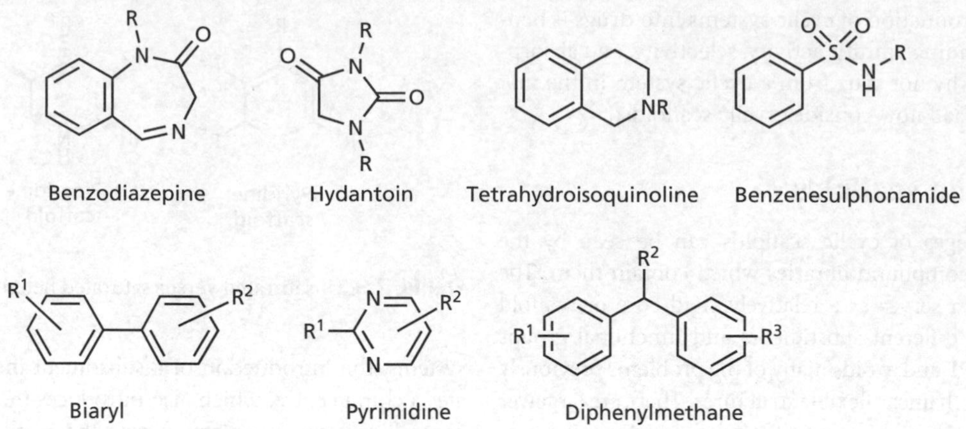

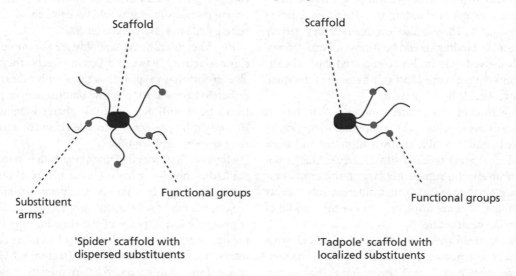

FIGURE 7.10 Examples of 'privileged' scaffolds.

FIGURE 7.11 'Spider-like' and 'tadpole-like' molecules.

which, in turn, is determined by the synthetic route used and the reagents available. Therefore, some thought has to be put into which synthetic route is most likely to give the widest diversity in the structures produced.

In general, it is best to synthesize 'spider-like' molecules—so called because the various 'arms' or substituents that include potential binding groups are spread evenly around the scaffold (Fig. 7.11). The chances of success are best if the 'arms' are evenly spread around the scaffold compared with the more localized arrangement of substituents in a 'tadpole-like' structure, as this allows a more thorough exploration of the 3D space (**conformational space**) around the molecule.

Ideally, we would also want to create structures where we could place different functional groups on each substituent arm, and at various distances along the arm. All the various structures could then be used as 'probes' to find complementary binding regions within the binding site (Fig. 7.6).

To conclude, the ideal scaffold should be rigid and small, in order to allow a wide variation of substituents. It should also have its substituents widely dispersed round its structure (spider-like) rather than restricted to one part of the structure (tadpole-like). Finally, a suitable synthesis has to be available that allows each of the substituents to be varied independently of each other.

7.3.5 Examples of scaffolds

Benzodiazepines, hydantoins, β-lactams and pyridines are examples of good scaffolds (Fig. 7.12). They all have small molecular weights, and there are various synthetic routes available which produce the substitution patterns required to fully explore the conformational space about them. For example, it is possible to synthesize benzodiazepines such that there are variable substituents round the whole structure (see also section 6.2.4).

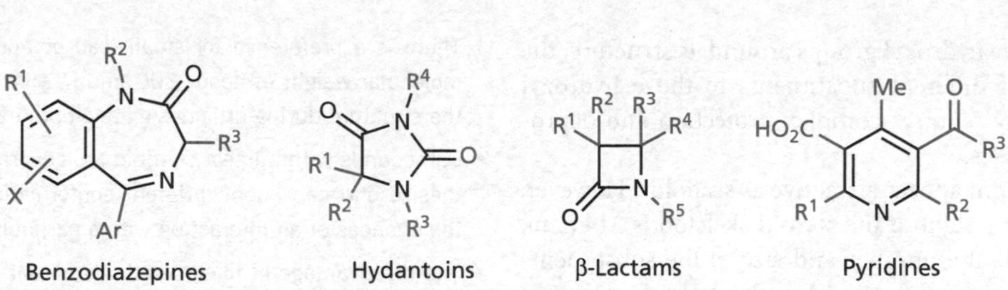

FIGURE 7.12 Examples of 'spider-like' scaffolds.

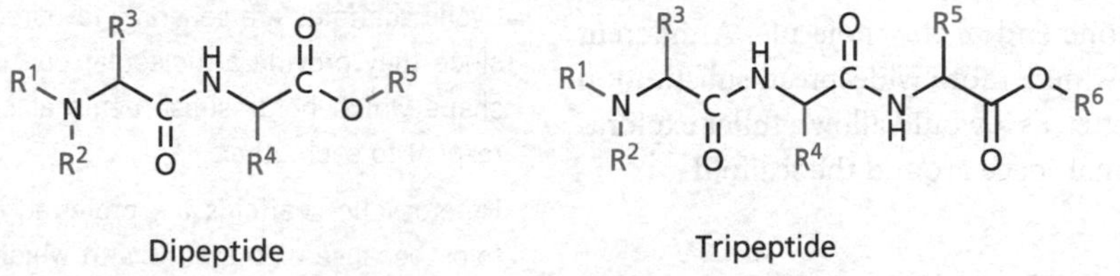

FIGURE 7.13 Peptide scaffolds.

Peptide scaffolds are flexible scaffolds which have the capacity to form hydrogen bonds with target binding sites. They are easy to synthesize and a large variety of different substituents are possible by using the amino acid building blocks. Further substitution is possible on the terminal amino and carboxylic acid functions. The substituents are widely distributed along the peptide chain, allowing a good exploration of conformational space. However, the peptide scaffold should ideally be restricted to di- and tripeptides in order to keep the molecular weight below 500 if it is intended to develop an orally active drug (Fig. 7.13). It is interesting to note that the antihypertensive agents **captopril** and **enalapril** are dipeptide-like and orally active, whereas larger peptides such as the **enkephalins** are not (Fig. 7.14). Oral activity has also been a problem with those HIV protease inhibitors having molecular weights over 500.

Some scaffolds have inherent disadvantages (Fig. 7.15). For example, although **glucose** has a small molecular

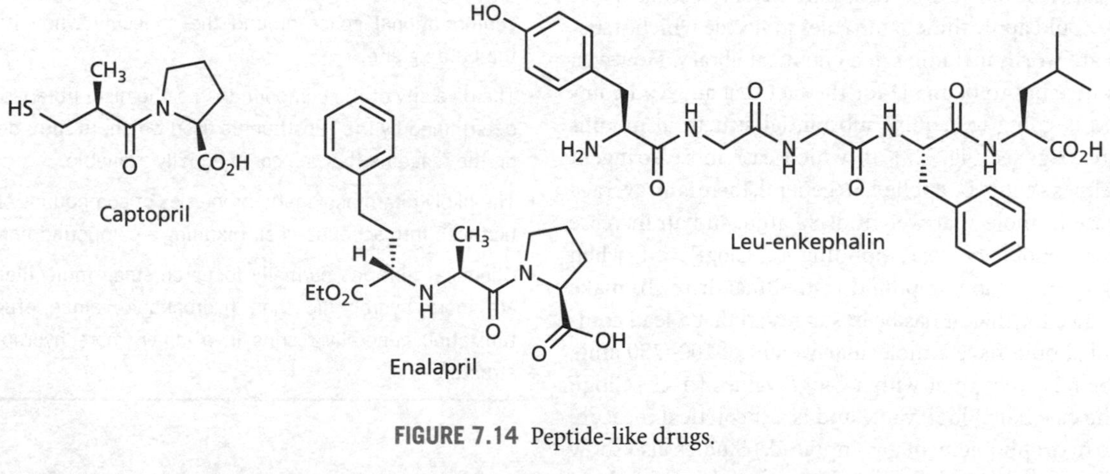

FIGURE 7.14 Peptide-like drugs.

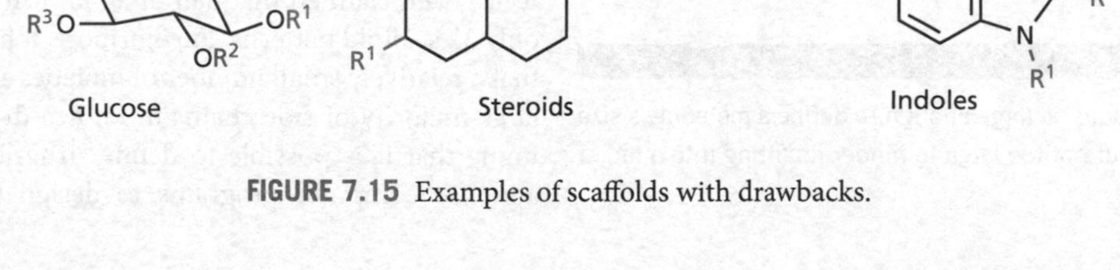

FIGURE 7.15 Examples of scaffolds with drawbacks.

weight and five hydroxyl groups around its structure, the attachment of different substituents to those hydroxyl groups usually requires complex protection and deprotection strategies.

Steroids might appear attractive as scaffolds. However, the molecular weight of the steroid skeleton is 314 amu, and this limits the number and size of the substituents which can be added if we wish to keep the overall molecular weight below 500.

Indoles are good scaffolds, but the substitution pattern shown in Figure 7.15 is not ideal. The variable substituents are all localized to one end of the molecule. A different synthetic route allowing a more widespread substitution pattern would be better, as it would allow a fuller exploration of conformational space around the scaffold.

7.4 Designing 'drug-like' molecules

The use of 'spider-like' structures increases the chances of finding hits that will interact with a target binding site, but it is also worth remembering that compounds with good binding interactions do not necessarily make good medicines. There are also pharmacokinetic issues to be taken into account (section 1.5.3), and so it is worthwhile introducing certain restrictions on the types of molecule that will be produced in order to increase the chances of developing an orally active final drug. In general, the chances of oral activity are increased if the final drug structure obeys Lipinski's rule of five or Veber's parameters (section 1.5.3).

We could apply these same rules to decide which structures are worth including in a chemical library. However, allowance has to be made for the fact that any lead compound is going to require substantial structural modifications (e.g. see Fig. 7.7), in which case more stringent guidelines should be applied. In general, there is an average increase in molecular weight of 80 amu, and an increase in hydrophobicity corresponding to $ClogP = 1$ when going from a lead compound to the final drug. To make allowance for that, it has been suggested that a lead compound should have a molecular weight of 100–350 amu, and be relatively polar with a $ClogP$ value of 1–3. ($ClogP$ is a the calculated $logP$ value and is a theoretical measure of the hydrophobicity of a compound.) Studies also show that drugs developed from lead compounds tend to have more aromatic rings and hydrogen bond acceptors.

We shall now look at how we can plan chemical libraries so that they contain structurally diverse compounds.

KEY POINTS

• Scaffolds should be large enough to define a molecule's size and shape, but not too large to hinder its fitting into a target binding site.

• There is a preference for small lead compounds having a molecular weight of about 300 amu to allow elaboration of the structure during drug design and optimization.

• Compounds with a linear scaffold are generally easy to synthesize and can adopt different conformations, increasing the chances of an interaction with a particular target.

• The disadvantage of linear scaffolds is that the active conformation may only be one of many conformations. This will result in low binding affinity and the possibility of interactions with undesired targets.

• Cyclic scaffolds are generally favoured over linear scaffolds since they provide a molecular core of a defined size and shape which holds substituents at specific positions with respect to each other.

• Heterocyclic scaffolds are preferred over carbocyclic scaffolds because of the ease with which heterocycles can be synthesized, as well as the possibility of additional binding interactions involving the heteroatoms.

• Heteroaromatic scaffolds have an advantage since the presence of substituents does not result in asymmetric centres. However, they result in drugs which are generally two-dimensional in shape.

• Drugs containing too many planar hydrophobic rings may suffer from poor solubility.

• A privileged scaffold is one that is common to a range of drugs having different pharmacological activities.

• Spider scaffolds have substituents positioned around the whole scaffold, thus allowing a thorough exploration of the conformational space around the molecule when it is in the binding site.

• The diversity of the compounds in a chemical library can be determined by the substituents used which, in turn, depend on the reagents that are commercially available.

• The likely pharmacokinetic properties of compounds should be taken into account when planning a compound library.

• Chemical libraries normally focus on small molecules that are more hydrophilic than hydrophobic, since drug optimization generally results in a larger, more hydrophobic structure.

7.5 Computer-designed libraries

7.5.1 Introduction

It has been claimed that half of all known drugs involve only 32 scaffold patterns. Furthermore, it has been stated that a relatively small number of moieties account for the large majority of side chains in known drugs. This may imply that it is possible to define 'drug-like molecules' and use computer programs to design more focused compound libraries. Descriptors used in this approach include log*P*, molecular weight, number of hydrogen bond donors, number of hydrogen bond acceptors, number of rotatable bonds, aromatic density, the degree of branching in the structure, and the presence or absence of specific functional groups. One can also choose to filter out compounds that do not obey the rules mentioned in section 7.4. Computer programs can also be used to identify which structures should be synthesized in order to maximize the number of different pharmacophores represented.

7.5.2 Planning compound libraries

Combinatorial and parallel syntheses (Chapter 6) are methods of rapidly creating a large number of compounds on a small scale using a set synthetic scheme. The compounds produced constitute a compound library which can be tested to find hits for a set target. A compound library could be created that would include all the possible compounds obtainable from the reaction scheme, using commercially available starting materials and reagents. However, the number of molecules that are possible can be immense, making such a goal unfeasible. Molecular modelling can help to provide a focus such that a smaller number of structures are made, whilst maintaining a high probability of finding hits.

One method of doing this is based on the identification of **pharmacophore triangles** (section 1.7). Let us assume that a synthesis is being carried out to generate 1000 structurally diverse compounds. The number of different pharmacophores generated from the 1000 compounds would be an indication of the structural diversity. Therefore, a library of compounds which generates 100 000 different pharmacophores would be superior to a library of similar size which produces only 100 different pharmacophores. Thus, an effective way of designing a more focused library is to carry out a pharmacophore search on all the possible products from a synthetic route in order to select those products that demonstrate the widest structural diversity. Those compounds would then be the ones chosen for inclusion in the library.

First, all the possible synthetic products are automatically ranked on their level of rigidity. This can be achieved by identifying the number of rotatable bonds. Pharmacophore searching then starts with the most rigid structure, and all the possible pharmacophore triangles are identified for that structure. If different conformations are possible, these are generated and the various pharmacophore triangles arising from these are added to the total. The next structure is then analysed for all of its pharmacophore triangles. Again, triangles are identified for all of the possible conformations. The pharmacophores from the first and second structures are then compared. If more than 10% of the pharmacophores from the second structure are different from those of the first, both structures are added to the list for the intended library. Both sets of pharmacophores are combined, and the next structure is analysed for all of its pharmacophores. These are compared with the total number of pharmacophores from the first and second structures, and if there are 10% new pharmacophores represented, the third structure is added to the list and the pharmacophores for all three structures are added together for comparison with the next structure. This process is repeated throughout all the target structures, eliminating all compounds which generate less than 10% new pharmacophores. In this way, it is possible to cut the number of structures which need to be synthesized by 80–90% with only a 10% drop in the number of pharmacophores generated.

There is a good reason for starting this analysis with a rigid structure. A rigid structure has only a few conformations and there is a good chance that most of these will be represented when the structure interacts with its target. Therefore, one can be confident that the associated pharmacophores are fairly represented. If the analysis started with a highly flexible molecule having a large number of conformations, there is less chance that all the conformations and their associated pharmacophores will be properly represented when the structure meets its target binding site. Rigid structures which express some of these conformations more clearly would not be included in the library, as they would be rejected during the analysis. As a result, some pharmacophores which should be present are actually left untested.

It is also possible to use modelling software to carry out a substituent search when planning a compound library. Here, one defines the common scaffold created in the synthesis, the number of substituents which are attached, and their point of attachment. Next, the general structures of the starting materials used to introduce these substituents are defined. The substituents which can be added to the structure can then be identified by having the computer search databases for commercially available starting materials. The program then generates all the possible structures which can be included in the library, based on the available starting materials. Once these have been identified, they can be analysed for pharmacophore diversity as described previously.

Alternatively, the various possible substituents can be clustered into similar groups on the basis of their structural similarity. This allows starting materials to be preselected choosing a representative compound from each group. The structural similarity of different substituents would be based on a number of criteria such as the distance between important binding groups and the types of binding group present.

KEY POINTS

- Molecular modelling can be used to identify the number of pharmacophores represented in different structures.

- The synthesis of a compound library can be made more efficient by identifying those structures having the greatest diversity of pharmacophores.

- Rigid structures represent a pharmacophore more efficiently than linear structures.

- Computer programs can be used to identify the structures that are feasible from a set synthetic scheme using commercially available reagents and starting materials.

7.6 Synthesis

7.6.1 Introduction

Before undertaking the synthesis of a compound library, it is necessary to decide what kind of scaffold will be used. Conventionally, all the compounds in the library will contain that scaffold, although, section 6.5.8.4 illustrates that this is not always the case

The choice of scaffold is up to the individual research team and could involve the use of a privileged scaffold that has been proved to be successful in other projects, or the use of a completely novel scaffold. There is plenty of opportunity for novel scaffolds. In a survey of known drugs, 1506 different scaffold patterns were identified, but half of the drugs involved only 32 of these patterns (Fig. 7.16). There are an almost unlimited number of novel possibilities still to be explored, let alone the number of variations that are possible within each scaffold pattern.

The advantage of using a known scaffold is that there may be several published synthetic routes towards it, whereas a synthetic route for a completely new scaffold will often have to be devised from scratch. The advantage

of a completely novel scaffold is that there is perhaps more chance of finding a novel lead compound, and it is less likely to fall foul of existing patents.

Having decided on the scaffold, it is then a case of deciding the best approach for synthesizing it. If several synthetic routes are available, these can be compared to see which route is the one most likely to result in the substituents and functional groups being evenly spread around the scaffold—the 'spider-like' scaffold. Moreover, the variation required in the different compounds will result from the choice of reagents used at various stages of the synthesis. Therefore, it is best to adopt a synthetic route which uses commercially available reagents that are as diverse in nature as possible. For example, there are a large number of different amines which are commercially available, and so a synthesis involving an amine as a reagent allows a large number of different substituents to be introduced at that stage. Finally, the synthetic route should be as short as possible, ideally with high yielding reactions that introduce diversity at each stage and avoid the need for purifications or protecting groups. Thus, the challenge is to create as diverse a range of compounds as possible using the shortest feasible synthetic route, involving reactions that are high yielding and which use as wide a range of reactants as possible—quite a challenging set of objectives.

Having identified the scaffold and the synthetic route, it is then a case of deciding what reagents are to be used. As already stated, the reagents will determine the variation in substituents between each structure within the library. One of the most important variations is the presence or otherwise of different functional groups at different locations on each substituent arm. However, there are other variations which can be considered, such as the length and/or bulk of the substituent, its hydrophobic character, its polarity, and whether it is electron-donating or electron-withdrawing. In any library, it is important to consider all these factors in order to obtain as diverse a selection of structures as possible. If the chemical library is being created to identify hits

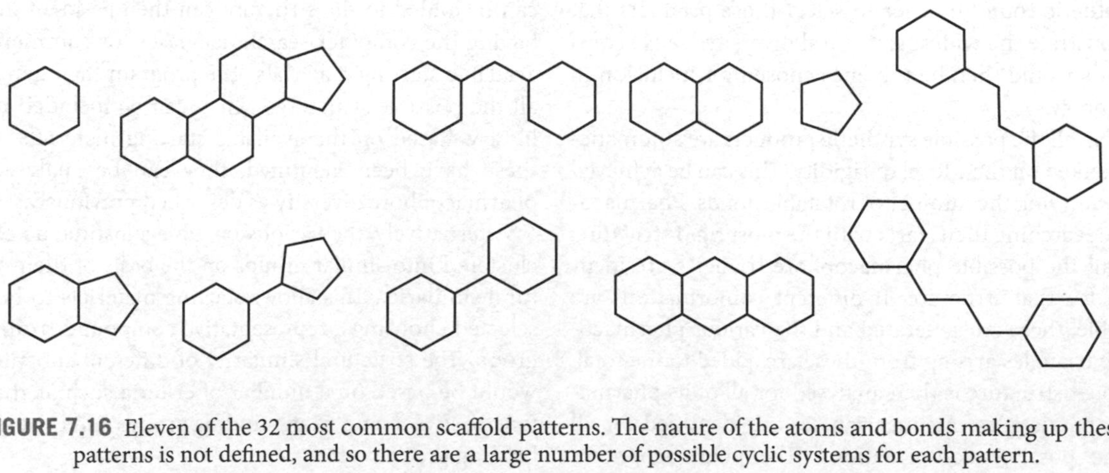

FIGURE 7.16 Eleven of the 32 most common scaffold patterns. The nature of the atoms and bonds making up these patterns is not defined, and so there are a large number of possible cyclic systems for each pattern.

for a specific target, then certain functional groups known to be important for binding ought to be included. For example, if the target is a zinc-containing protease (e.g. angiotensin-converting enzyme), a library of compounds containing a carboxylic acid or thiol group would be relevant.

Having decided on which reagents would be suitable, it is important to check whether they will actually react, especially if combinatorial or parallel synthesis is being used. For example, there is no point in using a weakly nucleophilic amine as a reagent if the reaction is known to require a strongly nucleophilic amine, so trial experiments have to be carried out to identify any 'problem' reagents that should be avoided. There is also no point in including reagents that will introduce toxicity in the form of a 'problem' functional group such as an alkyl halide or an aromatic nitro group. Similarly, it may be worth avoiding any substituents or functional groups that are likely to be prone to metabolism (e.g. an ester group).

In the next section, we shall consider an example of a mix and split combinatorial synthesis aimed at preparing a library of compounds with diverse substituents around a common scaffold.

7.6.2 An example of a combinatorial synthesis to create a chemical library

We shall now look at an example of how an asymmetric synthesis was used to produce a large chemical library of

over two million structurally diverse compounds using a small number of reactions and without the need for protecting groups. A novel saturated tricyclic ring system was chosen as a scaffold rather than a more planar heteroaromatic ring system. This was in order to achieve a greater 'penetration' of conformational space around the structure (see also section 7.7).

In order to produce such a vast number of structures, a mix and split combinatorial synthesis was devised using resin beads as a solid phase support. The beads contained a linker with a primary amine group, and so the first stage was to link the amine to a fairly complex looking chiral carboxylic acid (I) in the presence of a peptide coupling agent (Fig. 7.17). The chiral carboxylic acid had previously been prepared from the natural product **shikimic acid** and the chirality served to control the stereoselectivity of the subsequent reactions such that each final product would be present as a distinct diastereoisomer.

The coupling agent involved was PyBOP, which is commonly used in solid phase synthesis (see also Case Study 1). The mechanism of the coupling reaction is shown in Figure 7.18. and takes advantage of the very strong phosphorus oxygen double bond that is formed in the phosphorus by-product.

The opposite enantiomer of structure I was also prepared from shikimic acid and coupled to the resin beads before the resin beads from both couplings were combined and mixed together. Although both enantiomers

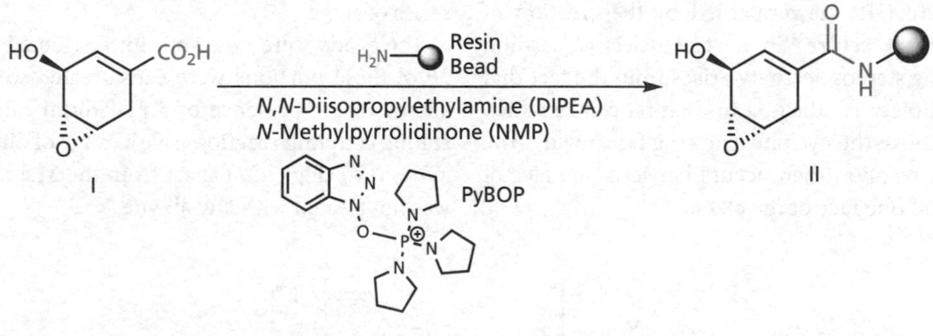

FIGURE 7.17 Attachment of the chiral carboxylic acid to the resin bead.

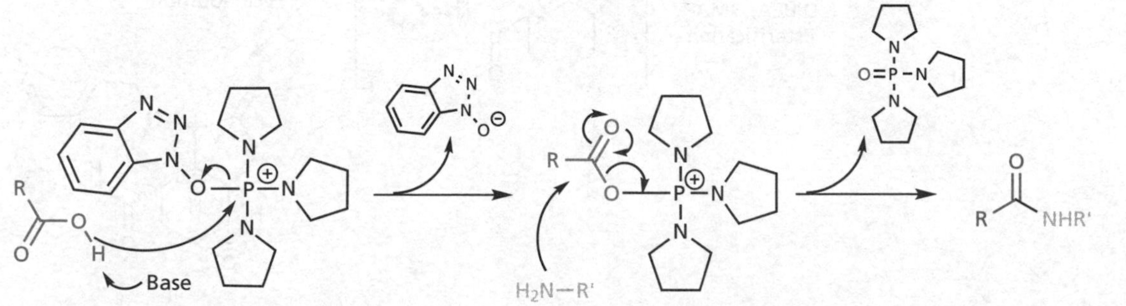

FIGURE 7.18 The coupling mechanism using PyBOP.

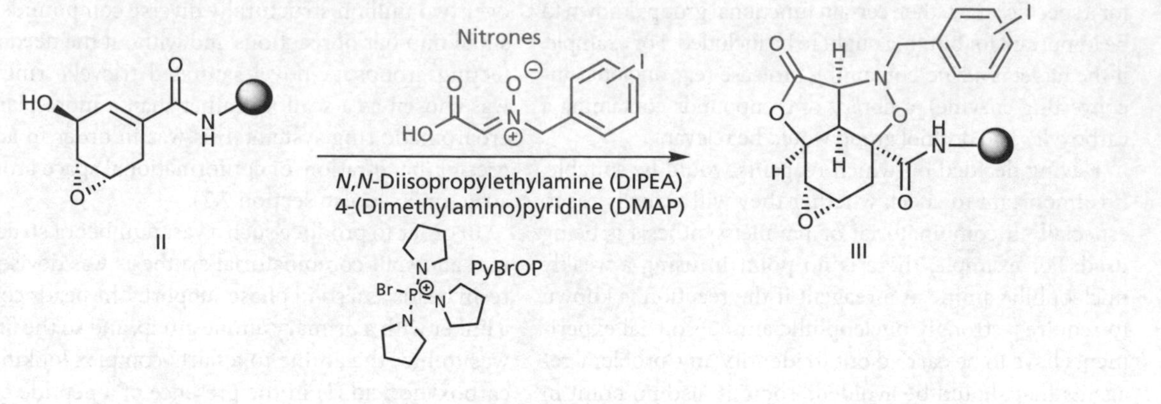

FIGURE 7.19 Reaction of the alcohol (II) with a nitrone in the presence of the coupling agent PyBrOP. Three different structures were formed with the iodine substituent at the *ortho*, *meta*, or *para* position on the aromatic ring.

were used throughout the rest of the synthesis, for clarity the diagrams will show only one of these.

The resin-bound amide (II) was divided into three portions, and each portion was treated with a different nitrone in the presence of the coupling agent (PyBrOP) (Fig. 7.19). Each of the nitrones contained the same carboxylic acid, but differed in the position of the iodine substituent in the aromatic ring.

This time the coupling was between the hydroxyl group on structure II and the carboxylic acid group on the nitrone to provide an ester. This was followed by a spontaneous 1,3-dipolar cycloaddition to give a tetracyclic structure (III), accompanied by the creation of three new chiral centres in a stereoselective fashion (Fig. 7.20). The stereoselectivity arises from the fact that the nitrone moiety is attached to a chiral centre which orientates it above the cyclohexene ring (as drawn). The cycloaddition reaction then occurs between one face of the nitrone and one face of the alkene.

The resulting tetracyclic scaffold is both rigid and has the capability of being functionalized in several positions without the need for protecting groups. The positions that can be functionalized are the iodophenyl ring, the lactone, the epoxide and the N–O link (Fig. 7.21). These groups are evenly spread around the structure and so the diversity introduced by subsequent reagents will also be well spread around the molecule.

Up until this stage, little structural diversity has been introduced other than the position of the iodine substituent on the aromatic ring. However, the molecule has been set up to allow diversity to be introduced in the next series of steps.

The beads were now split into 31 equal portions and 30 of these portions were each treated with a different alkyne in the presence of a palladium catalyst. The resulting coupling reaction gave a range of different aryl alkynes (IV) (Fig. 7.22) apart from the 31st portion which was not treated with any alkyne.

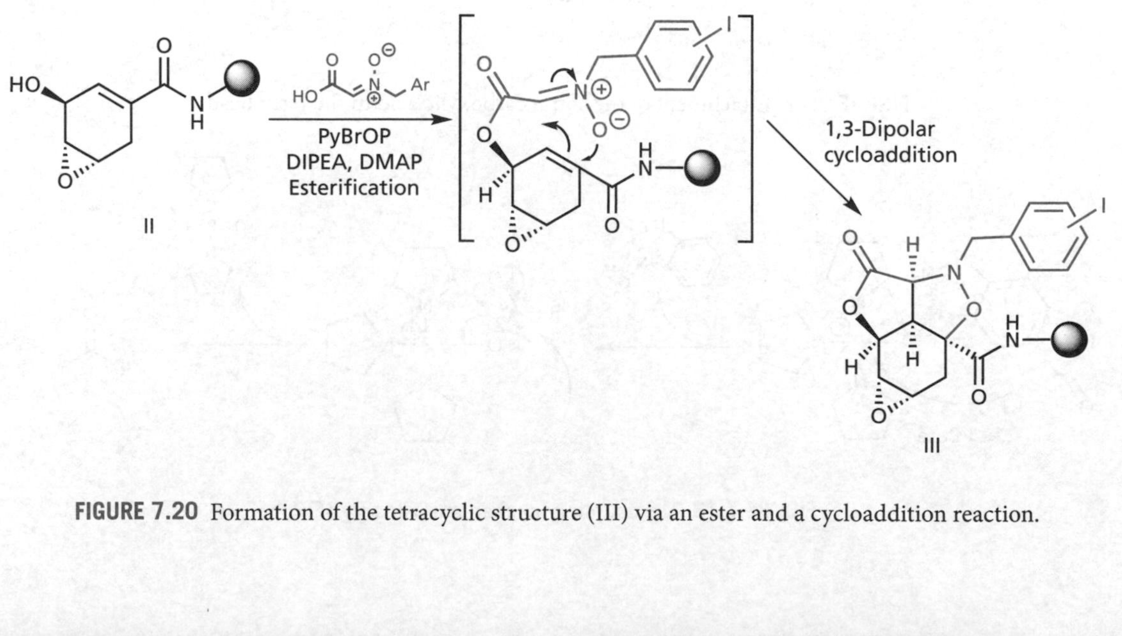

FIGURE 7.20 Formation of the tetracyclic structure (III) via an ester and a cycloaddition reaction.

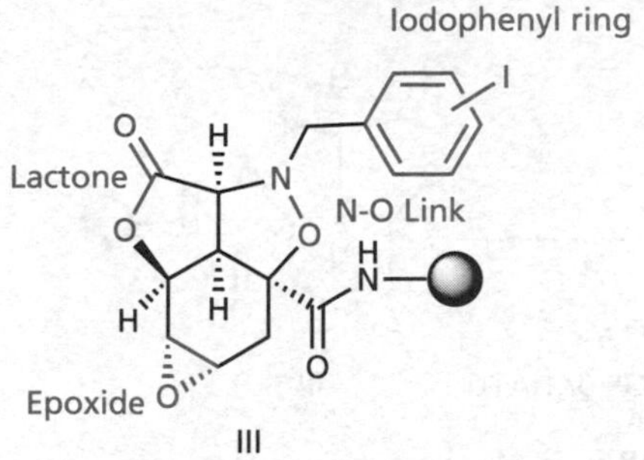

FIGURE 7.21 Positions in the tetracyclic ring that can be functionalized with reagents.

All the beads from this phase of the synthesis were mixed together. A small sample of the mixture was set aside, and the rest was split into 62 equal portions, with each portion being treated with a different primary amine (Fig. 7.23). The amine reacted with the lactone to form an amide, and in doing so revealed an alcohol, which was now available for further reaction.

Once again, the beads were mixed together and split into 63 different portions, and then each portion (except one) was treated with a different carboxylic acid to form an ester. (The carboxylic acids used were actually treated in advance with a coupling agent (DIC) and a base (DMAP) to prepare symmetrical anhydrides.) A photocleavable linker was used in the synthesis and so treatment of the resin-bound products with light was sufficient to release the compounds from the beads for analysis and testing.

The full synthesis is shown in Figure 7.24 and it was calculated that the complete library contained 2 180 106 different compounds, Despite this, it only took two

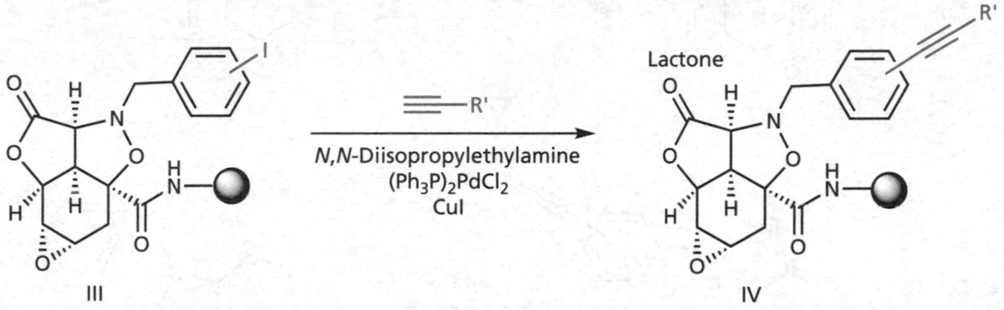

FIGURE 7.22 Coupling of an alkyne with an aryl iodide.

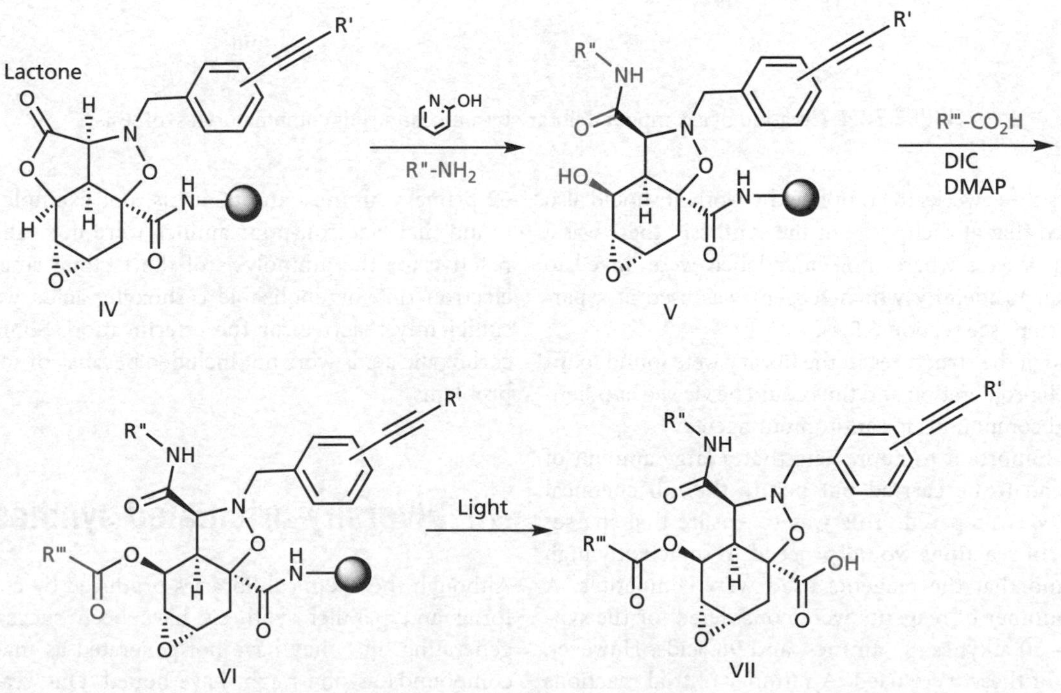

FIGURE 7.23 Latter stages of the synthesis.

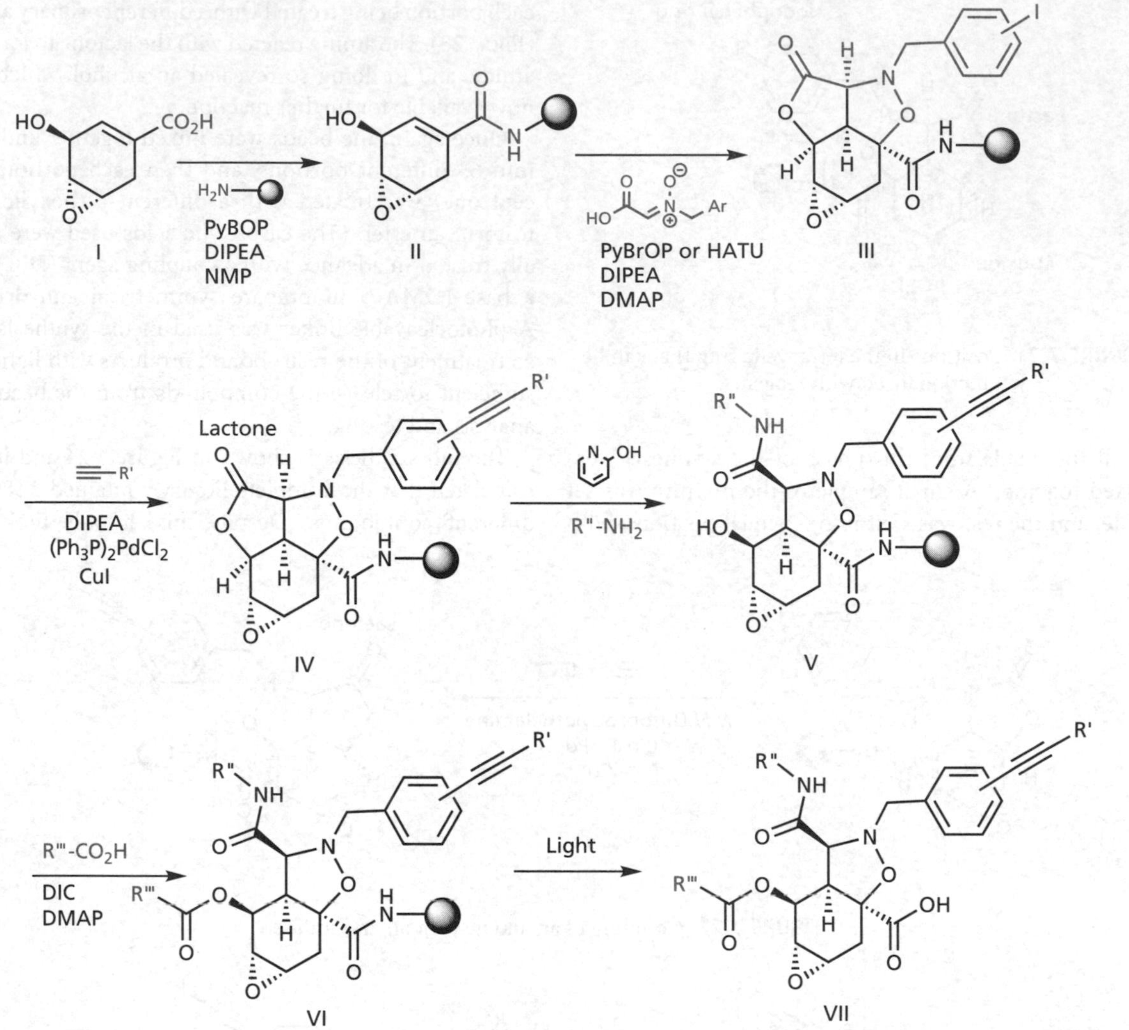

FIGURE 7.24 Creation of a compound library by a mix and split combinatorial synthesis.

workers three weeks to complete the work. It should also be noted that at each stage of the synthesis, there was a tagging process where molecular labels were linked to the resin to identify which reagent was used at a particular step (see section 6.5.4).

Some of the structures in the library were found to inhibit cell proliferation and thus could be viewed as potential lead compounds for antitumour agents.

It is important to appreciate that a large amount of work had to be carried out before the full chemical library was prepared. This was to ensure that the sequence of reactions would proceed in sufficiently high yield and that the reagents used were compatible. A large number of reagents were considered for the synthesis—50 alkynes, 87 amines, and 98 acids. However, not all of these were used. A number of trial reactions were carried out to test how effective the various reagents were, and only those resulting in high yields were used for the library synthesis—a total of 30 alkynes,

62 primary amines, and 62 acids. For example, it was found that electron-poor amines were not sufficiently reactive for the aminolysis of the lactone ring, while electron-rich or enolizable carboxylic acids were not sufficiently reactive for the esterification. Some other carboxylic acids were not included because of solubility problems.

7.7 Diversity-orientated synthesis

Although the chemical libraries produced by combinatorial and parallel synthesis have been successful in generating hits, they have not generated as many lead compounds as one might have hoped. One reason for this might be that the structures present in a particular library are not sufficiently diverse in nature. Traditionally, most chemical libraries have been synthesized to

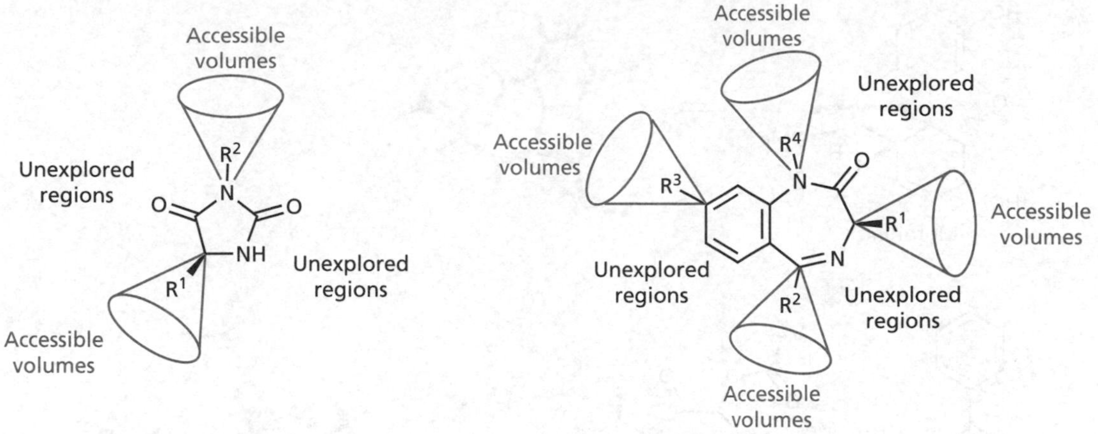

FIGURE 7.25 Accessible volumes and unexplored regions. Note that the space above and below these rings is completely unexplored.

generate compounds with the same scaffold, and the diversity has been achieved by using different reagents to introduce different substituents around the scaffold. However, this means that the structures in the library are essentially very similar in size and shape, since it is the scaffold that chiefly determines those attributes. Moreover, the substituents are at particular positions around the scaffold, and so the 'exploration' of space around the scaffold is restricted to those particular regions (Fig. 7.25). This is particularly the case with heteroaromatic ring systems where substituents are confined to the same plane as the ring system.

Furthermore, many of the privileged scaffolds that are so favoured in the synthesis of chemical libraries tend to be fairly simple cyclic systems which, although not always planar, are rather two-dimensional in nature (Fig. 7.26).

In contrast, many of the important clinical agents derived from the natural world have multicyclic ring systems which contain fewer aromatic and heteroaromatic rings, and are more three-dimensional in shape (Fig. 7.27). As a result, the scaffold and the substituents around it access a much wider region of space. For example, the multicyclic ring systems of morphine and paclitaxel are more three-dimensional or 'globular' in shape.

Therefore, it has been proposed that the discovery of lead compounds might be more successful if chemical libraries were prepared that contained structures with more complex multicyclic scaffolds having globular 3D shapes. That would allow substituents to access

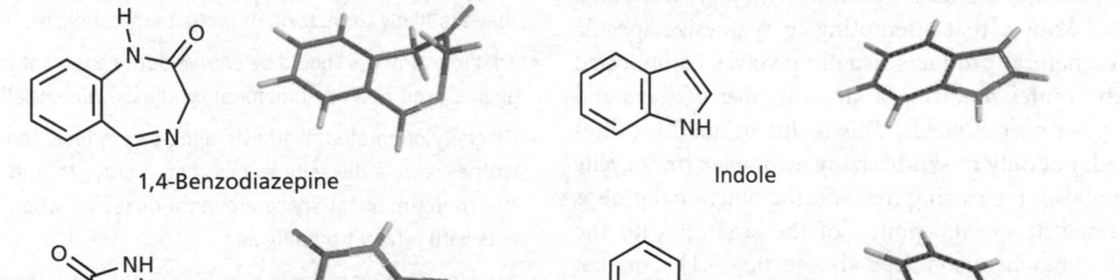

1,4-Benzodiazepine

Indole

Hydantoin

Pyridine

Biaryl

Diphenylmethane

FIGURE 7.26 Shapes of some common scaffolds.

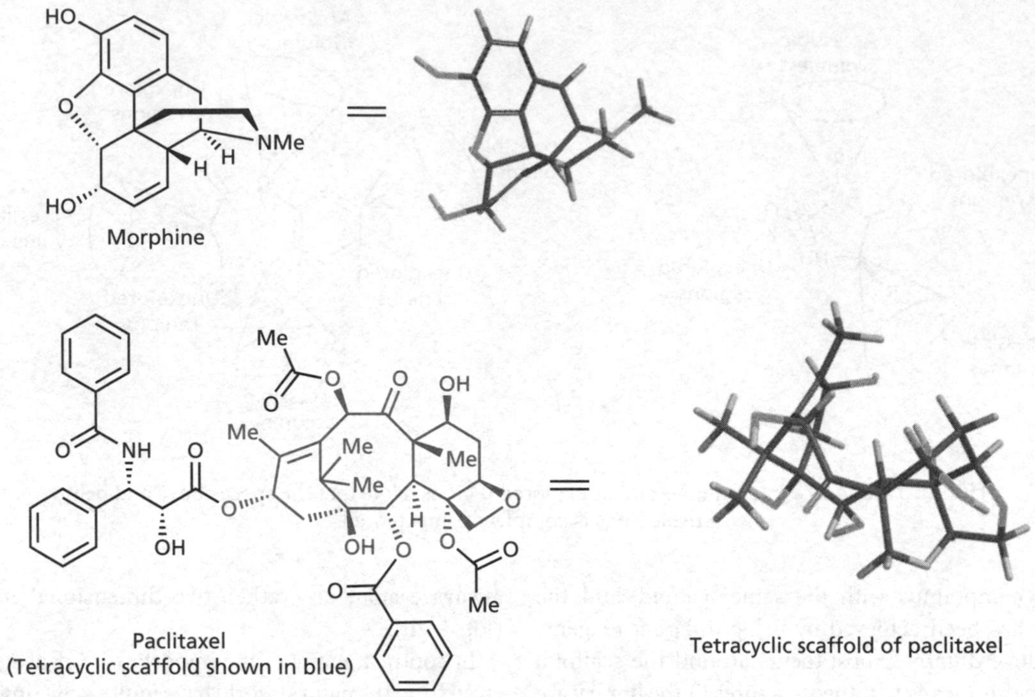

FIGURE 7.27 Shapes of scaffolds present in morphine and paclitaxel.

a greater volume of space around the scaffold than substituents around a more simple mono- or bicyclic ring system (Fig. 7.28). In theory, this should increase the chances of a lead compound being discovered that interacts effectively with different binding regions within the target binding site.

In general, chemical libraries have focused on compounds with relatively simple cyclic scaffolds for ease of synthesis and the desire to fit in with Lipinski's rules. It is well known that attempting to synthesize specific complex natural products usually involves complicated synthetic routes requiring a large number of steps and giving poor overall yields. This is due to the difficulties involved, not only in synthesizing a complex multicyclic ring, but also in ensuring that specific functional groups are located at specific points of the scaffold with the correct stereochemistry (see also section 9.1). For that reason, few researchers have contemplated the possibility of creating libraries of complex structures. However, in the last decade it has been shown that it is possible to synthesize complex multicyclic compounds in a small number of steps as long as there is no requirement to incorporate specific substituents at specific positions with specific stereochemistry (see also section 6.5.8).

KEY POINTS

• The synthesis of a compound library involves choosing a particular scaffold and identifying a feasible synthesis that will allow various substituents to be located at different parts of the scaffold.

• The synthesis should ideally involve a minimum number of reactions that proceed in high yield without the need for protecting groups.

• Reagents should be used that will result in a diverse range of substituents and functional groups around the scaffold.

• Certain substituents or functional groups are best avoided if they are likely to be toxic or prone to metabolism.

• Trial experiments should be carried out to ensure that the reagents used in a combinatorial synthesis will actually react.

• Diversity-orientated synthesis aims to produce compounds with as wide a diversity as possible in order to fully explore the conformational space around a molecule when it interacts with a target binding site.

• Simple heteroaromatic scaffolds are easy to synthesize, but the structures obtained are generally two-dimensional in shape, restricting the amount of conformational space that can be explored with different substituents.

• Saturated multicyclic systems are more 'globular' in shape. As a result, substituents around their structure can explore more of the surrounding conformational space.

• The efficient synthesis of rigid multicyclic saturated ring systems is feasible as long as there are no requirements for substituents to be located at specific positions of the scaffold with specific stereochemistry.

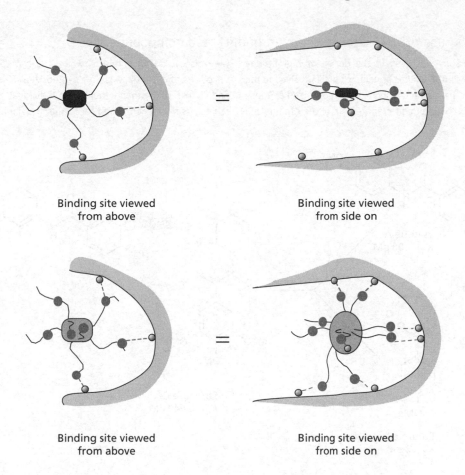

Binding site viewed
from above

Binding site viewed
from side on

Binding site viewed
from above

Binding site viewed
from side on

FIGURE 7.28 Comparison of (a) a planar scaffold with (b) a globular scaffold in probing a binding site. The black and light blue shapes represent the scaffolds. The wiggly lines and blue dots represent substituents and functional groups respectively. The grey spheres represent binding regions within the binding site.

7.8 Fragment-based lead discovery

So far we have described approaches that can be used to create chemical libraries of diverse structures, but, in order to detect any hits, the structures must have sufficient binding affinity or activity to show up in whatever assay is used. Unfortunately, there is no guarantee that this will be the case and assaying the structures in a chemical library may prove unproductive.

Recently, NMR spectroscopy has been used to *design* a lead compound rather than to discover one (see Box 7.1). In essence, the method sets out to find small molecules (**epitopes**) which will bind to specific, but different, regions of a protein's binding site. These molecules will only have a low measurable binding affinity since they only bind to a small part of the binding site, but if a larger molecule is designed which links these epitopes together, a compound may be created which binds to far more of the binding site and does have a measurable binding affinity (Fig. 7.29).

Lead discovery by NMR is also known as SAR by NMR (SAR = structure–activity relationships) and can be applied to proteins of known structure which are labelled with ^{15}N or ^{13}C such that each amide bond in the protein has an identifiable peak. About 200 mg of the purified protein is required for these studies, and it has to be soluble in order to make the NMR measurements. Therefore, the procedure is applicable to enzymes, but not to membrane-bound proteins such as receptors and ion channels.

A range of low molecular weight compounds are screened to see whether any of them bind to a specific region of the binding site. Binding can be detected by observing a shift in any of the amide signals, which will not only show that binding is taking place, but will also reveal which part of the binding site is occupied. Once a compound (or ligand) has been found that binds to one region of the binding site, the process can be repeated to find another ligand that will bind to a different region. This is usually done in the presence of the first ligand to

BOX 7.1 The use of NMR spectroscopy in finding lead compounds.

NMR spectroscopy was used in the design of high-affinity ligands for the **FK506 binding protein**, which is involved in the suppression of the immune response. Two optimized epitopes (A and B) which bound to different regions of the binding site were discovered. Structure C, where the two epitopes were linked by a propyl link, was then synthesized. This compound had higher affinity than either of the individual epitopes and represents a lead compound for further development.

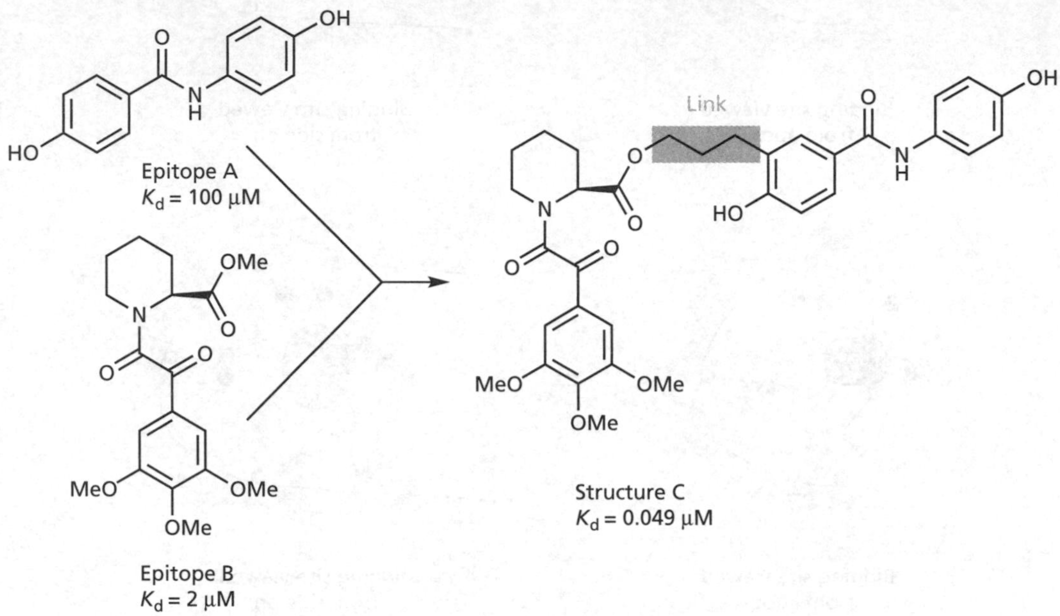

Epitope A
$K_d = 100\ \mu M$

Epitope B
$K_d = 2\ \mu M$

Structure C
$K_d = 0.049\ \mu M$

Link

FIGURE 1 Design of a ligand for the FK506 binding protein.

ensure that the second ligand does, in fact, bind to a distinct region.

Once two ligands (or epitopes) have been identified, the structure of each can be optimized to find the best ligand for each of the binding regions and then a molecule can be designed where the two ligands are linked together.

There are several advantages to this approach. Since the individual ligands are optimized for each region of the binding site, a lot of synthetic effort is spared. It is much

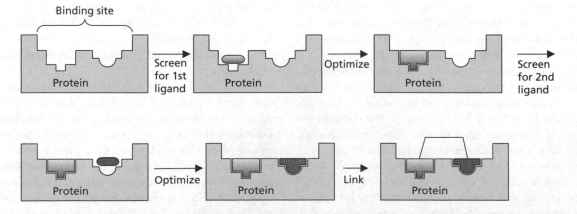

FIGURE 7.29 Epitope mapping.

easier to synthesize a series of low molecular weight compounds to optimize the interaction with specific parts of the binding site than it is to synthesize a range of larger molecules to fit the overall binding site. A high level of diversity is also possible, as various combinations of fragments could be used. A further advantage is that it is more likely to find epitopes that will bind to a particular region of a binding site than to find a lead compound that will bind to the overall binding site. Moreover, fragments are more likely to be efficient binders, having a high binding energy per unit molecular mass. Finally, some studies have demonstrated a 'super-additivity' effect where the binding affinity of the two linked fragments is much greater than one might have expected from the binding affinities of the two independent fragments.

The method described above involves the linking of fragments. Another strategy is to 'grow' a lead compound from a single fragment—a process called **fragment evolution**. This involves the identification of a single fragment that binds to part of the binding site, and then finding larger and larger molecules which contain that fragment but which bind to other parts of the binding site as well.

A third strategy is known as **fragment self-assembly** and is a form of dynamic combinatorial chemistry (section 6.5.6). Fragments are chosen that can bind to different regions of the binding site and then react with each other to form a linked molecule in situ. This could be a reversible reaction as described in section 6.5.6. Alternatively, the two fragments can be designed to undergo an irreversible linking reaction when they bind to the binding site. This has been called '**click chemistry in situ**' (section 7.9.2).

NMR spectroscopy is not the only method of carrying out fragment-based lead discovery. It is also possible to identify fragments that bind to target proteins using the techniques of X-ray crystallography, *in vitro* bioassays, and mass spectrometry. X-ray crystallography, like NMR, provides information about how the fragment binds to the binding site, and does so in far greater detail. However, it can be quite difficult to obtain crystals of protein-fragment complexes because of the low affinity of the fragments. Recently, a screening method called **CrystaLEAD** has been developed which can quickly screen large numbers of compounds, and detect ligands by monitoring changes in the electron density map of protein-fragment complexes, relative to the unbound protein.

Finally, it may be possible to use fragment-based strategies as a method of optimizing lead compounds that have been obtained by other means. The strategy is to identify distinct fragments within the lead compound and then to optimize these fragments by the procedures already described. Once the ideal fragments have been identified, the full structure is synthesized, incorporating the optimized fragments. This could be a much quicker method of optimization than synthesizing analogues of the larger lead compound. One potential problem with this approach is that the fragments might bind to different binding regions from those used when they are part of the larger lead compound.

A rule of three has been suggested for the fragments used for fragment-based lead discovery:

- a molecular weight less than 300
- no more than three hydrogen bond donors
- no more than three hydrogen bond acceptors
- ClogP = 3
- no more than three rotatable bonds
- a polar surface area = 60 Å^2.

7.9 Click chemistry in lead discovery

7.9.1 Click chemistry in the design of compound libraries

The term 'click chemistry' was invented by Barry Sharpless in 2001 to describe reactions having most of the following properties.

- They are cheap and efficient.
- They involve readily available reagents and starting materials, and involve simple reaction conditions.
- They have a high thermodynamic driving force which is greater than $20 \, \text{kcal mol}^{-1}$, and go to completion in a short time period.
- They are high yielding, showing regioselectivity and stereospecificity, with a high selectivity for a single product.
- They are chemoselective for a narrow range of functional groups.
- They create carbon–heteroatom bonds rather than carbon–carbon bonds.
- They generate non-toxic by-products.
- They should be feasible in a benign solvent such as water, and be insensitive to oxygen.
- The product should be easily isolated and purified, avoiding chromatography.

Such reactions are seen as being ideal for the creation of chemical libraries, with distinct advantages over reactions which do not proceed in high yields, or which involve tricky reagents or reaction conditions. This is because any lead compound identified from a click reaction should be inherently easy to synthesize and so there would be fewer synthetic problems in synthesizing the analogues needed to develop the lead compound into a clinically useful drug. Moreover, scaling up the synthesis of any potential drug candidate arising from click chemistry should prove easier, cheaper, and quicker to carry out than other types of reaction, allowing ready production of sufficient sample for preclinical and clinical trials.

Examples of reactions that fit the description of click chemistry are shown in Figure 7.30.

Water is seen as the preferred solvent for click reactions for a number of reasons.

- It is observed that click reactions tend to proceed more effectively when water is the solvent, even when the reactants and product appear insoluble.
- Poorly soluble organic reactants can be more reactive in water than in organic solvents where they are fully dissolved.
- Water is capable of forming hydrogen bonds as both a hydrogen bond donor and a hydrogen bond acceptor—features that can be useful in several click reactions

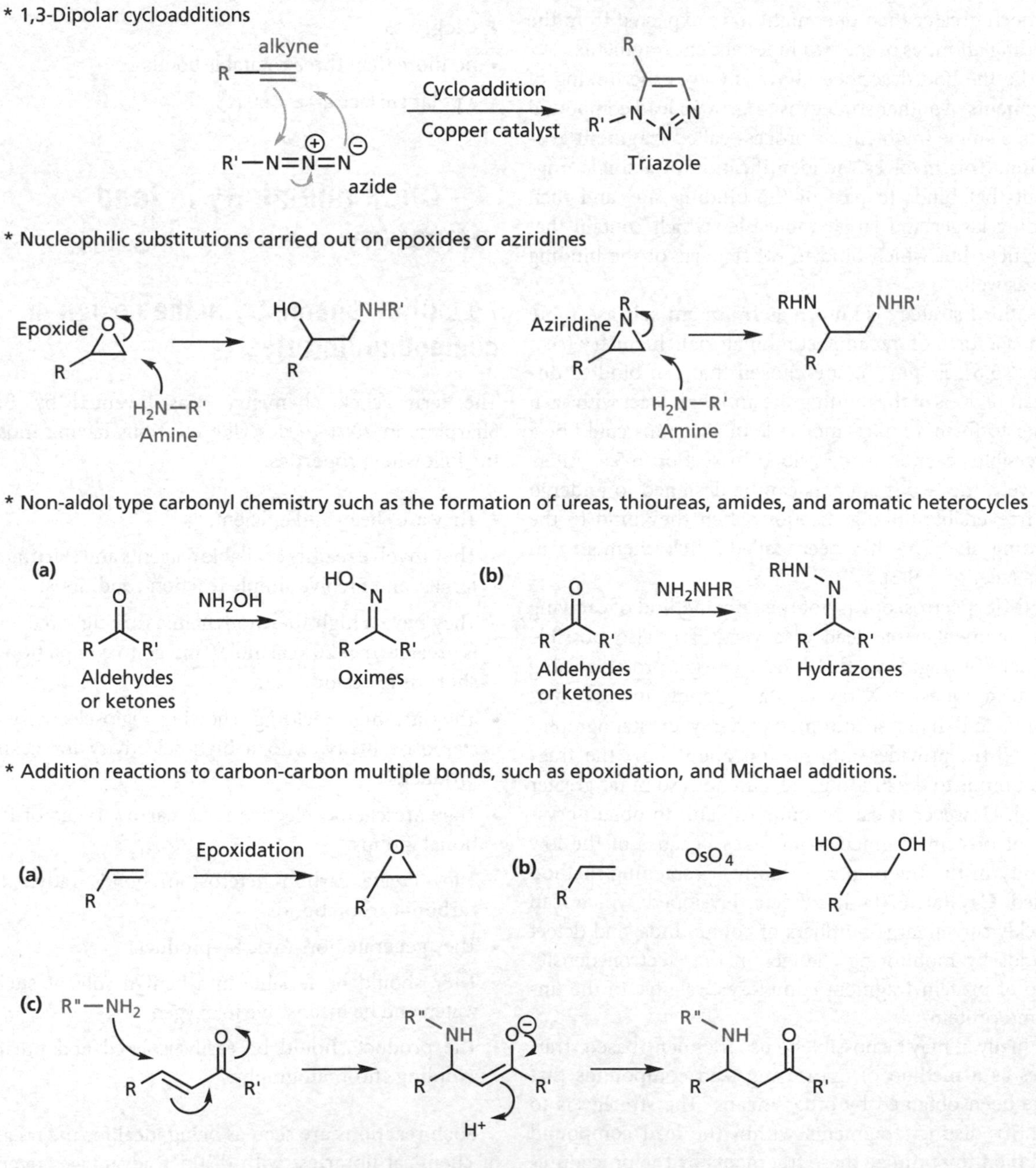
* 1,3-Dipolar cycloadditions
alkyne
Cycloaddition
Copper catalyst
azide
Triazole
* Nucleophilic substitutions carried out on epoxides or aziridines
Epoxide
Amine
Aziridine
Amine
* Non-aldol type carbonyl chemistry such as the formation of ureas, thioureas, amides, and aromatic heterocycles
(a)
Aldehydes or ketones
Oximes
(b)
Aldehydes or ketones
Hydrazones
* Addition reactions to carbon-carbon multiple bonds, such as epoxidation, and Michael additions.
(a)
Epoxidation
(b)
(c)

FIGURE 7.30 Examples of 'click' reactions.

that are aided by different types of hydrogen bonding interactions.

- Most OH and amide NH groups will not interfere with click reactions when the reactions are carried out in water, and so these groups do not need to be protected.

- Water acts as a good heat sink for exothermic reactions, but also has a convenient boiling point if reactions require heating.

- Water is an ideal solvent for production scale as it is cheap and environmentally friendly. Thus, scaling up a click reaction carried out in water is easier than for other types of reaction.

Alkene and alkyne functional groups are particularly crucial for many click reactions as they are high energy functional groups containing reactive pi bonds. For example, an alkene can be seen as standing at the top of a mountain representing a high energy peak (Fig. 7.31). Converting the alkene to an epoxide results in a functional group which has descended from the summit of the energy mountain, but is still on the upper slopes. Therefore, it is feasible to carry out a click reaction on the epoxide as well. The careful planning of a synthesis can result in a series of click reactions where the products are steadily descending the energy mountain to the stability of the valley floor. Such a synthesis would take full advantage of the energy that was inherent in the original unsaturated system. That is why reactions involving the creation of new carbon–heteroatom bonds are

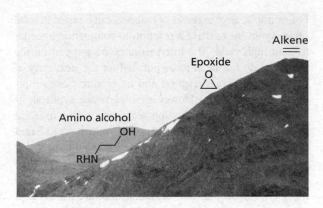

FIGURE 7.31 Relative energies of functional groups.

particularly important. A reaction that involves the formation of a new carbon–carbon bond is less favourable in thermodynamic terms and is equivalent to clambering over another energy hill, rather than descending directly to the valley.

An example of a multistep synthesis using click chemistry was devised by Sharpless and involved the conversion of an alkene to an epoxide, which was then ring-opened with hydrazine. From this key intermediate, a number of different scaffolds were obtained, some of which are shown in Figure 7.32. Any one of these routes could be developed to create a chemical library that could be used to search for hits for particular targets.

As a postscript, solid phase synthesis allows non-click reactions to occur as if they were click-reactions.

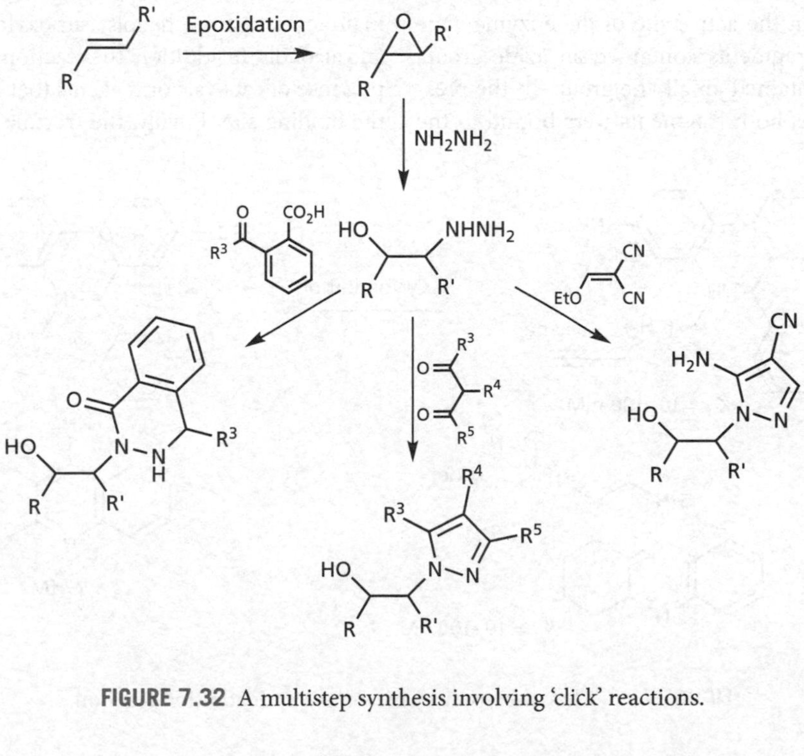

FIGURE 7.32 A multistep synthesis involving 'click' reactions.

For example, large excesses of reagents can be used in solid phase synthesis to drive a reaction to completion in order to gain high yield. The intermediates are trapped on the resin and do not need to be purified for the next stage of the reaction. Excess reagents and by-products can simply be washed through. However, solid phase synthesis is impractical on the production scale, and so the inherent difficulties of particular reactions will come back to haunt development chemists when the reactions have to be scaled up in the pilot plant. Moreover, the compounds prepared by solid phase tend to be hydrophobic in nature, as the number of hydrophilic functional groups are usually kept to a minimum to avoid the need for protecting groups.

7.9.2 **Click chemistry in fragment-based lead discovery**

Because click reactions occur so easily under mild conditions, they are ideal for 'clicking' molecules together within the binding sites of target proteins. In section 7.8, we discussed fragment-based drug discovery where small molecules are added to a target binding site to see whether they are capable of binding to specific regions of the overall site. If the fragments being used to 'probe' a binding site are chosen such that they can undergo a click reaction, that reaction will essentially be catalysed by the protein target if the two component fragments bind sufficiently close to each other within the binding site. Therefore, the products obtained should provide good lead compounds for that target.

For example, a femtomolar inhibitor for the acetylcholinesterase enzyme was obtained by fragment self-assembly within the active site of the enzyme. One of the molecular fragments contained an azide group, while the other contained an alkyne group. In the presence of the enzyme, both fragments were bound to the

active site, and were positioned close enough to each other for an irreversible 1,3-dipolar cycloaddition to take place, forming the inhibitor in situ (Fig. 7.33). This type of reaction has been called 'click chemistry in situ'. In this example, the two components have a binding affinity (K_d) of 10–100 nM, but the click product from the cycloaddition reaction has a K_d of 77 fM, demonstrating that the effects on binding affinity of combining the fragments are more than just additive. Part of the reason for this effect may well be due to the formation of the triazole ring. The nitrogen atoms in the ring are capable of acting as hydrogen bond acceptors, while the ring itself can form pi–pi stacking interactions with any heterocycles that may be present in the binding site, such as the indole ring system of tryptophan residues.

The 1,3-dipolar cycloaddition has been particularly popular for these kinds of study. This is because of a nice balance of reactivity as far as the azide and the alkyne are concerned. They are reactive enough to undergo rapid cycloaddition when they are held close together in the binding site of a target protein, but react much more slowly when they are free in solution. This is due to a large thermodynamic driving force for cycloaddition combined with a high kinetic barrier. The latter property means that the reaction is slow at room temperature when the reagents are free in solution. Therefore, one can be confident that the products obtained from in situ click chemistry are from reagents bound close together in the binding site.

The alkynes and azides involved are also described as 'benign' reactants because they are relatively unreactive when they encounter biological macromolecules and do not react with the various functional groups that are found in proteins, such as phenols, carboxylates, amines, amides, and alcohols. In addition, the reaction does not require the presence of catalysts or reagents that might interfere with the binding site. Finally, the triazole ring that is formed

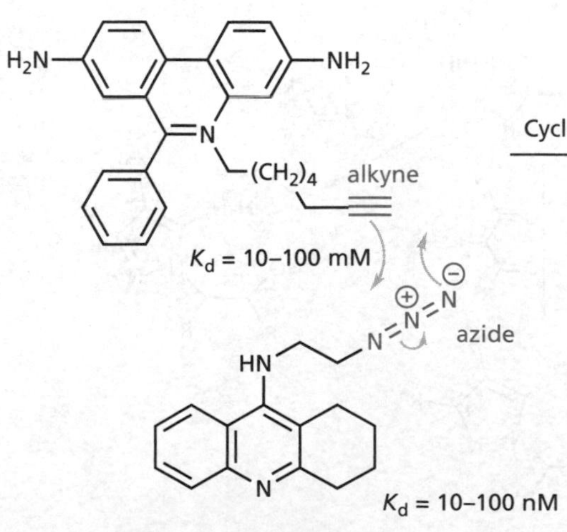

FIGURE 7.33 Click chemistry by means of a cycloaddition reaction.

is stable to acidic or basic hydrolysis, as well as to strong reducing and oxidizing agents. Therefore, any lead compounds obtained from these studies are amenable to many of the reactions used in drug optimization studies.

Because of these advantages, the dipolar cycloaddition reaction has been used in in situ click reactions to investigate lead compounds for a number of enzymes such as carbonic anhydrase, acetylcholinesterase, HIV protease, protein tyrosine phosphatases, and metalloproteases.

There are huge advantages in this approach, but there are also some disadvantages. One of these is the difficulty in carrying out these reactions on target proteins which are scarce and difficult to obtain. One possible solution to this has been to develop microfluidic devices which carry out the reactions on a much smaller scale (section 6.4.6).

7.9.3 Click chemistry in synthesizing bidentate inhibitors

The 1,3-dipolar cycloaddition reaction has been used to link together structures that interact with two different binding sites within a target enzyme. For example, kinase enzymes catalyse the phosphorylation of alcohol or phenol groups in protein substrates using ATP as a cofactor (Fig. 7.34). Thus, they contain a binding site for

ATP which acts as the phosphorylating agent, as well as a binding site for the protein substrate.

A number of kinase inhibitors have been designed that bind to the ATP binding site, but there is an interest in designing novel structures which bind to both the ATP binding site and the substrate binding site. One approach has been to use a cycloaddition reaction to create a library of compounds which link molecules having binding affinity for the ATP binding site with molecules binding elsewhere in the kinase active site.

Imatinib is a clinically important anticancer drug that binds to the ATP binding site of a kinase enzyme called Abl kinase. Two truncated imatinib structures were synthesized which incorporated the alkyne group required for the cycloaddition reaction and which would serve as the part of the molecule expected to bind to the ATP binding site (Fig. 7.35). The portion of the drug removed is present primarily to improve solubility and has little contribution to the binding interactions with the target binding site.

These 'clickable warheads' were then linked to 45 different azides to produce a library of 90 compounds. The truncated imatinib structure binds to the ATP binding site, and so the different moieties introduced by the azides acted as probes to see whether additional interactions were obtained with the rest of the active site. One of the structures obtained proved to be slightly more potent than imatinib itself (Fig. 7.36).

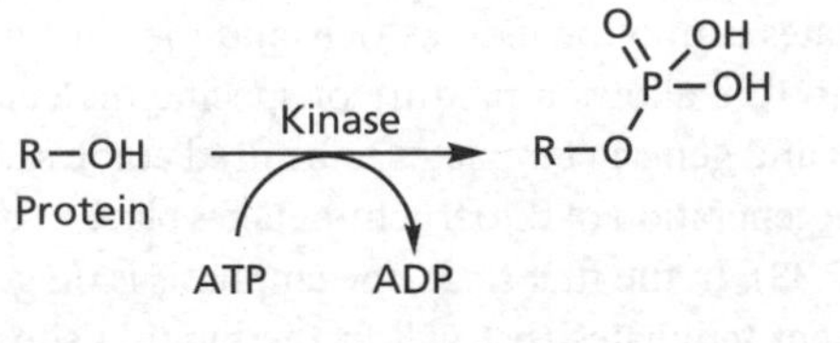

KEY POINTS

• NMR spectroscopy can be used to identify whether small molecules (epitopes) bind to specific regions of a binding site. Epitopes can be optimized and then linked together to give a potential lead compound.

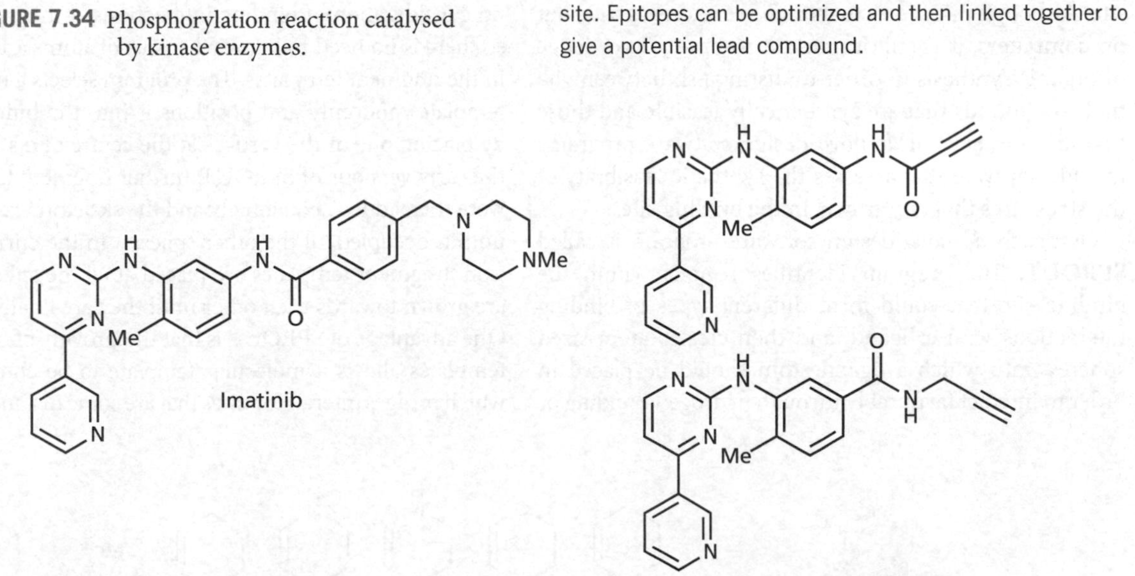

FIGURE 7.34 Phosphorylation reaction catalysed by kinase enzymes.

FIGURE 7.35 Imatinib and the truncated imatinib analogues.

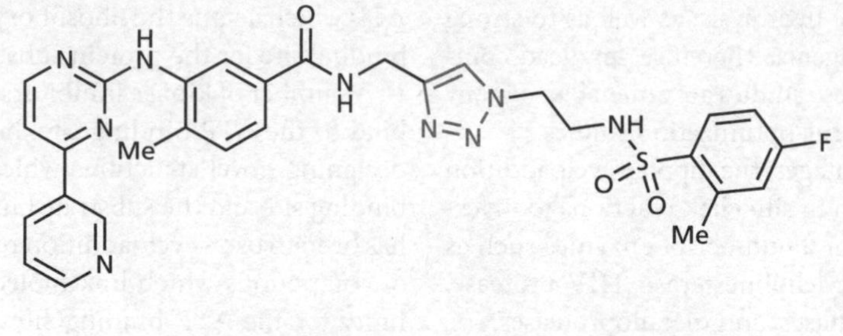

FIGURE 7.36 Active bidentate kinase inhibitor.

- Click chemistry involves reactions that are thermodynamically favoured and are completed in a short time period.

- Reactions involving click chemistry are more likely to be scaled up during chemical and process development.

- Click reactions are applicable to synthetic routes that start from high energy functional groups such as alkenes and alkynes.

- Click chemistry in situ involves a click reaction between two reagents that are bound close together within a binding site. The resulting structure has the potential to be a useful lead compound for further drug design and optimization.

7.10 *De novo* drug design

If the structure of a target binding site is known from X-ray crystallographic studies, it is possible to design potential lead compounds using dedicated software programs. This is known as *de novo* drug design, where compounds are designed *in silico* to fit and bind to the binding site. Although this is a virtual *in silico* exercise carried out on computers, it is still important to have a knowledge of organic synthesis in order to distinguish between virtual compounds that are synthetically feasible and those that are not. Several *de novo* design software programs include software that assesses the synthetic feasibility of the structures that are 'grown' in the binding site.

One such *de novo* design software program is called **SPROUT**. The program identifies regions within the binding site that could form different types of binding interactions with a ligand, and then creates atom-sized spheres, into which a ligand atom should be placed in order to interact favourably through hydrogen bonding or van der Waals interactions. A series of stored molecular fragments or building blocks are then fitted into the binding site to see whether they will interact with these spheres. These 'building blocks' are actually **templates** which are designed to represent several different molecular fragments. Each template is defined by vertices and edges rather than by atoms and bonds. A vertex represents a generalized sp, sp^2, or sp^3 hybridized atom, while an edge represents a single, double, or triple bond, depending on the hybridization of the vertices at either end. For example, the template shown in Figure 7.37 can represent a large number of different six-membered rings. This approach has the advantage that it radically cuts down the number of different fragments that have to be stored in the program, making the search for novel structures more efficient. However, there is no reason why specific templates cannot be used as well, and the current version of SPROUT allows a mixture of specific molecular fragments and generic templates to be used at the same time.

The generation of the structures takes place in two stages (Fig. 7.38). In the first stage, the emphasis is on generating fragment templates that will fit the binding site. There is no consideration of binding interactions at this stage, and so there is no need to know what sort of atoms are present in the fragment templates. The program selects a fragment template randomly and positions it into the binding site by placing one of the vertices at the centre of a sphere. In the early versions of SPROUT, further fragment templates were then added sequentially and the skeleton was 'grown' until it occupied all the other spheres. In the current version, fragment templates are placed at all the spheres and are grown towards each other until they are finally linked. One advantage of SPROUT is that the 'growth' of fragment templates allows a molecular template to be constructed which bridges interaction sites that are some distance apart.

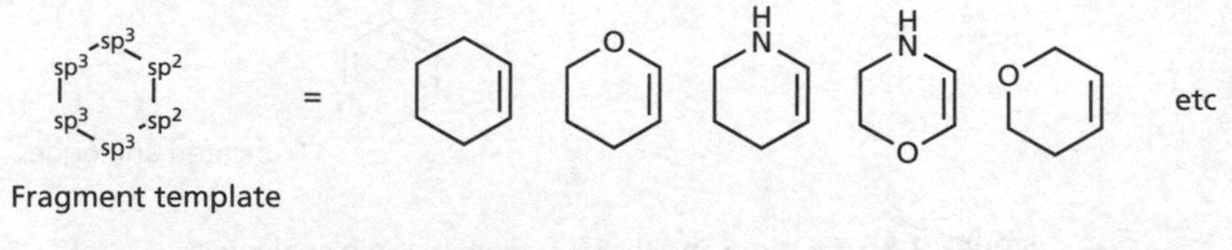

FIGURE 7.37 Examples of structures represented by a template used in SPROUT.

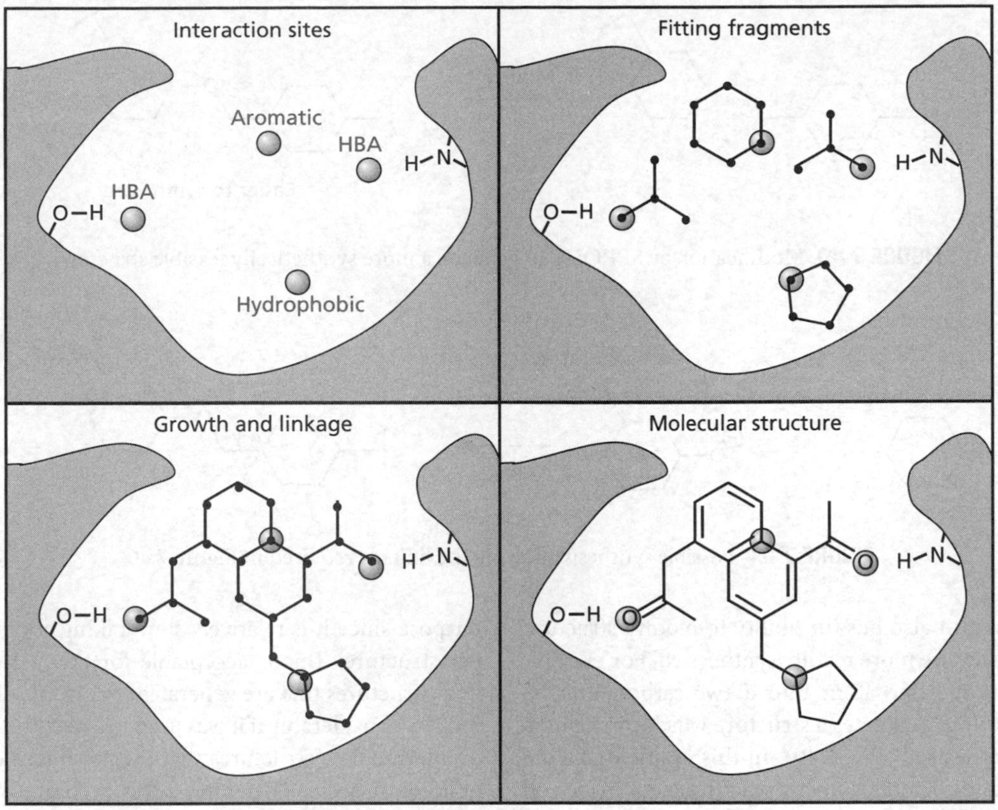

FIGURE 7.38 Generating structures using SPROUT.

The second stage in the process is to create specific molecules from the molecular templates that have been produced. This involves replacing the vertices with suitable atoms to allow favourable hydrogen bonding and van der Waals interactions with the binding site. For example, if a vertex is located within a sphere that requires a hydrogen bond acceptor, an oxygen or a nitrogen atom can be added at that position. Since generic templates have been used to generate each skeleton, a large variety of molecular structures can be generated from each molecular template.

SPROUT has the capacity to identify certain structural features that might be unrealistic and then modify them. For example, an OH might be generated during the second stage in order to introduce a hydrogen bond donor, but if the OH is linked to a double bond this results in an enol that would tautomerize to a ketone. The latter would not be able to act as a hydrogen bond donor. The program can identify an enol and modify it to a carboxylic acid which can still act as a hydrogen bond donor (Fig. 7.39).

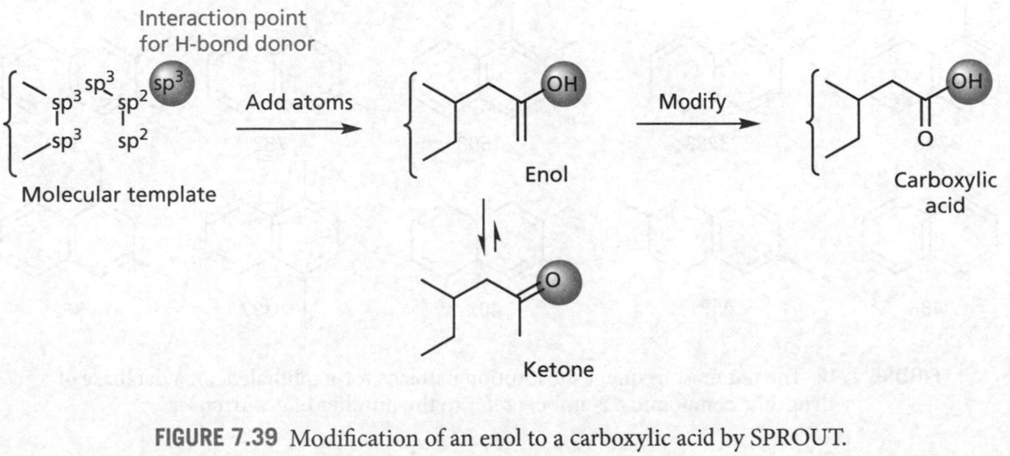

FIGURE 7.39 Modification of an enol to a carboxylic acid by SPROUT.

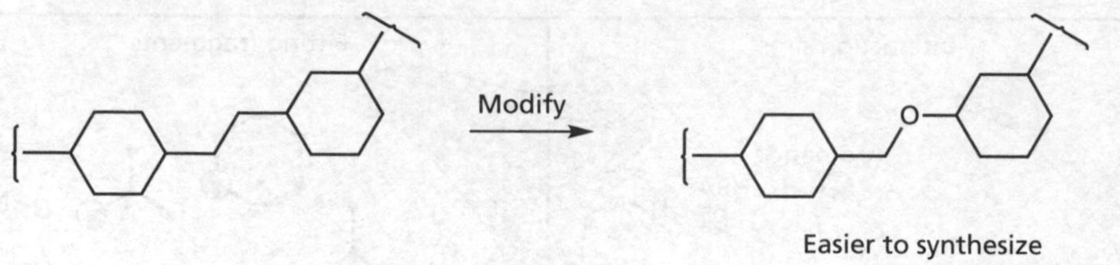

FIGURE 7.40 Modification by SPROUT to generate a more synthetically feasible structure.

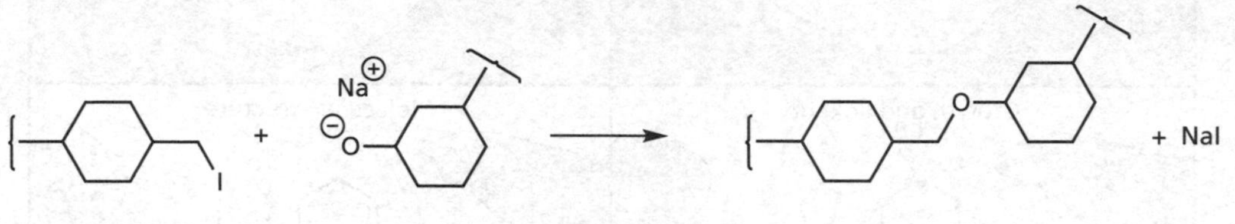

FIGURE 7.41 Possible synthesis allowing the linkage required in Figure 7.40.

The program also has the ability to modify structures such that they are more readily synthesized. For example, introducing a heteroatom into a two-carbon link between two rings generates a structure which can be more readily synthesized (Fig. 7.40). In this example, the link could be made synthetically by reacting an alkoxide with an alkyl halide (Fig. 7.41).

The structures that are finally generated by SPROUT are then evaluated *in silico* for a variety of properties including possible toxicity and pharmacokinetic properties. The program **CAESA** is used to evaluate how easily each structure can be synthesized and to give an indication of likely starting materials for the synthesis. The program does this by carrying out a retrosynthetic analysis of each structure.

More recently, SPROUT has introduced a method of assessing the synthetic feasibility of the partial structures created during the *de novo* construction process. Such an analysis is useful as it allows a pruning process to take place which rejects partial structures that are not easily synthesized, and directs the program to generate more suitable structures. CAESA itself cannot be used for this

purpose since it is relatively slow, taking about a minute per structure. This is acceptable for the analysis of the final structures that are generated, but would slow up the process considerably if it was used to assess the thousands of intermediate structures that are generated during the building process. Therefore, a less accurate but quicker method of analysis is carried out. The method involves the identification of molecular features within each partial structure and identifying how frequently they occur in known structures. The rationale is that if a particular feature commonly exists in known compounds, it is likely that that same feature should be capable of synthesis in the novel structures generated by *de novo* design.

The major structural features within a molecule can be defined as the rings of various sizes that are present, as well as any connecting chains. The synthetic feasibility of rings and chains generally depends on their substitution pattern. For example, the ten most frequent substitution patterns for a naphthalene ring amongst a database of known compounds are shown in Figure 7.42. The analysis of partial structures can be carried out such that structures with

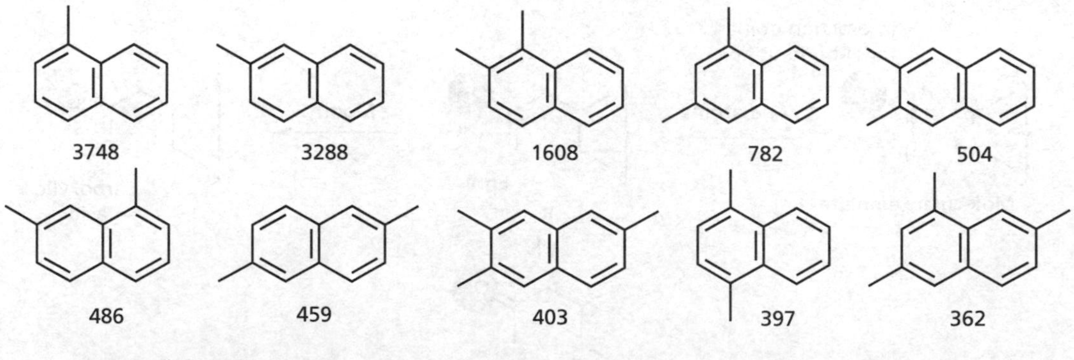

| 3748 | 3288 | 1608 | 782 | 504 |

| 486 | 459 | 403 | 397 | 362 |

FIGURE 7.42 The ten most frequent substitution patterns for naphthalene in a database of drug-like compounds. Numbers refer to the number of occurrences.

uncommon structural features (such as a tetrasubstituted naphthalene ring) are penalized and rejected. A measure of the drug-like character of the partial structures can also be gleaned if the database used in the analysis is restricted to active compounds from drug databases.

KEY POINTS

• *De novo* drug design involves the use of software programs which create novel structures *in silico* such that they fit and bind to binding sites.

• The SPROUT program stores a series of molecular templates which act as molecular fragments that are fitted to different parts of the binding site.

• The templates are linked together and then specific atoms are added to obtain relevant binding interactions.

• The synthetic feasibility of the molecules constructed is assessed automatically.

FURTHER READING

Bemis, G.W. and Murcko, M.A. (1996) 'The properties of known drugs. 1: Molecular frameworks', *Journal of Medicinal Chemistry*, **39**, 2887–93.

Ertl, P., et al. (2006) 'Quest for the rings: *in silico* exploration of ring universe to identify novel bioactive heteroaromatic scaffolds', *Journal of Medicinal Chemistry*, **49**, 4568–73.

Kalesh, K.A., Liu, K., and Yao, S.Q. (2009) 'Rapid synthesis of Abelson tyrosine kinase inhibitors using click chemistry', *Organic and Biomolecular Chemistry*, **7**, 5129–36.

Kolb, H.C., Finn, M.G., and Sharpless, K.B. (2001) 'Click chemistry: diverse chemical function from a few good reactions', *Angewandte Chemie, International Edition*, **40**, 2004–21.

Krasinski, A., et al. (2005) 'In situ selection of lead compounds by click chemistry: target-guided optimisation of acetylcholinesterase inhibitors', *Journal of the American Chemical Society*, **127**, 6686–92.

Lovering, F., Bikker, J., and Humblet, C. (2009) 'Escape from flatland: increasing saturation as an approach to improving clinical success', *Journal of Medicinal Chemistry*, **52**, 6752–6.

Lowe, D. (2009) 'The problems of favourite reactions', *Chemistry World*, October, p 22 (aryl–aryl scaffolds).

Manetsch, R., et al. (2004) 'In situ click chemistry: enzyme inhibitors made to their own specifications', *Journal of the American Chemical Society*, **126**, 12 809–18.

Pitt, W.R., et al. (2009) 'Heteroaromatic rings of the future', *Journal of Medicinal Chemistry*, **52**, 2952–63.

Ritchie, T.J. and Macdonald, S.J.F. (2009) 'The impact of aromatic ring count on compound developability—are too many aromatic rings a liability in drug design?', *Drug Discovery Today*, **14**, 1011–20.

Ritchie, T.J., et al. (2011) 'The impact of aromatic ring count on compound developability: further insights by examining carbo- and hetero-aromatic and -aliphatic ring types', *Drug Discovery Today*, **16**, 164–71.

Strang, D.R. (2003) 'Diversity-orientated synthesis; a challenge for synthetic chemists', *Organic & Biomolecular Chemistry*, **1**, 3867–70.

Tan, D.S., et al. (1998) 'Stereoselective synthesis of over two million compounds having structural features both reminiscent of natural products and compatible with miniaturised cell-based assays', *Journal of the American Chemical Society*, **120**, 8565–6.

Wang, J., et al. (2006) 'Integrated microfluidics for parallel screening of an in situ click chemistry library', *Angewandte Chemie, International Edition*, **45**, 5276–8.

8 Analogue synthesis in drug design

8.1 Introduction

In Chapter 7 we discussed how active compounds can be discovered from the natural world or synthesized in the chemistry laboratory. Such compounds may have some kind of binding affinity for a target biomolecule such as a protein or a nucleic acid. They may also have some pharmacological activity, but more often than not that activity is weak and unselective. Nevertheless, such compounds can serve as the starting point in the quest for a drug candidate that has:

- good activity and good binding interactions with the target
- selectivity of action for a selected target or targets
- minimal side effects
- lack of toxicity
- oral activity
- chemical and metabolic stability
- ease of synthesis.

This process can involve the synthesis of large numbers of analogues before an acceptable drug candidate can be put forward for preclinical and clinical trials. More often than not, the final drug candidate can look dramatically different from the original lead compound. For synthetic lead compounds, it is invariably the case that the final drug going forward to the clinic is larger and more elaborate (Fig. 8.1). In some cases, the structure of the original lead compound is still clearly visible, as in the anticancer drug gefitinib. In other drugs such as **imatinib**, the modifications are starting to disguise the original lead compound, while in some compounds the structure of the initial lead compound is barely recognizable, as in the antiviral drug **saquinavir**.

For complex lead compounds derived from the natural world, it is often the case that the final drug is simpler and has a smaller molecular weight (Fig. 8.2).

Having said that, if a natural lead compound has potent activity in itself, it is possible that clinically important analogues may be feasible which have structures that are similar to or even more complex than the original lead compound. For example, the opioid analgesics **hydromorphone** and **buprenorphine** are analogues of morphine (Fig. 8.3).

In the case of buprenorphine and hydromorphone, there is a clear structural relationship with the lead compound, but in some simplified analogues the comparison may be difficult to spot. An extreme case is that of neuromuscular agents such as **suxamethonium** and **rocuronium**, which are based on the lead compound **tubocurarine** (Fig. 8.4). Here, the key structural similarity is the separation between two charged quaternary ammonium centres.

Not all natural lead compounds have complex structures. For example, some of the body's own neurotransmitters have quite simple structures, in which case the drugs that are developed from them are likely to be more complex and have a higher molecular weight (Fig. 8.5).

Synthetic chemists have been preparing analogues of pharmacologically active lead compounds for well over 100 years in order to carry out studies into structure–activity relationships (SAR) (section 1.7). Initially, the analogues that were synthesized depended on which reagents and starting materials were available, and so it was very much a hit and miss affair. Nowadays, the medicinal chemist tends to have a much clearer idea which analogues are likely to have the best chance of success, especially if the structure of the target protein is known. However, it is still important to prepare a large number of analogues in order to achieve success.

There are two main reasons for synthesizing analogues of a lead compound:

a) to identify the structural features and functional groups that are important to the binding affinity and

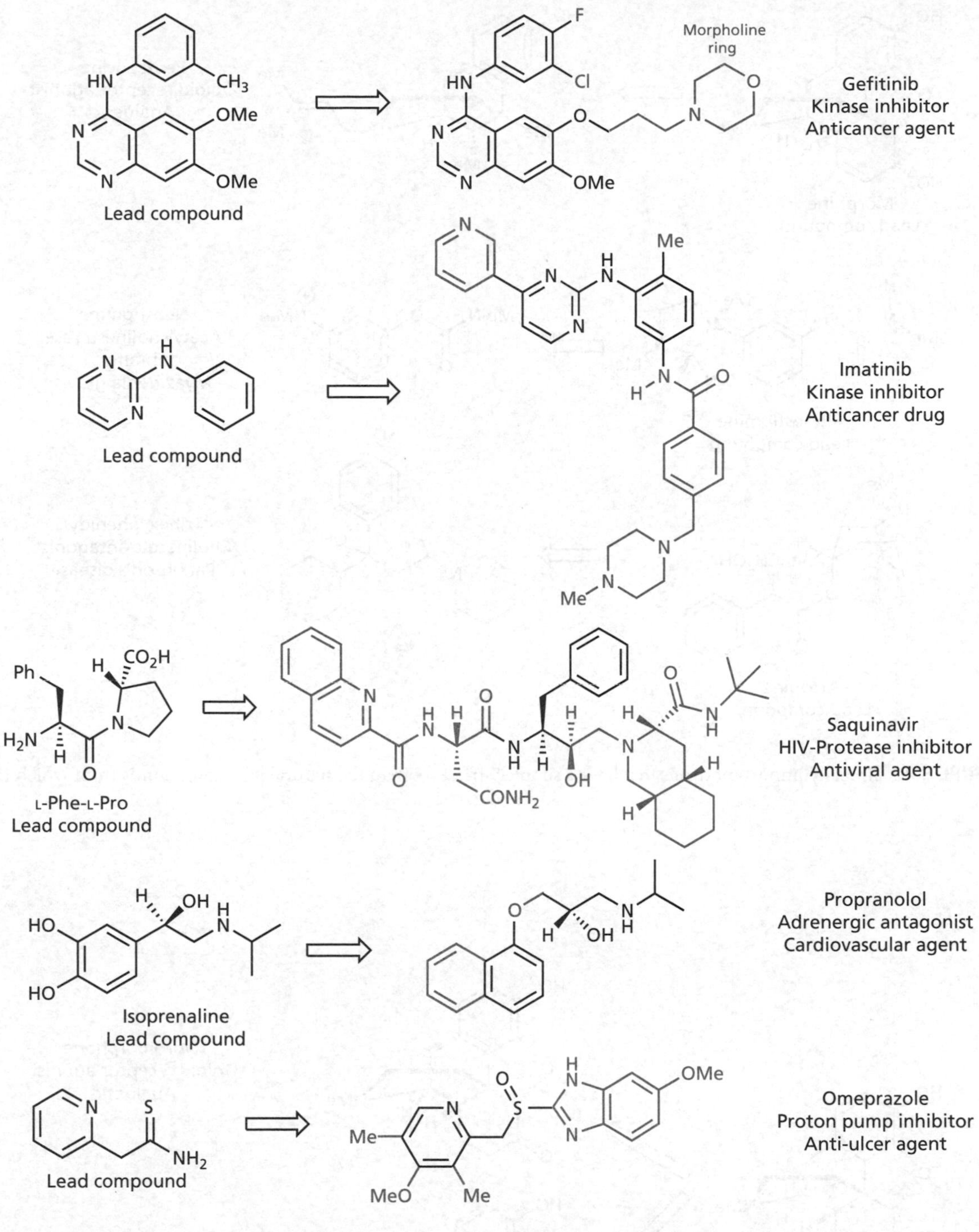

FIGURE 8.1 Clinically important drugs and the synthetic lead compounds from which they were derived.

activity of the lead compound—the identification of the pharmacophore;

b) to modify the structure of the lead compound to find analogues with improved pharmacodynamic and pharmacokinetic properties which can be obtained

by a practical and economic synthesis—this is known as **drug optimization**.

We shall first look at the need for analogues in order to carry out SAR studies. Such studies allow us to identify the pharmacophore that is responsible for a particular activity.

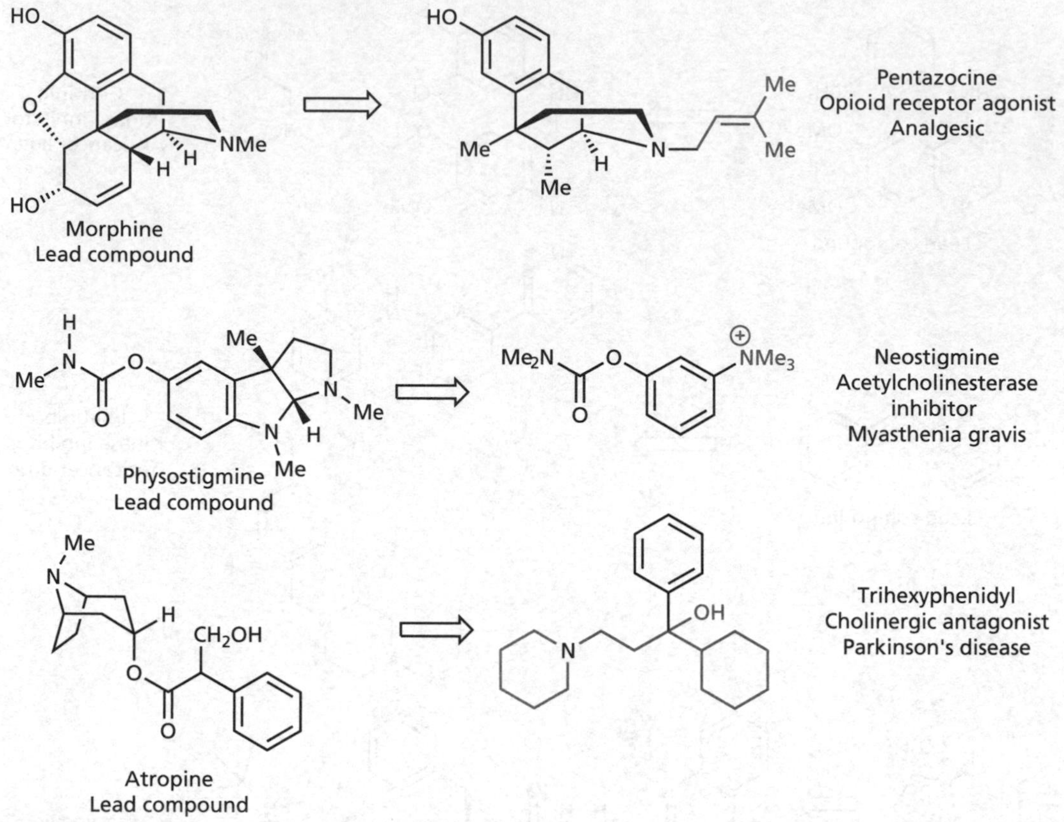

FIGURE 8.2 Clinically important drugs that have a simpler structure than the natural lead compounds from which they were derived.

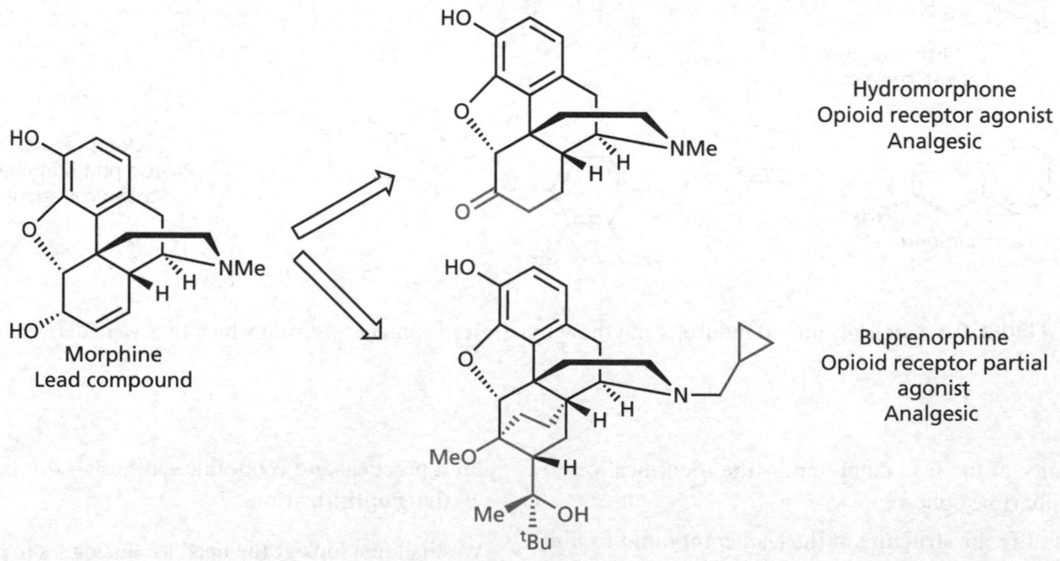

FIGURE 8.3 Clinically important analgesics that have either similar or more complex structures than morphine.

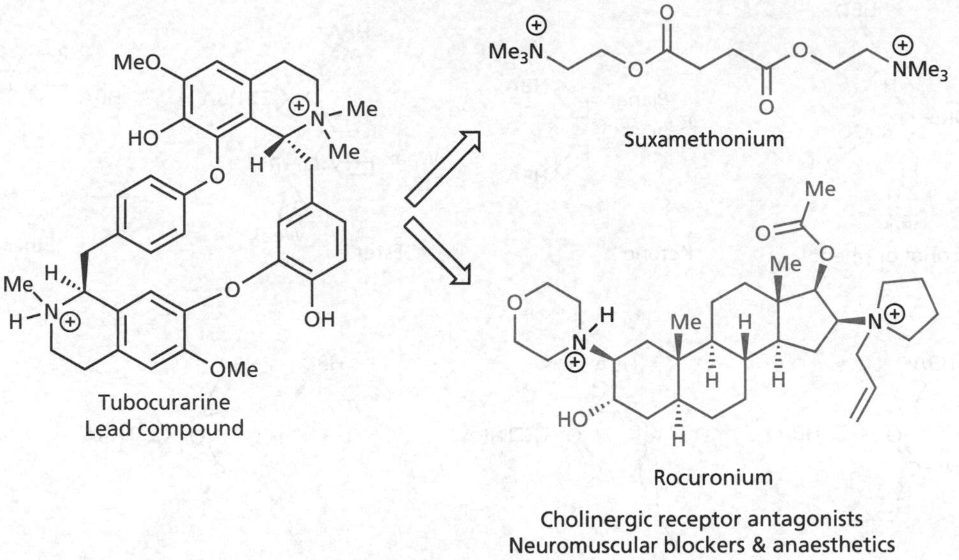

FIGURE 8.4 Clinically important drugs and the natural lead compounds from which they were developed.

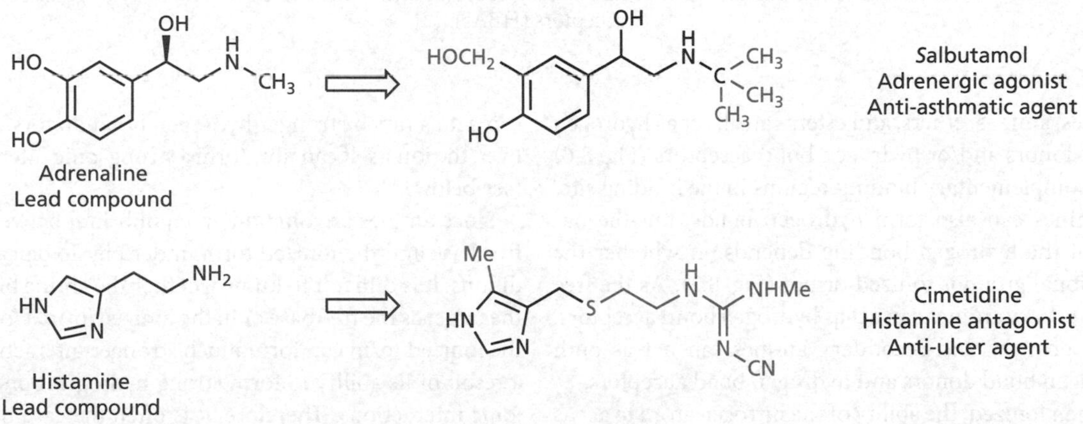

FIGURE 8.5 Clinically important drugs and the natural lead compounds from which they were derived.

8.2 Analogues for SAR studies and pharmacophore identification

8.2.1 Introduction

An important priority at the start of any drug research project is to define the functional groups that are involved in binding a lead compound to its macromolecular target, whether that be through hydrogen bonding, ionic interactions, van der Waals interactions, dipole–dipole interactions, cation–pi interaction, interactions with transition metal cofactors, or covalent bond formation (see also section 1.7). The principle is quite simple. A single functional group in the lead compound is modified, masked, or removed in a deliberate attempt to prevent it interacting with the target. The resulting analogue is then tested for pharmacological activity or binding affinity using an *in vitro* assay. If the activity/affinity of the analogue is not adversely affected, then clearly that functional group is not important. If activity/binding *is* significantly reduced or eliminated, then it implies that the functional group is important.

8.2.2 Binding roles played by functional groups

A number of functional groups are commonly found in lead compounds and have the potential to act as important binding groups. For example, alcohols, phenols,

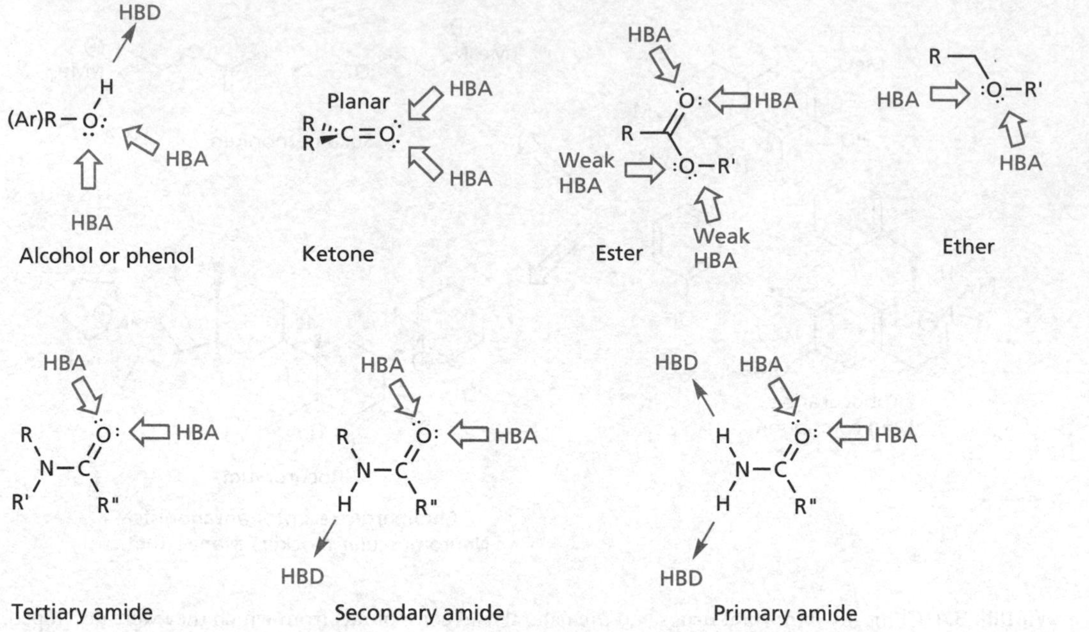

FIGURE 8.6 Common functional groups that can act as hydrogen bond donors (HBDs) and/or hydrogen bond acceptors (HBAs).

ketones, amides, ethers, and esters can all act as hydrogen bond donors and/or hydrogen bond acceptors (Fig. 8.6) with complementary binding regions in the binding site.

Amines can also form hydrogen bonds, but the nature of the hydrogen bonding depends on whether the functional group is ionized or not (Fig. 8.7). As the free base, tertiary amines can act as hydrogen bond acceptors, while primary and secondary amines can act as both hydrogen bond donors and hydrogen bond acceptors.

When ionized, the ability of the nitrogen atom to act as a hydrogen bond acceptor is lost, but the N^+–H protons can act as much stronger hydrogen bond donors. Moreover, the ion itself can now form a strong ionic interaction (see below).

Since amines are often in an equilibrium between the free base and the ionized form under physiological conditions, it is difficult to know whether the amine binds to the target as the free base or in the ionized form. However, the ionized form can form much stronger interactions as a result of its ability to form strong hydrogen bonds and ionic interactions. Therefore, it is often the case that active compounds containing an amine group bind to their

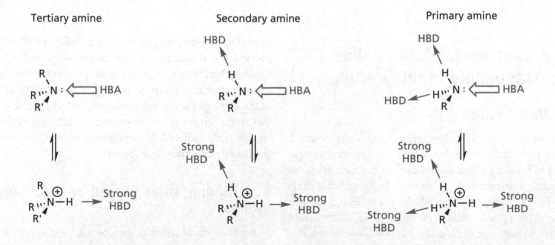

FIGURE 8.7 The hydrogen bonding capacity of amines as free bases or in the ionized form.

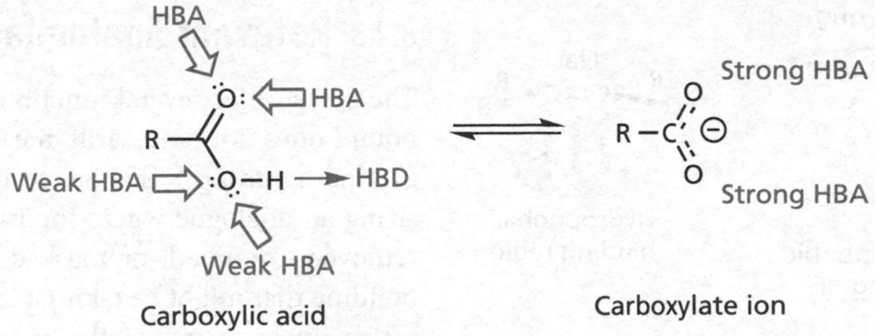

FIGURE 8.8 The hydrogen bonding capacity of a carboxylic acid as the free acid or in the ionized form.

targets in the ionized form. Indeed, the amine group is one of the most common functional groups present in both lead compounds and clinically important drugs.

Carboxylic acids are less commonly found in lead compounds and drugs, but they are still quite prevalent. They too can form hydrogen bonds, either as the free acid or as the ionized carboxylate ion (Fig. 8.8). The ion cannot act as a hydrogen bond donor, but it can act as a strong hydrogen bond acceptor. That, allied with the fact that it can form ionic interactions with the binding site, makes it a very strong binding group. If the group is playing a role in binding, then the odds are that it does so in the ionized form.

Finally, nitrogen-containing heterocycles are commonly present in lead compounds as they are often used as scaffolds (section 7.3). Many of these have the potential to act as hydrogen bond donors or hydrogen bond acceptors (Fig. 8.9). Note that not all nitrogen atoms in a heterocycle have the potential to act as hydrogen bond acceptors. If the nitrogen's lone pair of electrons can interact with a pi system, it is less available to participate in a hydrogen bond. Thus, the nitrogen of an aromatic amine makes a poor hydrogen bond acceptor. Similarly, if the nitrogen's lone pair of electrons is part of an aromatic system, that nitrogen is a poor hydrogen bond acceptor.

Compared with the number of functional groups capable of forming hydrogen bonds, there are far fewer capable of forming ionic or electrostatic interactions. Here, we are restricted to amines and carboxylic acids (Fig. 8.10). Aminium ions can form ionic interactions with charged carboxylate groups in the binding site. These would be provided by the amino acids aspartate or glutamate.

Carboxylate ions can form ionic interactions with charged aminium or guanidinium ions in the binding site. These would be provided by the amino acids lysine or arginine.

Hydrophobic regions of the lead compound can interact with hydrophobic regions of the binding site through van der Waals interactions. Although these interactions are weak, there can be far more of them than hydrogen bonding or ionic interactions, and so the cumulative effect of van der Waals interactions can be decisive in whether a compound binds strongly to its target. The hydrophobic regions in a lead compound include hydrophobic regions of the scaffold, as well as alkyl substituents and functional groups such as alkenes and aromatic rings (Fig. 8.11).

Other types of interaction include dipole–dipole interactions, pi–pi stacking, and pi–cation interactions. Pi–cation interactions can be important in drugs containing

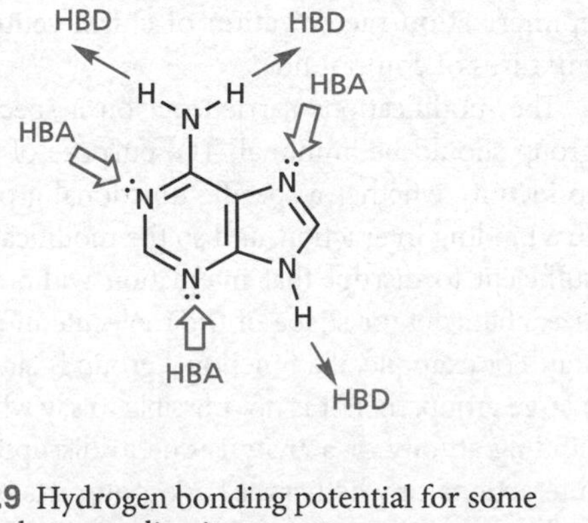

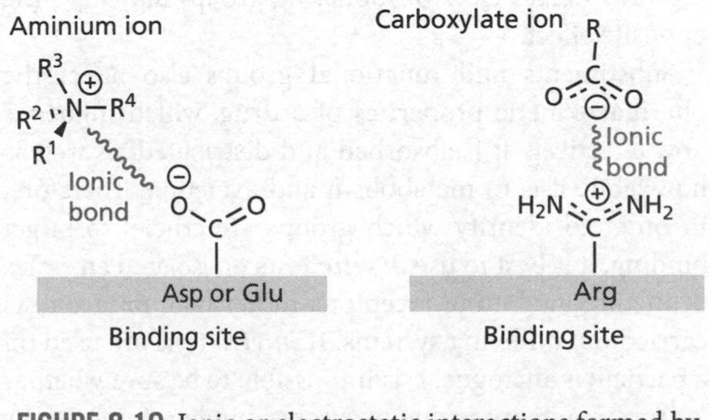

FIGURE 8.9 Hydrogen bonding potential for some heterocyclic nitrogen atoms.

FIGURE 8.10 Ionic or electrostatic interactions formed by aminium and carboxylate ions.

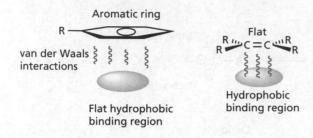

FIGURE 8.11 van der Waals interactions between hydrophobic groups.

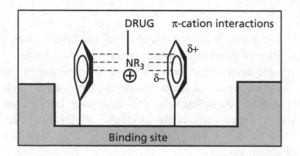

FIGURE 8.12 Pi–cation interactions.

an aminium or a quaternary ammonium group (Fig. 8.12). The charged group interacts with aromatic or heteroaromatic rings present in the side chains of amino acids such as tryptophan, tyrosine or phenylalanine.

Other functional groups may also be present in a lead compound, some of which have an indirect effect on its activity. For example, a rigid functional group such as an alkene, alkyne, or amide may play a role in rigidifying a structure so that it is more likely to be in the active conformation. In the absence of the group, the molecule is less conformationally restrained and activity will decrease.

Other groups may have an electronic effect that affects the binding strength of another group. For example, electron-withdrawing groups may affect a neighbouring amine such that it is less effective as a hydrogen bond acceptor, whereas electron-donating groups may have the opposite effect.

Substituents and functional groups also affect the pharmacokinetic properties of a drug, which influence how effectively it is absorbed and distributed, as well as how labile it is to metabolism and excretion. Therefore, in order to identify which groups are crucial to target binding, it is best to use *in vitro* tests on isolated enzymes or membrane-bound receptors, rather than *in vivo* tests carried out on living systems. If *in vivo* tests are used on a particular analogue, it is impossible to be sure whether a drop in activity is due to weaker binding interactions with the target, or less of the analogue reaching the target.

8.2.3 Relevant analogues for SAR studies

The presence of several functional groups in a lead compound does not necessarily mean that they are all acting as binding groups. This has to be established by synthesizing an analogue where the functional group has been removed, modified, or masked in order to disrupt any bonding that might be taking place. If the analogue is less active, this suggests that the functional group is important to binding either directly or indirectly. If activity is unaffected, the functional group is not essential to binding.

There are different approaches that can be used to synthesize analogues of a lead compound. Some approaches are particularly relevant to lead compounds from the natural world and are discussed in Chapter 9. Here, we concentrate on methods that are more relevant to fully synthetic lead compounds. There are two main approaches:

a) synthesis from the lead compound itself (also suitable for natural lead compounds) (see also section 8.5);

b) repeating the synthesis that was used to obtain the lead compound, but modifying the reagents or starting materials (section 8.6).

8.2.3.1 Analogues synthesized from a lead compound

Analogues can be prepared from a lead compound by carrying out reactions on the functional groups that are present. The reactions concerned should be chemoselective for a particular functional group, such that only one group is modified in the structure. That way, any change in activity can be attributed to the modification of that group. If chemoselectivity is a problem, it may be necessary to carry out protection and deprotection strategies in order to modify a specific group.

The reactions that are carried out on a lead compound usually need to be relatively mild, especially if the lead compound has several functional groups and chiral centres. Harsh reaction conditions can result in unwanted reactions leading to degradation of the structure or epimerization/racemization of chiral centres leading to mixtures of compounds.

The modifications carried out on a specific functional group should be 'minimal'. The purpose of the exercise is to identify whether a specific functional group is involved in a binding interaction, and so the modification should be sufficient to disrupt that interaction without affecting the size, character, or shape of the molecule in any significant way. For example, if a functional group is altered by adding a large group, then it is not possible to say whether a loss of binding affinity or activity is due to disruption of binding interactions, or whether the large group acts as a steric shield and prevents the analogue from fitting the binding site.

Typical reactions that can often be carried out on a lead compound are the following (Fig. 8.13):

- acetylation of an alcohol or phenol to form an ester
- acylation of an amine to form an amide
- methylation of an alcohol or phenol to form a methyl ether
- hydrogenation of an alkene to form an alkyl group
- reduction of a ketone to form an alcohol
- hydrolysis of an ester or an amide to form a carboxylic acid and an alcohol/amine.

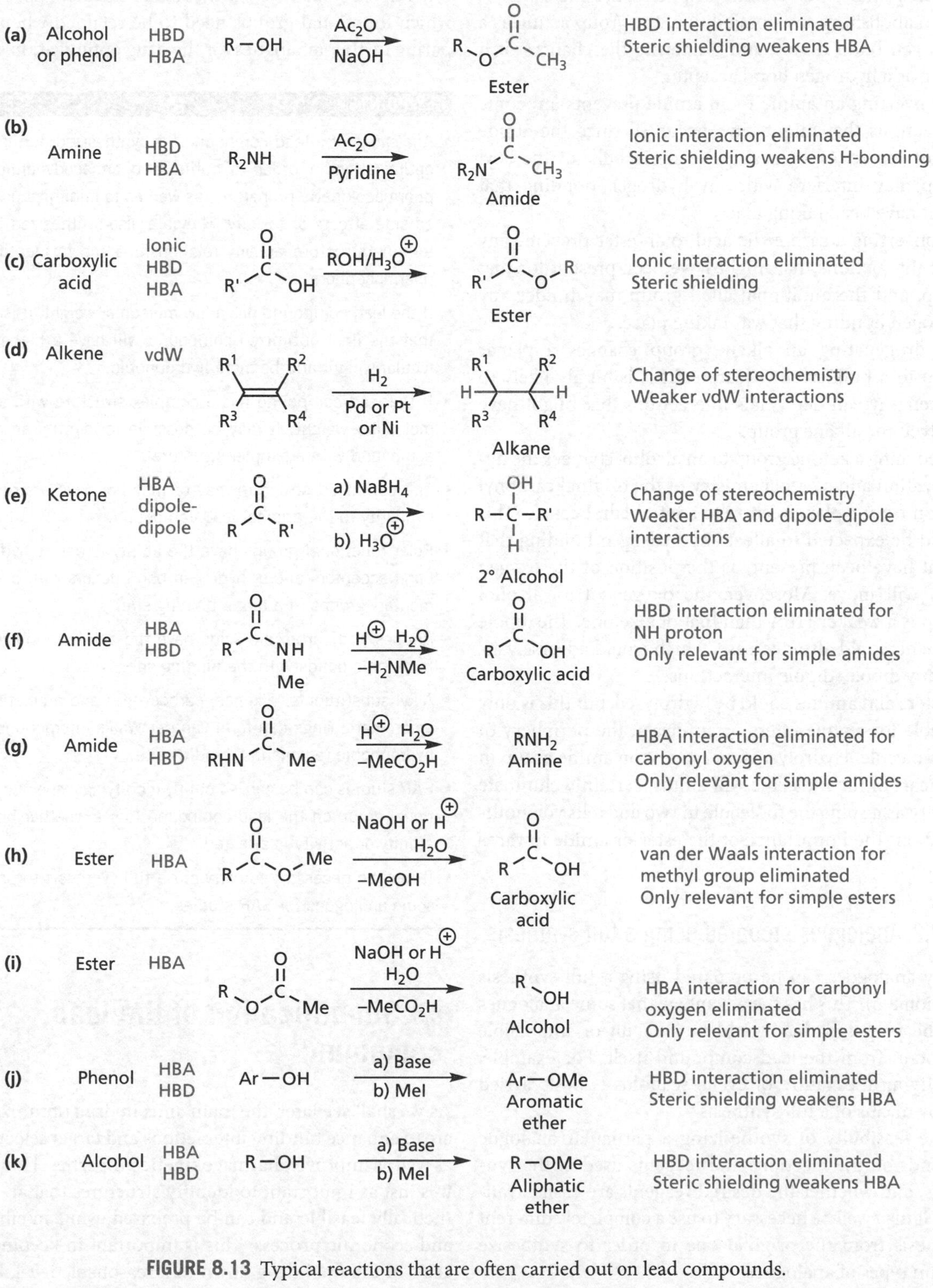

				Reagents	Product	Effect
(a)	Alcohol or phenol	HBD HBA	$R-OH$	Ac_2O / NaOH	Ester ($R-O-C(=O)-CH_3$)	HBD interaction eliminated. Steric shielding weakens HBA
(b)	Amine	HBD HBA	R_2NH	Ac_2O / Pyridine	Amide ($R_2N-C(=O)-CH_3$)	Ionic interaction eliminated. Steric shielding weakens H-bonding
(c)	Carboxylic acid	Ionic HBD HBA	$R'-C(=O)-OH$	$ROH/H_3O^{\oplus}$	Ester ($R'-C(=O)-O-R$)	Ionic interaction eliminated. Steric shielding
(d)	Alkene	vdW	$R^1R^3C=CR^2R^4$	H_2 / Pd or Pt or Ni	Alkane ($R^1R^3CH-CHR^2R^4$)	Change of stereochemistry. Weaker vdW interactions
(e)	Ketone	HBA dipole-dipole	$R-C(=O)-R'$	a) $NaBH_4$ b) $H_3O^{\oplus}$	2° Alcohol ($R-CH(OH)-R'$)	Change of stereochemistry. Weaker HBA and dipole–dipole interactions
(f)	Amide	HBA HBD	$R-C(=O)-NH-Me$	$H^{\oplus}$ H_2O / $-H_2NMe$	Carboxylic acid ($R-C(=O)-OH$)	HBD interaction eliminated for NH proton. Only relevant for simple amides
(g)	Amide	HBA HBD	$RHN-C(=O)-Me$	$H^{\oplus}$ H_2O / $-MeCO_2H$	Amine (RNH_2)	HBA interaction eliminated for carbonyl oxygen. Only relevant for simple amides
(h)	Ester	HBA	$R-C(=O)-O-Me$	NaOH or $H^{\oplus}$ H_2O / $-MeOH$	Carboxylic acid ($R-C(=O)-OH$)	van der Waals interaction for methyl group eliminated. Only relevant for simple esters
(i)	Ester	HBA	$R-O-C(=O)-Me$	NaOH or $H^{\oplus}$ H_2O / $-MeCO_2H$	Alcohol ($R-OH$)	HBA interaction for carbonyl oxygen eliminated. Only relevant for simple esters
(j)	Phenol	HBA HBD	$Ar-OH$	a) Base b) MeI	Aromatic ether ($Ar-OMe$)	HBD interaction eliminated. Steric shielding weakens HBA
(k)	Alcohol	HBA HBD	$R-OH$	a) Base b) MeI	Aliphatic ether ($R-OMe$)	HBD interaction eliminated. Steric shielding weakens HBA

FIGURE 8.13 Typical reactions that are often carried out on lead compounds.

Acetylation of an alcohol or a phenol results in an ester and eliminates any hydrogen bonding that might have occurred due to the OH proton acting as a hydrogen bond donor. Moreover, the larger acyl group is likely to interfere with any hydrogen bonding that might have involved the oxygen as a hydrogen bond acceptor.

Methylation of an alcohol or phenol has the same effect in abolishing any possibility of the group acting as a hydrogen bond donor, and hindering the chances of it acting as a hydrogen bond acceptor.

Converting an amine to an amide prevents any ionic interactions that might have occurred since the amide cannot ionize. Once again, the steric bulk of the acyl group may interfere with any hydrogen bonding that might have been taking place.

Converting a carboxylic acid to an ester prevents any ionic interactions that might have been present for that group, and the additional alkyl group may hinder any hydrogen bonding that was taking place.

Hydrogenating an alkene group changes a planar group to a bulkier alkyl feature. This is highly likely to weaken any van der Waals interactions that might have involved the alkene group.

Reducing a ketone group to an alcohol changes the hybridization and stereochemistry of the original carbonyl carbon from a planar centre to a tetrahedral centre. This would be expected to affect any hydrogen bonding that might have been present, as the position of the oxygen atom will move. Moreover, the oxygen of an alcohol group is a weaker HBA than that of a ketone. The dipole moment would also decrease, which would adversely affect any dipole–dipole interactions.

Esters and amides could be hydrolysed, but this is only sensible for simple esters or amides at the periphery of the molecule. Hydrolysing an ester or an amide group in the heart of the molecule will almost certainly eliminate activity as it splits the molecule in two and tells you nothing about the importance of the ester or amide towards binding.

8.2.3.2 Analogues prepared using a full synthesis

Many analogues can be prepared using a full synthesis (section 8.6). This has the advantage that some analogues can be prepared that would be difficult or impossible to obtain from the lead compound itself. For example, modifying the nature of cyclic scaffolds is best carried out by means of a full synthesis.

The feasibility of synthesizing a particular analogue depends on the reactions and reagents used in the synthesis, and whether the desired reagents are easily available. It may well be necessary to use a completely different synthesis from the original one in order to synthesize certain types of analogues.

8.2.4 Identification of a pharmacophore

Once the functional groups involved in binding interactions have been identified, a pharmacophore can be defined (section 1.7). This identifies the important functional groups, as well as their relative positions to each other in the structure. The medicinal chemist then knows which functional groups need to be retained when preparing further analogues for the drug optimization stage.

> **KEY POINTS**
>
> - Analogues of a lead compound are synthesized during drug optimization in order to optimize pharmacodynamic and pharmacokinetic properties, as well as to minimize unwanted side effects or toxicity. They are also synthesized to determine structure–activity relationships and to identify the pharmacophore.
> - If the lead compound has a low molecular weight, it is likely that the final optimized compound will have a higher molecular weight and be more hydrophobic.
> - If the lead compound has a complex structure with a high molecular weight, it may be possible to identify an active compound with a simpler structure.
> - The optimized structure may or may not have a structural similarity to the original lead compound.
> - Polar functional groups have the ability to act as hydrogen bond acceptors and/or hydrogen bond donors with complementary groups in a target binding site.
> - Amines and carboxylic acids have the ability to ionize and form ionic bonds with the binding site.
> - Alkyl substituents, alkenes, carbocyclics and aromatic/heteroaromatic rings can form van der Waals interactions with hydrophobic regions in a binding site.
> - SAR studies can be carried out by modifying a specific functional group on the lead compound to see whether binding affinity or activity is affected.
> - It may be necessary to carry out a full synthesis to prepare some analogues for SAR studies.

8.3 Simplification of the lead compound

As we shall see later, the main aims in drug optimization are to enhance binding interactions and target selectivity, as well as improve pharmacokinetic properties. However, it is just as important to identify structures that are synthetically feasible and can be prepared using an efficient and economic process. This is important in keeping the costs of the final drug down to a reasonable level, but it

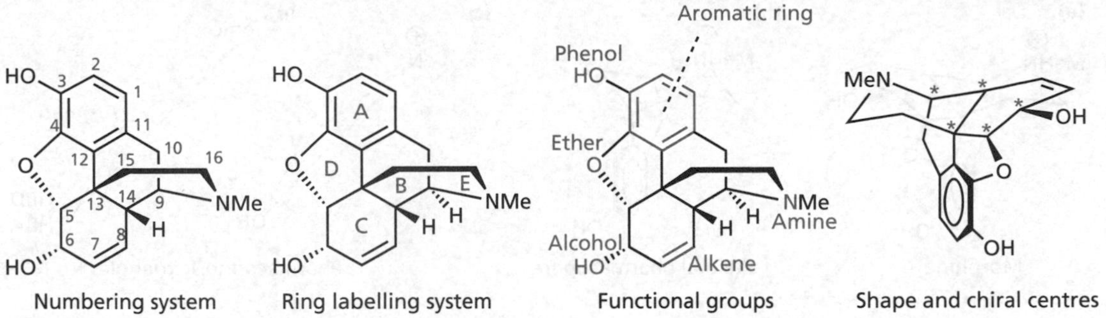

FIGURE 8.14 Structure of morphine.

is also crucial for the preparation of analogues aimed at finding that drug in the first place. If the synthesis of the lead compound is difficult, then generating analogues is also going to be difficult.

This is a particular consideration when looking at lead compounds with a complex structure, especially those from the natural world. Such compounds are likely to have several functional groups, rings, and chiral centres, all of which make synthesis extremely challenging.

Therefore, a common approach in drug design is to prepare simplified analogues that retain the desired pharmacophore, in the hope that they retain the desired activity. If such a compound can be found, that structure can be adopted as a new lead compound for drug optimization.

A knowledge of the pharmacophore is crucial in this respect as it identifies those regions of the original

complex structure that are crucial to binding affinity or activity and those regions which are not.

The development of opioid analgesics is a good example of this approach. Morphine is a potent analgesic, and is still one of the most important painkillers used in medicine. However, it has a relatively complex structure containing six functional groups, a pentacyclic ring system, and five chiral centres (Fig. 8.14). The first total synthesis of racemic morphine was achieved in 1952 and involved 29 steps with an overall yield of only 0.0014%. The synthesis started with a bicyclic structure representing rings A and B of the final structure (Fig. 8.15). From this, the structure was built up such that a Diels–Alder reaction could be used to form ring C. This also introduced the double bond that would eventually become the alkene in the final structure.

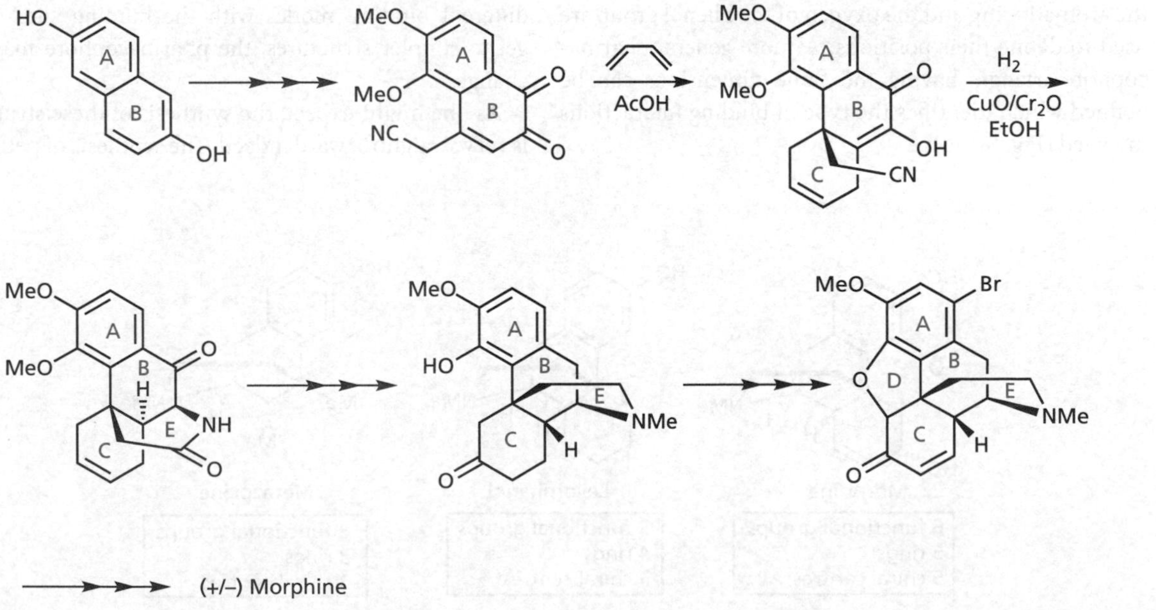

FIGURE 8.15 Key intermediates in the first synthesis of (+/–)morphine.

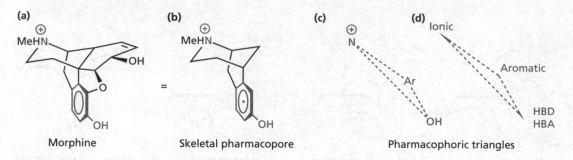

FIGURE 8.16 The analgesic pharmacophore for morphine.

Ring E was then created by means of a reductive amination, and the final ring (ring D) was formed towards the end of the synthesis.

There have been several other approaches to synthesizing morphine, but all of them result in low overall yields. Consequently, it is still more economic to extract morphine from opium than to synthesize it from scratch.

Therefore, it makes sense to design simpler analogues of morphine that retain the analgesic pharmacophore, and which will be much simpler to synthesize. SAR studies have demonstrated that the analgesic pharmacophore consists of the phenol, the aromatic ring, and the ionized amine (Fig. 8.16a). Since morphine is a rigid structure, the relative positions of these groups are well defined and can be represented by a skeletal pharmacophore which represents the minimal scaffold required to hold them in the same position (Fig. 8.16b). Alternatively, the pharmacophore can be defined by a pharmacophore triangle, where each corner of the triangle represents one of the important functional groups (Fig. 8.16c). The centre of the aromatic ring and the oxygen of the phenol group are used to define their positions. A more general pharmacophore triangle having the same dimensions can be defined which identifies the type of binding interactions involved (Fig. 8.16d).

Over the years, a large number of simplified morphine analogues have been synthesized that retain the pharmacophore and are found to have similar analgesic activity to morphine (Fig. 8.17).

The benefits of these simpler structures can be seen by the ease with which they are synthesized. For example, **levorphanol** can be synthesized in seven steps, compared with the 29 steps involved in the first synthesis of morphine (Fig. 8.18).

Even simpler analogues of morphine have been discovered such as **pethidine** and **methadone** (Fig. 8.19). However, in these structures the phenol group is no longer an important feature of the analgesic pharmacophore. This illustrates an important point. The analogues shown in Figure 8.17 are all rigid structures similar to morphine, and bind in the same way to the target binding site. The much simpler analogues shown in Figure 8.19 are smaller, more flexible structures, and have lost the rigidity that was present in the previous analogues. As a result, they adopt different conformations and can have different binding modes with the binding site. As we get to simpler structures, the pharmacophore may well change.

As one might expect, the synthesis of these structures is very straightforward indeed. The synthesis of pethidine

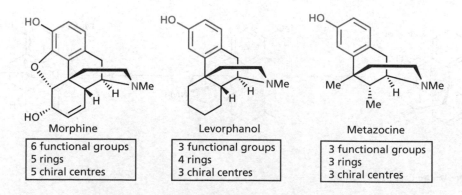

Morphine	Levorphanol	Metazocine
6 functional groups	3 functional groups	3 functional groups
5 rings	4 rings	3 rings
5 chiral centres	3 chiral centres	3 chiral centres

FIGURE 8.17 Examples of simplified morphine analogues.

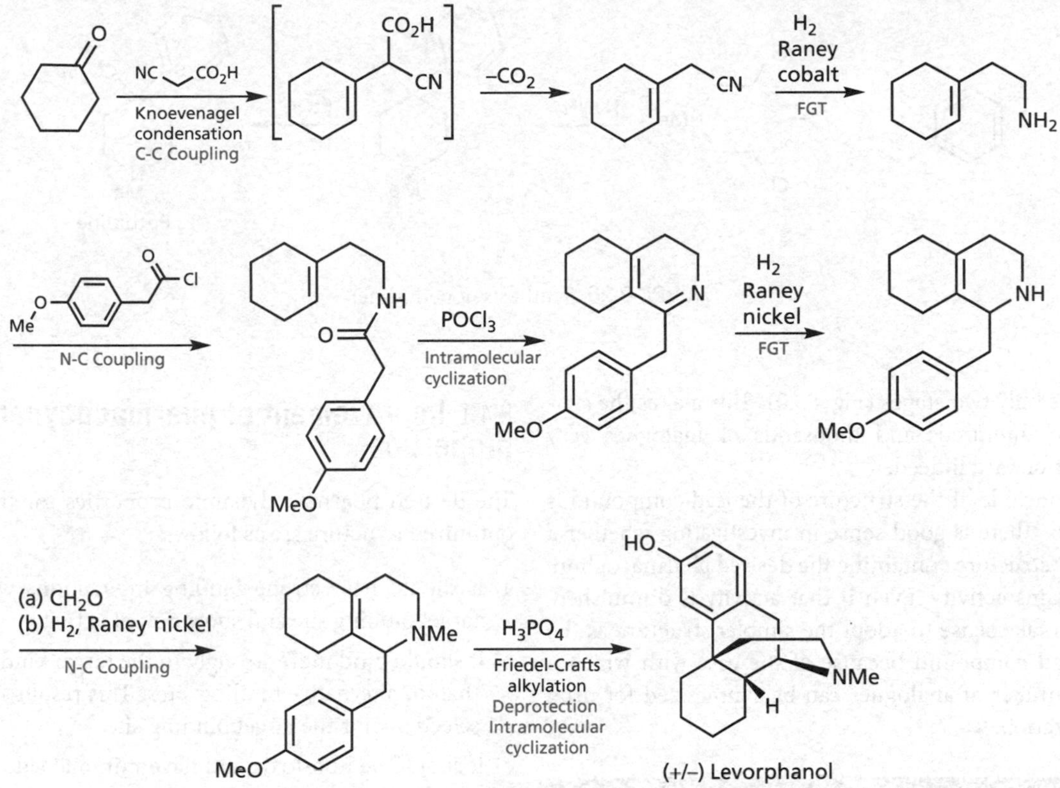

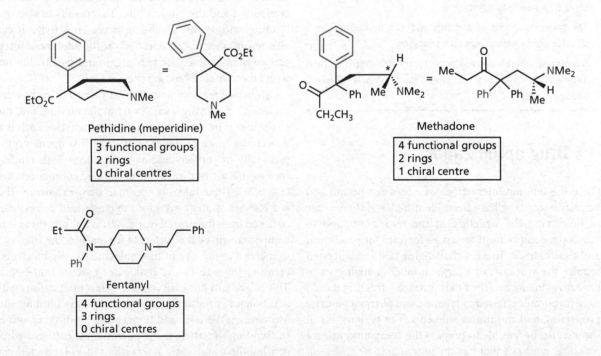

FIGURE 8.18 Synthesis of racemic levorphanol.

FIGURE 8.19 Simplified flexible analogues of morphine.

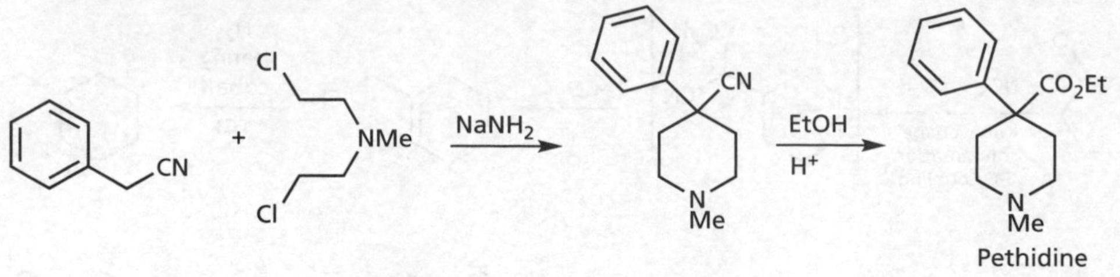

FIGURE 8.20 Synthesis of pethidine.

involves only two stages (Fig. 8.20). This makes the synthesis of hundreds and thousands of analogues very straightforward indeed.

To conclude, if the structure of the lead compound is complex, there is good sense in investigating whether a simpler structure containing the desired pharmacophore still retains activity. Even if that activity is diminished, it may make sense to adopt the simpler structure as the new lead compound because of the ease with which a large number of analogues can be synthesized for drug optimization.

KEY POINTS

- Complex lead compounds are difficult to synthesize, limiting the scope for analogue synthesis.

- The synthesis of simpler structures that retain the pharmacophore of the lead compound may lead to active compounds that are more easily synthesized. One of these may serve as a new lead compound.

- The pharmacophore of a compound can be defined as a structure or by pharmacophore triangles.

- Simpler structures are easier to synthesize, but may bind differently to the binding site, resulting in a different pharmacophore.

8.4 Drug optimization

Once the binding interactions of a lead compound and its pharmacophore have been identified, it is then a case of modifying the structure of the lead compound to develop a compound that can go forward for preclinical and clinical trials. This is a challenging task and will often require the synthesis of a large number of analogues in order to gain success. The final optimized structure should have favourable pharmacodynamic and pharmacokinetic properties, and minimum side effects or toxicity, but it should also be feasible to prepare the compound using a practical synthesis that has the potential to be scaled up to production level (Chapter 10).

8.4.1 Improvement of pharmacodynamic properties

The desired pharmacodynamic properties for the final optimized structure are as follows.

a) It should have strong binding interactions with the target binding site and show potent activity.

b) It should bind more strongly to the target binding site than to alternative binding sites. This results in good selectivity for the target binding site.

c) It should be non-toxic and have minimal side effects. This can usually be achieved if the compound shows increased binding and good selectivity for the target binding site.

So how can all this be achieved? One of the most important methods of drug optimization is to increase the number of binding interactions between the initial lead compound and the binding site. This results in stronger binding, and usually results in increased activity. It can also result in increased selectivity if the additional interactions are possible with the target binding site, but not with other types of binding site.

Synthetic lead compounds tend to be relatively small molecules occupying a small part of the binding site, and so the idea is to increase the size of the molecule such that it occupies more of the binding site. This opens up the possibility of further binding interactions with binding regions that are not accessible to the lead compound itself (Fig. 8.21). This strategy is known as **drug extension**. This is a relevant strategy even for the larger lead compounds obtained from the natural world, as it is unlikely that a lead compound will be the perfect fit for any binding site.

Increasing the size of the molecule usually means increasing the size of the molecule's carbon framework. This is likely to increase van der Waals interactions with additional hydrophobic regions within the binding site. Moreover, if there are additional polar binding regions in the binding site, there is the potential for further hydrogen bonding and ionic interactions if polar functional groups are positioned correctly in the drug 'extensions'.

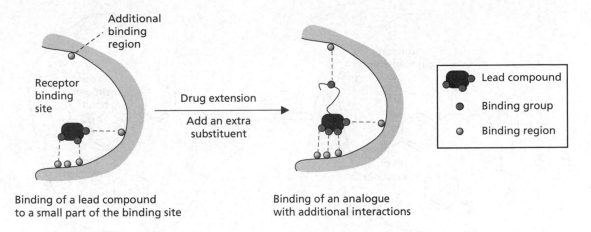

FIGURE 8.21 Drug extension to achieve additional binding interactions.

You might well ask how a lead compound should be increased in size. Should it be expanded in all directions like an inflated balloon, or should it only be extended in a particular direction, similar to a branch on a tree? Also, where should we place a polar functional group on the extension if we want to make interactions with a previously unused polar binding region?

These decisions can be helped enormously if it is possible to crystallize the target protein with the lead compound bound to the binding site. The structure of the ligand–protein complex can then be determined using X-ray crystallography, and studied using molecular modelling to identify where the vacant regions of the binding site are. This indicates where substituents should be added to the lead compound in order to access vacant parts of the binding site. A study of those vacant areas in the binding site can also lead to the identification of extra binding regions and indicate whether it is worth adding polar or non-polar binding groups to the additional substituent.

Unfortunately, such information is often unavailable, as many important drug targets (such as membrane-bound receptors) are extremely difficult to crystallize. In such cases, it is necessary to revert to traditional methods where one synthesizes a large number of analogues to probe the binding site for the additional binding interactions.

Analogues can be prepared with additional substituents at different parts of the lead compound's scaffold to see whether they have a beneficial effect on activity. Substituents varying in length and bulk provide clues about the space available around that region of the scaffold, while the nature of the substituent (whether it is hydrophilic or hydrophobic) provides information on possible hydrophobic and hydrophilic binding regions. However, it is important to appreciate that substituents cannot be added to every part of the lead compound's scaffold. Indeed, the number of positions where this is possible may

be quite limited and are really determined by the building blocks used in the synthesis of the lead compound, or the reactivity of the lead compound itself. Therefore, extending the lead compound is more akin to a tree growing branches, than a balloon being inflated.

Finally, there may well be additional binding regions situated in the part of the binding site occupied by the lead compound, but which do not interact with the lead compound itself. This is possible if the lead compound does not contain the relevant binding group for such an interaction. Varying the structure of the lead compound's core scaffold to introduce such a binding group may result in additional interactions without having to enlarge the structure (Figs 8.21 and 8.22).

8.4.2 Improvement of pharmacokinetic properties

As well as finding analogues with increased potency and selectivity for the intended target, it is important to develop structures that are capable of reaching the desired target in the body. For that to happen, the analogues must have good pharmacokinetic properties with respect to absorption, distribution, metabolism, and excretion (ADME) (section 1.5.3).

It is normally desirable to develop an orally active drug, although this is not always the case; for example, inhaled anti-asthmatic drugs should not be orally active as this might result in cardiovascular side effects. However, in most other fields of medicine, the preference is for an orally active compound which can be easily administered. This means that the final drug candidate must have the following properties.

a) It must be chemically stable to the environments encountered in the gastrointestinal tract.

b) It must be resistant to digestive enzymes.

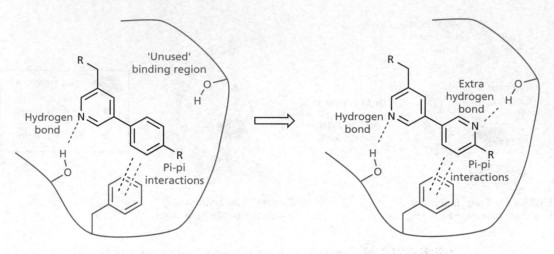

FIGURE 8.22 Modifying a scaffold to obtain an additional binding interaction.

c) It must be absorbed efficiently from the gut into the blood supply.

d) Since the blood supply from the gut passes through the liver, the drug must be capable of surviving the metabolic enzymes present in the liver—the **first pass effect**.

e) It must be resistant to metabolic enzymes in the blood supply and other tissues. However, it must not be so resistant that it 'lingers' in the body for prolonged periods of time.

f) It must have the correct physicochemical properties to be distributed to target tissues effectively.

g) It must be capable of crossing cell membranes if its target is within cells.

h) It must be capable of crossing the blood–brain barrier if its target is in the central nervous system.

i) It must be excreted at a predictable and reasonable rate.

As far as absorption from the digestive system is concerned, several studies have been carried out to identify why some drugs are orally active and some are not. **Lipinski's rules** and **Veber's parameters** (section 1.5.3) serve as useful guidelines on how well potential analogues might be absorbed, and help to prioritize which analogues are best synthesizing. Essentially, the analogues should have a good balance of hydrophobic and hydrophilic properties, and should not be too flexible in nature. In other words, too many rotatable bonds can be bad for absorption.

A knowledge of these properties helps in the drug optimization process. If an analogue has good activity at the target binding site, but is too hydrophobic or hydrophilic, then it may be possible to add substituents that will change those properties. For example, adding an alkyl substituent will increase hydrophobicity and may improve the pharmacokinetic properties of a drug which is too polar to cross hydrophobic cell membranes. Alternatively, polar functional groups such as alcohols and phenols can be masked as ethers. If a polar functional group is an essential part of the pharmacophore, a masking group can be used which is susceptible to metabolism and is removed once the drug has been absorbed. Such an analogue is known as a **prodrug** as it normally has poor activity, but is converted to the active drug in the body. A common method of masking polar carboxylic acids, phenols, or alcohols is to synthesize an ester derivative. Once absorbed, esterase enzymes catalyse the hydrolysis of the ester group.

On the other hand, a drug might not be sufficiently polar and show poor solubility in aqueous solution. A popular method of addressing this is to add an amine substituent. Since this is a weak base, it is possible to form a water-soluble salt for administration. Once administered, the amine acts as a weak base and can equilibrate between the ionized and free base form (Fig. 8.7), allowing it to cross cell membranes as the free base and to be more soluble in aqueous environments as the ionized form.

However, there is a potential problem with adding an amine group. Since the amine is polar in nature, it will be highly solvated with water molecules, which means that the amine has to be desolvated before the drug can access a relatively hydrophobic binding site. This involves an energy penalty without the compensation of additional binding interactions between the amine and the binding site. As a result, the binding affinity and activity of the analogue is likely to be decreased (Fig. 8.23).

One way round this problem is to place the amine group at the end of a long chain such that it sticks out from the binding site and does not need to be desolvated.

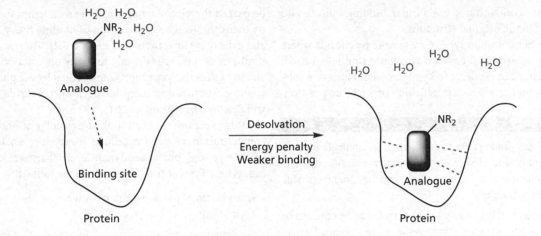

FIGURE 8.23 The energy penalty resulting from desolvating a polar group.

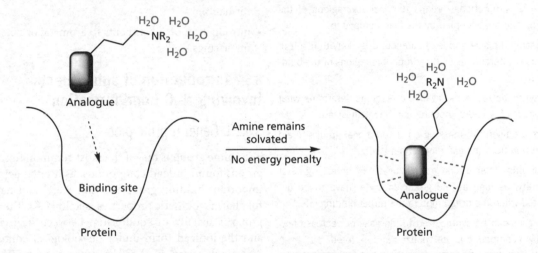

FIGURE 8.24 Positioning a solvating group such that it does not need to be desolvated.

If this is to work effectively, the substituent should be placed at the part of the drug scaffold that is closest to the binding site entrance (Fig. 8.24). For example, gefitinib contains a morpholine ring which was added to increase polarity and solubility (Fig. 8.1). The ring was placed at the end of an alkyl chain such that it stuck out of the binding site and did not need to be desolvated (section 8.6.7).

Another crucial part of drug optimization is to ensure that candidate drugs have a reasonable lifetime within the body. The length of time a drug is present in the system depends on both metabolism and excretion (section 1.5.3). Drugs which are metabolized or excreted too quickly will have to be administered at frequent intervals—possibly several times per day. Ideally, a drug should be administered no more than once per day, and so analogues may need to be found that are more resistant to metabolism and excretion. At the other end of the spectrum, there are

distinct disadvantages in drugs that are too resistant to metabolism and excretion, because their effects may be too prolonged. This complicates dosing regimes and may lead to patient non–compliance. Therefore, the ideal drug should be metabolized and excreted in a predictable manner and with a reasonable rate.

A rapidly metabolized compound usually contains a metabolically susceptible group. Suitable analogues could be made where that group is replaced with a different group so that it is similar in nature, but is more resistant to metabolism. For example, an exposed methyl substituent is often susceptible to oxidation by cytochrome P450 enzymes in the liver. A common solution is to replace the methyl group with a chloro substituent. The chlorine is similar in size and hydrophobic character to the methyl group, and so it can interact with the binding site in a similar fashion. However, it is resistant to oxidation. A group such as this is called a **bio-isostere** since the analogue containing it has similar binding affinity and activity to the original structure.

Another common type of metabolic reaction is where part of an aromatic or heteroaromatic ring is oxidized. Analogues can be designed which contain an extra substituent that acts as a steric shield to protect that position.

KEY POINTS

- Drug optimization aims to develop a drug candidate that can be synthesized efficiently, has optimum pharmacodynamic and pharmacokinetic properties, and has minimum side effects or toxicity.

- Improved binding affinity and selectivity can be obtained by synthesizing analogues that have increased binding interactions with the desired target binding site.

- Extra binding interactions can be achieved by adding substituents or functional groups that occupy regions of the binding site not occupied by the lead compound.

- Increasing the size of a lead compound is likely to result in increased interactions with hydrophobic regions of the binding site.

- The synthetic route and available reagents determine what analogues of a lead compound can be synthesized.

- Esters are useful in masking polar functional groups to aid absorption from the gastrointestinal tract.

- The addition of an amine group is useful in increasing water solubility, as long as the group does not have to be desolvated when the drug binds to the target binding site.

- Analogues can be synthesized to increase or decrease how rapidly a compound is metabolized and excreted.

8.5 Synthesis of analogues from a lead compound

8.5.1 Introduction

In general, there are two approaches to the synthesis of analogues. One is to prepare analogues using a full synthesis; this will be covered in section 8.6.

The other approach is to prepare analogues from the lead compound. This is particularly relevant for complex lead compounds, including many of the lead compounds isolated from the natural world. Such compounds can be very difficult to synthesize, and so a full synthesis of analogues would be equally difficult.

Since the lead compound is going to be used as the starting material for analogue synthesis, the accessible analogues will depend on the functional groups present in the lead compound and their reactivity. However, there is a complication. Several of these functional groups will

be part of the pharmacophore, and any attempt to modify them will lead to a disruption of binding interactions and a decrease in activity (see section 8.2). Therefore, the synthesis of analogues needs to focus on reactions that will not affect the key groups represented by the pharmacophore. There is one exception to this general rule, as we will discover in section 8.5.2.1.

In this section, we shall look at examples of how lead compounds have been modified to produce analogues with improved pharmacodynamic or pharmacokinetic properties. Typical reactions include the following:

- introduction of substituents involving N–C bond formation;

- introduction of substituents involving C–O bond formation;

- introduction of substituents involving C–C bond formation;

- introduction of substituents to aromatic or heteroaromatic rings.

8.5.2 Introduction of substituents involving N–C bond formation

8.5.2.1 General principles

The amine group is one of the most common functional groups found in lead compounds, as it often acts as an important binding group (section 8.2.1) and has useful pharmacokinetic properties (section 8.4). The amine group is usually in equilibrium between its free base and the ionized form under physiological conditions. As a result, it can cross cell membranes as the free base and has the potential to form a strong ionic bond with the binding site when it is ionized. Amines can be incredibly useful in generating analogues for the following reasons.

- Because the ionized amine can form such a strong interaction with the binding site, it is often possible to add or vary its substituents without seriously affecting its ability to ionize and interact with the binding site. Of course, it is important that alkyl substituents are used rather than acyl substituents. The latter change the amine to an amide, which would result in a large drop in activity since amides cannot ionize (section 8.2.2.1).

- It is relatively straightforward to introduce different amine substituents. The formation of a N–C bond is a favoured process involving a nucleophilic amine and an electrophilic reagent such as an alkyl halide, carboxylic acid, carboxylic acid derivative, aldehyde, ketone, or epoxide (Fig. 8.25).

- Many of the amines present in natural lead compounds are tertiary amines containing a methyl substituent.

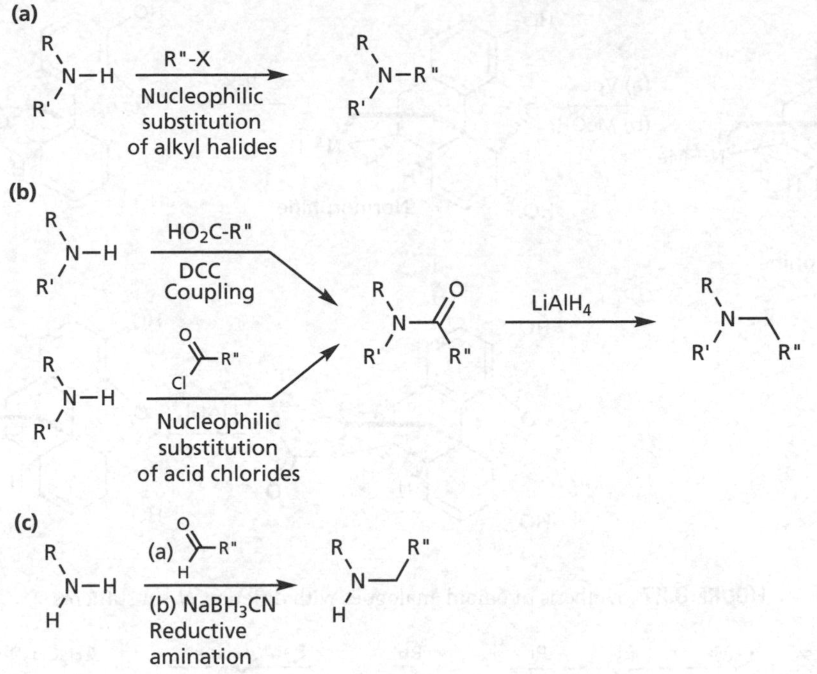

(a)

(b)

(c)

FIGURE 8.25 Commonly used reactions involving N–C bond formation to form analogues.

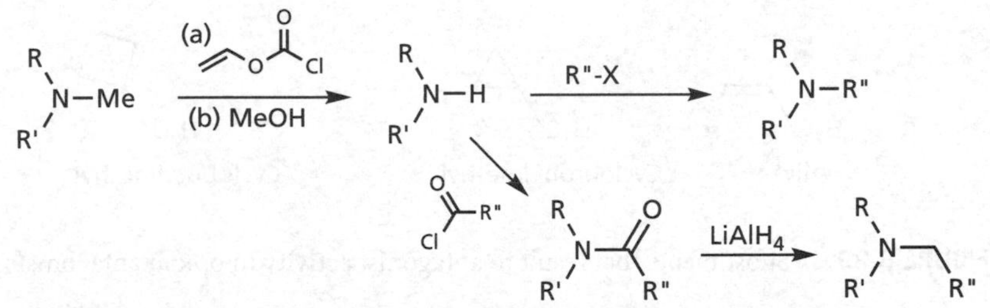

FIGURE 8.26 Removal of an *N*-methyl group to replace it with different substituents.

This methyl group can be removed under mild conditions and replaced with a different substituent (Fig. 8.26).

• Variation of *N*-substituents can be a good method of improving both pharmacodynamic and pharmacokinetic properties. The additional substituent can be added to obtain additional binding interactions, to modify the hydrophobic/hydrophilic character of the compound, to alter the pK_a of the amine group, or to add groups which will enhance solubility.

The following are some examples of how C–N formation has been used successfully to generate analogues with improved pharmacodynamic or pharmacokinetic properties.

8.5.2.2 Development of opioid analgesics

Morphine is a well-established opioid analgesic that is extracted from opium, and it has served as a lead compound for the development of other opioid analgesics. Because of the difficulty in preparing analogues by a full synthesis, analogues have been prepared from morphine itself (Fig. 8.27). The *N*-methyl substituent of morphine was removed by reaction with vinyloxycarbonyl chloride, followed by treatment with methanol (see also section 9.6.3). The resulting normorphine then served as the starting material for analogues containing different types of *N*-substituent. These can be produced by alkylating the amine with an alkyl halide. Alternatively, an acylation can be carried out with a range of acid chlorides to form an amide, followed by reduction to the amine. The latter method avoids the

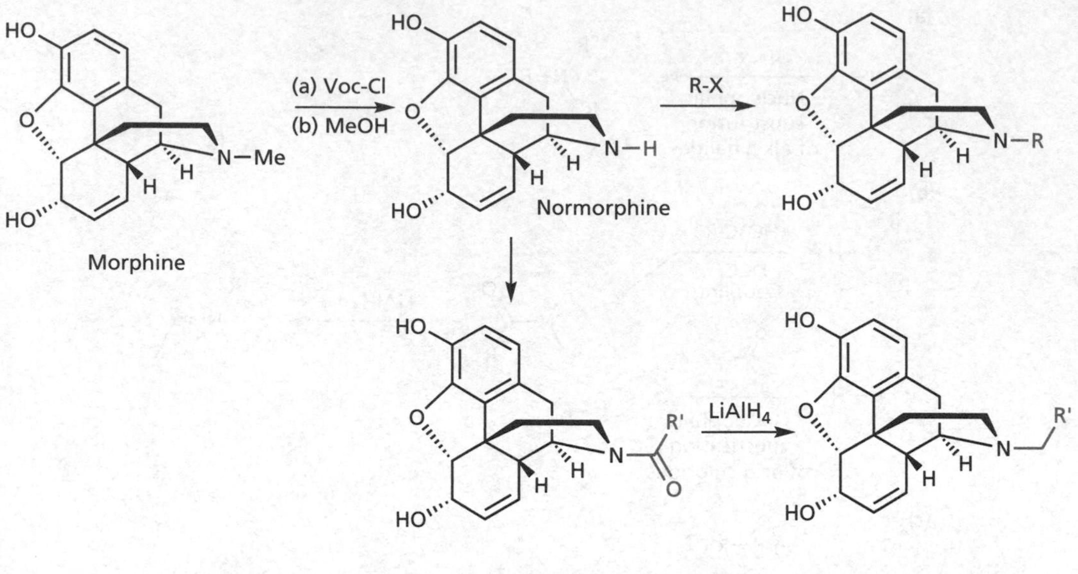

FIGURE 8.27 Synthesis of opioid analogues with different *N*-substituents.

R = Me, Et, Pr — Agonism decreases, Antagonism increases; Bu — Zero activity; Pentyl, Hexyl — Agonists; CH_2CH_2Ph — 14 x activity of morphine

FIGURE 8.28 Change in activity for different *N*-substituents.

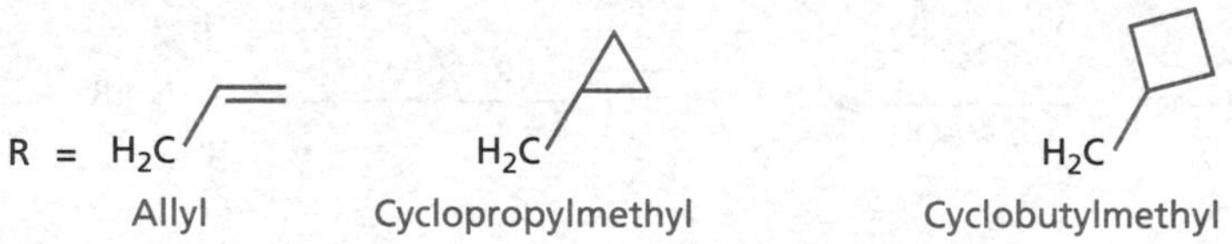

FIGURE 8.29 *N*-Substituents that result in antagonist activity in opioid antagonists.

problems of over-alkylation, leading to quaternary salts. The number and nature of possible analogues is limited only by the variety of alkyl halides and carboxylic acids that are commercially available or are easily synthesized.

An early study involved the addition of different lengths of *N*-alkyl substituent, varying from an ethyl group to a hexyl group. As the alkyl group increased in size from methyl to butyl, the activity dropped to zero (Fig. 8.28). With a larger group such as pentyl or hexyl, activity recovered slightly. None of this was particularly exciting, but when a phenethyl group was attached, the activity increased 14-fold relative to morphine—a strong indication that a hydrophobic binding region which interacts favourably with the new aromatic ring had been located within the binding site.

A particularly interesting result was the finding that certain types of substituent resulted in opioid analogues that acted as antagonists rather than agonists. This indicates that substituents can not only enhance a binding interaction, but can change the pharmacological effect. Allyl and cyclopropylmethyl substituents were chiefly responsible for antagonism, and it is thought that the additional interaction involving these substituents produces a different type of induced fit that fails to activate the receptor (Fig. 8.29).

8.5.2.3 Hydrophobic versus hydrophilic substituents

In general, greater success has been obtained by adding hydrophobic substituents to an amine group, especially if an aromatic or heteroaromatic ring is included. These rings have the potential to form pi–pi interactions with hydrophobic binding regions containing other aromatic/heteroaromatic rings. With a heteroaromatic ring, there is the additional possibility of one of the heteroatoms forming a hydrogen bond with the binding site.

Although hydrophobic substituents are generally more successful in finding additional interactions, the careful positioning of a polar group within a substituent can sometimes be beneficial if it is capable of forming a hydrogen bond with a polar binding region (Box 8.1).

BOX 8.1 Salmeterol

Salmeterol is an analogue of the anti-asthmatic agent **salbutamol**, but has a much longer duration of action due to the very long *N*-substituent that is present. Most of the chain is hydrophobic and forms van der Waals interactions with hydrophobic regions in the binding site, but the ether oxygen is able to act as a hydrogen bond acceptor and forms an important hydrogen bond with a tyrosine residue. It has been proposed that this not only increases binding interactions, but also helps to direct the remainder of the chain such that the terminal aromatic ring fits into a hydrophobic pocket for even better binding interactions.

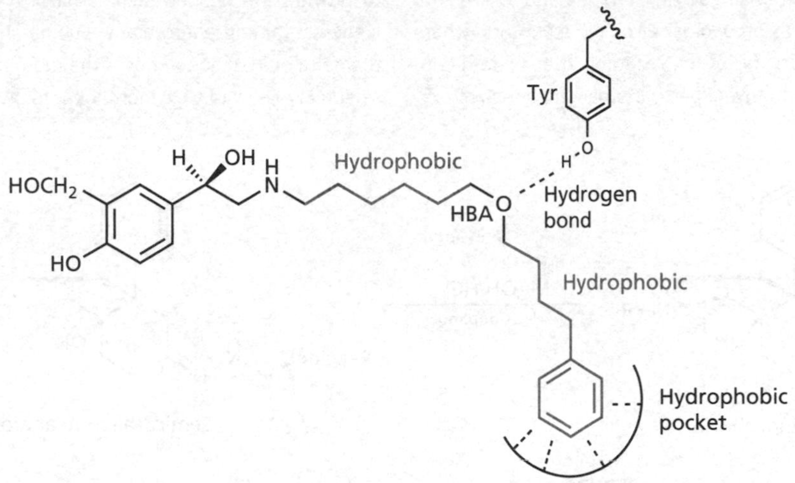

FIGURE 1 Additional interactions involving the *N*-substituent of salmeterol.

8.5.2.4 Modification of ketone groups

There are fewer examples of lead compounds containing ketone groups than of those containing amine groups. However, the presence of a ketone group can be extremely useful as it is possible to carry out C–N bond formations that result in imines, hydrazones, or semicarbazones This is achieved by reacting the ketone with amines, hydrazines, and hydrazinocarbonyl reagents, respectively (Fig. 8.30 and Box 8.2). If aminations of the ketone group are carried out under reducing conditions, the analogue obtained is an amine rather than an imine.

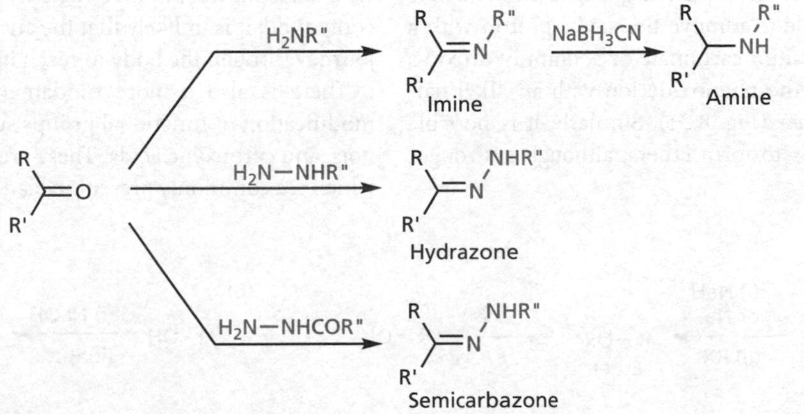

FIGURE 8.30 Methods of adding substituents to a ketone group by C–N bond formation.

BOX 8.2 Synthesis of oxymorphone analogues

Endorphins and enkephalins are endogenous polypeptides and act as the body's own analgesics by activating opioid receptors. The structures show a certain level of selectivity between the different types of opioid receptors, particularly for the delta receptor over the mu receptor. This appears to be due to specific amino acids within the peptide sequence. As a result, these amino acid sequences have been termed as 'addresses'.

Oxymorphone is an analogue of morphine and is a useful analgesic that acts as an agonist at opioid receptors, where it shows some selectivity for mu receptors over delta receptors. It was decided to investigate whether analogues of oxy-

morphone could be synthesized where the selectivity was reversed. This was achieved by adding peptide substituents to the opioid scaffold such that they would act as addresses for the delta receptor.

The ketone group at position C6 of oxymorphone was treated with the hydrazinocarbonyl derivative of the relevant peptide in a nucleophilic addition and elimination reaction to provide the desired semicarbazone products. The analogue containing phenylalanylleucine showed selectivity for the delta opioid receptor over the mu receptor, reversing the selectivity shown by oxymorphone itself.

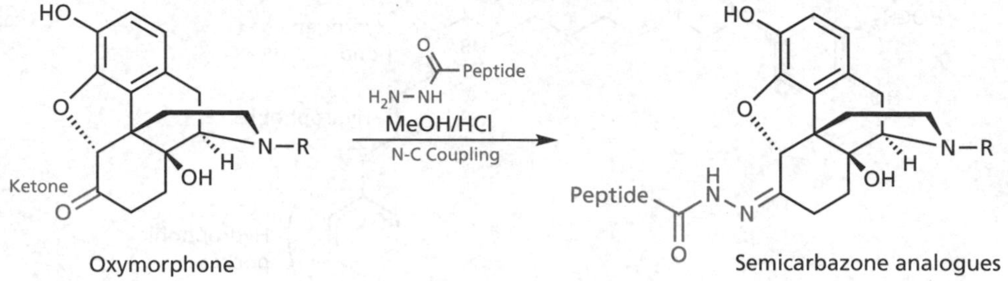

FIGURE 1 Synthesis of oxymorphone analogues.

8.5.3 **Introduction of substituents involving C–O bond formation**

8.5.3.1 Modification of alcohols, phenols, or carboxylic acids

Many lead compounds contain an alcohol, phenol, or carboxylic acid functional group. It is relatively easy to add substituents to these groups as a result of C–O bond formation. For example, a phenol group is mildly acidic and so it is possible to remove the acidic proton with a base such as potassium carbonate or sodium hydroxide, and then alkylate the phenoxide ion with an alkyl halide to form an ether (Fig. 8.31). Similarly, it is possible to alkylate alcohols to form ethers, although a stronger base is required.

Clearly, such reactions will only be feasible if the lead compound is stable to the base required.

If a carboxylic acid is present in the lead compound, it is possible to esterify the acid to introduce an alkoxy substituent. Similarly, it is relatively easy to esterify alcohols or phenols to prepare ester analogues. However, a particular problem with this approach is the presence of esterase enzymes in the blood supply which catalyse the hydrolysis of esters. Therefore, even if *in vitro* tests show that an ester analogue has improved activity over the original lead compound, it is unlikely that the analogue will survive its journey through the body to reach its target intact.

There is also a more fundamental problem in the modification of functional groups such as alcohols, phenols, and carboxylic acids. These are highly polar groups which are commonly present in lead compounds because

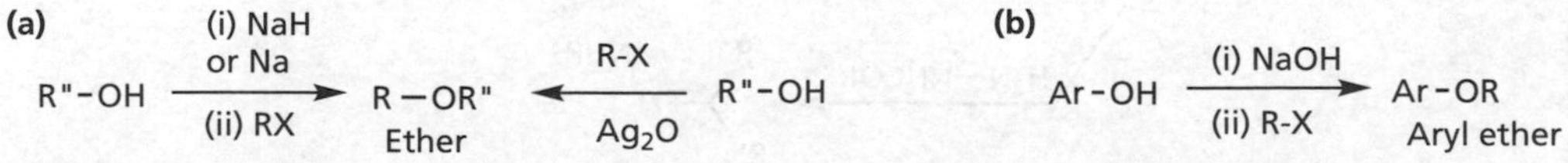

FIGURE 8.31 Synthesis of ethers from (a) alkyl halides and alcohols and (b) phenols and alkyl halides.

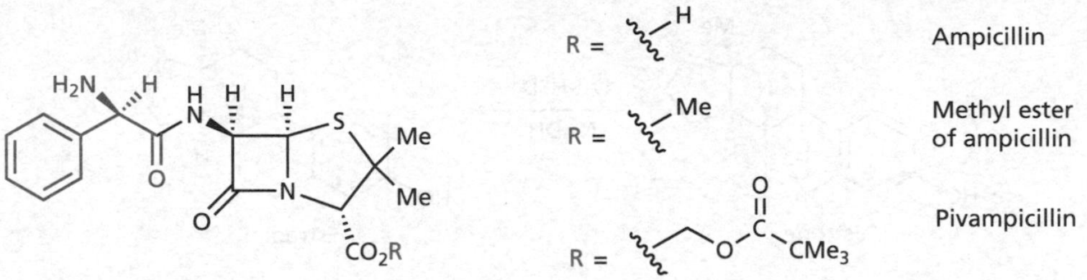

FIGURE 8.32 Ampicillin and ester analogues.

they form strong binding interactions with the binding site and are part of the pharmacophore. Any attempt to modify these groups in order to add substituents is likely to prevent those interactions taking place and result in a drop in activity (see also section 8.2.2).

Therefore, the addition of substituents to alcohol, phenol, and carboxylic acids is only likely to be useful if these groups are not part of the pharmacophore. Lead compounds that have been generated as a result of a simplification strategy are unlikely to have such groups, since one of the aims of that strategy is to remove functional groups that have no binding role.

Alcohols, phenols, and carboxylic acids that are part of the pharmacophore are generally 'off bounds' for modification, as explained above. However, there is one exception. There may well be mileage in esterifying these groups to produce prodrug analogues with improved pharmacokinetic properties. As stated earlier, alcohols, phenols, and carboxylic acids are all very polar groups and this may hinder absorption. If the groups are masked as esters, the resulting analogues will be far less polar and more likely to be absorbed from the gastrointestinal tract. Once the analogue reaches the blood supply, the esterase enzymes should hydrolyse the ester to reveal the active drug. Therefore, the synthesis of ester prodrugs is a popular strategy for improving the bioavailability of a polar drug.

This strategy has been used successfully on many occasions and it would be tempting to believe that all esters are susceptible to enzyme-catalysed hydrolysis. However, this is not always the case. Esters will only be hydrolysed if they can be accepted as substrates by esterase enzymes.

In some cases, the ester group may be in a rather 'crowded' region of the drug skeleton, in which case the enzyme is unable to accept it as a substrate. For example, the methyl ester of the antibacterial agent **ampicillin** was prepared as a potential prodrug, but proved resistant to human esterases (Fig. 8.32). This was because the ester group was linked directly to the bulky bicyclic ring system of penicillin, which acted as steric shield.

Consequently, a more complex ester prodrug called **pivampicillin**, which included an extended ester that was more accessible to esterase enzymes, had to be designed. After enzyme-catalysed hydrolysis took place, the resulting structure was chemically unstable and spontaneously broke down to provide the active penicillin (Fig. 8.33).

If a lead compound has a sterically hindered alcohol, phenol, or carboxylic acid that is not part of the pharmacophore, there may well be scope for synthesizing esters as analogues if the ester proves resistant to metabolic enzymes.

8.5.3.2 Modification of ethers

If an aromatic methyl ether is present in the lead compound, it may well be possible to remove the methyl group to form a phenol and replace it with different substituents. Even if the ether has a binding role as a hydrogen bond acceptor, there is still a chance that the interaction will be preserved as the modification is less 'drastic' than the conversion of an alcohol or a phenol to an ether. The modification of substituents on an ether played an important role in the development of the anticancer agent gefitinib (Fig. 8.1).

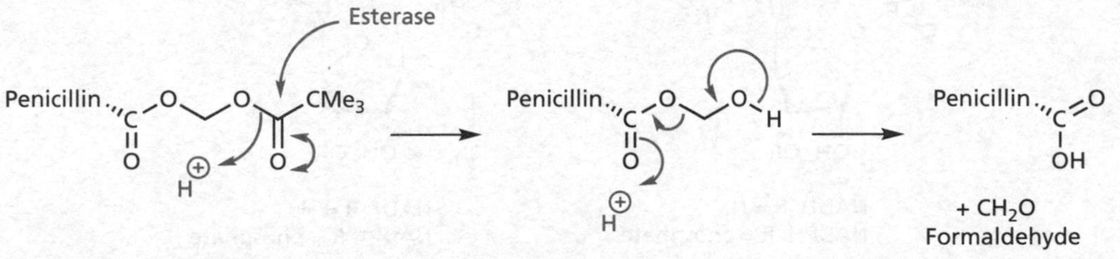

FIGURE 8.33 Mechanism by which an extended ester was hydrolysed.

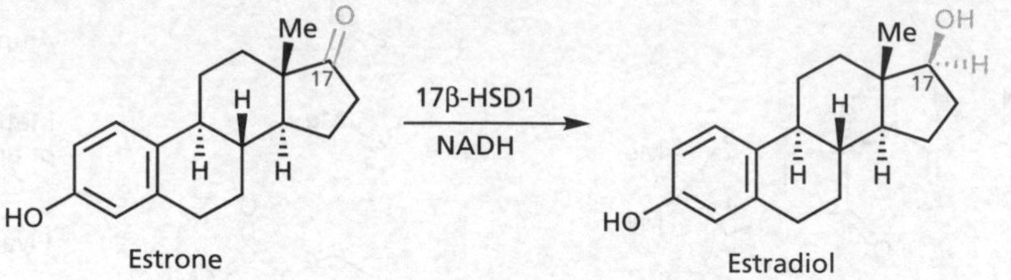

FIGURE 8.34 Enzyme-catalysed synthesis of estradiol from estrone.

The removal of a methyl group is feasible for aromatic methyl ethers, but attempting demethylation of an aliphatic methyl ether may lead to mixtures of products as the C–C bond on either side of the ether oxygen may be split.

8.5.4 Introduction of substituents involving C–C bond formation

8.5.4.1 General principles

There is generally less scope to introduce a substituent to a lead compound by C–C bond formation, compared with C–N or C–O bond formation. However, if the lead compound contains a ketone group, there are two feasible reactions:

- nucleophilic addition to the ketone group using an organometallic reagent;
- alkylation of the carbon atom at the α-position.

Examples of these approaches are provided in the following sections.

8.5.4.2 Synthesis of estrone analogues as potential anticancer agents

Estrone is a relatively inactive steroid that serves as the substrate for an enzyme called **17β-dehydroxysteroid dehydrogenase type 1** (17β-HSD1). This enzyme catalyses the reduction of the 17-keto group to form the alcohol group present in the female hormone **estradiol** (Fig. 8.34).

Although 17β-HSD1 catalyses this reduction, it is incapable of acting as the actual reducing agent. For reduction to take place, the body's own reducing agent has to be present. This is the enzyme cofactor known as **nicotinamide adenine dinucleotide** (NADH) (Fig. 8.35), which provides the equivalent of a hydride ion for the reduction process. NADH itself is converted to NAD+ during the process. The active site of the enzyme contains a binding

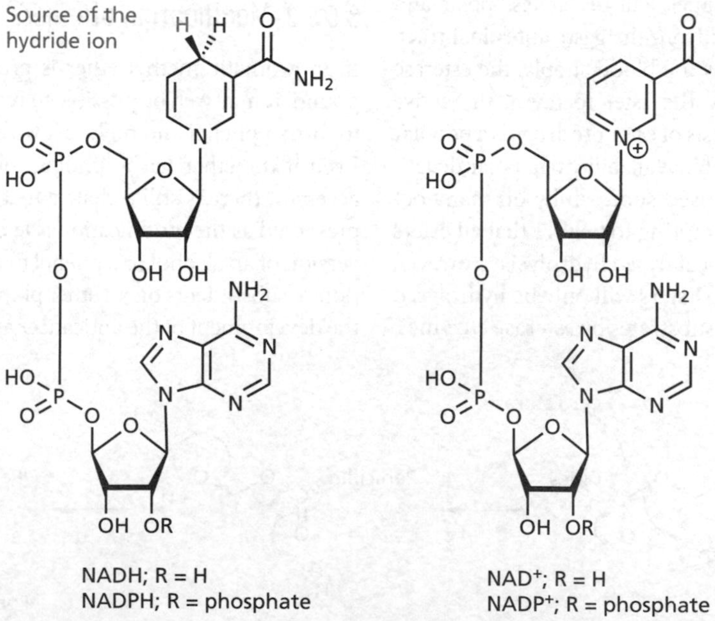

FIGURE 8.35 Nicotinamide adenine dinucleotide and nicotinamide adenine dinucleotide phosphate.

site for the estrone substrate, as well as a neighbouring binding site for the cofactor. The substrate and cofactor are positioned such that the 'hydride' ion of NADH is positioned close to the ketone group of the substrate, allowing an efficient transfer of the hydride ion during the reduction process.

Inhibition of 17β-HSD1 would lower estradiol levels in the body and would be a potentially useful method of slowing the development of estrogen-dependent tumours. Therefore, 17β-HSD1 inhibitors are potentially useful anticancer agents.

Since estrone binds to the target enzyme, it can be viewed as a lead compound for such inhibitors. Thus, the aim is to develop an analogue which binds more strongly than estrone, but does not undergo the enzyme-catalysed reaction. By binding more strongly than estrone, the inhibitor would block estrone from accessing the active site. By being resistant to the enzyme-catalysed reduction, there would be no chance of the inhibitor itself being converted to an active analogue of estradiol.

One project aimed at designing 17β-HSD1 inhibitors involved the synthesis of estrone analogues where substituents were added to C16, which is the α-carbon relative to the ketone group at position 17. There were several reasons for introducing a substituent at this position.

First, the protons at position 16 are slightly acidic due to the neighbouring ketone group. Therefore, it should be possible to form a carbanion at position 16 and react that with electrophiles to introduce the substituents.

Secondly, the cofactor has to be close to C17 in order to reduce the ketone, and so it must also be close to C16. A substituent at C16 should be close enough to the cofactor to allow possible intermolecular interactions.

Thirdly, since the substituent at C16 is close to the ketone at C17, there is a strong probability that it could act as a steric shield and prevent the transfer of the hydride ion from NADH to the ketone group.

A series of estrone analogues containing various substituents at C16 were prepared. As stated above, it is possible to form a carbanion at C16, which can then be alkylated with different alkyl halides. However, a seemingly more elaborate synthesis was used to produce the desired analogues (Fig. 8.36). First, the phenol group was protected as a benzyl ether to prevent it being alkylated in the next stage. Treatment with a strong base (LDA) then generated the carbanion at C16, and this was alkylated with ethyl bromoacetate. This reaction generates an extra chiral centre and, as the alkylation is not stereoselective, a mixture of diastereoisomeric esters was obtained. The resulting esters were hydrogenated to deprotect the phenol group, and then hydrolysed to convert the esters to carboxylic acids. The carboxylic acids were coupled with a series of amines to give the final products, which were then separated to isolate the two diastereoisomers.

You may well ask why substituents were not just added directly to position 16 instead of going through this longer process. In fact, there are distinct advantages in carrying out the synthesis in this manner.

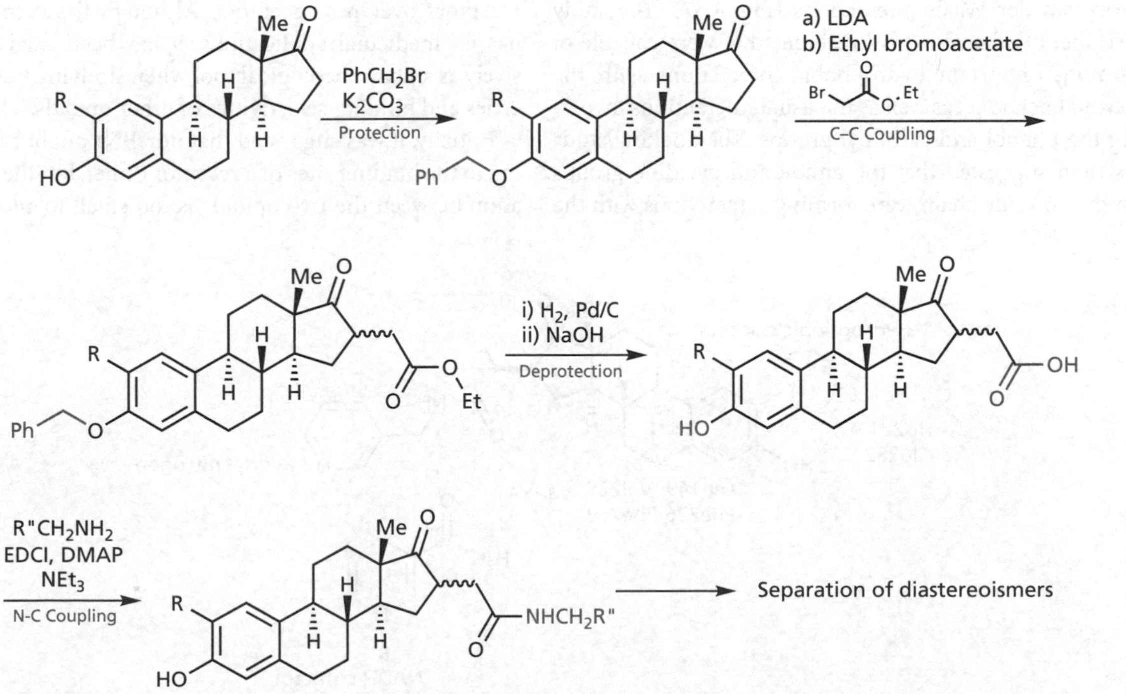

FIGURE 8.36 Synthesis of estrone analogues with substituents at C3 (R = H or Et) and C16.

First, most of the synthetic route can be carried out on a single batch on a large scale to provide a sizeable stockpile of the carboxylic acid. Once sufficient carboxylic acid has been obtained, the different analogues can be generated by a single reaction for each analogue.

Secondly, the final reaction involves the use of different amines. There are a large number of commercially available amines that can be used for this reaction.

Thirdly, the coupling reaction of an amine with an amide is easier to carry out than the alkylation of a carbanion. Moreover, the reaction is carried out under much milder conditions without the need for the phenol group to be protected. Therefore, it makes sense to create the various analogues using this reaction rather than alkylating the carbanion with a range of different alkyl halides, some of which would have to be synthesized themselves.

The amide group that appears in the final structure was included as a synthetic linker rather than as a deliberate attempt to find potential binding interactions. Nevertheless, it was hoped that the group might be capable of forming a hydrogen bonding interaction with the peptide backbone of the active site. As we shall see later, the amide group did indeed serve as a binding group, but not with the peptide backbone.

A promising analogue from these studies was one containing an ethyl substituent at C2 and a longer substituent at C16. This analogue was found to act as an inhibitor, and docking experiments indicated that the structure could bind to the active site such that the ethyl group at C2 fitted into a normally empty hydrophobic pocket to form van der Waals interactions (Fig. 8.37). The study also identified amino acid residues that were capable of forming important hydrophobic interactions with the steroid backbone, as well as the usual interactions involving the phenol and carbonyl groups. The docking studies then suggested that the amide and pyridine groups on the 16β side chain were forming interactions with the

bound cofactor. The amide was interacting with the primary amide of the nicotinamide ring, while the pyridine interacted with a phosphate group (not shown). Therefore, this structure represents a rather novel approach to inhibitor design where the inhibitor interacts with both the active site and a bound cofactor.

The inhibitor was also found to be resistant to enzyme-catalysed reduction, and further studies of the docked structure revealed that one of the methylene protons in the substituent is relatively close (2.63 Å) to the Pro-S hydrogen of the cofactor, as is the methine proton at C16 (1.81 Å). The proximity of these protons suggests that they are acting as steric shields to prevent the transfer of the hydride ion to the carbonyl group.

As stated above, adding a synthetic linker to the lead compound can make the generation of analogues easier and more efficient. Another example of this approach can be found in a synthesis of glycylcyclines, where an α-halo amide group was added to a tetracycline structure. Once formed, the halogen could then be substituted with a range of amines (see Fig. 12.19).

8.5.4.3 Synthesis of norbinaltorphimine (nor-BNI)

Norbinaltorphimine is an opioid dimer that was synthesized to see whether opioid dimers would bind to opioid receptors more strongly than opioid monomers. In fact, nor-BNI was found to be a highly potent and selective kappa antagonist, showing 100-fold selectivity for kappa receptors over mu receptors. Although the compound has no medicinal applications, it has been used extensively as a pharmacological tool when studying the properties and binding selectivities of other opioids.

Initially, it was suggested that nor-BNI might be binding to the binding sites of a receptor dimer, but the separation between the two opioids is too small to allow that

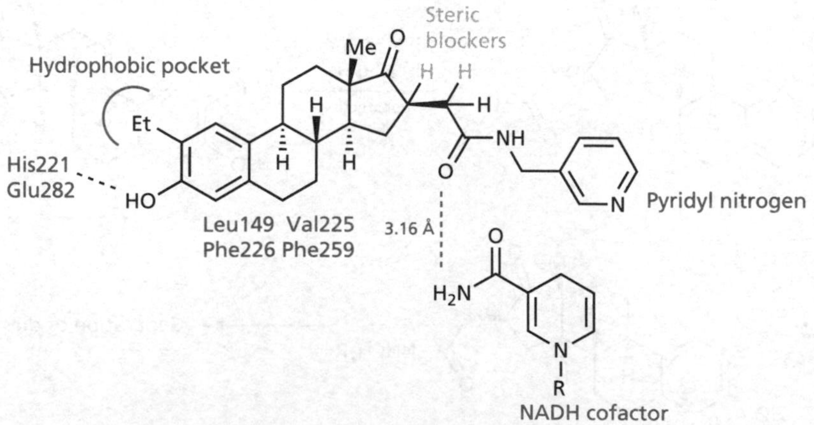

FIGURE 8.37 Proposed binding interactions.

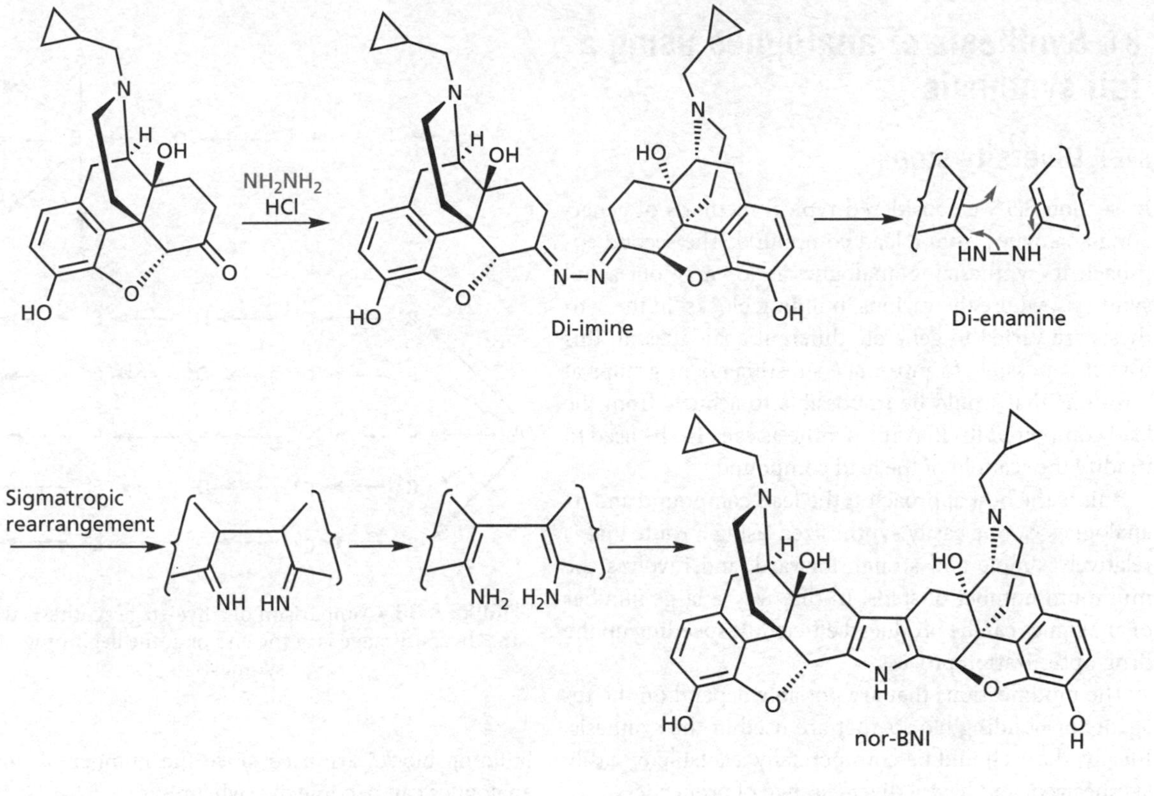

FIGURE 8.38 Synthesis of nor-BNI.

to happen. It is now felt that one half of the dimer binds to the receptor binding site, while the other half interacts with another region of the same receptor protein.

The synthesis of the opioid dimer nor-BNI takes advantage of the ketone group present in the opioid antagonist **naltrexone**, as well as the activated carbon at the alpha position, in order to create a pyrrole ring using a synthetic method called the **Piloty–Robinson synthesis** (Fig. 8.38). This involves heating naltrexone with hydrazine hydrochloride in DMF. The hydrazine reacts with the ketone groups of naltrexone by a nucleophilic addition and elimination mechanism to give a di-imine. In the presence of acid, tautomerism takes place to give the di-enamine intermediate, which then undergoes a [3,3] sigmatropic rearrangement, followed by ring closure.

8.5.5 Addition of substituents to aromatic rings

Substituents can be added to an aromatic ring by electrophilic substitution and then modified to generate different functional groups and substituents. Such substituents may result in enhanced binding interactions either directly or indirectly. Introducing different substituents at different positions of an aromatic ring has also been a popular approach to modifying the pharmacokinetic

properties of a compound. Substituents can be varied in terms of their hydrophobicity, size, and electronic properties, and this can have a significant effect on both the pharmacodynamic and pharmacokinetic properties of the resulting analogues. However, the feasibility of carrying out the reactions required to introduce these groups depends on the lead compound being stable to the reaction conditions used.

For example, **chlorotetracycline** is an antibiotic which is unstable to nitration, whereas other tetracyclines are more stable and can be nitrated under acid conditions. This has allowed the synthesis of several novel tetracycline structures (see section 12.4).

KEY POINTS

- The analogues that can be synthesized from a lead compound depend on the functional groups present.

- Reactions carried out on functional groups that are part of the pharmacophore are likely to result in analogues with poorer binding affinity and activity. Amine groups are an exception to the rule.

- Analogues can be easily synthesized from lead compounds by carrying out reactions at amines, alcohols, phenols, carboxylic acids, and ketones.

8.6 **Synthesis of analogues using a full synthesis**

8.6.1 **Diversity steps**

In section 8.5 we considered typical methods of generating analogues from a lead compound. The second approach to synthesizing analogues is to carry out a full synthesis, where the various 'building blocks' in the synthesis are varied to generate different analogues. In this way, it is possible to introduce substituents or groups at positions that would be impossible to achieve from the lead compound itself. A full synthesis can also be used to modify the scaffold of the lead compound.

This is the best approach if the lead compound and its analogues can be easily synthesized using a route that is relatively simple and straightforward, and involves the minimum number of steps. In that way, a large number of analogues can be produced efficiently, speeding up the drug optimization process.

The modifications that are possible depend on the reagents or building blocks that are used in the synthesis. Ideally, these should be commercially available or easily synthesized, and have a diverse range of properties.

The synthetic steps where different reagents/building blocks can be used are described as **diversity steps** as the synthesis branches at these steps to produce the different analogues desired. Ideally, the diversity step should come late on in the synthesis to minimize the number of reactions required to create each individual analogue. For example, if the full synthesis involves five steps, then the ideal situation is where the diversity step comes at the final stage (Fig. 8.39). In that way, the first four steps can be carried out on a large scale to prepare a stockpile of the penultimate structure (E). Only one synthetic step is then needed for the preparation of each analogue.

In contrast, a five-stage synthesis involving a diversity stage at step 1 would mean that every analogue would have to be synthesized using the complete synthetic route.

It is highly likely that a synthesis will have two or more diversity stages. Once again, it is best to use a synthesis where these steps are clustered towards the end of the synthesis rather than distributed throughout the synthetic route. The greater the number of diversity stages that are present in a synthesis, the greater the number of possible analogues that can be synthesized involving all kinds of substituent permutations. For example, a synthesis containing two diversity stages can generate 25 different analogues if five different reagents are used at each diversity stage (Fig. 8.40). This can be an advantage in providing access to large numbers of analogues, but it can also be a disadvantage when far more reagents/

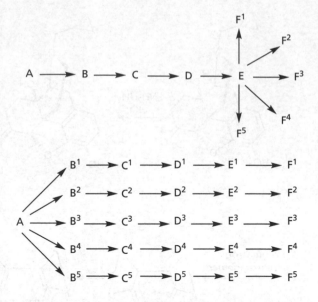

FIGURE 8.39 Comparison of a five-stage synthesis where the diversity stage is at the end or at the beginning of the synthesis.

building blocks are used since the number of possible analogues can become overwhelming.

Moreover, rationalizing why one analogue is more active than another can be difficult if the two structures have two or more structural differences. Therefore, the traditional approach has been to synthesize analogues where the structural variation is confined to one part of the molecule in order to find out which analogue has the optimum binding affinity or activity.

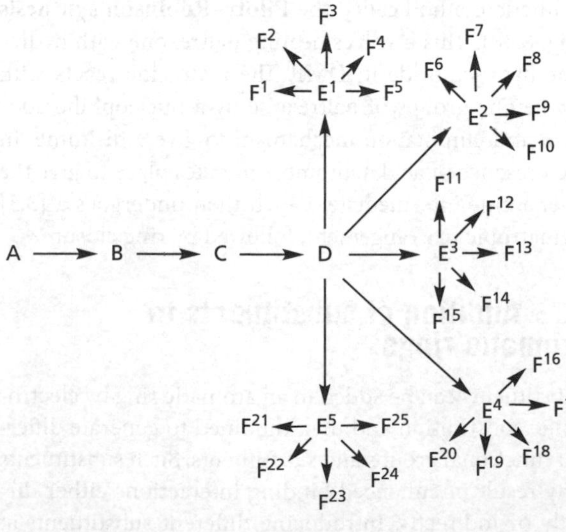

FIGURE 8.40 Number of possible analogues obtained by using five different reagents at two diversity stages.

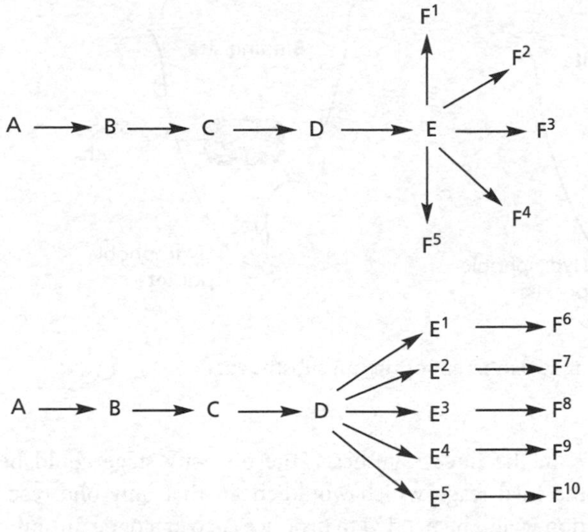

FIGURE 8.41 Preparation of analogues to identify the optimum modification at a specific region of the structure.

Another set of analogues are then prepared to identify the optimum analogue when modifications are made to another part of the molecule, and the process is repeated for each diversity stage in the synthesis.

For example, if the final two stages of a synthesis were diversity stages, a set of analogues could be prepared where different reagents are used in the final stage alone to find the most active analogue between analogues F^1–F^5. A second set of analogues are then prepared where the variation is introduced at the penultimate stage and the optimum structure from those analogues (F^6–F^{10}) is identified (Fig. 8.41). In this case, 10 analogues would be synthesized rather than 25 in order to identify which substituent is optimum for each diversity stage.

The synthesis of the second set of analogues will inevitably be less efficient than the synthesis of the first set, since two reactions are required to generate each analogue rather than one. If a large number of analogues are planned, there may well be a benefit in using a different synthetic route where the relevant diversity stage comes last.

This traditional approach of identifying the optimum substituent or feature at a specific position of the molecular scaffold has been extremely successful in drug optimization, but it is not foolproof. There are several instances where an analogue has been synthesized which contains all the optimum features identified from individual sets of analogues, and yet it shows disappointing activity. In some cases, the optimum feature in one set of analogues may actually be detrimental to activity when combined with the optimum feature in another set of analogues. In other words, the two features may 'oppose' each other in some way. For example, there may be some kind of steric interaction between the features that forces the molecule out of its active conformation or disrupts its binding interactions with the target binding site. One possible scenario is where a hydrogen bonding group may be beneficial at two different positions of the structure because it forms hydrogen bonding interactions with the same region of the target binding site. However, an analogue containing both groups shows poor activity because the groups interact with each other, rather than with the binding site (Fig. 8.42).

There may even be slightly different binding modes for different sets of analogues. Combining the optimum features into the one structure may result in one of the features having a steric clash with the binding site, regardless of which binding mode is used (Fig. 8.43).

Therefore, rather than synthesizing an analogue where all the optimum substituents have been included, it is

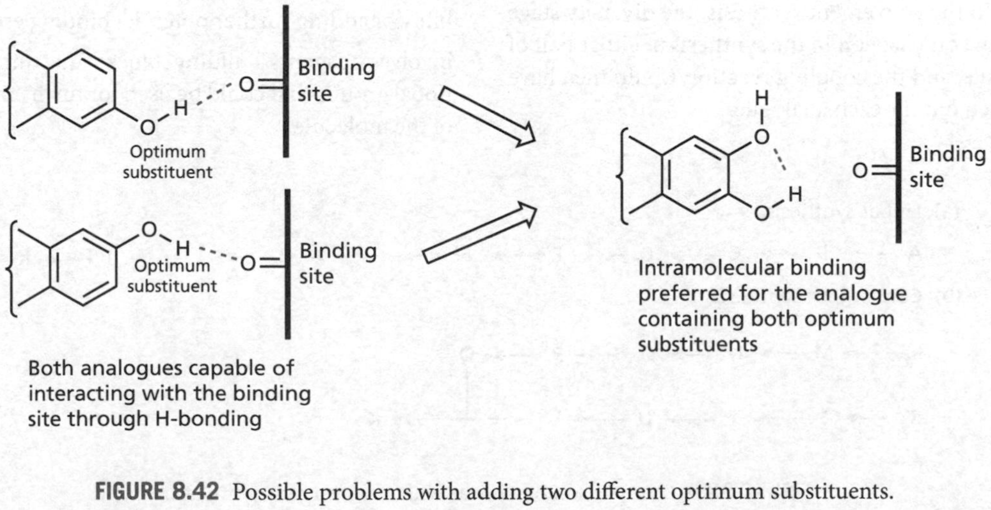

FIGURE 8.42 Possible problems with adding two different optimum substituents.

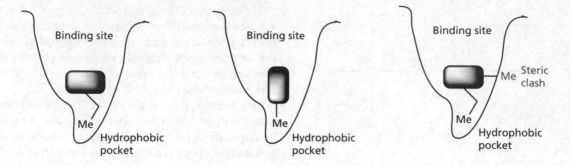

FIGURE 8.43 Possible problems with adding two different optimum substituents.

worth synthesizing a range of analogues where different combinations of optimum substituent are included to find the most active compound. In addition, the number of possible analogues could be increased by including substituents or modifications that were near optimum.

However, success is never guaranteed. It may be that the approaches described here fail to produce analogues with significantly improved activity. In such cases, it might be worth synthesizing analogues with random combinations of substituents in the hope of striking lucky (Box 8.3)!

8.6.2 **Linear versus convergent syntheses**

Synthetic routes can be described as being linear or convergent (Fig. 8.44). In a linear synthesis, the final structure is built up stage by stage using relatively simple reagents and building blocks. In a convergent synthesis, the two halves of the molecule are built up separately bit by bit, and then coupled together. A well-designed convergent synthesis will usually be more efficient in synthesizing a specific compound, as fewer synthetic steps are involved and the overall yield will generally be better.

Choosing between a linear and a convergent synthesis for the synthesis of a series of analogues may not be so clear-cut. In the convergent synthesis, the diversity stage will need to come late on in the synthesis of either half of the molecule, and the coupling reaction would then have to be carried out for each analogue.

In the linear synthesis, the diversity stage could be the final stage which would mean that only one reaction would be needed to produce each analogue. In some situations, it might make more sense to carry out the longer synthesis to stockpile the penultimate structure, rather than to use the convergent synthesis. The decision on which route to use will depend on how efficient each method is in generating the desired analogues.

8.6.3 **Preferred types of reaction for diversity steps**

There are clear advantages in using a synthesis where each diversity step involves a reaction that is:

- high yielding and can be carried out under mild conditions;
- involves reagents or building blocks that are commercially available or easily synthesized;
- involves reagents/building blocks that have a diverse range of physicochemical properties such as hydrophobicity, size, and electron-donating/electron-withdrawing properties;
- involves reagents/building blocks that allow the possibility of adding further potential binding groups;
- involves reagents/building blocks that include functional groups that could be used for further elaboration of the molecule.

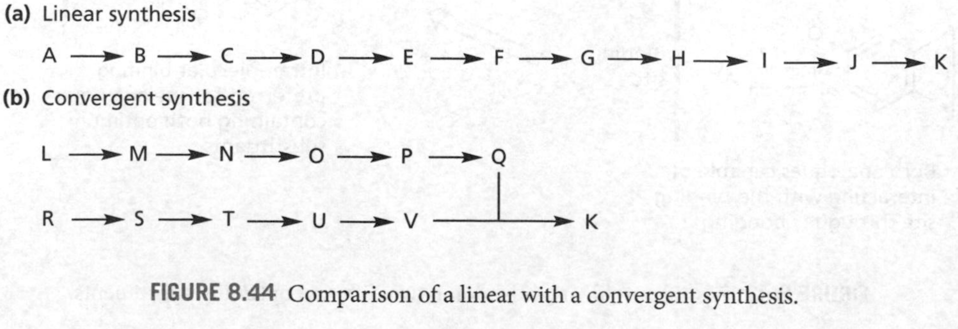

(a) Linear synthesis

A → B → C → D → E → F → G → H → I → J → K

(b) Convergent synthesis

L → M → N → O → P → Q

R → S → T → U → V ————— → K

FIGURE 8.44 Comparison of a linear with a convergent synthesis.

BOX 8.3 Design of the kinase inhibitor sorafenib

Sorafenib is a clinically approved anticancer drug which acts as a kinase inhibitor against a range of membrane-bound receptor kinases. The lead compound for sorafenib was discovered by high throughput screening of 200 000 compounds, leading to the identification of a urea (I) with micromolar activity (Fig. 1). Substituents and rings were altered in a systematic fashion and it was found that a *para*-methyl group on the phenyl ring resulted in a tenfold increase in activity. However, despite the synthesis of many more analogues, no further improvement in activity could be obtained. Up to this point, conventional medicinal chemistry strategies had been followed which involved altering one group at a time. This allows one to rationalize any alterations in activity that result from the change of any ring or substituent. It was now decided to use parallel synthesis to produce 1000 analogues having all possible combinations of the different substituents and rings that had been studied to date. This led to the discovery of a urea (IV) with slightly improved activity over structure II. The curious thing about this structure is that it deviates from the SAR results obtained by single-point modifications. Structure IV has a phenoxy substituent and an isoxazole ring, but neither of these groups would be considered good for activity based on the initial SAR. For example, structure III has the phenoxy substituent while structure VI has the isoxazole ring, but both structures have low activity compared with the lead compound. Conventionally, this would be taken to imply that neither group is good for activity. However, it does not take into account the synergistic effects that two or more modifications might have. The strategy of multiple-point modifications allows the identification of such synergistic effects, and demonstrates that there are limitations to simple SAR analyses.

Structure IV was now adopted as the new lead compound. Replacing the phenyl ring with a pyridine ring led to structure V and a fivefold increase in activity, as well as improving aqueous solubility and ClogP. Conventional optimization strategies then led to sorafenib which is 1000-fold more active than the original lead compound.

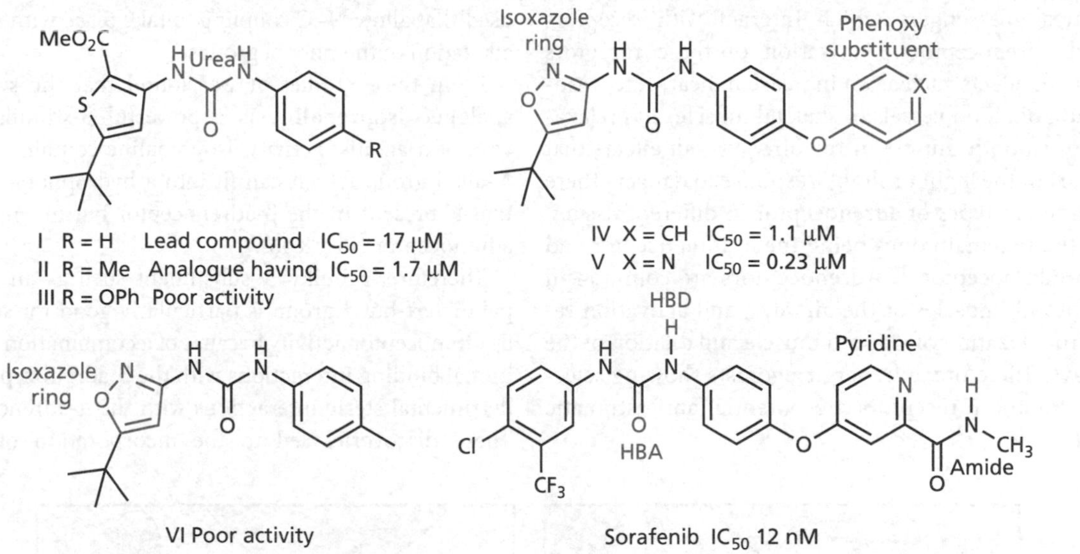

FIGURE 1 From lead compound to sorafenib.

Particularly suitable reactions are those defined as **click reactions** involving the formation of C–N bonds (section 7.9). Reactions involving amines have been particularly prevalent in preparing analogues. We have already noted this in section 8.5.2.

Amines can easily be reacted with alkyl halides, alkyl mesylates, alkyl tosylates, epoxides, aldehydes, ketones, acid chlorides, and acid anhydrides to form a wide range of analogues containing amine, imine, amide, and 1,2-amino alcohol functional groups. Therefore, many synthetic routes are designed to incorporate a reaction of this type (see also section 8.5.2).

There are also a huge number of commercially available reagents that can be used for these reactions, depending on which reagent is to be varied in the C–N bond formation. For example, there are large numbers

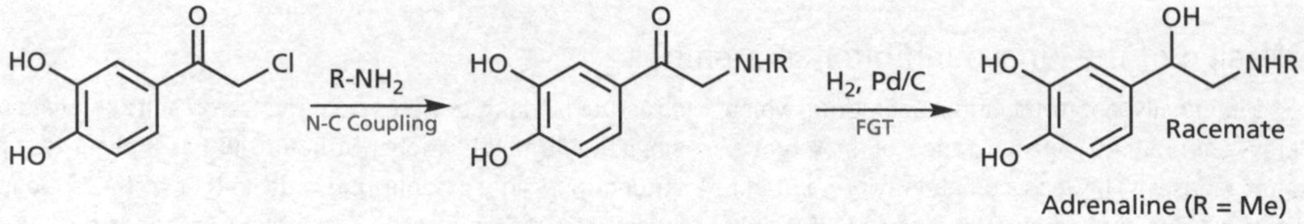

FIGURE 8.45 Synthesis of adrenaline (R = Me) and its analogues.

of structurally diverse amines that are commercially available, including amino acids. Equally, there are large numbers of commercially available alcohols, alkyl halides, aldehydes, ketones, and carboxylic acids which can be accessed and used for C–N bond formation, either directly or following activation.

The following sections provide examples of synthetic routes used to generate important analogues.

8.6.4 Development of the anti-asthmatic agent salbutamol

Adrenaline and noradrenaline are both endogenous chemical messengers which interact with receptors called adrenoceptors. Activation of these receptors produces effects such as an increase in heart rate, vasodilation of blood vessels to skeletal muscle, and relaxation of smooth muscle in the airways—all effects that are part of the 'fight or flight' response to danger. There are different types of adrenoceptor in different tissues, with the two main ones being the α-adrenoceptor and the β-adrenoceptor. β-Adrenoceptors are common in the smooth muscles of the airways, and activation results in relaxation of smooth muscle and dilation of the airways. Therefore, adrenergic agonists showing selectivity for those receptors are potential anti-asthmatic agents.

Since adrenaline naturally produces this effect, it was used as the lead compound to develop anti-asthmatic drugs that showed increased activity and selectivity for β-adrenoceptors over α-adrenoceptors. Adrenaline itself interacts equally well with both types of receptor, and so analogues were synthesized to test whether different amine substituents would make a difference. Adrenaline can be synthesized by reacting an amine with a chloro ketone, and the synthesis can easily be adapted to use different amines for the preparation of analogues (Fig. 8.45). Note that the two phenol groups are not protected during the synthesis. An amine group is more nucleophilic than a phenol group, and so reaction conditions can be used that allow N–C coupling to take place without self-alkylation of the phenol groups.

From these studies, it was found that the synthetic analogue **isoprenaline** is a powerful β-stimulant devoid of α-agonist activity. Isoprenaline contains a bulky *N*-alkyl group which can fit into a hydrophobic pocket that is present in the β-adrenoceptor but not in the α-adrenoceptor (Fig. 8.46).

Therefore, a bulky *N*-substituent such as an isopropyl or *tert*-butyl group is particularly good for selective β-adrenoceptor activity because of a combination of additional binding interactions with the β-adrenoceptor and detrimental steric interactions with the α-adrenoceptor. These discoveries led to the incorporation of bulky

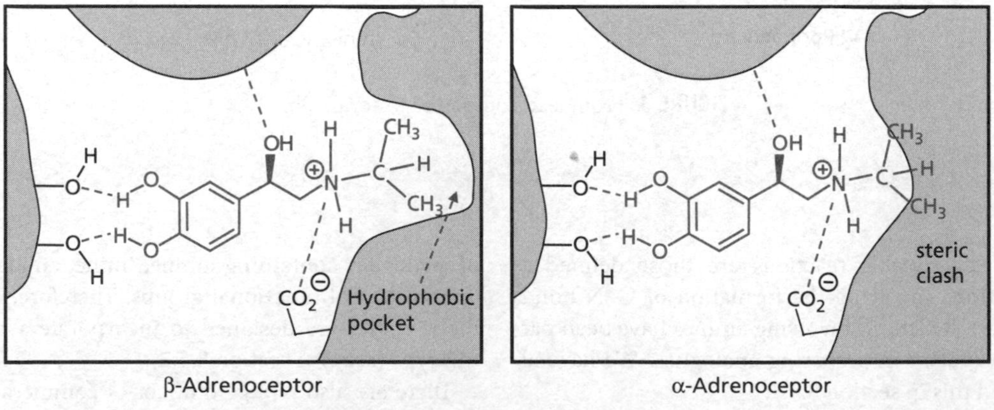

FIGURE 8.46 (a) Binding of isoprenaline with the β-adrenoceptor. (b) Steric clash preventing isoprenaline binding to the α-adrenoceptor.

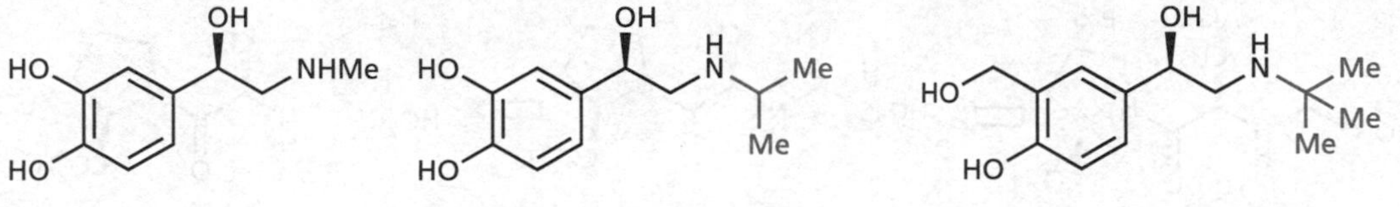

FIGURE 8.47 Comparison of adrenergic agonists.

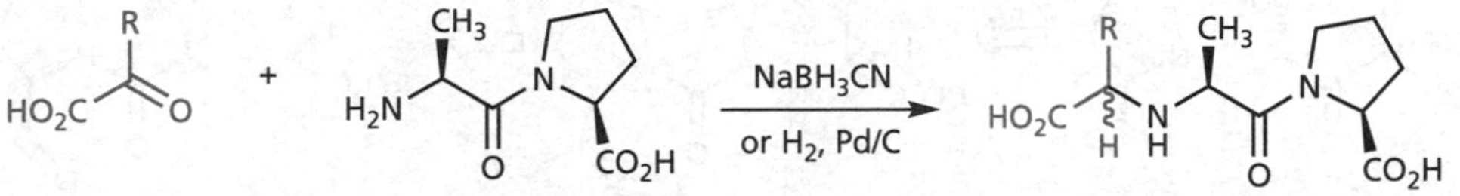

FIGURE 8.48 Synthesis of a series of N-carboxymethyldipeptides.

N-substituents into anti-asthmatic agents such as salbutamol (Fig. 8.47).

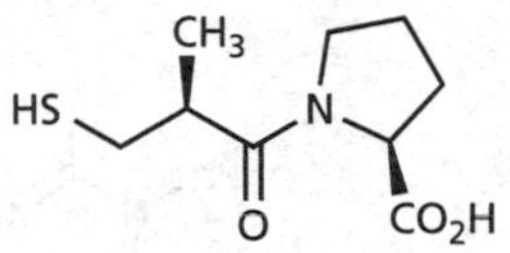

8.6.5 Development of the ACE inhibitors enalaprilate and enalapril

The **angiotensin converting enzyme** (ACE) is an important target for antihypertensive drugs that lower blood pressure. The enzyme is responsible for catalysing the conversion of a decapeptide called **angiotensin I** to the octapeptide **angiotensin II**. The latter is a potent vasoconstrictor, and so the inhibition of ACE lowers the levels of this hormone and results in dilation of blood vessels. This, in turn, leads to a lowering of blood pressure. An N-carboxymethyldipeptide (R = H in Fig. 8.50) showed weak activity as an ACE inhibitor and served as the lead compound for the development of more potent ACE inhibitors. SAR demonstrated that both carboxylic acid groups in the lead compound were ionized and involved in binding interactions with the enzyme's active site. However, if the activity was to be increased, it would be necessary to find additional binding interactions. It was reasoned that there must be pockets within the active site that are capable of accommodating amino acid side chains, since the substrate (angiotensin I) is a peptide. Therefore, it was decided to synthesize analogues with different substituents (R) at the position shown in Figure 8.48 to see whether a substituent could be found that would fit the predicted pocket. The synthesis involved a reductive coupling of an amine with a keto acid, and it was decided to keep the amine constant as this was similar to the first ACE inhibitor to be discovered—a structure called **captopril** (Fig. 8.49). Thus, a

FIGURE 8.49 Captopril.

variety of keto acids were used to introduce the various substituents (R).

Methyl and ethyl substituents increased activity, with the ethyl analogue proving as effective as captopril (Fig. 8.50). Activity dropped slightly with the introduction of a benzyl group, but a phenethyl substituent led to a dramatic increase in activity, indicating that the aromatic ring was interacting with a hydrophobic pocket.

The addition of a new substituent to structures II–V meant that a new asymmetric centre had been introduced, and all these structures had been tested as mixtures of diastereomers—the R,S,S and S,S,S diastereomers. Structure V was now separated by chromatography and the S,S,S diastereomer was found to be 700 times more active than the R,S,S diastereomer. This structure was named **enalaprilate**. **Enalapril** is its ethyl ester prodrug.

8.6.6 Development of fentanyl analogues as analgesics

Fentanyl (Sublimaze) (Fig. 8.51) was discovered in the 1960s and is still one of the most widely used opioid drugs in medicine today. It contains an aniline group linked to a piperidine ring, and is defined as a 4-anilinopiperidine

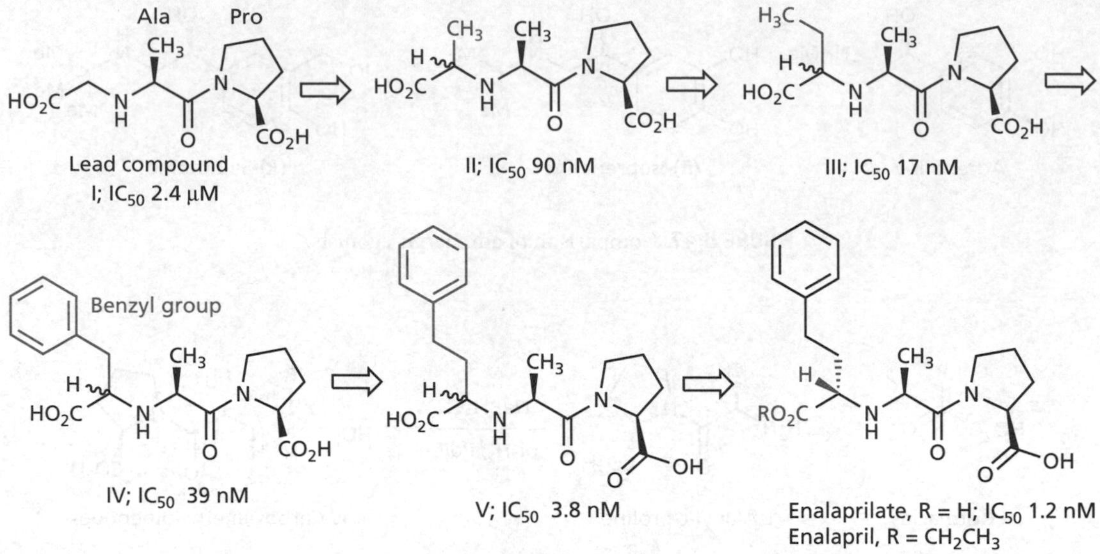

FIGURE 8.50 Development of enalaprilate.

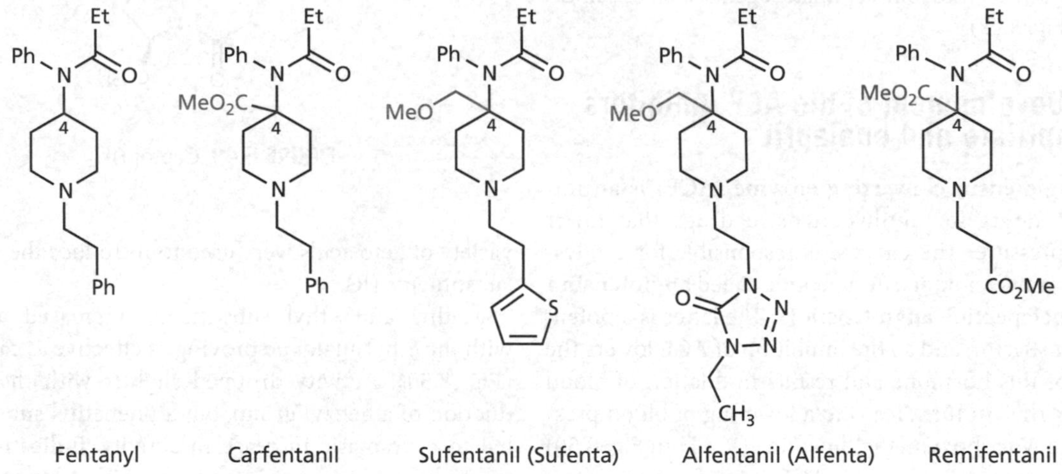

FIGURE 8.51 Fentanyl and its analogues.

structure. It is 100–300 times more potent than morphine (depending on the test used), and has a fast onset and a relatively short duration, making it an ideal intravenous analgesic for use in anaesthesia associated with surgical operations. Owing to the success of fentanyl, it has been used as a lead compound for the development of other successful agents. A key advance was the discovery that a polar ester or methoxymethyl substituent at position 4 of the piperidine ring resulted in an increase in activity. For example, carfentanil is 27 times more potent than fentanyl. Both the ester and methoxymethyl groups are capable of acting as hydrogen bond acceptors, implying that an additional binding interaction is taking place with a previously unused binding region in the binding

site. Following this discovery, carfentanil and the corresponding methoxymethyl analogue were adopted as new lead compounds to explore whether a variation in the N-substituent would enhance activity even further. This led to the discovery of **sufentanil** and **alfentanil**.

Fentanyl can be prepared from 4-piperidinone by a straightforward synthesis involving the reductive amination of an N-substituted piperidone (Fig. 8.52).

This synthesis allows the preparation of a variety of analogues where the acyl group and both N-substituents can be varied (Fig. 8.53).

However, the reductive amination does not allow for the introduction of an extra substituent at C4. Therefore, the synthesis has to be modified in order to produce

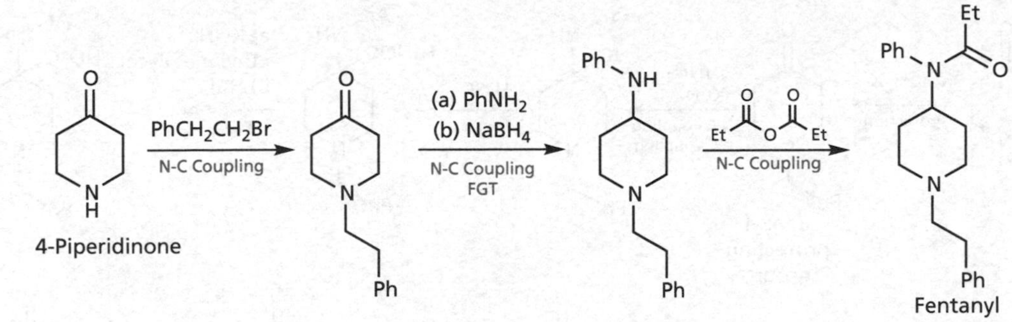

FIGURE 8.52 Synthesis of fentanyl.

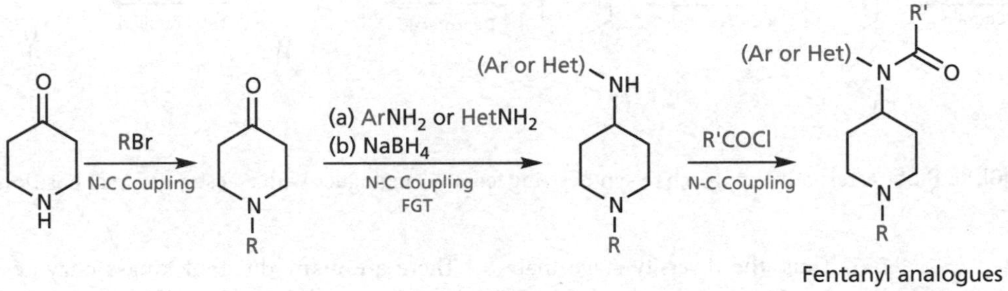

FIGURE 8.53 Synthesis of fentanyl analogues.

these kinds of compound. This can be achieved by replacing the reductive amination with a condensation reaction involving an aniline in the presence of a cyanide ion (corresponding to the Strecker synthesis). Thus, the aniline is added along with a cyanide group (Fig. 8.54). The cyanide group can then be modified to form ester substituents.

This synthesis allows analogues to be produced with variable substituents at four positions. Unfortunately, two of the diversity stages come at the very start of the synthesis. However, this can be avoided by first protecting the amine before introducing the ester group at position 4 and the N-acyl group. The protecting group can then be removed and different N-substituents added as

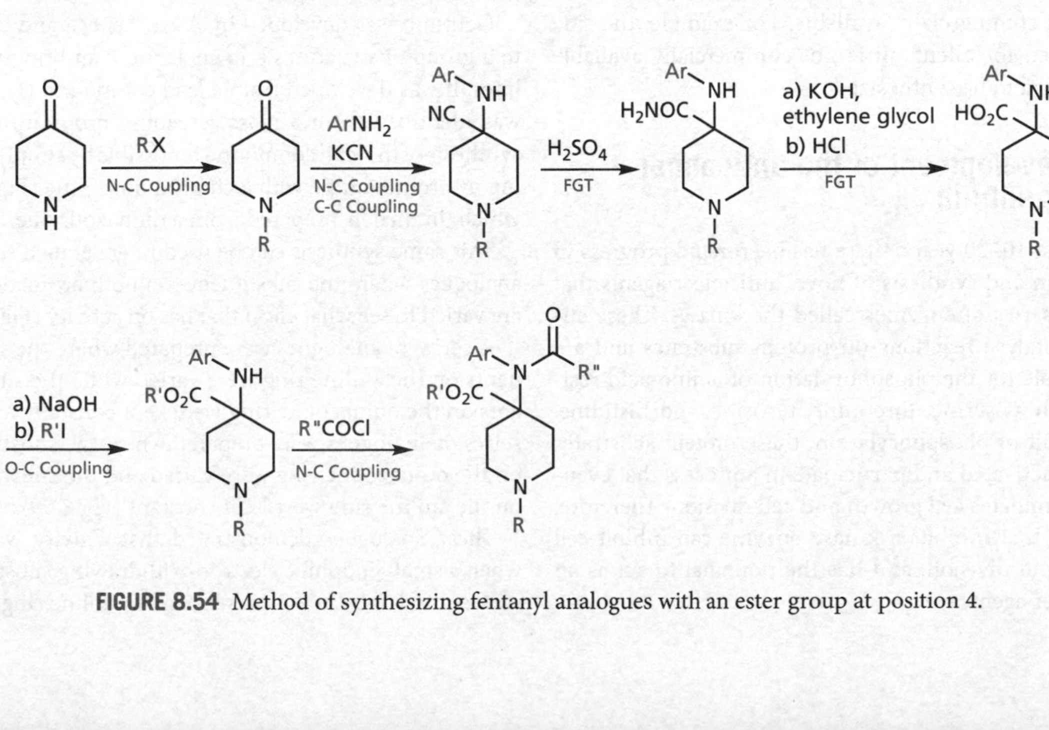

FIGURE 8.54 Method of synthesizing fentanyl analogues with an ester group at position 4.

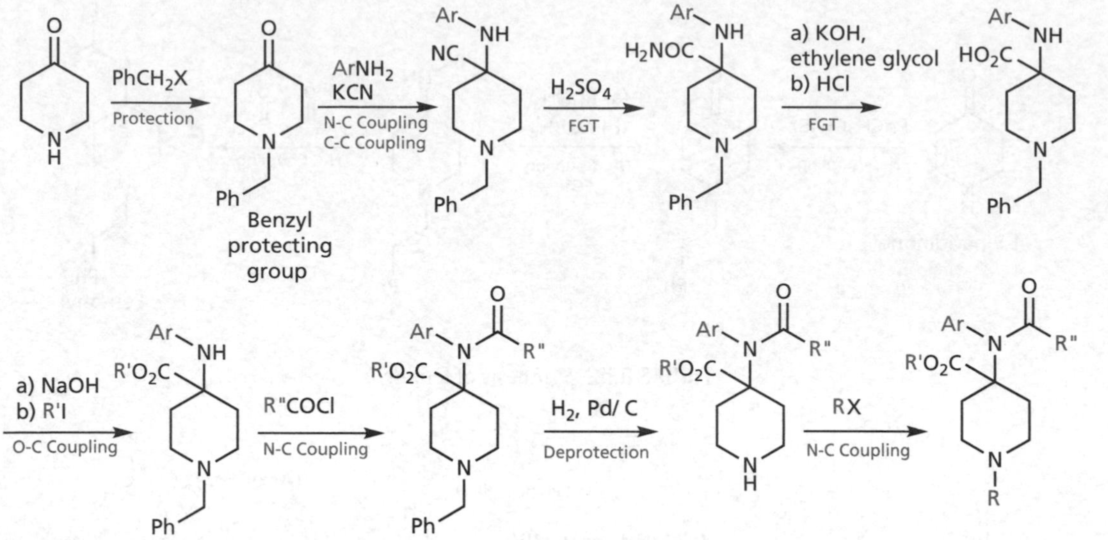

FIGURE 8.55 Alternative approach to synthesizing fentanyl analogues with an ester group at position 4.

the final step (Fig. 8.55). Thus, the diversity stage that appeared at step 1 in Figure 8.54 now becomes the final diversity stage in the synthesis shown in Figure 8.55.

The number of possible analogues that are possible from variation at four different positions is vast, and so analogues were synthesized where the group at C4 was varied whilst keeping both *N*-substituents constant. Once optimum substituents at C4 had been identified, the substituent at C4 was kept constant and the *N*-substituent on the piperidine ring was varied. This was a particularly efficient process as a stockpile of the penultimate structure in the synthesis could be built up. The synthesis of each analogue then involved a single reaction. This also made it practical to synthesize more complex alkyl halides that were not commercially available. For example, the side chain used for alfentanil is not commercially available and needed to be synthesized.

8.6.7 **Development of the anticancer agent gefitinib**

In the last 10–20 years, there has been rapid progress in the design and synthesis of novel anticancer agents that target a series of enzymes called the kinases. These enzymes catalyse reactions on protein substrates and are responsible for the phosphorylation of amino acid residues such as serine, threonine, tyrosine, and histidine. As a result of phosphorylation, these protein substrates become activated and participate in a process that eventually stimulates cell growth and cell division. Therefore, an agent that inhibits a kinase enzyme can inhibit cell growth and division, and has the potential to act as an anticancer agent.

There are many different kinase enzymes within the cell and one of these is the kinase enzyme associated with a membrane-bound receptor called the **epidermal growth factor receptor (EGF-R)**. The EGF receptor is so called because it is activated by a circulating hormone called the epidermal growth factor (EGF). Activation of the receptor results in the activation of an associated kinase enzyme, which is on the intracellular part of the cell membrane. The enzyme then catalyses the phosphorylation of tyrosine residues in protein substrates within the cell, eventually leading to cell growth and division.

Several agents have been studied as EGF-R kinase inhibitors and the first of these to reach the market was **gefitinib (Iressa)** (Fig. 8.56).

Gefitinib was developed by Astra Zeneca and belongs to a group of structures known as the 4-anilinoquinazolines. It was developed from a lead compound (I), which was identified from a mass screening programme. The synthesis of the lead compound is possible by simply reacting *m*-bromoaniline with a chloroquinazoline (Fig. 8.57) which, in turn, is prepared from a quinazolinone.

This same synthesis can be used to generate a series of analogues where the substituents on both aromatic rings are varied to see what effect that has on activity (Fig. 8.58).

A series of analogues were prepared where the substituents on the aniline ring were varied while the substituents on the quinazoline ring were kept constant Another series of analogues were prepared where the substituents on the quinazoline ring were varied and the substituents on the aniline ring were kept constant (Fig 8.59).

These analogues demonstrated that activity was best when a small lipophilic electron-withdrawing substituent, such as 3-Cl or 3-Br, was present on the aniline ring. It was

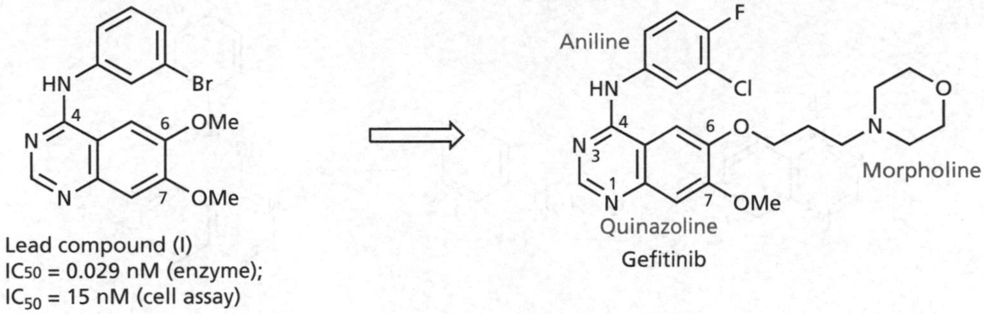

FIGURE 8.56 Structural comparison of gefitinib and its lead compound.

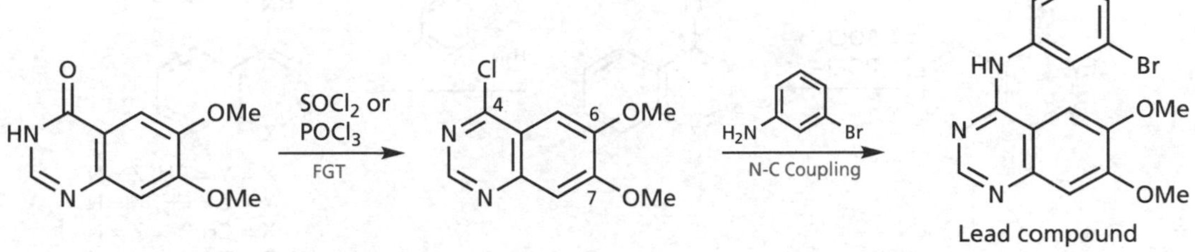

FIGURE 8.57 Synthesis of the lead compound (I).

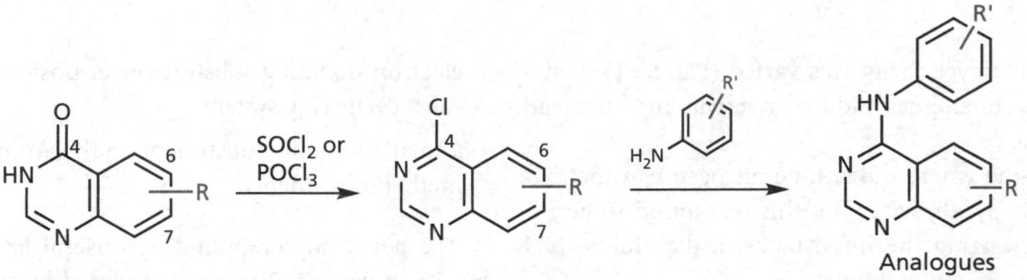

FIGURE 8.58 Synthesis of analogues to explore the effects of different substituents.

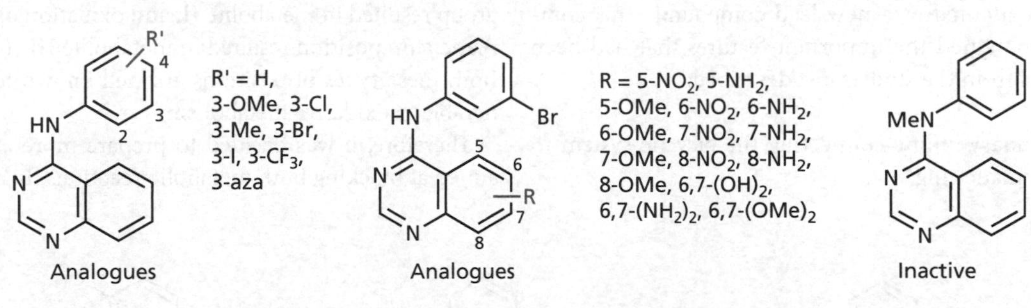

FIGURE 8.59 Analogues with different aromatic substituents.

also found that the best substituents on the quinazoline ring were the two electron-donating methoxy groups at positions 6 and 7, as found in the lead compound.

The same synthesis could be used to link amines to the quinazoline ring, rather than anilines. This allowed a series of analogues to be prepared where the separation between the quinazoline ring system and the single aromatic ring was increased (Fig. 8.60). However, activity fell, demonstrating that the aniline moiety needed to be directly linked to the quinazoline ring.

Different bicyclic systems were used as starting materials to produce analogues where the pattern of nitrogen

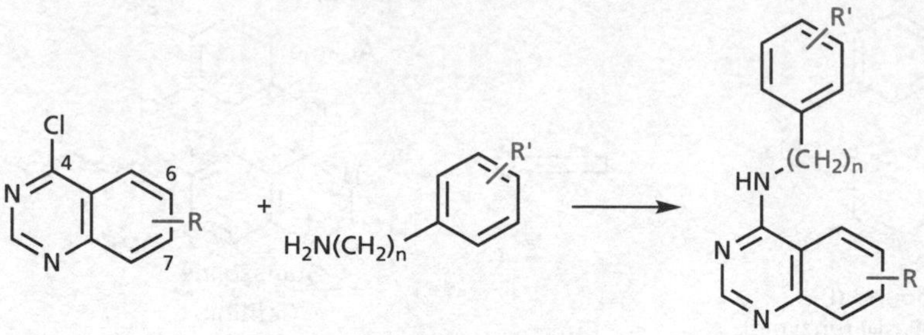

FIGURE 8.60 Analogues with different chain lengths and substituents.

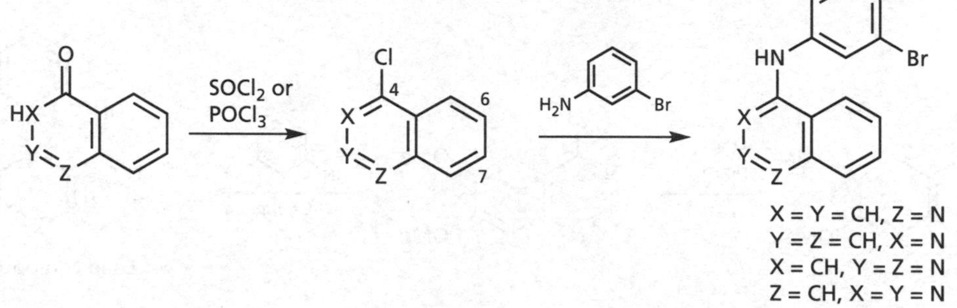

X = Y = CH, Z = N
Y = Z = CH, X = N
X = CH, Y = Z = N
Z = CH, X = Y = N

FIGURE 8.61 Analogues with different bicyclic scaffolds.

atoms in the bicyclic ring was varied (Fig. 8.61). However, all these analogues had lower activity than the lead compound.

An analogue where the aniline nitrogen was methylated was also synthesized, but this was found to be inactive, emphasizing the importance of the aniline NH group for binding interactions.

Following on from this work, the analogue containing a methyl substituent at the *meta* position of the aniline ring was adopted as a new lead compound. This compound contained the important features that had been identified from the studies just described:

- a secondary amine connecting the bicyclic system to the aromatic ring;

- electron-donating substituents at positions 6 and 7 of the bicyclic ring system;

- a small lipophilic substituent on the aromatic ring (the methyl substituent).

The new lead compound had useful *in vitro* activity, but its *in vivo* activity was hampered by the fact that it was rapidly metabolized by cytochrome P450 enzymes to give two metabolites. Oxidation of the aromatic methyl group resulted in metabolite II, and oxidation of the aromatic *para* position resulted in metabolite III (Fig. 8.62). Both these types of positions are well known to be vulnerable to oxidative metabolism.

Therefore, it was decided to prepare more analogues aimed at blocking both metabolic reactions. In structure

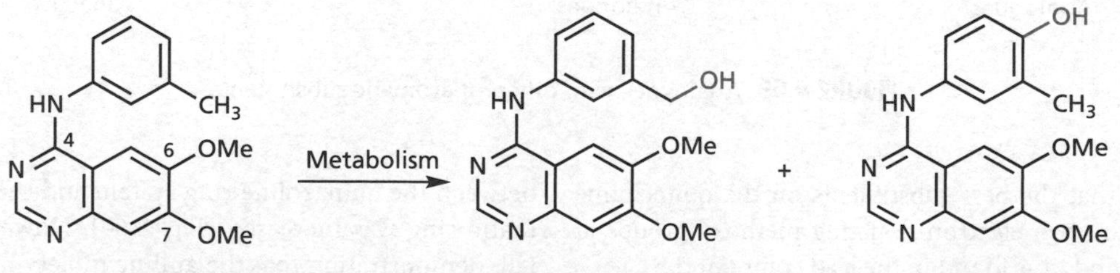

FIGURE 8.62 Main metabolites of the lead compound.

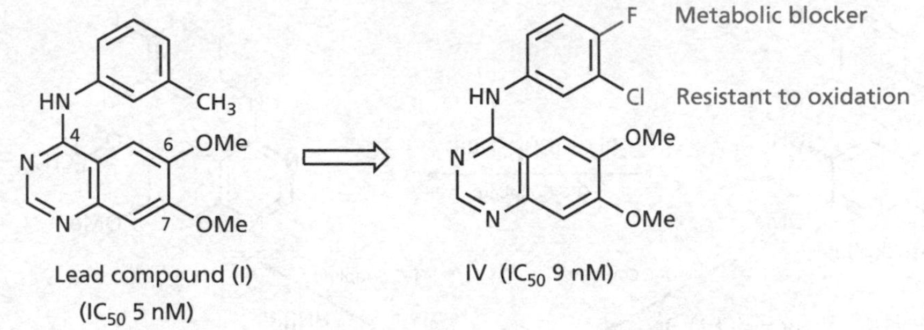

FIGURE 8.63 Introduction of substituents to block metabolism.

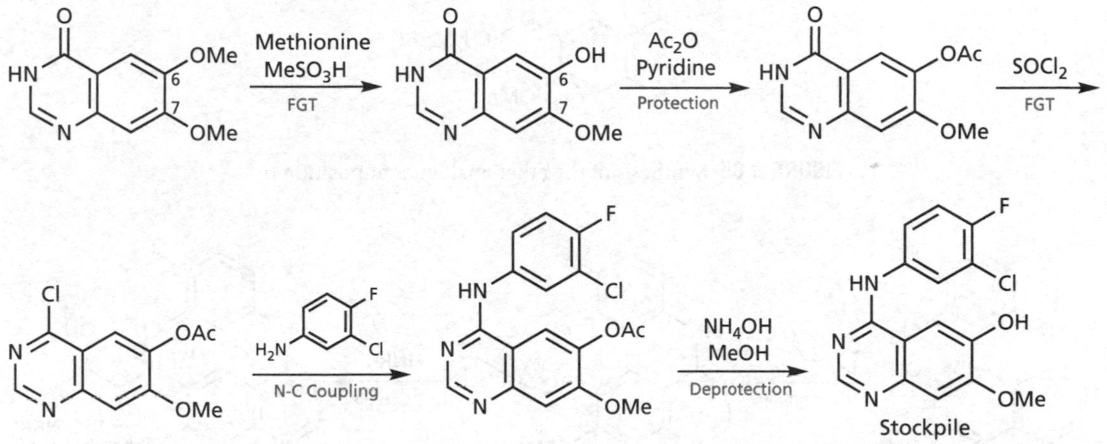

FIGURE 8.64 Synthesis of the starting material required to synthesize ether analogues at position 6.

IV (Fig. 8.63), the methyl group was replaced by a chloro substituent. This can be viewed as a bioisostere for the methyl group since it has a similar size and lipophilicity, allowing it to fit and bind into the same hydrophobic pocket. However, it has the advantage that it is resistant to oxidation by cytochrome P450 enzymes.

A fluoro substituent was introduced as a **metabolic blocker** to block oxidation at the *para* position of the aromatic ring. Fluorine is essentially the same size as hydrogen and so there is little risk of any adverse steric effects arising from its introduction.

Although the resulting compound was less active *in vitro* as an enzyme inhibitor, it showed better *in vivo* activity because of its resistance to metabolism. Unfortunately, the compound showed poor solubility and so further modifications were required. It was decided to modify one of the methoxy groups at position 6 in order to add polar substituents designed to enhance water solubility.

In order to synthesize these analogues, it was necessary to synthesize an analogue of structure IV with a phenol group at position 6 instead of a methyl ether. This was achieved by treating the quinazoline starting material with methionine and methanesulphonic acid to achieve a chemoselective demethylation of the methoxy group at

position 6 but not at position 7 (Fig. 8.64). It was then a straightforward case of protecting the resulting phenol group, introducing the chloro substituent at position 4, and then substituting the chlorine group with the aniline substituent. The acetyl protecting group was then removed to give the required starting material.

Having stockpiled the starting material required to generate the ether analogues, each analogue was efficiently synthesized by a one- or two-step process (Fig. 8.65). The one-step process involved alkylation of the phenol group directly with bromo amines. The two-step process involved addition of a linker chain using a dibromo alkane. The resulting structure could then be treated with a much wider range of amines than was possible using the one-step process.

A third route allowed the incorporation of an alcohol functional group into the side chain to produce even more polar substituents (Fig. 8.66).

All the analogues synthesized from these routes were found to be potent inhibitors, demonstrating the feasibility of modifying the 6-methoxy substituent.

The most effective agent was gefitinib, where the *O*-substituent contains a morpholine ring—a feature which is commonly added to structures to enhance water solubility. Because the morpholine ring includes a basic nitrogen, it is

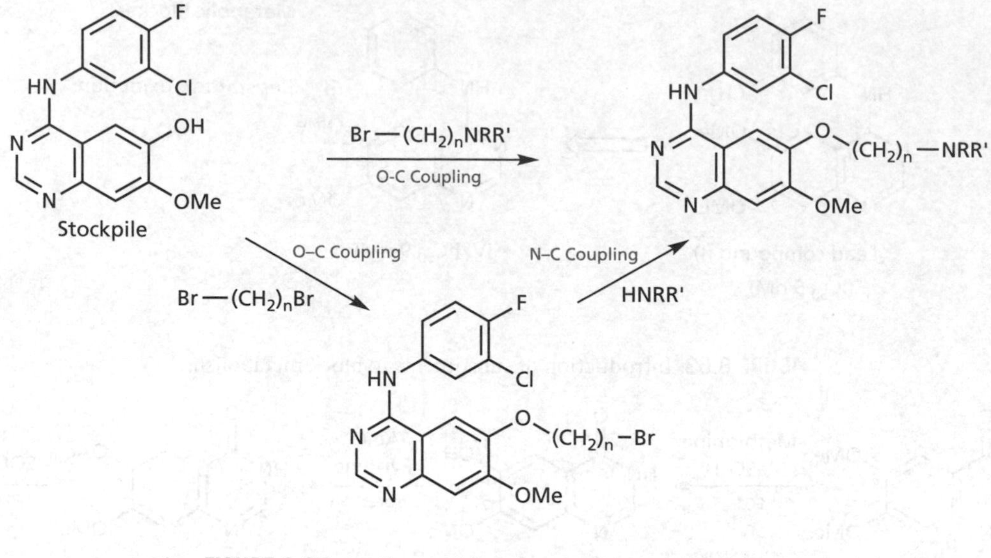

FIGURE 8.65 Synthesis of the ether analogues at position 6.

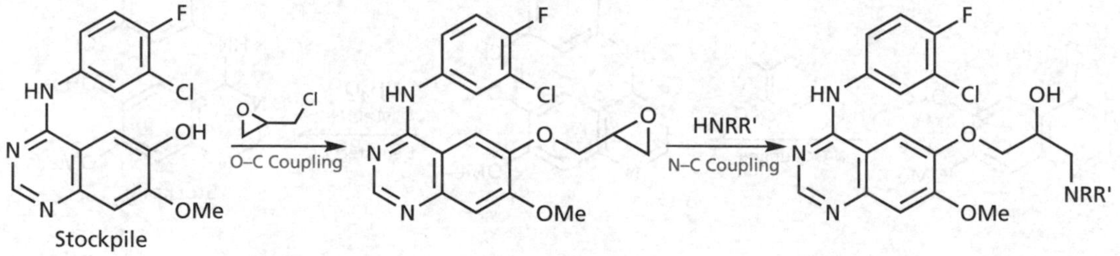

FIGURE 8.66 Synthesis of analogues with a 1,2-amino alcohol.

possible to protonate it and form water-soluble salts of the drug. The morpholine group plays no role in target binding, and it is important that it is linked to the drug scaffold such that it 'sticks out' of the target binding site when the drug is bound. This exposes the morpholine ring to the surrounding aqueous environment, which means that the group does not need to be desolvated when the drug binds to the binding site, thus avoiding any energy penalty involved in desolvation (section 8.4.2).

Binding studies indicate that the two nitrogens in the quinazoline ring form hydrogen bonds with the active site. One of these is a hydrogen bond to a methionine residue, while the other forms a hydrogen bond to a water molecule which acts as a hydrogen bonding bridge to a tyrosine residue. The chlorine substituent fits into a hydrophobic pocket, while the chain with the morpholine ring sits in a cleft that leads to the outside of the enzyme (Fig. 8.67).

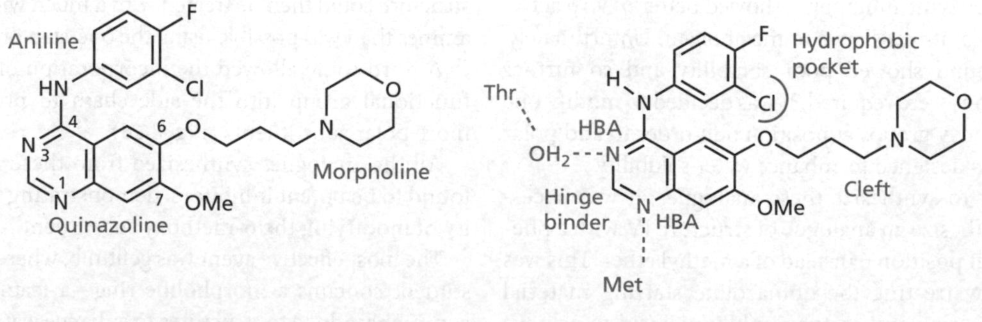

FIGURE 8.67 Structure and binding interactions of gefitinib (Iressa).

- A full synthesis is likely to generate a larger variety of analogues than synthesizing analogues from a lead compound.

- The analogues that can be synthesized using a full synthesis depend on the synthetic steps involved and the reagents that are available.

- Reagents/building blocks should be commercially available or easily synthesized.

- The synthetic steps involved in coupling molecular 'building blocks' are known as diversity steps.

- An efficient synthesis involves diversity steps which are clustered together at the end of a synthetic route.

- The more diversity steps present in a synthesis, the greater the number of possible analogues.

- The traditional approach to synthesizing analogues is to carry out a synthesis where modifications are limited to one diversity step at a time.

- Different synthetic routes may be used to synthesize different analogues, such that the relevant diversity step comes close to the end of the route.

- Combining the optimum features from each diversity step may not be beneficial for binding affinity or activity. It is often better to synthesize analogues using different combinations of the optimum features.

- Convergent syntheses are more efficient than linear syntheses at producing a specific target structure, but may be less useful when synthesizing a range of analogues.

FURTHER READING

General references

Patrick, G.L. (2013) *An introduction to medicinal chemistry* (5th edn). Oxford University Press, Oxford (Chapter 24, 'The opioid analgesics'; Chapter 23, 'Drugs acting on the adrenergic nervous system'; Section 19.5.1, 'Penicillins'; Section 21.6.2, 'Protein kinase inhibitors').

Specific syntheses

Alikhani, V., et al. (2004) 'Long chain formoterol analogues: an investigation into the effect of increasing amine substituent chain length on the β_2-adrenoceptor activity', *Bio-organic and Medicinal Chemistry Letters*, **14**, 4705–10 (salmeterol).

Bagley, J.R., et al. (1989) 'New 4-(heteroanilido)piperidines, structurally related to the pure opioid agonist fentanyl, with agonist and/or antagonist properties', *Journal of Medicinal Chemistry*, **32**, 663–71 (fentanyl analogues).

Barker, A.J., et al. (2001) 'Studies leading to the identification of ZD1839 (IressaTM): an orally active, selective epidermal growth factor receptor tyrosine kinase inhibitor targeted to the treatment of cancer', *Bioorganic and Medicinal Chemistry Letters*, **11**, 1911–14C (gefitinib).

Colapret, J.A., et al. (1989) 'Synthesis and pharmacological evaluation of 4,4-disubstituted piperidines', *Journal of Medicinal Chemistry*, **32**, 968–74 (fentanyl analogues).

Gates, M. and Tschudi, G. (1956) 'The synthesis of morphine', *Journal of the American Chemical Society*, **78**, 1380–93 (morphine).

Janssens, F., et al. (1986) 'Synthetic 1,4-disubstituted-1,4-dihydro-5H-tetrazol-5-one derivatives of fentanyl: Alfentanil (R 39209), a potent, extremely short-acting narcotic analgesic', *Journal of Medicinal Chemistry*, **29**, 2290–7 (alfentanil).

Lawrence, H.R., et al. (2005) 'Novel and potent 17β-hydroxysteroid dehydrogenase type I inhibitors', *Journal of Medicinal Chemistry*, **48**, 2759–62 (estrone analogues).

Lipkowski, A.W., et al. (1986) 'Peptides as receptor selectivity modulators of opiate pharmacophores', *Journal of Medicinal Chemistry*, **29**, 1222–5 (oxymorphone analogues).

Lunts, L.H.C. (1994) 'Salbutamol: a selective β2-stimulant bronchodilator', C.R. Ganellin and S.M. Roberts (eds), *Medicinal chemistry—the role of organic research in drug research* (2nd edn). Academic Press, London, Chapter 4 (salbutamol).

Patchett, A.A., et al. (1980) 'A new class of angiotensin-converting enzyme inhibitors', *Nature*, **288**, 280–3 (*N*-carboxymethyldipeptides).

Portoghese, P.S., et al. (1987) 'Bimorphinans as highly selective, potent κ opioid receptor antagonists', *Journal of Medicinal Chemistry*, **30**, 238–9 (norbinaltorphimine).

Rewcastle, G., et al. (1995) 'Tyrosine kinase inhibitors. 5. Synthesis and structure–activity relationships for 4-[(phenylmethyl)amino]- and 4-(phenylamino)quinazolines as potent adenosine 5'-triphosphate binding site inhibitors of the tyrosine kinase domain of the epidermal growth factor receptor', *Journal of Medicinal Chemistry*, **38**, 3482–7 (gefitinib).

Vardanyan, R. and Hruby, V. (2006) *Synthesis of essential drugs*. Elsevier, Amsterdam, pp 19–56 (levorphanol).

9 Synthesis of natural products and their analogues

9.1 Introduction

The natural world has been a rich source of clinically important compounds that are still in use today. Important drugs have been isolated from plants, trees, micro-organisms, animals, and even our own bodies. These include the antimalarial drugs **quinine** and **artemisinin**, the analgesic **morphine**, the anticancer agent **paclitaxel**, the antibiotic **penicillin G**, and the anti-inflammatory agent **cortisol** (Fig. 9.1).

The synthesis of many of these agents has proved to be a huge challenge, and although synthetic routes have been devised, they are often impractical for large-scale synthesis because of the large number of steps required, low overall yields, and high cost. Why should the synthesis of these compounds be so difficult? There are four main reasons:

- the number and variety of functional groups that are present;
- the presence of multicyclic or unusual ring systems;
- the number of asymmetric centres;
- the number and variety of substituents.

Although functional groups are essential to synthesis, their presence in a final product also poses a challenge. How do you incorporate them into the structure and how do you ensure that they are not altered during the synthesis? There are a number of ways of getting round this problem.

- Introduce the functional group at a late stage of the synthesis, ideally as a result of a carbon–carbon bond formation.
- Introduce a functional group that is relatively unreactive early on in the synthesis, and then convert it to the required functional group(s) at the end of the synthesis. Such a group can be called a latent group for the required functional group.
- Introduce the required functional group early on in the synthesis and add a protecting group that is resistant to the reagents used in the synthesis. Deprotect the functional group at the end of the synthesis.

All these tactics have been used successfully in the full synthesis of natural products, but the more functional groups that are present, the more complicated the situation becomes. Synthetic design requires intelligent use of a variety of tactics and strategies in order to build up the target molecule, but this inevitably increases the number of synthetic steps required. Another problem is the presence of particularly labile functional groups. For example, penicillins have a reactive beta-lactam ring, while there is a reactive trioxane ring system in artemisinin. This usually requires coming up with a strategy where particularly reactive functional groups are introduced or unmasked at the very end of the synthesis.

Multicyclic ring systems are often present in natural products, some of which may never have been synthesized before and will require careful synthetic planning. The more complicated the ring system, the more complicated the required synthesis, especially if there are several substituents present. In general, the more substituents on a ring system, the more difficult the synthesis.

Asymmetric centres are another complicating factor. If there are n asymmetric centres present in a structure, there are 2^n possible stereoisomers. For example, cortisol has seven asymmetric centres, which means that there are 2^7 possible stereoisomers, amounting to a total of 128 structures. Any synthesis of cortisol would have to be designed such that only one of these 128 structures is produced.

Since the full synthesis of many of these agents is too complicated and costly to be economically viable, other methods of obtaining them have become necessary.

KEY POINTS

- The full synthesis of a complex natural product is often impractical on a production scale because of the number of synthetic steps required and the low overall yield obtained.

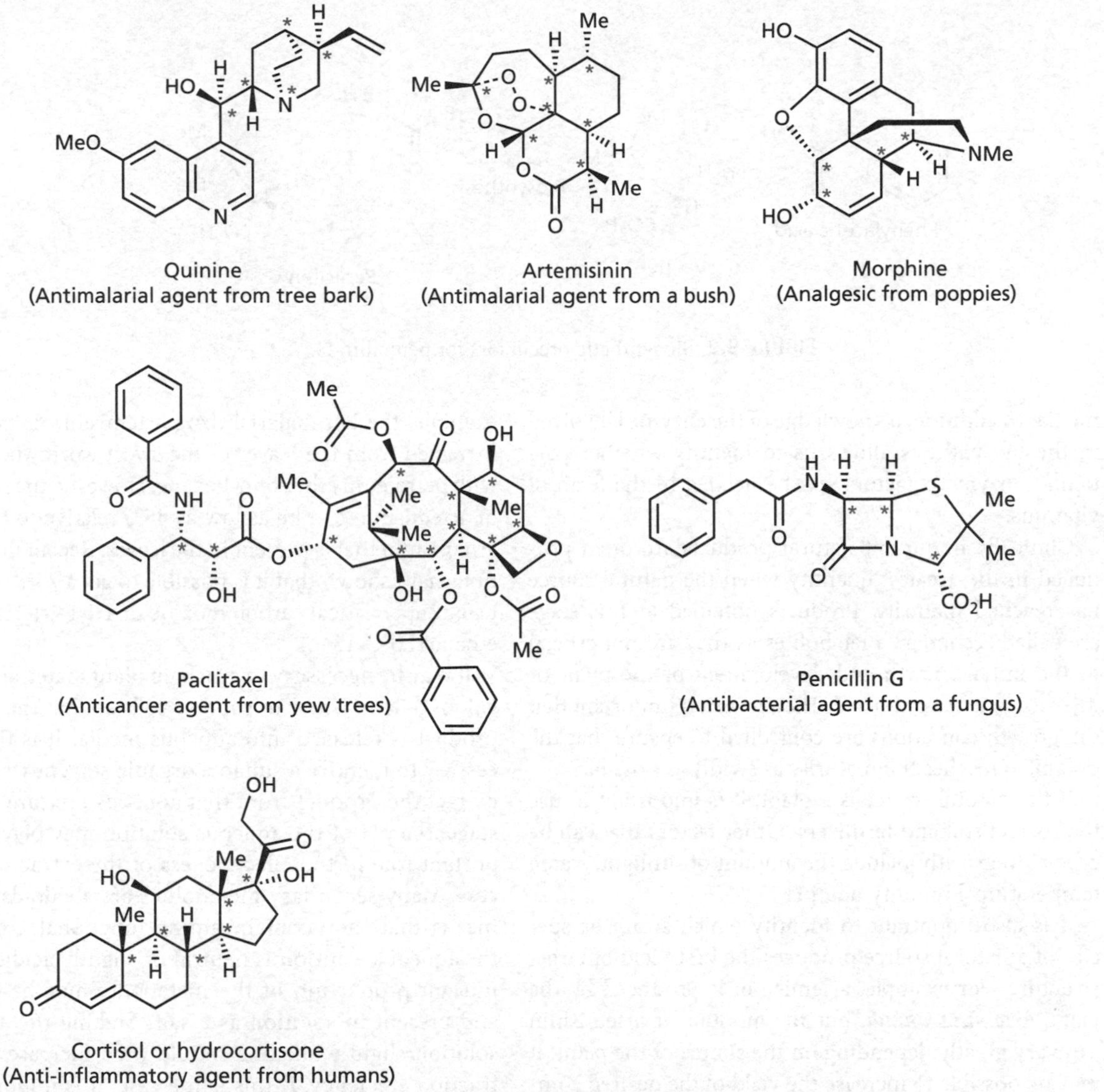

Quinine
(Antimalarial agent from tree bark)

Artemisinin
(Antimalarial agent from a bush)

Morphine
(Analgesic from poppies)

Paclitaxel
(Anticancer agent from yew trees)

Penicillin G
(Antibacterial agent from a fungus)

Cortisol or hydrocortisone
(Anti-inflammatory agent from humans)

FIGURE 9.1 Examples of clinically important drugs isolated from the natural world. Asterisks indicate asymmetric centres. The pharmacological activity and natural source are in brackets.

• Factors affecting the difficulty of a synthesis include complex ring systems and the number of asymmetric centres, substituents, and functional groups present.

9.2. Extraction from a natural source

The most effective method of obtaining many clinically important natural products is to extract them from their natural source. For example, morphine is extracted from opium obtained from the seed pods of poppy plants. Penicillin G is obtained by growing cultures of a fungal Penicillium species and extracting the penicillin from the fermentation medium. Artemisinin is extracted from the leaves of a plant called *Artemisia annua* (sweet wormwood). There are several things one can do to maximize the yield of the desired natural product.

• **Optimize the growing conditions to favour biosynthesis of the natural product.**

There may well be key nutrients that are required for the biosynthesis of the desired natural product. In the case of penicillin G, the presence of phenylacetic acid, valine, and cysteine are crucial as biosynthetic precursors if good yields are to be obtained (Fig. 9.2). Thus, an understanding of the biosynthetic route and the biosynthetic precursors involved can be extremely important in identifying key nutrients that should be added to the growth

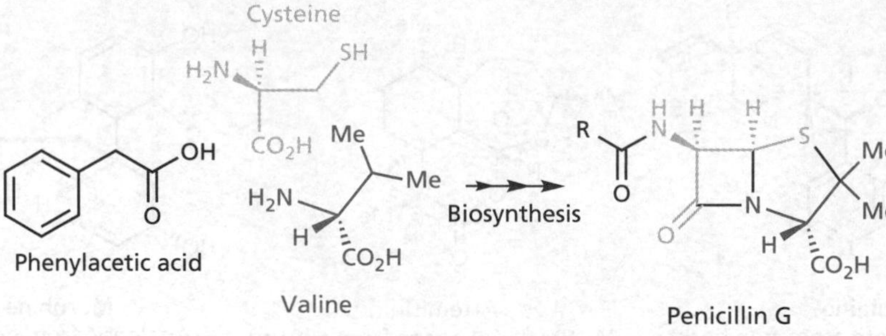

FIGURE 9.2 Biosynthetic precursors for penicillin G.

media. In addition, a knowledge of the enzymes involved in the biosynthesis allows us to identify whether particular enzyme cofactors will be needed in the form of vitamins.

Clinically important natural products are often produced in the greatest quantity when the natural source has reached maturity. Products obtained at this stage are called secondary metabolites as they are not crucial to the initial growth and development of the plant or microorganism concerned. Therefore, it is important that the growth conditions are controlled to ensure that the organism reaches full maturity as swiftly as possible.

If the natural source is a plant, it is important to use the correct soil and fertilizers. Other factors that can be crucial to growth include the amount of sunlight, water, temperature, humidity, and pH.

It is also important to identify which strain or species of a natural source produces the best yield of target structure. For example, artemisinin is produced by the plant *Artemisia annua*, but the amount of artemisinin can vary greatly depending on the source of the plant. It may be possible to increase the yield of the desired compound from a plant by conventional hybridization procedures. The market leader for artemisinin production is Artemis—a particular plant hybrid that resulted from a cross between different parent plants.

• **Optimize the harvesting procedures.**

As already stated, many of the important natural products used in medicine are **secondary metabolites** and are more likely to be present when the organism has reached a certain maturity. Therefore, identifying when the levels of natural product reach a maximum is crucial in determining when to harvest a plant crop or a microbiological culture.

Having identified the best time to harvest a crop or culture, it is also important to identify the optimum extraction methods to ensure that most of the natural product is obtained. If the natural source is a plant, it is also important to know whether the desired product is produced in the leaves, roots, stems, flowers, or seeds. For

example, the antimalarial drug artemisinin is normally extracted from the leaves of the sweet wormwood plant with petroleum ether or n-hexane. However, the recovery of artemisinin can be as low as 45% relative to the total artemisinin that is present in the leaves. Recent investigations have shown that it is possible to get a 75% recovery using supercritical carbon dioxide or 1,1,1,2-tetrafluoroethane (HFC-134a).

It may be necessary to crush up plant material or fungal mycelia in order to rupture cells so that the desired product is released into aqueous media. It is then necessary to identify a suitable organic solvent which can extract the product from that aqueous mixture. At this stage, the pH of the aqueous solution may play an important role in the effectiveness of the extraction process. Many secondary metabolites are alkaloids, which means that they contain amine functional groups. If the aqueous solution is neutral or slightly acidic, a significant proportion of the metabolite may be ionized and present in solution as a salt. Making the aqueous solution slightly alkaline should then increase the extraction efficiency. If this is the case, it is important to check the pH levels of the aqueous solution after each extraction. For example, if the first extraction is successful in taking up a certain percentage of the alkaloid present, the pH of the aqueous solution may well swing back towards neutral because of the removal of the basic compound. More base would have to be added to the solution to compensate for this. On the other hand, it is important not to make the aqueous solution too alkaline. The desired product might be unstable under strong alkaline conditions and decompose. Alternatively, the product may contain a functional group such as a phenol or a carboxylic acid which will ionize under alkaline conditions, thus preventing it from being extracted. For example, morphine has an amine group that ionizes under acid conditions, but it also contains a phenol group that ionizes under alkaline conditions. In order to extract this product efficiently from an aqueous solution, it is necessary to identify a pH when neither group is ionized (Fig. 9.3).

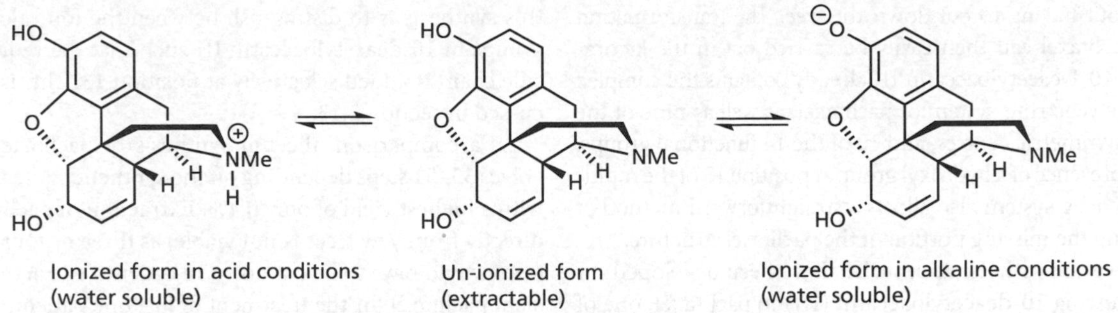

FIGURE 9.3 Different forms of morphine at different pHs.

Assuming that an extraction process has been successful, it then becomes necessary to purify the compound, ideally as a crystalline salt.

KEY POINTS

- Many clinically important natural products are obtained from their natural source.

- Growing conditions that optimize the amount of natural product produced should be identified.

- Harvesting should be carried out at a time when the levels of natural product are high. The optimum extraction procedures need to be determined.

- The parts of the plant that contain the most quantity of natural product should be identified.

9.3 Semi-synthetic methods

In certain cases, it is impractical to extract a clinically important natural product from its natural source, especially if the compound is present in low quantities, or the natural source is rare or difficult to cultivate. An alternative approach is to extract a structurally related natural product which is more readily obtained, and then convert that synthetically to the desired compound. This is known as a semi-synthetic approach to distinguish it from a full synthesis. In a full synthesis, a drug is prepared from simple commercially available compounds. In the semi-synthetic approach, the chemist makes use of nature to provide a suitable starting material which already has most of the complex features of the target compound included in its structure.

The most likely natural starting materials for such a semi-synthetic approach are biosynthetic intermediates of the target compound. These should be towards the end of the biosynthetic route, since these will be closer in nature to the final structure and have most of the 'difficult features' already present—for example, asymmetric centres, rings, functional groups, and substituents. Clearly, a knowledge of the biosynthetic route to a particular compound is extremely useful in identifying potential intermediates.

A successful example of this approach is the semi-synthetic method used to produce the anticancer agent **paclitaxel** (Taxol). **10-Deacetylbaccatin III** (Fig. 9.4) has been identified as a late intermediate in the biosynthesis of paclitaxel and can be extracted from yew tree needles in six to ten times greater yield than paclitaxel itself,

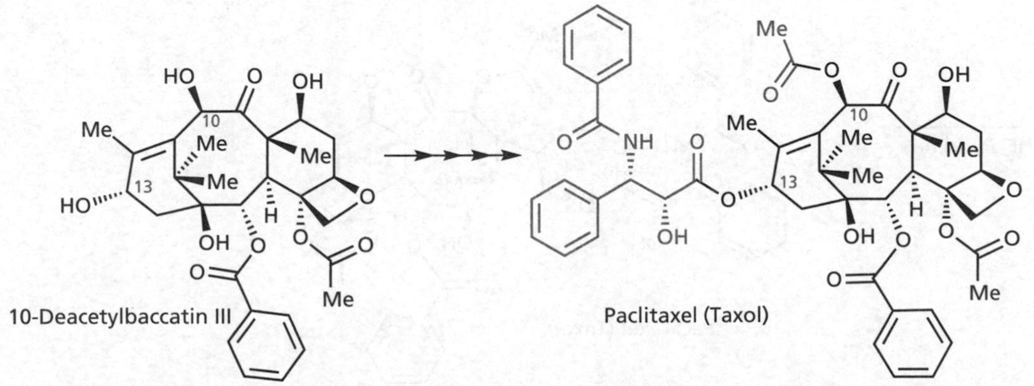

FIGURE 9.4 Semi-synthetic approach to paclitaxel.

without having to cut down the tree. The transformation to paclitaxel can then easily be carried out in the laboratory. 10-Deacetylbaccatin III already contains the complex multicyclic ring system of paclitaxel, as well as nine of the 11 asymmetric centres, and 10 of the 14 functional groups. The presence of a hydroxyl group at position 13 of the multicyclic ring system also allows a straightforward method of linking the missing portion of the paclitaxel structure.

Several synthetic approaches have been developed for converting 10-deacetylbaccatin (III) to paclitaxel, one of which is shown in Figure 9.5. This synthesis was approved by the FDA in 1994 and has been used to produce paclitaxel in multi-kilogram quantities. A key requirement of

this synthesis is to distinguish between the four alcohol groups in 10-deacetylbaccatin III such that the required side chain is added selectively at position 13. This is discussed in section 2.13.

As a comparison, the full synthesis of paclitaxel involves 35–51 steps depending on the synthetic route used, with a highest yield of only 0.4%. Extraction of paclitaxel directly from yew trees is not viable, as three or four yew trees would have to be cut down in order to extract sufficient paclitaxel for the treatment of just one patient.

A semi-synthetic approach to producing the antimalarial agent **artemisinin** is also being developed, since a full synthesis of artemisinin is not financially viable and

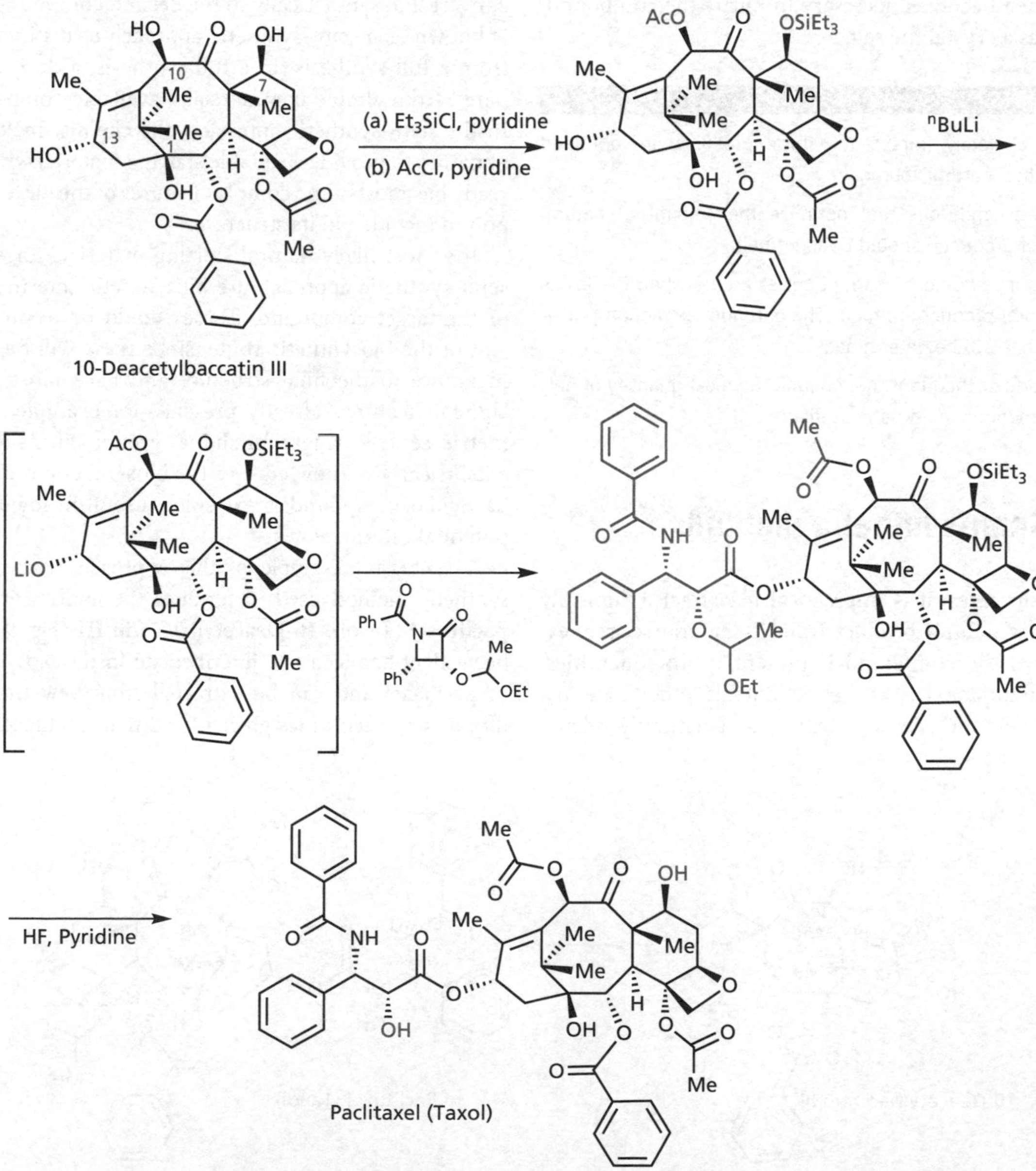

FIGURE 9.5 Synthesis of paclitaxel from 10-deacetylbaccatin III.

extraction of the compound from its natural source (*Artemisia annua* or sweet wormwood) can give yields as low as 0.1%. Moreover, the extracted product is impure and has to be 'cleaned up' using chromatography.

It is possible to extract from the plant a molecular precursor of artemisinin called **artemisinic acid**, which is ten times more abundant than artemisinin. Moreover, it can be extracted without the need for chromatographic purification. A genetically engineered strain of a yeast called *Saccharomyces cerevisiae* that can produce the metabolite has also been developed (section 9.5).

Artemisinic acid can be reduced to **dihydro-artemisinic acid** (Fig. 9.6). This compound is then treated with singlet oxygen to promote an ene reaction to give a hydroperoxide structure. Treatment with TFA results in cleavage of the hydroperoxide followed by a rearrangement reaction to give an enol, which is then treated with triplet oxygen to give another hydroperoxide. This undergoes a series of spontaneous cyclization reactions to give the final product in 30% yield from artemisinic acid.

One of the barriers to scaling up this synthesis relates to the practical problems associated with the use of singlet oxygen, which has to produced in situ using specialized equipment. Another issue is that light is required, but the light can only penetrate a limited distance into the solution as it is absorbed by a photosensitizer. The bigger the reaction flask, the less efficient the process is. Work is currently being carried out to identify a cost-effective method of converting dihydroartemisinic acid to artemisinin by means of a continuous flow system. This essentially involves the reactant flowing through narrow diameter tubing wrapped round the light source such that illumination of the reaction

solution is far more efficient. Increasing the duration of the reaction is also a more practical method of increasing yields than increasing the size of the reaction vessel. Further work has shown that it is possible to carry out the ene reaction, Hock cleavage, and subsequent condensations as a single continuous flow chemical process that does not require the isolation of any of the intermediates. This can result in a 39–46% yield of artemisinin from dihydroartemisinic acid, with production of 200g of artemisinin per day. It has been estimated that about 1500 of these photoreactors would be required to meet current needs.

The above examples illustrate how a natural product can be prepared from a biosynthetic intermediate, but it may also be possible to synthesize a natural product from another natural product that has some structural similarity to the desired compound. An example of this approach involved the use of a plant steroid to produce **progesterone**—an important mammalian sex hormone. Before the Second World War, the synthesis of progesterone involved long, complicated, multi-stage routes which inevitably meant that the final product was extremely expensive and was produced in low overall yield. A much quicker synthesis involving only five steps was developed using a hexacyclic steroid called **diosgenin** (Fig. 9.7), which is produced by the Mexican yam (*Dioscorea mexicana*). The procedure started with an oxidative degradation of the spiro ring system present in diosgenin to produce a pentacyclic steroid, where the atoms that had made up the sixth ring were now present as an acyclic substituent. Oxidation with CrO_3 cleaved the substituent and opened up the fifth ring to provide a tetracyclic steroid that was very similar to the target structure of progesterone. Functional group transformations

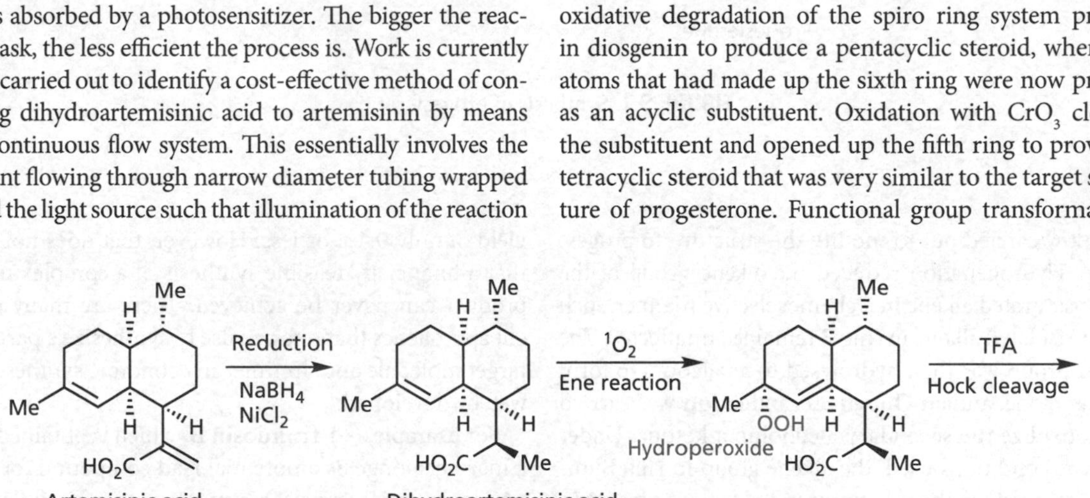

FIGURE 9.6 Semi-synthetic route to artemisinin.

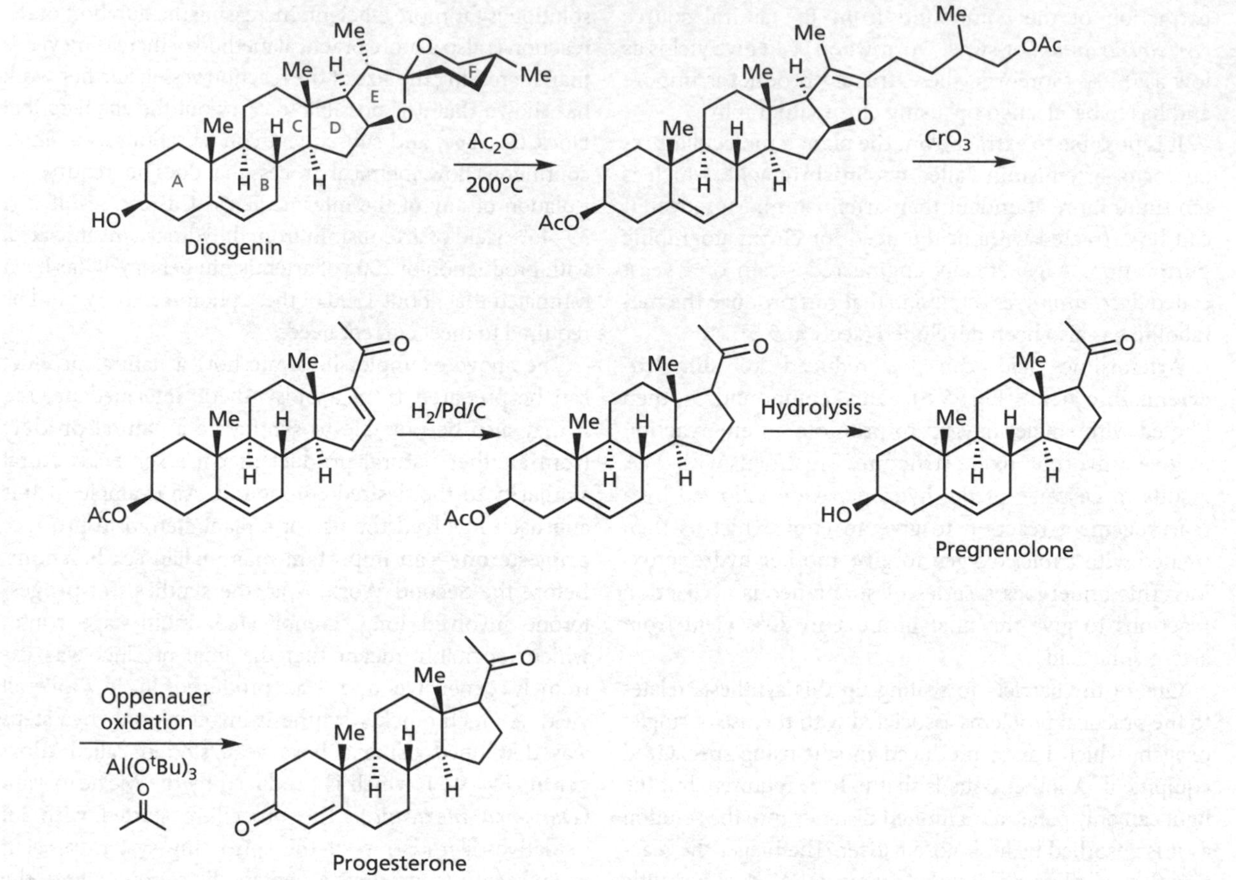

FIGURE 9.7 Synthesis of progesterone.

were now carried out to modify the structure to progesterone. Hydrogenation reduced the alkene group of the α,β-unsaturated alkene in a chemoselective manner, such that the isolated alkene in ring B remained unaffected. The acetate group was then hydrolysed to an alcohol to form pregnenolone, and an **Oppenauer oxidation** was carried out to oxidize the secondary alcohol to a ketone. Under the basic conditions used, the alkene group in ring B migrated to produce the α,β-unsaturated ketone present in progesterone. The company Syntex was set up to take advantage of this process and went on to devise syntheses of testosterone and various estrogens from diosgenin.

9.4 Full synthesis

In many cases, the full synthesis of a natural product is not financially feasible because of the number of steps involved and the low overall yield. We have already seen that a full synthesis of paclitaxel takes 35–51 steps with a

yield of only 0.4% or less. However, that does not mean that a financially feasible synthesis of a complex natural product can never be achieved. There are many different approaches that one can use to synthesize a particular target molecule and, in time, an economic synthesis may well be developed.

For example, (+)-**frondosin B**, which is obtained from a marine sponge, is a potential lead compound for novel agents that might act as treatments for cancer, inflammatory diseases, or HIV. The compound has a tetracyclic structure with only one chiral centre, but synthesizing the pure enantiomer has proved a challenge. The first successful synthesis of the product involved 17 steps with only a 1% overall yield. A second synthesis developed the following year involved 20 steps, and succeeded in obtaining a yield of 7%. In 2009, a synthetic route was devised involving 10 steps with a 13% yield. However, in 2010, a synthetic route was devised that involved only three steps with a 50% overall yield (Fig. 9.8).

The first stage of the reaction sequence was an enantioselective Friedel–Crafts alkylation of a benzofuran ring

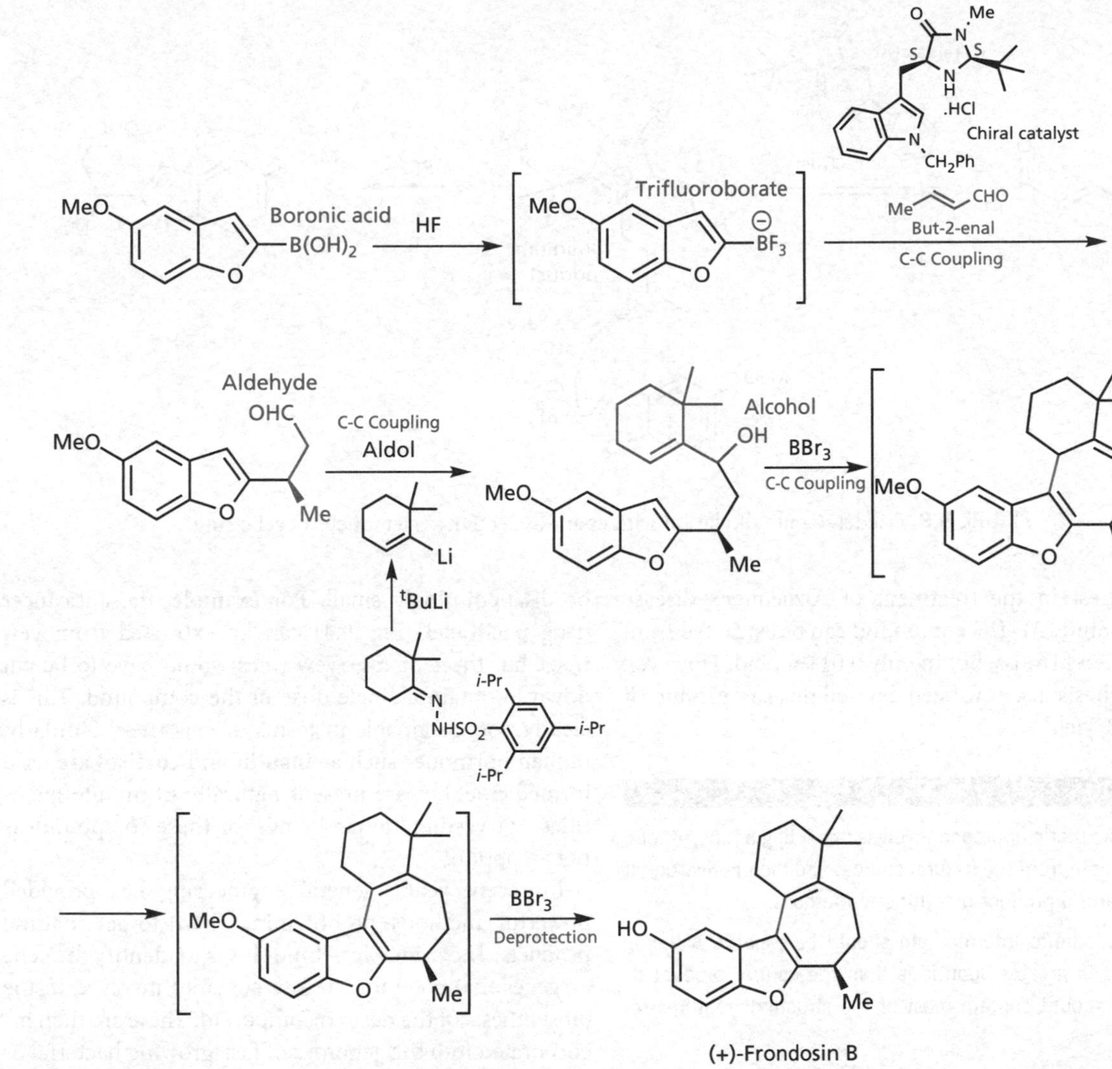

FIGURE 9.8 Three-step synthesis of (+)-frondosin B.

system with but-2-enal in the presence of a chiral amine catalyst and hydrogen fluoride. But-2-enal reacts with the catalyst to form an iminium adduct, which serves as the alkylating agent for the reaction (Fig. 9.9). At the same time, the boronic acid group of the benzofuran structure is converted to a trifluoroborate group with HF. The trifluoroborate group then acts as an activating and directing group for the Friedel–Crafts reaction on the benzofuran ring.

The stereoselectivity and enantioselectivity is determined by the iminium adduct. First, the tertiary butyl group that is present ensures that the configuration of the iminium adduct is as shown in Figure 9.9, in order to minimize steric interactions. Secondly, the heterocyclic substituent (Het) acts as a steric shield to prevent the fluoroborate approaching the *si* face of the alkene group. Therefore, the reaction takes place selectively at the *re* face. In other words, the fluoroborate approaches

from behind the alkene as drawn in Figure 9.9. Once the alkylation has taken place, the iminium ion hydrolyses to give the product and restore the chiral catalyst. The alkylation reaction was carried out with an 84% yield and an enantiomeric excess of 91–97%.

The second step involved a vinyl lithium reagent which was generated in situ from a hydrazone and a base. The vinyl lithium reagent reacts with the aldehyde group of the starting material in an aldol reaction to give a secondary alcohol.

The final step involved three separate processes of allylic alkylation, olefin isomerization, and demethylation of the methyl ether. These were originally carried out as separate steps, but it was found that BBr$_3$ could perform all three operations in a single reaction.

Another recent success story has been the full synthesis of a natural product called huperzine A, which

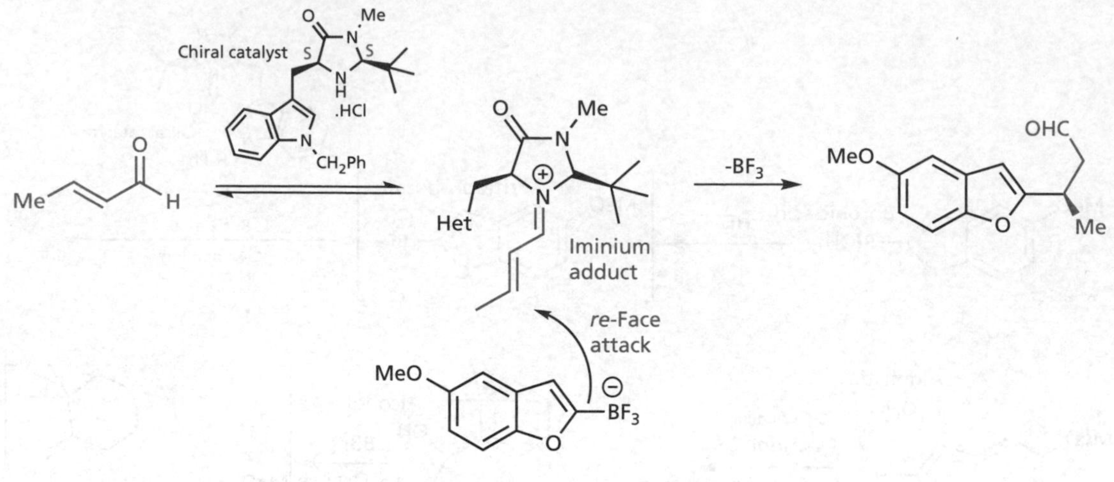

FIGURE 9.9 Friedel–Crafts alkylation with enantioselectivity (Het = heterocyclic ring).

is of interest in the treatment of Alzheimer's disease (see Case Study 3). The compound can be extracted from a herb called firmoss, but in only 0.011% yield. However, a full synthesis has now been devised that can produce it in 25–45% yield.

KEY POINTS

- A semi-synthetic approach involves extracting a biosynthetic intermediate from the natural source, and then converting it to the natural product by synthetic methods.

- The biosynthetic intermediate should be capable of being extracted in greater quantities than the natural product itself and should contain most of the structural features required.

- Full syntheses of a number of natural products have been developed which produce the compounds in good overall yields.

9.5 Cell cultures and genetic engineering

9.5.1 Modification of microbial cells

In section 9.2, we discussed how a natural product might be harvested from its natural source. This may seem the most obvious method of isolating it, but it is not necessarily the best method. It can take a long time to grow the source crop, and the harvesting/extraction procedures may prove costly and 'messy'. Moreover, poor weather conditions, droughts, floods, and earthquakes can adversely affect the crops concerned, resulting in worldwide shortages of important drugs. To add to these problems, the amount of natural product obtained may be disappointingly small. For example, the anticancer drug **paclitaxel** (Fig. 9.4) can be extracted from yew trees, but three or four yew trees would have to be cut down to obtain a single dose of the compound. This is clearly not sustainable in terms of yew trees! Similarly, human hormones such as **insulin** and **cortisol** are used in medicine, but are present naturally in minute quantities. Harvesting human beings for these compounds is not an option!

In recent years, genetic engineering has provided powerful methods of obtaining 'hard-to-get' natural products. The principle behind this is to identify the gene or genes that code for the key enzymes involved in the biosynthesis of the desired compound. These are then incorporated into the genomes of fast-growing bacterial or fungal strains. Because microbial cells grow quickly, the rate at which the product is produced will be greater than in the natural source. An additional advantage is that the harvesting and extraction processes are likely to be cheaper, simpler, and more efficient.

For example, the supplies of **artemisinin** depend on growing sufficient crops of *Artemisia annua* (sweet wormwood), which makes worldwide supplies of the drug vulnerable to any natural disasters that might devastate those crops. Therefore, other methods of producing the drug are highly desirable to provide an alternative supply. A total synthesis of artemisinin has been devised, but it is too costly to produce artemisinin in that fashion. A more promising approach is to carry out a semi-synthetic conversion of a biochemical precursor called **artemisinic acid** (section 9.3). Artemisinic acid can be extracted from *Artemisia annua* in greater quantities than artemisinin itself. However, the extraction process is still relatively slow and there would be huge advantages in genetically modifying microbial cells to produce the intermediate more speedily. In order to carry out genetic transformations, an understanding of the biosynthetic pathway leading to artemisinin is required so that the key enzymes in that process can be identified.

Artemisinin is a sesquiterpene lactone endoperoxide. The initial stages of the biosynthesis involve a biosynthetic pathway that is common to many organisms and is known as the **farnesyl pyrophosphate (FPP) pathway**—a pathway which provides the necessary biosynthetic precursors for a whole range of terpenes and sterols (Fig. 9.10). The synthesis of farnesyl diphosphate involves linking together isoprene C5 units (coloured blue in Fig. 9.10). Once farnesyl diphosphate has been formed, specialized enzymes that are unique to *Artemisia annua* catalyse the various cyclizations and modifications that eventually result in artemisinin.

The first of these is an enzyme called **amorphadiene synthase**, which catalyses the cyclization of farnesyl diphosphate to form a bicyclic structure called amorpha-4,11-diene. A cytochrome P450 enzyme (**CYP71AV1/CPR**) then oxidizes a methyl group in

three stages to form artemisinic acid. Further enzymes catalyse the conversion of artemisinic acid to artemisinin.

The FPP pathway is present in both plants and fungal cells, but the products generated from farnesyl diphosphate are different. For example, the yeast *Saccharomyces cerevisiae* uses farnesyl diphosphate as the starting material for the biosynthesis of **squalene** and the sterol structure **ergosterol**, which is an important constituent of yeast cell membranes (Fig. 9.11).

However, the fact remains that the yeast cell can produce farnesyl diphosphate, and so it is conceivable that yeast cells could produce artemisinic acid if they were supplied with amorphadiene synthase and the necessary cytochrome P450 enzyme. Therefore, the plant gene that codes for amorphadiene synthase was inserted into the yeast's DNA such that the enzyme would be produced by the cell.

Further genetic engineering was carried out to modify the control of the FPP pathway in order to increase FPP production, whilst inhibiting the production of fungal sterols. This would guarantee that sufficient quantities of

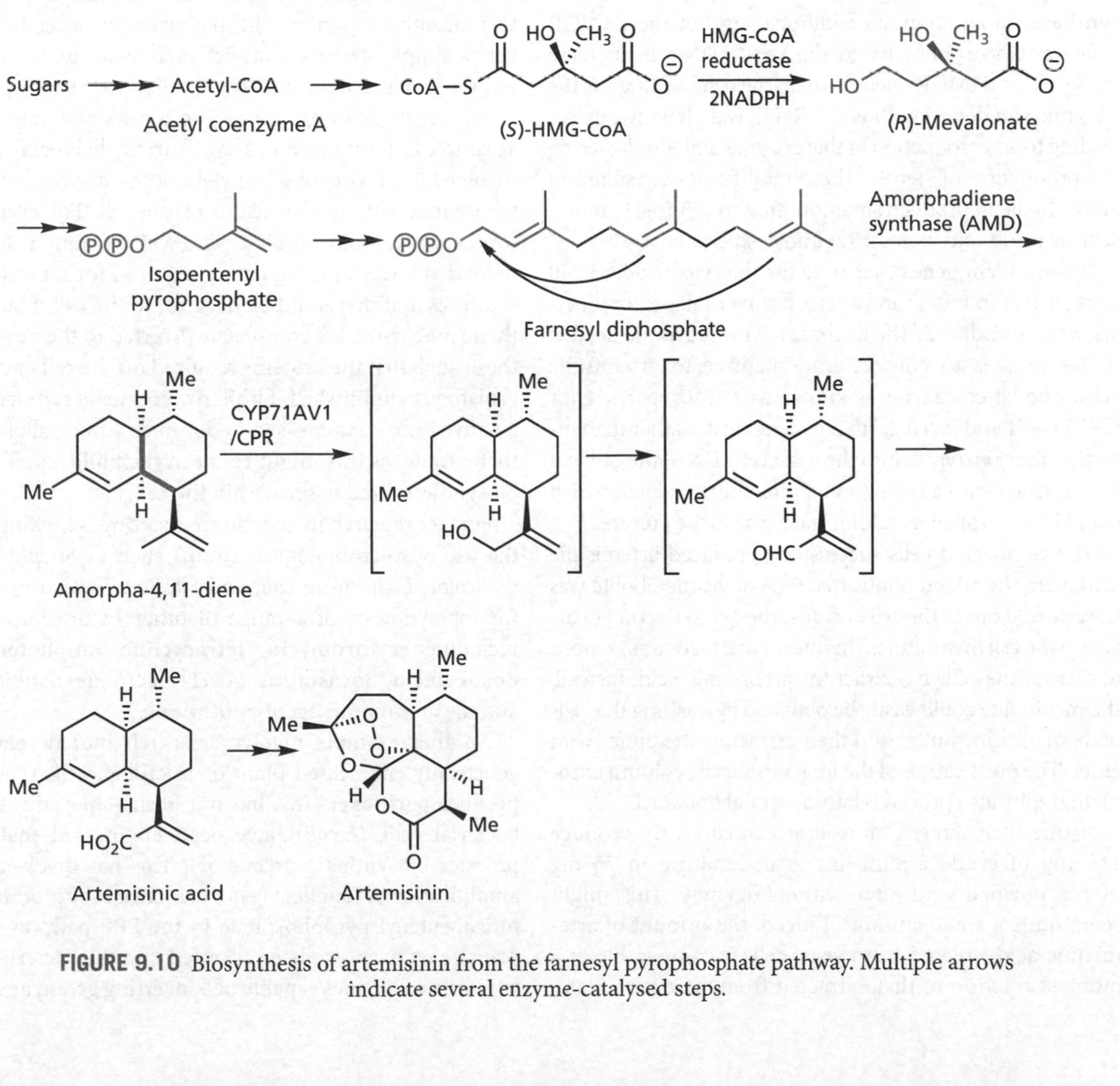

FIGURE 9.10 Biosynthesis of artemisinin from the farnesyl pyrophosphate pathway. Multiple arrows indicate several enzyme-catalysed steps.

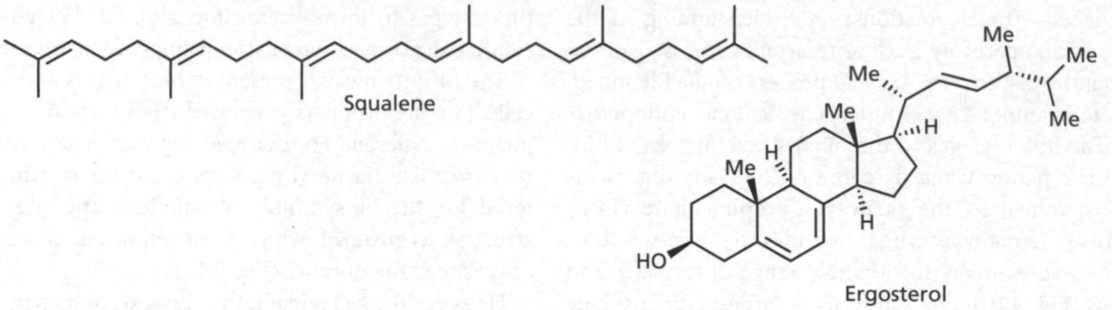

FIGURE 9.11 Squalene and ergosterol.

FPP would be available as starting material for the synthesis of amorpha-4,11-diene. The necessary control was achieved by upregulating the expression of two genes that code for enzymes involved in FPP synthesis, whilst downregulating one of the genes that codes for an enzyme in sterol synthesis. To be specific, upregulation of the genes tHMGR and ERG20 led to increased production of **3-hydroxy-3-methylglutaryl-coenzyme A reductase** (tHMGR) and **FPP synthase**. In addition, an additional copy of the tHMGR gene was incorporated into the yeast's DNA to increase the levels of tHMGR even further. The gene coding for the enzyme **squalene synthase** (ERG9) was downregulated, leading to less production of that enzyme, and thus lowering the production of sterols. These modifications resulted in amorphadiene production amounting to 153 mg/L, representing a 500-fold increase on initial experiments.

Finally, plant genes that code for the cytochrome P450 enzyme (CYP71AV1) and a cytochrome P450 oxidoreductase were introduced. The former enzyme is the one needed if the yeast is to convert amorphadiene to artemisinic acid. The latter enzyme is known as a redox partner for CYP71AV1 and is crucial in regenerating it. Rather than inserting these enzymes into the yeast cell's DNA, the cell was transformed with a vector containing the two genes, which could be controlled by adding galactose to the culture.

The engineered cells successfully produced artemisinic acid, with the added bonus that 95% of the metabolite was transported out of the cell and adsorbed onto the outer surface of the cell membrane. This meant that there was no need to disrupt the cells to extract the artemisinic acid. Instead, the metabolite could easily be obtained by washing the cells with an alkaline buffer and then extracting the buffer with ether. The purification of the final product by column chromatography also proved relatively straightforward.

A one-litre aerated bioreactor can currently produce 115 mg of crude artemisinic acid, resulting in 76 mg of the purified acid after chromatography. This might seem quite a small amount. Indeed, the amount of artemisinic acid isolated from yeast cells is comparable as a biomass fraction to that extracted from the native plant.

However, the process is much faster. In other words, the quantity of artemisinic acid isolated from yeast over four to five days would take several months to obtain if it was extracted from the plant.

In all, about a dozen genes were inserted into the yeast cell in order to modify its metabolism, which is why this project has been described as an example of **synthetic biology**. Studies of this nature are designed to radically alter metabolic pathways in order to convert a simple starting material such as acetyl CoA into a complex natural product that is alien to the host cell.

Introducing foreign genes into a cell's genome is not as simple as it might sound, and a tricky balancing act is involved in maximizing the yield of the desired product whilst maintaining the health of the cell. For example, overexpression of a foreign gene will certainly result in increased levels of the enzyme required for a particular synthesis, but this could be harmful to the cell if biosynthetic precursors are completely diverted to the new synthesis such that the proteins required for the cell's normal growth are diminished. Similarly, problems can occur if a native gene is suppressed to diminish a normal biosynthetic route, as this might result in the build-up of toxic biosynthetic precursors within the cell.

Further research in genetic engineering is looking into the use of microbiological strains such as *Streptomyces coelicolor*, *Escherichia coli*, and *Saccharomyces cerevisiae* for the synthesis of a range of other natural products including **erythromycin, tetracycline, amphotericin, doxorubicin, lovastatin**, novel macrolide antibiotics, and the antitumour agent **epothilone**.

Another example involves research into developing genetically engineered plant or microbial cells that will produce **paclitaxel**. This has not been achieved yet, but bacterial cells (*E.coli*) have been engineered that can produce biosynthetic precursors. This has involved the amplification of four host genes to increase the production of isopentenyl pyrophosphate in the FPP pathway. Two plant genes have also been inserted into the bacterial cell to code for enzymes capable of converting geranylgeranyl

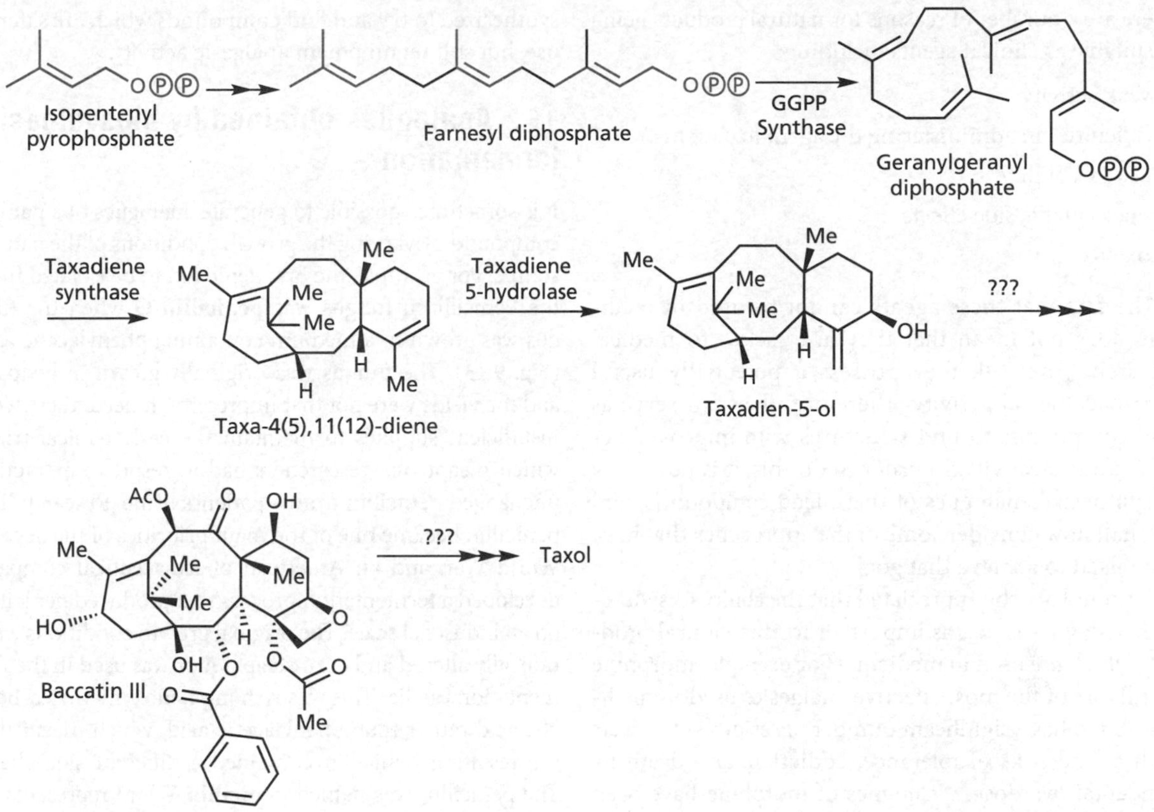

FIGURE 9.12 Production of taxadien-5-ol in genetically modified bacterial cells. Question marks indicate stages that have still to be developed.

diphosphate (a biosynthetic intermediate leading to terpenes) to taxadiene and taxadien-5α-ol (Fig. 9.12).

9.5.2 Modification of normal host cells

Another approach which is designed to increase the yield of a medicinally important natural product is to genetically modify the cells which naturally produce it. This can be achieved by carrying out random mutagenesis studies. The natural cells are subjected to random mutations to see whether any of the mutated cells produce the desired product in increased yields. For example, the antibiotic **erythromycin** is normally extracted from the fungus *Saccharopolyspora erythraea*. These fungal cells were subjected to multiple cycles of random mutations and screening to develop a mutated strain which produced 8 g of erythromycin per litre of culture. This corresponded to a 50–100 times increase in yield compared with the original wild type strain.

More traditional breeding programmes can be carried out on plants with the aid of genetic mapping to identify genes that are important in achieving high yields. For example, breeding programmes have successfully resulted in sweet wormwood plants that can produce over twice as much artemisinin as the original source.

KEY POINTS

- Genetic engineering has been used to modify microbological cells such that they produce natural products of interest.

- The ability to generate natural products using microorganisms allows those products to be obtained more quickly and more reliably than from the natural source.

- The genetic modifications involved aim to maximize the amount of desired product produced without affecting the viability of the microbiological cell.

- Genetic engineering and traditional breeding programmes have resulted in modified strains of a natural source that result in increased yields of natural product.

9.6 Analogues of natural products

9.6.1 Introduction

The natural products that have been described so far are all useful clinical agents, but they are vastly outnumbered by the number of natural products which have a useful pharmacological activity but cannot be used in medicine.

There are a number of reasons for natural products being unsuitable as clinical agents including:

- weak activity
- difficulties in administering the agent to patients
- lack of stability
- unacceptable side effects
- toxicity.

The fact that these agents cannot be used in medicine does not mean that they are useless to medical research. After all, they possess a potentially useful pharmacological activity. Therefore, they can serve as lead compounds to find structures with improved activity and selectivity. In order to do this, it is necessary to synthesize analogues of these lead compounds, and we shall now consider some of the approaches that have been used to achieve that goal.

It should also be appreciated that the ability to synthesize analogues is just as important for the natural products which *are* used in medicine. For example, morphine is still one of the most effective analgesics used in medicine, but it has a significant number of serious side effects such as the risks of tolerance, addiction, and death by accidental overdose. Analogues of morphine have been synthesized to try and find compounds which are safer to use, but still retain potent analgesic activity.

9.6.2 Analogues obtained by biosynthesis/fermentation

It is sometimes possible to generate analogues of a natural compound by varying the growth conditions of the natural source. For example, the first penicillin to be isolated from the Penicillium fungus was **penicillin G** when the fungus was grown in a medium containing phenylacetic acid (Fig. 9.13). The fungus was originally grown in bedpans and the yields were not that impressive. Indeed, there were insufficient supplies to maintain the early clinical trials, which meant that researchers had to resort to extracting unchanged penicillin from a patient's urine. Research into penicillin became one of the main priorities of the Second World War, and an American pharmaceutical company developed a fermentation process that produced penicillin on an industrial scale. The original growth conditions were radically altered and corn steep liquor was used in the fermentation media. This was rich in phenoxymethylcarboxylic acid rather than phenylacetic acid, which meant that the resulting penicillin contained a different side chain. The penicillin was named **penicillin V** and represents an

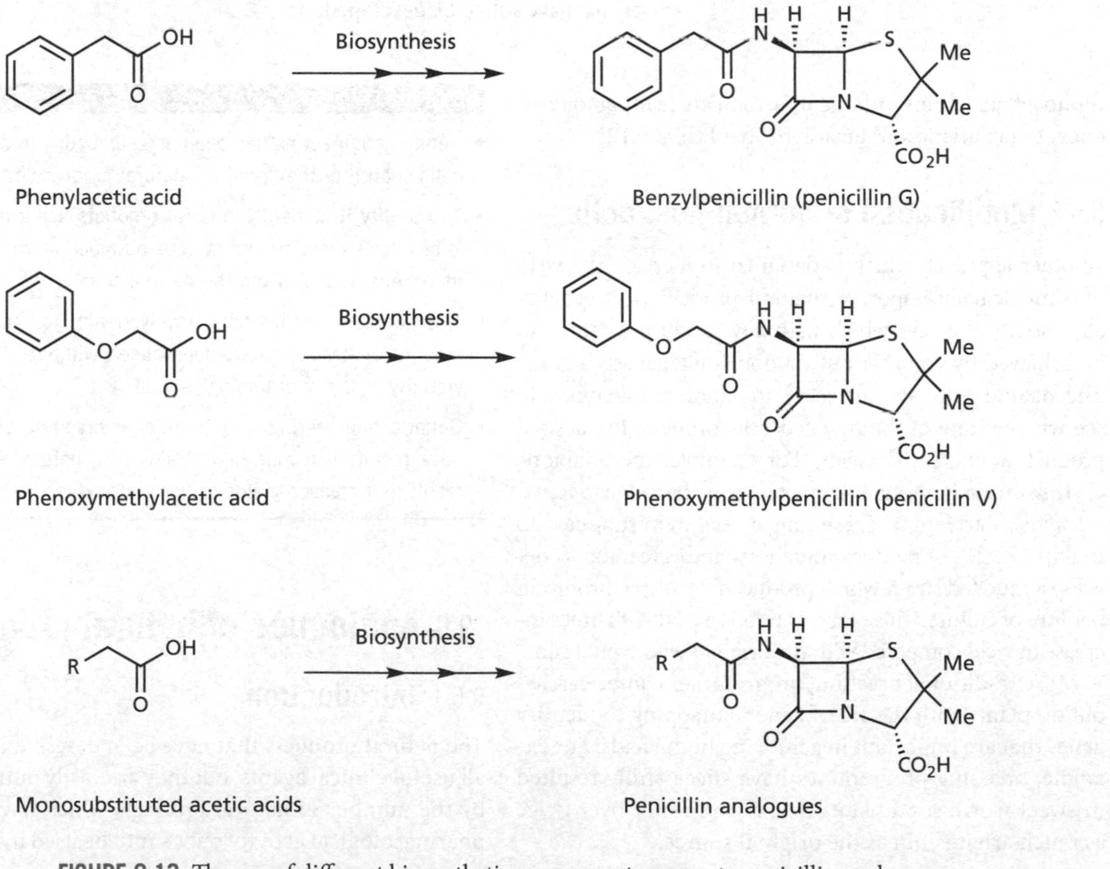

FIGURE 9.13 The use of different biosynthetic precursors to generate penicillin analogues.

early analogue of penicillin G (Fig. 9.13). The advantages of synthesizing analogues were soon demonstrated by the fact that penicillin V was more stable to the acidic conditions of the stomach and could be administered orally, whereas penicillin G had to be administered by injection.

Having identified that it was possible to generate different analogues by varying the fermentation conditions, a variety of other analogues were produced by adding different monosubstituted carboxylic acids to the fermentation medium. This proved successful, but suffered severe limitations. For example, it takes time to grow the fungus, extract it, and then purify the resulting analogue. Therefore, the generation of analogues was relatively slow. There are also limitations on the types of analogue that can be prepared. This is related to the substrate selectivity of the enzymes present in the biosynthesis. For example, penicillin analogues with branching at the α-carbon cannot be produced in this manner. As we shall see later, many of the most important penicillins used in medicine today are branched at that position.

9.6.3 Analogues synthesized from the natural product itself

If it is possible to extract a natural product in good yield, then the simplest method of synthesizing analogues is to make use of the functional groups that are present in the structure. The analogues obtained can thus be described as semi-synthetic analogues, as they have been synthesized from the natural product. Many of the important drugs introduced into medicine in the first half of the twentieth century were generated in this manner. For example, morphine has six functional groups, and so it was possible to carry out functional group transformations on all these groups to generate a large range of analogues, many of which proved to be useful analgesics. The structures shown in Figure 9.14 are examples of morphine analogues which are still used as analgesics today.

Functional groups such as phenols, alcohols, amines, and carboxylic acids are commonly found in natural

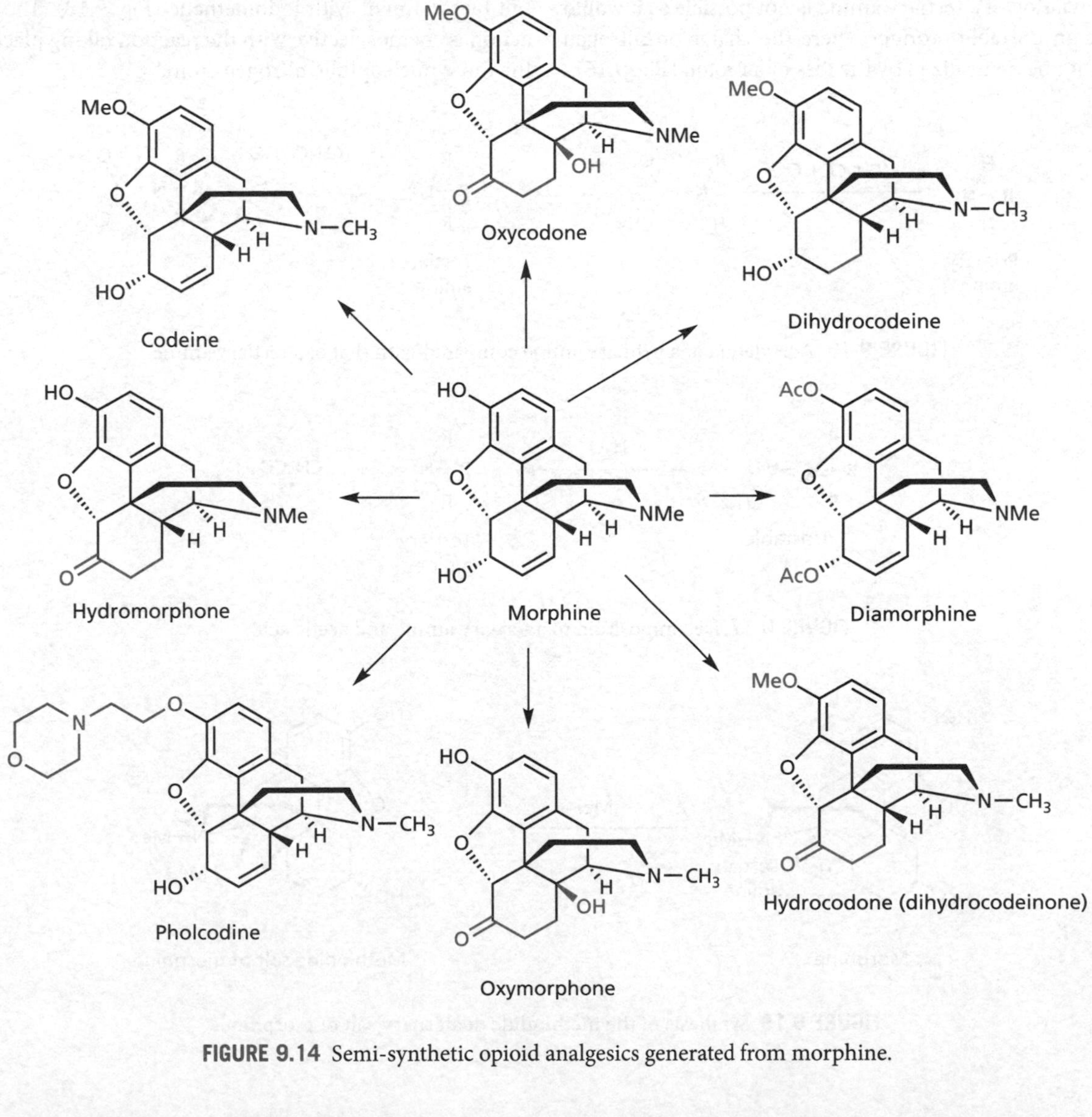

FIGURE 9.14 Semi-synthetic opioid analgesics generated from morphine.

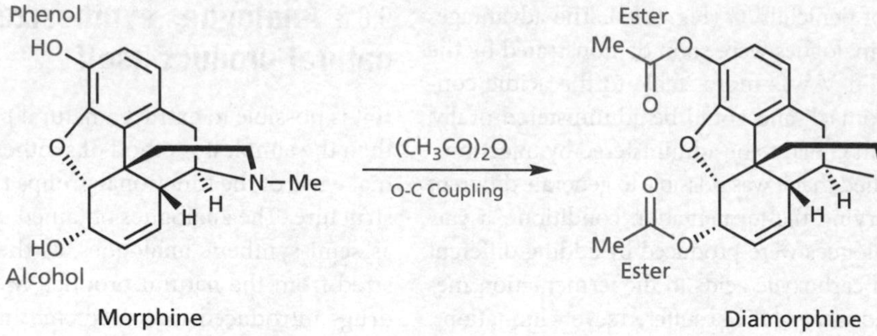

FIGURE 9.15 Synthesis of diamorphine from morphine.

products and can be used to add substituents. For example, morphine is easily converted to diamorphine (heroin) by acetylation of both the alcohol and phenol groups (Fig. 9.15). One might ask why the amine is not acetylated, since an amine is more nucleophilic than either an alcohol or a phenol. The reason lies in the fact that the amine present in morphine is a tertiary amine. Acetylation of a tertiary amine is not possible as it would give an unstable product where the charge on nitrogen cannot be neutralized by the loss of a proton (Fig. 9.16).

The charged product resulting from acetylation of a tertiary amine will rapidly decompose back to the tertiary amine and acetic acid in the presence of water during the work-up procedure (Fig. 9.17).

Although it is not possible to acetylate a tertiary amine, it is possible to alkylate it to form a stable salt. For example, morphine can be converted to its methiodide salt by treatment with iodomethane (Fig. 9.18). The reaction is chemoselective with the reaction taking place at the more nucleophilic nitrogen atom.

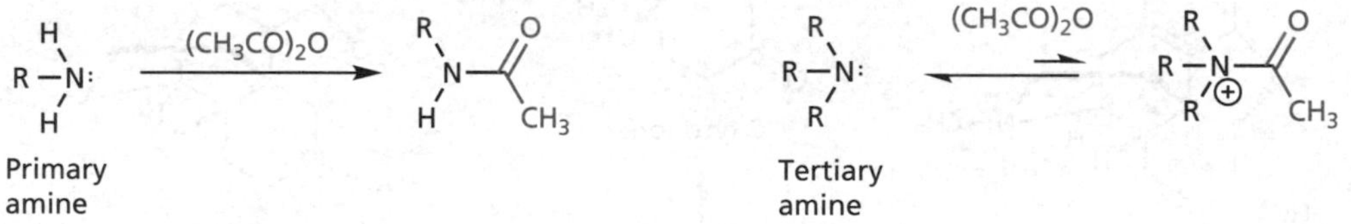

FIGURE 9.16 Acetylation of a primary amine compared with that of a tertiary amine.

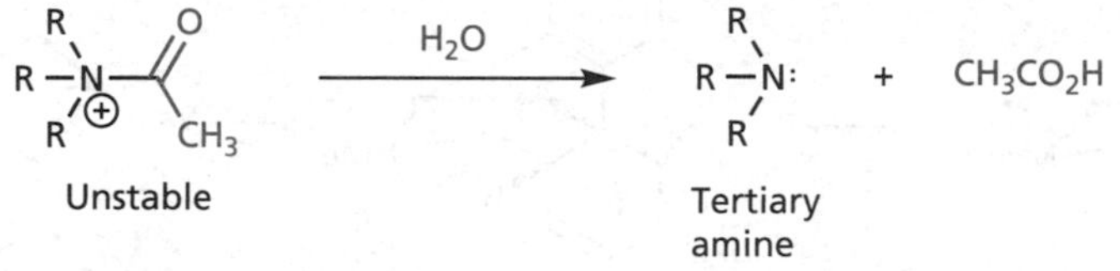

FIGURE 9.17 Decomposition to a tertiary amine and acetic acid.

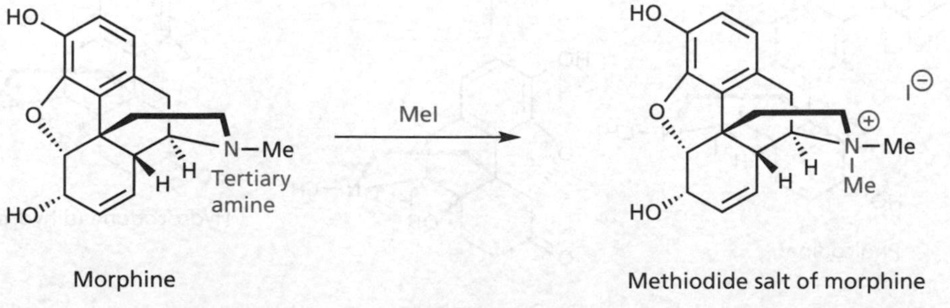

FIGURE 9.18 Synthesis of the methiodide quaternary salt of morphine.

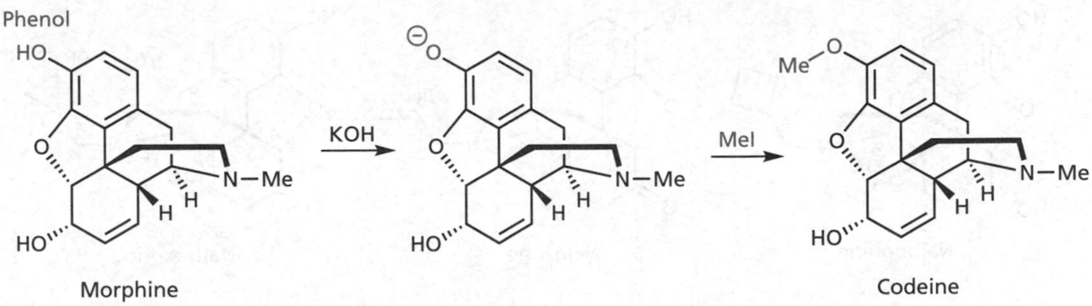

FIGURE 9.19 Synthesis of codeine.

It is possible to synthesize codeine by treating morphine with iodomethane in the presence of potassium hydroxide (Fig. 9.19). How can this occur, when we have already seen that nitrogen is more nucleophilic than oxygen? The key factor here is the presence of KOH, which can remove the slightly acidic proton of a phenol group to form a phenoxide ion. A negatively charged oxygen is now more nucleophilic than nitrogen and so a chemoselective reaction is possible on the phenol oxygen to obtain codeine. It should be noted that a certain amount of methylation still occurs on the nitrogen atom to form the methiodide salt, and this reduces the yield of codeine obtained. Therefore, the reaction is chemoselective rather than chemospecific. Note also that there is no methylation of the alcohol group in morphine. A much stronger base than KOH would be needed to remove this proton.

9.6.4 Analogues synthesized by fragmenting the natural product

With some natural products, it may be possible to cleave or remove a group to produce a simpler analogue. That analogue could then be used as a starting material for the synthesis of other analogues. For example, it is possible to remove the methyl group from the tertiary amine present in **morphine** to give the analogue **normorphine**. Once that has been obtained, a large number of alternative alkyl groups can be added to the nitrogen (Fig. 9.20) (see also section 8.5.2.1). Direct alkylation can be carried out, but this can be difficult to control and over-alkylation may lead to mixtures of products. For that reason, it is often better to acylate the nitrogen to form an amide, as that will only go once. The amide is then reduced to the amine.

In this way, it has been possible to synthesize a number of important opioids, such as those shown in Figure 9.21. **Nalbuphine** is used clinically as an analgesic, whereas **naloxone** and **naltrexone** act as opioid antagonists. In other words, they block the actions of opioid analgesics. Such agents are useful in the treatment of opioid overdose or addiction. They also demonstrate that the synthesis of analogues can result in important differences in pharmacological activity.

If a natural product contains a simple ester or amide substituent, it is often feasible to hydrolyse the functional group. Analogues can then be made by re-forming the ester or amide with a different moiety (Fig 9.22).

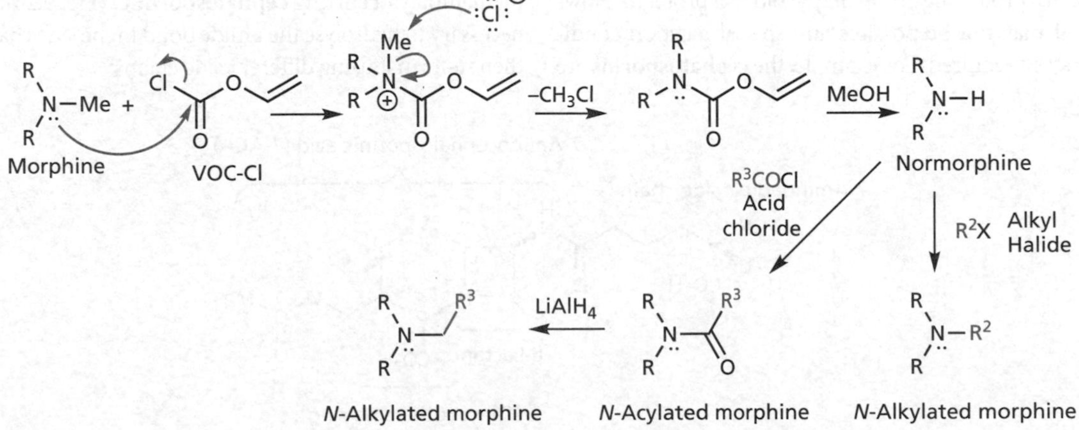

FIGURE 9.20 Demethylation and alkylation of the basic centre.

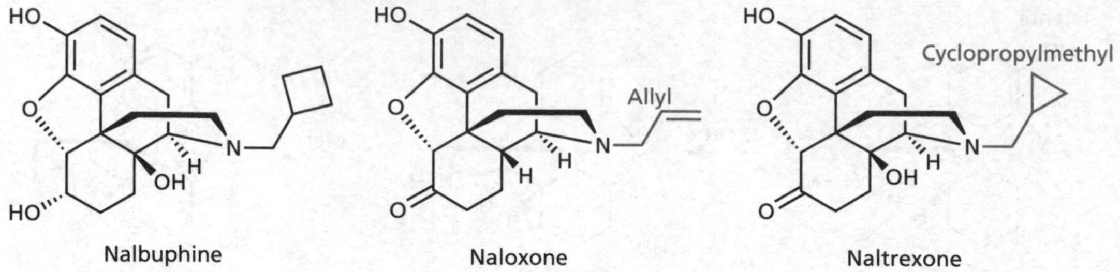

Nalbuphine   Naloxone   Naltrexone

FIGURE 9.21 Clinically important opioids obtained by replacing the *N*-methyl group of morphine.

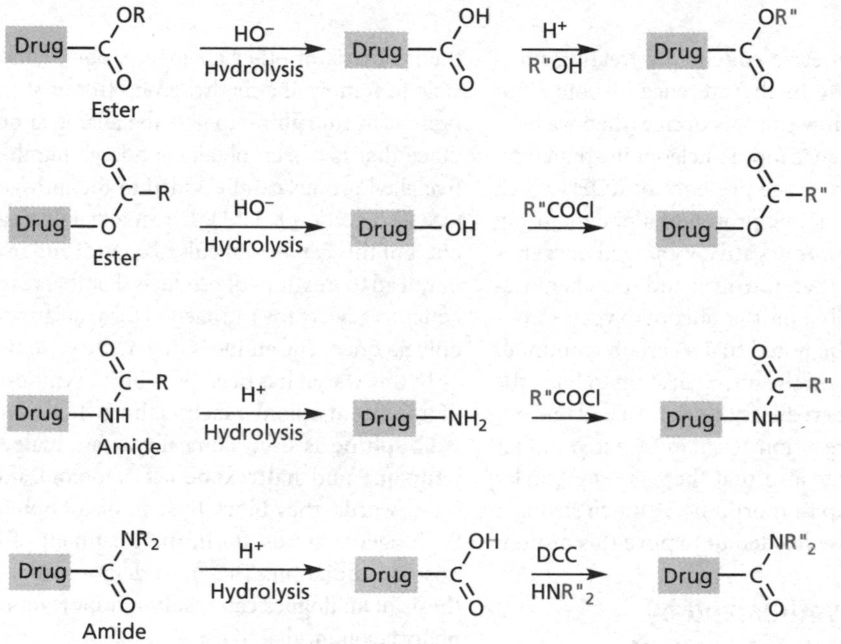

FIGURE 9.22 Synthesis of analogues resulting from the hydrolysis and re-forming of an ester or amide group.

The hydrolysis of an ester or amide by standard procedures may prove problematic if there are other functional groups present that would be susceptible to the reaction conditions. The use of a protecting group may avoid that problem. However, that may not be possible and special reaction conditions may be required. For example, the **cephalosporins** are important antibacterial agents which are structurally related to penicillins in having a β-lactam ring that is crucial to the activity of the compound. In order to produce analogues of the naturally occurring **cephalosporin C** (Fig. 9.23), it was necessary to hydrolyse the amide bond in the side chain and then re-form it using different side chains.

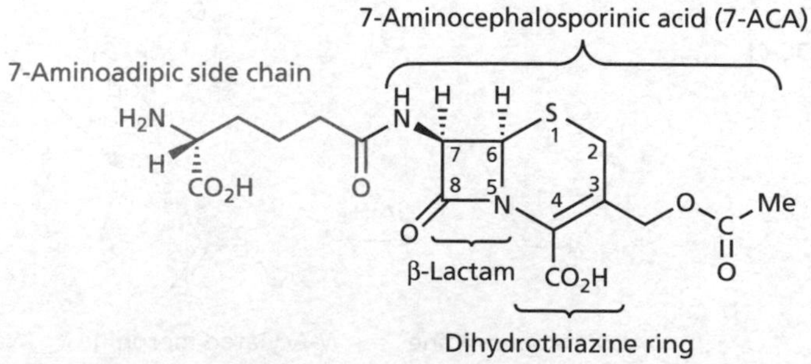

FIGURE 9.23 Cephalosporin C.

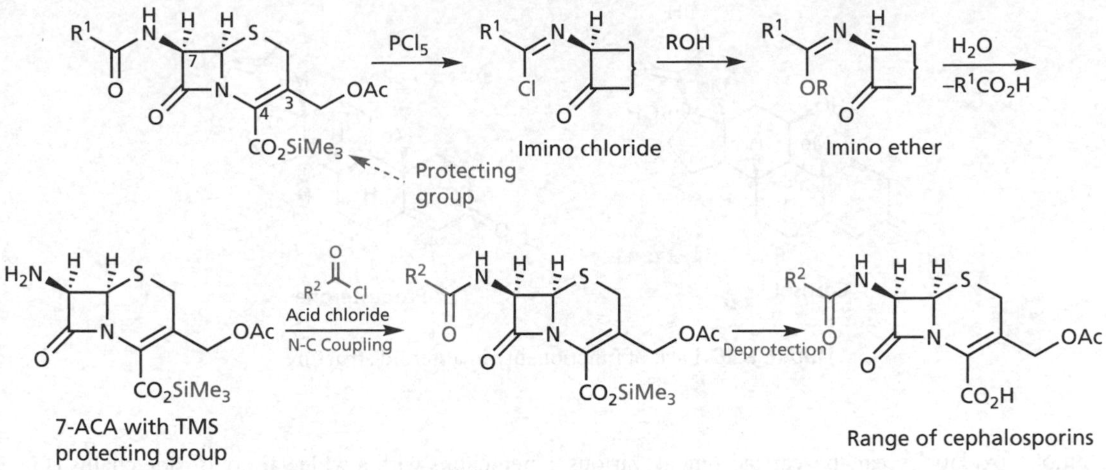

FIGURE 9.24 Synthesis of 7-ACA and cephalosporin analogues.

However, the normal conditions used to hydrolyse a peptide bond would also hydrolyse the β-lactam ring, resulting in an inactive compound. Indeed, owing to ring strain, the β-lactam ring is more likely to react than the secondary amide. Therefore, reaction conditions were required which would activate the amide side chain such that it would be more reactive than the β-lactam ring (Fig. 9.24).

The strategy employed involved reaction with phosphorus pentachloride, and took advantage of the fact that a β-lactam nitrogen is unable to share its lone pair of electrons with the neighbouring carbonyl group. The first step of the mechanism requires the formation of a double bond between the nitrogen on the side chain and its neighbouring carbonyl group (Fig. 9.25). This is only possible for the secondary amide group, since ring constraints prevent the β-lactam nitrogen forming a double bond within the β-lactam ring. A chlorine atom is now introduced to form an imino chloride which can then be treated with an alcohol to give an imino ether. This functional group is now more susceptible to hydrolysis than

the β-lactam ring and so treatment with aqueous acid successfully gives the desired 7-ACA, which can then be acylated to give a range of analogues (Fig. 9.25).

The normal hydrolysis conditions used to cleave esters or amides can also result in the racemization or epimerization of asymmetric centres. In such situations, it may be feasible to use enzymes (esterases or peptidases) in order to cleave the bond. An example of this can be seen in section 11.11.6 where methyl pseudomonate was hydrolysed to pseudomonic acid by esterase enzymes present in yeast cells. Another example is the enzyme-catalysed hydrolysis of the amide side chain in penicillins to produce 6-aminopenicillanic acid (6-APA) (Fig. 9.29).

In some natural products, a lack of functional groups may restrict the number of analogues that can be synthesized and may also limit variation to specific parts of the structure. This was particularly the case with some steroids where the majority of the tetracyclic skeleton was a barren desert as far as functionality is concerned (Fig. 9.26). In such situations, it may be possible to introduce functional groups by taking advantage of enzymes.

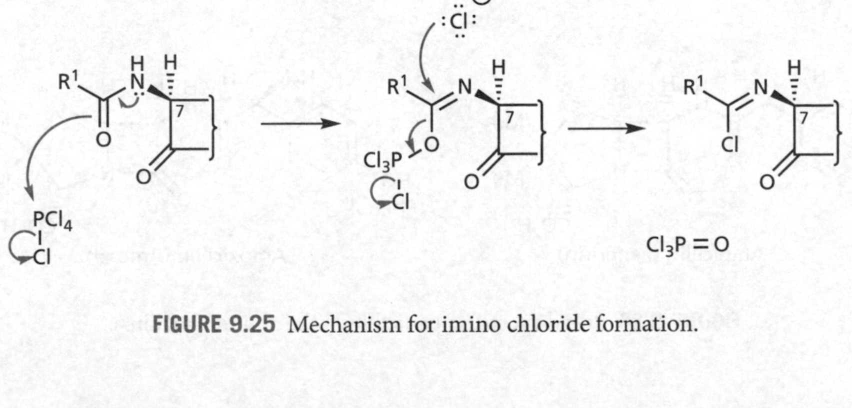

FIGURE 9.25 Mechanism for imino chloride formation.

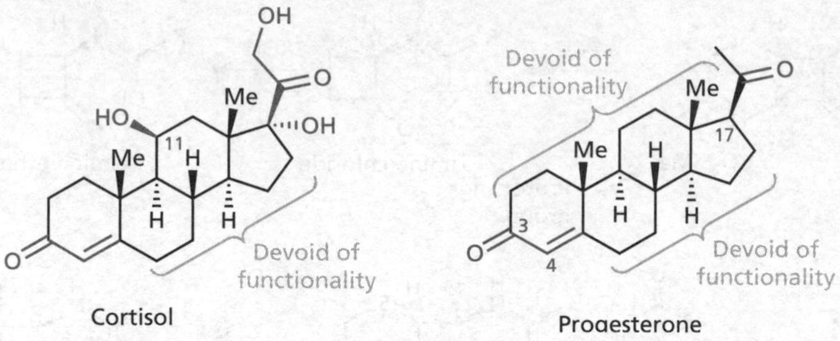

FIGURE 9.26 Lack of functionality in a steroid structure.

For example, oxidations can be carried out at various positions of the steroid skeleton by performing microbiological oxidations.

9.6.5 **Analogues synthesized from biosynthetic intermediates**

If a biosynthetic intermediate can be extracted from its natural source in good yield, it provides a method of generating analogues as well as the natural compound itself (section 9.3). For example, one of the big breakthroughs in the penicillin area occurred in the 1960s when Beecham Pharmaceuticals isolated a biosynthetic intermediate called **6-aminopenicillanic acid** (6-APA) from a fungal culture. This was achieved by using a fermentation medium that lacked the carboxylic acid precursor required for the normal side chain of penicillins. Having obtained this intermediate, it was a simple matter of reacting 6-APA with different acid chlorides to produce

penicillins with a wide variety of side chains (Fig. 9.27). This had huge advantages over the fermentation procedures described in section 9.6.2 as it was easier and quicker to produce the analogues. Moreover, it was also possible to synthesize penicillins that could not be made by fermentation.

For example, the important broad spectrum penicillins **ampicillin** and **amoxicillin** (Fig. 9.28) are orally active compounds that are commonly used as a first line of defence against infection. It is not possible to produce these penicillins by fermentation because of the substrate selectivity of the enzymes involved in biosynthesis, but they can easily be prepared from 6-APA.

6-APA is now produced more efficiently by hydrolysing penicillin G or penicillin V with an enzyme called **penicillin acylase** (Fig. 9.29) or by a chemical method that allows the hydrolysis of the side chain in the presence of the highly strained β-lactam ring (compare Fig. 9.24).

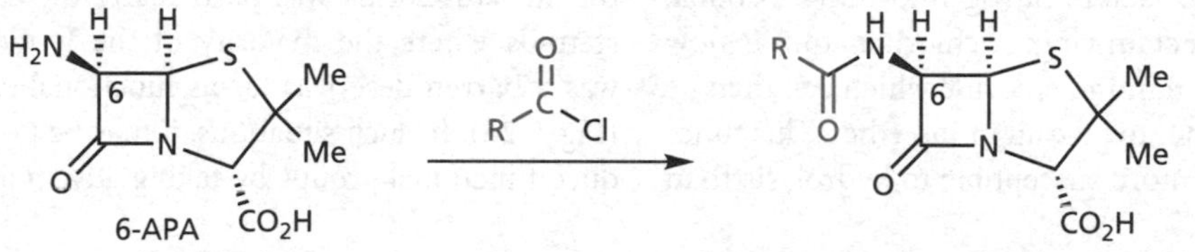

FIGURE 9.27 Penicillin analogues synthesized by acylating 6-APA.

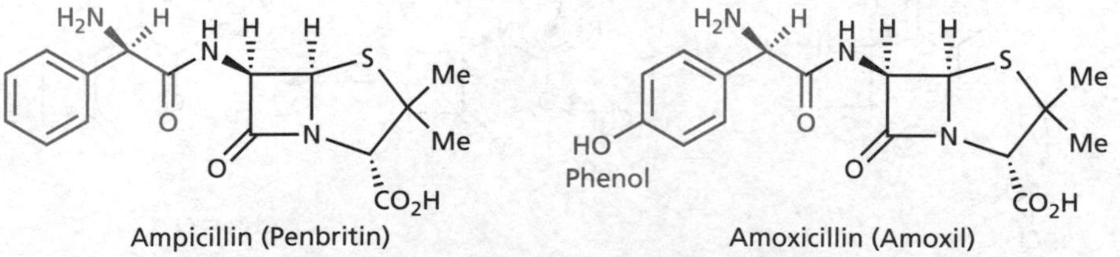

FIGURE 9.28 Broad spectrum penicillins—the aminopenicillins.

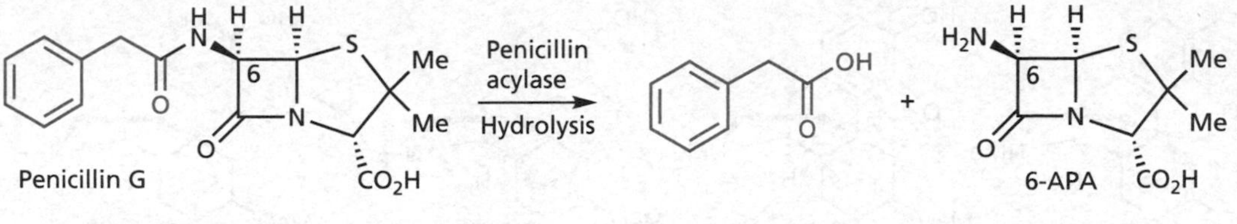

FIGURE 9.29 Synthesis of 6-APA from penicillin G.

9.6.6 Analogues from related natural products

In certain situations, the extraction of a natural source may lead to the isolation of a natural product that has a similar structure to a clinically important agent. These can be studied in their own right as analogues. Alternatively, they provide another method of creating analogues. For example, opium contains the opioid structures **codeine** and **thebaine**, as well as **morphine** (Fig. 9.30).

Both compounds are structural analogues of morphine. Codeine is a useful analgesic in its own right, whereas thebaine is devoid of analgesic activity. It might have been tempting to dismiss thebaine as an uninteresting structure. However, chemists at Reckitt & Colman recognized that it showed potential as a starting material for the synthesis of more complex analogues of morphine

(Fig. 9.31). The basis for this belief lay in the presence of a diene functional group, which is well known to undergo the Diels–Alder reaction with electron-deficient alkenes. Indeed, reaction with methyl vinyl ketone produced a structure called **thevinone**, which could undergo the Grignard reaction to produce a range of potent analgesic structures called the **thevinols** and **orvinols** (see section 4.6.1). The orvinols are particularly potent as they contain a phenol group which is an important part of the analgesic pharmacophore (Fig. 9.32). In the thevinols, this group is masked as a methyl ether.

9.6.7 Analogues from genetically modified cells

We have already described how genetic engineering can be used to synthesize natural products in fast-growing

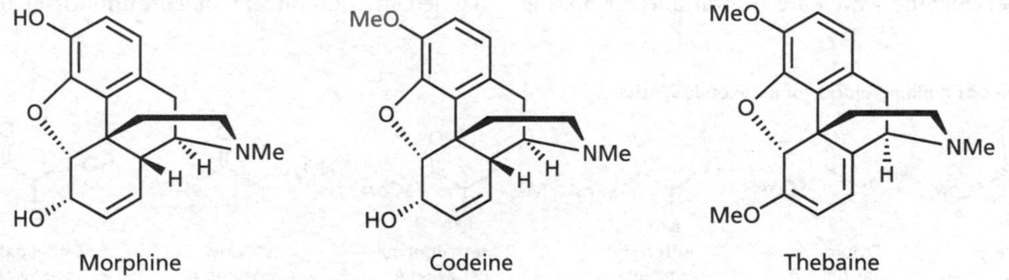

FIGURE 9.30 Opioids extracted from opium.

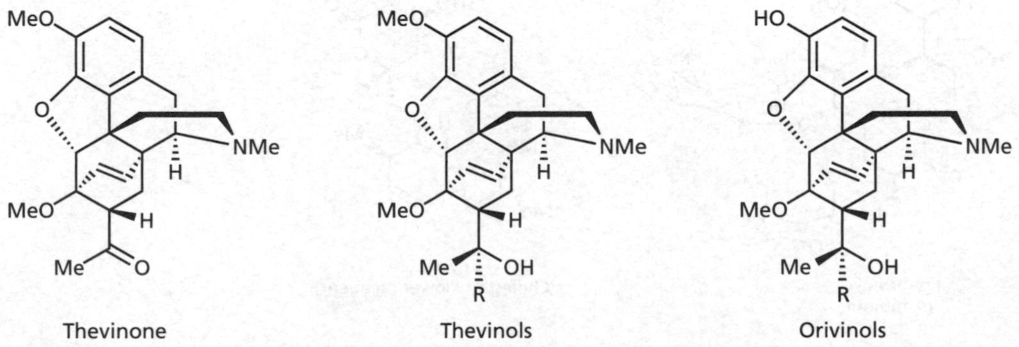

FIGURE 9.31 Thevinone, thevinols, and orvinols.

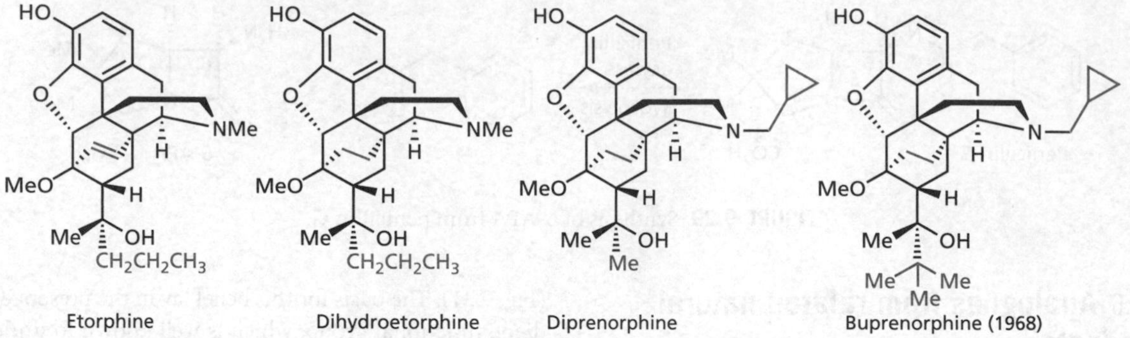

Etorphine   Dihydroetorphine   Diprenorphine   Buprenorphine (1968)

FIGURE 9.32 Important orvinols used in human and veterinary medicine.

cells (section 9.5). Genetically engineered cells have also been used to synthesize analogues of natural products that would be difficult to obtain by synthetic or semi-synthetic means. A particular area of research has been the study of 'super-enzymes' that are responsible for the biosynthesis of clinically important natural products produced in microbial cells. These super-enzymes are extremely large proteins that contain several active sites within their structure and provide an ordered 'assembly line' for linking a series of molecular building blocks together to form a linear structure, which then undergoes one or more cyclizations to form complex ring systems.

The **polyketide synthases** (PKSs) (type I) are one class of super-enzyme which are responsible for linking various acyl coenzyme A building blocks together to form polyketides, which are then released and cyclized to form the biosynthetic intermediates that are the basis for structures such as **erythromycin** and **lovastatin** (Fig. 9.33).

Genetic engineering has been used to modify the various catalytic sites within such super-enzymes. For example, various modifications have been carried out on the polyketide synthase responsible for creating the macrocyclic ring of erythromycin. As a result, these modified synthases have successfully generated erythromycin analogues with different substitution patterns around the ring (see Chapter 13).

This ability to genetically modify a polyketide synthase has led to the concept of **combinatorial biosynthesis**

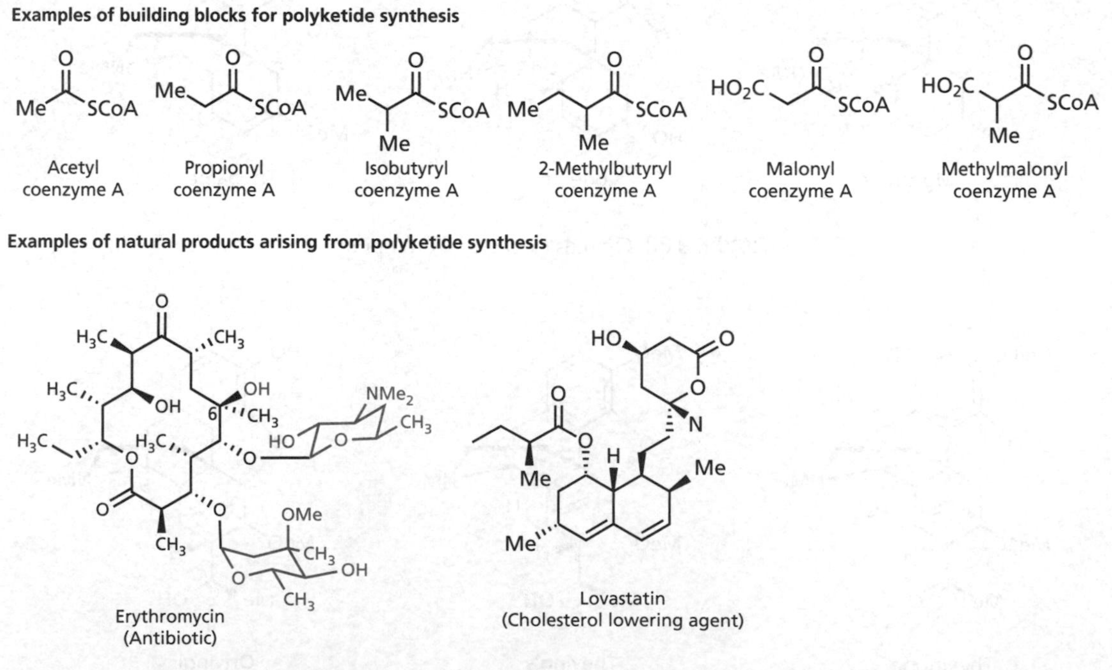

Examples of building blocks for polyketide synthesis

Acetyl coenzyme A   Propionyl coenzyme A   Isobutyryl coenzyme A   2-Methylbutyryl coenzyme A   Malonyl coenzyme A   Methylmalonyl coenzyme A

Examples of natural products arising from polyketide synthesis

Erythromycin (Antibiotic)   Lovastatin (Cholesterol lowering agent)

FIGURE 9.33 Important building blocks for polyketide synthases and examples of natural products arising from such syntheses.

where genetic modifications are carried out to generate a library of polyketide synthases which differ in the number, sequence, and nature of the different catalytic regions within their structure. This, in turn, results in a library of novel analogues which can potentially be generated from fast-growing microbial cells or even under cell-free conditions.

Similar studies have been carried out on another class of super-enzymes called the **non-ribosomal peptide synthases** (NRPSs). These catalyse the synthesis of the peptide precursors used for the biosynthesis of natural products such as penicillin and vancomycin. Like the PKSs, the NRPSs contain several catalytic sites and construct peptides in a similar assembly line process using amino acids as building blocks. Genetically engineering NRPSs opens up the possibility of preparing novel analogues using biosynthesis rather than synthesis.

The biosynthesis of some natural products involves both a PKS and an NRPS. For example, **epothilone C** (Fig. 9.34) is one such structure produced by the myxobacterium *Sorangium cellulosum*. The biosynthesis involves an NRPS which catalyses the synthesis of a heterocyclic ring from malonyl SCoA and cysteine. The resulting structure is then transferred to the PKS which catalyses the synthesis of a polyketide chain. Finally, there is a cyclization reaction which produces the macrocyclic lactone and releases the compound from the PKS enzyme complex (Fig. 9.35).

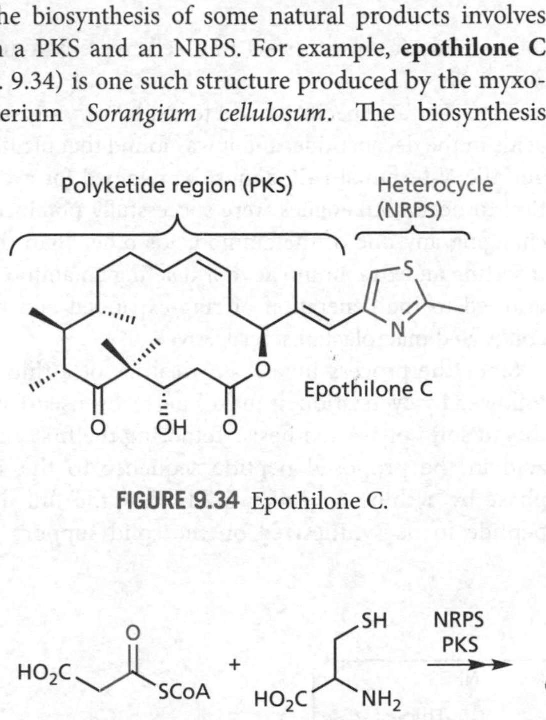

Epothilone C has anticancer activity and is of interest as a lead compound in developing novel anticancer agents. Genetic modifications of the biosynthetic pathway may well be one approach to producing analogues.

9.6.8 Synthesis of analogues with the aid of genetically modified enzymes

Genetic engineering has been used to generate enzymes that can be used to catalyse synthetic procedures in the chemistry laboratory. For example, thioesterase enzymes have been generated by excising the final catalytic site from PKS and NRPS super-enzymes. All the PKSs and NRPSs have this catalytic site since it serves to split the final product from the enzyme. In the case of a PKS, the thioesterase catalytic site uses a serine residue to 'capture' the completed polyketide chain by hydrolysing the thioester group which tethers the polyketide to another part of the PKS complex (Fig. 9.36). As a result of this reaction, the polyketide chain is transferred to the serine residue. The thioesterase then catalyses an intramolecular cyclization, whereby a hydroxyl group within the polyketide structure hydrolyses the acyl link with serine to release the polyketide as a macrocyclic lactone.

A similar process takes place with NRPS enzymes, where the thioesterase captures the final peptide

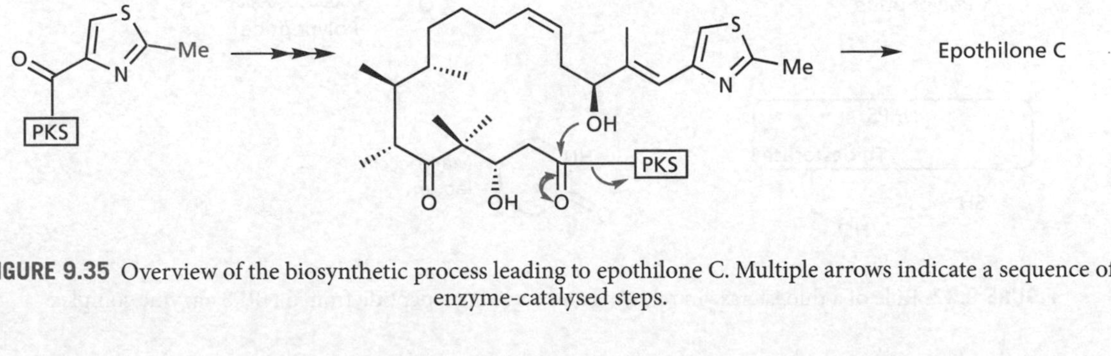

FIGURE 9.34 Epothilone C.

FIGURE 9.35 Overview of the biosynthetic process leading to epothilone C. Multiple arrows indicate a sequence of enzyme-catalysed steps.

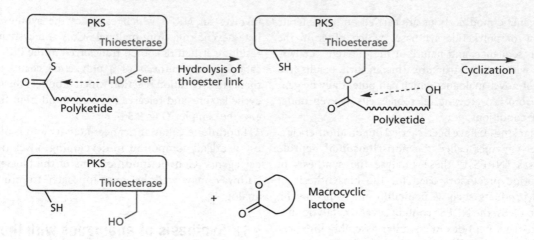

FIGURE 9.36 Role of a thioesterase in releasing and cyclizing a polyketide from a PKS enzyme complex.

structure. The thioesterase then catalyses the cyclization such that an amine group hydrolyses the acyl link to generate a macrocyclic lactam (Fig. 9.37).

These enzyme-catalysed cyclizations would be immensely useful if they could be transferred to cell-free reactions in the chemistry laboratory since macrocyclic cyclizations are quite tricky synthetic operations to carry out.

Therefore, genetic engineering has been used to generate the portion of an NRPS that contains the thioesterase catalytic site as a single isolated protein. The NRPS that was used in this work was **tyrocidine synthetase**, which is responsible for the biosynthesis of a cyclic decapeptide antibiotic called **tyrocidine A** (Fig. 9.38).

The thioesterase enzyme produced from this research was found to catalyse the cyclization of the same linear decapeptide (as a thioester) in a cell-free environment (Fig. 9.39).

The substrate specificity was tested by varying amino acids in the decapeptide, and it was found that ornithine and the N-terminal D-Phe must be retained for cyclization to occur. Analogues were successfully obtained by changing any one of the amino acids other than these. Inserting an extra amino acid or deleting an amino acid also led to the generation of ring-expanded and ring-contracted macrolactams, respectively.

Since the process involves hydrolysis of a thioester followed by cyclization, it makes perfect sense to apply this to solid phase synthesis. Tethering the first amino acid in the proposed peptide sequence to the solid phase by a thioester link would allow the full decapeptide to be synthesized on the solid support. The

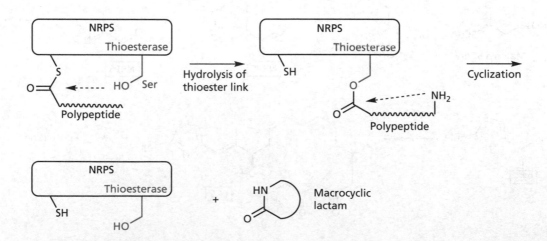

FIGURE 9.37 Role of a thioesterase in releasing and cyclizing a peptide from a NRPS enzyme complex.

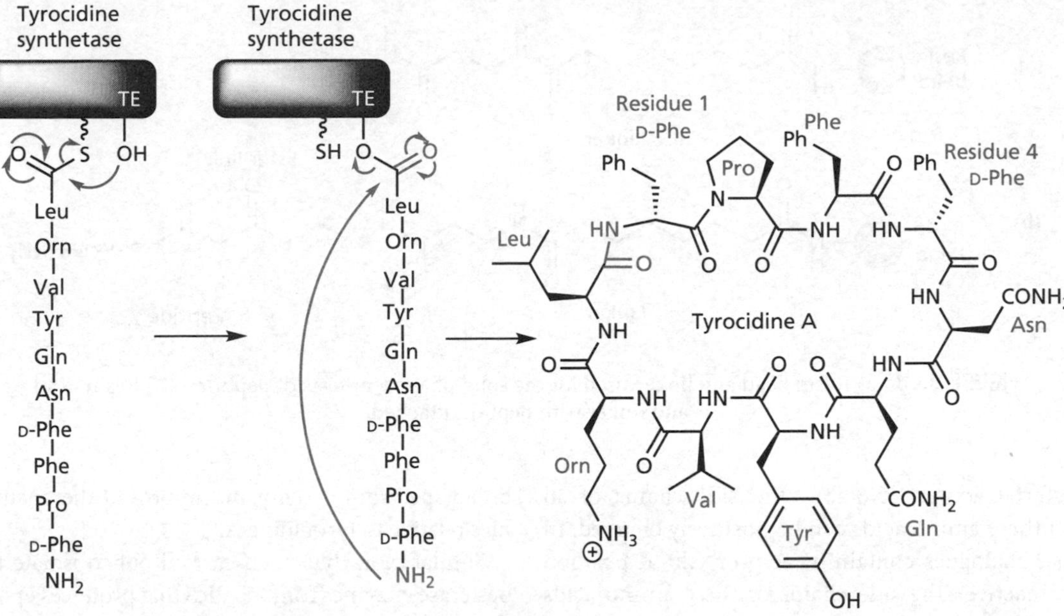

FIGURE 9.38 The final thioesterase-catalysed cyclization of tyrocidine A from tyrocidine synthetase (TE = thioesterase active site).

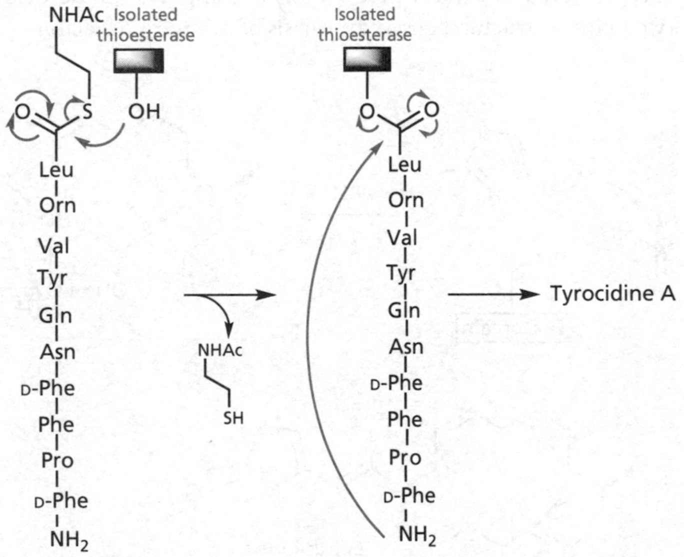

FIGURE 9.39 Thioesterase-catalysed hydrolysis and cyclization in a cell-free environment.

thioesterase enzyme could then be used to release the final peptide by cleaving the thioester link whilst simultaneously catalysing the cyclization. In fact, it has been found that a thioester linkage is not essential and that the thioesterase-catalysed reaction will work with a normal ester.

The solid phase synthesis of peptides was carried out on resin beads carrying a biomimetic linker that mimics the linker used by the NRPS super-enzyme (Fig. 9.40).

Once the peptide synthesis was completed, the resin beads were incubated with the thioesterase enzyme to release the peptide chain from the solid phase and induce cyclization to form tyrocidine A.

The solid phase procedure was successfully used to generate a series of tyrocidine analogues where the D-Phe residue normally present at position 4 was replaced with different amino acids. It was found that the introduction of D-Arg, D-Lys, or D-Orn all increased

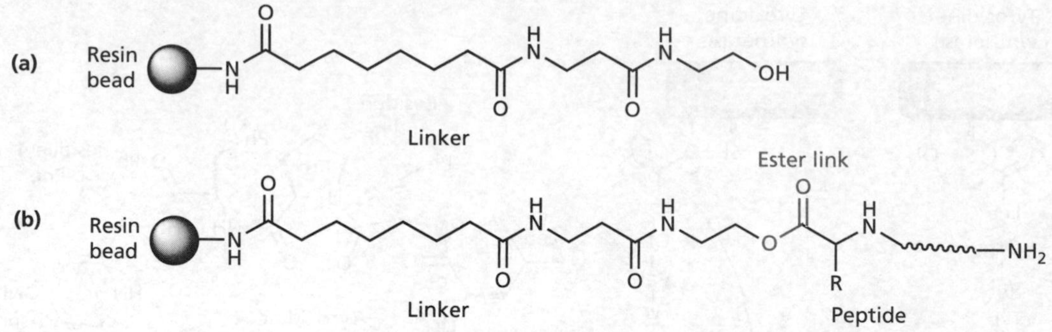

FIGURE 9.40 (a) Resin bead and linker used for the solid phase synthesis of peptides. (b) Resin bead and linker with peptide attached.

antibacterial activity. Notably, the side chains of all three of these amino acids can be positively charged. In contrast, analogues containing Asp or Glu at position 4 were inactive. The side chains of these amino acids contain negatively charged carboxylate groups, thus emphasizing the importance of a side chain containing a positive charge.

Modifications at other positions were also carried out and a total of 300 solid phase tethered substrates were synthesized and cyclized. Two of these structures showed

broad spectrum activity and improved therapeutic profiles relative to tyrocidine A.

Similar work has been carried out to isolate the thioesterase enzyme from the PKS that produces epothilone C (Fig. 9.41).

The isolated enzyme has proved successful in cyclizing the relevant polyketide precursor in cell-free conditions (Fig. 9.42). The use of this enzyme opens up the possibility of using solid phase synthetic methods for the synthesis of macrocyclic lactones.

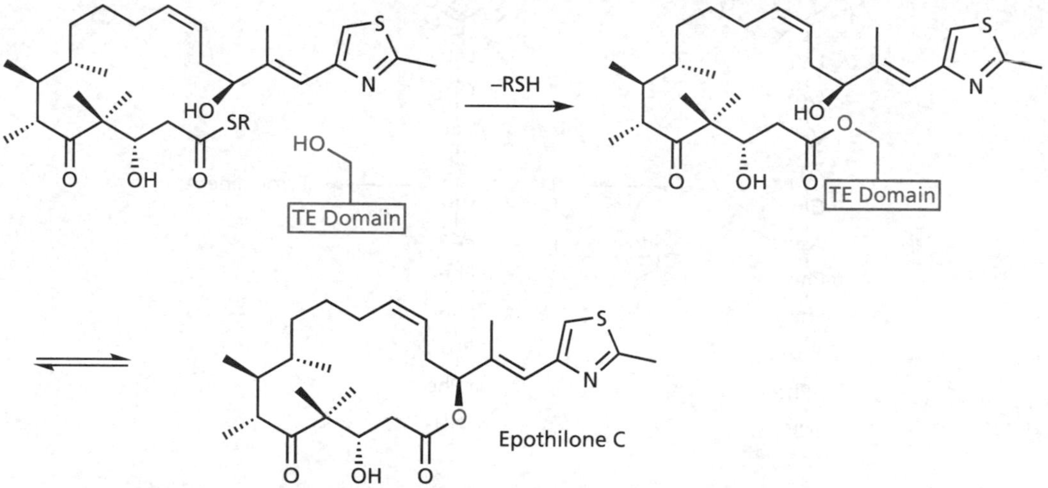

FIGURE 9.41 Release and cyclization of epothilone C.

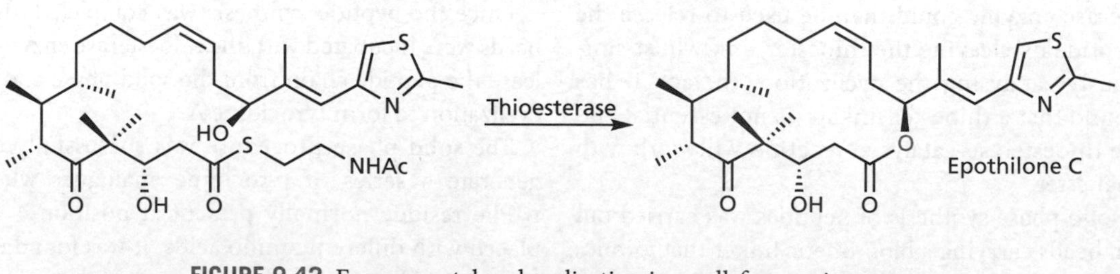

FIGURE 9.42 Enzyme-catalysed cyclization in a cell-free environment.

FURTHER READING

General reading

Cane, D.E., Walsh C.T., and Khosla, C. (1998) 'Harnessing the biosynthetic code: combinations, permutations and mutations', *Science*, **282**, 63–8.

Khosla, C. and Keasling, J.D. (2003) 'Metabolic engineering for drug discovery and development', *Nature Reviews Drug Discovery*, **2**, 1019–25.

Mootz, H.D., Schwarzer D., and Marahiel, M.A. (2002) 'Ways of assembling complex natural products on modular nonribosomal peptide synthetases', *ChemBioChem*, **3**, 490–504.

Pfeifer, B.A. and Khosla, C. (2001) 'Biosynthesis of polyketides in heterologous hosts', *Microbiology and Molecular Biology Reviews*, **65**, 106–18.

Wu, N., et al. (2001) 'Assessing the balance between protein–protein interactions and enzyme–substrate interactions in the channeling of intermediates between polyketide synthase modules', *Journal of the American Chemical Society*, **123**, 6465–74.

Specific compounds

Ajikumar, P.K., et al. (2010) 'Isoprenoid pathway optimization for Taxol precursor overproduction in *Escherichia coli*', *Science*, **330**, 70–74 (taxanes).

Boddy, C.N., et al. (2003) 'Epothilone C macrolactonisation and hydrolysis are catalyzed by the isolated thioesterase domain of epothilone polyketide synthase', *Journal of the American Chemical Society*, **125**, 3428–9 (epothilone C).

Davies, J. (2010) 'Cultivating the seeds of hope', *Chemistry World*, June, pp 51–53 (artemisinin).

Frense, D. (2007) 'Taxanes: perspectives for biotechnological production', *Applied Microbiology and Biotechnology*, **73**, 1233–40 (taxanes).

Holton, R.A., Biediger, R.J., and Boatman, P.D. (1995) 'Semisynthesis of Taxol and Taxotere', in M. Suffness (ed.), *Taxol: science and applications*. CRC Press, London, Chapter 7, pp 97–119 (paclitaxel).

Kohli, R.M., Walsh, C.T., and Burkart, M.D. (2002) 'Biomimetic synthesis and optimization of cyclic peptide antibiotics', *Nature*, **418**, 658–61 (tyrocidin A analogues).

Levesque, F. and Seeberger, P.H. (2012) 'Continuous-flow synthesis of the anti-malaria drug artemisinin', *Angewandte Chemie, International Edition*, **51**, 1706–9 (artemisinin).

McDaniel, R., et al. (1999) 'Multiple genetic modifications of the erythromycin polyketide synthase to produce a library of novel "unnatural" natural products', *Proceedings of the National Academy of Sciences of the USA*, **96**, 1846–51 (erythromycin analogues).

Reiter, M., et al. (2010) 'The organocatalytic three-step total synthesis of (+)-frondosin B', *Chemical Science*, **1**, 37–42 (frondosin B).

Ro, D.-K., et al. (2006) 'Production of the antimalarial drug precursor artemisinic acid in engineered yeast', *Nature*, **440**, 940–3 (artemisinic acid).

Roth, R.J. and Acton, N. (1991) 'A facile semisynthesis of the antimalarial drug qinghaosu', *Journal of Chemical Education*, **68**, 612–13 (artemisinin).

Trauger, J.W., et al. (2000) 'Peptide cyclization catalysed by the thioesterase domain of tyrocidine synthetase', *Nature*, **407**, 215–18 (tyrocidine A).

Xue, Q., et al. (1999) 'A multiplasmid approach to preparing large libraries of polyketides', *Proceedings of the National Academy of Sciences of the USA*, **96**, 11 740–5 (erythromycin analogues).

Zhang, Y., et al. (2008) 'The molecular cloning of artemisinic aldehyde $\Delta11(13)$ reductase and its role in glandular trichome-dependent biosynthesis of artemisinin in *Artemisia annua*', *Journal of Biological Chemistry*, **283**, 21 501–8 (artemisinin).

10 Chemical and process development

10.1 Introduction

10.1.1 Chemical development

Once a compound has been identified as a drug candidate and has progressed to preclinical trials, it is necessary to start the development of a large-scale synthesis as soon as possible (section 1.10). This is known as chemical development and is carried out in specialist laboratories. To begin with, a quantity of the drug may be obtained by scaling up the synthetic route used by the research laboratories. In the longer term, however, such routes often prove unsuitable for large-scale manufacture. There are several reasons for this. During the drug discovery/design phase, the emphasis is on producing as many different compounds as possible in the shortest period of time. The yield is unimportant as long as sufficient material is obtained for testing. The reactions are also done on a small scale, which means that the cost is trivial even if expensive reagents or starting materials are used. Hazardous reagents, solvents, or starting materials can also be used because of the small quantities involved.

The priorities in chemical development are quite different. A synthetic route has to be devised which is straightforward, safe, cheap, efficient, high yielding, has the minimum number of synthetic steps, and will provide a consistently high quality product which meets predetermined specifications of purity.

During chemical development, the conditions for each reaction in the synthetic route are closely studied and modified in order to obtain the best yields and purity. Different solvents, reagents, and catalysts may be tried. The effects of temperature, pressure, reaction time, excess reagent or reactant, concentration, and method of addition are studied. Consideration is also given to the priorities required for scale-up. For example, the original synthesis of **aspirin** from salicylic acid involved acetylation with acetyl chloride (Fig. 10.1). Unfortunately, the by-product of this is hydrochloric acid, which is corrosive and environmentally hazardous. A better synthesis involves acetic anhydride as the acylating agent. The by-product formed here is acetic acid, which is safer than hydrochloric acid and can also be recycled.

Therefore, the final reaction conditions for each stage of the synthesis may be radically different from the original conditions, and it may even be necessary to abandon the original synthesis and devise a completely different route (Box 10.1).

Once the reaction conditions for each stage have been optimized, the process needs to be scaled up. The priorities here are cost, safety, purity, and yield. Expensive or hazardous solvents or chemicals should be avoided and replaced by cheaper and safer alternatives. Experimental procedures may have to be modified. Several operations carried out on a research scale are impractical on large scale. These include the use of drying agents, rotary evaporators, and separating funnels. Alternative large-scale procedures for

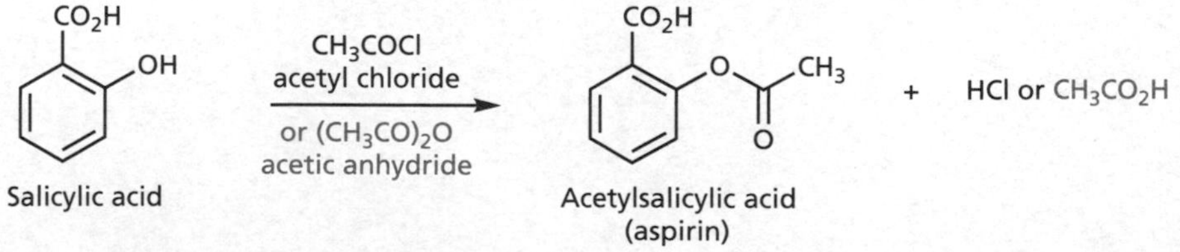

FIGURE 10.1 Synthesis of aspirin.

BOX 10.1 Synthesis of ebalzotan

Ebalzotan is an antidepressant drug produced by Astra which works as a selective serotonin (5-HT$_{1A}$) antagonist. The original synthesis from structure I involved six steps and included several expensive and hazardous reagents, resulting in a paltry overall yield of 3.7% (Fig. 1). Development of the route involved the replacement of 'problem' reagents and optimization of the reaction conditions, leading to an increase in the overall yield to 15%. Thus, the expensive and potentially toxic reducing agent sodium cyanoborohydride was replaced with hydrogen gas over a palladium catalyst. In the demethylation step, BBr$_3$ is corrosive, toxic, and expensive, and so it was replaced with HBr which is cheaper and less toxic.

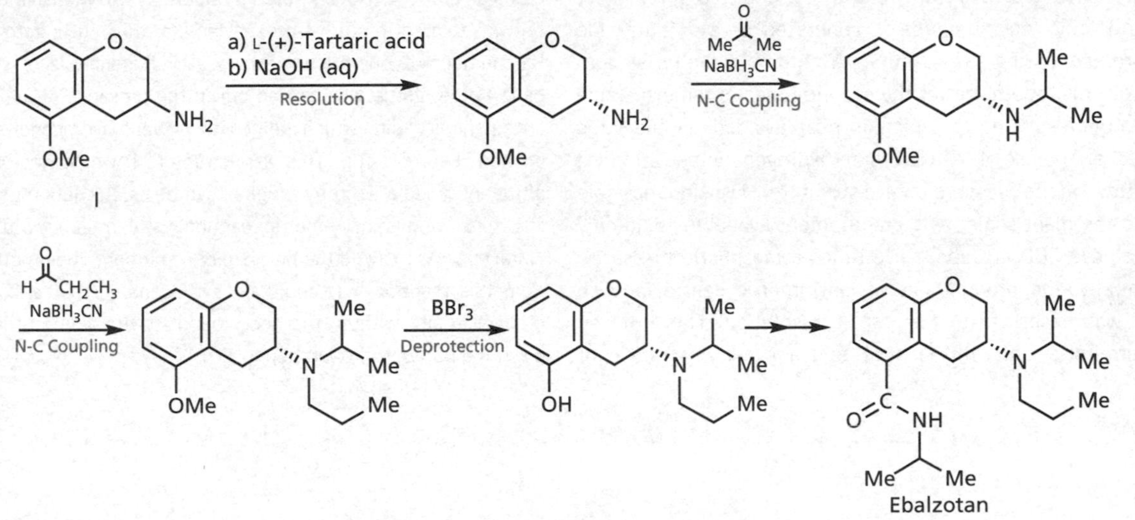

FIGURE 1 Synthesis of ebalzotan.

these operations are, respectively, removing water as an azeotrope, distillation, and stirring the different phases.

There are several stages in chemical development. In the first stage, about a kilogram of drug is required for short-term toxicology and stability tests, analytical research, and pharmaceutical development. Often, the original synthetic route will be developed quickly and scaled up in order to produce this quantity of material, as time is of the essence. The next stage is to produce about 10 kg of the drug for long-term toxicology tests, as well as for formulation studies. Some of the material may also be used for phase I clinical trials. The third stage involves a further scale-up to the pilot plant, where about 100 kg is prepared for phase II and phase III clinical trials.

Because of the time scales involved, the chemical process used to synthesize the drug during stage 1 may differ markedly from that used in stage 3. However, it is important that the quality and purity of the drug remain as constant as possible for all the studies carried out. Therefore, an early priority in chemical development is to optimize the final step of the synthesis and to develop a purification procedure which will consistently give a high quality product. The **specifications** of the final product are defined and determine the various analytical tests and purity standards required. These define predetermined limits for a range of properties such as melting point, colour of solution, particle size, polymorphism, and pH. The product's chemical and stereochemical purities must also be defined, and the presence of any impurities or solvent should be identified and quantified if they are present at a level greater than 1%. Acceptable limits for different compounds are proportional to their toxicity. For example, the specifications for ethanol, methanol, mercury, sodium, and lead are 2%, 0.05%, 1 ppm, 300 ppm, and 2 ppm respectively. Carcinogenic compounds such as benzene or chloroform should be completely absent, which in practice means that they must not be used as solvents or reagents in the final stages of the synthesis.

All future batches of the drug must meet these specifications. Once the final stages have been optimized, future development work can look to optimizing or altering the earlier stages of the synthesis (Box 10.2).

As a result they are known as **promoters**. One common use for promoters is to remove impurities from commercial solvents and reagents. In the research lab, solvents are routinely distilled before use, but this would be too costly to carry out at production level. Therefore, promoters can be useful in removing any troublesome impurities that might be present.

As an example, RedAl is used as a promoter in the cyclopropanation of alkenes with zinc (Fig. 10.11a). It serves to remove any zinc oxides from the surface of the zinc, as well as remove any trace moisture or peroxides present in the solvent. Chemicals such as 1,2-dibromoethane, magnesium bromide, and methylmagnesium bromide have been used to promote Grignard reactions. It is thought that they may act by removing a magnesium hydroxide coating from the magnesium metal used in the reaction.

10.7 Methods of increasing the yield of an equilibrium reaction

In an equilibrium reaction, the yield can be increased if one of the reactants is added in excess. This will push the equilibrium towards products. However, this

is only feasible if the reactant used in excess is cheap, readily available, and easily removed from the reaction product.

Equilibrium reactions can also be shifted towards products if one of the products is removed from the reaction by precipitation, distillation, or crystallization. A well-known example of this is the formation of ketals, where the reaction can be carried out under distillation conditions to remove the water produced (Fig. 10.22).

10.8 Troublesome intermediates

An intermediate in a synthetic route may prove problematic for a variety of reasons; for example, it might be toxic which is clearly a safety issue, or it might be unstable resulting in poor yields. For example, the synthesis of the antipsychotic agent **quetiapine** involved an imino chloride intermediate which proved unstable and was easily hydrolysed (Fig. 10.23).

The purpose of the imino chloride was to allow a substitution reaction to take place which would introduce a piperazine ring into the structure. To avoid the problem

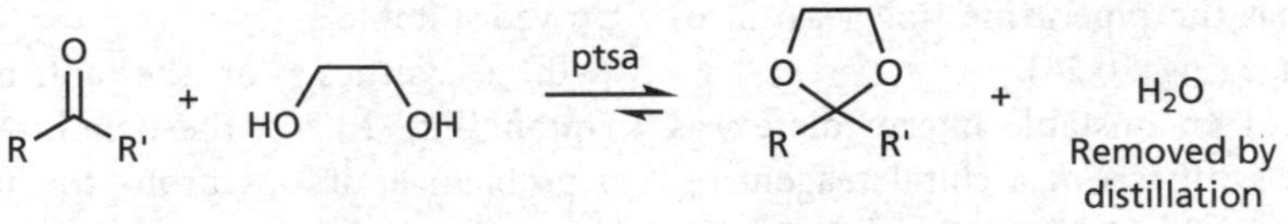

FIGURE 10.22 Removal of water to encourage ketalization.

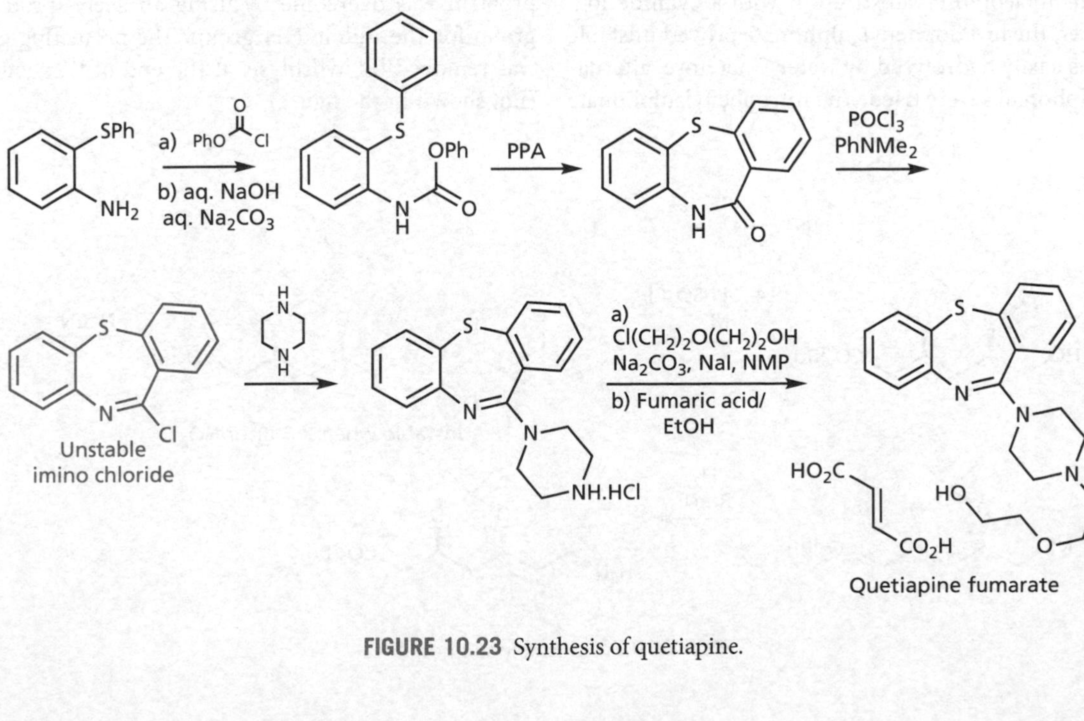

FIGURE 10.23 Synthesis of quetiapine.

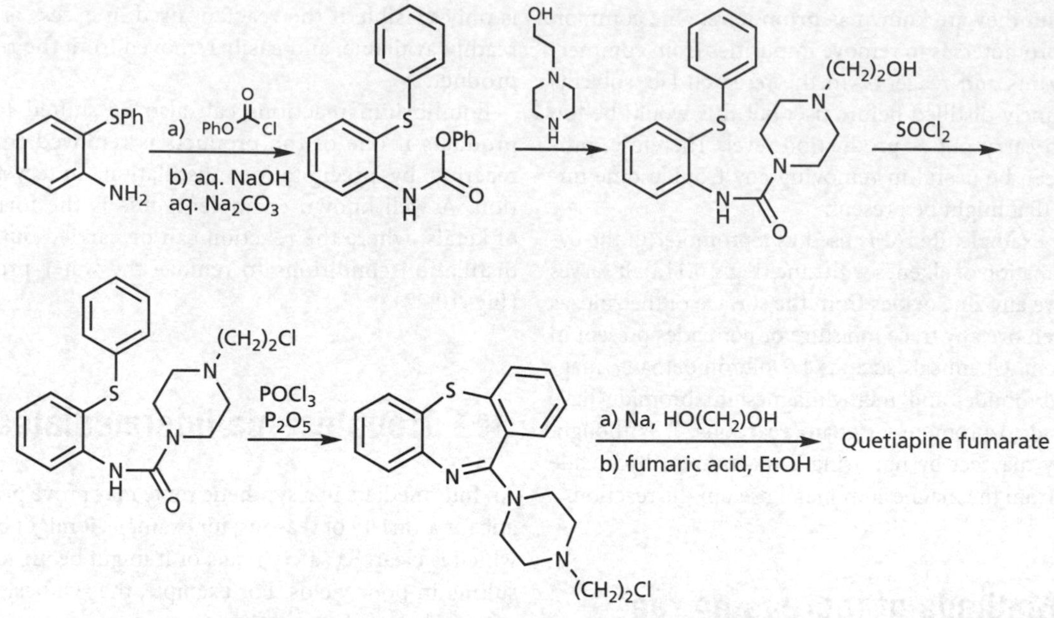

FIGURE 10.24 Alternative synthesis of quetiapine.

of the unstable imino chloride, the order of the reactions was changed such that the piperazine ring was introduced at an earlier stage (Fig. 10.24).

Another example of an unstable intermediate was observed during the synthesis of a chiral reagent required for the synthesis of **atorvastatin** (Fig. 10.25). A bromophenylsulphonate (X = Br) was synthesized such that it could act as a good leaving group for a subsequent nucleophilic substitution with a cyanide ion. However, the bromophenylsulphonate proved unstable and was easily hydrolysed by water. Therefore, alternative sulphonates were tried. The nitrophenylsulphonate

was also unstable, but the chlorophenylsulphonate proved suitable.

In a synthesis of the anti-migraine agent **eletriptan** (Fig. 10.26), the penultimate structure proved a problem as it was prone to dimerization. This was due to the indole nitrogen in one molecule adding to the vinyl sulphone of another. Unsurprisingly, this resulted in reduced yields and purification problems. The problem was overcome by using an acetyl protecting group for the indole NH group. The protecting group was removed by hydrolysis at the end of the synthesis (not shown in the figure).

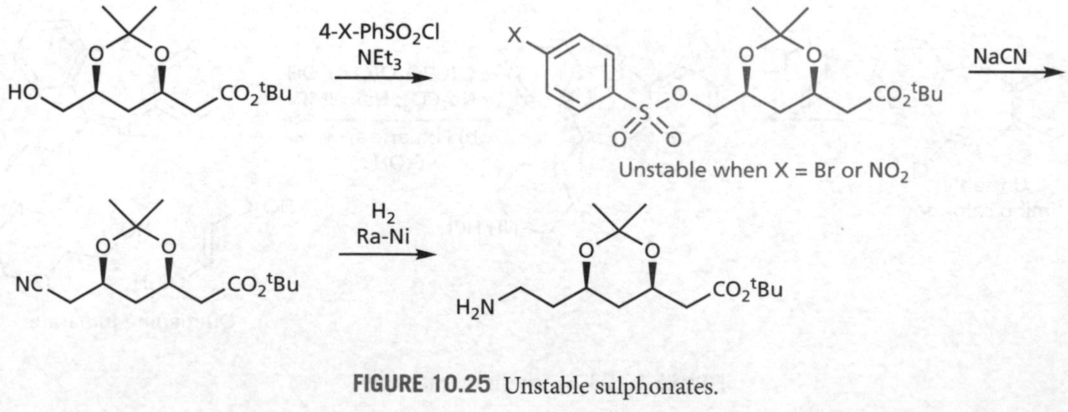

FIGURE 10.25 Unstable sulphonates.

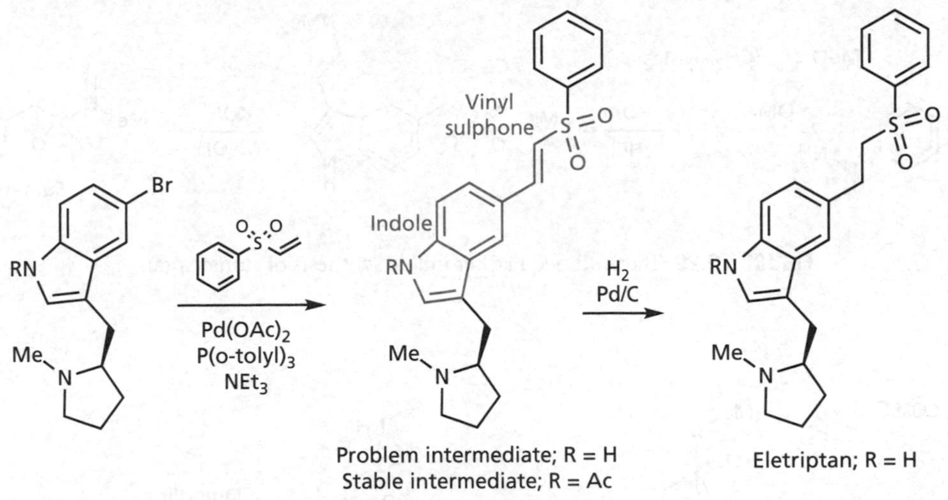

FIGURE 10.26 Synthesis of eletriptan.

10.9 Avoiding impurities

There is always the likelihood of impurities being present in a reaction product. In some cases, these may be due to side reactions involving either the starting material or the product. For example, a Fischer indole synthesis was used to synthesize the anti-migraine agent **sumatriptan** (Fig. 10.27). However, the yield was low due to the formation of an impurity resulting from a reaction occurring between two molecules of the product (Fig. 10.28a). This reaction involved nucleophilic substitution of the sulphonamide group in one of the structures.

To tackle this problem, the reagent was modified to include a protecting group on the sulphonamide group (Fig. 10.29), which prevented the formation of the first impurity shown in Figure 10.28a.

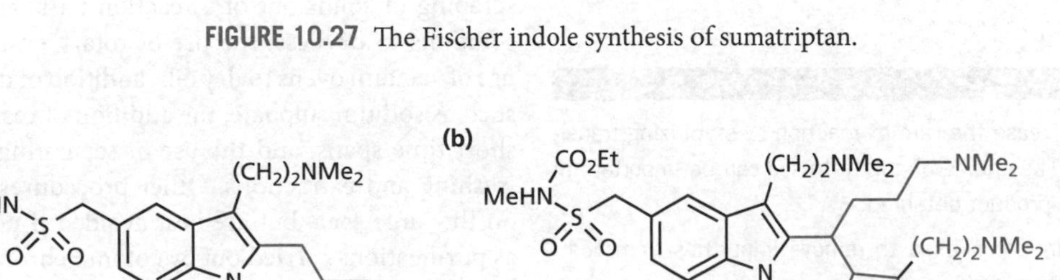

FIGURE 10.27 The Fischer indole synthesis of sumatriptan.

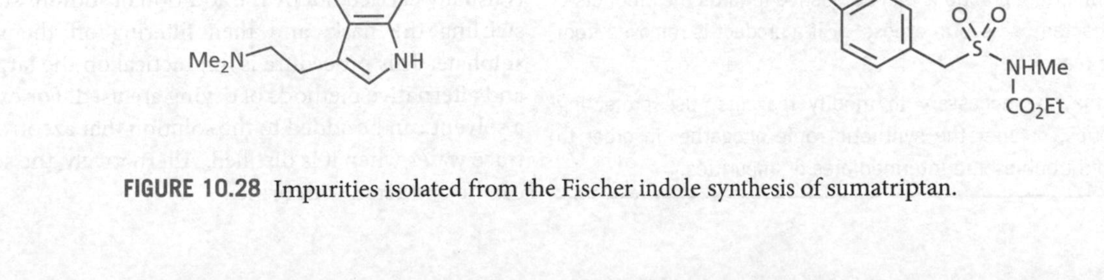

FIGURE 10.28 Impurities isolated from the Fischer indole synthesis of sumatriptan.

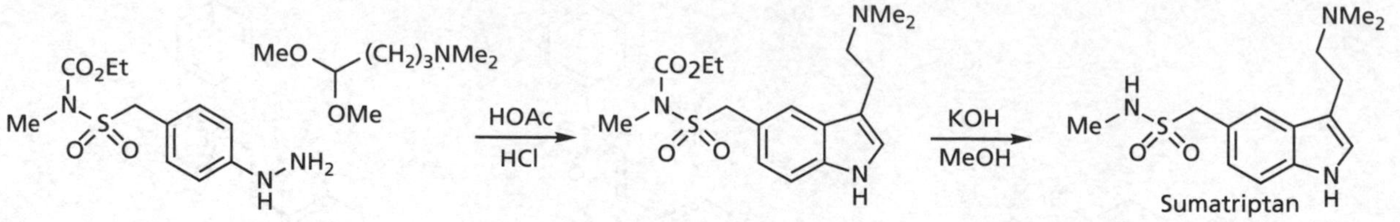

FIGURE 10.29 The modified Fischer indole synthesis of sumatriptan.

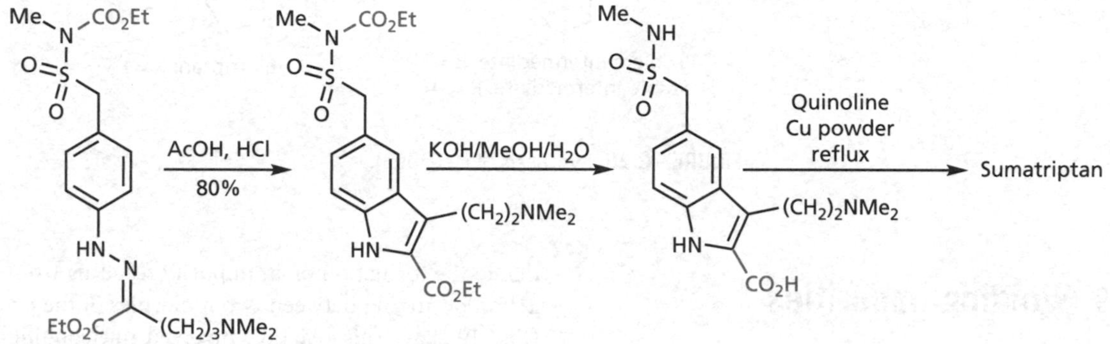

FIGURE 10.30 Modified reagent used in the Fischer indole synthesis leading to sumatriptan.

However, a new impurity was formed, as shown in Figure 10.28b. This was prevented by synthesizing a hydrazone which contained a second ester group on the hydrazone side chain (Fig. 10.30). This ester group acted as a deactivating group in the final indole product and the desired product was obtained in 80% yield without any evidence of the impurity. The protecting and deactivating groups were subsequently removed in high yield.

KEY POINTS

- Catalysts increase the rate of reaction by stabilizing transition states. The choice of catalyst used can be important in defining the product obtained.

- Promoters are often used to remove impurities from commercial solvents and reagents.

- Equilibrium reactions can be pushed towards the products if a reactant is used in excess, or if a product is removed from the reaction.

- It may be necessary to modify reagents, use protecting groups, or alter the synthetic route altogether in order to avoid troublesome intermediates or impurities.

10.10 Experimental and operational procedures on the large scale

10.10.1 Experimental procedures

Several common experimental procedures are totally impractical in the production plant. These include the scraping of solids out of a reaction flask, concentrating solutions to dryness, the use of rotary evaporators, the use of vacuum ovens to dry oils, addition of drying agents such as sodium sulphate, the addition of reagents within short time spans, and the use of separating funnels for washing and extractions. Other procedures are feasible on the large scale but are best avoided if possible, such as purifications carried out by column chromatography.

Drying an organic solution in the research laboratory is usually carried out by the addition of sodium sulphate, swirling the flask, and then filtering off the sodium sulphate. This procedure is impractical on the large scale and alternative methods of drying are used. For example, a solvent can be added to the solution that azeotropes off trace water when it is distilled. Alternatively, the solution can be extracted with brine.

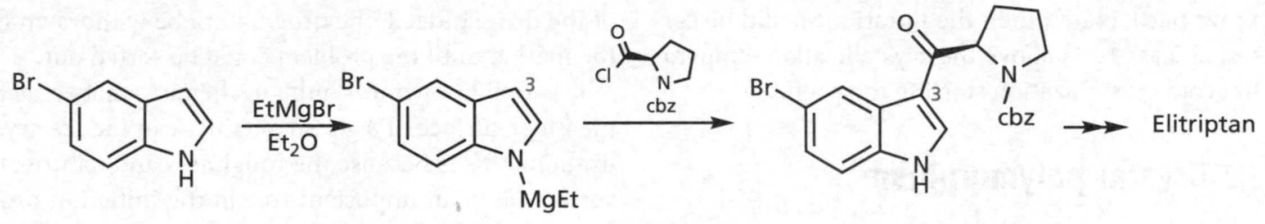

FIGURE 10.31 Acylation of an indole ring system at position 3.

Removing a solvent in order to concentrate a reaction solution can only be done by normal distillation procedures as a large-scale equivalent of a rotary evaporator does not exist. Washing and extracting solutions has to be carried out by stirring the two phases in large reaction vessels. Clearly, the process is less efficient than shaking a separating funnel, and so the stirring may have to be carried out for much longer. Another approach is to use **countercurrent extraction** where the two phases are in contact with each other, but flowing in opposite directions. Purifications are best carried out by crystallization, preferably at the final stage.

10.10.2 **Physical parameters**

A number of physical parameters play an increasingly important role as reactions are scaled up to production level. These include the stirring efficiency of the system, the surface-to-volume ratio of the reactor vessel, the rate of heat transfer from the outside of the reaction vessel to the body of the solution held within it, and the temperature gradient between the centre of the reactor and the walls.

Reactions can behave differently when they are carried out using large-scale reaction vessels, and reactions may have to be modified to take account of this. For example, **eletriptan** is a conformationally restrained analogue of the anti-migraine agent **sumatriptan**. Part of the synthesis involved the use of a Grignard reagent to form the magnesium salt of an indole structure, which can then be acylated at position 3 of the indole ring. However, when this was carried out on a large scale it resulted in a disappointing yield of only 50% (Fig. 10.31). The reaction conditions were changed such that the Grignard reagent and the acid chloride were added simultaneously to opposite sides of the reaction vessel over 2–3 hours at a controlled temperature. A much better yield of 82% was obtained on a multi-kilogram scale.

10.10.3 **The number of operations in a process**

One of the priorities at the production level is to decrease the number of operations in a process to a minimum. For example, an ideal process would be one where there is

no need to isolate or purify the intermediates involved in the synthesis until the crystallization of the final product. Ideally, after each reaction, the reaction solution would be transferred to the next reaction vessel for the next synthetic stage. The efficiency of this process is further enhanced if the same solvent can be used for a series of reactions.

10.10.4 **Clean technology**

An important aspect of process development is to minimize the impact of the production process on the environment. Therefore, there is a clear benefit in developing processes that use non-toxic reagents and solvents as much as possible.

Clean technology refers to experimental procedures involving techniques such as electrochemistry, photochemistry, ultrasound, and microwaves. This also helps with respect to environmental issues (see also section 10.15).

10.10.5 **Minimizing costs**

To minimize costs, it is important to minimize the cost of raw materials and to maximize the overall yield from each batch. Therefore, it is advisable to produce large batches on each run as this also minimizes the cost of labour and overheads.

10.11 **Crystallization**

10.11.1 **Introduction**

Crystallization is the preferred method of purification on large scale, especially if it only needs to be carried out on the final product. The crystallization conditions need to be carefully controlled to ensure consistent purity, crystal form, and crystal size. This involves careful monitoring of the rates of cooling and stirring. If the rate of cooling is too fast, it may result in very fine crystals which clog up filters. If the cooling rate is too slow, it may lead to large crystals which trap solvent within the crystal form. If a hot filtration is required prior to crystallization in order to remove particulates, then the filtration should be carried out at least 15°C above the crystallization temperature to avoid crystallization starting too soon.

10.11.2 Crystal polymorphism

It is possible to obtain different crystal forms (polymorphs) of a final product. For example, progesterone has five known crystal forms, while aspirin has two. Part of a drug's specifications is that it must be produced as a particular polymorph. The reason for this is that polymorphs differ in stability and properties. For example, the polymorphs of **aspirin** are known as polymorphs I and II. The latter is a metastable polymorph which is transformed to the more stable polymorph I under mechanical stresses such as milling and tableting. Both crystal forms consist of layers of aspirin molecules, and there is a subtle difference in the relative positions of the molecules in one layer relative to the molecules in adjacent layers. The transformation from polymorph II to polymorph I results from a shearing action where the layers 'slip' sideways with respect to each other. This affects the interactions between the molecules in different layers and can have important consequences on properties such as solubility. This, in turn, can affect the drug's activity and effectiveness. Therefore, it is important to identify the most stable polymorph and ensure that it is produced consistently in the crystallization process.

The polymorph obtained from crystallization depends on conditions such as the solvent used, rate of cooling, etc. The types of impurity that are present can also play a role. For example, a particular crystal form of **progesterone** can no longer be made, and studies suggest that the presence of certain types of impurity might hold the key to this mystery. The most stable crystal form of progesterone has been shown to have 11 impurities present, although these have not been identified.

It is possible that the crystal form produced may change despite careful controls. For example, the antiviral agent **ritonavir** was marketed as a particular polymorph but, over time, a different crystal form evolved which proved less soluble and reduced the effectiveness of the drug. Indeed, the drug had to be withdrawn from the market until the problem could be sorted out.

It is well known to synthetic chemists that scratching the inner surface of a glass container can induce crystallization. This is because the roughness and nature of the surface plays an important role in the initiation process for crystallization. The nature of the container surface can also affect the type of polymorph that crystallizes, and research is being carried out to design surfaces with nanopores or indentations of different shapes and sizes. One study showed that a polymer film with hemispherical nanopores 15–120 nm in diameter hindered aspirin crystallization, whereas square/box-shaped nanopores promoted it. Further studies may lead to methods that will ensure more efficient crystallization of a desired polymorph.

10.11.3 Examples of crystal polymorphs

Choosing the crystal form with the optimum pharmaceutical properties is important. For example, it was found that different crystal forms of the antipsychotic agent **risperidone** were obtained depending on the conditions used in the final synthetic step (Fig. 10.32).

The original patented synthesis of risperidone involved a final stage using DMF as solvent and DMF/isopropanol as a mixed crystallization solvent. However, DMF is difficult to remove from the final product, and so the reaction was carried out in isopropanol or acetonitrile instead, followed by crystallization using isopropanol or acetone. Crystal form A was obtained under these conditions.

A different crystal form (crystal form B) was obtained if the product was crystallized using chloroform/cyclohexane. The same crystal form was obtained if the product was dissolved in aqueous HCl and then precipitated with aqueous sodium carbonate. A third form of crystal (form E) was obtained by dissolving the product in isopropanol and precipitating with water.

Crystals can trap solvent within their crystal form, and the amount of trapped solvent can vary. For example, the antipsychotic agent **olanzapine** can be crystallized as the monohydrate or dihydrate, depending on the work-up

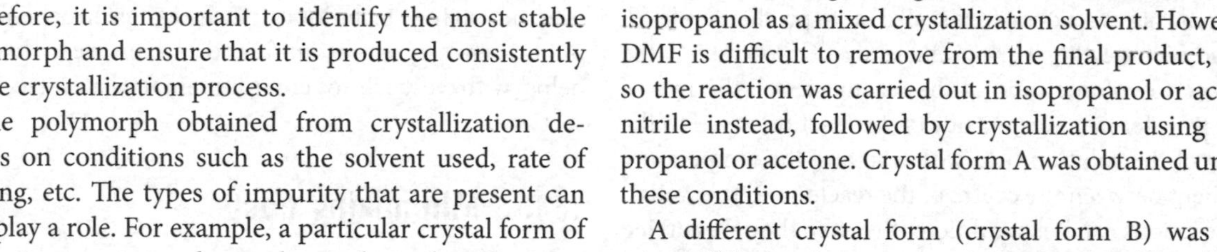

FIGURE 10.32 Synthesis of risperidone.

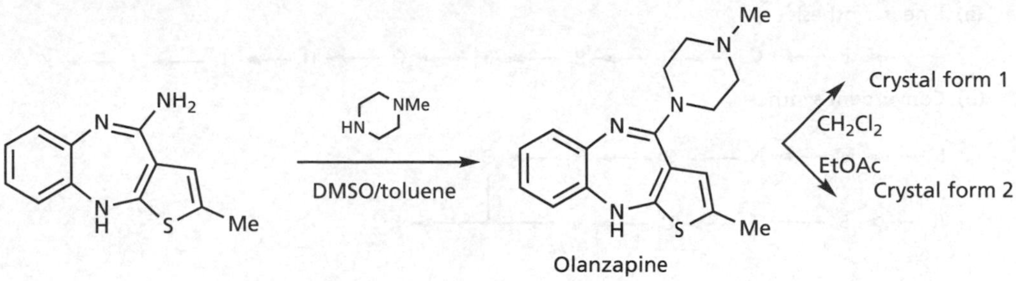

FIGURE 10.33 Different crystal forms of olanzapine

conditions after the final stage (Fig. 10.33). Water was added to precipitate the product, which was washed and then dried under vacuum. The dihydrate was obtained by warming (30–50°C) and the monohydrate was obtained at room temperature. Recrystallization from dichloromethane gave crystal form 1, while crystallization from ethyl acetate gave crystal form 2.

10.11.4 Co-crystals

Pharmaceutical co-crystals contain the pharmaceutical agent of interest along with a different molecular structure in a specific stoichiometric ratio. The properties of different co-crystals can have a significant effect on solubility which, in turn, affects bioavailability. Therefore, if a drug is being produced as a co-crystal, it is important that it is produced in a consistent fashion. There is some dispute about what qualifies as a co-crystal. Crystals that consist of an ionized drug along with an organic counterion, such as a maleate ion, are not considered co-crystals. However, a crystal containing the un-ionized drug along with maleic acid would be considered a co-crystal. Crystals containing solvent of crystallization are not normally considered to be co-crystals.

KEY POINTS

- Several experimental procedures that are carried out routinely in the research laboratory are not feasible on the production scale.
- Ideally, purifications should be avoided on the large scale and restricted to a crystallization of the final product.
- The use of large-scale reactors leads to significant differences in physical parameters compared with the use of ordinary glassware in the research laboratory.
- In process development, the number of operations should be kept to a minimum.
- The conditions of crystallization need to be carefully controlled to ensure that consistent crystals are obtained for each batch.

- If different crystal forms are possible for the final product, it is essential to produce the crystal form defined by the specifications.

10.12 Synthetic planning in chemical and process development

10.12.1 Introduction

One of the most satisfying aspects of chemical and process development is being able to develop a synthesis that is shorter, more efficient, and more cost effective than the original research route. This can best be achieved by reducing the number of reactions involved in the synthesis. There are a number of ways in which that can be done, such as carrying out reactions in a different order, starting from a different starting material, or combining two reactions into one.

Another successful strategy is to design a convergent synthesis rather than a linear synthesis (Fig. 10.34). In a linear synthesis, the structure is built up using relatively simple reagents and building blocks. In contrast, a convergent synthesis involves building up both halves of the target structure as separate processes and then linking the two halves towards the end of the synthesis. A convergent synthesis is generally more efficient than a linear synthesis with the same number of steps, and should give better overall yields. For example, a linear synthesis involving 10 reactions would give an overall yield of 10.7%, assuming an average yield of 80% for each reaction. In contrast, a convergent synthesis involving the same number of reactions would give a much higher overall yield. For example, assuming an average of 80% yield per reaction, the overall yield of structure K would be 26.2% from structure L, and 32.8% from structure R (Fig. 10.34).

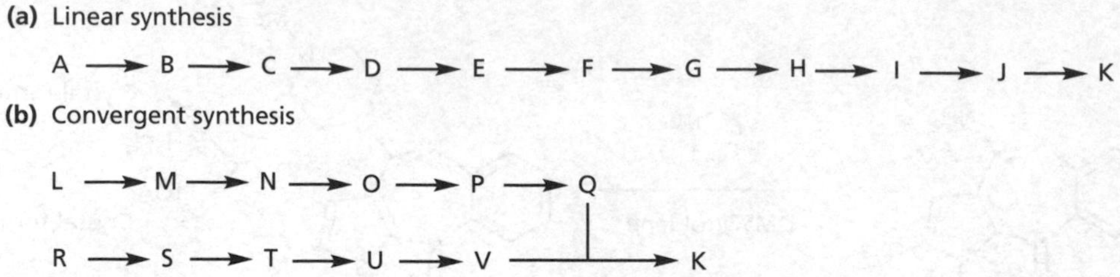

(a) Linear synthesis

A ⟶ B ⟶ C ⟶ D ⟶ E ⟶ F ⟶ G ⟶ H ⟶ I ⟶ J ⟶ K

(b) Convergent synthesis

L ⟶ M ⟶ N ⟶ O ⟶ P ⟶ Q

R ⟶ S ⟶ T ⟶ U ⟶ V ⟶ K

FIGURE 10.34 Comparison of a linear versus a convergent synthesis.

10.12.2 Cutting down the number of reactions in a route

The fewer the number of reactions in a synthetic process the better, and so it may be possible to modify a synthetic route to make it more efficient.

For example, a four-step synthesis of **naratriptan** (an anti-migraine agent) was modified to a three-step process by cutting out a step in the early part of the synthesis. The original route involved two hydrogenation reactions at stages 2 and 4 (Fig. 10.35).

An improved synthesis missed out the first hydrogenation reaction such that both double bonds were reduced at the same time in the final stage (Fig. 10.36).

Another way of shortening a synthesis is to use a different starting material. For example, a five-stage process starting from an aromatic ketone was involved

in producing an important tetralone intermediate used in the synthesis of the antidepressant **sertraline** (Fig. 10.37). However, this only went in an overall yield of 8%.

By using a different starting material, it was possible to use an alternative route to the intermediate involving only three stages (Fig. 10.38). The first stage was a highly regioselective Friedel–Crafts acylation of 1,2-dichlorobenzene with succinic anhydride, followed by a chemoselective reduction of the ketone group to give an alcohol which, on heating, reacted in an intramolecular cyclization reaction with the carboxylic acid to form a lactone. The final stage was a combined Friedel–Crafts alkylation and acylation. The first alkylation was of benzene to give an intermediate, which then underwent an intramolecular cyclization by means of a Friedel–Crafts acylation. The overall yield of the

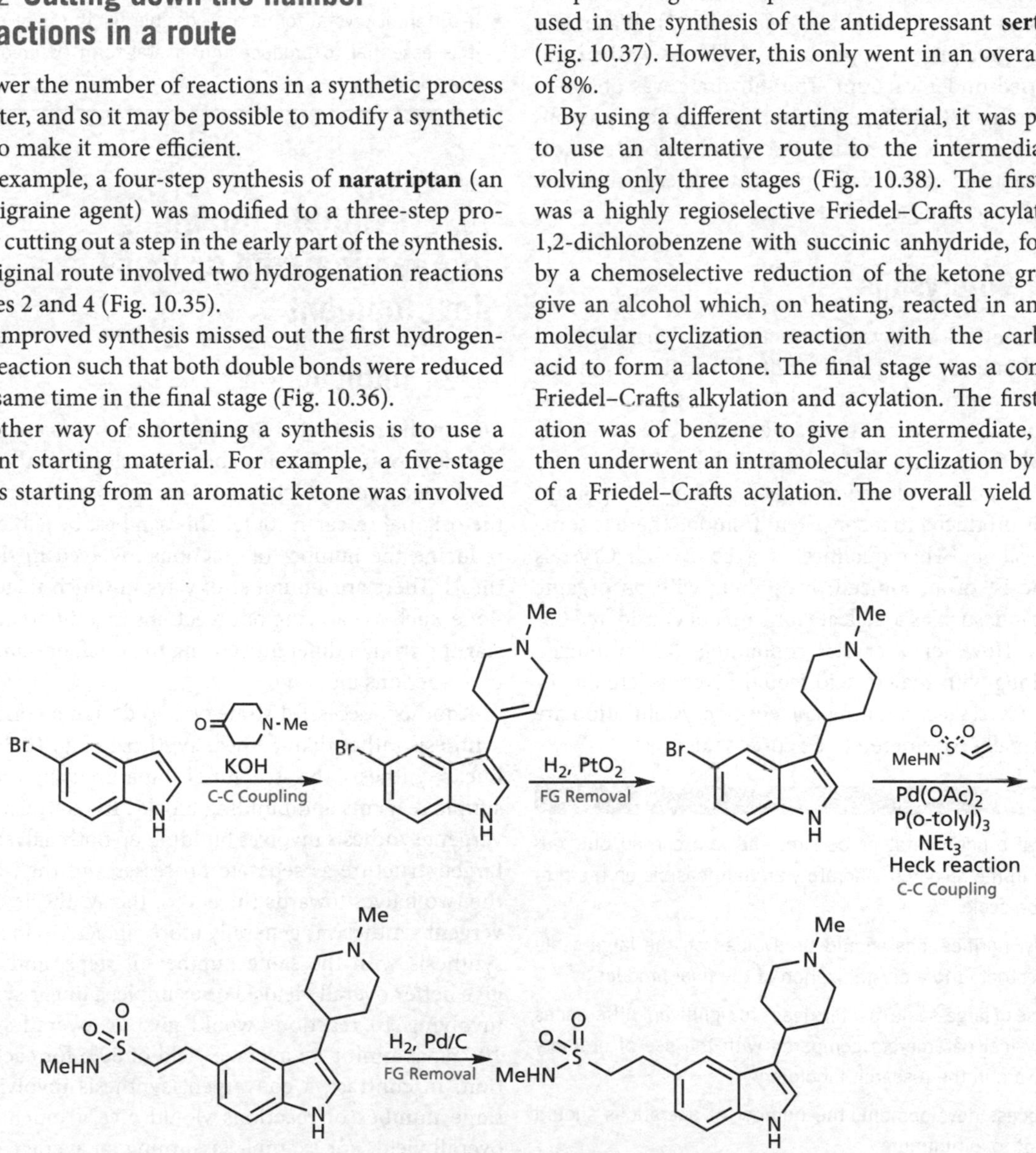

FIGURE 10.35 Four-step synthesis of naratriptan.

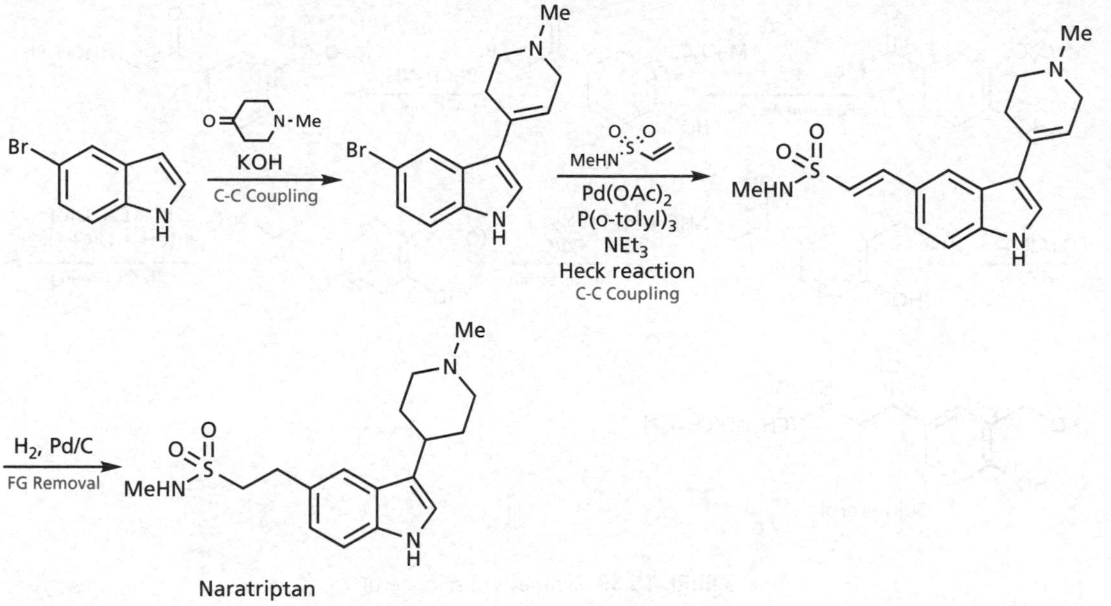

FIGURE 10.36 Three-stage synthesis of naratriptan.

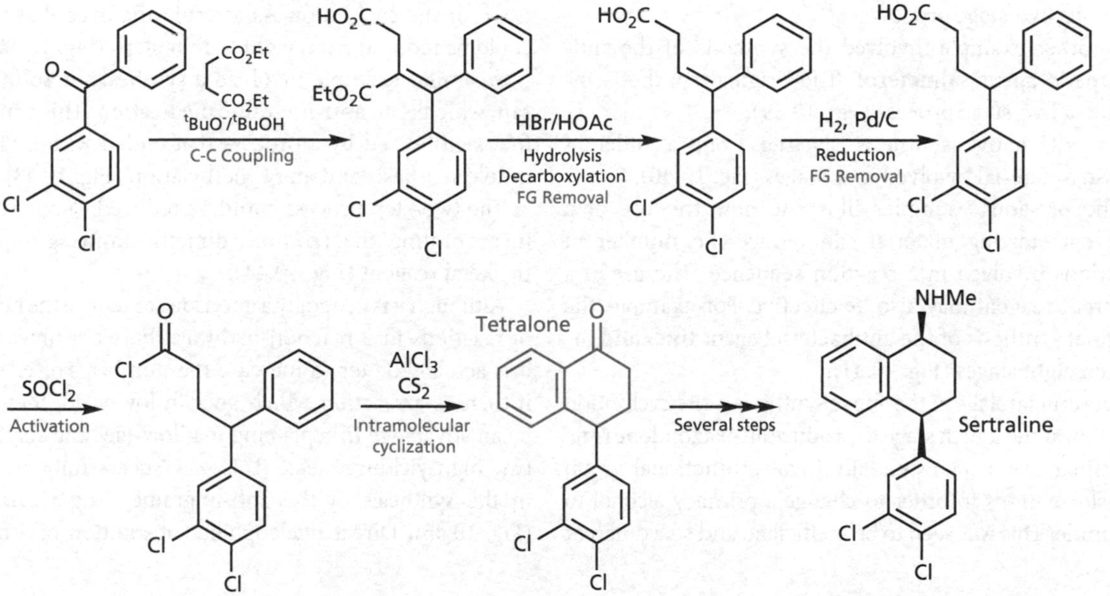

FIGURE 10.37 Synthesis of the tetralone ring system of sertraline.

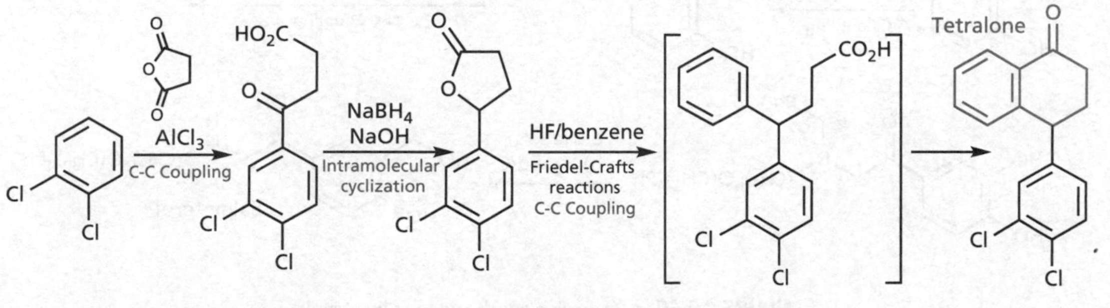

FIGURE 10.38 Alternative synthesis to the tetralone ring system.

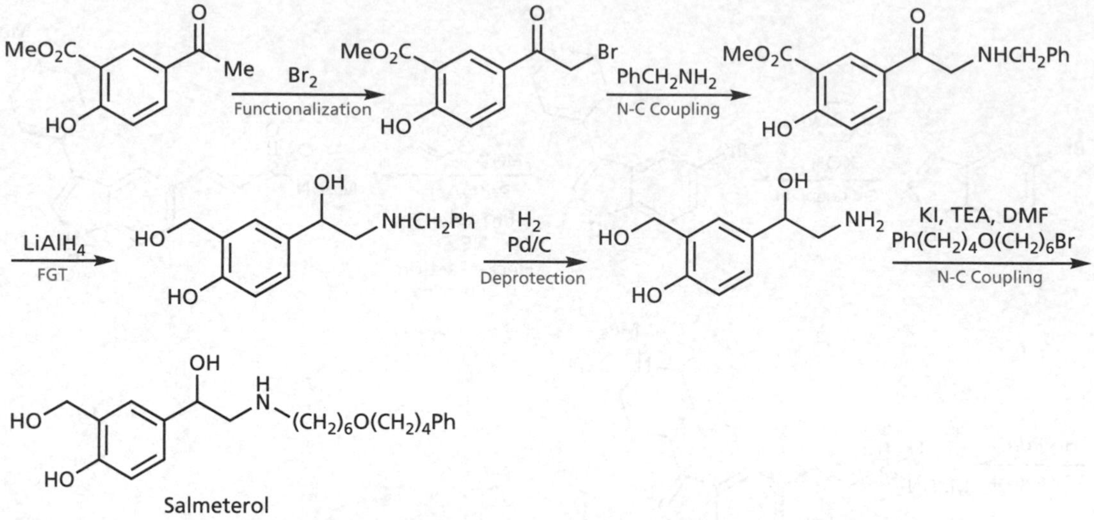

FIGURE 10.39 Synthesis of salmeterol.

three-stage process was 82% compared with 8% for the previous five-stage process.

Another example involved the synthesis of the anti-asthmatic agent **salmeterol**. The original synthesis involved a five-stage process (Fig. 10.39).

An alternative synthesis starting from a different starting material involved four stages (Fig. 10.40).

The previous examples illustrate how the use of a different starting material can reduce the number of reactions involved in a reaction sequence. The use of a different reagent may also be effective. For example, the original synthesis of the antibacterial agent **linezolid** involved eight stages (Fig. 10.41).

The crucial stage in the above synthesis is the cyclization reaction at the fourth stage to produce an oxazolidone ring. The final five steps essentially involve functional group transformations in order to change a primary alcohol to an amide. This was seen to be inefficient, and so a different

reagent which already contained the amide group, was used for the cyclization As a result, the overall synthesis could be reduced from eight to four steps (Fig. 10.42).

A similar example involved a synthesis of **zolmitriptan**, which is an anti-migraine medication. This drug was first synthesized by a process that ended with a Fischer indole synthesis and an *N*-methylation (Fig. 10.43).

The two-step process could be reduced to one step by incorporating the required dimethylamine group into the ketal reagent (Fig. 10.44).

Although it is generally a good idea to reduce the number of reactions in a reaction pathway, there are times when it is actually better to increase the number. For example, if there is a reaction which goes in low yield, there may be an advantage in replacing that low-yielding stage with two high-yielding steps. This was successfully achieved in the synthesis of the anti-migraine drug **rizatriptan** (Fig. 10.45). Direct nucleophilic substitution of a benzyl

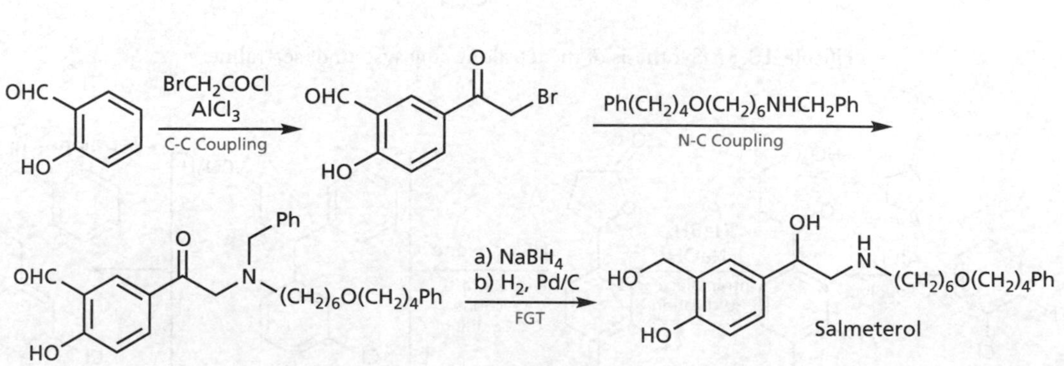

FIGURE 10.40 Alternative synthesis of salmeterol.

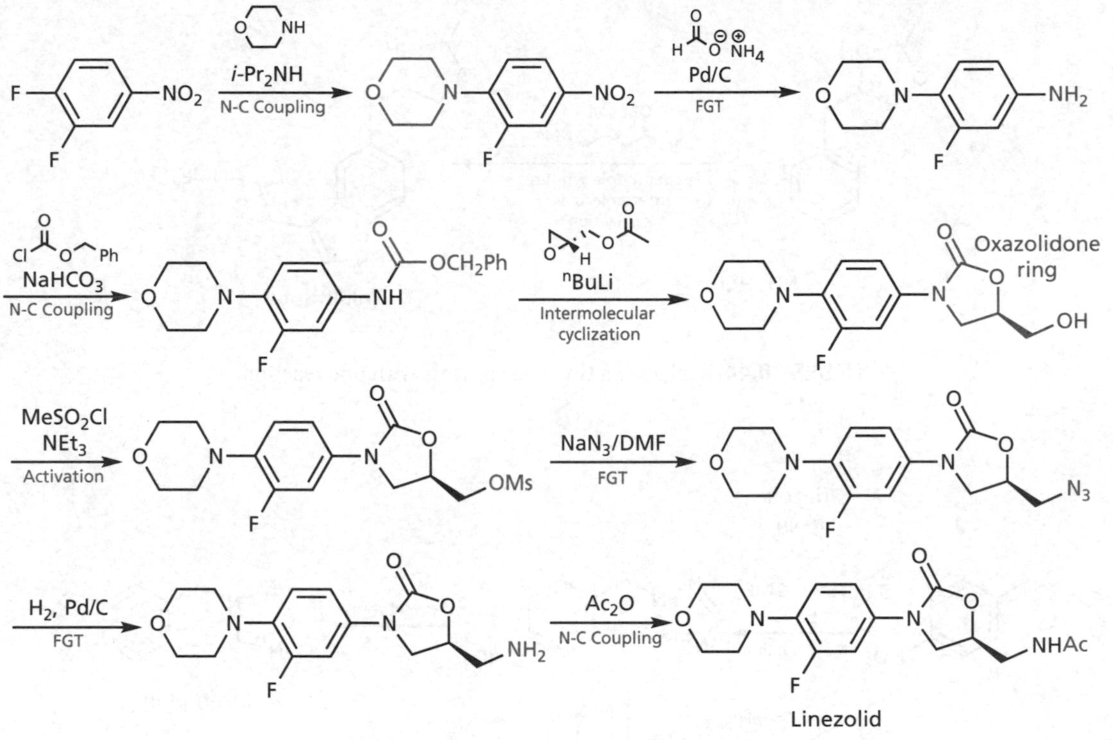

FIGURE 10.41 Original synthesis of linezolid.

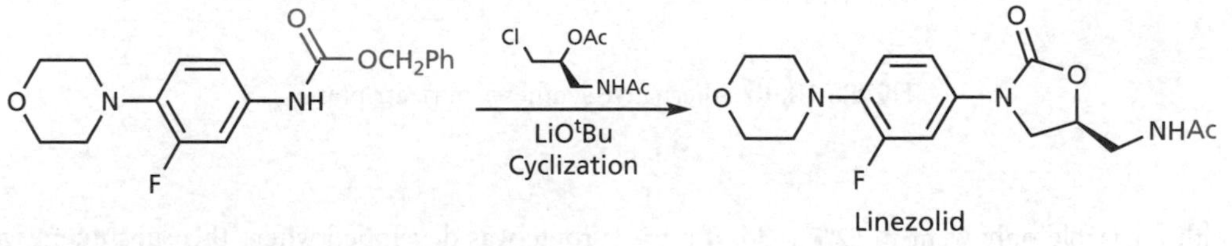

FIGURE 10.42 Cyclization reaction with a modified reagent to produce linezolid directly.

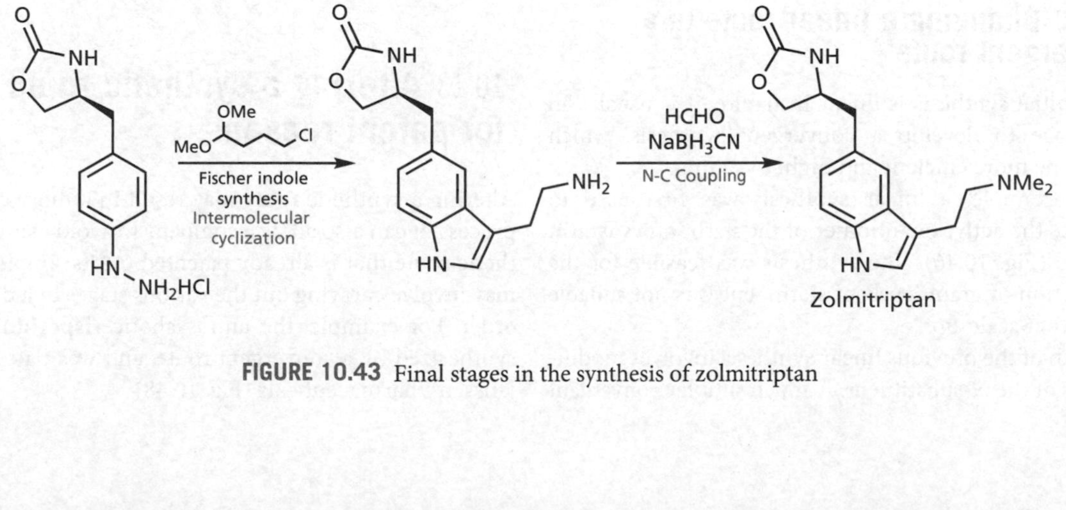

FIGURE 10.43 Final stages in the synthesis of zolmitriptan

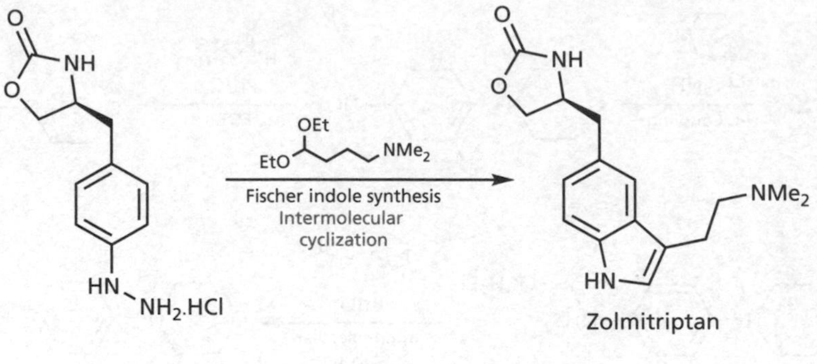

FIGURE 10.44 Replacing a two-stage process with one reaction.

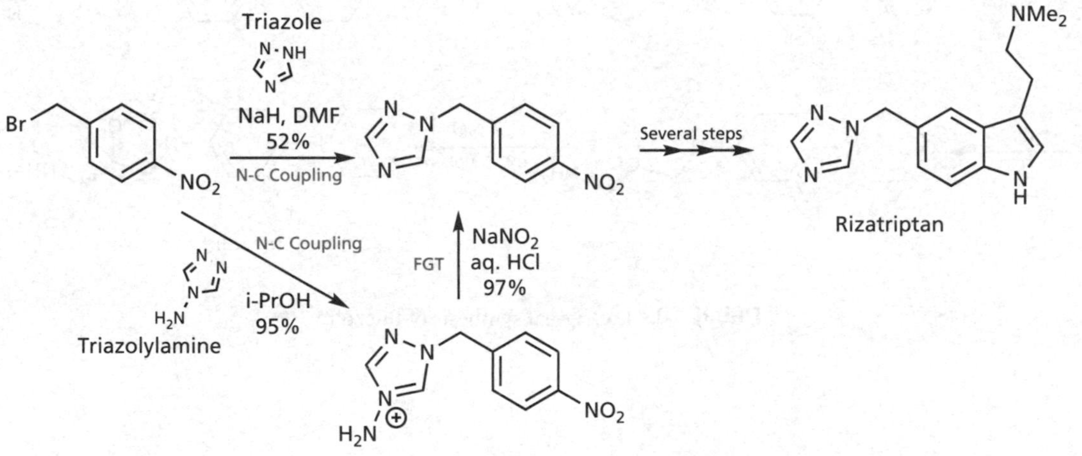

FIGURE 10.45 Alternative syntheses of rizatriptan.

bromide with a triazole only went in 52% yield. If a triazolylamine was used instead, the reaction went in 95% yield. The amino group was then removed in 97% yield, which meant that the overall yield for the two-stage process was 92% which was far superior to the one-stage process.

10.12.3 Changing a linear route to a convergent route

If an initial synthesis is linear in nature, it is usually an advantage to develop a convergent synthesis, which should be more efficient and higher yielding.

For example, a linear synthesis was first used to produce the active enantiomer of the statin **atorvastatin lactone** (Fig. 10.46). The synthesis was feasible for the production of gram levels of statin, but was not suitable for further scale-up.

Much of the previous linear synthesis involves modifications of the *N*-substituent. A much simpler convergent

route was developed where this substituent was synthesized in a three-step process in advance and then incorporated during the Paal–Knorr pyrrole synthesis. An equivalent of pivalic acid was added to the reaction as a catalyst (Fig. 10.47).

10.13 Altering a synthetic route for patent reasons

Altering a synthetic route may result in a more efficient process. It can also allow a company to avoid using a synthetic route that is already patented. At its simplest, this may involve carrying out the various stages in a different order. For example, the antipsychotic **risperidone** was synthesized by a convergent route, and was patented by Janssen Pharmaceuticals (Fig. 10.48).

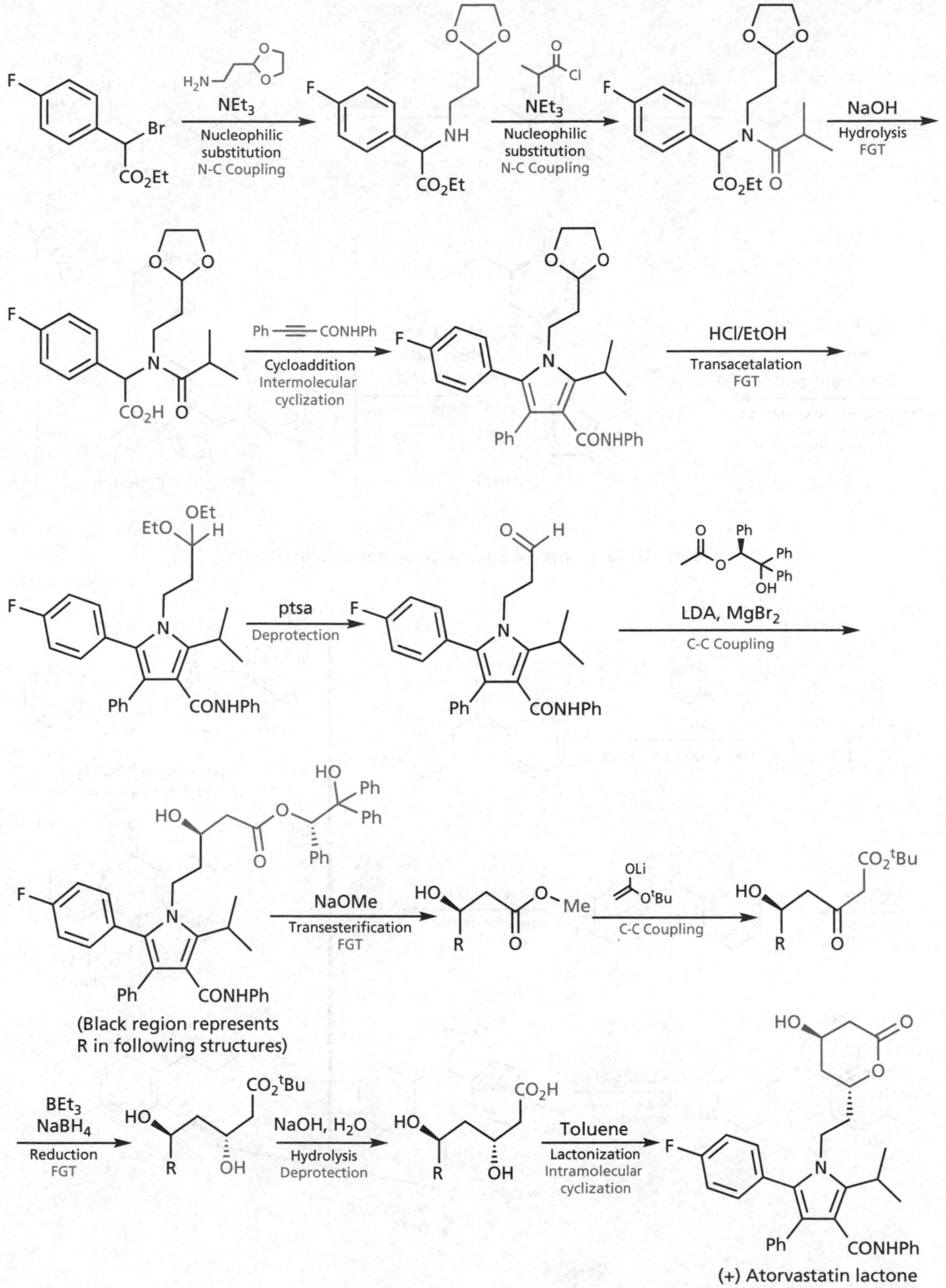

FIGURE 10.46 Linear synthesis of atorvastatin lactone.

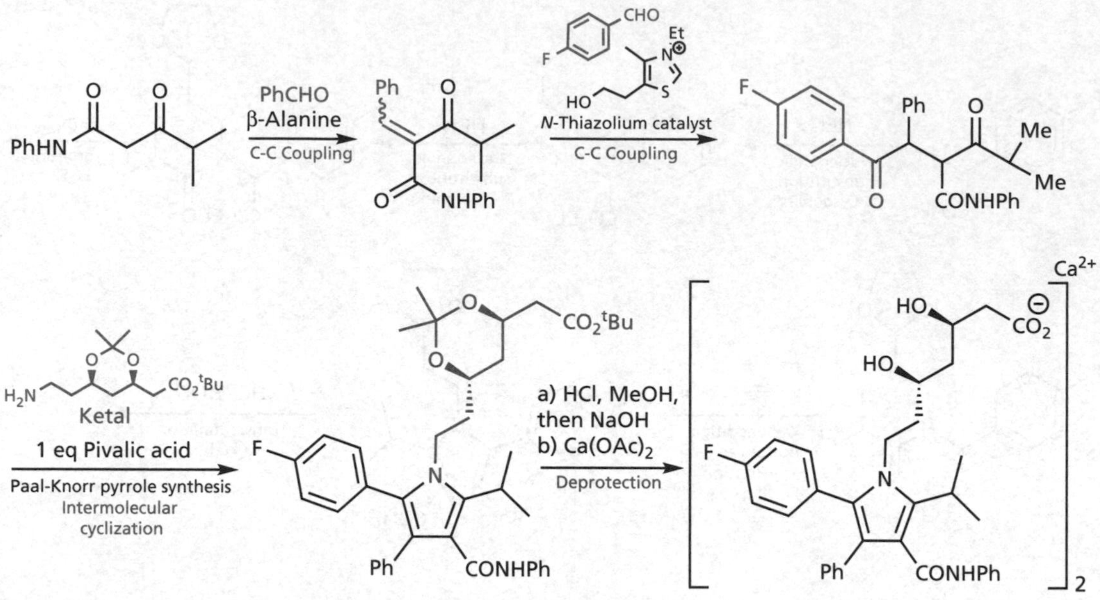

FIGURE 10.47 Convergent synthesis of atorvastatin lactone.

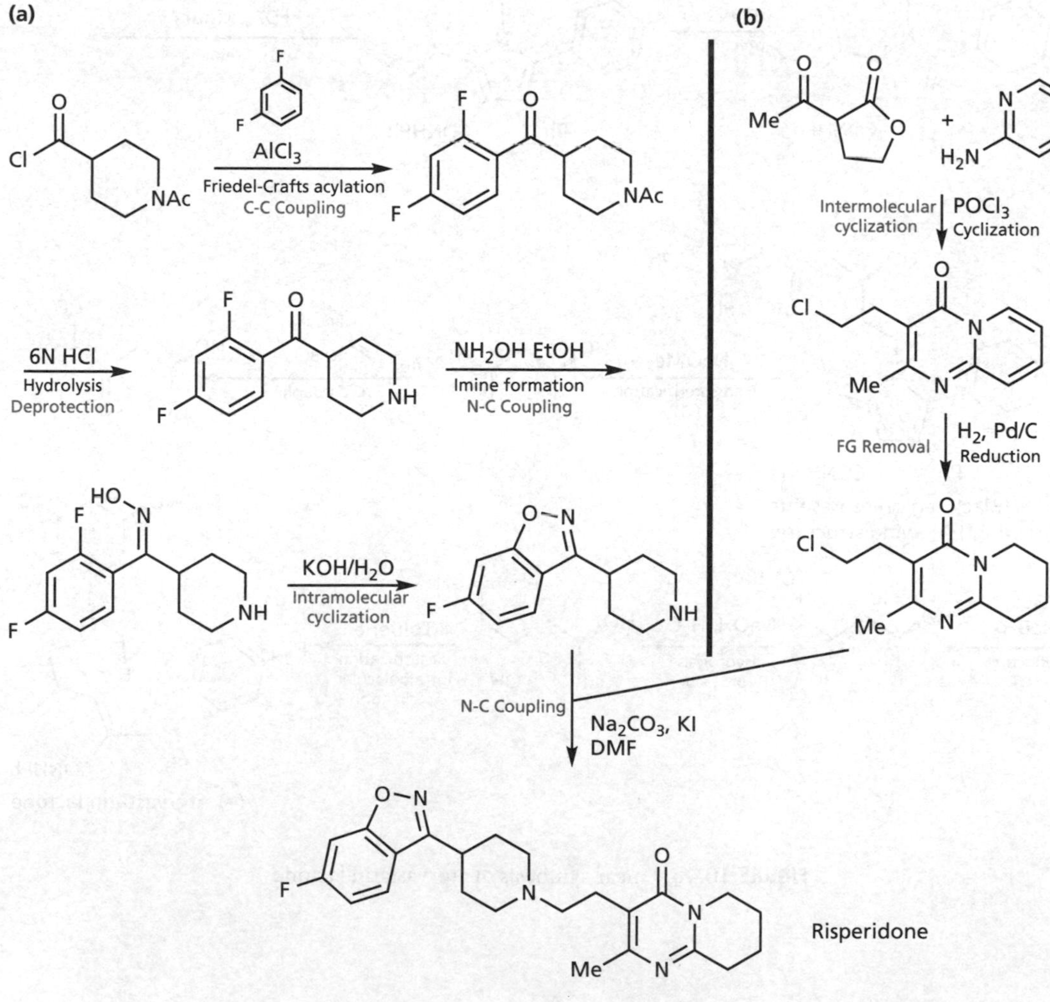

FIGURE 10.48 Synthesis of risperidone.

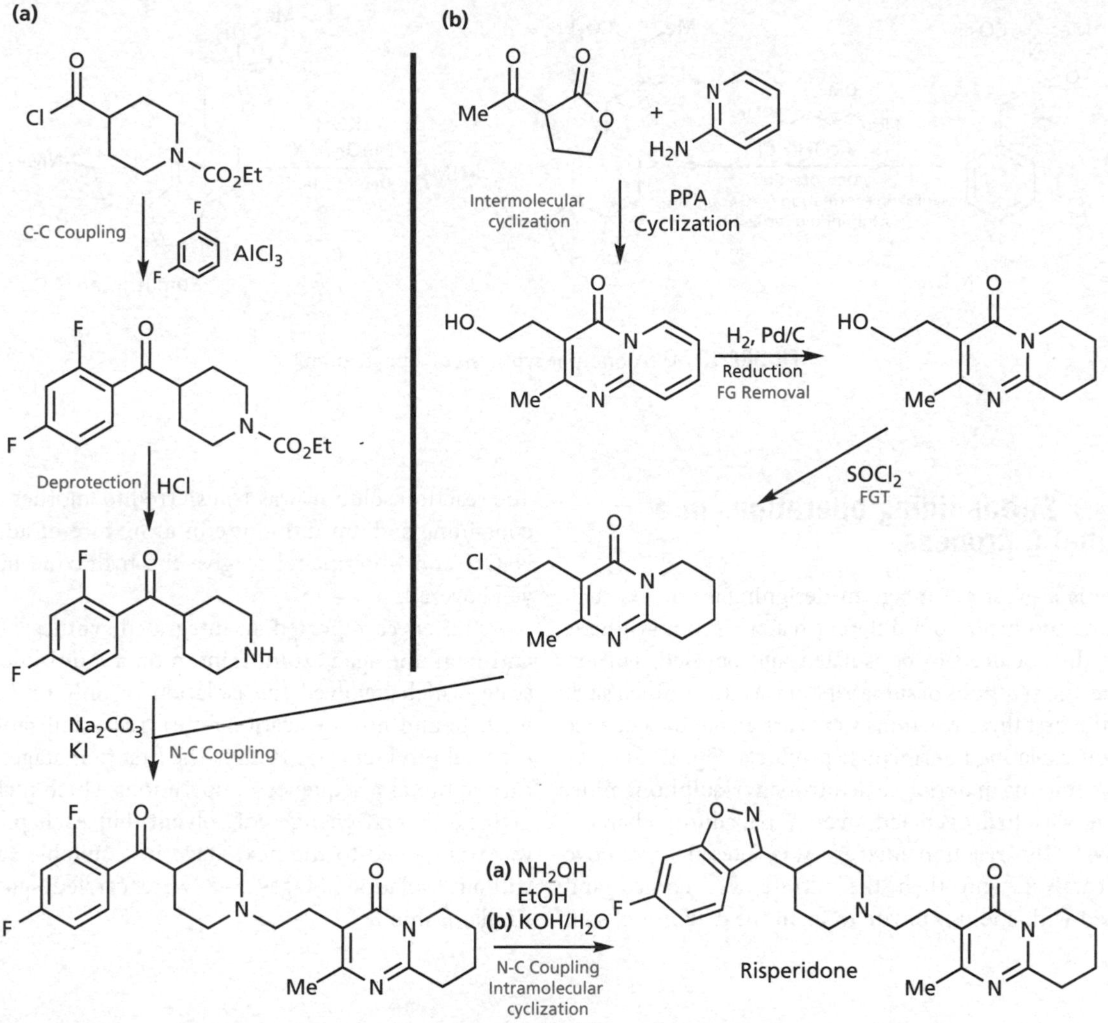

FIGURE 10.49 Alternative synthesis of risperidone.

A rival company modified the order of the reactions as shown in Figure 10.49.

10.14 Minimizing the number of operations in a synthesis

10.14.1 Introduction

There is an economy of effort and expense at the production level if the number of operations involved in a synthetic route can be cut to a minimum. One of the most effective ways of doing this is to develop a synthesis where several reactions can be carried out in sequence without having to isolate the products from each reaction. Ideally, the reactions involved would be carried out in the same solvent, allowing the reaction product from one reaction to be transferred in solution directly to the next reaction vessel. Alternatively, one might be able to carry out a one-pot synthesis where a sequence of reactions is carried out in one reaction vessel without the need for work-ups or the isolation of the intermediates involved—a process described as **telescoping**.

10.14.2 One-pot reactions

A one-pot operation involving three reactions was carried out as part of a synthesis of the anti-migraine drug **sumatriptan** (Fig. 10.50). Reaction of a hydrazine starting material with a dimethyl acetal was carried out under catalytic acid conditions, resulting in deprotection of the aldehyde group, hydrazone formation, and Fischer indole cyclization in 50% yield. A further hydrolysis then gave sumatriptan.

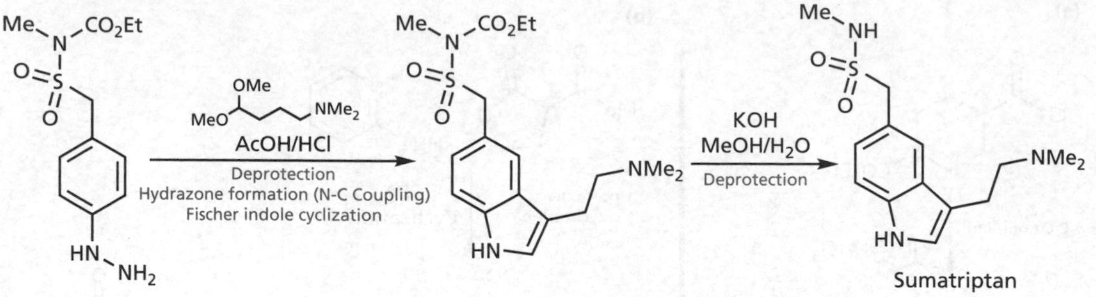

FIGURE 10.50 A one-pot synthesis of sumatriptan.

10.14.3 Streamlining operations in a synthetic process

There is a great advantage in designing a process such that the products from different reactions in a synthetic route do not need to be isolated and purified. For example, the synthesis of sumatriptan was streamlined such that the first three reactions were carried out in sequence without isolating intermediate products (Fig. 10.51).

The starting material was a nitrobenzyl sulphonamide, which was hydrogenated over a palladium charcoal catalyst. The reaction mixture was filtered to remove the catalyst, and then the filtrate was cooled and treated with sodium nitrite to form the diazonium salt.

The reaction solution was transferred to another vessel containing sodium dithionite in a mixture of aqueous NaOH and isopropanol to give the hydrazine in 77% yield overall.

AstraZeneca reported an integrated synthesis of the anti-migraine agent **zolmitriptan** on a multi-kilogram scale, which involved the isolation of only one intermediate and no purifications until recrystallization of the final product (Fig. 10.52). The first four stages were carried out as a sequence of operations which included extractions and changes of solvent, but each product was transferred to the next stage in a suitable solvent without isolation. Stages 5–8 were carried out in a similar fashion.

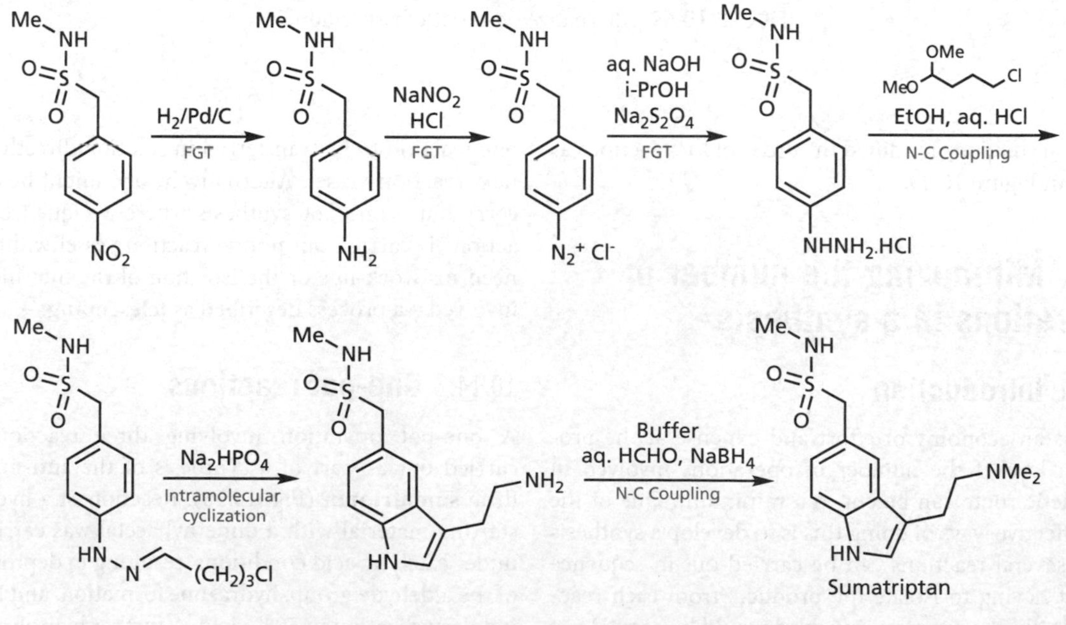

FIGURE 10.51 Synthesis of sumatriptan.

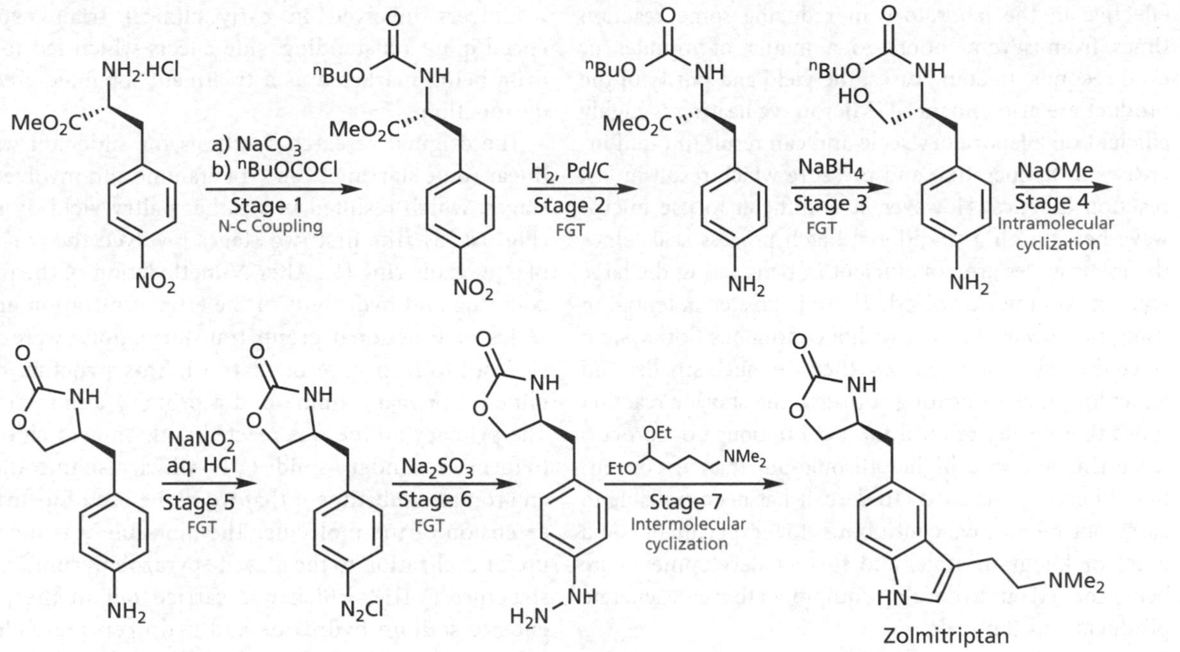

FIGURE 10.52 Integrated process for the synthesis of zolmitriptan.

10.15 Continuous flow reactors

The conventional approach to chemistry is to mix reagents in a flask or a reactor and to carry out a reaction until it has gone to completion or has reached equilibrium. In contrast, continuous flow chemistry involves a series of pumps and tubes which carry a flow of the different reagents required for a reaction. When the tubes meet, the reagents are mixed together and the reaction mixture continues flowing at an optimal rate through a tubular system until the reaction is complete. However, not all reactions can be completed in a relatively short period of time, in which case the reaction mixture may have to be stored or incubated until the reaction is complete. Continuous flow experiments can be carried out on an extremely small scale as described in section 6.4.6. However, continuous flow systems can also be used on a production scale and have a number of advantages over the traditional approach of preparing compounds as a series of batches (see also section 9.3). In particular, they can prove to be safer, more environmentally friendly, and create less waste.

For example, with large batch reactors, a number of factors are inevitably less efficient compared with the laboratory scale. These include heat transfer, stirring efficiency, and the rapid addition of reagents. These problems become particularly crucial when it comes to exothermic reactions, which often means that relatively high levels of solvent are required to act as a heat sink in order to prevent the reaction going out of control. Continuous flow chemistry can solve a lot of these problems. The mixing process involved when the two reagents meet is faster and more efficient since it is on a much smaller scale. The volumes and size of tubing involved are also much smaller which allows more efficient temperature control. Moreover, less solvent is required which reduces the amount of waste that is likely to be generated. Indeed, it has proved possible to carry out many reactions without using any solvent. The rate of flow is crucial. The faster the rate, the more product is produced per hour. However, the rate must also be slow enough to allow the reaction to go to completion before the final product is collected.

The use of continuous flow techniques also allows experimental conditions to be used that would be impractical or dangerous on a batch process. In some cases, this results in increased reaction rates whilst maintaining reaction selectivity. This has been termed **novel process windows**. The use of high temperatures and pressures can be particularly effective in increasing reaction rates.

Another advantage of continuous flow reactors is the ability to use green energy sources that would be impossible to use with batch reactors. For example, photochemical reactions can be carried out using continuous flow systems (section 9.3). The technique is also being developed to allow microwave heating, which has proved effective in the laboratory in reducing some reaction times from several hours to a matter of minutes or even seconds. In some cases, the yield and purity of the product are also enhanced. Microwave heating is highly efficient on a laboratory scale and can result in rapid increases in temperature and pressure which result in fast reaction kinetics. However, it is difficult to use microwave heating on a traditional batch process scale since the microwaves are not efficient in penetrating the large reaction volumes involved. There is greater potential in using microwave heating with a continuous flow system since the volumes mixed together are much smaller and allow for efficient heating. In turn, the shorter reaction times that result are ideal for a continuous flow process since this allows a higher throughput than if conventional heating was used. To date, it has been possible to carry out microwave continuous flow experiments on a gram or kilogram scale, and further developments are being carried out to develop equipment that can generate products on a ton scale.

KEY POINTS

• The number of reactions in a process synthesis should be minimized.

• A convergent synthesis is more efficient than a linear synthesis.

• A synthetic route can be shortened by carrying out reactions in a different order or by using a different starting material or reagent.

• A synthesis to a particular pharmaceutical might be modified for patent reasons.

• The number of operations in a production synthesis can be minimized by carrying out a series of reactions without isolating the intermediates involved.

• Carrying out a sequence of reactions using the same solvent cuts down the number of operations required.

• Efficiency is enhanced by carrying out a sequence of reactions in one reactor.

• Continuous flow systems offer advantages over traditional batch processes and allow the use of green energy sources such as photochemistry and microwave heating.

10.16 Case Study—Development of a commercial synthesis of sildenafil

Sildenafil is an enzyme inhibitor that targets an enzyme called **phosphodiesterase 5**. It was originally designed as a potential antihypertensive agent, but

volunteers involved in early clinical trials experienced quite 'outstanding' side effects which led to the drug being marketed as a treatment for male erectile dysfunction.

The original research synthesis for sildenafil was a linear route starting from 2-pentanone and involved 11 stages, which resulted in a rather paltry yield of 4.2% (Fig. 10.53). The first two stages involved the creation of a pyrazole ring (I). After N-methylation of the pyrazole ring and hydrolysis of the ester, a nitration and a series of functional group transformations were carried out to form intermediate VI. This structure contained a primary amine and a primary amide group. The primary amine was reacted with an acid chloride to form a secondary amide (VII) which also introduced an aromatic substituent that would be used for further extension of the molecule. The molecule was now set up for cyclization to the bicyclic pyrazolopyrimidinone structure (VIII), which was carried out in the presence of sodium hydroxide and hydrogen peroxide. A sulphonation reaction introduced a sulphonyl chloride group (structure IX) which was treated with N-methylpiperazine to form the final sulphonamide structure (sildenafil).

Apart from the low yield, there were a number of problems with the research synthesis which made it unattractive as a large-scale manufacturing route.

• The penultimate sulphonyl chloride structure (IX) used in the final stage is toxic and would be a probable impurity in the final product. It would be crucial to reduce the levels of this compound to an absolute minimum.

• Multiple crystallizations of the final product were required to achieve a satisfactory purity, and to reduce the levels of structure IX to an acceptable level.

• It was not easy to scale up the chlorosulphonation reaction required to produce structure IX. As one scales up the reaction, it takes longer to quench. This results in increased levels of a hydrolysis side reaction which produces increased levels of a sulphonic acid structure rather than the chlorosulphonate. As the sulphonic acid fails to react with N-methylpiperazine, it results in a lower yield for the final step.

• The quenching process itself requires large quench volumes and an increase in the volume of aqueous waste, which is environmentally unsound.

• The cyclization reaction to give the pyrimidinone (VIII) was an effective method of cyclization, but a certain amount of the primary amide was hydrolysed, resulting in the acid (X) as an impurity (Fig. 10.54).

Early development work established that it was possible to carry out the pyrimidinone cyclization more

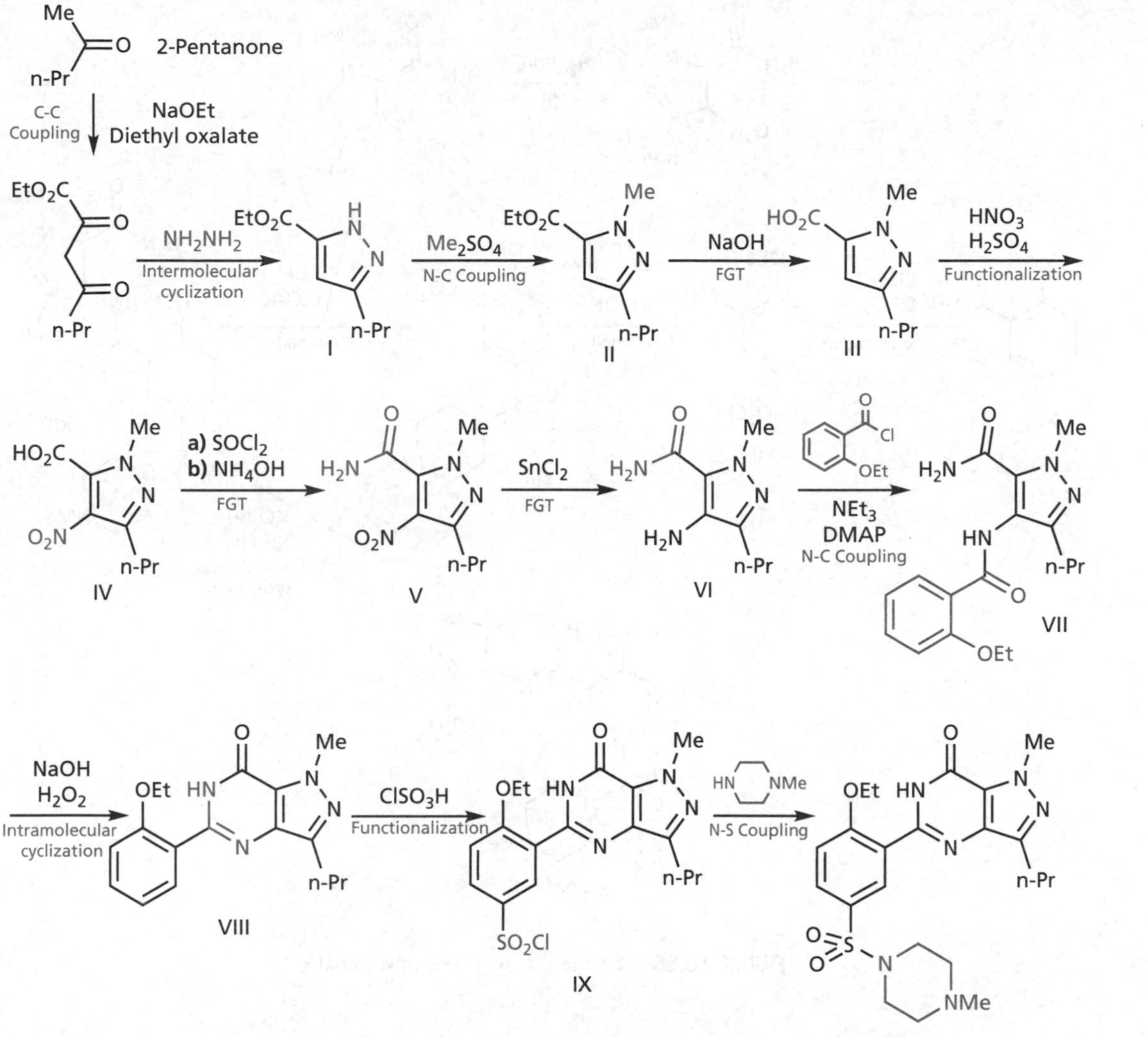

FIGURE 10.53 Research synthesis of sildenafil.

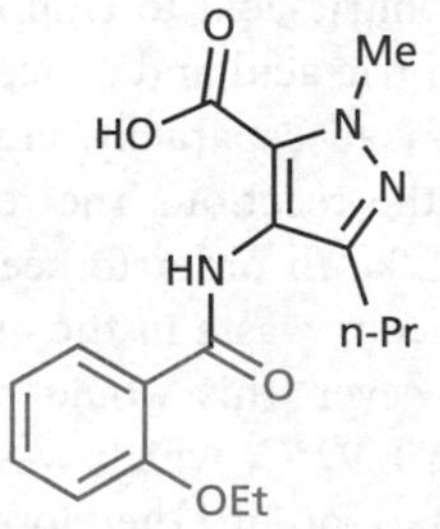

FIGURE 10.54 Impurity (X) formed during the cyclization reaction as a result of hydrolysis.

efficiently if non-aqueous conditions were used, namely potassium *tert*-butoxide in tertiary butanol. The reaction went in quantitative yield with no sign of the hydrolysed impurity (X), or any other impurity for that matter.

Considering the quality of product obtained, it was decided to redesign the synthesis such that the pyrimidone cyclization would come at the end of the synthesis rather than halfway through (Fig. 10.55). Assuming that the cyclization was equally effective, that would result in easier purification of the final product. Moreover, since the chlorosulphonation would come much earlier in the synthesis, there was little risk of toxic intermediates being present in the final product. The newly designed route was also convergent in nature to reduce the number of stages involved and to boost yields. Thus, instead of coupling the amine (VI) to 2-ethoxybenzoyl chloride, and then building up the rest of the structure, it was coupled to a benzoic acid structure (XIII) which already contained the required substituents for the final sildenafil structure.

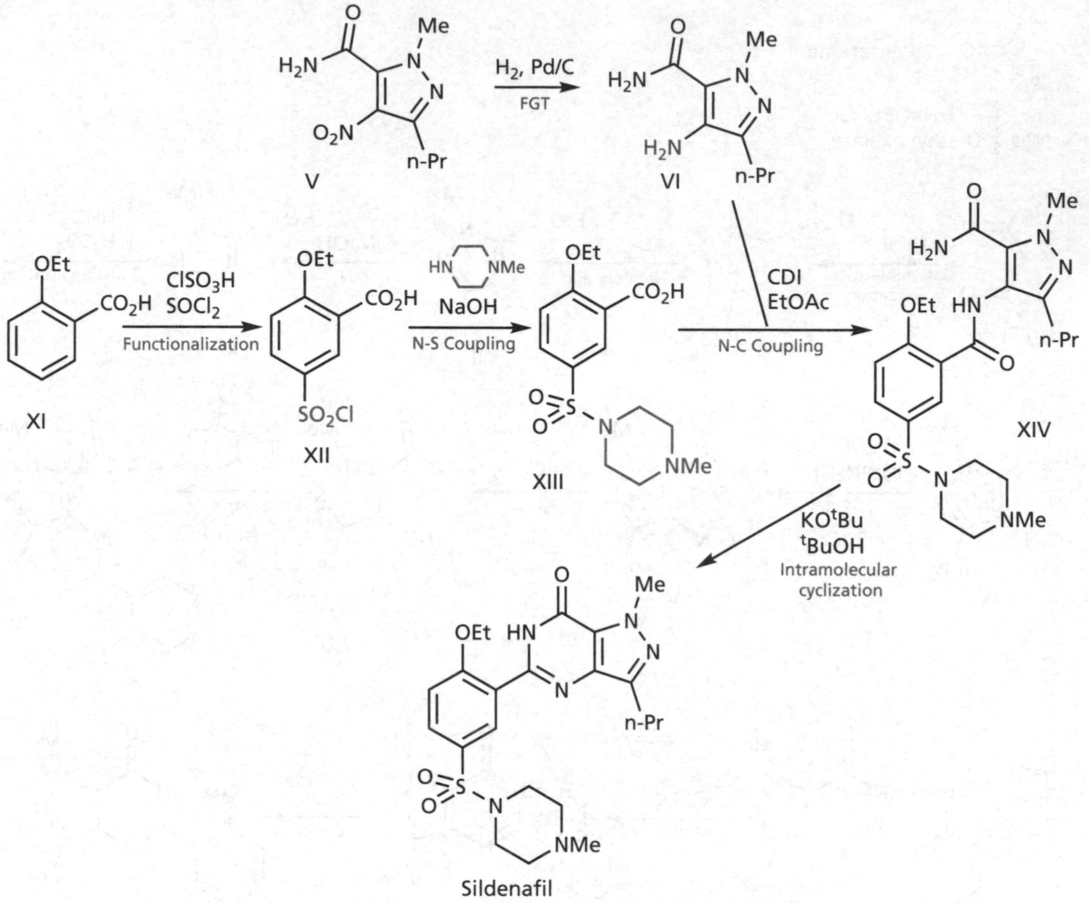

FIGURE 10.55 Commercial synthesis of sildenafil.

Studies were carried out on the various stages. The nitro-substituted pyrazole (V) was synthesized as before (Fig. 10.53), but the nitration step was identified as a potentially hazardous operation for large-scale production. The nitro pyrazole was found to have two exotherms, one of which evolved 16.2 kJ/mol at 130°C due to decarboxylation. The resulting release of carbon dioxide also caused a pressure increase. The other exotherm was more serious and evolved 294kJ/mol at 295°C. Furthermore, under the acidic conditions used in the reaction, the decarboxylation reaction could occur at 100°C. Therefore, it was important to reduce the chances of these reactions taking place as there could be a risk of an uncontrolled reaction becoming self-sustaining and violent, leading to large increases in temperature and pressure.

The experimental method used in the research procedure involved adding the pyrazole (III) to fuming nitric acid and oleum at 50°C to liberate 249 kJ/mol of heat. It was calculated that this would be sufficient to

raise the temperature of the large-scale process to 127°C and trigger the decarboxylation reaction. To avoid this possibility, the pyrazole (III) was initially dissolved in concentrated sulphuric acid to eliminate 67 kJ/mol of energy. Fuming nitric acid and concentrated sulphuric acid were also mixed separately, eliminating another 44 kJ/mol from the reaction. The reaction was to be carried out at 6 L/kg in order to keep the throughput high and the aqueous waste in the quench reaction to a minimum. However, this would still cause a temperature rise of 50–92°C, which was too close to the decomposition start point. Therefore, only a third of the nitrating acid mixture was charged at any one time. The reaction was monitored by HPLC to ensure that the reaction was proceeding to plan before the next third was added. The maximum adiabatic temperature rise was calculated to be 21°C.

The reduction of the nitro group in structure V was carried out using hydrogenation over a palladium catalyst, rather than by $SnCl_2/HCl$.

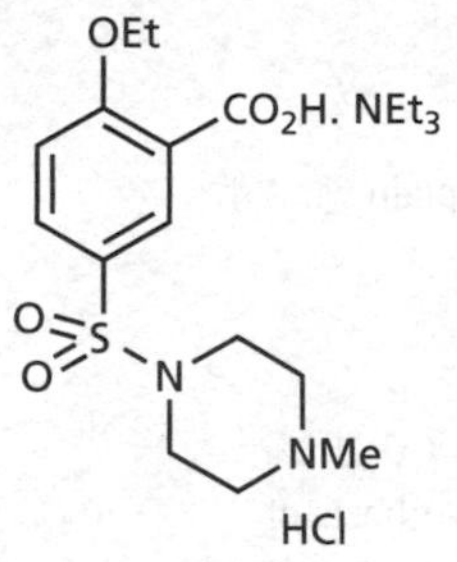

FIGURE 10.56 The double salt (structure XV).

The chlorosulphonation reaction was now carried out at the start of the synthesis on a cheaper, lower molecular weight starting material (structure XI) which minimized the problems associated with scale-up and environmental issues. There were some initial problems with the formation of the sulphonamide (XIII). Early experiments were carried out using NaOH in triethylamine, and this resulted in a double salt (XV) (Fig. 10.56) which proved difficult to crystallize and was very insoluble for further reactions. Fortunately, the addition of water to the double salt eventually led to its dissociation and crystallization of structure XIII. Following this discovery, the reaction was carried out in water and, on completion, the solution was adjusted to the isoelectric point to precipitate the product (Fig. 10.57).

The coupling reaction requires activation of the carboxylic acid (XIII) which can be carried out by converting the acid to an acid chloride using thionyl chloride or oxalyl chloride. However, a coupling agent was used instead (N,N'-carbonyldiimidazole (CDI)). Although this was a more expensive method, it meant that reduction of the nitro group, activation of the carboxylic acid, and the subsequent coupling reaction could all be carried

out in the same solvent (ethyl acetate) without needing to isolate any of the products. Thus, three individual stages were condensed into one. It also meant that the environmental impact was substantially reduced since there would be no aqueous waste to deal with and it would be relatively easy to recover the ethyl acetate. Moreover, the coupled product crystallized from the reaction solution with the main byproduct (imidazole) remaining in solution. One disadvantage of using CDI is that it is sticky and hygroscopic, and there were problems in charging the reagent into the reaction vessels.

The final cyclization reaction was carried out with heating for several hours and then the reaction was diluted with water and the pH adjusted to 7.5 (the isoelectric point) to precipitate a product which met the required specifications. Further purification stages were not required.

The work carried out by Pfizer in developing this process resulted in the company being awarded the Crystal Faraday Award for Green Chemical Technology by the Institute of Chemical Engineers as a recognition that the manufacturing process was environmentally friendly.

A comparison of the research synthesis with the commercial synthesis showed that the total aqueous volume was reduced by a factor of 5.

The organic waste from the research synthesis would have involved 125 000 L of solvents such as pyridine, 2-butanone, acetone, ethyl acetate, dichloromethane, and toluene. The commercial route involved 13 500 L of solvents consisting only of toluene and ethyl acetate. Organic solvents were used as single solvents, making their recovery easier.

The final route was described as safe and robust, resulting in a high-yielding process of 76% from the nitro acid pyrazole (IV), compared with the original yield of 7.5%.

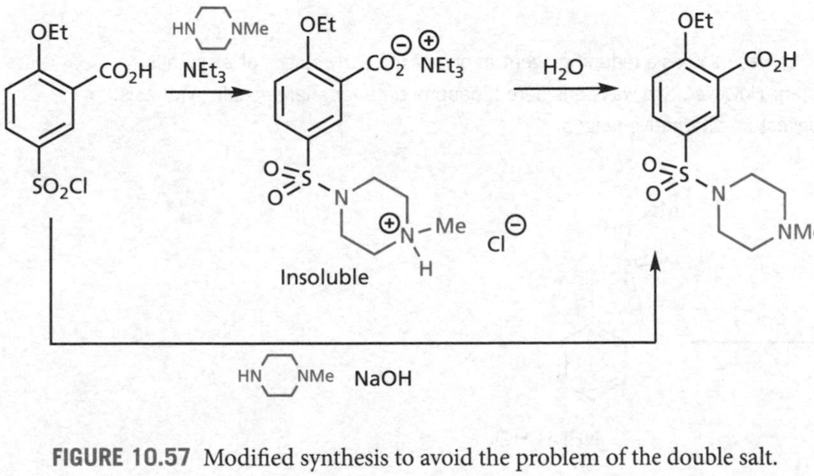

FIGURE 10.57 Modified synthesis to avoid the problem of the double salt.

QUESTIONS

1. Usually, a 'balancing act' of priorities is required during chemical development. Explain what this means.

2. Discuss whether chemical development is simply a scale-up exercise.

3. The following synthetic route was used for the initial synthesis of fexofenadine (R = CO$_2$H), an analogue of terfenadine (R = CH$_3$). The synthesis was suitable for the large-scale synthesis of terfenadine, but not for fexofenadine. Suggest why not. (Hint: consider the electronic effects of R)

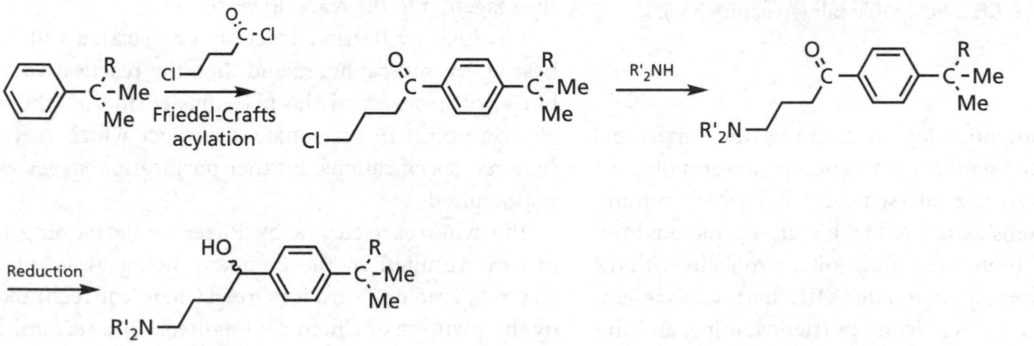

4. The following reaction was carried out with heating under reflux at 110°C. However, the yield was higher when the condenser was set for distillation. Explain why.

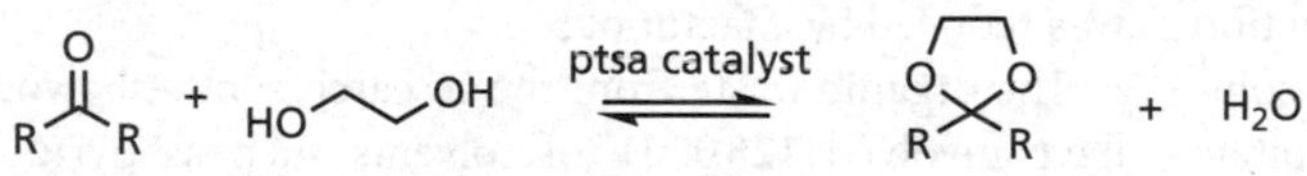

5. What considerations do you think have to be taken into account when choosing a solvent for scale-up? Would you consider diethyl ether or benzene as a suitable solvent?

6. Phosphorus tribromide was added to an alcohol to give an alkyl bromide, but the product was contaminated with an ether impurity. Explain how this impurity might arise and how the reaction conditions could be altered to avoid the problem.

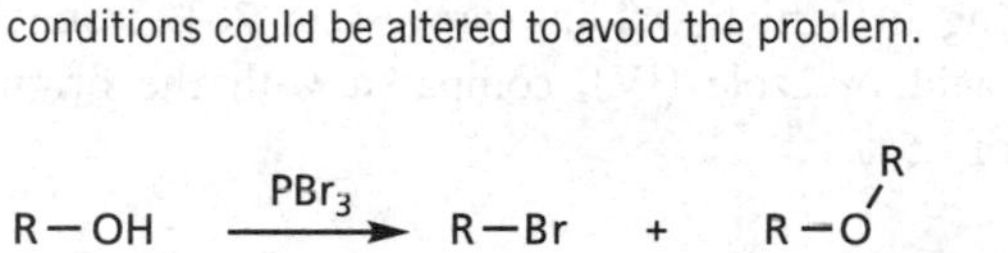

7. Tin(II) chloride (SnCl$_2$) was used as a reducing agent in one of the early stages of a synthesis leading to sumatriptan. However, this was considered inappropriate for a large-scale synthesis. Explain why and suggest an alternative reagent.

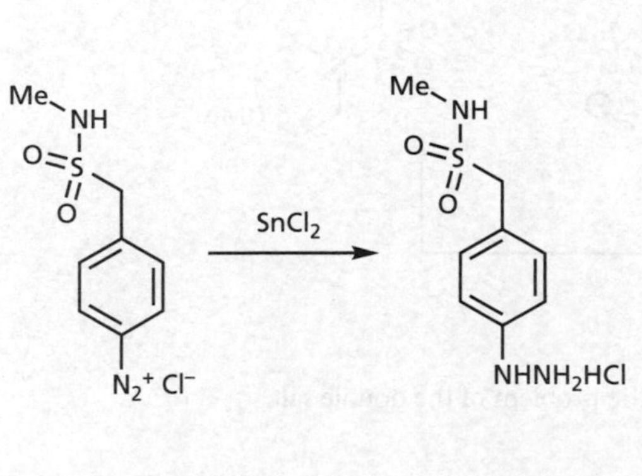

8. The Fischer indole synthesis has been used to synthesize the anti-migraine agent **sumatriptan** (see Figure 10.27). Propose a mechanism by which the two impurities shown in Figure 10.28 might have been formed, then suggest how the two ester groups used in Figure 10.29 hinder the production of these impurities.

9. Propose a mechanism by which the product from the Fischer indole synthesis shown in Figure 10.41 is formed.

FURTHER READING

General reading

Anastas, P.T. and Kirchhoff, M.M. (2002) 'Origins, current status, and future challenges of green chemistry', *Accounts of Chemical Research*, **35**, 686–94.

Diao, Y., et al. (2011) 'The role of nanopore shape in surface-induced crystallization', *Nature Materials*, **10**, 867–71.

Lowe, D. (2011) 'In the pipeline', *Chemistry World*, August, p 21 (polymorphs).

Morschhauser, R., et al. (2012) 'Microwave-assisted continuous flow synthesis on industrial scale', *Green Processing & Synthesis*, **1**, 218–90.

Wiles, C. and Watts, P. (2012) 'Continuous flow reactors: a perspective', *Green Chemistry*, **14**, 38–54.

Specific compounds

Baker, R., et al. (1993) 'The sulphate salt of a substituted triazole, pharmaceutical compositions thereof, and their use in therapy', EP Patent 573,221 (rizatriptan).

Blatcher, P., et al. (1995) 'Process for the preparation of *N*-methyl-3-(1-methyl-4-piperidinyl)-14-indole-5-ethanesulphonamide', WO Patent 95/09166 (naratriptan).

Bozsing, D., et al. (2001) 'A process for the preparation of quetiapine and intermediates thereof', WO Patent 01/55,125 (quetiapine).

Dale, D.J., et al. (2000) 'The chemical development of the commercial route to sildenafil: a case history', *Organic Process Research and Development*, **4**, 17–22 (sildenafil).

Federsel, H.-J. (2000) 'Development of a process for a chiral aminochroman antidepressant: a case story', *Organic Process Research and Development*, **4**, 362–9 (ebalzotan).

Glen, R.C., et al. (1995) 'Computer-aided design and synthesis of 5-substituted tryptamines and their pharmacology at the $5HT_{1D}$ receptor: discovery of compounds with potential anti-migraine properties', *Journal of Medicinal Chemistry*, **38**, 3566–80 (zolmitriptan).

Haning, H., et al. (2002) 'Imidazo[5,1-f][1,2,4]triazin-4(3H)-ones, a new class of potent PDE 5 inhibitors', *Bioorganic and Medicinal Chemistry Letters*, **12**, 865–8 (vardenafil).

Holman, N.J. and Friend, C.L. (2001) 'Processes for the preparation of sumatriptan and related compounds', WO Patent 01/34561 (sumatriptan).

Howard, H.R., et al. (1996) '3-Benzisothiazolylpiperazine derivatives as potential atypical antipsychotic agents', *Journal of Medicinal Chemistry*, **39**, 143 (ziprasidone).

Kennis, L.E.J. and Vandenberk, J. (1989) '3-Piperidinyl-substituted 1,2-benzisoxazoles and 1,2-benzisothoazoles', US Patent 4,804,663 (risperidone).

Koprowski, R., et al. (2002) 'Process for preparation of hydrates of olanzapine and their conversion into crystalline forms of olanzapine', WO Patent 02/18390 (olanzapine).

Krochmal, B., et al. (2002) 'Preparation of risperidone', WO Patent 02/14,286 (risperidone).

Larsen, S.D., et al. (2001) 'Olanzapine dihydrate D', US Patent 6,251,895 (olanzapine).

Nowakowski, M., et al. (2002) 'Method for producing sulphonamide-substituted imidazotriazinones', WO Patent 02/50076 (vardenafil).

Ogilvie, R.J. (2002) 'New process', WO Patent 02/50063 (eletriptan).

Oxford, A.W., et al. (1991) 'Indole derivatives', US Patent 4,997,841 (naratriptan).

Patel, R. (2000) 'One pot synthesis of 2-oxazolidinone derivatives', US Patent 6,160,123 (zolmitriptan).

Perkins, J.F. (2001), 'Process for the preparation of 3-acyl-indoles', EP Patent 1,088,817 (eletriptan).

Perrault, W.R., et al. (2002), 'Process to prepare oxazolidinones', WO Patent 02/085,849 (linezolid).

Pete, B., et al. (1998) 'Synthesis of 5-substituted indole derivatives. I: An improved method for the synthesis of sumatriptan', *Heterocycles*, **48**, 1139–49 (sumatriptan).

Radhakrishnan, T.V., et al. (2001) 'A process for the preparation of antipsychotic 3-[2-[4-(6-fluoro-1,2-benzisoxazol-3-yl)-1-piperidinyl]ethyl]-6,7,8,9-tetrahydro-2-methyl-4H-pyrido[1,2,-a]pyrimidin-4-one', WO Patent 01/85731 (risperidone).

Robertson, A.D., et al. (1995) 'Indolyl compounds for treating migraine', US Patent 5,466,699 (zolmitriptan).

Robertson, A.D., et al. (1999) 'Therapeutic heterocyclic compounds', US Patent 5,863,935 (zolmitriptan).

Schaeffer, H.J. (1981) 'Purine derivatives', US Patent 4,294,831 (aciclovir).

Thuresson, B. and Pettersson, B.G. (1960) 'Synthesis of *N*-alkyl-piperidine and *N*-alkyl-pyrrolidine-α-carboxylic acid amides', US Patent 2,955,111 (bupivacaine).

Welch, W.M., et al. (1985) 'Antidepressant derivatives of *cis*-4-phenyl-1,2,3,4-tetrahydro-1-naphthalenamine', US Patent, 4,536,518 (sertraline).

Williams, M. and Quallich, G. (1990) 'Sertraline: development of a chiral inhibitor of serotonin uptake', *Chemistry and Industry*, May, 315–19 (sertraline).

11 Synthesis of isotopically labelled compounds

11.1 Introduction

Isotopes are elements which differ in the number of neutrons they have in their nucleus. They can be categorized as stable/heavy isotopes or radioactive isotopes. The synthesis of isotopically labelled compounds plays an important role in various aspects of the drug development process, either directly or indirectly. The following are examples of where they are of use:

- drug metabolism studies and the identification of drug metabolites;
- determination of biosynthetic pathways and biosynthetic intermediates;
- binding assays in pharmacological tests to determine the affinity of drugs for different protein targets;
- radioactive anticancer drugs;
- radiolabelled compounds used in diagnostic tests;
- drugs containing stable isotopes.

It is possible to label drugs with stable or radioactive isotopes. Obviously, stable isotopes are safer and easier to work with. However, radioactive isotopes are easier to detect and quantify, allowing the detection of labelled structures even when they are present in very small quantities.

11.2 Radioisotopes used in the labelling of compounds

The typical radioisotopes used to label pharmaceuticals and related compounds are tritium (^{3}H or T), carbon (^{14}C), sulphur (^{35}S), and iodine (^{125}I).

11.2.1 Types of radioactive decay

Radioisotopes decay exponentially to form more stable isotopes, and release radiation as they do so. The radiation released can be alpha, beta, or gamma depending on the radioisotope concerned.

Alpha decay (α-decay) involves the release of an α-particle which corresponds to the nucleus of a helium atom ($^4_2$He^{2+}). Therefore, the nucleus that results from α-decay has a mass number that has decreased by four units and an atomic number that has decreased by two units. The positive charges on the α-particle are not normally shown in the fragmentation equation. α-Decay generally only occurs for the heavier radioactive isotopes. The resulting α-particles are relatively heavy compared with β-particles and are easily stopped by paper or skin. They can also only travel a few centimetres through air.

$$^X_Y A \longrightarrow {}^{X-4}_{Y-2} B + {}^4_2 He$$

Beta decay (β-decay) involves the release of an electron (β^- or e$^-$) or a positron (β^+ or e$^+$) depending on the radioisotope involved. If an electron is released, an electron antineutrino ($\bar{\gamma}_e$) is also released. If a positron is released, an electron neutrino (γ_e) is also released. The mass number of the resulting nucleus remains unchanged, but the atomic number increases or decreases by one, depending on whether β^+ or β^- decay has taken place. In the case of β^- decay, a neutron is converted into a proton and an electron, which accounts for the released electron and the increased atomic number.

The energy of the radiation depends on the isotope involved, but can usually be blocked by silver foil.

$$\beta^- \text{-Decay } {}^X_Y C \longrightarrow {}^X_{Y+1} D + e^- + \bar{\gamma}_e \qquad \beta^+ \text{-Decay } {}^X_Y E \longrightarrow {}^X_{Y-1} F + e^+ + \gamma_e$$

Gamma decay (γ-decay) is associated with some decay processes, and involves the release of high energy electromagnetic radiation which can be hazardous to tissues. The radiation travels through barriers such as skin, paper, or foil, but can be stopped by lead shields.

11.2.2 Radioactive and biological half-life

Radioactive decay is exponential, which means that the rate of decay is not a linear process. Therefore, the lifetime of a radioisotope is defined by its half-life ($t_{1/2}$). For example, tritium has a half-life of 12.3 years, which means that half the tritium will have decayed in 12.3 years. Half of the remaining tritium will then decay over the next 12.3 years, and so it continues. Therefore, radioactivity can 'linger' for a much longer period of time than the half-life quoted. For example, it would take 61.5 years for a sample of tritium-labelled drug to decay to 3% of its original radioactivity.

The biological half-life is the half-life of a radiolabelled compound in the body and is not related to the radioactive half-life, as it depends on the rate at which the radioactivity is excreted, rather than its decay. Moreover, the biological half-life depends on the molecule involved. For example, tritiated water has a biological half-life of 12 days, while tritiated thymidine has a half-life of 190 days.

Radioactivity can be detected and measured in units called curies (Ci), where 1 curie corresponds to 2.22×10^{12} disintegrations per minute (dpm). The intensity of radiation varies depending on the isotope involved. The specific activity is a measure of the radioactivity per millimole of source, whether that be the element itself or the molecule in which it has been incorporated.

11.2.3 Commonly used radioisotopes

Tritium (^{3}H or T) is only present in trace amounts in the natural world and is generated by cosmic rays. It has one proton and two neutrons in its nucleus, giving it a mass of 3, and it has a half-life of 12.3 years. It decays to helium and emits a weak form of beta radiation, which is not picked up by standard Geiger counters. Indeed, the radiation can only travel through 5–6 mm of air, and cannot pass through skin. However, it is sensitively and accurately picked up by liquid scintillation counters. It should be pointed out that tritium is potentially dangerous if inhaled or ingested, and if it combines with oxygen to produce tritiated water, it can be absorbed through pores in the skin. Tritium has a maximum specific activity of 29.1 Ci/mmol.

$$^3_1\text{H} \longrightarrow \ ^3_2\text{He}^+ + \text{e}^- + \bar{\gamma}_\text{e}$$ Mean energy = 0.006 Mev
Max energy = 0.019 Mev

Carbon-14 (^{14}C) has six protons and eight neutrons, and decays with a half-life of 5730 years to form nitrogen. A weak form of beta radiation is emitted with a maximum specific activity of 62.4 mCi/mmol. ^{14}C is easier to work with than tritium since it is easily detected by normal hand-held Geiger counters, allowing one to monitor for any spills. Liquid scintillation counters are also sensitive and quantitative in detecting compounds that are labelled with the isotope. The radiation has a range of about 22 cm through air and about 0.27 mm in tissue.

$$^{14}_6\text{C} \longrightarrow \ ^{14}_7\text{N} + \text{e}^- + \bar{\gamma}_\text{e}$$ Mean energy = 0.049 Mev
Max energy = 0.156 Mev

Sulphur-35 (^{35}S) has a half-life of 87.5 days and decays to chlorine with the emission of weak energy beta radiation. The radiation energy is very similar to that of ^{14}C and so it is difficult to distinguish between the two forms. The radiation only travels through 24–30 cm of air and cannot pass through skin. The safety precautions are similar to those used when working with tritium or ^{14}C.

$$^{35}_{16}\text{S} \longrightarrow \ ^{35}_{17}\text{Cl} + \text{e}^- + \bar{\gamma}_\text{e}$$ Mean energy = 0.0487 Mev
Max energy = 0.167 Mev

Phosphorus-32 and phosphorus-33 are both radioactive isotopes that emit beta radiation with half-lives measured in days. The half-lives of ^{32}P and ^{33}P are 14 days and 25 days, respectively. Both isotopes decay to a sulphur isotope, and are useful in labelling nucleotides and nucleic acids. The radiation from ^{32}P is stronger than that from ^{33}P, and can travel through 1–6 m of air compared with 46 cm for ^{33}P. A Perspex shield may be necessary when working with ^{32}P.

$$^{32}_{15}\text{P} \longrightarrow \ ^{32}_{16}\text{S}^+ + \text{e}^- + \bar{\gamma}_\text{e}$$ Mean energy = 0.5 Mev
Max energy = 1.71 Mev

$$^{33}_{15}\text{P} \longrightarrow \ ^{33}_{16}\text{S}^+ + \text{e}^- + \bar{\gamma}_\text{e}$$ Mean energy = 0.085 Mev
Max energy = 0.249 Mev

Iodine-125 has 72 neutrons and 53 protons. It has a half-life of 59–60 days and decays by electron capture to give an excited state of tellurium-125, which decays immediately to the stable form of ^{125}Te with the release of gamma radiation. Electron capture involves one of the protons in the nucleus 'capturing' an inner-shell electron to produce a neutron. As a result, the atomic number decreases by one and the mass number remains the same. An electron neutrino is also released.

The isotope is used to tag proteins such as antibodies. Since gamma radiation is involved, more stringent precautions are required than with the β-emitting isotopes described earlier. For example, workers should wear film badges to detect the level of exposure, and use lead foil shields. Double gloves should also be worn with the outer pair being regularly renewed. The isotope can be particularly risky to the thyroid, and so there should be periodic thyroid monitoring as well as regular measurement of urine samples.

Electron
capture
$$^{125}_{53}\text{I} \xrightarrow{\ \ \ } \ ^{125}_{52}\text{Te}^* \longrightarrow \ ^{125}_{52}\text{Te} + \gamma$$ Mean energy = 0.028 Mev
Max energy = 0.035 Mev
γ_e
Electron
neutrino

Fluorine-18 has a half-life of 110 minutes and mostly decays to a heavy isotope of oxygen (^{18}O), with release of a positron and an electron neutrino. A smaller proportion (3.1%) decays to ^{18}O by electron capture. Each positron produced has a range of 2.4 mm in water and tissue before it meets an electron, whereupon both the positron and the electron are annihilated with release of two gamma ray photons. These photons can be detected, and this is the basis for diagnostic tests involving PET scans (section 11.10.1).

$$^{18}_{9}F \longrightarrow \; ^{18}_{8}O \; + \; e^+ \; + \; \gamma_e \quad \text{Max energy = 0.635 Mev}$$

Carbon-11 is another isotope that decays with the release of a positron and electron neutrino, this time to form ^{11}B. Structures labelled with the isotope are also used diagnostically for PET scans. The half-life of the isotope (20 minutes) is much shorter than for ^{18}F. A small proportion (0.2%) decays by electron capture to ^{11}B.

$$^{11}_{6}C \longrightarrow \; ^{11}_{5}B \; + \; e^+ \; + \; \gamma_e \quad \text{Max energy = 0.96 Mev}$$

Iodine-123 has a half-life of about 13 hours and decays by electron capture to give tellurium-123 with emission of gamma radiation of 159 keV and an internal conversion that causes an electron emission of 127 keV. It is a useful isotope for diagnostic tests involving single photon emission computed tomography (SPECT), which has some advantages over PET imaging (section 11.10).

$$^{123}_{53}I \xrightarrow{\text{Electron capture}} \; ^{123}_{52}Te^* \longrightarrow \; ^{123}_{52}Te \; + \; \gamma \quad \text{Mean energy = 159 kev}$$

Electron neutrino γ_e

11.2.4 Commonly used stable isotopes

The most commonly used stable isotopes in labelling studies are deuterium (^2H or D) and carbon-13 (^{13}C). Deuterium has one proton and one neutron in its nucleus, compared with hydrogen which has only one proton. Structures labelled with a deuterium isotope will have a molecular weight that is 1 amu greater than normal. Therefore, it is possible to detect the presence of such labelled structures in a mass spectrum.

Carbon-13 has six protons and seven neutrons present in its nucleus, as opposed to the naturally abundant isotope of carbon (^{12}C) which has six protons and six neutrons. The relative abundance of ^{13}C is 1.1%, which means that there is an approximately 1% chance of finding a molecule of methane containing a ^{13}C isotope. However, as the number of carbon atoms in a molecule increases, the chance of finding a molecule containing a ^{13}C isotope also increases. For example, there is an 11% chance of finding a molecule of decane with a ^{13}C isotope present. Drug molecules generally have a molecular weight of 300–500, and this equates to a 20–30% chance of finding a molecule with one ^{13}C isotope present. The evidence for this can be seen in the fragmentation pattern of a mass spectrum where one normally sees a peak which is 1 amu greater than the molecular weight. The chances of finding a molecule with two isotopes of ^{13}C present are only 1% of 1%, which means that it is negligible for drug-sized molecules. Introducing a ^{13}C isotope into a molecule will result in a significantly larger fragmentation peak in the mass spectrum.

The ^{13}C isotope also has a spin of ½ which means that it can be detected by NMR spectroscopy. Therefore, ^{13}C NMR spectra can be run on unlabelled organic compounds to detect signals for all the carbon atoms in the skeleton. However, it is important to appreciate that each signal is coming from a different molecule in the sample, as it is highly unlikely that one specific molecule will have more than one ^{13}C isotope present. When a compound is labelled with a ^{13}C label, then the position of the label can be detected on the NMR spectrum since the intensity of the relevant signal is dramatically increased compared with the spectrum of the unlabelled structure.

The signals in a ^{13}C spectrum appear as singlets, since it is extremely unlikely for two ^{13}C molecules to be linked together in the same molecule. Coupling is possible to any protons that are attached to the carbon concerned, but these protons are normally decoupled when the spectrum is being taken in order to prevent this. However, it is possible to carry out labelling experiments where the labelled structure contains two neighbouring ^{13}C isotopes. In such cases, the signals will show up as doublets.

Nitrogen-15 is a stable heavy isotope of nitrogen which can be used to label amino acids and proteins. Compared with the naturally abundant ^{14}N, the natural abundance of ^{15}N is only 0.4%. Nitrogen-15 can be detected using mass spectrometry. Since the isotope has a spin of ½, it can also be detected using NMR spectroscopy. This method of detection has proved useful in determining whether small molecules bind to a protein binding site, since the NMR signals for labelled nitrogens in the binding site are shifted when a molecule is bound close by (section 11.9).

Oxygen-18 has a natural abundance of only 0.2% and is a stable heavy isotope of the naturally occurring oxygen-16. It has been used in kinetic studies in both chemistry and biology, and can be incorporated into peptides when a larger protein is hydrolysed by protease enzymes in ^{18}O-enriched water. Labelled molecules can be detected by mass spectrometry, but NMR spectroscopy is not possible since the isotope has a spin of zero. Oxygen-18 is also used to generate the isotope ^{18}F in cyclotrons.

KEY POINTS

- Isotopes are elements which differ in the number of neutrons present in their nuclei.

- Radiolabelled and heavy isotopes are useful in a variety of pharmaceutical applications.

- The three types of radioactive decay are known as alpha, beta, and gamma decay.
- The half-life of a radioisotope is the time taken for half of its radioactivity to decay.
- The biological half-life of a radiolabelled drug is its half-life in the body and is not related to the radioactive half-life.
- Radioactivity is measured in curies or microcuries, which are related to the number of disintegrations per minute. The specific activity is a measure of the radioactivity per mole or millimole of the radiolabelled molecule.
- Commonly used radioisotopes include tritium (^{3}H or T), ^{14}C, ^{35}S, ^{32}P, ^{33}P, ^{125}I, ^{18}F, ^{11}C, and ^{123}I.
- Commonly used heavy isotopes include deuterium (^{2}H or D), ^{13}C, ^{15}N, and ^{18}O.
- Molecules containing heavy isotopes can be detected using mass spectrometry.
- Molecules containing ^{13}C or ^{15}N can be detected by NMR spectroscopy.

11.3 The production of isotopes and labelled reagents

The production of isotopes and labelled reagents is a specialized field and beyond the scope of a typical pharmaceutical research laboratory. Instead, there are specialized industries which generate the isotopes and incorporate them into common reagents for organic synthesis. Pharmaceutical research laboratories can then purchase these reagents and incorporate the label into drugs or other molecules of interest.

11.3.1 Synthesis of radioisotopes and labelled reagents

The synthesis of radioisotopes involves the use of cyclotrons and nuclear reactors. The typical radioisotopes used in pharmaceutical research are generated by nuclear reactions where a stable isotope is subjected to high energy protons or neutrons (Fig. 11.1). Once radioisotopes have been generated, they are incorporated into simple organic molecules that are likely to be useful in organic synthesis.

11.3.2 Generation of stable, heavy isotopes and labelled reagents

Stable heavy isotopes are generated by enrichment processes which are designed to increase the proportion of molecules containing the heavy isotope. For example, heavy water is water (H_2O) that is enriched with HDO.

11.4 The synthesis and radiosynthesis of labelled compounds

11.4.1 Radiosynthesis

The radiosynthesis of pharmaceuticals and other organic molecules is a specialized field of synthetic organic chemistry which involves synthesizing the target structure with a radiolabel inserted at a defined position(s) of the molecular skeleton. In order to do this effectively and efficiently, it is important to design a practical synthetic route which incorporates the label in high yield. Radiolabelled reagents are extremely expensive. For example, the cost of L-[3-^{14}C] serine in 2013 was £648 for 35 μg/50 μCi. Therefore, the emphasis in radiosynthesis is to ensure that as much of the radiolabel as possible ends up in the final product, and does not end up in side products or unreacted intermediates. Moreover, it is preferable to keep the number of reactions involving radioactive materials to a minimum. This means that the radiolabelled reagent should be used as late on in the synthesis as possible, which can often mean changing the normal synthetic route to a different one altogether.

In order to synthesize a labelled drug, it is necessary to use a simple organic molecule or reagent that already contains the radioisotope and is commercially available. When labelling a drug or target structure, not every atom in the molecule has to be labelled. Indeed, it is actually preferable to label only one position in the molecule. Nor is it necessary for every molecule in the sample to be labelled. Detection methods are accurate enough to detect labelled molecules, even if they represent only a fraction of the total molecules present. The measure of how many radiolabelled molecules are present in a particular

$$^6_3\text{Li} + {}^1_0\text{n} \longrightarrow {}^3_1\text{H} + {}^4_2\text{He}$$

$$^{14}_7\text{N} + {}^1_0\text{n} \longrightarrow {}^{14}_6\text{C} + {}^1_1\text{H}^+$$

$$^{14}_7\text{N} + {}^1_1\text{H}^+ \longrightarrow {}^{11}_6\text{C} + {}^4_2\text{He}$$

$$^{18}_8\text{O} + {}^1_1\text{H}^+ \longrightarrow {}^{18}_9\text{F} + {}^1_0\text{n}$$

FIGURE 11.1 Examples of nuclear reactions used to generate radioisotopes.

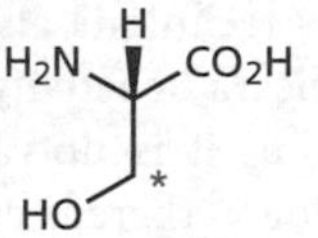

FIGURE 11.2 Structure of L-[3-^{14}C] serine. The asterisk indicates the position of the ^{14}C radiolabel.

sample is given by the specific activity of the compound. This is normally in the order of µCi/mmol.

The method used to incorporate a label into a drug structure will be determined by the labelled compounds that are commercially available. The amount of compound present in commercially available radiolabelled reagents is extremely small. For example, 50 µCi of L-[3-^{14}C] serine (Fig. 11.2) corresponds to approximately 35 µg. This is far too small a quantity to convert to a target structure, nor is there any need to do so. Instead, the labelled material is mixed with an excess of unlabelled compound (~100 mg) and the diluted mixture is used in the first reaction of the reaction sequence. This means that the number of labelled structures present in the starting material and product is very small in comparison to the number of unlabelled structures. For example, if a sample of ^{14}C-labelled product was found to give a radioactive count of 60 000 dpm (or 0.027 µCi), the number of radiolabelled molecules will be 2.42×10^{14}, which corresponds to 0.402 nmol. For a molecule with a molecular weight of 300, this corresponds to 120 ng. This does not really matter since the detection methods for radioactivity are so sensitive. However, it does have an important consequence when it comes to measuring chemical versus radiochemical purity (section 11.4.3).

11.4.2 Practical issues in radiosynthesis

Before any radiosynthesis is attempted, it is important to devise a suitable synthetic route, and then carry out a dress rehearsal of the full synthesis with unlabelled reagents to practice all the procedures, to establish whether the different reactions actually work, and to ensure that products can be isolated and purified.

The synthesis of radiolabelled drugs containing weak β-emitting isotopes such as tritium and ^{14}C can feasibly be carried out without the need for physical barriers, since the radiation is not sufficiently strong to affect the worker unless it is inhaled or ingested. Therefore, normal lab coats, safety glasses, and protective gloves are perfectly suitable. There is also no need for special shields or monitoring badges. However, a Geiger counter should be on hand to regularly check for any radioactivity that might have contaminated glassware, working surfaces, or gloves. All work can be carried out in a normal fume

cupboard. However, it is best to carry out all reactions and work-up procedures in a metal tray such that any spills that might occur are safely contained and are not spread over the whole fume cupboard.

Disposable protective gloves should be worn for all operations, and removed whenever the worker has finished an operation or wishes to leave the fume cupboard for any reason. The gloves should be checked for radioactive contamination before they are removed, and placed in a waste container within the fume cupboard. Normal lab coats can be worn, but it is good practice to restrict the use of that lab coat to the radiosynthetic lab and to remove it whenever the worker leaves the laboratory. The lab coat should also be coloured or marked in such a way that it is clearly reserved for radiosynthetic work. It should also be regularly checked for radioactive contamination. Plastic overshoes are another sensible precaution and must only be worn in the radiosynthetic laboratory.

In general, reactions are carried out on a semi-microscale (typically 100 mmol scale) using semi-micro apparatus. Normal experimental procedures may have to be modified. For example, rather than using separating funnels to carry out extractions or washings, these can be carried out in small flasks or vials, using Pasteur pipettes to mix the phase and then separate them. If a filtration is needed to remove an insoluble impurity or drying agent, the solution can be passed through a Pasteur pipette with a cotton wool plug, rather than a separating funnel. If solvent is to be removed, then it can be removed using a gentle stream of nitrogen rather than with a rotary evaporator. Radiosynthesis requires the chemist to be extremely precise and tidy in all the procedures used and to be constantly checking for radioactive contamination.

Normal laboratories can be used for low level radiosynthetic operations, but a specialized laboratory should be used if radiosynthesis is being carried out on a regular basis, or if high levels of radioactivity are being used. Dedicated radiosynthetic laboratories should have sealed floors and walls which allow the lab to be thoroughly cleaned if there is an accidental spill. There should also be an anteroom with a Geiger counter to check lab coats, hands, clothes, and feet for any accidental contamination before the worker leaves the area. The lab itself should be monitored on a regular basis. The spread of radioactive contamination can be rapid and insidious. If a single crystal of a radioactive compound escapes from the fume hood and drops onto the floor of the lab, the researcher only has to stand on it and his/her progress around the lab will result in invisible radioactive footprints everywhere he/she goes.

It is also important to account for all the radioactivity used in a particular reaction and to identify where it ends up. For example, as well as measuring the amount of radioactivity that ends up in the final product, it is necessary to measure how much of it ends up in washings, extracts, and filtered material. If there is a discrepancy, it is possible that the radioactivity has been lost as a result of volatile or gaseous by-products, in which case traps should be used if the synthesis is repeated.

The measurement of radioactivity is carried out by taking measured aliquots (typically 10 μl in a microcapillary tube) of a radioactive solution, and then adding it to plastic vials containing toluene/methanol solutions and a scintillant. A liquid scintillation counter is then used to measure the radiation in disintegrations per minute. Two or three measurements should be taken to obtain an average result. The total amount of radioactivity in the solution can then be determined by knowing the total volume of the solution.

11.4.3 **Chemical and radiochemical purity**

It has to be appreciated that there is a clear difference between chemical purity and radiochemical purity when carrying out a radiolabelled synthesis. The level of radioactively labelled molecules is very small compared with the total number of molecules present in a product, and so it is possible for the product to have a very good chemical purity but a very poor radiochemical purity, and vice versa. Therefore, it is important to monitor reactions closely for both chemical and radiochemical purity.

Thin layer chromatograms (TLCs) can be run and then monitored under ultraviolet light or by iodine staining to gain a qualititative measure of chemical purity.

Radioactive products can be revealed by autoradiography where an X-ray film is placed over a silica plate. After a suitable exposure time, the film is developed and the radioactive products are shown up as dark spots or bands on the film, providing a qualitative measure of the number of different radiolabelled products present and their relative intensity.

A quantitative measure of radiochemical purity can be obtained by using a thin layer radioactivity scanner to scan a TLC or preparative plate for the distribution of radioactivity. This technique also provides a quantitative measure of the radioactivity for different bands on the TLC plate, but it is not as accurate as liquid scintillation counting. Therefore, a more accurate measurement of radiochemical purity is to scrape off the relevant bands from the TLC plate and add the silica to plastic vials. The silica can then be digested with a mixture of water (0.8 ml) and 48% hydrofluoric acid (0.8 ml) for about 3–4 hours before adding a scintillant and measuring the radioactivity with a scintillation counter. Another method of obtaining a measure of radiochemical purity is to pass a solution of the product down an HPLC column and to collect fractions for counting.

11.4.4 **Radiodilution analysis**

Radiodilution analysis is another means of identifying radiochemical purity. A sample of the radiolabelled product is diluted with a solution of the pure unlabelled product. Aliquots of the solution are taken to measure the radioactivity, and the specific activity is calculated for the diluted sample. The solution is concentrated to obtain the solid product, which is then crystallized. A sample of the crystallized product is then measured for activity. If the specific activity of the crystallized product is the same as the original diluted solution, the radiochemical purity is 100%. If it has dropped, the product is recrystallized until a constant specific activity is obtained. Relating the initial specific activity to the final specific activity then identifies the radiochemical purity of the initial sample.

Radiodilution analysis can also be used to measure the ratio of two different radiolabelled compounds within a mixture. For example, a product was known to contain a mixture of cyclo-(D-phenylalanyl-D-[3-14C]seryl) and cyclo-(D-phenylalanyl-L-[3-14C]seryl) (Fig. 11.3). Two samples of the mixture were taken. One sample was diluted with the unlabelled D,L-diastereoisomer and the other sample was diluted with the unlabelled D,D-diastereoisomer. Each sample was crystallized several times until the specific activity remained constant. This

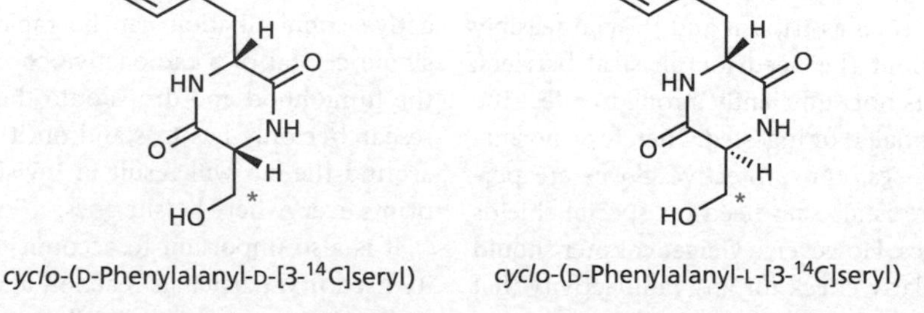

cyclo-(D-Phenylalanyl-D-[3-14C]seryl)   cyclo-(D-Phenylalanyl-L-[3-14C]seryl)

FIGURE 11.3 Structures of radiolabelled cyclic dipeptides. The asterisk indicates the position of the 14C radiolabel.

established that the original mixture contained 68%:28% of D,L to D,D radiolabelled diastereoisomers, plus 4% unidentified labelled material.

If the final product of a radiolabelled synthesis is diluted with pure product and then crystallized to constant specific activity, it ensures that the final crop of crystals has good chemical and radiochemical purity.

11.4.5 Chemical and radiochemical purity of radiolabelled starting materials

It is important to appreciate that the chemical and radiochemical purities of commercially available reagents are unlikely to be 100%, and the purities should be assessed before committing the reagent to a synthesis. For example, a commercial sample of radiolabelled *ortho*-cresol was found to have a radiochemical purity of only 93% and had to be purified before it was used in the synthesis (section 11.5.5).

11.4.6 Stability of radiolabelled compounds

The shelf life of ^{14}C-labelled compounds is relatively short and they should be used as soon as possible after they have been synthesized. The isotope has a long half-life of about 5730 years, and so this does not explain the short lifespan. However, the release of β-particles (electrons) and their interaction with molecules in the sample can create free radicals which lead to decomposition. This is known as **radiolysis**. The lifetime of samples can be increased by keeping them in solution or freezing them.

11.4.7 Practical considerations when synthesizing products labelled with heavy isotopes

Although many of the precautions required when synthesizing radiolabelled compounds are unnecessary when synthesizing a product containing a stable heavy isotope, the principles involved in designing and carrying out suitable synthetic routes are very similar. Commercially available reagents containing heavy isotopes can be purchased and vary in price. For example, deuterated water costs about £16.00 for 10 g. However, other labelled starting materials are much more costly. For example ^{13}C-labelled serine costs about £500 for 100 mg.

11.4.8 Isotope effects

Isotopes have different numbers of neutrons in their nuclei which affects their atomic mass. However, other differences may be observed in the properties of isotopes.

For example, bond strengths can be significantly different. As a result, isotope effects may be observed in chemical and biochemical reactions, where the reaction rates differ between the labelled and the unlabelled compounds.

The isotope effects for carbon, oxygen, and nitrogen are small, and it can be assumed that the labelled molecule behaves in the same way as an unlabelled molecule. However, the isotope effects are far more significant with deuterium and tritium, and can affect reaction rates and equilibrium constants by as much as 50%. It is even possible to separate some tritium-labelled compounds from the unlabelled compound by chromatography. Although the isotope effect may be seen as a problem in some contexts, it can be very useful in others. For example, it has been put to very good use in designing analogues which are more resistant to metabolism (section 11.8.2).

> **KEY POINTS**
> • Radioisotopes are generated by bombarding stable isotopes with high energy protons or neutrons.
> • Stable heavy isotopes are generated by enrichment processes.
> • Radiolabelled compounds can be synthesized using a commercially available radiolabelled reagent. The radiolabel should be incorporated as late on in the synthesis as possible.
> • Only a small fraction of the molecules present in a radiolabelled sample contain the radiolabelled isotope.
> • The chemical purity of a radiolabelled compound is a measure of how pure the overall sample is, whereas the radiochemical purity indicates the purity of the radiolabelled molecules within the sample.
> • Radiochemical purity can be measured by TLC, HPLC, or radiodilution analysis.
> • Radiolysis is where radioactive decay causes chemical decomposition.
> • Isotope effects may be observed in molecules labelled with deuterium or tritium.

11.5 The use of labelled drugs in drug metabolism studies

11.5.1 The importance of studying drug metabolism

The body has an arsenal of metabolic enzymes that can modify the structure of a drug to produce different compounds known as drug metabolites. In general, these metabolites are more polar than the original compound, and are more rapidly excreted as a result. Ideally, any metabolites that are formed should be inactive and quickly excreted. However, it is quite likely that some will have some form of biological activity. One aspect of preclinical testing is to identify the metabolites formed from a pharmaceutical agent and to assess what kind of activity they have. This is important for a number of reasons.

- The parent drug may appear relatively safe and potent when *in vitro* tests are carried out, but if a toxic metabolite is formed then the drug may fail toxicity tests carried out *in vivo*.

- It is important to establish whether any metabolites have a similar pharmacological activity to the parent drug. It may be that the parent drug has a defined biological half-life, but if active metabolites are formed, the observed activity may last longer than anticipated. This will affect the level and frequency of dosing during clinical trials.

- Metabolites may interact with other target proteins in the body, and result in side effects.

- Metabolites may inhibit or enhance the activity of the metabolic enzymes which create them. This can lead to drug–drug interactions. In other words, a metabolite that inhibits a metabolic enzyme may prolong the activity of a different drug that is normally metabolized by that enzyme.

- It is possible that the parent drug is not actually the active compound at all, but is converted by metabolic enzymes to the active compound. In other words, the agent is acting as a prodrug. There are several examples where drug metabolism studies have identified this phenomenon and allowed medicinal chemists to design improved agents. For example, **oxamniquine** is an antiprotozoal drug that was discovered following drug metabolism studies. These demonstrated that a methyl group in the original drug candidate was oxidized by cytochrome P450 enzymes (Fig. 11.4).

Drug metabolites are likely to be present in very small quantities in biological fluids or tissues, and so a sensitive method of detecting them is required. This is why radiolabelling is so useful. Introducing a radioactive isotope into a candidate drug allows one to identify the presence of metabolites far more easily, since the metabolites should also be radioactive and easily detectable.

The radiolabelled drug can be administered to test animals during preclinical trials, or to human volunteers in phase I clinical trials. Blood and urine samples can then be taken at regular intervals to detect the levels of radioactivity present, allowing one to assess the rates of absorption, distribution, and excretion. The samples can also be studied by HPLC to see whether the parent drug has been metabolized, and to assess how many different metabolites have been formed. These metabolites would be revealed by a number of radioactive peaks appearing in the HPLC trace. The relevant fractions could then be collected and analysed to try and identify the structure of the metabolite present.

11.5.2 Synthetic priorities for radiolabelling

There are several priorities to consider when introducing a radiolabel into a drug structure. The first of these is to ensure that it is incorporated into a defined position within the molecule in good yield. The isotope should also be stable in that position and not easily lost as a result of chemical or metabolic reactions.

11.5.3 Incorporation of tritium

Tritium can be very easily incorporated into any organic molecule containing an exchangeable proton, such as those present on an alcohol, carboxylic acid, or phenol. This is done by simply shaking a solution of the drug with T_2O (Fig. 11.5). Unfortunately, the label is just as easily lost as a result of proton exchange with water once the compound is administered to a test animal or patient.

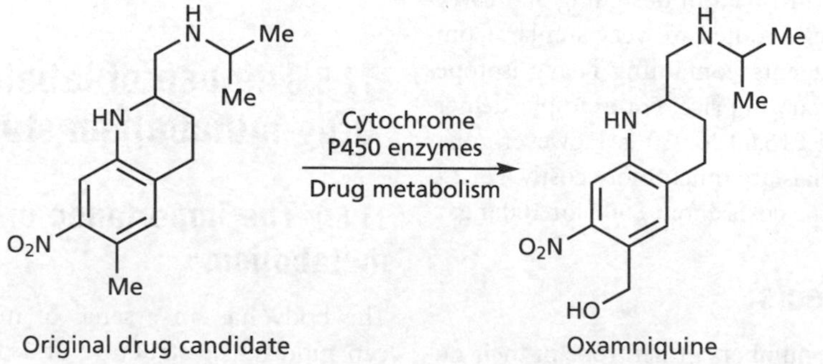

FIGURE 11.4 Metabolism of a drug candidate for antiprotozoal activity.

Therefore, it is best to carry out a synthesis that places the label on the carbon skeleton of the drug. The label will then be stable to proton exchange with water. However, there are certain parts of a carbon skeleton where the label should not be incorporated. For example, there is always the possibility that deuterium or tritium could be lost through a metabolic reaction affecting the part of the molecule where the label is situated. If the label has been incorporated to detect drug metabolites, the metabolite resulting from that metabolic reaction would not be detected. Therefore, the label should not be placed on groups or positions which are known to be prone to metabolic reactions.

One possible way of inserting tritium onto the carbon skeleton of a drug is to identify a weakly acidic proton in the structure which is stable at pH 7.4, but could be removed under more basic conditions. The protons on an α-carbon atom relative to a carbonyl group are likely candidates as it is possible to remove them in the presence of strong base to form an enolate, which can then be quenched with deuterated or tritiated water (Fig. 11.6). The D or T label introduced in this way will remain attached to the molecule under normal physiological conditions. However, it is important to ensure that the basic conditions used to introduce the label do not degrade the compound or racemize any asymmetric centres that are present.

If the drug does not contain a suitably acidic proton, there may be a synthetic intermediate which does. The label could be introduced at that stage and the labelled intermediate converted to the final product. However, it is important to ensure that the label remains attached throughout the synthesis and does not become 'scrambled' (i.e. moved to different positions due to isomerizations).

Some common methods of introducing a tritium label as part of a synthesis are shown in Figure 11.7. There is no guarantee that the standard synthetic route used to

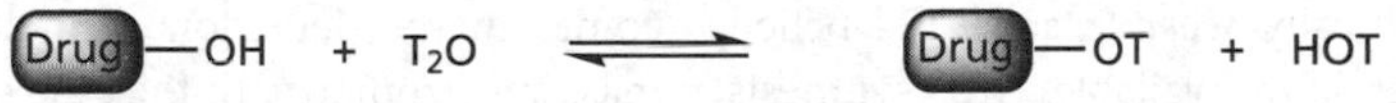

FIGURE 11.5 Replacing an exchangeable proton with tritium.

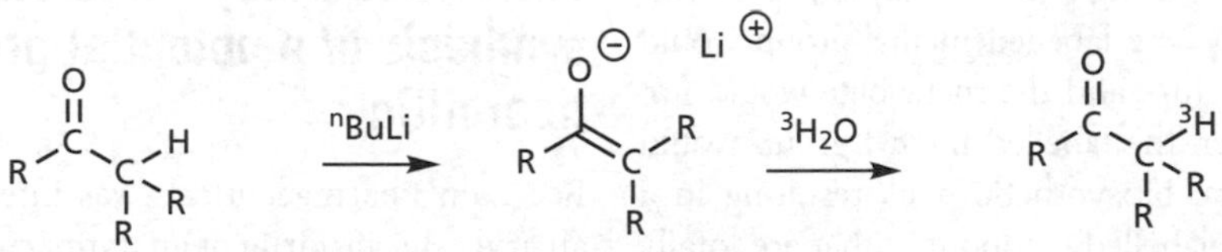

FIGURE 11.6 Incorporating tritium by reaction with an enolate ion.

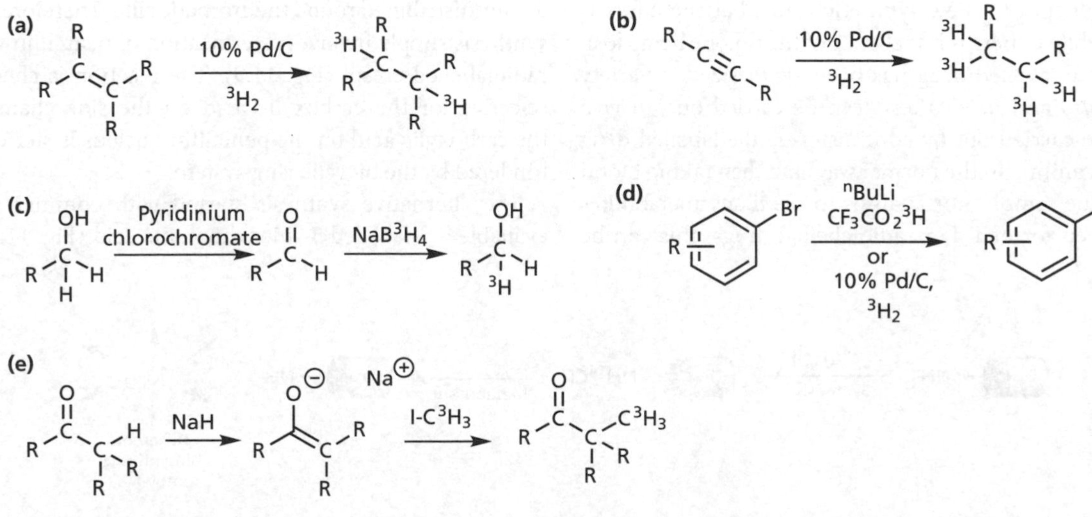

FIGURE 11.7 Reactions by which a tritium label can be incorporated into a carbon skeleton.

produce the target compound contains a suitable reaction by which the label can be incorporated, in which case it may be necessary to design a totally different synthesis.

11.5.4 Incorporation of carbon-14

There are several advantages in using ^{14}C rather than ^{3}H for drug metabolism studies. Primarily, carbon labels that are incorporated into the drug's skeleton do not run the risk of being exchanged or 'scrambled', and there is less of an isotope effect. The disadvantage of using a carbon label is the difficulty that might arise in incorporating it into the molecule. Indeed, a completely new synthesis may need to be devised in order to be successful.

When designing a labelled synthesis, it is important to carry out a retrosynthesis in order to identify a simple reagent that is likely to be commercially available in the labelled form. This will usually restrict the positions of the drug which can be labelled. However, it is also important to find a biologically stable position for the label. For example, consider a drug which contains an *N*-methyl group. An *N*-methylation is synthetically feasible and ^{14}C-labelled iodomethane is commercially available. The synthesis could also be designed such that *N*-methylation is the final stage in order to maximize the overall radiochemical yield. However, it is known that *N*-methyl groups are susceptible to drug metabolism through *N*-demethylation. If this happened to the drug, the labelled methyl group would be lost from the structure and the metabolite would not be detected. Moreover, the labelled methyl group would enter the cell's general biosynthetic pool, resulting in a large number of radiolabelled compounds that are totally unrelated to the drug (Fig. 11.8).

Therefore, it is important to avoid placing the ^{14}C label at any position on the drug scaffold that is likely to be susceptible to drug metabolism. This may require the development of a new synthetic route, but the effort is worthwhile if there is less chance of the isotope being lost.

Once a labelled drug has been synthesized, a variety of *in vitro* and *in vivo* tests can be carried out. *In vivo* tests are carried out by administering the labelled drug to a test animal in the normal way, and then taking blood and urine samples for analysis to see if any metabolites have been formed. For radiolabelled drugs, this can be done by using high performance liquid chromatography (HPLC) with a radioactivity detector. It is important to choose the correct animal for these studies, as there are significant metabolic differences across different species. *In vivo* drug metabolism tests are also carried out as part of phase I clinical trials to see whether the drug is metabolized differently in humans from any of the test animals.

In vitro drug metabolism studies can also be carried out using perfused liver systems, liver microsomal fractions, or pure enzymes. Many of the individual cytochrome P450 enzymes that are so important in drug metabolism are now commercially available.

11.5.5 Incorporation of stable heavy isotopes

There may be situations where it is desirable to synthesize a drug containing a stable heavy isotope such as deuterium or carbon-13. If so, the methods and strategies involved are the same as those described for tritium and ^{14}C, respectively (sections 11.5.2–11.5.4). The difference comes in the detection of metabolites, which requires mass spectrometry. In the case of ^{13}C labelled structures, ^{13}C NMR spectroscopy can also be used.

11.5.6 Case Study—Radiolabelled synthesis of a potential prodrug for ticarcillin

Beecham Pharmaceuticals was interested in developing an aryl ester prodrug of an antibacterial penicillin called **ticarcillin**, and wished to carry out drug metabolism studies to determine the metabolic fate of both parts of the ester. The aryl part of the ester is derived from *ortho*-cresol which can be purchased with the ^{14}C label uniformly distributed round the aromatic ring. Therefore, the synthesis simply involves esterification of ticarcillin with radiolabelled cresol (Fig. 11.9). The reaction is chemoselective for the carboxylic acid on the side chain, as the carboxylic acid on the penicillin nucleus is sterically hindered by the bicyclic ring system.

The alternative synthesis started with commercially available ^{14}C-labelled 3-thienylmalonic acid (Fig. 11.10),

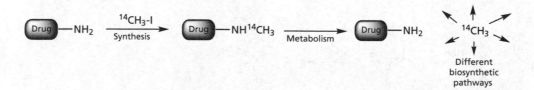

FIGURE 11.8 Disadvantages of labelling an *N*-methyl group.

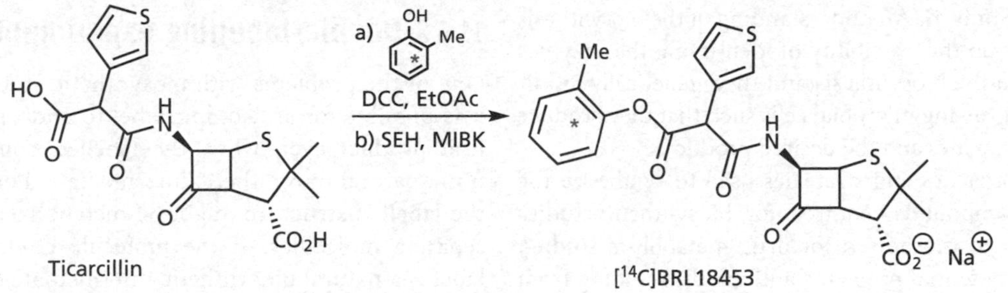

FIGURE 11.9 Synthesis of a ^{14}C-labelled prodrug of ticarcillin.

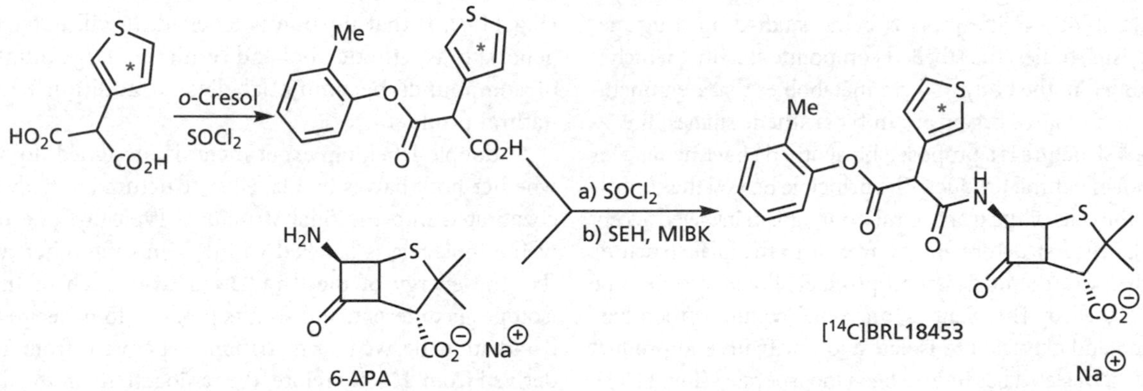

FIGURE 11.10 Synthesis of a ^{14}C-labelled prodrug of ticarcillin.

which was diluted with 1 mmol of unlabelled sample and then reacted with *ortho*-cresol. The product was treated with thionyl chloride to generate the acid chloride and then linked to 6-aminopenicillanic acid. The coupled product was then converted to its sodium salt.

As with all radiosynthetic operations, it was important to assess where all the radioactivity ended up. The reaction started with 5.5 mCi of 3-thienylmalonic acid, and 3.52 mCi ended up in the final product. The remainder of the radioactivity could be accounted for by the radioactivity measured in the aqueous and organic washings carried out during the synthesis, as well as the mother liquors from the crystallizations carried out on the final product. The chemical and radiochemical purities were consistent and were determined by HPLC.

KEY POINTS

- Radiolabelled drugs are used in drug metabolism studies. They generate radiolabelled metabolites which can be detected and isolated.

- Radioisotopes should be incorporated into a chemically and metabolically stable position within a structure.

- It is feasible to radiolabel a drug with tritium if a carbonyl group is present in the structure. Weakly acidic protons on an α-carbon can be removed with base and replaced with tritium.

- Compounds can be radiolabelled at specific positions by using a synthesis which involves a suitable radiolabelled reagent.

11.6 The use of labelled compounds in biosynthetic studies

11.6.1 Introduction

Labelled compounds are essential to biosynthetic studies aimed at identifying how natural products are produced and the biosynthetic intermediates involved. An understanding of a biosynthetic process can be important to the pharmaceutical industry for a number of reasons. For example, the identification of biosynthetic intermediates leading to a clinically important natural product may allow that intermediate to be isolated and then converted to the final product by a semi-synthetic process. Alternatively, it may be possible to produce analogues of the natural product by feeding the natural source with analogues of the biosynthetic intermediate (see Case Study 4). An understanding of the biosynthesis also opens up the possibility of identifying the enzymes involved in the biosynthesis and then genetically modifying fast-growing microbial cells such that they produce both the enzymes and the desired product.

The approaches and strategies used to synthesize the labelled compounds required for biosynthetic studies are similar to those used for drug metabolism studies, as are the potential problems and pitfalls resulting from chemical or metabolic susceptibility of the isotopic label. Tritium and ^{14}C radioisotopes are commonly used, as are stable heavy isotopes such as deuterium and ^{13}C.

One difference between drug metabolism and biosynthetic studies is the process being studied. In drug metabolism studies, the labelled compound is a drug which is modified in the body to form metabolites that are unnatural to the source organism. In biosynthetic studies, the labelled structure is a proposed biosynthetic intermediate for a known natural product. The principle behind the study is that the cells of the plant or microorganism involved accept the labelled structure and convert it to the final structure. In this way, the final natural product should also become radiolabelled. The plant or microbial culture is then harvested and extracted to isolate and purify the end product to see if the label has indeed been incorporated (Fig. 11.11). If the label is present, it suggests that the labelled compound is a biosynthetic intermediate, but it is not absolute proof. Further experiments have to be carried out in order to establish whether the label was incorporated as a result of the biosynthetic pathway, or whether the label ended up in the final structure by a more indirect route (section 11.6.2).

11.6.2 **Double-labelling experiments**

One of the problems with biosynthetic studies is that it is possible for an isotopic label to end up in a natural product even when the labelled compound is not a natural biosynthetic intermediate. For example, the labelled structure might be metabolized into two separate molecules. If the molecule containing the label *is* a natural biosynthetic intermediate or starting material, then the label will still be incorporated into the final structure (Fig. 11.12). A similar possibility is where a metabolic reaction cleaves off a small portion of the test compound, for example a methyl group (Fig 11.8). If that portion is labelled, it will enter the general biosynthetic pool and result in a large number of compounds becoming labelled in addition to the natural product.

A double-labelling experiment is designed to test whether both halves of a labelled structure are truly incorporated into the final structure. Typically, one part of the molecule is labelled with ^{14}C and the other with ^{3}H. The energy of the β-radiation from each of these isotopes is different, and so it is possible to measure the β-radiation derived from tritium separately from that derived from ^{14}C. Therefore, the radioactivity in the final product can be measured to identify whether both tritium and ^{14}C are present (Fig. 11.13). If both are present, this is further evidence that the test compound has been incorporated, but it is still not absolute proof. It is possible that the test compound might have been split into two halves, and that both halves were incorporated into

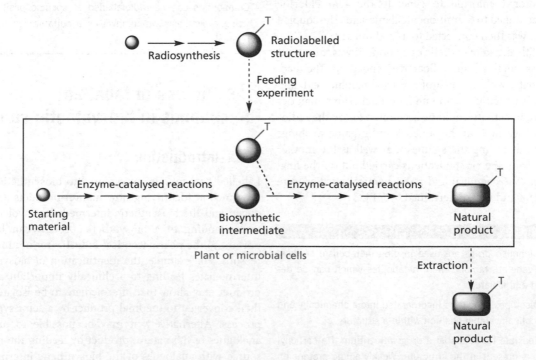

FIGURE 11.11 Principle behind labelling studies in a biosynthetic study.

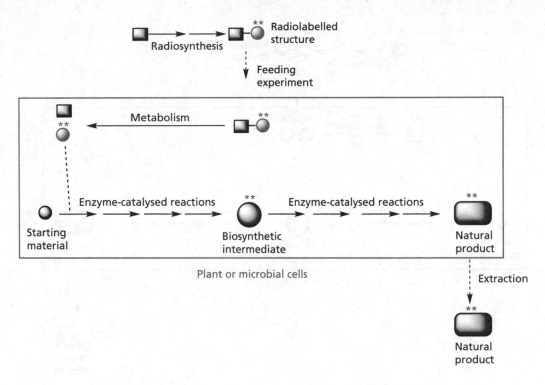

FIGURE 11.12 Incorporation of a label as a result of metabolic degradation of the labelled structure.

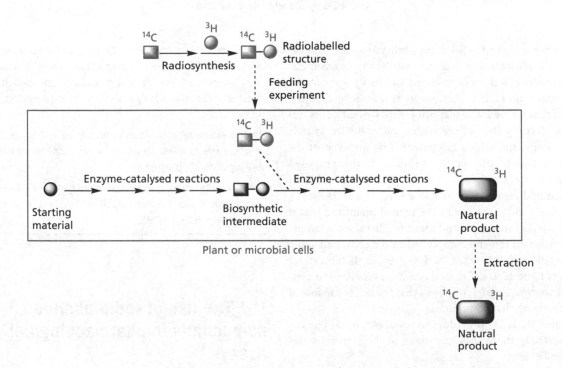

FIGURE 11.13 Double-labelling experiment.

the final product at different stages of the biosynthetic route (Fig. 11.14).

This possibility can be discounted by measuring the ratio of radioactivity due to tritium and ^{14}C in both the original test compound and the final product. If they are the same, this is good evidence that the intermediate was incorporated intact. If the molecule had been split into two halves which had been incorporated separately, it is highly unlikely that they would have been incorporated to the same extent.

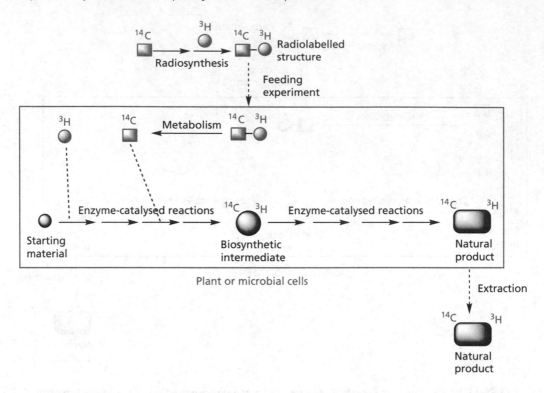

FIGURE 11.14 Possible route by which both labels of a double-labelled structure could be incorporated into a natural product, despite metabolic breakdown.

Even so, there is still a possibility that the double-labelled structure is not an authentic biosynthetic intermediate, but was converted to one by a metabolic reaction (Fig. 11.15). This can be ruled out by carrying out an **intermediate trapping experiment**. This involves feeding the culture with a radiolabelled sample of a compound which is known to be involved in the biosynthesis of the natural product. If the proposed intermediate is authentic, it should be formed in the cell and should also be radiolabelled (Fig. 11.16). However, it is likely to be present in such small quantities that it is impossible to extract and identify. Therefore, a sample of unlabelled intermediate is fed to the cells, and some time later it is extracted back out again. If the extract proves to be radioactive and can be crystallized to constant specific activity, it proves that the cell is capable of synthesizing that compound.

Case Study 4 provides an example of a biosynthetic study that illustrates many of the principles described above.

- Labelled compounds are crucial to biosynthetic studies.
- A knowledge of a biosynthetic process can help to identify intermediates and enzymes that could be used in the synthesis of a natural product or its analogues.

- Labelling studies involve feeding a plant or culture with a proposed biosynthetic intermediate which is labelled with an isotope. The natural product is extracted to detect whether the isotope has been incorporated into the structure.
- Double-labelling experiments are carried out to establish whether the labelled structure is incorporated intact into the biosynthetic product.
- Intermediate trapping experiments are carried out to establish whether a cell is capable of synthesizing a proposed biosynthetic intermediate.

11.7 The use of radiolabelled compounds in pharmacological assays

Radiolabelled compounds are used routinely in pharmacological studies to detect whether novel compounds are interacting with a target receptor, and to carry out quantitative measurements of binding affinities. They are also used to determine the location of different types of receptors in different tissues.

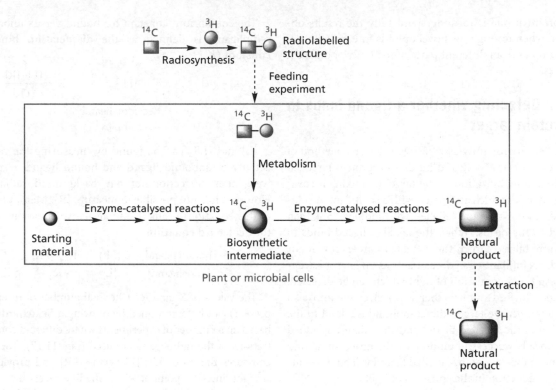

FIGURE 11.15 Route by which a double-labelled structure could be incorporated intact into a natural product despite its not being an authentic biosynthetic intermediate.

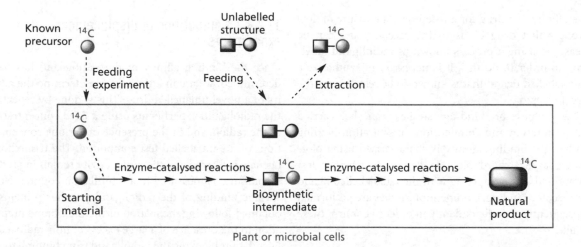

FIGURE 11.16 Intermediate trapping experiment.

11.7.1 Detection of receptor distribution in different tissues

Radioligand studies can be carried out on frozen tissue sections using autoradiography to determine the distribution of different types of receptors. In order to do this, the radioligand used in the experiments must show selectivity for the receptors of interest. For example, the distribution of different types of opioid receptor was determined in CNS tissue. These studies demonstrated that the three main opioid receptors (mu, delta, and kappa) are not distributed evenly throughout the CNS. Certain regions of the CNS are rich in one type of opioid receptor, while other regions are rich in another. Moreover, the pattern of receptor distribution varies from one animal species to another. These observations are important if one is to understand fully the results obtained when testing how novel opioids interact with tissue samples from different parts of the body or different species.

11.7.2 **Detecting whether a ligand binds to a protein target**

Scintillation proximity assay (SPA) is a visual method of detecting whether a ligand binds to a protein target. It involves immobilization of the target by linking it covalently to beads which are coated with a scintillant. A solution of a known ligand labelled with iodine-125 is then added to the beads. When the labelled ligand binds to the immobilized target, the ^{125}I acts as an energy donor and the scintillant-coated bead acts as an energy acceptor, resulting in emission of light which can be detected. In order to find out whether a novel compound also interacts with the target, the compound is added to the solution of the labelled ligand and the mixture is added to the beads. Successful binding by the novel compound will mean that less of the labelled ligand will bind, resulting in a reduction in the emission of light.

11.7.3 **Measurement of binding affinities with a receptor**

The affinity of a drug for a receptor is a measure of how strongly that drug binds to the receptor, and can be measured using a process known as **radioligand labelling**. In order to do this, it is necessary to synthesize a radiolabelled ligand that is known to be selective for the target receptor.

Two types of binding studies can be carried out—saturation and competition. In saturation binding studies, the binding affinity is characterized using plots known as Scatchard plots to determine whether test compounds are competing with the radiolabelled ligand for receptor binding. Competition studies are performed subsequently or independently in order to confirm these results.

11.7.3.1 Saturation binding studies

In saturation binding studies, tissue homogenates or genetically modified cells are incubated with various concentrations of the radiolabelled ligand. Once an equilibrium has been reached, the unbound ligands are removed by washing, filtration, or centrifugation. The extent of binding can then be measured by detecting the amount of radioactivity present in the cells or tissue, and the amount of radioactivity that was removed. Thus, the amount of bound ligand can be measured as well as the amount of free unbound ligand.

The equilibrium constant for bound versus unbound radioligand is defined as the **dissociation binding constant** (K_d).

$$[L] + [R] \quad \rightleftharpoons \quad [LR] \qquad K_d = \frac{[L]+[R]}{[LR]}$$

Receptor-ligand
complex

[L] and [LR] can be found by measuring the radioactivity of unbound ligand and bound ligand, respectively, after correction for any background radiation. However, it is not possible to measure [R]. Manipulation of this equation results in an alternative equation called the **Scatchard equation**.

$$\frac{[\text{Bound ligand}]}{[\text{Free ligand}]} = \frac{[LR]}{[L]} = \frac{R_{tot}-[LR]}{K_d}$$

The values of K_d and R_{tot} (the total number of receptors present) can be determined by drawing a **Scatchard plot** based on a number of experiments where different concentrations of the radioligand are used (Fig. 11.17). The plot compares the ratio [LR]/[L] versus [LR] and provides a straight line. The point at which the line meets the x-axis represents the total number of receptors available (R_{tot}) (line A in Fig. 11.17). The slope is equivalent to $-1/K_d$ and provides a measure of the radioligand's affinity for the receptor.

11.7.3.2 Competition or displacement experiments

Once the binding affinity of the radioligand has been determined, we are in a position to determine the affinity of a novel unlabelled drug. This is done by repeating the radioligand experiments using a fixed concentration of the radioligand in the presence of various concentrations of the unlabelled test compound. The tissue homogenate is filtered and the radioactivity remaining in the membranes is measured in a scintillation counter. Nonspecific binding of the radioligand to other proteins is obtained following incubation of the membrane preparation in the presence of a large excess of non-radioactive ligand, thus blocking the radioligand from binding to the target receptors. This measure can then be subtracted from the previous results to get the true binding of radioligand to the receptor binding sites.

The test compound competes with the radioligand for the receptor's binding sites and is called a **displacer**. The stronger the affinity of the test compound, the more effectively it will compete for binding sites and the less radioactivity will be measured for [LR]. This will result in a different line in the Scatchard plot.

If the test compound competes directly with the radiolabelled ligand for the same binding site on the receptor, the slope is decreased but the intercept on the x-axis remains the same (line X in Fig. 11.17). In other words, if the radioligand concentration is much greater than the test compound it will bind to all the receptors available.

If the slope remains the same and R_{tot} decreases, the radiolabelled ligand and test compound are binding at different sites on the receptor.

The data from these displacement experiments can be used to plot a different graph which plots the percentage of the radioligand that is bound to a receptor versus the concentration of the test compound. This results in a sigmoidal curve termed the **displacement curve** or **inhibition curve**, which can be used to identify the **IC$_{50}$ value** for the test compound (i.e. the concentration of compound that prevents 50% of the radioactive ligand from being bound). The lower the value of IC$_{50}$ or K_d, the stronger is the binding affinity.

Affinities for a test compound are sometimes defined by K_i, which is the inhibitory or affinity constant. This is the same as the IC$_{50}$ value if non-competitive interactions are involved. For compounds that *are* competing with the radioligand for the binding site, the inhibitory constant depends on the level of radioligand present and is defined as

$$K_i = \frac{IC_{50}}{1 + [L]_{tot}/K_d}$$

where K_d is the dissociation constant for the radioactive ligand and $[L]_{tot}$ is the concentration of radioactive ligand used in the experiment.

It is important to appreciate that these studies determine a drug's ability to bind to a receptor. They do not reveal whether that agent acts as an agonist or an antagonist. In order to determine this, other radiolabelling experiments need to be carried out to detect a measurable response in an intact cell.

11.7.4 Radioligand binding studies in the study of opioid receptors

Radioligand binding studies have proved crucial in assessing the binding affinity of opioids to different opioid receptors, as well as identifying the location and distribution of these receptors in different body tissues. Agents that have known selectivity for a certain type of receptor are synthesized such that they contain a radioactive isotope. They are then added to a homogenized tissue sample such that the ligand binds to the binding site of any target receptor that is present. Unfortunately, most drugs bind in a non-specific fashion to a whole range of different proteins that are present in tissue samples, which means that some of the radiolabelled sample will bind to proteins other than the target receptor. Fortunately, the interactions involved in this non-specific binding are weaker than the binding interactions to a target binding site, and so any radiolabelled opioid that is not bound to the target receptor can be removed by thoroughly washing the tissue. Moreover, it is possible to detect radioactivity using a very small quantity of sample, which, in itself, favours selective binding to target receptor binding sites.

There are three opioid receptors which are defined as mu (μ), kappa (κ), and delta (δ). Various tritiated structures showing selectivity for these receptors have been used in pharmacological tests (Fig. 11.18). Tritiated [D-Ala^2,MePhe^4,Glyol^5]enkephalin (DAGO or DAMGO) is a modified enkephalin that is commonly used as a selective radiolabelled ligand for mu receptors and has a 200-fold higher affinity for the mu receptor over the delta receptor. The tritiated cyclic enkephalin c[D-Pen^2, D-Pen^5]enkephalin or Tyr-c(D-Pen-Gly-Phe-D-Pen) (DPDPE) has been used as a selective ligand for delta receptors, while tritiated U-69593 is a selective ligand for the kappa receptor.

Another method of carrying out radioligand binding studies on opioid receptors is to use a non-selective radiolabelled opioid such as **etorphine** or **diprenorphine** (Fig. 11.19) along with a selective unlabelled opioid. The selective opioid binds selectively to its target opioid receptor, which results in the labelled opioid binding to two out of the three opioid receptors.

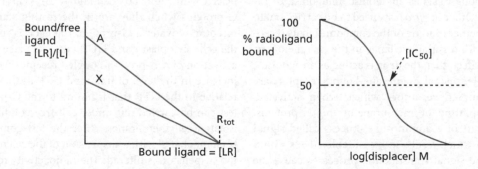

FIGURE 11.17 Scatchard plot. A = radioligand only; X = radioligand + competitive ligand.

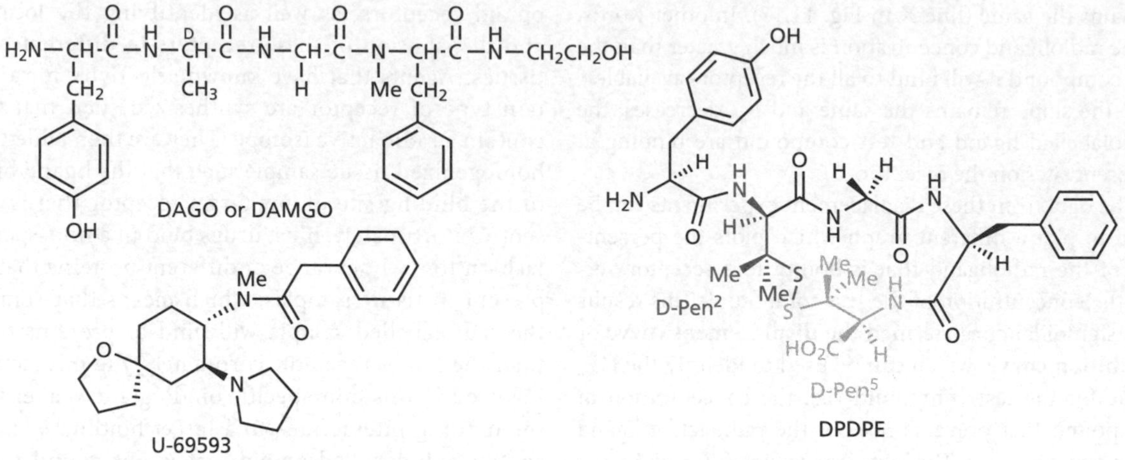

DAGO or DAMGO

U-69593

DPDPE

FIGURE 11.18 Structures showing selectivity for different types of opioid receptor.

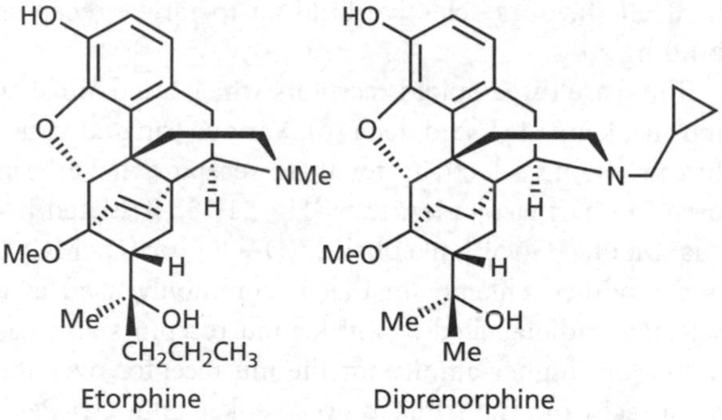

Etorphine   Diprenorphine

FIGURE 11.19 Etorphine and diprenorphine.

11.7.5 Radiolabelling experiments to determine whether a receptor ligand acts as an agonist or an antagonist

Binding affinities measure how strongly a ligand binds to a receptor, but do not indicate whether the ligand activates the receptor. In other words, they do not tell you whether the test compound acts as an agonist, antagonist, or inverse agonist. Other *in vitro* tests need to be carried out to determine whether binding of the compound to its receptor results in a measurable change in the cell chemistry. If there is an effect, then the drug is acting as an agonist.

For example, several membrane-bound receptors are **G-protein-coupled receptors** which, when activated, trigger the splitting of membrane-bound G-proteins. This is the start of a domino-like process called **signal transduction** which results in measurable effects within the cell. Some signal transduction processes cause the levels of a secondary messenger called **cyclic AMP** to increase or decrease depending on the types of membrane-bound receptor that are activated.

In order to measure the amount of cyclic AMP present, cells are fed with tritiated adenine such that any cyclic AMP produced becomes radiolabelled. The levels of cyclic AMP within the cell can then be determined by scintillation counting. If activation of a G-protein-coupled receptor normally increases the levels of cyclic AMP within the cell, the levels of radiolabelled cyclic AMP should increase when an agonist is being tested.

If activation of the receptor normally diminishes cyclic AMP levels, an agonist will diminish the levels of radiolabelled cyclic AMP. This test is carried out by first administering a natural product called **forskolin** (Fig. 11.20) which stimulates a receptor that increases the production of cyclic AMP. The amount of radiolabelled cyclic AMP is measured to determine how much the levels have been raised. Test compounds are now added to the cells to see whether they counter this effect. If they do, the compounds are acting as agonists. As a further check, an antagonist can be added to see whether it reverses the effect of the agonist on cyclic AMP levels.

Another method of testing whether G-protein-coupled receptors are being activated is to use an analogue of GTP, where one of the oxygen atoms has been replaced with radioactive sulphur (^{35}S). GTP is bound to G-proteins when they are in the resting state. When a receptor activates a G-protein, the GTP is released into the cell's cytoplasm and GDP takes its place. Therefore, activation of a G-protein-coupled receptor would see an increase in the level of unbound GTP in the cytoplasm, relative to the GTP that is bound to the G-protein. The test can be carried out under cell-free conditions using isolated cell membranes. After the test compound has been incubated with a suspension of the cell membranes, the suspension is filtered. The radioactivity measured in the cell membranes indicates the amount of GTP bound to the G-protein, while the radioactivity of the filtrate indicates the amount of unbound GTP.

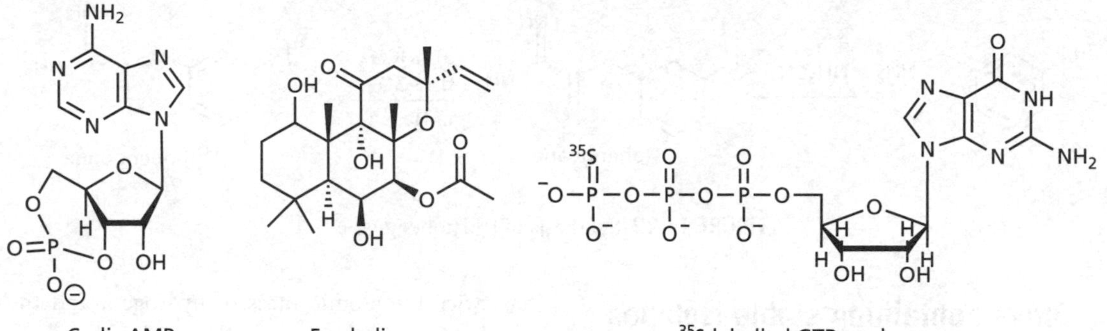

FIGURE 11.20 Cyclic AMP, forskolin, and a ^{35}S-labelled analogue of GTP.

- Radiolabelled ligands are used to identify the distribution of different receptor types in various tissues.
- Ligands for a protein target are labelled with ^{125}I and used in scintillation proximity assays to determine whether a test compound binds to the same target.
- Affinity is a measure of how strongly a drug binds to a receptor.
- Affinity can be measured from Scatchard plots derived from radioligand displacement experiments.
- Radiolabelled ligands are used to determine a compound's selectivity for different receptor types.
- Radiolabelled compounds can be used to detect the increase or decrease of enzyme-catalysed reactions within the cell. This can help to identify receptor agonists and antagonists if the enzymes are affected by receptor-mediated signal transduction processes.
- Radiolabelled analogues of GTP are used to detect whether a G-protein receptor is activated by a test compound. Activation of the receptor will result in an increased level of GTP in the cell cytoplasm.

11.8 Isotopically labelled drugs

11.8.1 Radioactive drugs

Radioactivity is harmful to cells and one approach to anticancer therapy has been to link radioactive isotopes to proteins called antibodies. Antibodies are part of the body's immune system and are able to recognize and bind to a molecular feature (called an antigen) on specific proteins. Normally, antibodies identify foreign proteins revealed by infecting microbes. By binding to these proteins, the antibodies direct the cell's defences to attack and eliminate the foreign invaders and defeat disease.

It is possible to produce antibodies in the lab that recognize specific proteins that are overproduced in tumour cells. If those antibodies are administered, they seek out and bind selectively to any tumour cells that are overexpressing that particular protein. If a radioisotope is attached to those antibodies, the radioisotope is carried to those tumours and held there such that the radiation kills the tumour cells in preference to normal cells. Such antibodies are called **conjugated antibodies** because they are linked to the radioisotope.

Ibritumomab and **tositumomab** are two such conjugated antibodies which contain the radioactive isotopes ^{90}Y and ^{131}I, respectively. The antibodies recognize a protein called CD20 which is present on the surface of B-lymphocytes—cells that grow uncontrollably in non-Hodgkin's lymphoma. The antibodies carry the radioactive isotopes to the cancerous cells and bind to the CD20 protein such that the radiation from the isotopes proves fatal to the tumour cell.

Iobenguane is a compound that binds to adrenergic receptors. A radiolabelled version of the compound ([^{131}I]iobenguane) has proved useful in targeting and eradicating tumours that have an affinity for noradrenaline. The iodine isotope has a half-life of 8 days and decays with the emission of beta particles, antineutrinos, and gamma rays (Fig. 11.21), which result in DNA mutation and cell death.

Na^{131}I is used in the final stage of the synthesis to incorporate the radiolabel. This involves an exchange process where the unlabelled iodine in iobenguane is replaced with the labelled iodine (Fig. 11.22).

$$^{131}_{53}\text{I} \longrightarrow \,^{131}_{54}\text{Xe}^* + \bar{\nu} + 606.3 \text{ keV } \beta^-$$

$$^{131}_{54}\text{Xe}^* \longrightarrow \,^{131}_{54}\text{Xe} + 364.3 \text{ keV } \gamma$$

FIGURE 11.21 Decay process for iodine-131.

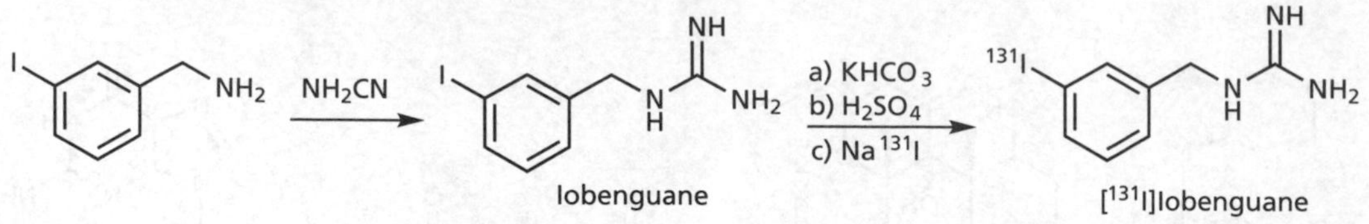

FIGURE 11.22 Synthesis of [^{131}I]iobenguane.

11.8.2 Drugs containing stable isotopes

Deuterium-modified drugs are being investigated by firms such as Concert Pharmaceuticals, GlaxoSmith Kline, and Auspex as a means of improving the biological half-life of known drugs. For example, **CTP-518** is a deuterium-modified version of the HIV-protease inhibitor **atazanavir** (Fig 11.23). Deuteration blocks metabolism at key positions of the drug and thus causes a 43–50% increase in plasma half-life without affecting activity. Currently, atazanavir is administered with another protease inhibitor called **ritonavir**. The latter drug is mainly added because it inhibits the enzyme that normally metabolizes atazanavir. The deuterated analogue may be sufficiently stable to metabolism that the addition of ritonavir is no longer necessary. This would simplify the dosing regime and remove the side effects associated with ritonavir.

This approach to drug design has several advantages. Deuterium is very similar in size to hydrogen and so it is unlikely that replacing hydrogen with deuterium will affect the binding interactions of the modified drug with the target binding site. Metabolic reactions can be blocked because there is a significant isotope effect when a proton is replaced with deuterium. Deuterium

has twice the atomic mass of hydrogen, and the C–D bond is 6–10 times stronger than a C–H bond. This has a dramatic effect on the rate at which a C–D bond can be cleaved compared with a C–H bond, and is known as the **kinetic isotope effect** (KIE).

Another example of a deuterated drug that has entered clinical trials is **SD 254**, which is a deuterated analogue of the antidepressant **venlafaxine** (Fig. 11.24). Normally, the N-methyl and O-methyl groups in venlafaxine are demethylated by metabolic enzymes to give polar metabolites, but the CD_3 groups in the deuterated analogue are resistant to this reaction and the analogue is metabolized at half the rate of the undeuterated drug, resulting in a longer biological half-life.

C20081 (Fig. 11.24) is a deuterated analogue of the antibacterial agent **linezolid**. There is a 43% increase in plasma half-life for the deuterated analogue as a result of blocked metabolism.

Some drugs are found to inhibit cytochrome P450 enzymes. These enzymes play an important role in drug metabolism, and so their inhibition decreases the rate at which other drugs are metabolized. This is an example of a drug–drug interaction, and it means that drugs that inhibit cyctochrome P450 enzymes should not be taken at the same time as drugs which are metabolized

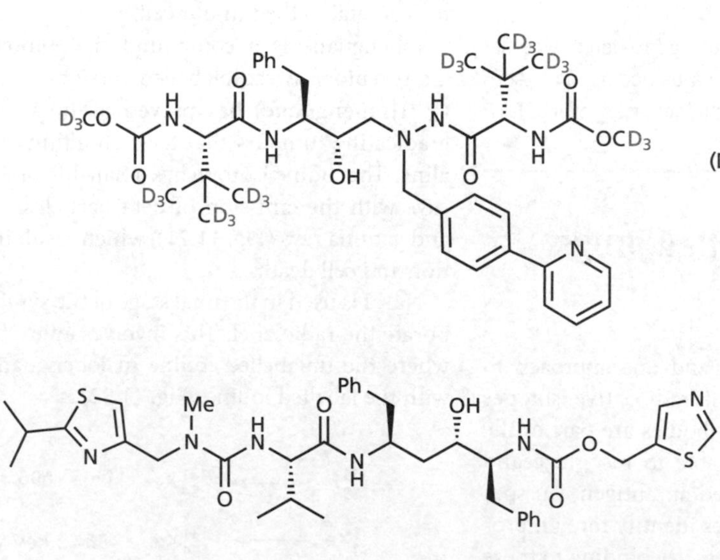

CTP 518
(Deuterated analogue
of atazanivir)

Ritonavir

FIGURE 11.23 CTP518 and ritonavir.

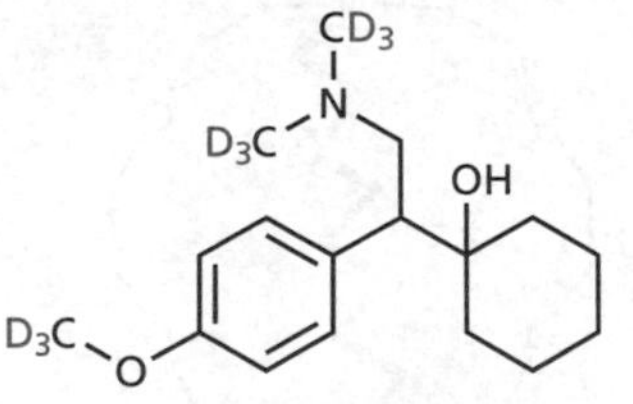

SD 254
(Deuterated
analogue of
venlafaxine)

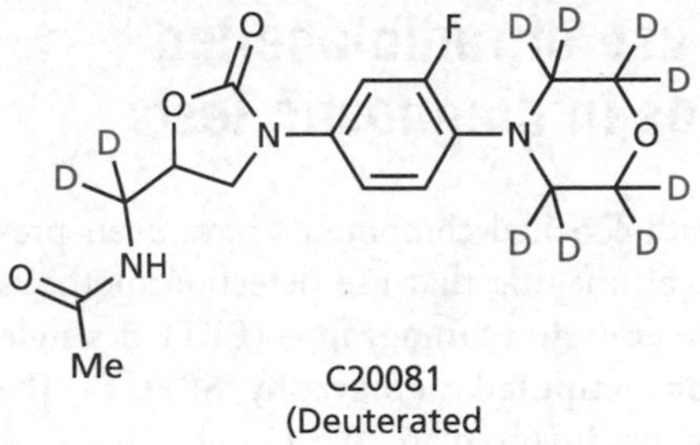

C20081
(Deuterated
analogue of
linezolid)

FIGURE 11.24 SD254 and C20081.

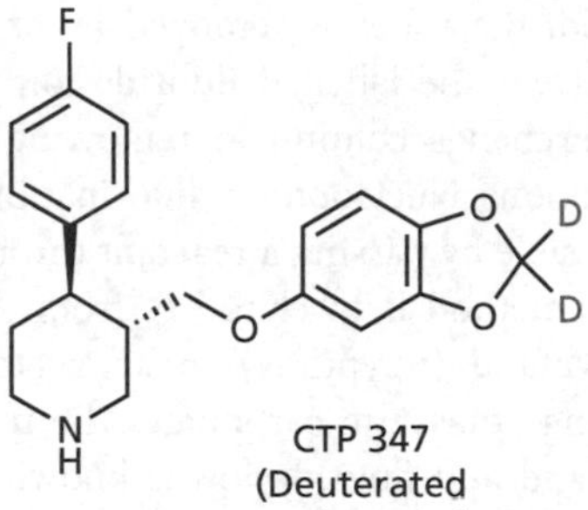

CTP 347
(Deuterated
analogue of
paroxetine)

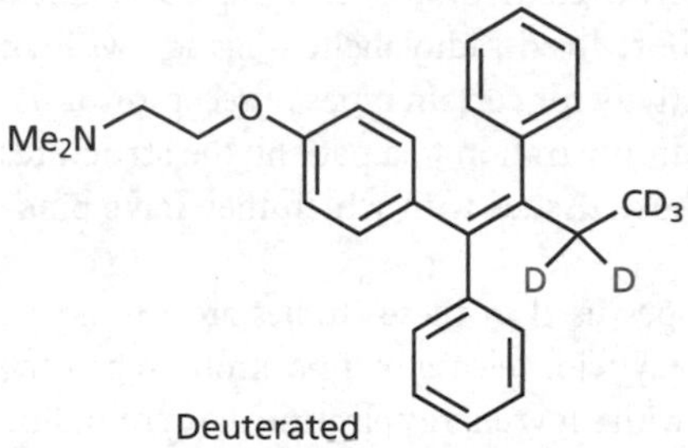

Deuterated
analogue of
tamoxifen

FIGURE 11.25 CTP 347 and a deuterated analogue of tamoxifen.

by those enzymes. Deuterium-modified drugs are a possible way round this problem. For example, the anti-depressant **paroxetine** inhibits the activity of a cyto-chrome P450 enzyme called CYP2D6. **CTP347** (Fig. 11.25) is a deuterium-modified version of paroxetine which has similar activity but far less inhibitory effect on the cytochrome P450 enzyme, thus removing the problems of interactions with other drugs. Using CTP347 would be an advantage over paroxetine as a greater number of other drugs would become feasible in combination therapies.

The inhibition of CYP2D6 by paroxetine is due to a metabolic reaction catalysed by the enzyme, which results in oxidation of the CH_2 group on the bicyclic ring. This results in the formation of a carbene which reacts with the enzyme and leads to irreversible inhibition. Thus, paroxetine acts as a **suicide substrate**. Replacing the protons of this methylene group with deuterium atoms prevents this process from taking place.

Tamoxifen is an antitumour drug, but suffers from toxic side effects. It has been found that toxicity can be reduced if the protons of the allylic methylene group are replaced with deuterium (Fig. 11.25). Oxidation at the allylic position is prevented as a result.

KEY POINTS

- Antibodies containing a radioisotope have been used as anti-cancer agents.

- Replacing metabolically susceptible protons with deuterium has resulted in drugs having a longer duration of action.

- Deuterium-modified drugs have the potential to prevent drug–drug interactions resulting from the inhibition of cyto-chrome P450 enzymes.

- Deuterium-modified drugs have the potential to reduce the toxicity of some drugs.

11.9 Fragment-based lead discovery

In recent years, NMR spectroscopy has been used to design lead compounds by identifying whether small molecules called **epitopes** bind to specific but different regions of a protein's binding site. The method involves labelling target proteins with ^{15}N or ^{13}C in order to detect such binding and is described in more detail in section 7.8.

11.10 The use of radiolabelled compounds in diagnostic tests

A number of radiolabelled compounds have been prepared for medical imaging that use detection methods such as **positron emission tomography** (**PET**) or **single photon emission computed tomography** (**SPECT**). The typical radioisotopes involved are ^{18}F, ^{11}C, and ^{123}I.

11.10.1 Medical imaging with positron emission tomography (PET)

PET is a commonly used method of carrying out medical imaging using short-lived radiolabelled ligands with an affinity and selectivity for certain types of receptor or tissue. Following administration to a patient, the structures concentrate in those tissues with which they have binding selectivity.

The radioisotopes used in these studies are ^{18}F or ^{11}C. Both isotopes decay with release of a positron. When the positron collides with a low energy electron in surrounding body tissue, both particles are annihilated and two gamma ray photons are created which move in roughly opposite directions. The detection of these photons is the basis for PET which allows 3D medical imaging of body tissues.

Owing to the short radiochemical half-life of ^{18}F and ^{11}C, it is important that the radiolabel is incorporated into a pharmaceutical compound as soon as it has been generated. Therefore, the synthesis of these radiopharmaceuticals is only feasible if a cyclotron is available on site.

Since the compounds involved are binding with certain types of tissue, it is important that the incorporation of a radiolabel does not interfere with that interaction. Replacing ^{12}C with ^{11}C is ideal as it will have no effect on the size or character of the molecule. When ^{18}F is incorporated, a proton has been replaced with ^{18}F to create a fluoro-substituted analogue of the original molecule. This is likely to affect the properties of the structure to some extent, but not to the extent that the analogue will lose its ability to bind to target tissues. A fluoro-substituent is slightly larger than a proton, but not so large that it will create a significant steric hindrance to binding interactions.

11.10.2 Synthesis of radiopharmaceuticals incorporating ^{18}F

^{18}F can be introduced into an organic molecule by a nucleophilic or an electrophilic fluorination reaction. Nucleophilic fluorination involves a labelled fluoride ion ($^{18}F^-$) and is the preferred option as the reaction goes in high yield with substitution of good leaving groups such as the triflate, mesylate, or tosylate groups. The products obtained have a high specific activity.

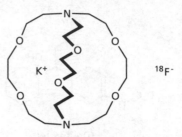

FIGURE 11.26 [(crypt-222)K]$^{+18}F^-$

$^{18}F^-$ is generated as a fluoride ion by proton bombardment of ^{18}O-enriched water within a cyclotron. Once generated, the water is removed by azeotroping it off. Alternatively, the labelled fluoride ion is trapped onto an ion exchange column to remove it from the aqueous solution. Nucleophilic fluorination reactions are then possible by passing a reactant through the column. Another method is to elute $^{18}F^-$ from the column with 2,2,2-cryptand (Kryptofix) in an acetonitrile solution containing potassium carbonate. The resulting complex of cryptand and fluoride ion is known as [(crypt-222) K]$^{+18}F^-$ and is used to fluorinate suitable substrates in an aprotic organic solvent. The cryptand serves to enhance the solubility of the potassium fluoride by complexing the potassium ion (Fig. 11.26).

Whichever method is used, the molecule undergoing nucleophilic fluorination has to be a good electrophile and contain good leaving groups such as the triflate, mesylate, or tosylate groups. The substitution reaction should also be carried out as soon as possible using an automated process. Unreacted fluoride ion can be removed using column chromatography (Sep-Paks), and the labelled structure can be purified using HPLC. Final products can be generated with a high specific activity in the range of 1000–10 000 Ci/mmol (37 370 GBq/mmol) depending on the ratio of ^{18}F to ^{19}F that is present.

The reactant for the nucleophilic fluorination reaction has to be a good electrophile, but this is not possible in some cases. If the reactant is nucleophilic in nature, it may be necessary to incorporate the ^{18}F in an electrophilic fluorination reaction using radiolabelled fluorine ($[^{18}F]F_2$). One way of doing this is to use $[^{18}F]F_2$ in a fluorodestannylation reaction (Fig. 11.27).

However, labelled fluorine is very reactive and requires specialist equipment. Moreover, the reaction can be unselective, which results in fluorination occurring at more

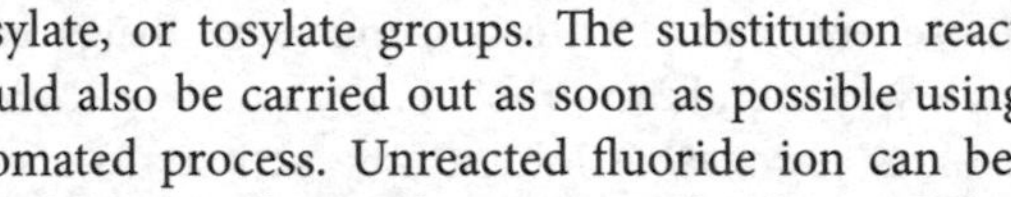

$$Ar-SnR_3 \xrightarrow[\text{Fluorodestannylation}]{[^{18}F]F_2} Ar-^{18}F$$

FIGURE 11.27 Fluorodestannylation reaction.

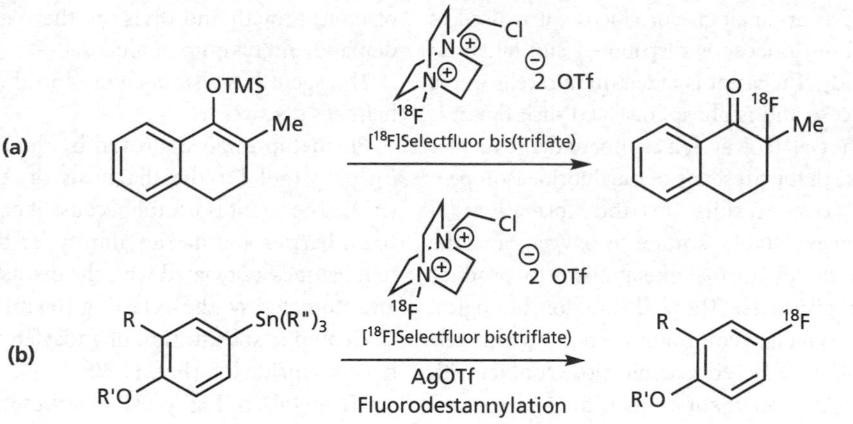

FIGURE 11.28 Introduction of ^{18}F using [^{18}F]Selectfluor bis(triflate).

than one position in the reactant. This complicates the separation and purification process, and decreases yields.

Therefore, other electrophilic fluorination agents have been developed (Case Study 5). For example, [^{18}F] Selectfluor bis(triflate) is an alternative electrophilic reagent which can be synthesized from labelled fluorine. It is milder and more selective than fluorine and is more easily handled. The reagent can be used with TMS enol ethers or in a fluorodestannylation reaction (Fig. 11.28).

Whichever electrophilic method is used, the amount of ^{18}F that can be incorporated into the target molecule is cut by at least 50%, since only one of the radiolabelled fluorine atoms in fluorine can be incorporated directly into a reactant or any of the alternative fluorinating agents that are derived from fluorine. Therefore, specific activities are limited to less than 10 Ci/mmol. Nevertheless, electrophilic fluorination has proved a useful method of introducing ^{18}F into electron-rich aromatic rings, for example FDOPA (see section 11.10.4).

Because of the short half-life of ^{18}F, the labelled product should be used on the day that it has been synthesized and needs to be transported to hospitals as quickly as possible. However, hospitals can now acquire on-site cyclotrons which incorporate the required chemistry setups in order to produce radiopharmaceuticals on site.

[^{18}F]Labelled compounds are often described as being in the 'carrier-added' or the 'no-carrier-added' form. The former term refers to labelled compound that has been diluted with unlabelled ^{19}F at some point in the process, whether that be dilution with the unlabelled compound itself or the use of diluted fluorinated structures during the synthesis. The 'no-carrier-added' form of the product has not been diluted with unlabelled structures either during or after the synthesis, but that does not mean that every fluorine in the structure is ^{18}F rather than ^{19}F.

11.10.3 Examples of ^{18}F-labelled radiopharmaceuticals prepared by nucleophilic substitution

[^{18}F]**Fluorodeoxyglucose** (^{18}F-FDG) has been used for medical imaging for many years and is the most commonly used PET radiopharmaceutical in the world. It is synthesized by reacting the [^{18}F]fluoride ion with a triflate derivative of mannose (Fig. 11.29). Nucleophilic substitution means that the incoming fluoride ends up in the equatorial position rather than the axial position to generate the glucose structure. The acetate protecting groups on the other alcohol groups are then removed. Protection is necessary in order to prevent the fluoride ion acting as a base with the OH groups to generate H^{18}F. This process has been automated to run on computer-controlled apparatus at sites where the radiopharmaceutical is used on a routine basis. The product can be obtained in over 60% yield in under 26 minutes.

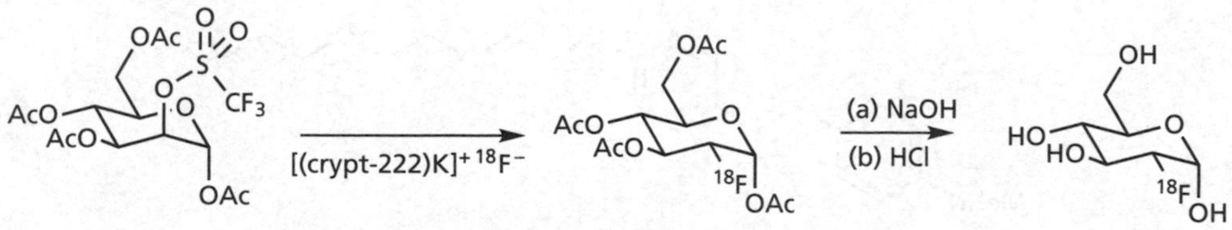

FIGURE 11.29 Synthesis of [^{18}F]fluorodeoxyglucose.

Since ^{18}F-FDG is an analogue of glucose, it is used as an indicator of how glucose is distributed and taken up by cells in the body. The agent is taken up into cells in the same way as glucose and is phosphorylated such that it is trapped within the cell. However, it cannot undergo further glycolysis because of the presence of the fluorine isotope.

Radioactive decay results in the formation of glucose with a heavy stable isotope of oxygen, and the molecule can undergo normal metabolism to produce non-radioactive products. The half-life for biological elimination of the labelled compound is only 16 minutes. However, patients are advised to avoid close contact with infants, children, and pregnant women for 12 hours—a period which corresponds to seven radioactive half-lives.

The agent is useful for assessing glucose metabolism in vital organs such as the heart, lungs, and brain. It has also proved useful in imaging tumours. Many tumours contain kinase enzymes which have been over-expressed, or are more active than in normal cells. Kinases are responsible for phosphorylating protein substrates, and so the phosphorylation of ^{18}F-FDG results in increased concentrations of the agent in tumour cells compared with normal cells. Moreover, as many tumours are in a process of rapid growth and division, there is a high metabolic demand and take-up of glucose.

The agent has also been used in the diagnosis of Alzheimer's disease.

Florbetapir was approved by the FDA in 2011 as an imaging tool for the diagnosis of Alzheimer's disease (AD). The agent is useful because it can cross the blood–brain barrier and has an affinity for the β-amyloid protein plaques associated with the disease. It is produced in an automated synthesis lasting 105 minutes, and involves nucleophilic substitution of a tosylate group with the labelled fluoride ion (Fig. 11.30).

Florbetaben (Fig. 11.31) is structurally related to florbetapir and is being used as a diagnostic tool to study whether amyloid plaques are being broken down as a result of AD treatments. This may also help to establish whether the plaques are the main culprit for AD or whether soluble amyloid oligomers may be causing more harm. The development of diagnostics alongside therapeutics is an emerging trend.

Another agent that can be used to study Alzheimer's disease is a [^{18}F]fluoropropyl analogue of **curcumin**, which can cross the blood–brain barrier and bind to

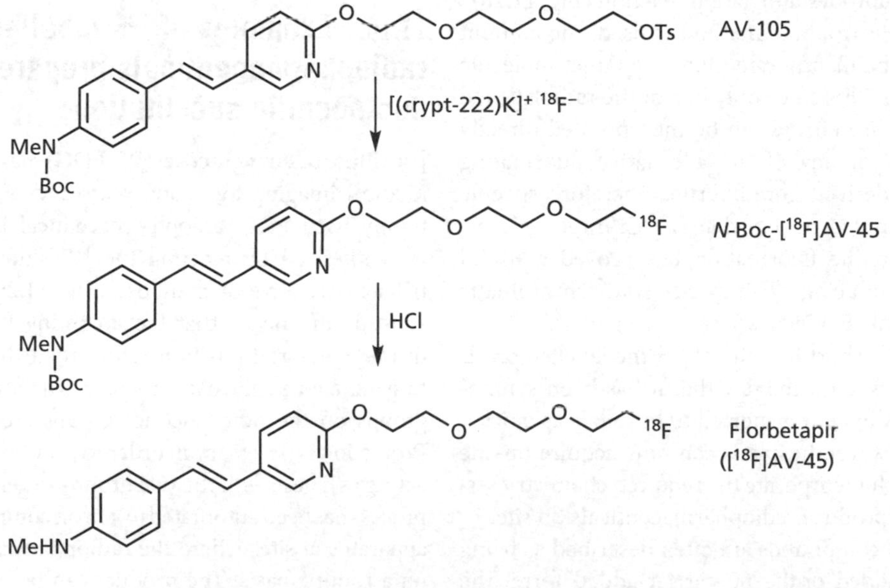

FIGURE 11.30 Synthesis of florbetapir.

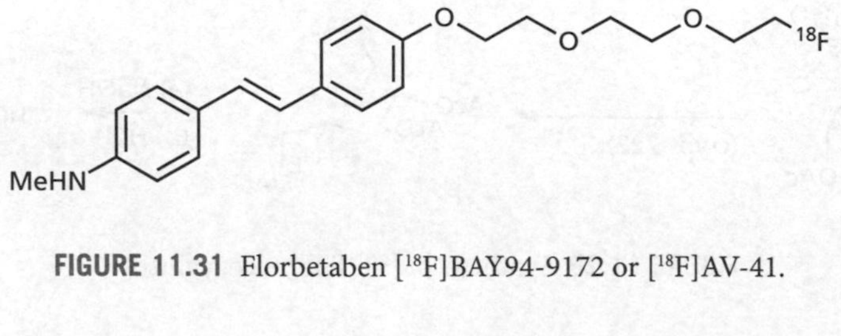

FIGURE 11.31 Florbetaben [^{18}F]BAY94-9172 or [^{18}F]AV-41.

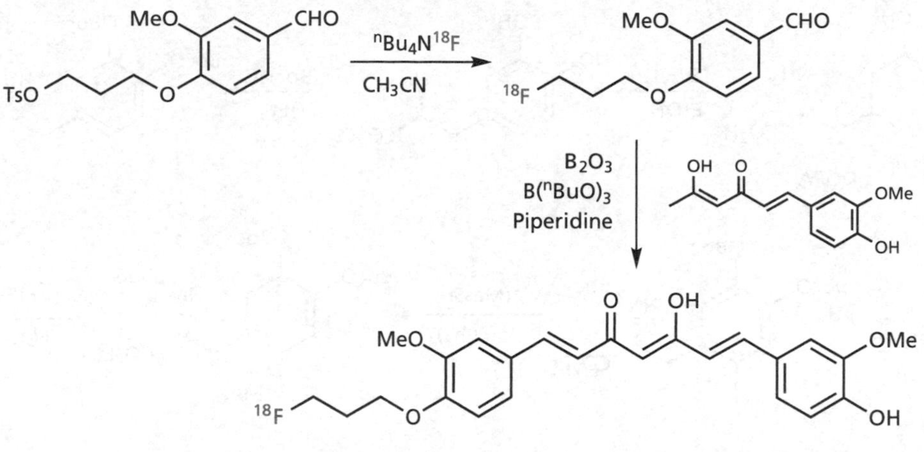

FIGURE 11.32 Synthesis of an ^{18}F- labelled analogue of curcumin.

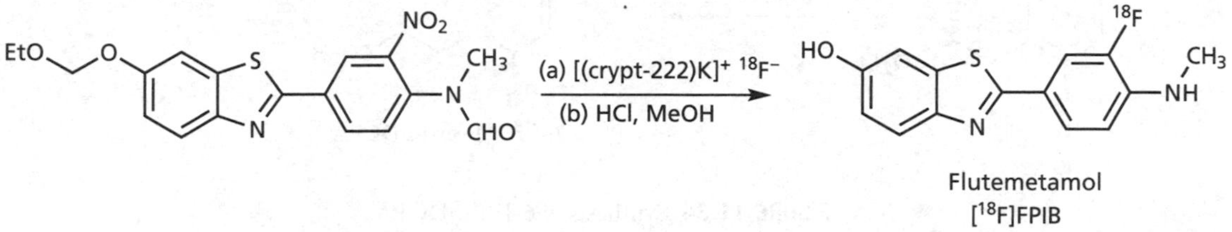

FIGURE 11.33 Synthesis of an ^{18}F- labelled analogue of Pittsburgh compound B.

β-amyloid plaques. The labelled compound has been synthesized in a two-stage process using tetrabutylammonium [^{18}F]fluoride as the fluoride source (Fig. 11.32). The structure was synthesized in a two-stage process as the one-stage process from the relevant tosylate resulted in low yields.

The previous examples illustrate nucleophilic substitutions on mesylates, tosylates, and triflates. It is also possible to substitute a nitro group on an aromatic ring as long as the aromatic ring has another electron-withdrawing substituent present, such as an aldehyde, nitro, ketone, nitrile, or ester group. Other groups that can be substituted in this manner are a halogen or a trimethylammonium ion.

For example, **flutemetamol** (FPIB) is the fluoroanalogue of a structure known as **Pittsburgh compound B** (Fig. 11.33). It has also been used in the study of Alzheimer's disease. Here, an aromatic nitro group has been replaced with the labelled fluorine in a nucleophilic substitution reaction.

11.10.4 Synthetic approaches to 6-[^{18}F] FDOPA

Labelling with [^{18}F]F$_2$ results in compounds with lower specific activity. However, this is acceptable in cases where high specific activities are not required. For

example, radiolabelled samples of DOPA do not need to have high specific activity as they can be administered to the patient in high doses. Despite the high doses administered, the compound is non-toxic and does not saturate the transport processes involved in its uptake into cells.

^{18}F-labelled 6-fluoroDOPA has been synthesized as a diagnostic tool for Parkinson's disease to study dopamine biosynthesis and the degeneration of dopamine neurons in the brain. Like Dopa itself, FDOPA is taken up by transport proteins into dopaminergic neurons and then decarboxylated to form 6-[^{18}F]dopamine. Imaging can identify whether there is a decrease in the uptake of dopamine by transport proteins. The agent also has potential for the diagnosis of neuroendocrine tumours.

A common method of synthesizing the compound is by an electrophilic destannylation reaction, which can be carried out with remote-controlled synthesis units (Fig. 11.34). However, the use of an organostannane requires rigorous analysis and purification of the radiopharmaceutical to ensure that no toxic tin by-products are present in the final material. One way round this problem would be to use resin bound organostannanes, which are only released following destannylation.

Unfortunately, the use of electrophilic fluorinations is expensive and is only possible on a small scale. Therefore, there are potential advantages in designing a

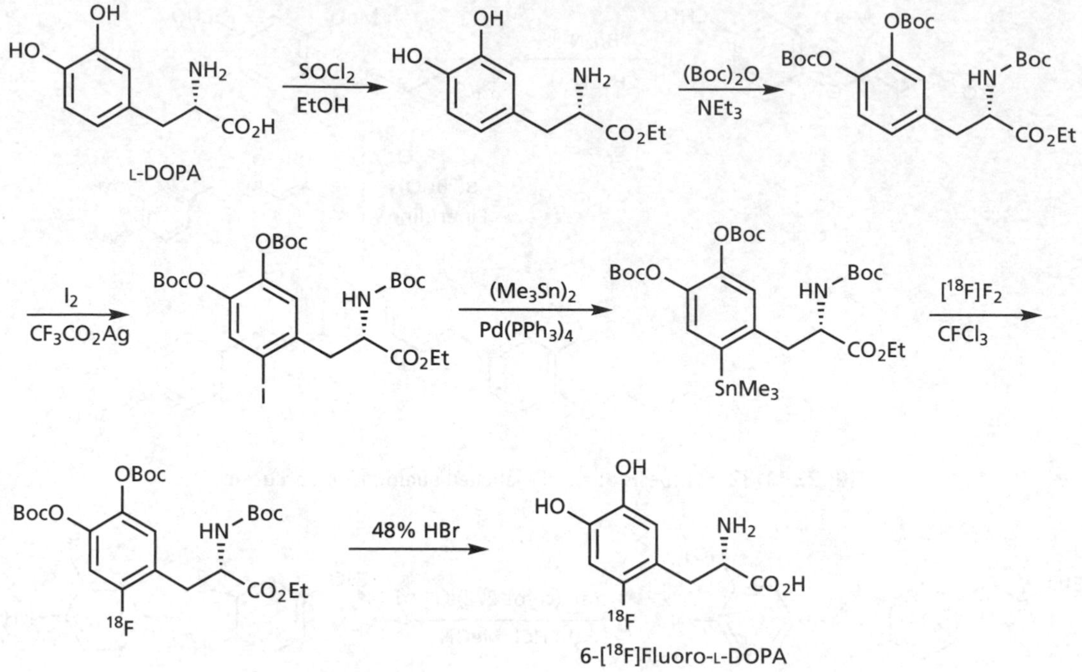

FIGURE 11.34 Synthesis of 6-[^{18}F]FDOPA.

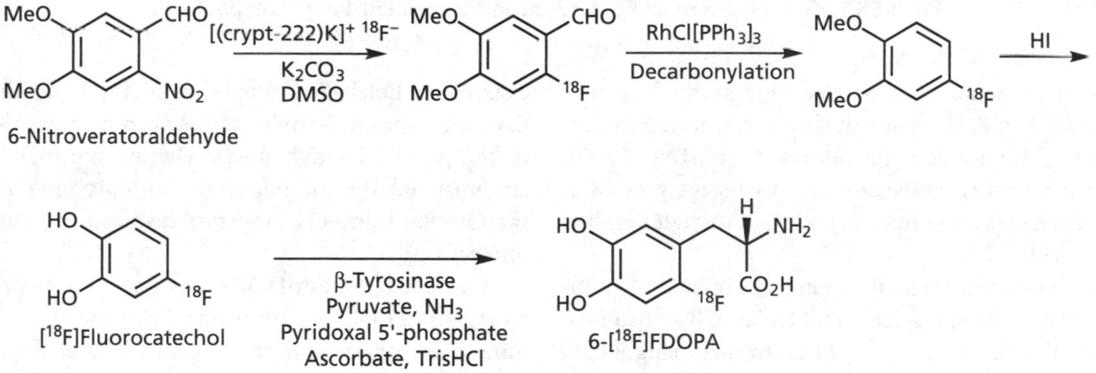

FIGURE 11.35 Synthesis of 6-[^{18}F]FDOPA using a nucleophilic fluoride ion.

synthesis of 6-[^{18}F]FDOPA where a nucleophilic fluoride ion is used instead. One approach is to start from 6-nitroveratoraldehyde, which can undergo a nucleophilic substitution reaction where the nitro group is replaced with ^{18}F (Fig. 11.35). The aldehyde group is then removed in the presence of a rhodium catalyst and the methyl ethers are demethylated with HI to give a fluorocatechol. This acts as the substrate for the enzyme β-tyrosinase which serves to catalyse the attachment of the amino acid side chain. The overall process takes about 150 minutes. Unfortunately, this procedure is not considered feasible for automation because of the number of steps involved and the relatively long time period required. In general, a maximum time of 60 minutes is considered the ideal for a full fluorination synthesis.

Another approach that has been considered involves a fluorine exchange reaction on a protected amino acid precursor (Fig. 11.36).

11.10.5 Synthesis of radiopharmaceuticals incorporating ^{11}C

^{11}C has a half-life of about 20 minutes and has been incorporated into several agents that have been used to image the brains of Alzheimer's sufferers. Such agents provide a non-invasive method of diagnosing the disease and evaluating the effectiveness of drug therapy. The most commonly used agent is **Pittsburgh compound B**, where the ^{11}C label is incorporated as an N-methyl group using [^{11}C]iodomethane (Fig. 11.37).

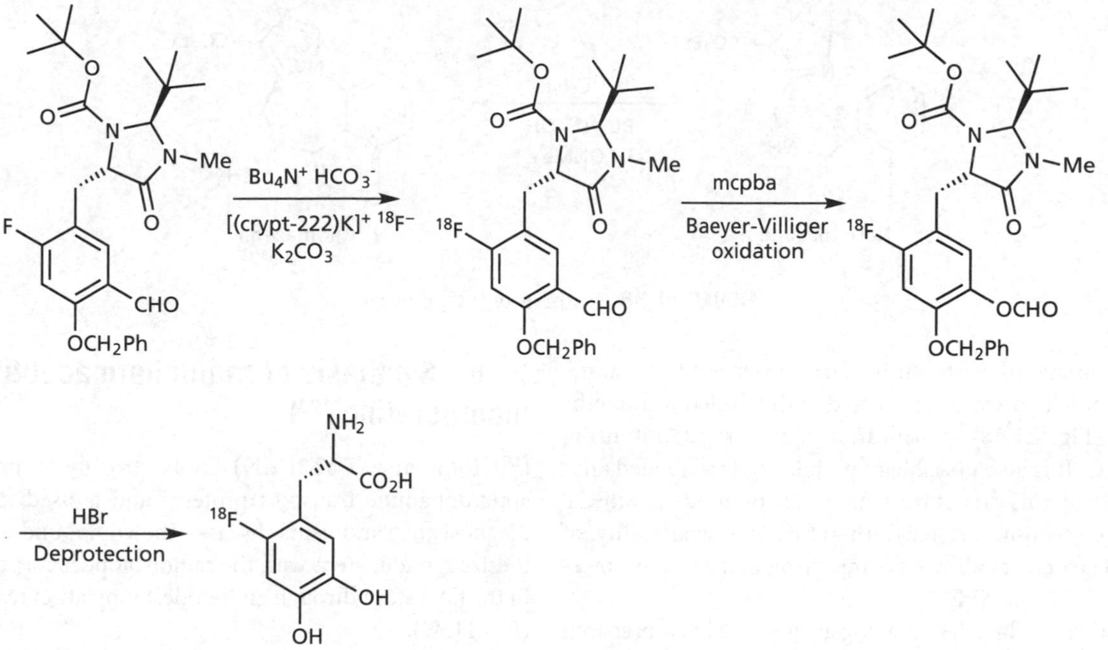

FIGURE 11.36 Synthesis of 6-[^{18}F]FDOPA by fluorine exchange.

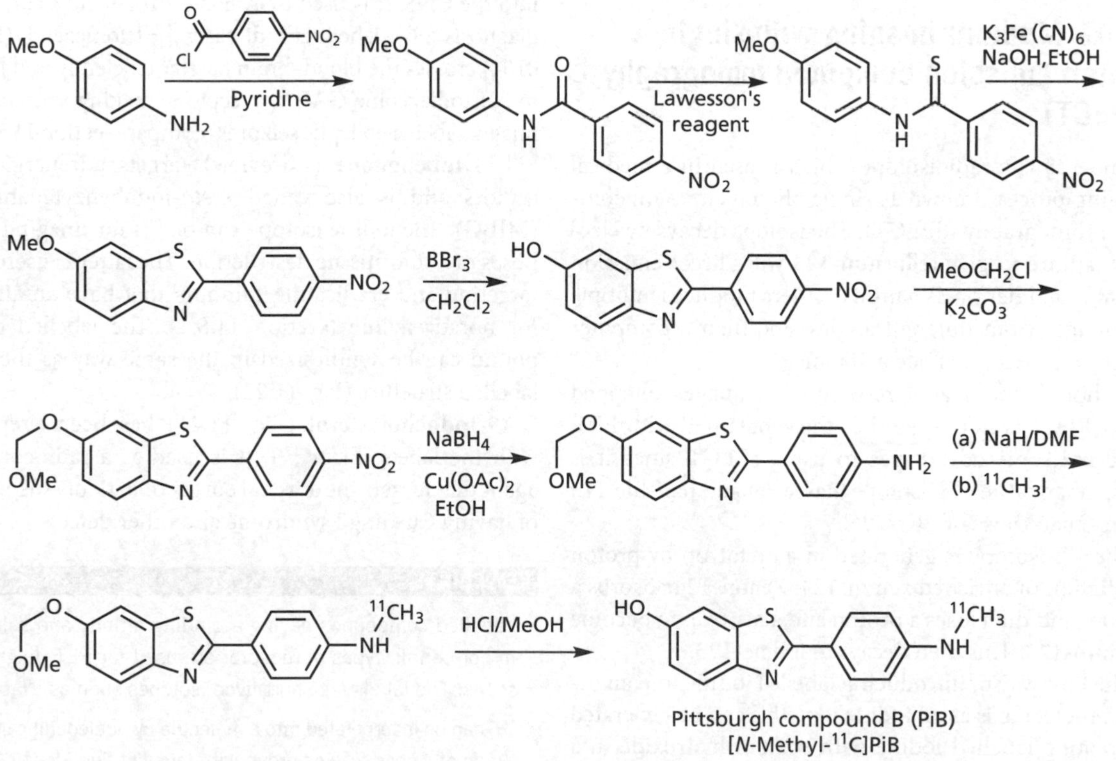

FIGURE 11.37 Synthesis of ^{11}C-labelled Pittsburgh compound B.

Despite the number of ^{11}C- and ^{18}F-labelled compounds that have been produced for the study of AD, there is still a need for improved agents with better signal-to-noise ratios. This would be possible by developing an agent which shows increased binding selectivity for amyloid plaques.

^{11}C-labelled compounds have been produced for purposes other than the study of AD. For example, [^{11}C]iomazenil binds to the benzodiazepine receptor GABA$_A$ and has been used to image the foci of epileptic seizures. It has been synthesized by treating

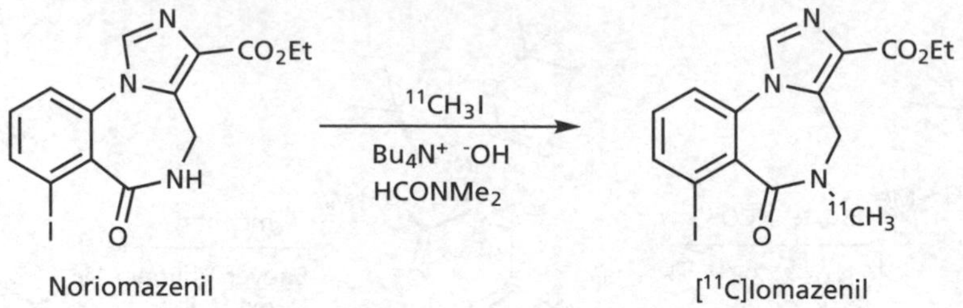

FIGURE 11.38 Synthesis of [^{11}C]iomazenil.

noriomazenil with $Bu_4N^{+-}OH$ to generate an amide anion, which was then treated with labelled iodomethane (Fig. 11.38). Purification was carried out using HPLC. It is also possible to prepare an ^{123}I-labelled analogue of this structure which can be used in SPECT scans (section 11.10.6). Therefore, the availability of both structures allows a comparison of the relative merits of PET and SPECT.

Other ^{11}C-labelled imaging agents have been prepared that can bind to the dopamine transport protein. These include ^{11}C-labelled cocaine and its analogues.

11.10.6 Medical imaging with single photon emission computed tomography (SPECT)

Iodine-123 is a radioisotope which is used in a medical imaging process known as single photon emission computed tomography (SPECT). The isotope decays by electron capture to give tellurium-123 with direct emission of gamma radiation. A gamma camera produces multiple 2D images from different angles, and then a computer program is used to create a 3D image.

Although the spatial resolution of images obtained with SPECT is not as good as those obtained with PET, there are some advantages to using SPECT since it is cheaper and the ^{123}I isotope has a longer half-life (13 hours) than those of ^{18}F or ^{11}C.

The ^{123}I isotope is generated in a cyclotron by proton irradiation of enriched xenon-124. Xenon-124 absorbs a proton, and then loses a proton and a neutron to become caesium-123. This then decays to iodine-123.

The best way of introducing labelled iodine into an organic molecule is as an iodide ion. This can be generated by treating labelled iodine with sodium hydroxide in a disproportionation reaction as follows:

$$3^{18}I_2 + 6HO^- \rightarrow 5^{18}I^- + {}^{18}IO_3^- + 3H_2O$$

A number of methods have been used to incorporate the iodide ion, and examples are shown in section 11.10.7.

11.10.7 Synthesis of radiopharmaceuticals incorporating ^{123}I

[^{123}I]**Ioflupane** (DaTSCAN) binds strongly to presynaptic dopamine transport proteins and is used for the diagnosis of Parkinson's disease. The compound is synthesized in four steps with the radioisotope incorporated in the final stage through an iododestannylation reaction (Fig. 11.39).

Other imaging agents include [^{123}I]**iofetamine** and [^{123}I] **iomazenil** (Fig. 11.40). [^{123}I]Iofetamine has a similar structure to amphetamine and can cross the blood–brain barrier into the CNS. It is used to evaluate stroke and as an early diagnosis of Alzheimer's disease. [^{123}I]Iomazenil (Ro16 0154) crosses the blood–brain barrier efficiently and binds to benzodiazepine GABA$_A$ receptors, making it useful for imaging foci of epileptic seizures (compare section 11.10.5).

[^{123}I]**Iobenguane** (Adreview) targets adrenergic receptors and is also called *meta*-iodobenzylguanidine (MIBG). The iodine isotope can be ^{123}I for imaging purposes or ^{131}I for tissue destruction. The latter is useful for targeting and eradicating tumours that have an affinity for noradrenaline (section 11.8.1). The labelled compound can be synthesized in the same way as the ^{131}I-labelled structure (Fig. 11.22).

19-Iodocholesterol (Fig. 11.41) has been prepared with the labels ^{125}I or ^{131}I. It is used as a radiocontrast agent that targets the adrenal cortex of patients suspected of having Cushing's syndrome and other defects.

> **KEY POINTS**
>
> - Labelled compounds having a binding affinity and selectivity for certain types of receptor are used for PET diagnostic scans. The labels are short-lived isotopes such as ^{18}F or ^{11}C.
>
> - ^{18}F can be incorporated into a molecule by nucleophilic substitution of a good leaving group with labelled fluoride ion (^{18}F$^-$).
>
> - ^{18}F can be incorporated into a molecule by electrophilic substitution using [^{18}F]F$_2$. However, the specific activity of the labelled compound is lower than that achieved using nucleophilic substitution.
>
> - Radiolabelled compounds should be used quickly because of the short half-life of the radioisotope.

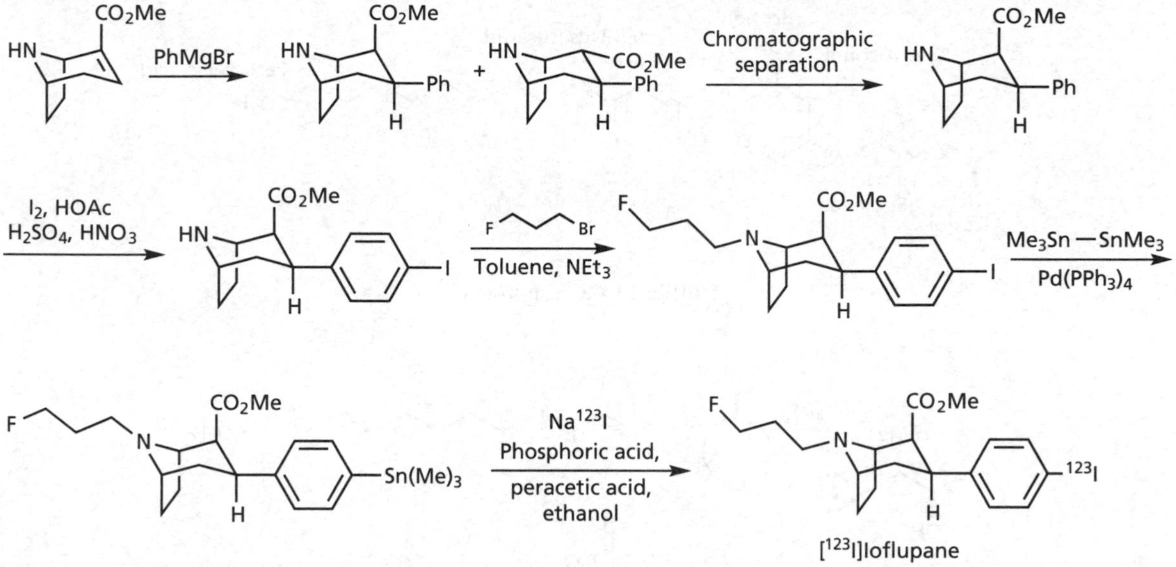

FIGURE 11.39 Synthesis of [^{123}I]ioflupane.

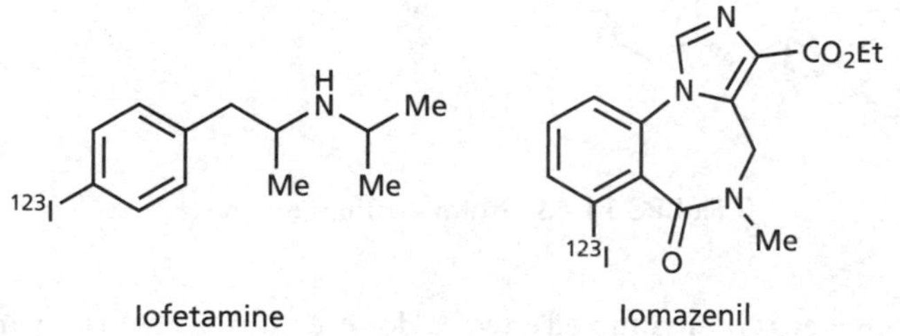

Iofetamine   Iomazenil

FIGURE 11.40 [^{123}I]Iofetamine and [^{123}I]iomazenil.

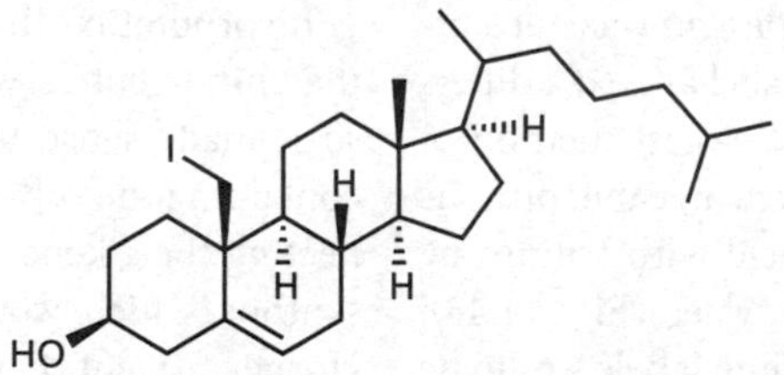

FIGURE 11.41 19-Iodocholesterol.

- On-site cyclotrons are available to synthesize specific labelled compounds in hospitals.

- Introducing ^{18}F by electrophilic substitution is acceptable if the labelled compound can be administered in high doses.

- ^{11}C can be incorporated into compounds by using a simple reagent that contains the isotope.

- ^{123}I is used to label compounds that are used in a medical imaging process known as single photon emission computed tomography.

11.11 Case Study—Synthesis of [2-^{14}C]mupirocin

11.11.1 Introduction

Mupirocin (formerly known as **pseudomonic acid**) is an antibiotic produced by the microorganism *Pseudomonas fluorescens*. The structure contains eight asymmetric centres and a variety of functional groups, namely an

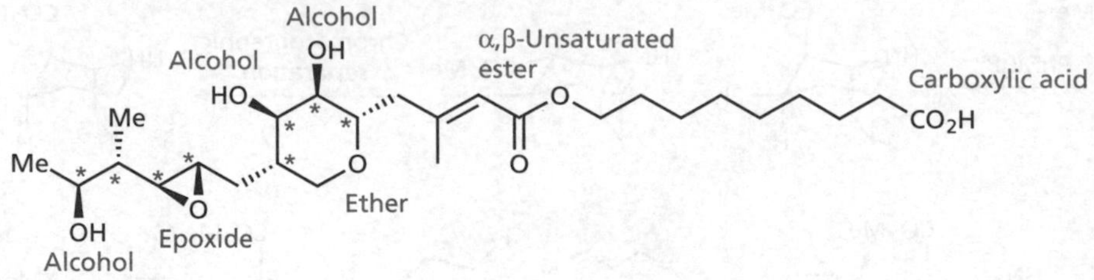

FIGURE 11.42 Mupirocin.

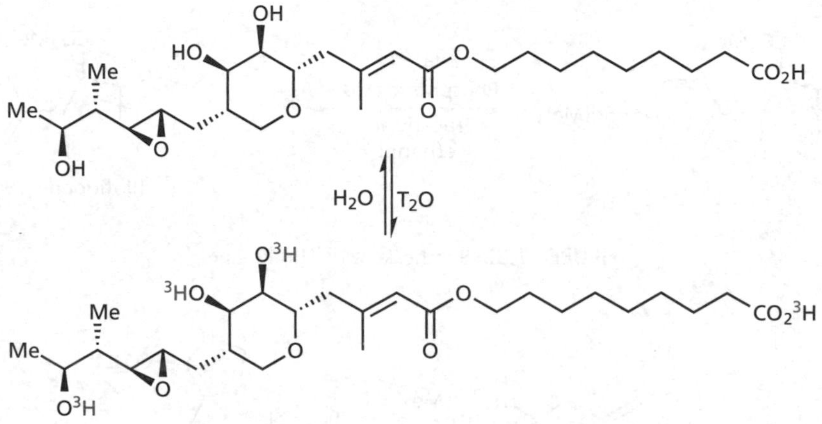

FIGURE 11.43 Proton–tritium exchange.

epoxide, a cyclic ether, an alkene, an α,β-unsaturated ester, carboxylic acid, and three alcohol groups (Fig. 11.42).

Beecham Pharmaceuticals was interested in studying how mupirocin was metabolized and wished to synthesize a radiolabelled sample of the antibiotic. However, the established synthetic route to mupirocin was not suitable for the incorporation of such a label and so a different synthesis had to be devised. The easiest method of labelling mupirocin is to label the exchangeable protons of the alcohols and the carboxylic acid with tritium by treating the molecule with tritiated water (Fig. 11.43), but this would be pointless as the tritium labels would be rapidly lost *in vivo* by proton exchange with water.

Therefore, it was necessary to devise a synthesis that would incorporate a ^{14}C label into the carbon skeleton of the structure.

11.11.2 Retrosynthetic analysis

A retrosynthetic analysis was carried out on the structure to identify a simple reagent that could be used to incorporate the label (Fig. 11.44).

An obvious disconnection is the ester, which would result in an α,β-unsaturated carboxylic acid and methyl 9-hydroxynonanoate. The labelling of methyl 9-hydroxynonanoate looks more straightforward, since it only has two functional groups and no chiral centres.

However, it is likely that hydrolysis of the ester will take place *in vivo*, and it would be interesting to track what happens to the more complex half of the molecule. Therefore, the challenge was to devise a method of labelling the left-hand portion of the structure.

Incorporating the label into the region containing all the chiral centres would be extremely challenging, and so it made sense to incorporate it into the side chain containing the α,β-unsaturated carboxylic acid. Disconnecting the alkene double bond results in a two-carbon synthon which would correspond to a simple Wittig–Horner reagent. Fortunately, a ^{14}C-radiolabelled version of this reagent was commercially available.

Therefore, it was proposed that this reagent could be used to introduce the ^{14}C label by means of a Horner–Wadsworth–Emmons reaction (Fig 11.45).

However, this all depended on whether the other components of the synthesis, namely methyl 9-hydroxynonanoate and the complex ketone required for the Horner–Wadsworth–Emmons reaction, could be readily prepared.

11.11.3 Synthesis of methyl 9-hydroxynonanoate

Oleic acid was identified as a suitable starting material for the synthesis of methyl-9-hydroxynonanoate as it is

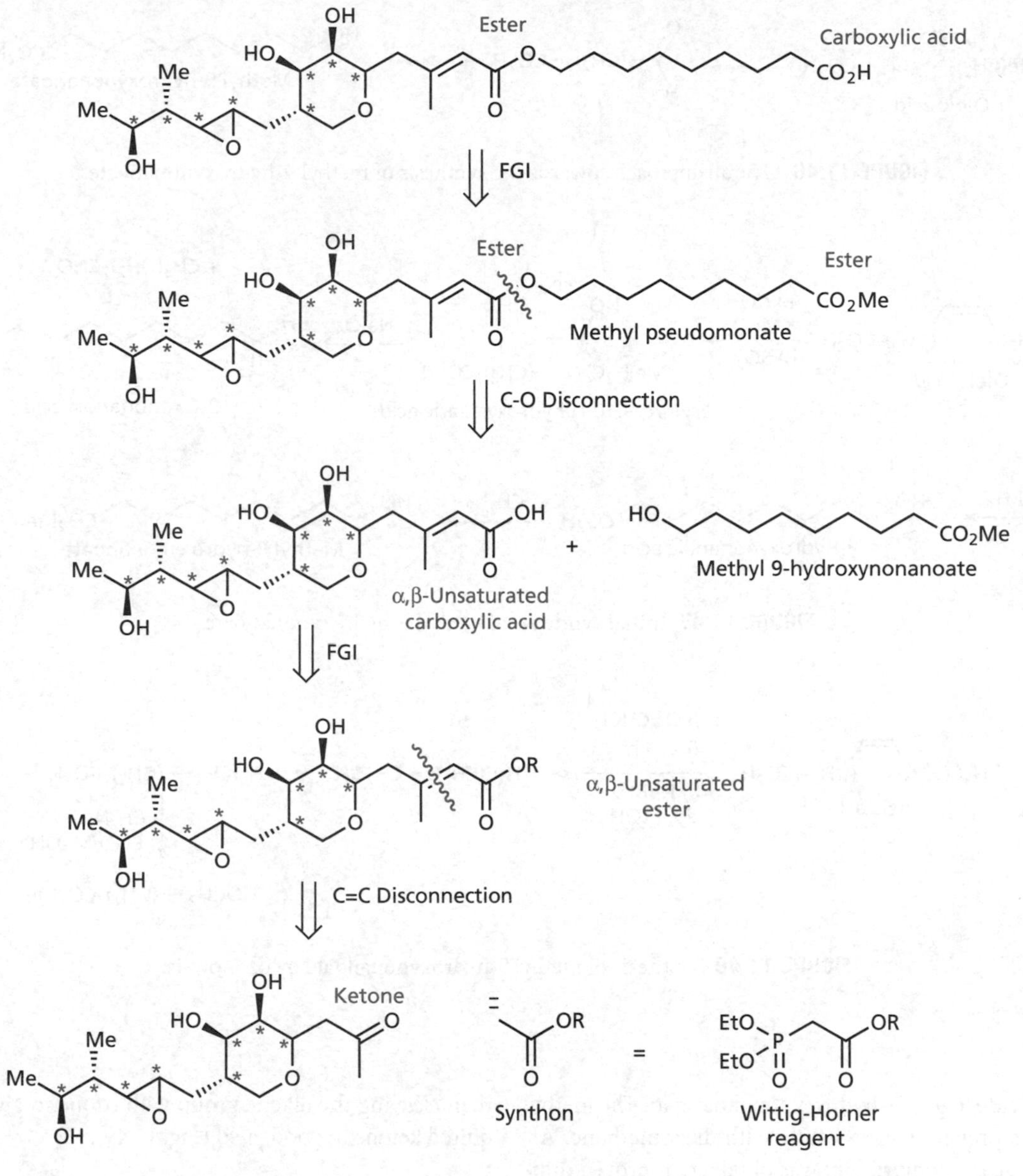

FIGURE 11.44 Retrosynthetic analysis of mupirocin (pseudomonic acid).

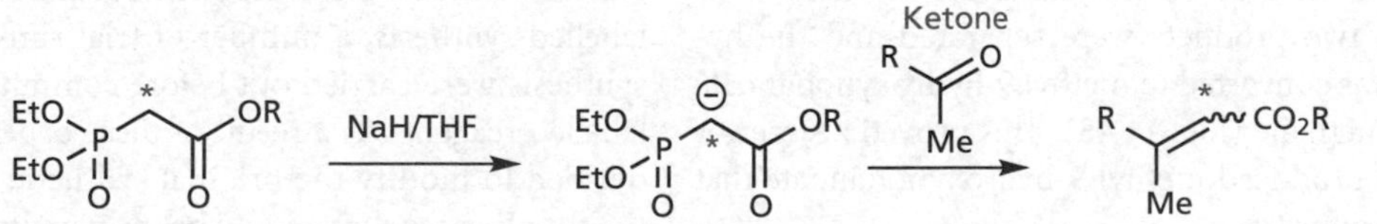

FIGURE 11.45 Horner–Wadsworth–Emmons reaction.

a readily available long-chain carboxylic acid with an alkene in the centre of the structure. Cleaving the alkene would produce a nine-carbon chain with a functional group at each end which could be modified to methyl 9-hydroxynonanoate (Fig. 11.46).

The initial approach was to oxidize the alkene group of oleic acid with potassium permanganate in alkaline solution to give a diol (Fig. 11.47). Treatment of the diol with sodium periodate split the molecule in two to give 9-oxononanoic acid, which was reduced by sodium

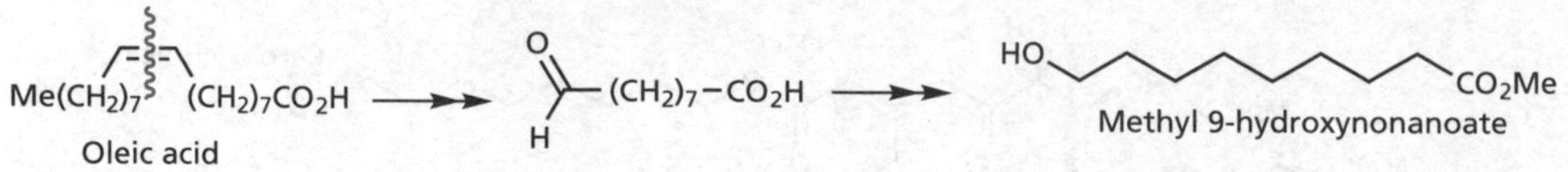

FIGURE 11.46 Overall approach towards the synthesis of methyl 9-hydroxynonanoate.

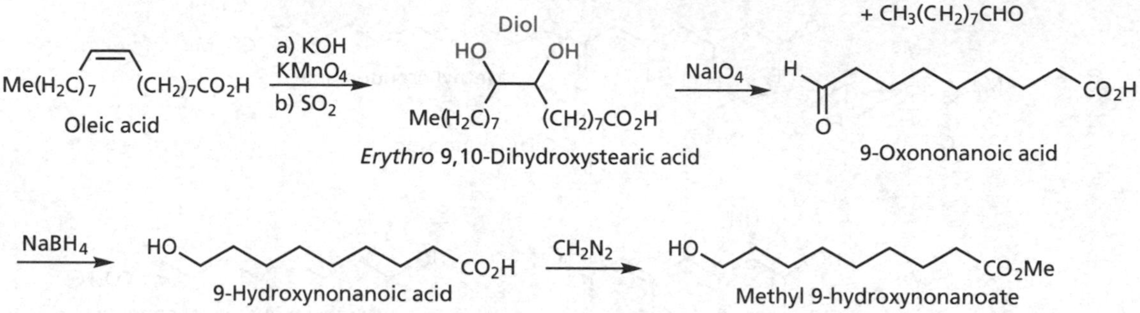

FIGURE 11.47 Initial synthesis of methyl 9-hydroxynonanoate.

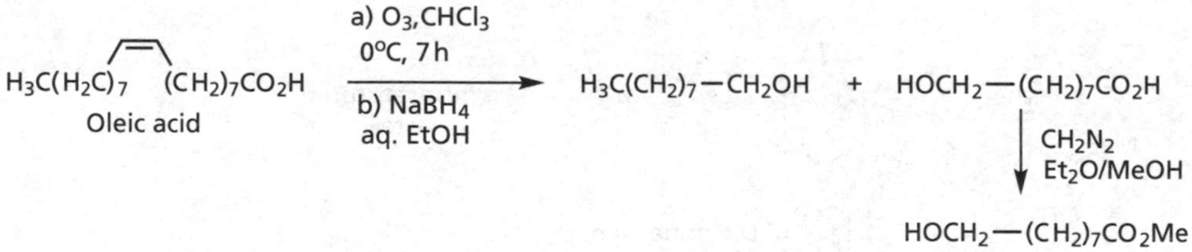

FIGURE 11.48 Synthesis of methyl 9-hydroxynonanoate by ozonolysis.

borohydride to give 9-hydroxynonanoic acid. The methyl ester was prepared successfully with diazomethane. Although the desired product was obtained, it proved difficult to remove side products and so purify the compound to an acceptable level.

An alternative approach involved cleaving the alkene group of oleic acid by ozonolysis under reducing conditions. The two products were separated and the hydroxy acid was converted to methyl 9-hydroxynonanoate using diazomethane (Fig. 11.48). This proved a superior method and produced methyl 9-hydroxynonanoate that was free of impurities.

11.11.4 Synthesis of the complex ketone

A full synthesis of the complex ketone would not be feasible because of the number of chiral centres and functional groups present. Therefore, it made more sense to synthesize this compound from mupirocin itself. This was achieved by converting mupirocin to methyl pseudomonate, and

then cleaving the alkene group with ozone to give the required ketone in good yield (Fig. 11.49).

11.11.5 Trial runs of the synthesis of mupirocin

Having obtained the components required for the radiolabelled synthesis, a number of trial runs of the overall synthesis were carried out before committing the radiolabelled reagent. As a result of these experiments, it was decided to modify the original synthetic plan such that the coupling reaction would be the penultimate step. This corresponds to the retrosynthesis shown in Figure 11.50.

11.11.6 Radiolabelled synthesis of mupirocin

[14C]Triethyl phosphonoacetate (25 mCi) was first diluted with 1 mmol of unlabelled triethyl phosphonoacetate and transferred to a 5 ml vial where the methyl ester

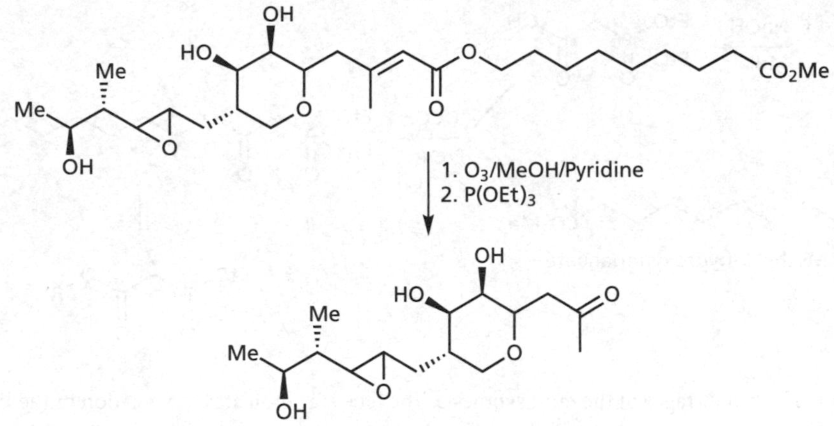

FIGURE 11.49 Synthesis of the ketone.

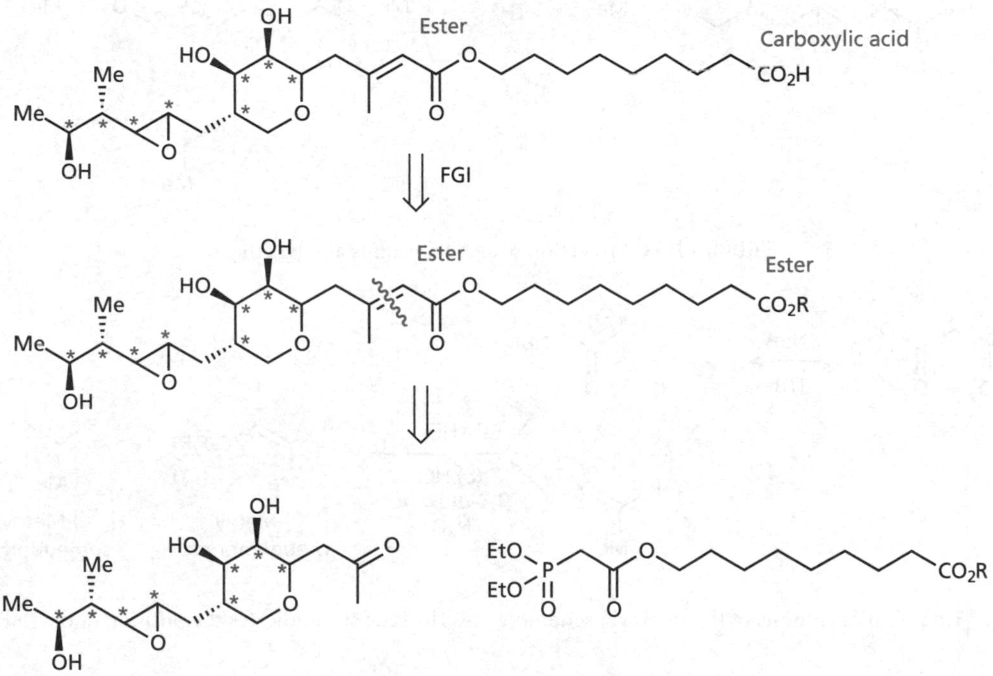

FIGURE 11.50 Modified retrosynthesis.

was hydrolysed with sodium hydroxide to give the carboxylic acid (Fig. 11.51). This was then esterified with methyl 9-hydroxy nonanoate using dicyclohexylcarbodiimide as a coupling agent.

Meanwhile, the alcohol groups of the ketone were protected as silyl ethers using bis(trimethylsilyl) acetamide (Fig. 11.52).

The Horner–Wadsworth–Emmons reaction was now carried out to give radiolabelled methyl pseudomonate plus the Z-isomer in a 2:1 ratio (Fig. 11.53). A radioactive scan of a TLC revealed that the ratio of radioactivity in the isomers was also 2:1. The specific activity for methyl

pseudomonate was measured as 19.6 mCi/mmol, while the specific activity for the Z-isomer was 19.2 mCi/mmol.

The two isomers were separated by preparative TLC and it was then a case of hydrolysing the methyl ester. Owing to the number of chiral centres present, the reaction was carried out using hog liver esterase at pH 7.2 (Fig. 11.54). Purification was then carried out using preparative HPLC followed by crystallization.

For the metabolic studies, samples of the radiolabelled mupirocin were taken up in solution and diluted with unlabelled mupirocin, and were then concentrated and crystallized until the specific activity remained constant.

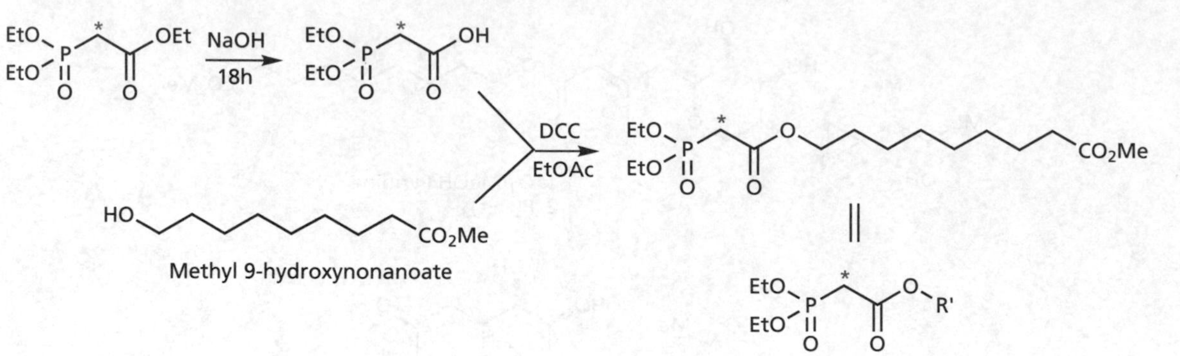

FIGURE 11.51 Initial stages of the radiosynthesis. The blue star indicates the position of the ^{14}C label.

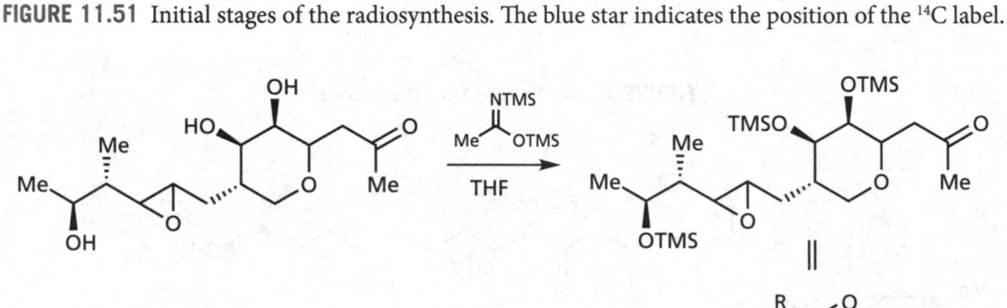

FIGURE 11.52 Protection of alcohol groups as silyl ethers.

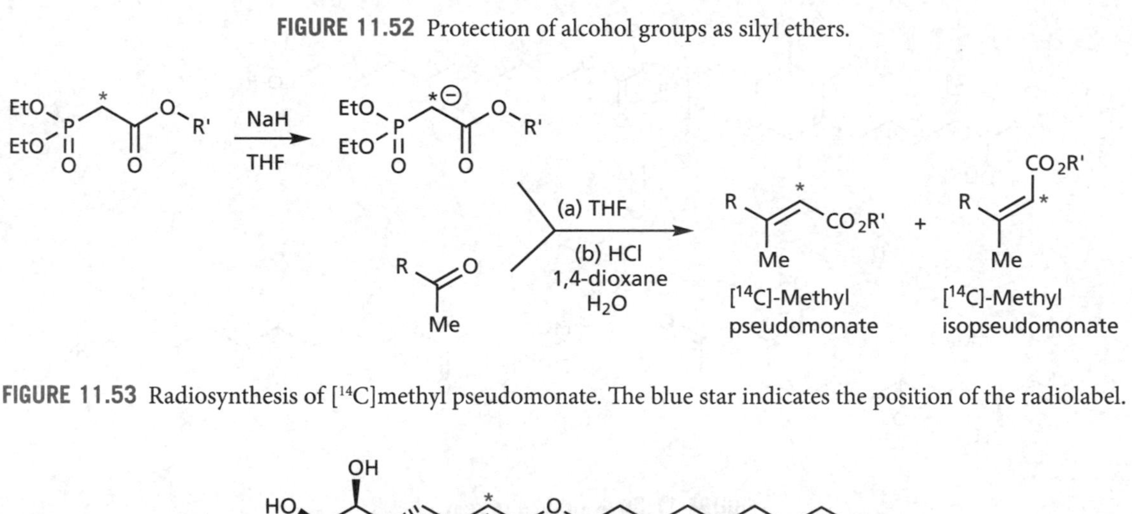

FIGURE 11.53 Radiosynthesis of [^{14}C]methyl pseudomonate. The blue star indicates the position of the radiolabel.

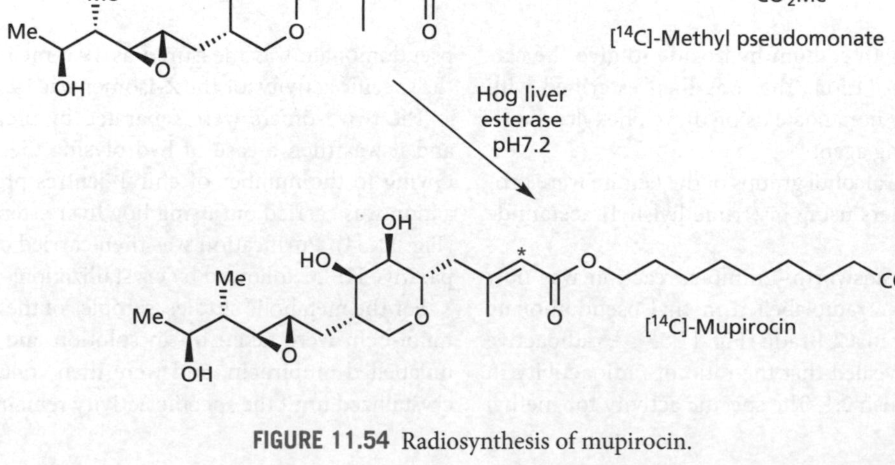

FIGURE 11.54 Radiosynthesis of mupirocin.

QUESTIONS

1. Discuss whether the doubly labelled atropine molecule shown is suitable for drug metabolism studies.

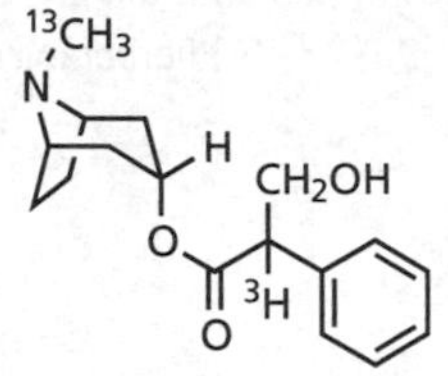

Atropine

2. Suggest how AY31660A could be synthesized from ^{14}C radiolabelled naphthalene.

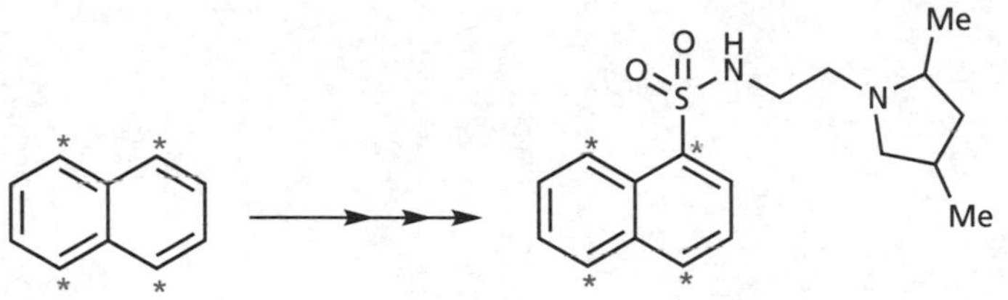

AY31660A

FURTHER READING

General reading

Buteau, K.C. (2009) 'Deuterated drugs: unexpectedly nonobvious?', *Journal of High Technology Law*, **10**, 22–74.

Kilbourn, M.R. and Shao, X. (2009) 'Fluorine-18 radiopharmaceuticals', in I. Ojima,(ed.), *Fluorine in Medicinal Chemistry and Chemical Biology*. Blackwell, Chichester, pp 361–88.

Specific compounds

Baldwin, R.M., et al. (1995) 'Synthesis and PET imaging of the benzodiazepine receptor tracer *N*-methyl[^{11}C] iomazenil', *Nuclear Medicine and Biology*, **22**, 659–65 (iomazenil).

Dolle, F., et al. (1998) '6-[^{18}F]Fluoro-L-dopa by radiofluorodestannylation: a short and simple synthesis of a new labelling precursor', *Journal of Labelled Compounds and Radiopharmaceuticals*, **41**, 105–14 (fluoro-L-dopa).

Hess, E., et al. (2002) 'Synthesis of 2-[^{18}F]fluoro-L-tyrosine via regiospecific fluoro-de-stannylation', *Applied Radiation and Isotopes*, **57**, 185–91 (Fluoro-DOPA).

Klunk, W.E., et al. (2004) 'Imaging brain amyloid in Alzheimer's disease with Pittsburgh compound-B', *Annals of Neurology*, **55**, 306–19 (Pittsburgh compound B).

Lemaire, C., et al. (1990) 'No-carrier-added regioselective preparation of 6-[^{18}F]fluoro-L-dopa', *Journal of Nuclear Medicine*, **31**, 1247–51 (fluoro-L-dopa).

Mathis, C.E., et al. (2003) 'Synthesis and evaluation of ^{11}C-labeled 6-substituted 2-arylbenzothiazoles as amyloid imaging agents', *Journal of Medicinal Chemistry*, **46**, 2740–54 (Pittsburgh compound B).

Neumeyer, J.L., et al. (1994) '*N*-ω-Fluoroalkyl analogs of (1*R*)-2β-carbomethoxy-3β-(4-iodophenyl)-tropane (β-CIT): radiotracers for positron emission tomography and single photon emission computed tomography imaging of dopamine transporters', *Journal of Medicinal Chemistry*, **37**, 1558–61 (ioflupane).

Ryu, E.K., et al. (2006) 'Curcumin and dehydrozinerone derivatives: synthesis, radiolabeling, and evaluation for β-amyloid imaging', *Journal of Medicinal Chemistry*, **49**, 6111–19 (curcumin).

Taylor, P. (2009) 'Tracing amyloid in Alzheimer's', *Chemistry World*, December, p 17 (florbetaben).

Teare, H., et al. (2010) 'Radiosynthesis and evaluation of [^{18}F] selectorfluor bis(triflate)', *Angewandte Chemie International Edition*, **49**, 6821–4 ([^{18}F]selectorfluor bis(triflate)).

Vasdev, N., et al. (2012) 'Synthesis and PET imaging studies of [^{18}F]2-fluoroquinolin-8-ol ([^{18}F]CABS13) in transgenic mouse models of Alzheimer's disease', *Medicinal Chemistry Communications*, **3**, 1228–30 (Pittsburgh compound B).

Wagner, F.M., Ermert, J., and Coenen, H.H. (2009) 'Three-step, "one-pot" radiosynthesis of 6-fluoro-3,4-dihydroxy-L-phenylalanine by isotopic exchange', *Journal of Nuclear Medicine*, **50**, 1724–9 (6-[^{18}F]fluoro-DOPA).

Wang, H., et al. (2011) 'Facile and rapid one-step radiosynthesis of [18F]BAY 94–172 with a new precursor', *Nuclear Medicine and Biology*, **38**, 121–7 (florbetaben).

Yao, C.-H., et al. (2010) 'GMP-compliant automated synthesis of [^{18}F]AV-45 (florbetapir F 18) for imaging β-amyloid plaques in human brain', *Applied Radiation and Isotopes*, **68**, 2293–7 ([^{18}F] florbetapir).

CASE STUDY 4
Studies on gliotoxin biosynthesis

CS4.1 Introduction

Biosynthetic studies are carried out in order to identify the enzymes and intermediates involved in the biosynthesis of natural products. A knowledge of a biosynthetic pathway can be useful for a number of reasons, especially if the final product has useful pharmacological activity. Knowing what enzymes and starting materials are involved can help to identify key nutrients that are required for the biosynthetic process, such as amino acids or enzyme cofactors. This is important when optimizing the growth conditions of the natural source in order to harvest the product in good yield. It may also be possible to genetically modify fast-growing microbial cells such that the natural product or its analogues can be produced more quickly and in greater quantities (see also section 9.6)

Knowing what intermediates are involved in the biosynthesis of a natural product can also be useful. For example, it might be possible to extract a key intermediate and then use it as the starting material for the semi-synthetic preparation of analogues. Alternatively, analogues of the intermediate could be fed to the natural source to see whether the biosynthetic route converts these to analogues of the natural product within the host cells themselves.

One of the difficulties in identifying biosynthetic routes is the fact that the intermediates involved may be present in very small quantities. Therefore, it is unlikely that any attempts to extract proposed intermediates from host cells will be successful. This means that more subtle approaches are required. A particularly effective method is to carry out radiolabelling experiments where a proposed intermediate including a radioisotope such as ^{14}C or ^{3}H is synthesized in the laboratory. The radiolabelled structure is then fed to the natural source to see whether it is converted to the natural product. If it is, the final product will also be radiolabelled. Moreover, since the detection of radioactivity is extremely sensitive, it is also possible to detect other radiolabelled structures extracted from host cells which might be biosynthetic intermediates or metabolites.

In this case study, we will look at an example of a biosynthetic study involving gliotoxin (Fig. CS4.1), a natural product produced by a fungus called *Trichoderma viride* (originally called *Gliocladium deliquescens*). Although gliotoxin is too toxic to be used clinically, it has interesting immunosuppressive properties. It is also an inhibitor of **farnesyl transferase** and can trigger cell death (apoptosis) in certain cells, making it an interesting lead compound for the development of novel anticancer agents. Other useful properties are antibiotic, antifungal, and antiviral activities. A ring-opened metabolite of gliotoxin which is formed in *Gliocladium virens* has also been found to be an inhibitor of **platelet-activating factor** (Fig. CS4.1).

Thus, an understanding of gliotoxin's biosynthesis may permit the production of analogues which have less toxicity to normal cells and might be useful in a number of medicinal fields.

Gliotoxin contains a dioxopiperazine ring with a disulphide bridge, which makes it an epidithiodioxopiperazine structure. It also contains an unusual cyclohexadiene containing an alcohol group that looks as if it should rapidly dehydrate to give an aromatic ring. However, the structure is perfectly stable.

Initial labelling studies using radiolabelled amino acids revealed that most of the structure is derived from the amino acids L-phenylalanine and L-serine. The methyl substituent on the nitrogen atom is derived from L-methionine and the sulphur atoms are provided by cysteine.

FIGURE CS4.1 Gliotoxin and a ring-opened metabolite.

FIGURE CS4.2 *cyclo*-(L-Phenylalanyl-L-seryl)—a proposed biosynthetic intermediate to gliotoxin.

Following on from these results, it was proposed that the cyclic dipeptide *cyclo*-(L-phenylalanyl-L-seryl) could be a crucial biosynthetic intermediate (Fig. CS4.2).

CS4.2 Synthesis of *cyclo*-(L-[U-¹⁴C] Phe-L-Ser)

In order to establish whether the cyclic dipeptide was an authentic biosynthetic intermediate or not, it was necessary to synthesize the structure with a radiolabel

incorporated. The synthesis involved initial protection of the two amino acids L-serine and L-phenylalanine. The carboxylic acid of phenylalanine was protected as a methyl ester, and the amino group of serine was protected with a benzyloxycarbonyl group (Fig. CS4.3). The alcohol group of serine was left unprotected as it was not expected to compete in the subsequent coupling reaction.

The two protected amino acids were coupled in the presence of the coupling agent dicyclohexylcarbodiimide (DCC) to give the protected dipeptide. The benzyloxycarbonyl protecting group was removed by hydrogenolysis

FIGURE CS4.3 Synthesis of *cyclo*-(L-phenylalanyl-L-seryl).

and the resulting structure was treated with ammonia in methanol to promote cyclization to the final product.

A trial run of the synthesis was carried out on a small scale with unlabelled compounds to ensure that all the reactions proceeded as anticipated, and to practice all the experimental techniques required. Having established that the route was effective, the radiosynthesis was carried out starting with unlabelled L-serine and radiolabelled L-[U-¹⁴C]phenylalanine. The 'U' indicates that all the carbon atoms in L-[U-¹⁴C]phenylalanine are universally labelled. Although the radioactivity of commercially available L-[U-¹⁴C]phenylalanine is high, the amount of compound present is extremely small and so the sample was taken up in solution and diluted with unlabelled L-phenylalanine to provide a manageable scale for the synthesis.

The radiosynthesis went according to plan, and the final product was subjected to **radiodilution analysis** to identify its radiochemical purity and to ensure that the radiochemistry resided in the correct product and not in some impurity. It is important to do this analysis as it is quite possible for the chemical purity of the final product to be high, but the radiochemical purity to be poor. This is because there is far more unlabelled compound present in the final structure than radiolabelled compound, and so the presence of a radiochemical impurity would be insignificant in chemical terms, but not in radiochemical terms.

The radiodilution procedure involved taking a sample of the radiolabelled product and dissolving it with pure unlabelled *cyclo*-(L-phenylalanyl-L-seryl). Aliquots were taken to measure the specific activity of the diluted sample, and then the sample was concentrated and the product was crystallized several times until the specific activity remained constant. Comparing the specific activity of the initial diluted sample with the specific activity of the final crystallized sample indicated a high level of radiochemical purity.

Radiodilution analysis was also carried out by diluting samples with pure unlabelled D,L and L,D-diastereoisomers of the cyclic dipeptides to determine whether any epimerization had taken place during the synthesis. The specific activities of the final crystallized products proved insignificant, indicating an absence of those diastereoisomers in the sample.

CS4.3 Synthesis of other isomers of *cyclo*-(L-Phe-L-Ser)

Although *cyclo*-(L-phenylalanyl-L-seryl) seems the most likely biosynthetic intermediate, it is quite possible that an epimerization could take place during the biosynthesis, and so one cannot rule out the possibility of another stereoisomer of the cyclic peptide being an intermediate. Therefore, the same synthetic route was used to prepare radiolabelled samples of the other three possible isomers of the cyclic dipeptide (Fig. CS4.4). As before, trial runs were carried out on unlabelled material to ensure that each stereoisomer could be obtained in good yield before committing radiolabelled starting material.

L-[U-¹⁴C]Phenylalanine was the radiolabelled starting material for the synthesis of *cyclo*-(L-[U-¹⁴C]Phe-D-Ser), while racemic DL-[3-¹⁴C]serine was the starting material for the diastereoisomers *cyclo*-(D-Phe-D-[3-¹⁴C]Ser) and *cyclo*-(D-Phe-L-[3-¹⁴C]Ser). Since the labelled serine was a racemate, the final product from the radiosynthesis was a mixture of the two diastereoisomers. Radiodilution analysis using each of the unlabelled diastereoisomers established that the ratio of the radiolabelled diastereoisomers in the mixture was 68%:28% in favour of the D,D-diastereoisomer.

CS4.4 Labelling studies with labelled isomers of *cyclo*-(Phe-Ser)

Each labelled structure (typically 32 mg, 6 µCi, dissolved in DMSO (2 ml)) was fed to one-day-old fungal cultures of *Trichoderma viride*, and the cultures were allowed to grow for another four days before they were harvested. The gliotoxin was extracted and purified to give a typical

cyclo-(L-[U-¹⁴C]Phe-L-Ser)   cyclo-(L-[U-¹⁴C]Phe-D-Ser)   cyclo-(D-Phe-D-[3-¹⁴C]Ser)   cyclo-(D-Phe-L-[3-¹⁴C]Ser)

FIGURE CS4.4 Radiolabelled dioxopiperazines. The blue star indicates the position of the radiolabel. The radiolabel in phenylalanine was distributed around all the carbon atoms.

yield of about 200 mg. The extracted gliotoxin was recrystallized to constant specific activity and from this it was possible to calculate the incorporation of radiolabel.

The results of these feeding experiments showed that the L,L-stereoisomer was incorporated to a level of 40–50%, whereas the other stereoisomers were poorly incorporated at only 0.5–1.5%. One has to be careful with negative results, since it is quite possible that the fungus might fail to produce gliotoxin on that particular run, or fail to take up the biosynthetic intermediate. So whenever the L,D-, D,D-, or D,L-stereoisomers were being tested, a control flask was used that was fed with the L,L-stereoisomer to ensure that that stereoisomer was still being taken up and incorporated into the final product.

The evidence clearly suggested that the L,L-cyclic dipeptide was being incorporated into gliotoxin, but there are many potential pitfalls in biosynthetic studies. For example, it could be argued that the cyclic dipeptide was actually being hydrolysed by fungal peptidases to the individual amino acids L-phenylalanine and L-serine, and that the amino acids were then being incorporated into gliotoxin via a different biosynthetic intermediate altogether (see section 11.6).

CS4.5 Double-labelling studies with *cyclo*-(L-[4'-³H]Phe-L-[¹⁴C]Ser)

As stated in section CS4.4, it is possible that the labelled cyclic dipeptide fed to the fungal culture could have been hydrolysed to its constituent amino acids, and that these were then incorporated into gliotoxin. To address that issue, a double-labelled compound was synthesized starting with L-[4'-³H]phenylalanine and L-[3-¹⁴C]serine, such that the cyclic dipeptide contained tritium at the

FIGURE CS4.5 Double-labelled *cyclo*-(L-[4'-³H]Phe-L-[3-¹⁴C]Ser).

para-position of the aromatic ring and a ¹⁴C label at the carbon bearing the hydroxyl group (Fig. CS4.5). Since the β-emissions from ³H and ¹⁴C have different energies, it is possible to measure these emissions separately and obtain a ratio of the radioactivity emitted from tritium versus the radioactivity emitted by ¹⁴C. The ratio of radioactivity for T:¹⁴C was found to be 10.9:1.

Double-labelled compounds are extremely useful in biosynthetic studies because a measurement of the isotope ratio for both the intermediate and the final product provides an indication of whether the intermediate has been incorporated intact. If this is the case, the isotope ratio should remain unchanged. However, if the intermediate has been degraded into different molecules which are incorporated separately, then one would expect the ratio to change. For example, if the cyclic dipeptide used in this research was hydrolysed to L-phenylalanine and L-serine, both labelled molecules would enter the cell's general pool of amino acids (Fig. CS4.6). It is highly unlikely that the concentration of these amino acids would be identical and so the labelled amino acids would be diluted to different extents. Moreover, L-phenylalanine and

FIGURE CS4.6 Possible hydrolysis of the L,L-cyclic dipeptide.

L-serine are used in the synthesis of all the cell's proteins, and so it is extremely unlikely that the same proportion of L-Phe and L-Ser would be taken up for gliotoxin biosynthesis. Finally, L-serine is known to be interconvertible with glycine, and so some of the labelled L-serine would be converted to glycine with loss of the ^{14}C label. The ^{14}C label would then enter the cell's general 1-C pool and be incorporated into a wide variety of different molecules.

As it turned out, the ratio of isotopes (T:^{14}C) in gliotoxin was found to be 11.1:1, which was essentially unchanged from the 10.9:1 measured for the cyclic dipeptide.

CS4.6 Degradation studies on double-labelled gliotoxin

Some degradation studies were now carried out to identify whether the radioisotopes were at their expected positions within the structure or had been scrambled. The radiolabelled gliotoxin obtained from the feeding experiments was converted to **anhydrodesthiogliotoxin**, with the isotope ratio remaining the same as expected (Fig. CS4.7). Treatment of anhydrodesthiogliotoxin with ozone cleaved the ^{14}C label from the structure as [^{14}C]formaldehyde, which was trapped as its dimedone derivative. The resulting trioxopiperazine structure contained tritium but no ^{14}C, demonstrating that all the ^{14}C label had been at the expected position.

Another experiment involved heating gliotoxin in basic solution to cause degradation and release of methylamine gas derived from the N-methyl moiety of gliotoxin. This was trapped in acid solution as the hydrochloride salt and converted to a phenylthiourea derivative which proved to be inactive. If the cyclic dipeptide had been degraded *in vivo*, the resulting serine would have been converted to glycine with release of a one carbon unit that would have found its way into methionine and thence to the N-methyl group of gliotoxin.

CS4.7 Intermediate trapping study

Even with all this evidence, it is still not fully established that the cyclic dipeptide is a true biosynthetic intermediate. For example, it could be argued that that the cyclic dipeptide is ring-opened by a peptidase enzyme to a linear dipeptide and that this is the true biosynthetic intermediate, not the cyclic dipeptide. In order to rule out this possibility, it was important to establish whether the fungus could actually produce the proposed cyclic dipeptide structure. This involved an **intermediate trapping experiment** which was carried out as follows. The fungal culture was fed with 40 mg of unlabelled cyclic dipeptide, and two hours later a sample of L-[U^{14}C]-phenylalanine (9 µg, 25 µg Ci) was added.

Instead of allowing the culture to grow for the normal four days, it was harvested two hours later and a continuous extraction of the aqueous filtrate was carried out with EtOAc for six days in order to recover the cyclic dipeptide. If the fungus was capable of producing the cyclic dipeptide, one would expect some of the labelled phenylalanine to have been used in its synthesis, and so the recovered cyclic dipeptide should include labelled material that had been produced by the cell.

The crude cyclic dipeptide that was recovered was indeed radioactive, but it is possible that the radioactivity could be due to other cellular products, and so the sample was diluted with unlabelled cyclic peptide and crystallized several times until the specific activity was constant. This indicated that there had been a 1–2% conversion of the radiolabelled amino acid into the cyclic

FIGURE CS4.7 Degradation experiments on labelled gliotoxin.

dipeptide, and confirmed that the fungus was capable of synthesizing the proposed cyclic intermediate.

CS4.8 Labelling studies with *cyclo*-(L-[U-¹⁴C]Phe-(NMe)-L-Ser)

Having confirmed *cyclo*-(L-Phe-L-Ser) as a biosynthetic intermediate, further labelling studies were carried out to test whether the *N*-methylated cyclic dipeptide (*cyclo*-(L-phenylalanyl-*N*-methyl-L-seryl)) was also a biosynthetic intermediate. In order to synthesize this structure, it was necessary to first prepare *O*-*t*-butyl-*N*-methyl-L-serine methyl ester (Fig. CS4.8). The alcohol group had to be protected in this synthesis since methylation of the amine group would also have methylated the alcohol group.

Having obtained the required serine derivative, the synthesis of *cyclo*-(L-phenylalanyl-*N*-methyl-L-seryl)

was rehearsed (Fig. CS4.9). The amine group of L-phenylalanine was protected with a benzyloxycarbonyl group, and then coupled with the *N*-methylated serine derivative. Following hydrogenolysis of the benzyloxycarbonyl group, the product spontaneously cyclized to give mainly *cyclo*-(L-phenylalanyl-*O*-*t*-butyl-*N*-methyl-L-seryl), as well as a smaller quantity of *cyclo*-(L-phenylalanyl-*O*-*t*-butyl-*N*-methyl-D-seryl) in a ratio of 9:1. The two diastereoisomers could be separated by carrying out a crystallization where the L,D-diastereoisomer crystallized preferentially. The pure L,L-diastereoisomer could be obtained by column chromatography followed by crystallization. However, recoveries were low.

The final deprotection step for the L,L-diastereoisomer proved problematic and various methods were tried until 45% HBr/HOAc proved effective. A small amount of the acetylated product *cyclo*-(L-phenylalanyl-*O*-acetyl-*N*-methyl-L-seryl) was also obtained, but this could be avoided by minimizing the reaction time for the deprotection.

FIGURE CS4.8 Synthesis of protected *O*-*t*-butyl-*N*-methyl-L-serine methyl ester.

FIGURE CS4.9 Synthesis of *cyclo*-(L-phenylalanyl-*N*-methyl-L-seryl).

Since the separation and purification of the two cyclized diastereoisomers had resulted in poor recoveries, the crude mixture was deprotected first before attempting the separation and purification process, in the hope of obtaining improved recoveries.

Before attempting this, a pure sample of *cyclo*-(L-phenylalanyl-*N*-methyl-D-seryl) was prepared as a standard. Unfortunately, it was found that the retention factors of the L,D-and the L,L-diastereoisomers were virtually identical. However, a separation could be achieved on preparative layer chromatography if the plate was eluted four times with the eluant.

A trial synthesis was carried out on a 1 mmol scale where the crude cyclized product containing both isomers was taken forward for deprotection. Fortunately, preparative layer chromatography and crystallization proved effective in purifying the desired L,L-diastereoisomer.

Having optimized the reaction conditions, the radiosynthesis was carried out starting from L-[U-^{14}C] phenylalanine as the labelled starting material to provide *cyclo*-(L-[U-^{14}C]phenylalanyl-*N*-methyl-L-seryl) with a specific activity of 83 μCi/mmol, corresponding to 98.8% of the initial specific activity for L-[U-^{14}C]phenylalanine.

Radioscanning and autoradiography of TLC plates run in three different solvent systems showed only one radiolabelled product. Radiodilution analysis with unlabelled *cyclo*-(L-Phe-L-(NMe)Ser) also indicated that the sample was pure.

The labelled *N*-methylated cyclic dipeptide was fed to a culture of *Trichoderma viride* as before, alongside a control flask which was fed with *cyclo*-(L-[U-^{14}C]

phenylalanyl-L-seryl). The resulting gliotoxin from the experiment was essentially inactive (1.8 × 10^{-3} μCi/mmol) as opposed to the control flask which provided gliotoxin with a specific activity of 2.91 μCi/mmol, representing a 55% incorporation. In other words, the specific activities for the two crops of gliotoxin differed by a factor of more than 1500. Therefore, it appears unlikely that the *N*-methylated dipeptide is an authentic biosynthetic intermediate.

CS4.9 Double-labelling studies with linear dipeptides

Labelling studies were also carried out with radiolabelled linear dipeptides. L-[3-^{14}C]Seryl-L-[4′-^{3}H]phenylalanine was prepared by coupling the *N*-benzyloxycarbonyl derivative of L-[3-^{14}C]serine with the benzyl ester of L-phenylalanine. Hydrogenolysis of the protected dipeptide removed both protecting groups to provide the required dipeptide (Fig. CS4.10).

The other possible dipeptide, L-[4′-^{3}H] phenylalanyl-L-[3-^{14}C]serine, was synthesized in a similar manner (Fig. CS4.11).

The gliotoxin extracted from the feeding experiments involving these labelled dipeptides was certainly radioactive. However, the isotopic ratio in the gliotoxin was significantly different from that in the dipeptides, indicating that tritium had been incorporated five times more effectively than ^{14}C. This demonstrated that the linear dipeptides had been hydrolysed by fungal enzymes

FIGURE CS4.10 Synthesis of L-[3-^{14}C]Seryl-L-[4′-^{3}H]phenylalanine.

FIGURE CS4.11 Synthesis of L-[4′-³H]Phenylalanyl-L-[3-¹⁴C]serine.

and that the two amino acids had been incorporated separately, with phenylalanine incorporated more efficiently than serine.

However, this does not rule out the possibility of the dipeptides being intermediates. Indeed, if the intermediates are fed to more mature cultures, the isotope ratios become more similar.

CS4.10 Generating gliotoxin analogues from cyclic dipeptides

Having established that cyclic dipeptides are biosynthetic intermediates of gliotoxin, it is possible that gliotoxin analogues could be generated by feeding different cyclic dipeptides to the fungal culture. However, this depends on whether the enzymes involved in the

biosynthetic pathway will accept the 'alien' cyclic dipeptides as substrates. The principle for this approach was established when *cyclo*-(L-alanyl-L-[U-¹⁴C]phenylalanyl) was fed to the fungal culture and resulted in a new metabolite that was detected on TLC at a higher retention factor by radioscanning and autoradiography. The analogue was isolated in about 10% yield and identified as **3a-deoxygliotoxin**, having an identical specific activity to the radiolabelled precursor (Fig. CS4.12).

The biosynthesis of natural product analogues is potentially useful when:

- the organism is easy to grow;
- the modified precursor is simple to synthesize;
- the biosynthetic conversion is efficient.

The results obtained from these labelling experiments certainly fitted these three criteria.

cyclo-(L-Alanyl-L-seryl)   Gliotoxin analogue

FIGURE CS4.12 Conversion of *cyclo*-(L-alanyl-L-seryl) to an analogue of gliotoxin.

QUESTIONS

1. Gliotoxin contains a hexadiene ring with an alcohol group present. Dehydration should result in a stable aromatic ring, yet this does not happen. Explain why this is the case.

2. Consider the intramolecular cyclization reaction shown in Figure CS4.3. Analyse the cyclization using Baldwin's rules (section 4.11) and state whether it is favoured or not favoured.

3. In section CS4.8, it was stated that the radiolabelled *N*-methylated dipeptide was not converted to gliotoxin. However, it was also stated that it was an 'unlikely' intermediate. Can you suggest why these results do not conclusively rule out this structure as a biosynthetic intermediate.

4. Considering the successful biosynthesis of a gliotoxin analogue described in section CS4.10, can you identify any other analogues that might be successfully generated in this way? Which analogues would have the best chance of success?

FURTHER READING

Kirby, G.W., Patrick, G.L., and Robins, D.J. (1978) 'cyclo-(L-Phenylalanyl-L-seryl) as an intermediate in the biosynthesis of gliotoxin', *Journal of the Chemical Society, Perkin Transactions 1*, 1978, 1336–8.

Kirby, G.W. and Robins, D.J. (1976) 'Analogue biosynthesis in *Trichoderma viride*: the formation of 3a-deoxygliotoxin', *Journal of the Chemical Society, Chemical Communications*, 1976, 354–5.

■ CASE STUDY 5
Fluorine in drug design and synthesis

5.1 Introduction

Several important drugs contain one or more fluorine atoms within their structure. In many cases, these atoms have been introduced to block the occurrence of a metabolic reaction. For example, **CGP 52411** is a protein kinase inhibitor and was put forward for clinical trials as an anticancer agent. However, metabolic studies revealed that cytochrome P450 enzymes catalysed the oxidation of the aromatic rings to form a polar metabolite that was rapidly excreted (Fig. CS5.1). Fluoro substituents were successfully introduced as metabolic blockers in the analogue **CGP 53353**. There are a number of reasons for choosing fluorine as a metabolic blocker. First, it is only slightly larger than hydrogen, and so introducing it is unlikely to have any detrimental steric effects that might block the analogue fitting the target binding site. Secondly, the C–F bond is strong and so it is unlikely that the fluorine will be easily removed from the structure by any chemical reaction. Thirdly, the enzyme-catalysed reactions responsible for replacing a metabolically susceptible hydrogen proton with an OH group require the departure of a positively charged proton. When a fluorine

atom is present, the enzyme-catalysed reaction would require the loss of a positively charged fluorine ion. This is a highly unstable species and so the enzyme-catalysed reaction cannot take place.

This tactic of blocking drug metabolism has also been applied successfully in the design of the protein kinase inhibitor **gefitinib**, which is used to treat lung cancers (Fig. CS5.2). Similarly, a number of anti-inflammatory steroids such as **flumetasone pivalate** contain a fluoro substituent that blocks metabolism at position 6.

The presence of fluorine in place of an enzymatically susceptible hydrogen can also disrupt some enzymatic reactions. For example, the antitumour drug **5-fluorouracil** acts as an inhibitor of an enzyme known as thymidylate synthase, which normally catalyses the conversion of dUMP to dTMP (Fig. CS5.3). Inhibiting this reaction means that there are fewer building blocks for DNA synthesis, leading to inhibition of cell growth and division.

The normal reaction involves the substrate dUMP reacting with a cysteine residue in the enzyme, as well as the cofactor N^5,N^{10}-methylene FH_4 (Fig. CS5.4). The cofactor is responsible for providing the methyl substituent that will appear in the product dTMP. The mechanism

FIGURE CS5.1 Metabolic oxidation of CGP 52411.

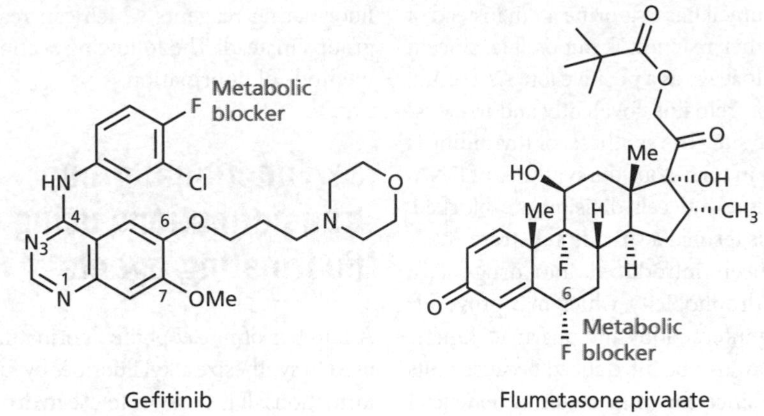

FIGURE CS5.2 Gefitinib and flumetasone pivalate.

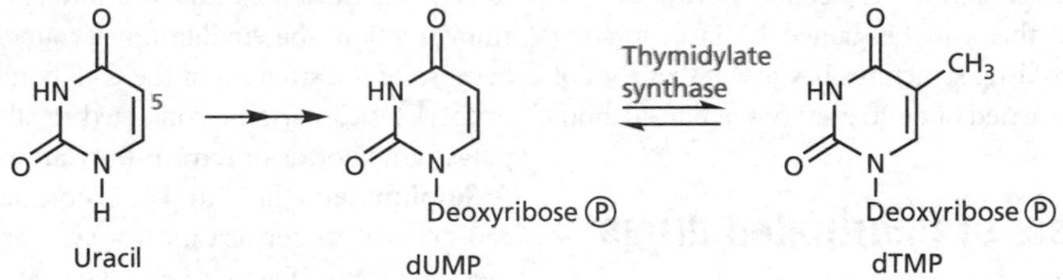

FIGURE CS5.3 Biosynthesis of dTMP (Ⓟ = phosphate).

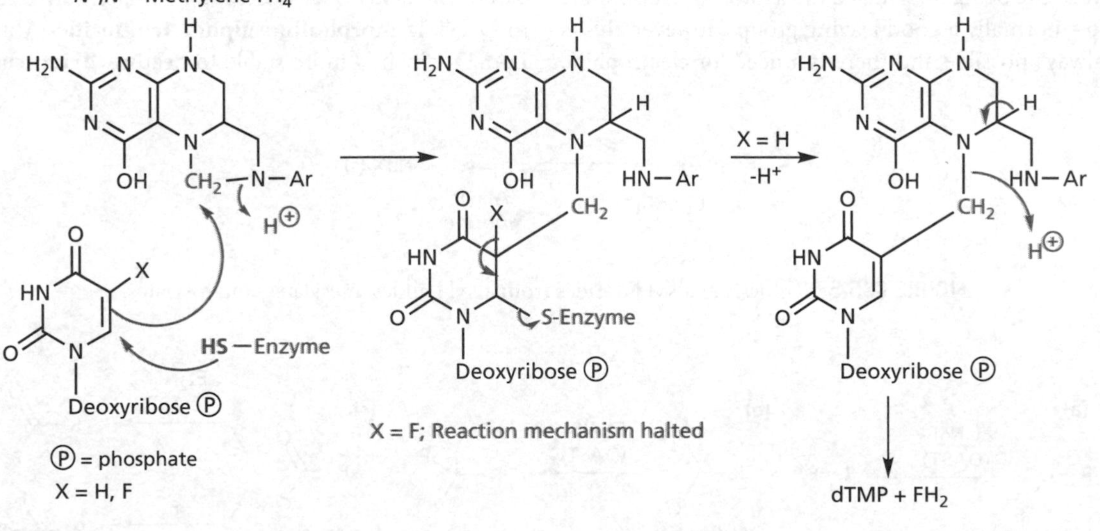

FIGURE CS5.4 Use of 5-fluorouracil as a prodrug for a suicide substrate.

requires the loss of a positively charged proton (X = H in Fig. CS5.4).

5-Fluorouracil is actually a prodrug and is converted in the body to the fluorinated analogue of deoxyuridylic acid monophosphate, which is mistaken for the normal substrate and combines with the enzyme and the cofactor. The first part of the mechanism proceeds normally. The tetrahydrofolate cofactor forms a covalent bond to the uracil skeleton via the methylene unit, but now things start to go wrong. In the second stage, it is not possible to lose a proton. 5-Fluorouracil has a fluorine atom instead of hydrogen (X = F). Further reaction is impossible, since it would require fluorine to leave as a positive ion. As a result, the fluorouracil skeleton remains covalently and irreversibly bound to the active site. The synthesis of thymidine is now terminated, which in turn stops the synthesis of DNA. Consequently, replication and cell division are blocked. The structure FdUMP is termed a **suicide substrate**.

Fluorine has also been introduced into drugs as a means of increasing hydrophobicity, which may prove advantageous for binding interactions and pharmacokinetic properties. Fluorine may also be introduced because of its electronegative nature since this may result in beneficial electronic effects that enhance binding to the target binding site. Finally, introducing the radioactive isotope of fluorine (^{18}F) has been very useful in preparing compounds that are used for diagnostic scans (section 11.10).

The advantages that can be gained by incorporating fluorine atoms into drug structures has meant that a lot of research has been carried out into methods of fluorination.

5.2 Synthesis of fluorinated drugs

The easiest method of introducing a fluorine substituent is to carry out a nucleophilic substitution using a fluoride ion. However, other nucleophilic reagents are available. In order to use a nucleophilic fluorinating reagent, the structure being fluorinated must have an electrophilic group—normally a good leaving group. However, this is not always possible and there is a need for electrophilic fluorinating reagents which can react with nucleophilic groups instead. The following sections describe different methods of fluorination.

5.3 Functional group transformations using nucleophilic fluorinating agents

A number of nucleophilic fluorinating reagents have been used to synthesize alkyl fluorides by functional group transformations. It is possible to use **hydrogen fluoride**, but this is a highly toxic agent which eats into glass and damages bone if it gets on the skin. **Potassium fluoride** is a more practical reagent and can be used to convert alkyl halides to alkyl fluorides (Fig. CS5.5). Although this is an equilibrium reaction, the equilibrium favours the alkyl fluoride because of the strength of the C–F bond. Alkyl mesylates or tosylates can also be converted to alkyl fluorides using potassium fluoride or **tetra-n-butylammonium fluoride**.

Sulphur tetrafluoride is a fluorinating agent which can be used to convert a carboxylic acid to a trifluoromethyl group (Fig. CS5.6c). However, the reagent is a highly toxic gas and has to be handled with care. **Diethylaminosulphur trifluoride (DAST)** is a more selective fluorinating agent and is easier to handle. It can be used to fluorinate alcohols, aldehydes, ketones, and carboxylic acids (Fig. CS5.6a–c). An alternative reagent to DAST is **morpholinosulphur trifluoride (Morpho-DAST)** which is more stable to heating. **Triethylamine**

FIGURE CS5.5 Synthesis of alkyl fluorides from alkyl halides, mesylates, and tosylates.

FIGURE CS5.6 Fluorination with nucleophilic reagents.

FIGURE CS5.7 Conversion of an aromatic methyl group to a trifluoromethyl group.

trihydrofluoride (Et$_3$N.3HF) has also been used as an alternative non-corrosive reagent to DAST for a large-scale fluorination reaction (Fig. CS5.6d). Perfluorosulphonyl fluorides have been used to activate alcohol groups which are then susceptible to nucleophilic substitution by the fluoride ion that is generated (Fig. CS5.6e).

A methyl group on an aromatic ring can be converted to a trifluoromethyl group by chlorinating the methyl group and then reacting it with **antimony(V) fluoride** (Fig. CS5.7).

5.4 Fluorination by C–C bond formation

A variety of fluorinated reagents are commercially available that can be used to synthesize fluorinated compounds by means of carbon–carbon bond formation.

A particularly useful reaction in drug synthesis is the trifluoromethylation of aromatic and heteroaromatic rings because of the pharmacokinetic and pharmacodynamic advantages that such a group might bring. Exposed methyl groups are commonly oxidized by cytochrome P450 enzymes to carboxylic acids, and replacing such a group with a CF$_3$ group improves the metabolic stability. Other pharmacokinetic and pharmacodynamic advantages can arise from the increased lipophilicity of a CF$_3$ group. For example, increased lipophilicity may improve absorption and distribution, or enhance binding interactions with a hydrophobic binding region in a target binding site.

We have already seen in Figures CS5.6c and CS5.7 how a CF$_3$ group can be introduced. Here, we discuss methods of introducing the CF$_3$ moiety by means of carbon–carbon bond formation. A common method of achieving this reaction is to carry out a cross-coupling reaction where an aryl iodide, aryl bromide, or vinyl halide is treated with CuCF$_3$ generated in situ (Figs CS5.8 and CS5.9). A number of reagents can be used for this purpose, such as CF$_3$SiEt$_3$, FSO$_2$CF$_2$CO$_2$Me, ClCF$_2$CO$_2$Me and CF$_3$CO$_2$Na. The first of these reagents is converted to $^-$CF$_3$ by adding a fluoride ion to the reaction. The fluoride ion forms a strong bond to silicon, thus breaking the bond between silicon and the trifluoromethyl group to generate the $^-$CF$_3$ carbanion (Fig. CS5.8a). The other three reagents decompose in the presence of CuI to form a carbene (:CF$_2$), which then combines with a fluoride ion to form $^-$CF$_3$ (Fig. CS5.8b–d).

A disadvantage with the last three reactions is the need for strong heating and the presence of one equivalent of cuprous iodide, and so more recent research has developed methods of carrying out the reaction under milder conditions.

An important step forward was the discovery that catalytic quantities of CuI could be used in trifluoromethylation reactions if 1,10-phenanthrene was included in the reaction.

New reagents have also been developed. Potassium (trifluoromethyl)trimethoxyborate is a relatively new agent that generates $^-$CF$_3$ under mild non-basic conditions using catalytic quantities of CuI and 1,10-phenanthrene (Fig. CS5.9a). A trifluoromethylcopper(I) reagent ligated with 1,10-phenanthroline has also been isolated and found to react under mild conditions (Fig. CS5.9b).

Further work has shown that it is possible to react aryl and alkenylboronic acids with Me$_3$SiCF$_3$ under mild conditions to introduce a trifluoromethyl group, thus extending the reaction to a range of readily available boronic acids (Fig. CS5.9c).

Trifluoromethylation of a carbonyl group is also possible using CF$_3$SiMe$_3$ in the presence of tetrabutylammonium fluoride (TBAF). TBAF provides the fluoride ion needed to release $^-$CF$_3$ from the reagent (Fig. CS5.9d).

FIGURE CS5.8 Methods used to introduce a CF$_3$ group.

FIGURE CS5.9 Methods used to introduce a CF₃ moiety (phen = 1,10-phenanthroline).

FIGURE CS5.10 Palladium-catalysed trifluoromethylation of aryl chlorides.

FIGURE CS5.11 Examples of trifluoromethylation reactions under radical conditions.

Palladium-mediated reactions have been tried out as an alternative to copper mediation. A particularly important application is the palladium-catalysed trifluoromethylation of aryl chlorides, as these compounds do not react well in copper-mediated reactions (Fig. CS5.10). However, high temperatures are required for this reaction.

The preceding reactions all require the starting material to contain a halogen or boronic acid, which is substituted with the trifluoromethyl group. Further studies have been carried out to investigate methods of using electrophilic reagents to trifluoromethylate unactivated positions in heteroaromatic rings. One way of doing this is to use CF_3I, but this reagent is a gas and is not easy to work with. A more convenient method is to use **Langlois reagent** (sodium trifluoromethanesulphinate), which is decomposed in the presence of *tert*-butyl peroxide to release a CF_3 radical. This radical then forms a bond to heterocyclic rings (Fig. CS5.11). The reaction normally takes place at the most nucleophilic position of the heterocyclic ring, but the regioselectivity can be altered by varying the solvents used.

An alternative way of incorporating a CF_3 group into a molecule is to take advantage of the large number of simple fluorinated molecules that are now commercially available (Fig. CS5.12). These can be used as reagents or starting materials for conventional synthetic reactions without having to use the specialized reagents or reaction conditions that would otherwise be required to introduce CF_3 and other fluorinated features.

Finally, the reagent CBr_2F_2 has been used to add a CF_2 group to an aldehyde by means of a Wittig reaction (Fig. CS5.13). The Wittig reagent is generated in situ.

Fluorination at the α-carbon of aldehydes and ketones is possible using an electrophilic fluorinating agent. Fluorine (F_2) itself has been used for this purpose, but fluorine is very reactive, non-selective, and difficult to handle. Moreover, toxic HF is formed as a by product of the reaction. A more convenient reagent is **Selectfluor** which is sufficiently stable to be handled safely and reacts more selectively (Fig. CS5.14). *N*-**Fluorobenzenesulphonimide (NFSi)** is another electrophilic fluorinating reagent that is commonly used (Fig. CS5.15).

FIGURE CS5.12 Examples of commercially available compounds containing fluorine.

FIGURE CS5.13 Addition of a CF_2 moiety through the Wittig reaction.

FIGURE CS5.14 Selectfluor.

FIGURE CS5.15 Fluorination of a ketone.

FURTHER READING

Ball, N.D., Kampf, J.W., and Sanford, M.S. (2010) 'Aryl-CF(3) bond-forming reductive elimination from palladium(IV)', *Journal of the American Chemical Society*, **132**, 2878–9.

Chen, Q.-Y. and Wu, S.-W. (1989) 'Methyl fluorosulphonyldifluoroacetate; a new trifluoromethylating agent', *Journal of the Chemical Society, Chemical Communications*, 1989, 705–6.

Chu, L. and Qing, F.-L. (2010) 'Copper-mediated oxidative trifluoromethylation of boronic acids', *Organic Letters*, **12**, 5060–3.

Grushin, V.V. and Marshall, W.J. (2006) 'Facile Ar-CF$_3$ bond formation at Pd. Strikingly different outcomes of reductive elimination from [(Ph$_3$P)$_2$Pd(CF$_3$)Ph] and [(xantphos)Pd(CF$_3$) Ph]', *Journal of the American Chemical Society*, **128**, 12 644–5.

Hadlington, S. (2011) 'Trifluormethylation' *Chemistry World*, October, p 28.

Ji, Y., et al. (2011) 'Innate C-H trifluoromethylation of heterocycles', *Proceedings of the National Academy of Sciences of the USA*, **108**, 14 411–15.

Knauber, T., et al. (2011) 'Copper-catalysed trifluoromethylation of aryl iodides with potassium (trifluoromethyl)trimethoxyborate', *Chemistry: A European Journal*, **17**, 2689–97.

Langlois, B.R. and Roques, N. (2007) 'Nucleophilic trifluoromethylation of aryl halides with methyl trifluoroacetate', *Journal of Fluorine Chemistry*, **128**, 1318–25.

Morimoto, H., et al. (2011) 'A broadly applicable copper reagent for trifluoromethylations and perfluoroalkylations of aryl iodides and bromides', *Angewandte Chemie International Edition*, **50**, 3793–8.

Neumeyer, J.L., et al. (1994) 'N-ω-Fluoroalkyl analogs of (1*R*)-2β-carbomethoxy-3β-(4-iodophenyl)-tropane (β-CIT): radiotracers for positron emission tomography and single photon emission computed tomography imaging of dopamine transporters', *Journal of Medicinal Chemistry*, **37**, 1558–61 (ioflupane).

Oishi, M., et al. (2009) 'Aromatic trifluoromethylation catalytic in copper', *Chemical Communications*, **14**, 1909–11.

Ojima, I. (ed.) (2009) *Fluorine in medicinal chemistry and chemical biology*. Blackwell, Chichester.

Qiu, X.-L. and Qing, F.-L. (2002) 'Practical synthesis of Boc-protected *cis*-4-trifluoromethyl and *cis*-4-difluoromethyl-L-prolines', *Journal of Organic Chemistry*, **67**, 7162–4.

Qiu, X.-L. and Qing, F.-L. (2002) 'Synthesis of Boc-protected *cis*- and *trans*-4-trifluoromethyl-D-prolines', *Journal of the Chemical Society, Perkin Transactions 1*, 2052–7.

Roy, S., et al. (2011) 'Trifluoromethylation of aryl and heteroaryl halides', *Tetrahedron*, **67**, 2161–95.

Sankar Lal, G., Pez, G.P., and Syvret, R.G. (1996) 'Electrophilic NF fluorinating agents', *Chemical Reviews*, **96**, 1737–55.

Senecal, T.D., et al. (2011) 'Room temperature aryl trifluoromethylation via copper-mediated oxidative cross-coupling', *Journal of Organic Chemistry*, **76**, 1174–6.

Takamatsu, S. (2001) 'Improved synthesis of 9-(2,3-dideoxy-2-fluoro-β-D-*threo*-pentofuranosyl)adenine (FddA) using triethylamine trihydrofluoride', *Tetrahedron Letters*, **42**, 2321–4.

Takamatsu, S. (2002) 'Convenient synthesis of fluorinated nucleosides with perfluoroalkanesulfonyl fluorides', *Nucleosides, Nucleotides and Nucleic Acids*, **21**, 849–61.

Teare, H., et al. (2010) 'Radiosynthesis and evaluation of [^{18}F] selectorfluor bis(triflate)', *Angewandte Chemie International Edition*, **49**, 6821–4.

Wang, X. (2010) 'Pd(II)-catalysed *ortho*-trifluoromethylation of arenes using TFA as a promoter', *Journal of the American Chemical Society*, **132**, 3648–9.

Wiehn, M.S., et al. (2010) 'Electrophilic trifluoromethylation of arenes and *N*-heteroarenes using hypervalent iodine reagents', *Journal of Fluorine Chemistry*, **131**, 951–7.

Xiao, J.-C., et al. (2005) 'Bipyridinium ionic liquid-promoted cross-coupling reactions between perfluoroalkyl or pentafluorophenyl halides and aryl iodides', *Organic Letters*, **7**, 1963–5.

Yamazaki, T., Taguchi, T., and Ojima, I. (2009) 'Unique properties of fluorine and their relevance to medicinal chemistry and chemical biology', in I. Ojima (ed.), *Fluorine in Medicinal Chemistry and Chemical Biology*. Blackwell, Chichester, pp 3–46.

PART C

Design and synthesis of selected antibacterial agents

Part C contains three chapters which focus on the design and synthesis of three different classes of antibacterial agent.

Chapter 12 covers the tetracyclines. The first tetracyclines to be discovered were natural products, and analogues were prepared by biosynthetic and semi-synthetic approaches. More recently, the full synthesis of tetracycline analogues has become feasible.

Chapter 13 covers erythromycin and macrolide antibacterial agents. Erythromycin is a natural product with a complex structure. As such, a full synthesis of erythromycin

or erythromycin analogues is impractical. The chapter describes how early analogues were synthesized from erythromycin by semi-synthetic procedures. Another approach to the preparation of erythromycin analogues is through biosynthesis, and the chapter describes how genetic engineering has been used to alter the biosynthetic pathway in order to produce novel analogues.

Chapter 14 covers quinolones and fluoroquinolones which are fully synthetic antibacterial agents. The chapter demonstrates that it is possible to use different synthetic routes to obtain similar types of structure.

12.1 Naturally occurring tetracyclines

Tetracyclines are clinically important broad-spectrum antibacterial agents that gain their name from the tetra-cyclic ring system which they all share (Fig. 12.1). The rings are labelled A–D and are bristling with a generous number of substituents, some of which are found in all the naturally occurring tetracyclines that have useful activity. These include a primary amide, a phenol, two enols, two α,β-unsaturated ketones, an allylic alcohol, and a tertiary amine. A number of other substituents (R^1–R^5), which vary from one tetracycline to another, are present. These include a chlorine atom, another tertiary amine group, a methyl group, and an alcohol.

Chlorotetracycline (Aureomycin) was the first of the tetracyclines to be isolated in 1948 from a

Streptomyces species. This was followed in the next few years by the discovery of **oxytetracycline (Terramycin)** and **tetracycline** itself (Fig. 12.2). All these structures showed good antibacterial activity, demonstrating that neither the 7-chloro group of chlorotetracycline nor the 5-OH group of oxytetracycline was crucial for activity.

6-Demethyltetracycline and **7-chloro-6-demethyl-tetracycline** (Fig. 12.3) were isolated from a mutant strain of *Streptomyces aureofaciens*, depending on the nature of the fermentation medium. If chloride ion was present in the medium, the chlorinated structure was produced. If absent, the non-chlorinated metabolite was formed. It is notable that the absence of the methyl group from ring C increases the stability of the tetracyclines to acid or base. Another mutant strain produced **7-bromotetracycline** when the fermentation medium contained bromide ion rather than chloride ion. This agent also reached the clinic.

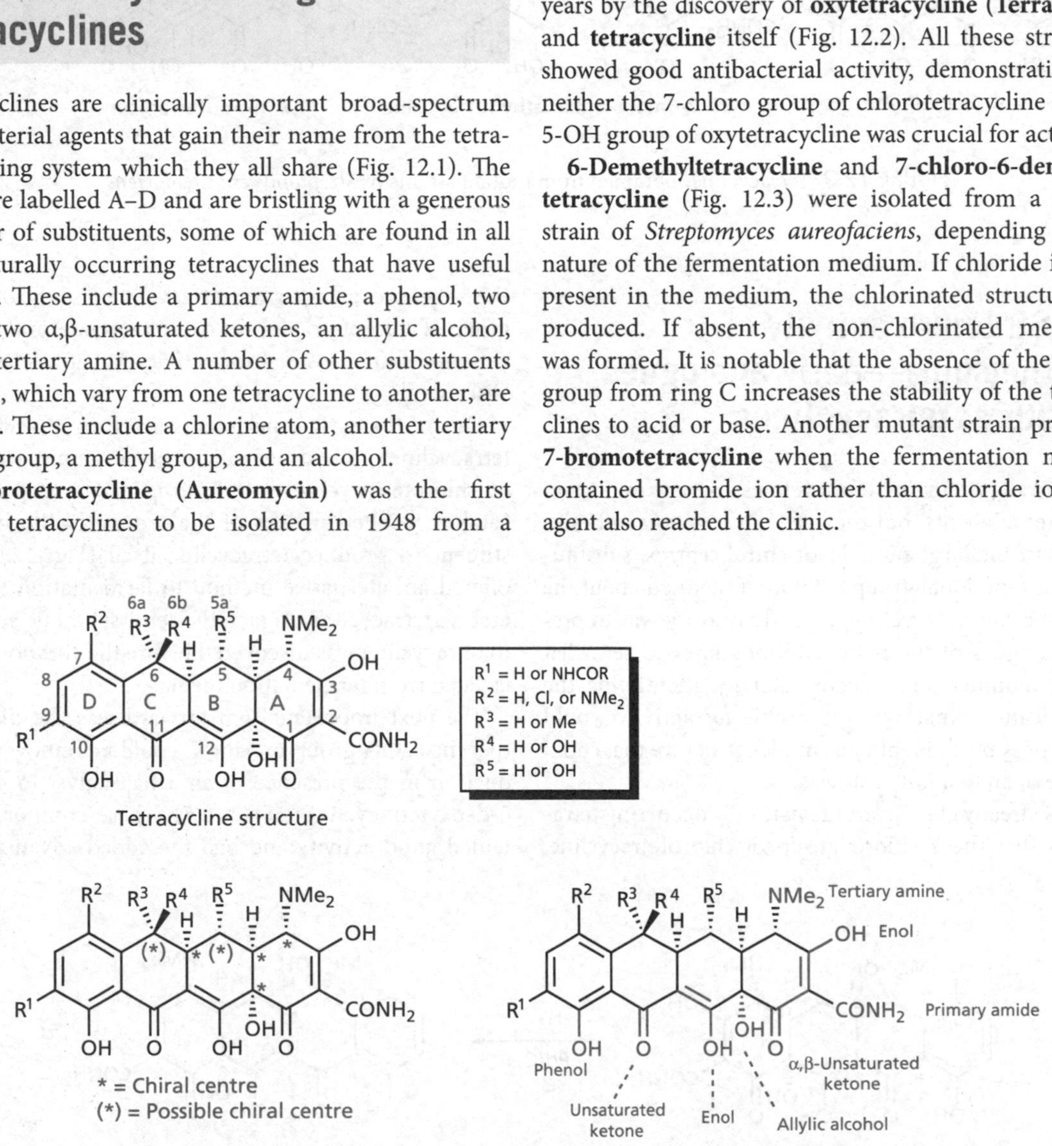

R^1 = H or NHCOR
R^2 = H, Cl, or NMe$_2$
R^3 = H or Me
R^4 = H or OH
R^5 = H or OH

Tetracycline structure

* = Chiral centre
(*) = Possible chiral centre

FIGURE 12.1 General structure of clinically important tetracyclines obtained from the natural world. Chiral centres are represented by a star. Stars in brackets represent chiral centres found in some but not all tetracyclines.

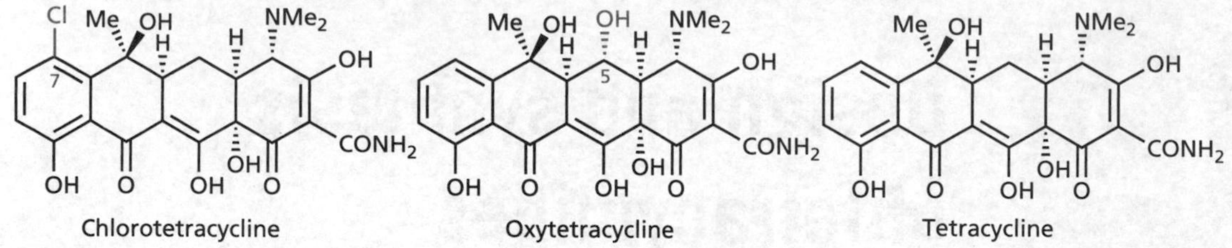

FIGURE 12.2 Naturally occurring tetracyclines.

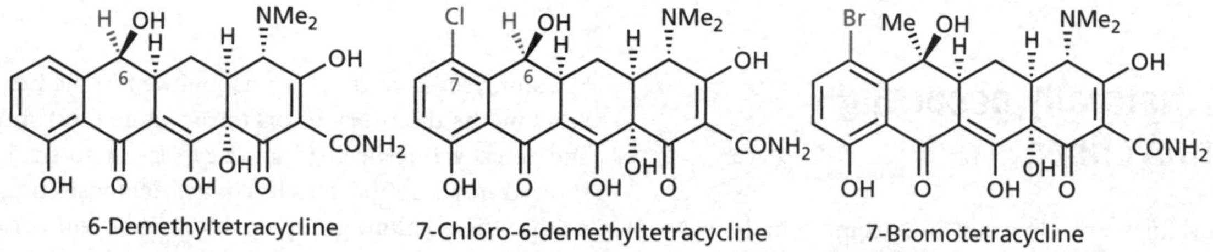

FIGURE 12.3 Tetracyclines obtained from mutant strains of *Streptomyces aureofaciens*.

12.2 Structure activity relationships—early analogues of natural tetracyclines

The naturally occurring tetracyclines are very important antibacterial agents, but they are not easy to synthesize because of the large number of chiral centres, substituents, and functional groups that are festooned about the tetracyclic rings. Therefore, an early priority was to prepare analogues of the natural tetracyclines to see what kind of modifications affected activity. Identifying the crucial features that are responsible for activity would make it possible to identify a simpler structure that could be more open to a full synthesis.

It was already clear from the naturally occurring tetracyclines that the 7-chloro group of chlorotetracycline, the 5-hydroxy group of oxytetracycline, and the 6-methyl group of chlorotetracycline, oxytetracycline, and tetracycline were not essential for activity. What else could be 'discarded'?

One of the early experiments involved subjecting the tetracyclines to hydrogenation. For example, treatment of chlorotetracycline with hydrogen over a palladium catalyst resulted in loss of the aromatic chlorine substituent to produce tetracycline itself (Fig. 12.4). This offered an alternative method to fermentation for generating tetracycline. In fact, the semi-synthetic approach to tetracycline was discovered before the compound was isolated from fermentation media.

The next important step forward was the discovery that the 6-OH group on ring C could be removed by reduction in the presence of an acid catalyst to produce 6-deoxytetracyclines (Fig. 12.5). These compounds retained good activity and had the added advantage that

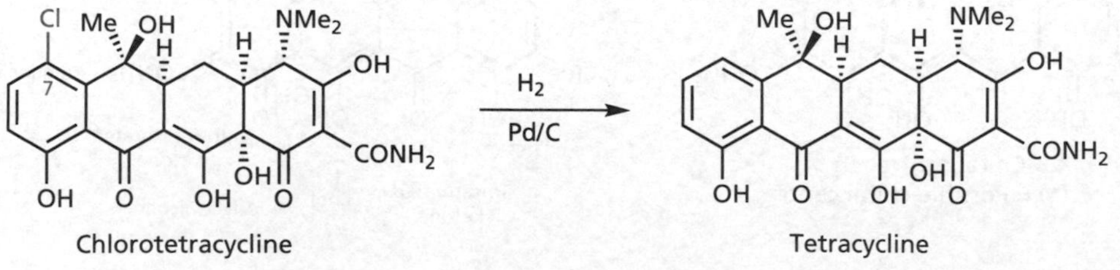

FIGURE 12.4 Catalytic hydrogenolysis of chlorotetracycline to form tetracycline.

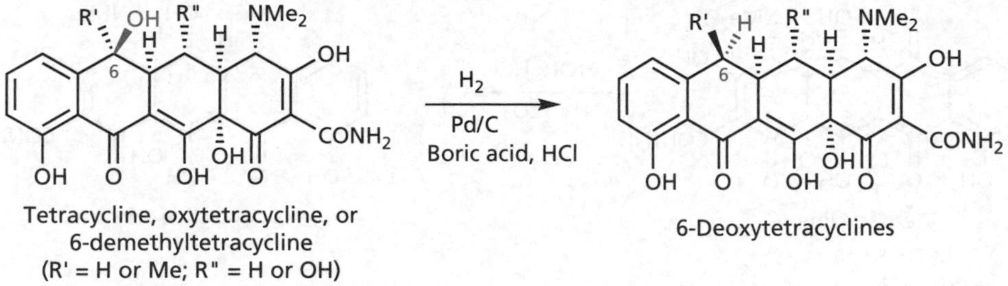

FIGURE 12.5 Removal of the 6-OH group by catalytic hydrogenation.

they were chemically more stable than the original tetracyclines. These results demonstrated that the 6-OH group was not essential for activity. Moreover, the chiral centre at position 6 has been inverted, indicating that the absolute configuration at C6 is not crucial either.

The compounds obtained from the above hydrogenation undergo inversion of configuration at C6, but it was possible to produce both possible isomers by a less direct route via 6-methylenetetracyclines (Fig. 12.6).

The 6-methylenetetracyclines have similar activity to the parent tetracyclines from which they are derived, emphasizing again that modifications can be carried out at C6 without adversely affecting activity. **Methacycline**

was one such agent that reached the clinic, and was synthesized from oxytetracycline (Fig. 12.6). It was also used as the starting material for the synthesis of the clinically important antibacterial agent **doxycycline** (Fig. 12.7). Hydrogenation reduced the methylene group to provide both possible isomers at C6, with the more active isomer being the one where the methyl group has retained its original orientation. An alternative two-stage process furnished doxycycline in better yield and purity (Fig. 12.7).

When 6-demethyltetracycline was reduced, the product was **6-deoxy-6-demethyltetracycline (sancycline)**, which was actually 70% more active than tetracycline

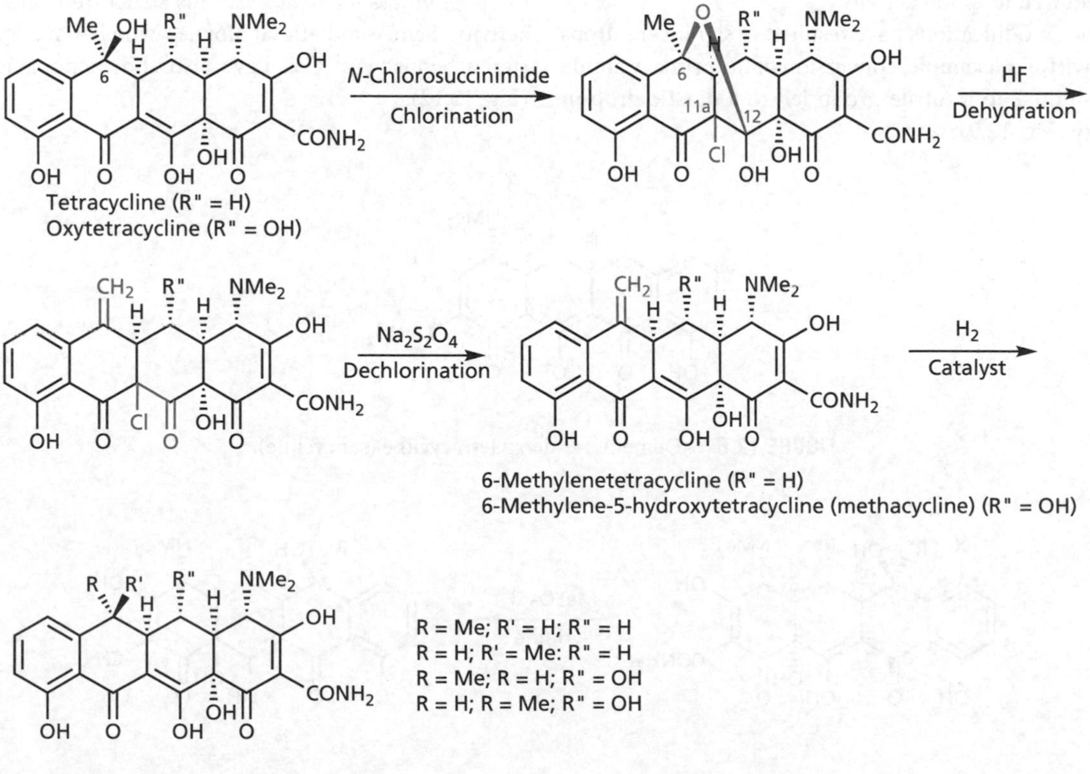

FIGURE 12.6 Synthesis and reduction of 6-methylenetetracyclines.

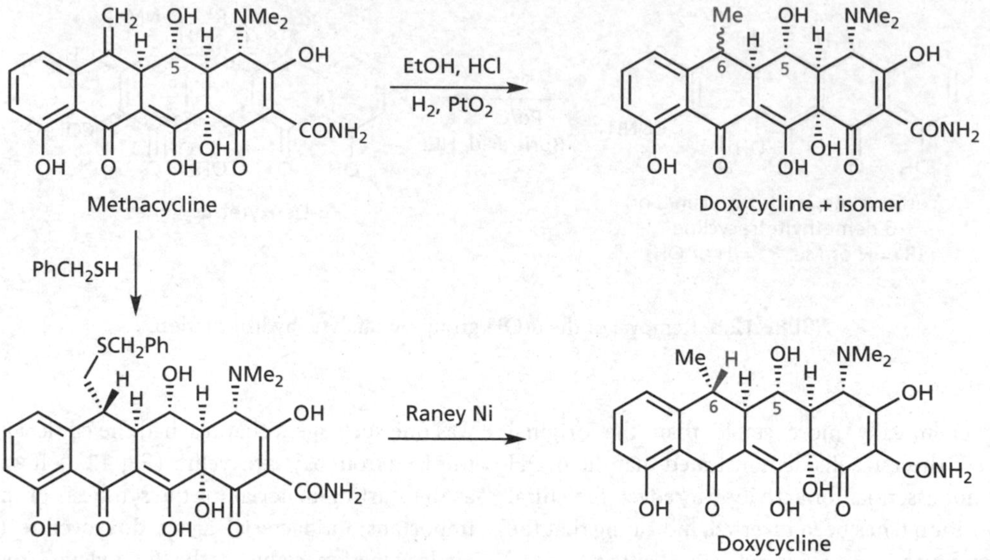

FIGURE 12.7 Synthesis of doxycycline.

itself (Fig. 12.8). This demonstrated that both the methyl group and the hydroxyl group could be removed from position 6, and also meant that the chiral centre at that position was not important to activity. The fact that the OH group at C5 is not crucial either indicates that two of the six chiral centres present in natural tetracyclines are not required for good activity.

Other modifications have resulted in significant drops in activity. For example, conversion of the primary amide at position 2 to a nitrile group led to a drastic drop in activity (Fig. 12.9).

Similarly, modifications of the 4-dimethylamino group adversely affected activity (Fig. 12.10).

Modifications affecting the chiral centre at C5a have also been bad for activity. For example, a tetracycline analogue which includes a double bond between C5a and C11a (Fig. 12.11) has been isolated from a mutant strain of *Streptomyces aureofaciens*. This structure is devoid of activity. Semi-synthetic analogues which have lost the stereochemistry at C5a have also shown poor activity (Fig. 12.12).

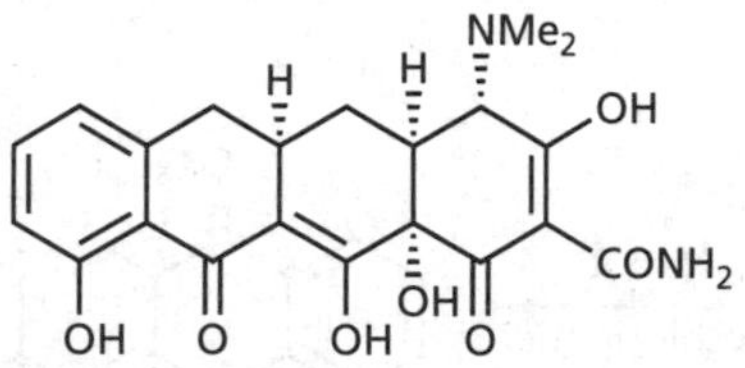

FIGURE 12.8 6-Demethyl-6-deoxytetracycline (sancycline).

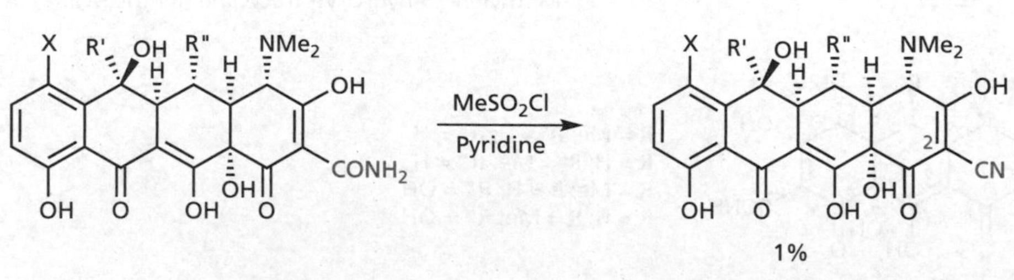

FIGURE 12.9 Modification of the primary amide group. Percentage values are levels of antibacterial activity relative to tetracycline.

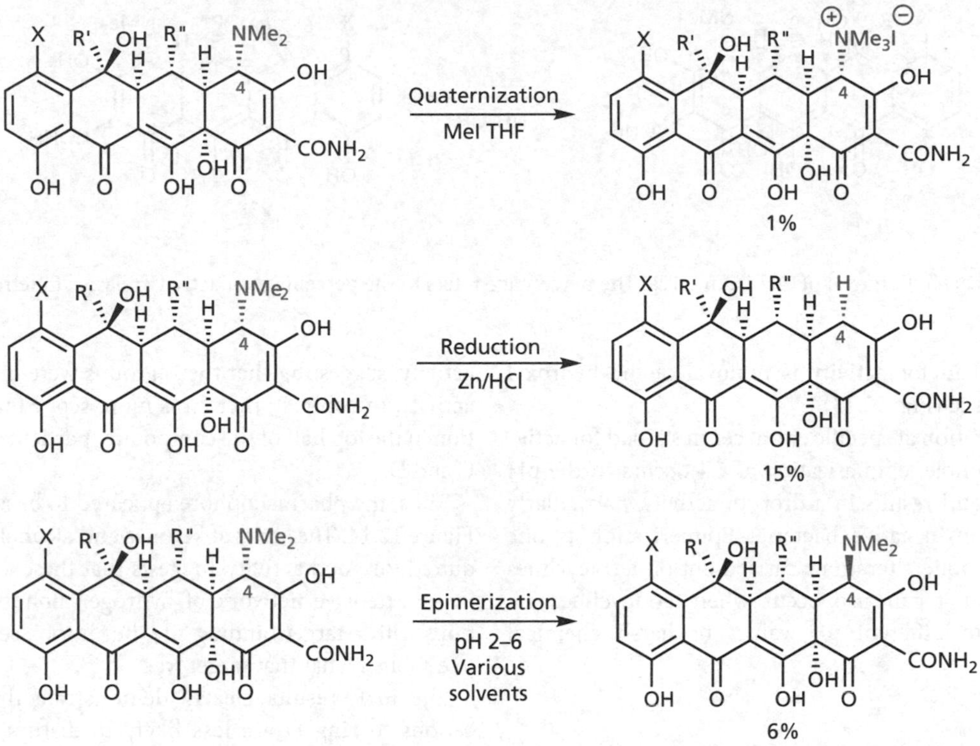

FIGURE 12.10 Analogues with poor activity. Percentage values are levels of antibacterial activity relative to tetracycline.

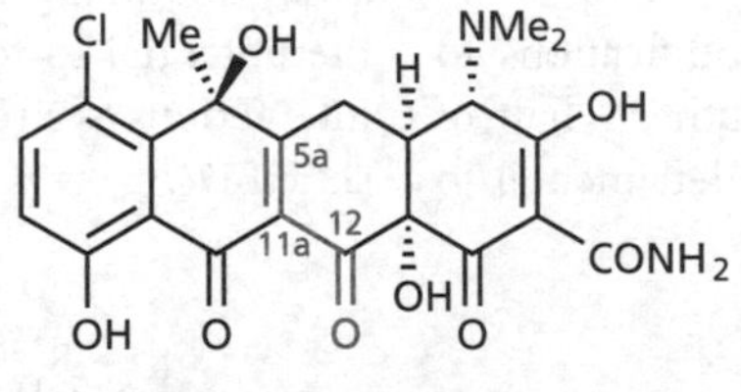

7-Chloro-5a(11a)-dehydrotetracycline

FIGURE 12.11 Fermentation product lacking activity.

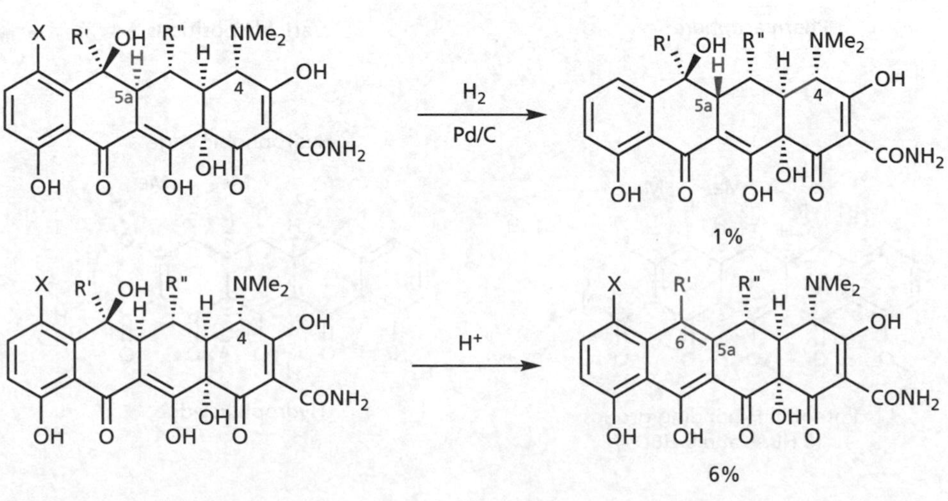

FIGURE 12.12 Analogues with poor activity. The percentage refers to the percentage of activity relative to tetracycline.

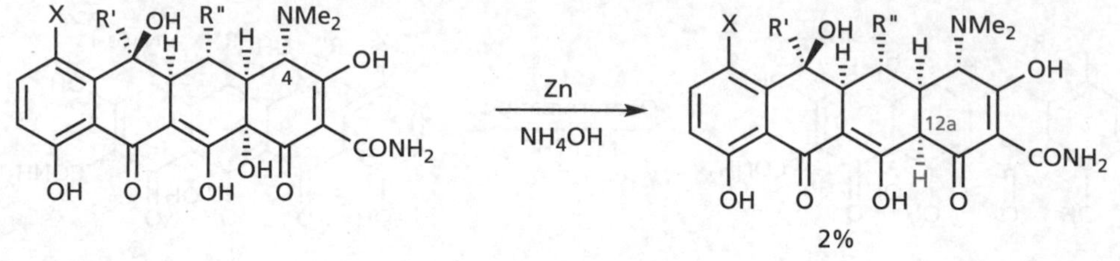

FIGURE 12.13 Removal of OH from C12a. The percentage refers to the percentage of activity relative to tetracycline.

Equally bad for activity is removal of the hydroxyl group at C12a (Fig. 12.13).

Epimerization at specific chiral centres is bad for activity. For example, epimerization at C4 occurs in the pH range 2–6 and results in a drop in activity, particularly against Gram-negative bacteria. Epimerization is one of the metabolic alterations carried out on tetracyclines *in vivo*, and it can also occur when tetracyclines are subjected to different pH values or harsh chemical conditions.

12.3 **Pharmacophore and mechanism of action**

In general, it was observed that any modifications to functional groups situated along the 'southern' half of the tetracyclic ring system were usually detrimental to activity, suggesting that these groups were important to activity. In contrast, there was more scope for modification in the top half of the compound, particularly in rings C and D.

Thus, the pharmacophore appeared to be as shown in Figure 12.14. The array of ketones and alcohol groups required for good activity suggests that these are involved in an extensive network of hydrogen bonding interactions with a target binding site, but it took several years to establish what that target was.

The SAR results clearly demonstrate that modifications to ring D are less likely to disrupt important binding interactions than modifications to rings A–C, implying that the majority of functional groups present in rings A–C are involved in binding interactions with a target binding site. In particular, tetracyclines have the potential to form a network of hydrogen bonding interactions as a result of having at least eight HBAs and six HBDs.

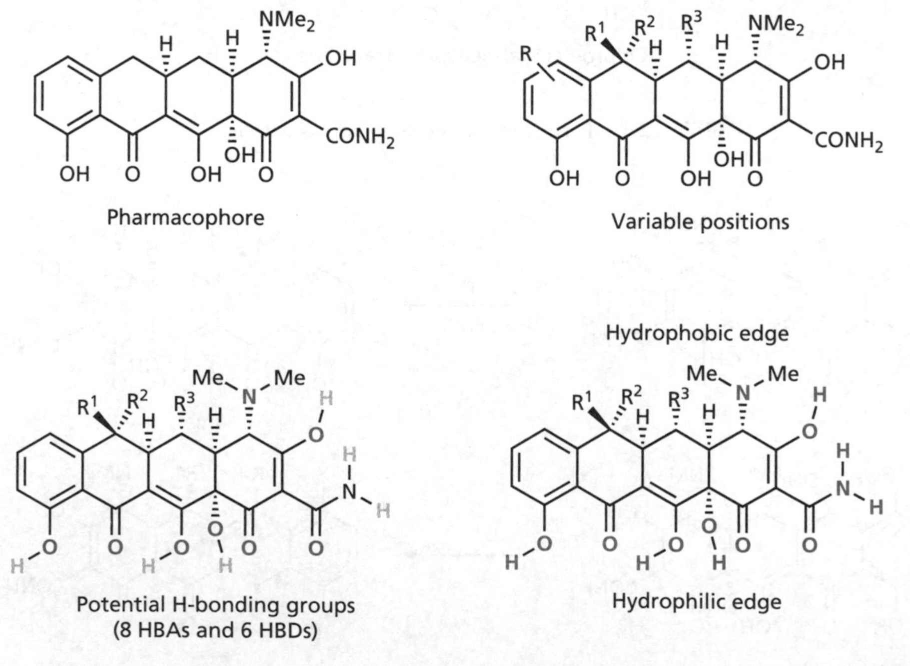

FIGURE 12.14 Pharmacophore and positions where substituents have been varied.

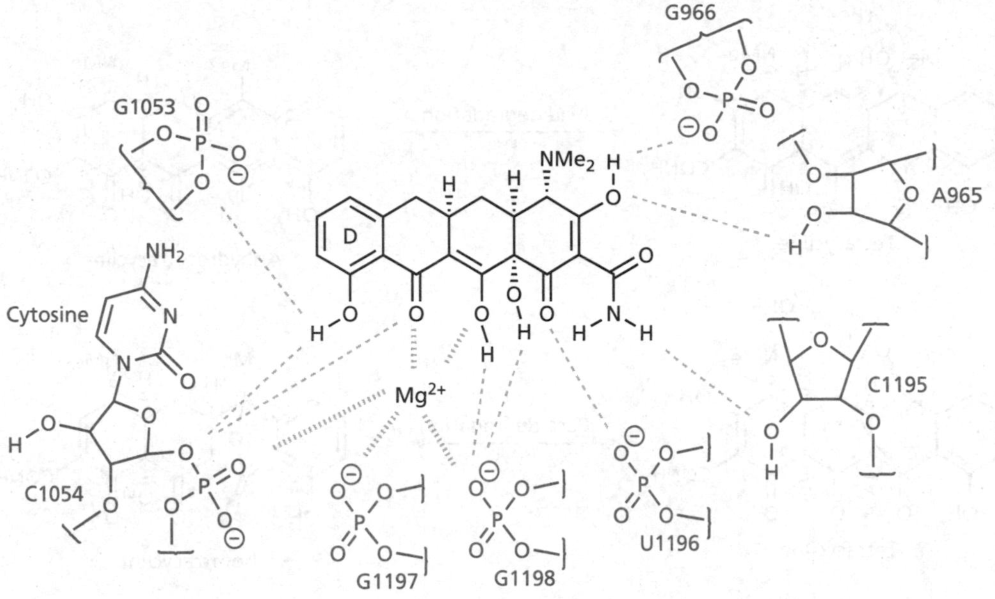

FIGURE 12.15 Binding interactions of tetracycline with RNA in bacterial ribosomes.

A better understanding of the targets for tetracyclines and their mechanism of action has now confirmed the importance of these functional groups. The targets for tetracyclines are bacterial ribosomes—structures which act as the cell's factory for producing proteins. In 2001, a crystal structure of tetracycline bound to a bacterial ribosome was obtained which revealed that the drug binds to the 30S subunit of the ribosome. The binding interactions observed also explain the SAR results that were obtained 40–50 years earlier (Fig. 12.15).

The tetracycline structure has an array of polar functional groups along one edge of its roughly planar tetracyclic structure which are clearly involved in important interactions. The other edge is hydrophobic in nature, and so any binding interactions associated with that part of the molecule involve van der Waals interactions. It has been found that tetracycline can bind to two different binding sites on ribosomes. The more significant binding site involves a series of hydrogen bonds to the sugar phosphate backbone of ribosomal RNA, as well as interactions involving a bridging magnesium ion. A cytosine ring is also thought to be involved in a pi–pi stacking interaction with ring D of the tetracycline. The binding interactions confirm the importance of the southern edge of the structure to binding and activity, and also help to explain why modifications to the top left region of the molecule can be carried out without losing activity. Therefore, modifications to ring D offer the best potential for generating novel tetracycline analogues, as it is not involved in crucial binding interactions.

12.4 Synthesis of semi-synthetic tetracyclines

Sancycline (Fig. 12.8) has a simpler structure than the natural tetracyclines and shows good activity. Therefore, it serves as a good lead compound for the development of novel tetracycline analogues which might be active against a broader range of infections. Such analogues are important because several infections have acquired resistance to the natural tetracyclines. Sancycline also serves as a good starting material for the synthesis of novel analogues as it has greater stability than the natural tetracyclines. Many of the reactions which have been carried out on the structure are not possible with the natural tetracyclines. For example, the natural tetracyclines are prone to degradation under acidic or basic conditions because of the presence of the 6-hydroxy group, resulting in the formation of inactive anhydrotetracyclines or isotetracyclines (Fig. 12.16).

The greater stability of sancycline means that it is possible to carry out electrophilic substitutions on the aromatic ring D under strongly acidic conditions to introduce a nitro group at either the 7- or the 9-position. Once the nitro group is introduced, it can be reduced to an amine, which can then be converted in turn to a diazonium salt, allowing halogens to be introduced.

However, the most useful modification has been the reductive alkylation of the nitro group at position 7 with formaldehyde to introduce a dimethylamino

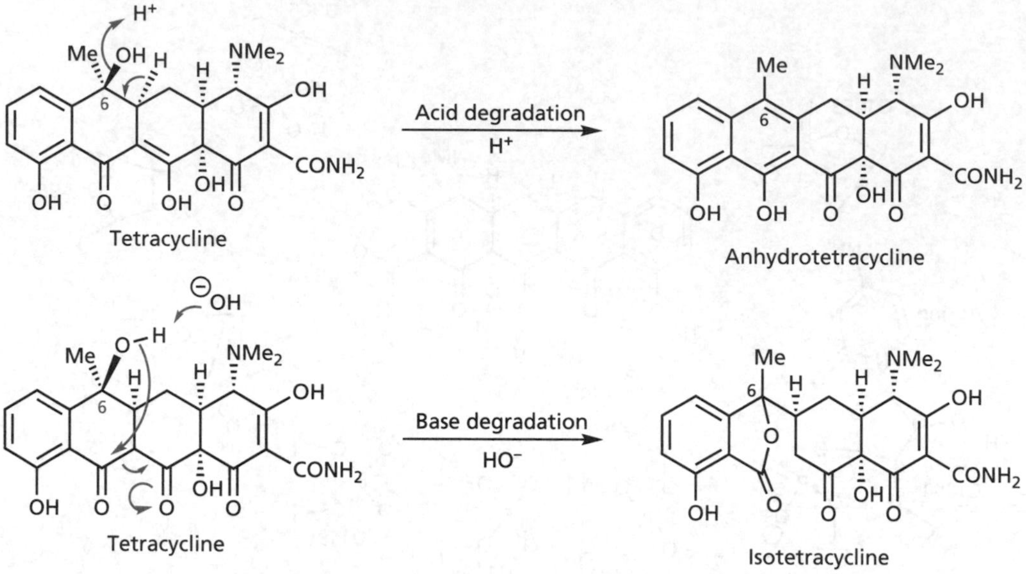

FIGURE 12.16 Common degradation pathways for tetracycline.

group (Fig. 12.17). This resulted in a structure called **minocycline** which proved to be a more potent compound than any of the previously discovered tetracyclines. Moreover, it has a broader range of activity than the parent tetracyclic structure, including activity against tetracycline-resistant staphylococci strains. Minocycline was added to the arsenal of clinically useful tetracyclines in the 1970s. In general, the presence of the 7-dimethylamino group appears to improve activity against bacterial strains that have gained tetracycline resistance as a result of efflux pumps. These are proteins

within cell membranes that 'capture' tetracyclines that have entered the cell and transport them back out again.

Unfortunately, the most abundant isomer from the nitration reaction in Figure 12.17 is the 9-nitro isomer rather than the 7-nitro isomer, by a ratio of 2:1. This means that the overall yield of minocycline is relatively low. Fortunately, it is not all bad news as it is possible to convert the 9-nitro structure to minocycline in 34% overall yield (Fig. 12.18). This means that the combined overall yield of minocycline from 6-demethyldeoxytetracycline via both nitro isomers is 33%.

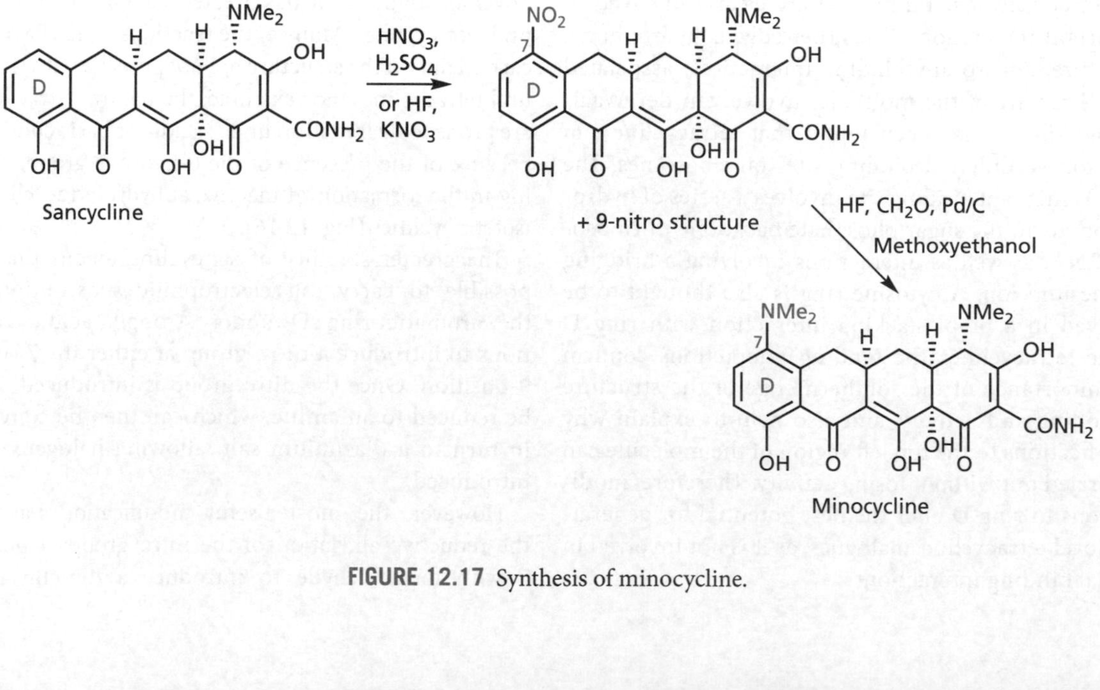

FIGURE 12.17 Synthesis of minocycline.

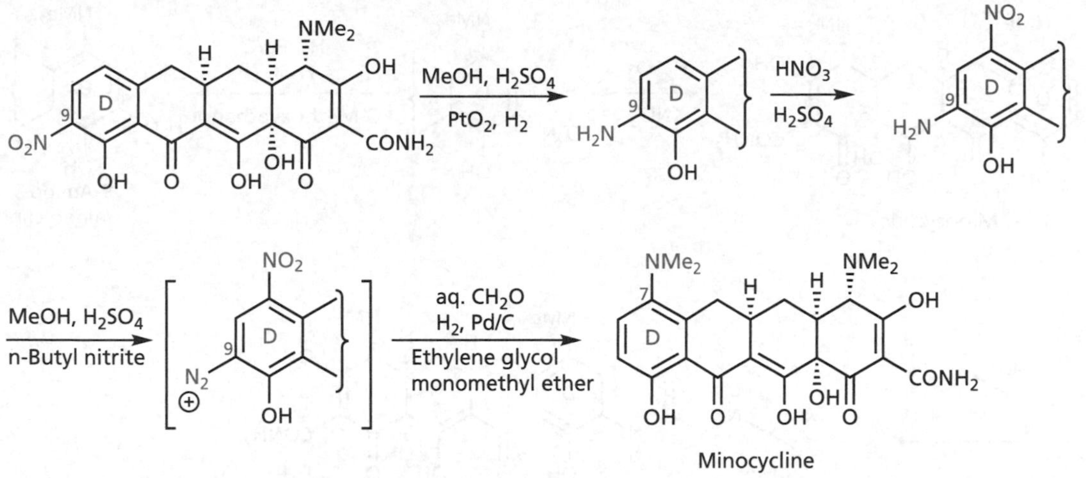

FIGURE 12.18 Synthesis of minocycline from the 9-nitro isomer.

These studies demonstrated that it is possible to prepare analogues which are substituted at both the 7- and the 9-positions of the aromatic D-ring—an important attribute which was crucial to the design and synthesis of the next important semi-synthetic tetracycline. This resulted from studies carried out in the 1990s where a variety of side chains were introduced at the 9-position of minocycline.

Minocycline was nitrated to introduce a nitro group at C-9, and then reduced to give 9-aminominocycline (Fig. 12.19). The 9-amino group was acylated with an α-bromo acid bromide to give an α-bromo amide. The bromo substituent could then be substituted with a wide range of primary and secondary amines to give a series

of compounds called the **glycylcyclines** which had useful activity.

The most important of these proved to be **tigecycline** (Tigacyl) which was approved in the USA in 2005 (Fig. 12.20). This is a broad-spectrum agent which is useful in treating many infections that have developed resistance to the older tetracyclines. A more efficient synthesis of tigecycline from minocycline involves acylation of the amine with an acid chloride that contains the full side chain required (Fig. 12.20).

Tigecycline is active against tetracycline-resistant bacterial strains that have become resistant because of mutations to the bacterial ribosome. These result in weaker binding interactions with older tetracyclines. Molecular

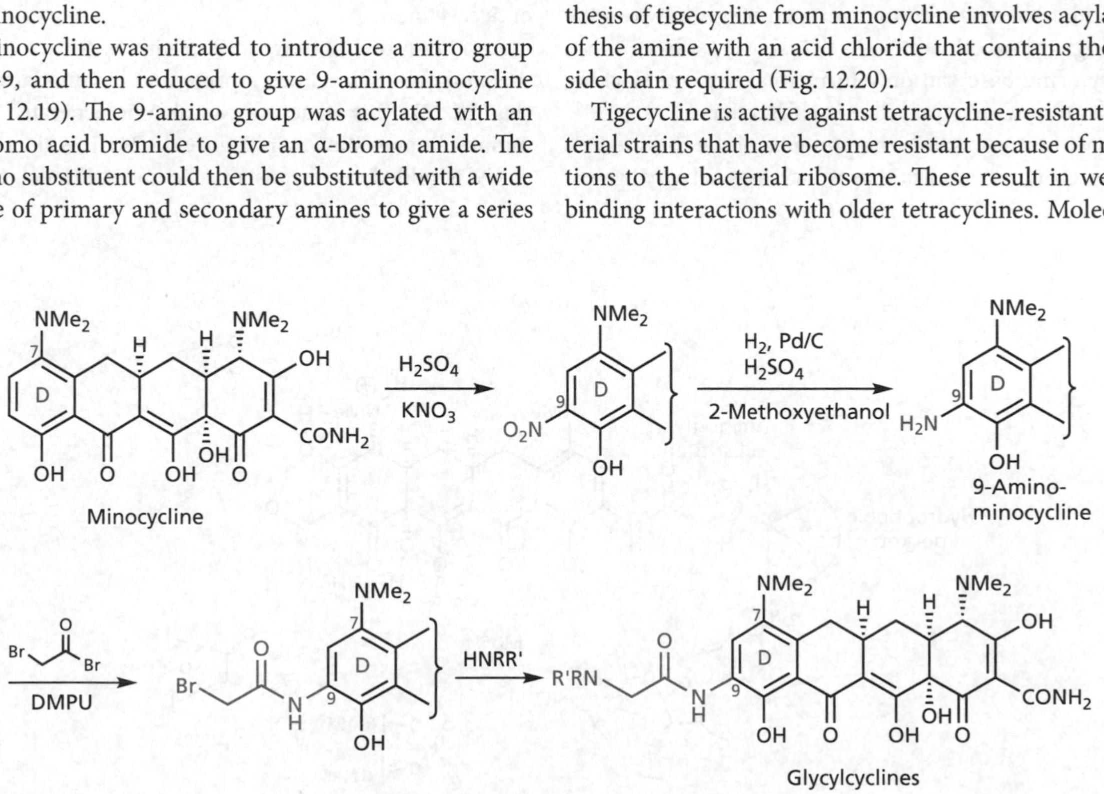

FIGURE 12.19 Synthesis of glycylcyclines from minocycline.

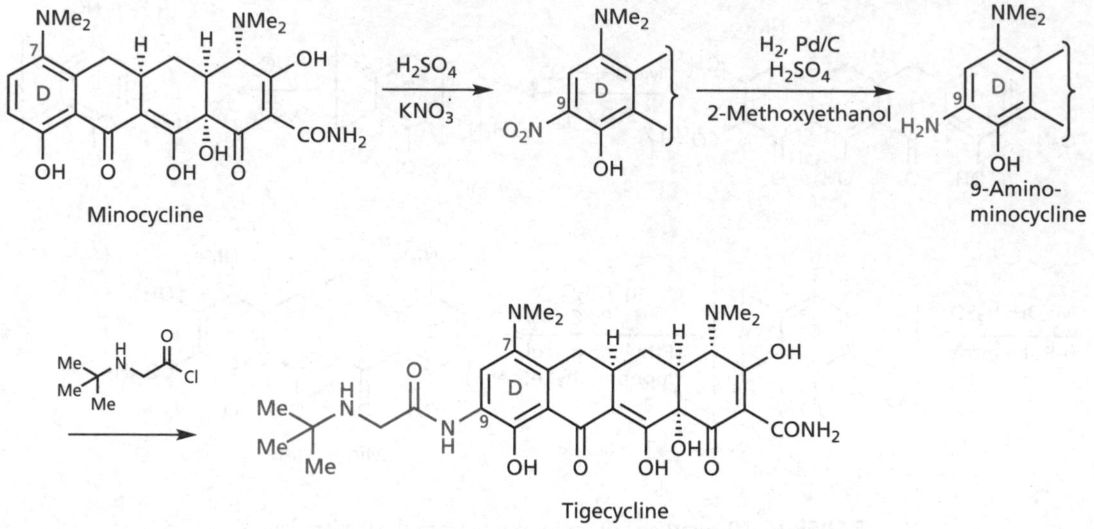

FIGURE 12.20 Synthesis of tigecycline (GAR-936) from minocycline.

modelling studies have suggested that tigecycline is effective against these strains because it has a different binding mode with the ribosome, which involves the side chain forming important binding interactions that are not possible with tetracycline itself (Fig. 12.21). In particular, there are strong hydrogen bonds between the aminoglycyl substituent and a cytosine base. In contrast, the cytosine base can only form weaker pi–pi stacking interactions with tetracycline itself (section 12.3).

As with minocycline, the presence of the 7-dimethyl-amino group is important as it increases activity against bacterial strains that have gained resistance to tetracycline as a result of efflux pumps. The polarity and electronegativity of the C7 substituent plays an important factor in this. Therefore, relevant substituents at positions 7 and 9 are both important to the effectiveness of tigecycline.

Another group of tetracyclines with useful activity are the aminomethylcyclines represented by **amadacycline** (Fig. 12.22). This structure can also be generated from minocycline as shown in Figure 12.23. The side chain at C9 is added by means of a form of Friedel–Crafts

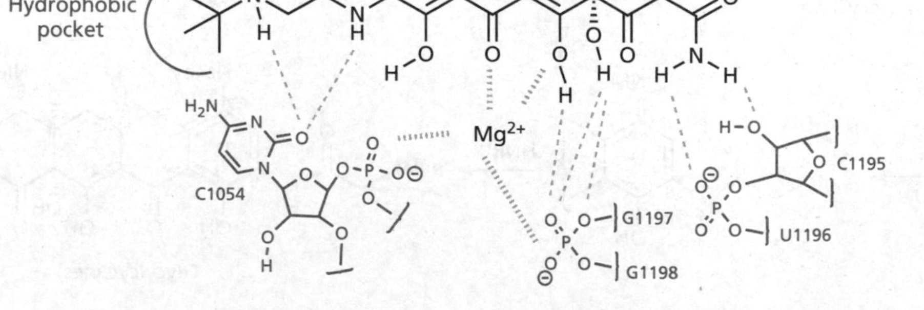

FIGURE 12.21 Binding interactions between tigecycline and the RNA binding site.

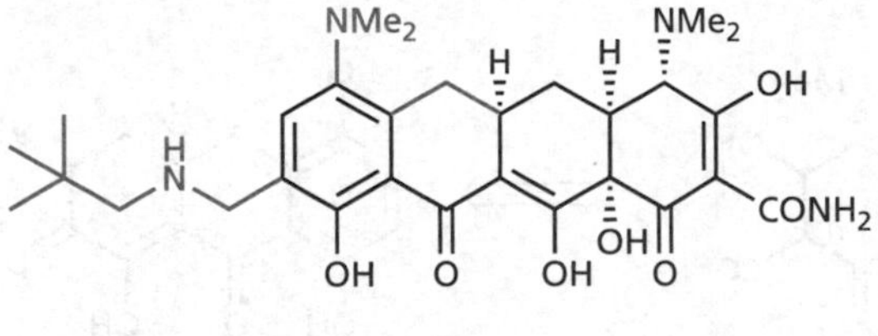

FIGURE 12.22 Amadacycline or omadacycline.

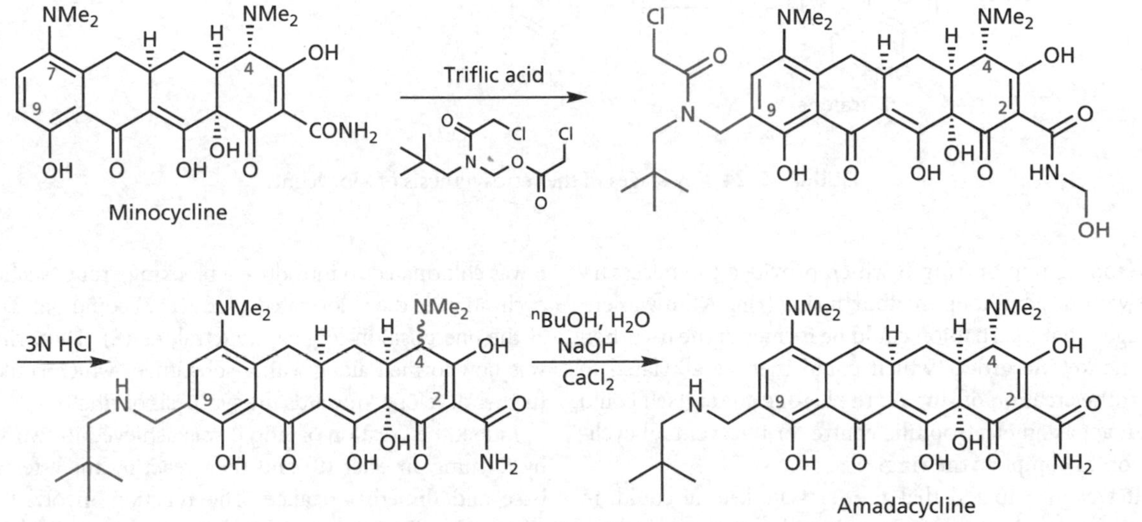

FIGURE 12.23 Synthesis of amadacycline.

reaction called the **Tscherniac–Einhorn reaction**. The primary amide group at C2 also reacts under these conditions and is hydroxymethylated. Heating under acid conditions reverts the latter group back to a primary amide, and deprotects the side chain on C9 to generate amadacycline, but epimerization occurs at C4 to produce an epimer as well. However, further heating in the presence of butanol, calcium chloride, and sodium hydroxide leads to another epimerization process which strongly favours amadacycline.

Further research into analogues containing suitable substituents at C7 and C9 is still relevant. Until recently, semi-synthetic methods have been used to obtain these agents by carrying out electrophilic substitution reactions on the aromatic D ring (halogenation and nitration). However, this limits the range of substituents that can be inserted at those positions. Ideally, a practical full synthesis would open the door to more varied analogues. New tetracyclines are always going to be important because of the ever present problem of acquired resistance to currently used tetracyclines.

12.5 Full synthesis of tetracyclines

A full synthesis of tetracyclines is a challenging concept because of the number of rings involved, the presence of four to six chiral centres, and the large number of functional groups and substituents that are present around the whole structure. Even the simplest of the early clinically important tetracyclines (sancycline) contains four chiral centres and eight functional groups. Sancycline (6-demethyl-6-deoxytetracycline) was, in fact, the first of the important tetracyclines to be fully synthesized. This was achieved by Woodward's team and involved 25 steps.

The overall strategy was to construct the tetracyclic ring starting from ring D and working towards ring A (from left to right). There was a good reason for this approach as ring A was the most complex part of the structure, involving five functional groups/substituents and three chiral centres.

Therefore, the strategy was to synthesize a tricyclic structure, on which the final ring could be built (Fig. 12.24). The intended tricyclic ring system contained

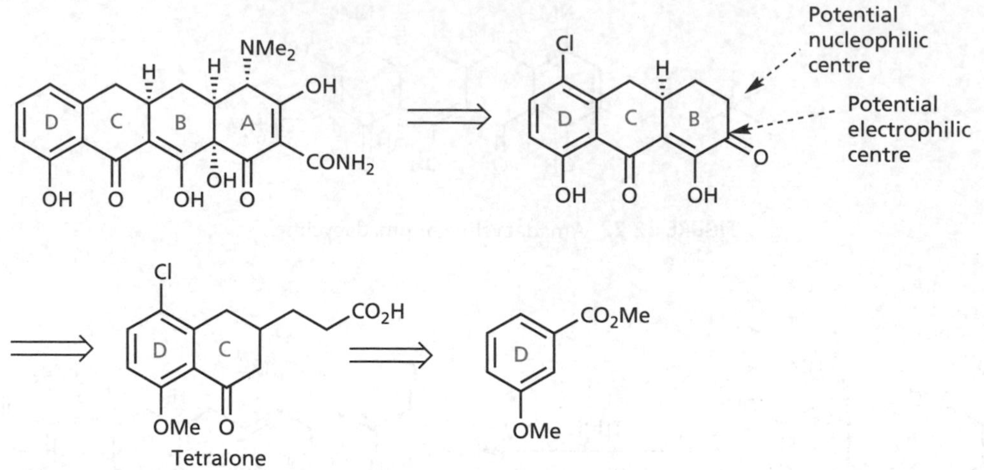

FIGURE 12.24 Key stages in the retrosynthesis of sancycline.

a ketone group on ring B which provided the necessary activation to build up the fourth ring (ring A). It was envisaged that a carbanion could be formed at the α-carbon to the ketone group, which could then be alkylated to start the creation of ring A. The ketone group itself could then act as an electrophilic centre for the eventual cyclization to complete the ring.

It was anticipated that the tricyclic ketone could, in turn, be synthesized from a bicyclic tetralone, which would be synthesized from a simple aromatic compound.

Therefore, the first phase of the synthesis was to create the tricyclic ring system (Fig. 12.25)—a process that involved a heavy emphasis on classical carbanion chemistry, such as the Claisen condensation and Michael addition, along with an important Friedel–Crafts reaction.

The initial starting material was an aromatic ester where the aromatic ring would end up as ring D in the final tetracyclic ring system. The next sequence of 7 reactions involved the building up of ring C to form a bicyclic tetralone.

The first reaction involved a Claisen condensation where the carbanion of methyl acetate underwent a nucleophilic substitution of the methyl ester to give a keto ester (1). This was then alkylated in situ to form the keto diester (2). The keto diester was treated with an α,β-unsaturated ester in the presence of Triton B as a catalyst, resulting in a Michael addition to form the keto triester (3).

Hydrolysis of the triester was accompanied by decarboxylation to give a keto diacid (4) which was esterified to a keto diester (5). The ketone group was now removed by hydrogenation to give a diacid (6), which was now set up for a Friedel–Crafts acylation to create ring C. However, there was a problem at this point since two products were possible. Cyclization could occur at either of the two positions that were *ortho* to the side chain. Therefore, structure

6 was chlorinated to introduce a blocking group, such that cyclization of the chlorinated product (7) could only occur at the one position to give the tetralone (8). Thus, ring C was now formed along with a substituent which could be further developed towards the synthesis of ring B.

In fact the creation of ring B was achieved in two steps by forming an ester (9) and then reacting the ester with base and dimethyl oxalate. The reaction involved two Claisen condensations, which 'spliced' in the two carbons required to complete ring B. Isomerization then created a dienol system (10). Acid conditions were then used to hydrolyse the ester and promote decarboxylation to give the crucial tricyclic ketone (11) required for the second crucial phase where the 'difficult' ring A would be created.

Having created the tricyclic ring system (11), the next part was the tricky process of building up ring A (Fig. 12.26). Carbanion chemistry was again used to introduce a suitable substituent at the α-position by means of an aldol condensation to give a single geometrical isomer (12).

Treatment with dimethylamine resulted in conjugate additions to the α,β-unsaturated ketone to give the amine (13). The relatively bulky amine added stereoselectivity such that the protons at the chiral centres shown ended up being *cis* to each other, as required.

One of the ketone groups was now reduced selectively with sodium borohydride to give an alcohol (14), which reacted with the ester group under acid conditions to give a lactone ring (15).

Reduction of the lactone ring then generated an amino acid (16) which was hydrogenated to remove the chlorine blocking group (17). It was now a case of completing ring A by adding the remaining portion of the ring using an organomagnesium reagent to give structure 19, and then inducing cyclization with sodium hydride to give structure 20. Hydrobromic acid was used as a dealkylating

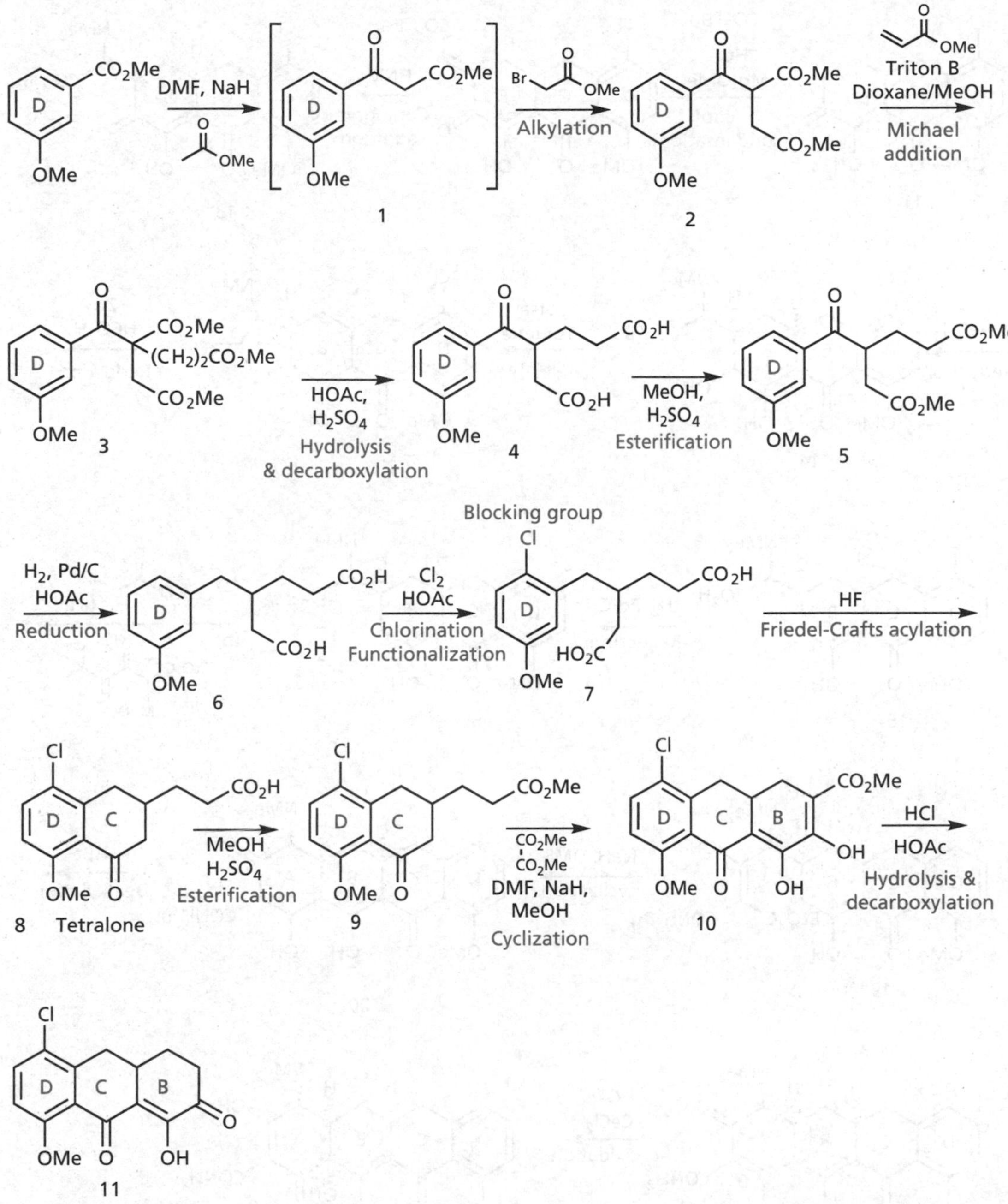

FIGURE 12.25 Synthesis of the tricyclic intermediate required in the synthesis of sancycline.

agent to reveal the required phenol on ring D and the primary amide on ring A (21). The final step was an oxidation using oxygen gas to introduce the alcohol group at the junction between rings A and B.

A number of other synthetic routes were subsequently developed to synthesize tetracyclines such as **5-oxytetracycline** (Terramycin) and tetracycline itself, one of which involved a bicyclic starting material representing rings C and D, but none of them was particularly practical

for synthesizing naturally occurring tetracyclines, or as a method of creating potentially novel tetracyclines. For a start, the number of steps involved inevitably meant a very low yield overall. For example, Woodward's synthesis of sancycline (Fig. 12.24) involved 25 steps and an overall yield of about 0.002%. A synthesis of 5-oxytetracycline (Terramycin) took 22 steps with an overall yield of 0.06%, and a synthesis of tetracycline itself took 34 steps with an overall yield of 0.002%. Therefore, it was more practical to

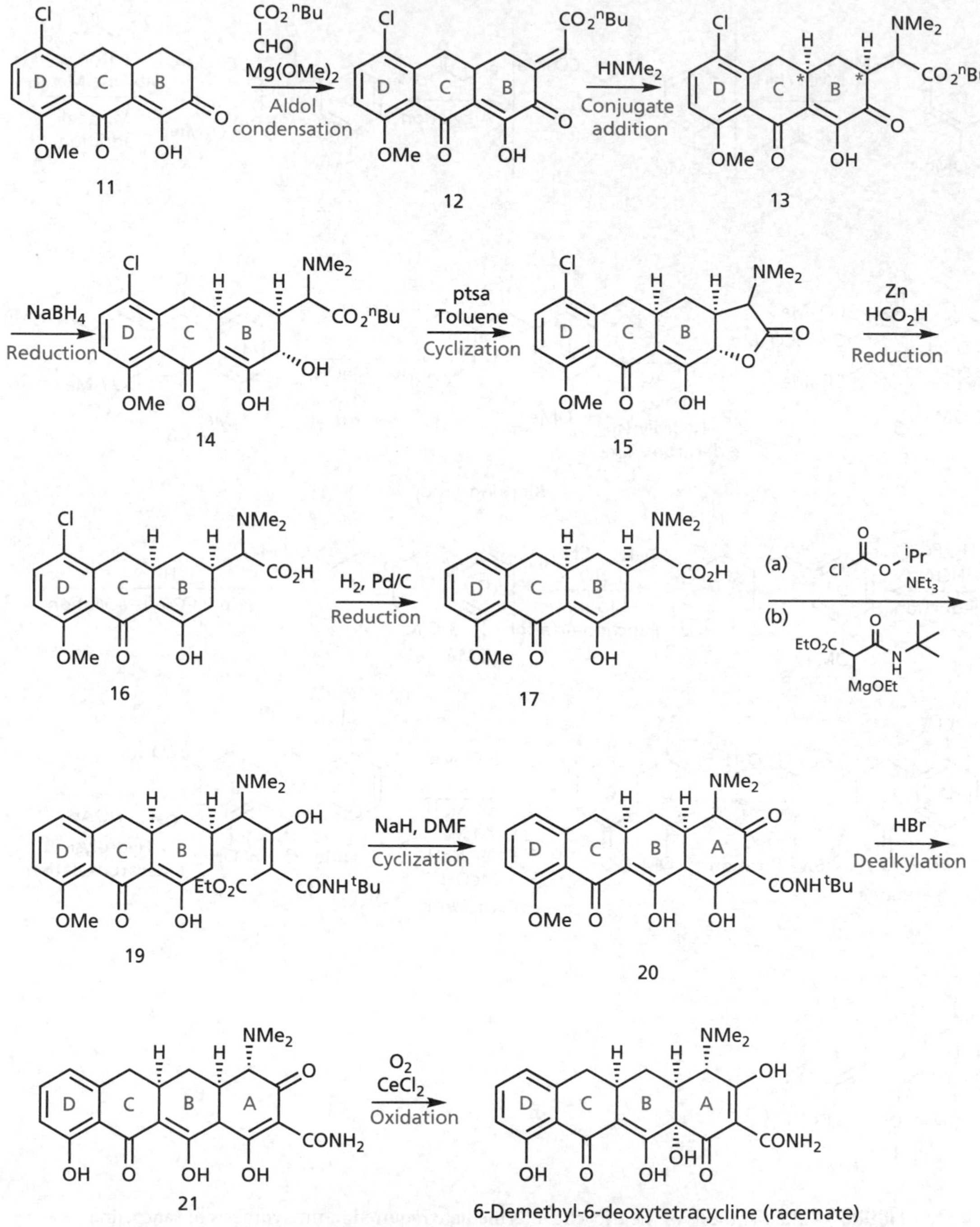

FIGURE 12.26 Final stages in the synthesis of sancycline.

obtain tetracyclines from fermentation or by semi-synthetic routes starting from naturally occurring tetracyclines.

Another major problem with these synthetic routes was the fact that they all built the molecules from the left to the right and finished off with the construction of ring A. However, SAR results have shown that ring D is the best ring to make variations in order to obtain novel active compounds (section 12.3). Variation in the other rings is more likely to result in a loss of activity. With the synthetic routes devised so far, every new analogue would have to be synthesized from scratch using the whole of the synthetic route. A more practical approach would be to build the structure from right to left such that the introduction of ring D comes at a late stage in the synthesis. That would

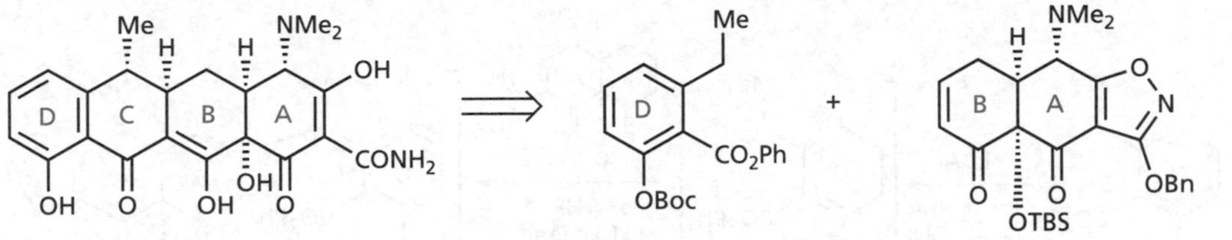

FIGURE 12.27 Retrosynthetic analysis for a convergent tetracycline synthesis. (TBS = *tert*-butyldimethylsilyl.)

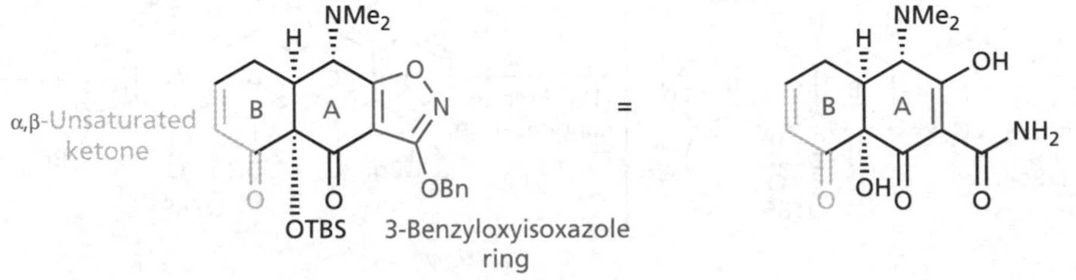

FIGURE 12.28 The tricyclic structure representing rings A and B. (TBS = *tert*-butyldimethylsilyl.)

make the synthesis far more practical for synthesizing novel structures with variability in ring D, since most of the synthesis would be common to all the analogues.

Recently, a synthesis has been devised that achieves that goal, and involves an aromatic precursor representing ring D being linked with a tricyclic structure that acts as a precursor for rings A and B (Fig. 12.27). The linkage of the two cyclic systems inevitably creates ring C in a convergent synthesis—another advantage over the previous linear synthesis carried out by Woodward.

The tricyclic compound used in this approach might seem unusual as it contains a 3-benzyloxyisoxazole ring which appears superfluous. However, this ring serves to 'wrap up' and protect the enol and primary amide groups that are normally present in ring A (Fig. 12.28). Therefore, it can be viewed as a dual-acting protecting group. The tricyclic ring also contains an α,β-unsaturated ketone that can undergo a Michael addition as part of the process needed to form ring C.

Only three chemical steps are actually required to synthesize a tetracycline from the tricyclic compound and the aromatic precursor for ring D (Fig. 12.29).

The first of these steps is cyclization to form the tetracyclic framework. The aromatic precursor is treated with a strong base (LDA) to form a benzylic carbanion which then adds to the α,β-unsaturated ketone of the tricyclic structure in a Michael addition. This produces two new chiral centres which means that four diastereoisomers are possible. Fortunately, the reaction is highly diastereoselective and produces the required diastereoisomer. This can be rationalized by proposing that the carbanion reacts at the less hindered face of the enone.

It is believed that the bulky TBS protecting group on the tricyclic ring acts as a steric shield to one face of the enone system. The addition product is then deprotonated by LDA to form another carbanion, which undergoes a Dieckmann condensation with the phenyl ester to effect cyclization in 81% yield to form the desired diastereoisomer (Fig. 12.29). The slower rate-determining step is the Dieckmann condensation.

The remaining two steps are deprotection steps. HF removes the *tert*-butoxycarbonyl protecting group from the phenol group on ring D as well as the TBS protecting group from the alcohol group at C12a. Reduction with hydrogen 'unzips' the benzylisoxazole ring to reveal the enol and the primary amide.

This route allows the synthesis of novel tetracycline analogues which would have been impossible using previous routes, allowing a greater variety of substituents to be added to the D ring, as well as the potential to modify the nature of ring D itself.

For example, over fifty 6-deoxytetracycline structures have been synthesized using this route, including known structures such as doxycycline, 6-deoxytetracycline, and minocycline.

Examples of novel structures synthesized in this manner include a range of novel tetracyclines with an aryl or heteroaryl substituent at position 6, which have been found to have useful activity. Similarly, it has been possible to synthesize a tetracycline with an 8-fluoro substituent, but even more remarkable modifications have been possible. For example, ring D has been modified to introduce pyridine and pyrazole rings, and a new class of **pentacyclines** have been synthesized (Fig. 12.30). Of

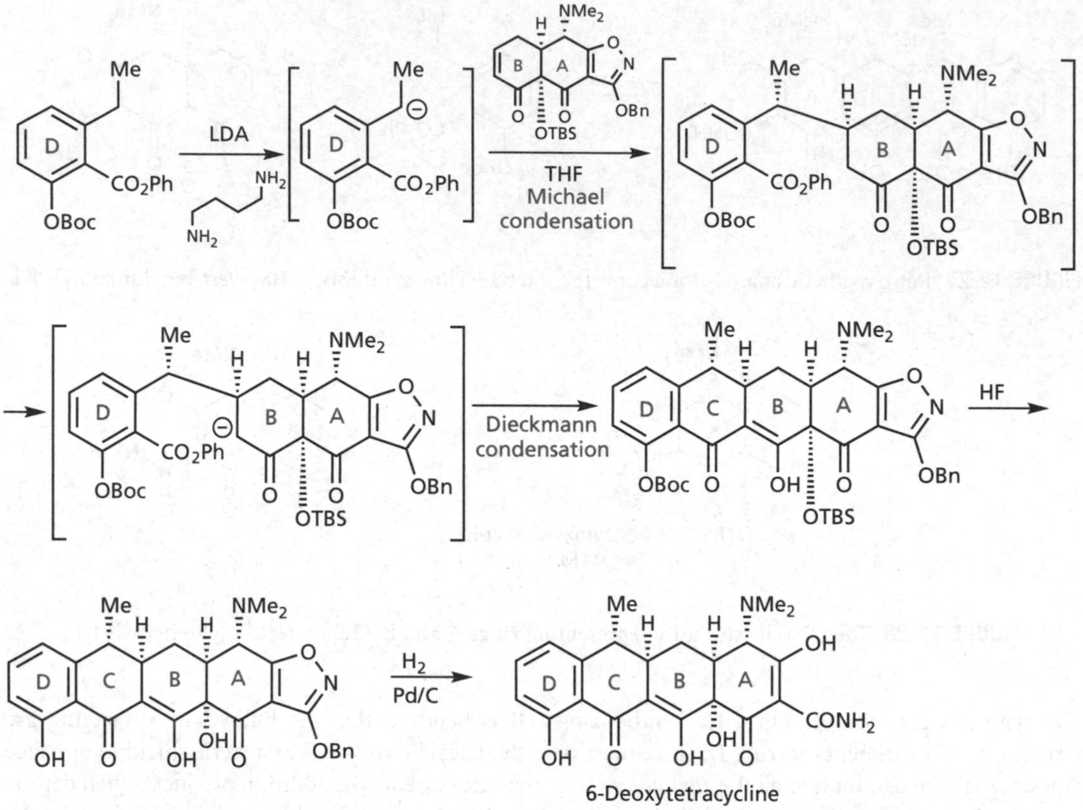

FIGURE 12.29 Synthesis of 6-deoxytetracycline (TBS = the *tert*-butyldimethylsilyl protecting group).

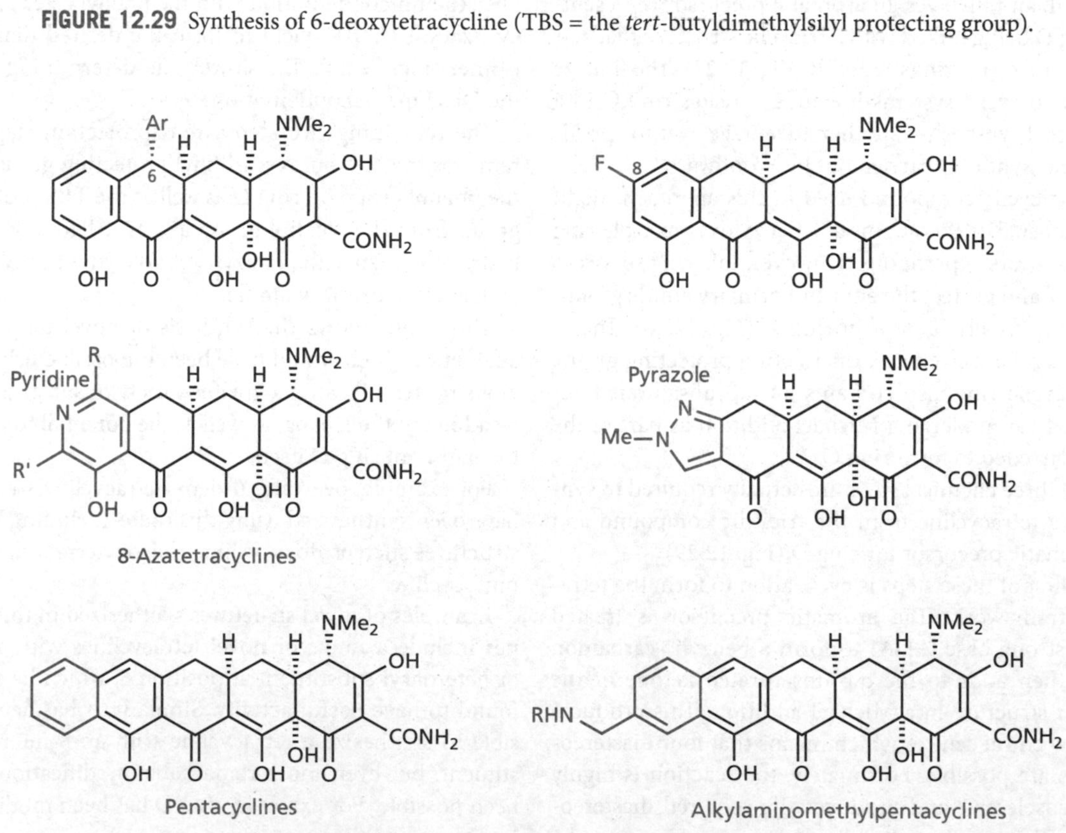

8-Azatetracyclines

Pentacyclines

Alkylaminomethylpentacyclines

FIGURE 12.30 Examples of novel tetracycline analogues.

these, the alkylaminomethylpentacyclines are of special interest as they show good antibacterial activity.

However, the most exciting structures to have come out of these studies so far has been a group of structures that have been called the **fluorocyclines**, which contain a fluorine substituent at position 7 and an amide group at position 9. The most important of these is **eravacycline** (Fig. 12.31). The results of phase II clinical trials have proved highly successful and the agent shows great promise as a novel broad-spectrum antibacterial agent.

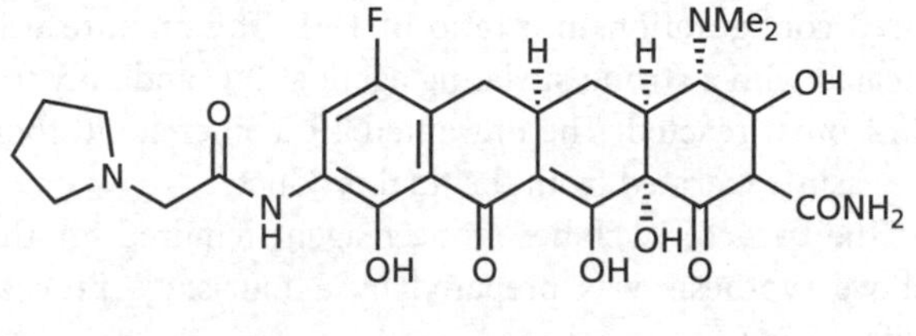

FIGURE 12.31 Eravacycline (TP434).

The full synthesis of eravacycline is shown in Figure 12.32 and involves a trisubstituted aromatic ring as starting material. This is methylated by deprotonating the aromatic ring regioselectively and then treating with iodomethane. The carboxylic acid group is then esterified to give a phenyl ester. The methyl ether is demethylated with BBr_3 to give a phenol which is then protected with the Boc protecting group. The key Michael–Dieckmann annulation is then carried out with the tricyclic precursor that represents the A and B rings. Two deprotection steps are then carried out to reveal the alcohols and the primary amide group. Nitration of ring D introduces a nitro group at C7, which is then reduced to an amine. Treatment of the amine with an acid chloride gives the final structure.

Eravacycline contains a pyrrolidine ring on its side chain at C7. A range of analogues have also been synthesized, demonstrating that a small secondary or tertiary amine is best. Weakly basic amines resulted in decreased potency, as did aromatic amines. Cyclic alkylamines varied in activity depending on the size of the ring.

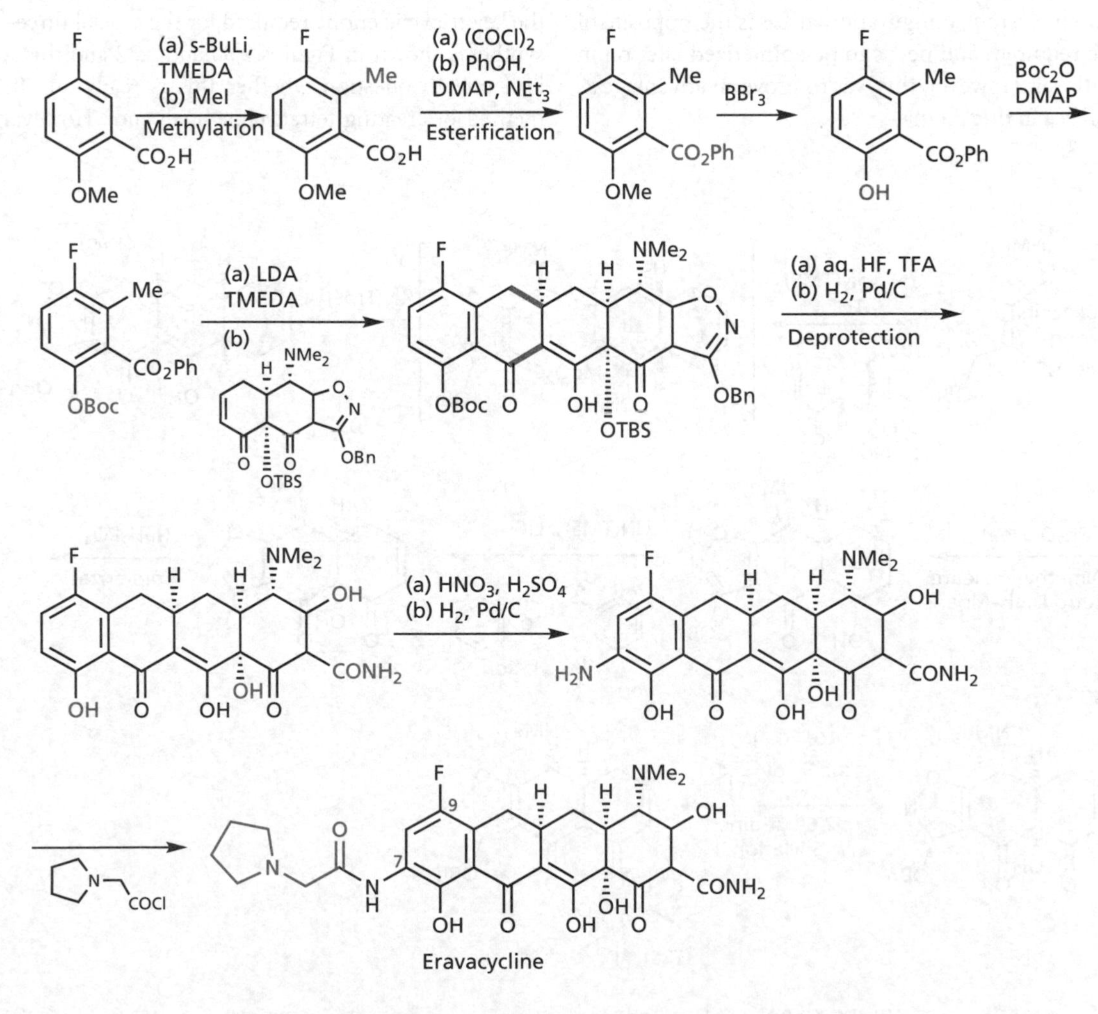

FIGURE 12.32 Synthesis of eravacycline.

The synthesis of eravacycline illustrates that a number of synthetic steps are required to prepare the aromatic reagent required for the cyclization step necessary to prepare the tetracycline ring system. This is also the case for the tricyclic enone reagent. Indeed, its preparation is by no means straightforward and a number of different methods have been used to synthesize it.

The most recent approach involves a convergent synthesis where a tricyclic cyclohexenone and an isoxazole ester are linked together by another diastereoselective Michael–Dieckmann condensation (Fig. 12.33). The isoxazole starting material is treated with base to form a carbanion at –78°C, which then reacts by a Michael addition with the enone of the cyclohexenone structure. This results in an enolate ion which undergoes the Dieckmann cyclization when the solution is allowed to warm up to –20°C. The reaction results in two new chiral centres and is highly diastereoselective. As before, the reaction proceeds with the carbanion adding to the least sterically hindered face of the enone system. In this case, the multicyclic feature of the enone acts as a steric shield and blocks addition to one face of the enone. Unfortunately, the configuration at C4 is the opposite of the one required, and needs to be epimerized later on in the synthesis. However, this was to prove an advantage as we shall see in due course.

A retro Diels–Alder reaction was now carried out to remove the 'steric shield' (dimethyl maleate was added to the reaction to trap the cyclopentadiene produced). The resulting product was then hydroxylated to introduce the OH group at what will be position 12a in the final tetracycline. It was at this stage that the wrong configuration of the 4-dimethylamino group proved advantageous since it acted as a steric shield and forced the hydroxylating agent to approach the α-face of the ring system (from below as drawn) rather than the β-face. Treatment with sodium dihydrogen phosphate in THF and MeOH then epimerized the chiral centre at C4 to give the desired configuration in a ratio of 11:1. The mixture was treated with a strong silylating agent at 0°C and only the 4S-isomer reacted. The unwanted 4R-isomer could then be easily separated from the final product.

The tricyclic cyclohexenone reagent required for the above synthesis was prepared by a four-stage process (Fig. 12.34).

Similarly, the isoxazole ester starting material required for the synthesis was prepared as shown in Figure 12.35.

Clearly, a lot of synthetic effort is required to prepare the key tricyclic enone required for the crucial three-stage syntheses shown in Figures 2.30 and 2.32 and this might lead one to question whether this is really an efficient method of creating tetracycline analogues. However, the

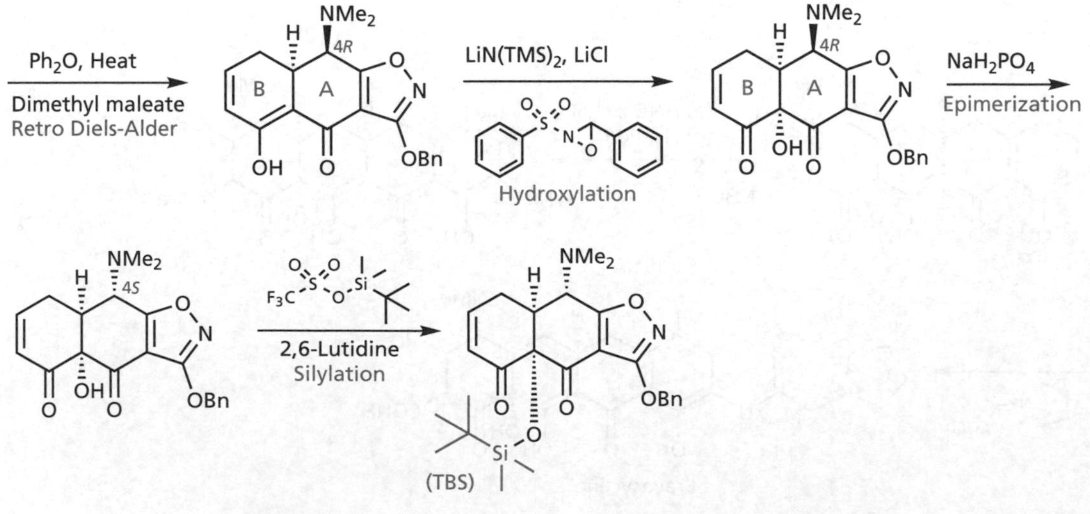

FIGURE 12.33 Synthesis of the tricyclic enone reagent (Bn = CH₂Ph).

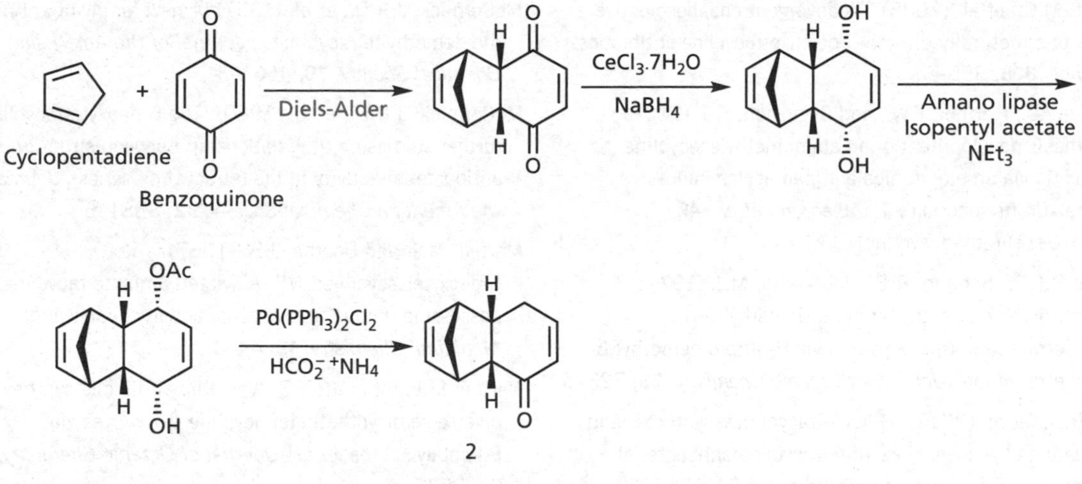

FIGURE 12.34 Synthesis of the tricyclic cyclohexenone reagent.

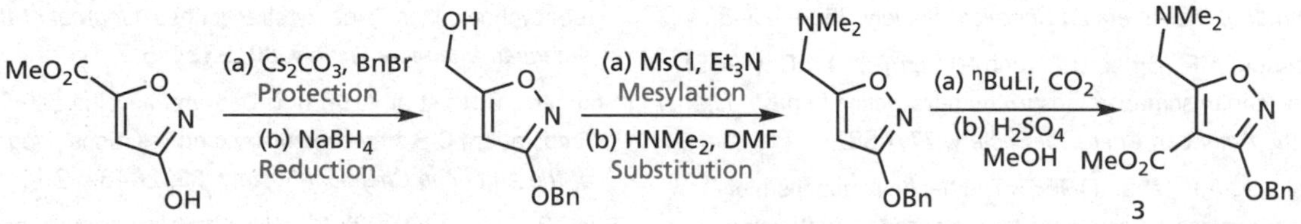

FIGURE 12.35 Synthesis of the isoxazole ester starting material.

same enone can be used to create a vast number of differ-ent analogues, and so if the synthesis of the enone can be carried out on a sufficiently large scale and in reasonable yield, it only has to be carried out once to provide sufficient material for the generation of several analogues. For ex-ample, it has been possible to scale up a previous synthesis of this compound to produce over 40 g of the precursor in one batch. This is still a huge advantage over a process like the original Woodward synthesis where every single novel structure would have to be generated by going through the whole process described in Figures 2.25 and 2.26.

12.6 Conclusions

Tetracyclines have been useful antibacterial agents for many years. However, their effectiveness has been di-minished by the appearance of resistant bacterial strains. Novel tetracyclines have been prepared with substituents at positions 7 and 9, and have proved effective against tetracycline-resistant strains. In the past, it was only practical to prepare such analogues by semi-synthetic ap-proaches, but novel synthetic approaches are now allow-ing the full synthesis of a new generation of tetracyclines.

FURTHER READING

Blackwood, R.K., et al. (1961) '6-Methylenetetracyclines. I: A new class of tetracycline antibiotics', *Journal of the American Chemical Society*, **83**, 2773–5.

Boothe, J.H., et al. (1953) 'Tetracycline', *Journal of the American Chemical Society*, **75**, 4621.

Broderson, D.E., et al. (2001) 'The structural basis for the action of the antibiotics tetracycline, pactamycin and hygromycin N on the 30S ribosomal subunit', *ChemBioChem*, **2**, 612–27.

Charest, M.C., et al. (2005) 'A convergent enantioselective route to structurally diverse 6-deoxytetracycline antibiotics', *Science*, **308**, 395–8.

Chung, J.Y.L., Hartner, F.W., and Cvetovich, R.J. (2008) 'Synthesis development of an aminomethylenecycline antibiotic via an electronically tuned acylaluminium Friedel–Crafts reaction', *Tetrahedron Letters*, **49**, 6095–6100 (omadacycline).

Church, R.F.R., Schaub, R.E., and Weiss, M.J. (1971) 'Synthesis of 7-dimethylamino-6-demethyl-6-deoxytetracycline (minocycline) via 9-nitro-6-demethyl-6-deoxytetracycline', *Journal of Organic Chemistry*, **36**, 723–5.

Clark, R.B., et al. (2011) '8-Azatetracyclines: synthesis and evaluation of a novel class of tetracycline antibacterial agents', *Journal of Medicinal Chemistry*, **54**, 1511–28.

Clark, R.B., et al. (2012) 'Fluorocyclines. 2: Optimization of the C-9 side-chain for antibacterial activity and oral efficacy', *Journal of Medicinal Chemistry*, **55**, 606–22.

Conover, L.H., et al. (1953) 'Terramycin. XI: Tetracycline', *Journal of the American Chemical Society*, **75**, 4622–3.

Doerschuk, A.P., Bitler, B.A., and McCormick, J.R.D. (1955) 'Reversible isomerisations in the tetracycline family', *Journal of the American Chemical Society*, **77**, 4687.

Doershuck, A.P., et al. (1956) 'The halide metabolism of *Streptomyces aureofaciens* mutants: the biosynthesis of 7-chloro-, 7-chloro[36]- and 7-bromotetracycline and tetracycline', *Journal of the American Chemical Society*, **78**, 1508–9.

Hlavaka, J.J., et al. (1962) 'The 6-deoxytetracyclines. III: Electrophilic and nucleophilic substitution', *Journal of the American Chemical Society*, **84**, 1426–30.

Korst, J.J., et al. (1968) 'The total synthesis of *dl*-6-demethyl-6-deoxytetracycline', *Journal of the American Chemical Society*, **90**, 439–57.

Kummer, D.A., et al. (2011) 'A practical, convergent route to the key precursor to the tetracycline antibiotics', *Chemical Science*, **2**, 1710–18.

McCormick, J.R.D., et al. (1957) 'Studies of the reversible epimerisation occurring in the tetracycline family: the preparation, properties and proof of structure of some 4-*epi*-tetracyclines', *Journal of the American Chemical Society*, **79**, 2849–58.

McCormick, J.R.D., et al. (1957) 'A new family of antibiotics: the demethyltetracyclines', *Journal of the American Chemical Society*, **79**, 4561–3.

McCormick, J.R.D., et al. (1960) 'The 6-deoxytetracyclines: further studies on the relationship between structure and antibacterial activity in the tetracycline series', *Journal of the American Chemical Society*, **82**, 3381–6.

Martell, M.J. and Boothe, J.H. (1967) 'The 6-deoxytetracyclines. VII: Alkylated aminotetracyclines possessing unique antibacterial activity', *Journal of Medicinal Chemistry*, **10**, 44–6.

Nelson, M.L., et al. (2003) 'Versatile and facile synthesis of diverse semisynthetic tetracycline derivatives via Pd-catalyzed reactions', *Journal of Organic Chemistry*, **68**, 5838–51.

Olson, M.W., et al. (2006) 'Functional, biophysical, and structural bases for antibacterial activity of tigecycline', *Antimicrobial Agents and Chemotherapy*, **50**, 2156–66.

Stephens, C.R., et al. (1958) Hydrogenolysis studies in the tetracycline series: 6-deoxytetracyclines', *Journal of the American Chemical Society*, **80**, 5324–5.

Stephens, C.R., et al. (1963) '6-Deoxytetracyclines. IV. Preparation, C-6 stereochemistry, and reactions', *Journal of the American Chemical Society*, **85**, 2643–52.

Sum, P.-E., et al. (1992) 'Glycylcyclines. 1: A new generation of potent antibacterial agents through modification of 9-aminotetracyclines', *Journal of Medicinal Chemistry*, **37**, 184–8.

Sum, P.E. and Petersen, P. (1999) 'Synthesis and structure–activity relationship of novel glycylcycline derivatives leading to the discovery of GAR-936', *Bioorganic and Medicinal Chemistry Letters*, **9**, 1459–62.

Sun, C., et al. (2008) 'A robust platform for the synthesis of new tetacycline antibiotics', *Journal of the American Chemical Society*, **130**, 17 913–27.

Sun, C., et al. (2011) 'Synthesis and antibacterial activity of pentacyclines: a novel class of tetracycline analogs', *Journal of Medicinal Chemistry*, **54**, 3704–41.

Xiao, X.-Y., et al. (2012) 'Fluorocyclines. 1: 7-Fluoro-9-pyrrolidinoacetamido-6-demethyl-6-deoxytetracycline—a potent, broad spectrum antibacterial agent', *Journal of Medicinal Chemistry*, **55**, 597–605.

Erythromycin and macrolide antibacterial agents

13.1 Introduction

Erythromycin and related macrolide structures are important antibiotics, and are the third most widely used class of antibiotics (Fig. 13.1). Erythromycin itself is a natural product that is produced by a bacterial strain called *Saccharopolyspora eythraea* (formerly *Streptomyces erythreus*) found in soil samples in the Philippines. The antibiotic disrupts protein synthesis in microbial cells by binding to the ribosomes that are responsible for their synthesis.

The structure of erythromycin contains a core 14-membered macrolactone ring which includes ten chiral centres, three alcohol groups, one ketone, six methyl substituents, one ethyl substituent, and two carbohydrate ethers. The carbohydrates are **cladinose** and **desosamine**, which add a further eight chiral centres, two alcohols, one amine, one methyl ether, and three methyl substituents to the overall structure. The structure without the carbohydrate rings (the aglycone portion of the structure) is called **erythronolide A. 6-Deoxyerythronolide** is the corresponding structure lacking the alcohol group at C6.

Erythromycin is unstable to stomach acids, but can be taken orally in tablet form. The formulation of the tablet involves a coating that is designed to protect the tablet during its passage through the stomach, but which is soluble once it reaches the intestines (enterosoluble). The acid sensitivity of erythromycin is due to the presence of the ketone at C9 and the two alcohol groups at C6 and C11 which are set up for an acid-catalysed intramolecular formation of a ketal (Fig. 13.2). One of the main aims in studying analogues of erythromycin is to find structures which are resistant to this acid-catalysed ketal formation.

13.2 Synthesis of erythromycin

Considering the presence of the macrocyclic ring and the number of chiral centres, substituents, and functional groups present in erythromycin, it comes as no surprise that a full synthesis of the structure posed a huge challenge. A full synthesis of erythromycin was finally achieved by Woodward in 1981 (Fig. 13.3), but it involves

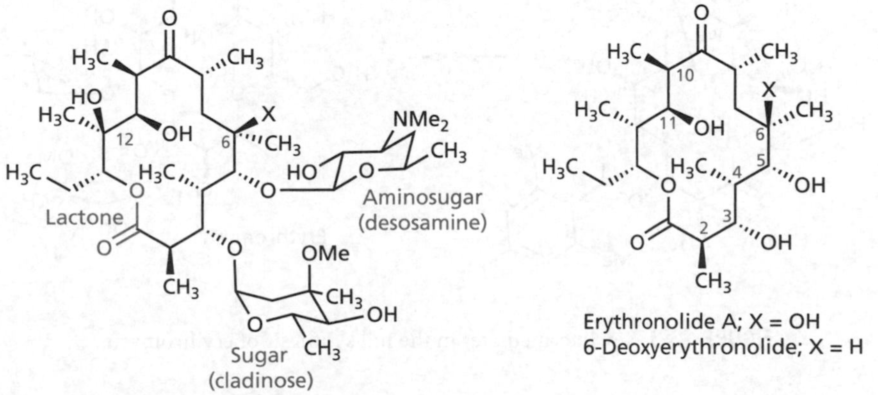

Erythromycin; X = OH
Clarithromycin; X = OMe

Erythronolide A; X = OH
6-Deoxyerythronolide; X = H

FIGURE 13.1 Erythromycin, clarithromycin, and 6-deoxyerythronolide.

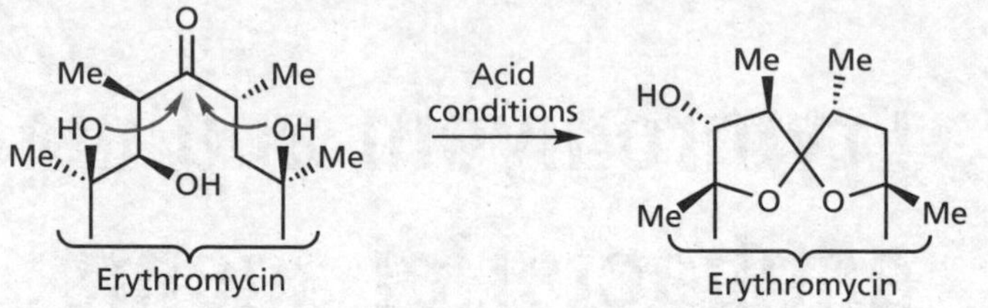

FIGURE 13.2 Intramolecular ketal formation in erythromycin.

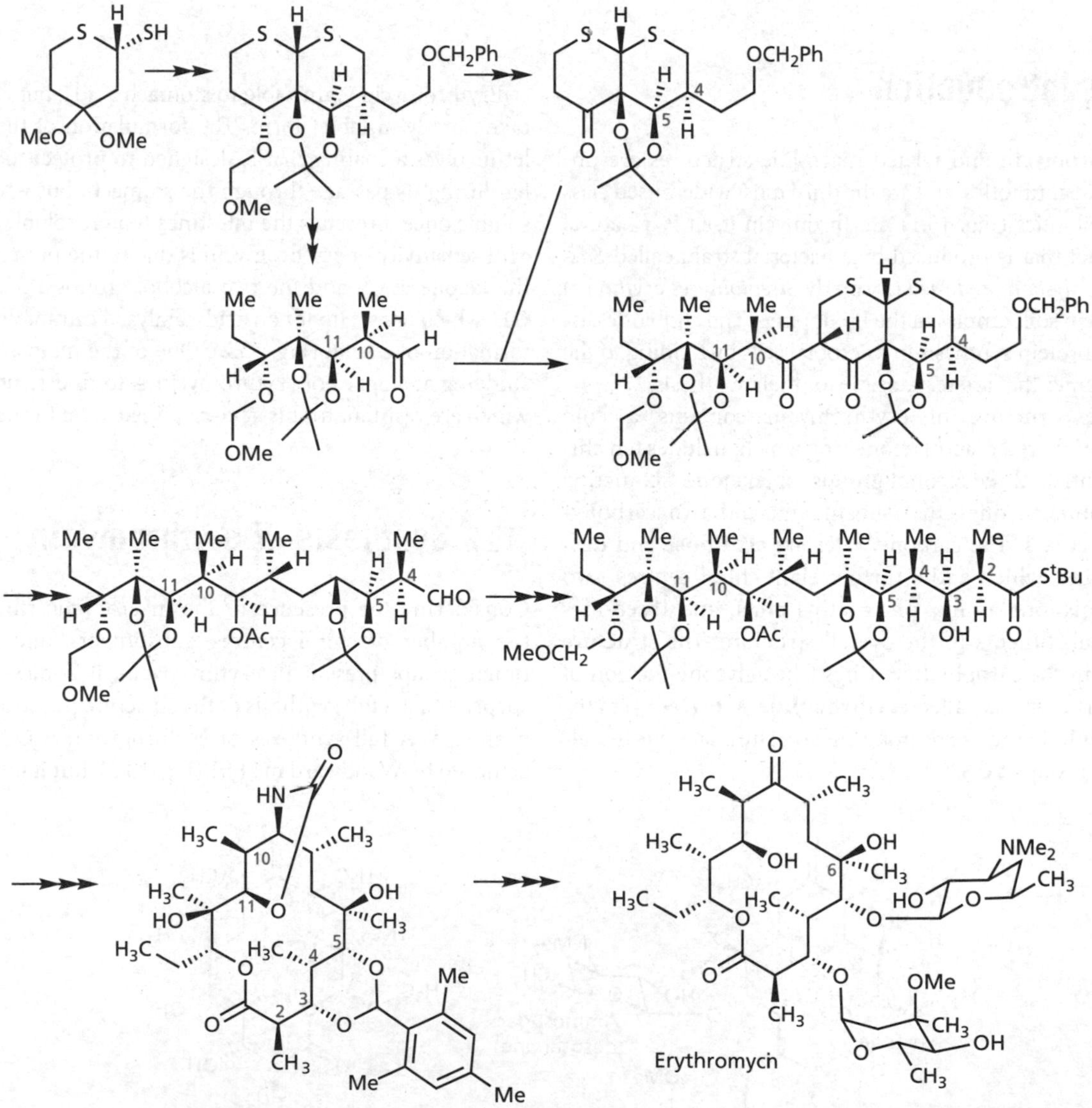

FIGURE 13.3 Key intermediates in the full synthesis of erythromycin.

50 steps and so it is totally impractical as an economic synthesis of erythromycin or its analogues.

13.3 Erythromycin analogues obtained from semi-synthetic methods

Since a full synthesis of erythromycin and its analogues is not really practical, semi-synthetic approaches appear to be a better option. Fortunately, erythromycin is readily extractable from cultures of the bacterium and has been used as the starting material for the preparation of analogues such as **roxithromycin**, **clarithromycin**, **azithromycin**, and **telithromycin** (Fig. 13.4). However, the scope of possible analogues from such semi-synthetic methods is restricted to the position and nature of the functional groups that are present in the starting material. Moreover, there is a difficulty in attaining regioselectivity because of the number of functional groups present.

Clarithromycin has a methyl ether at C6 instead of the alcohol group. The C6 alcohol is one of the groups which is involved in the acid-catalysed ketal formation that deactivates erythromycin, and so masking it makes the analogue more stable to gastric juices, resulting in improved oral absorption.

The synthesis of clarithromycin from erythromycin requires more than just methylating the alcohol group at position 6. The ketone at position 9 needs to be protected, as does the alcohol group of the amino sugar (Fig. 13.5). However, the alcohol groups at positions 11 and 12 do not need to be protected as long as there is a bulky *O*-substituent present on the oxime group. This allows the methylation reaction to show regioselectivity for the C6 alcohol.

Another method of increasing acid stability is to remove the ketone group. There have been two approaches to this. One is to convert the ketone to an alkoxyimine as seen in the structure of **roxithromycin**. The synthesis of roxithromycin involves reacting erythromycin with hydroxylamine, such that the ketone group is converted to an oxime (Fig. 13.6). O-Alkylation of the hydroxyl group can then be carried out in the presence of a mild base (sodium bicarbonate) to produce the final

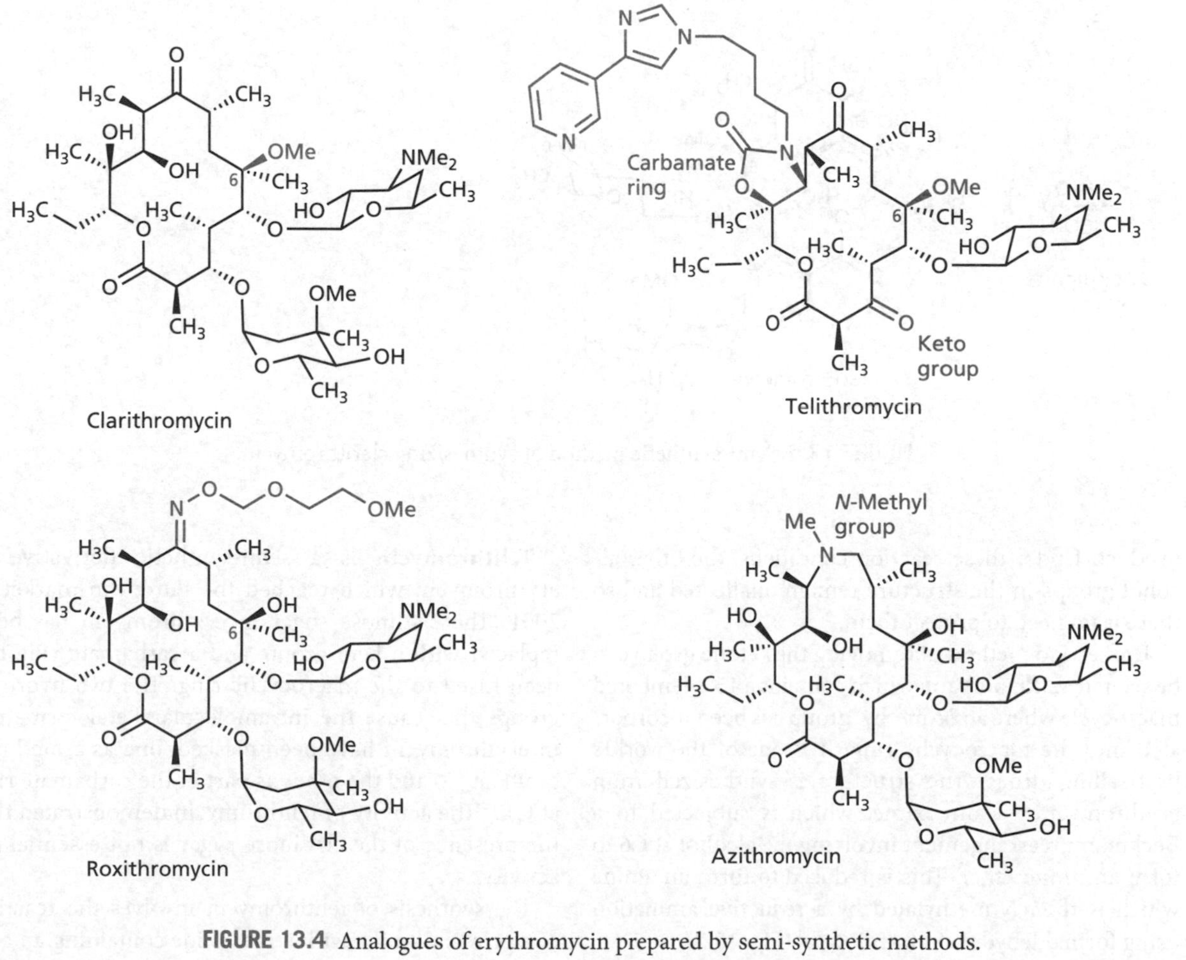

FIGURE 13.4 Analogues of erythromycin prepared by semi-synthetic methods.

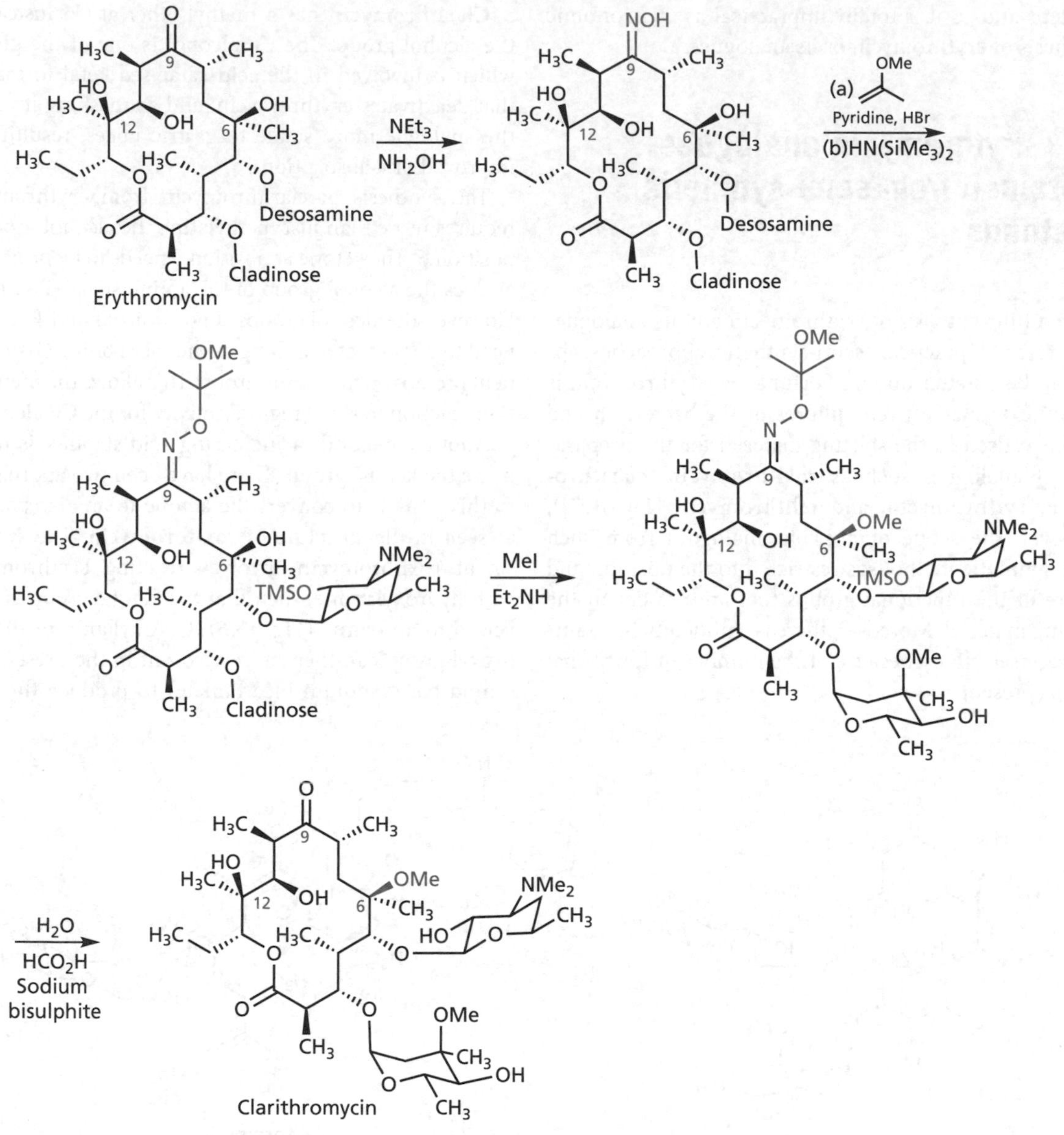

FIGURE 13.5 Semi-synthetic method of synthesizing clarithromycin.

product. Under these reaction conditions, the other alcohol groups in the structure remain unaffected and so there is no need to protect them.

The second method of removing the ketone group can be seen in **azithromycin**, which contains a 15-membered macrocycle where an *N*-methyl group has been incorporated into the macrocyclic ring. It is one of the world's best-selling drugs. The structure is synthesized from erythromycin via an oxime, which is subjected to a Beckmann rearrangement involving the alcohol at C6 to form an imino ether. This is reduced to form an amine which is then *N*-methylated by a reductive amination using formaldehyde and formic acid (Fig. 13.7).

Telithromycin is a semi-synthetic derivative of erythromycin which reached the European market in 2001. The cladinose sugar in erythromycin has been replaced with a keto group, and a carbamate ring has been fused to the macrocyclic ring. The two hydroxyl groups that cause the intramolecular ketal formation in erythromycin have been masked, one as a methoxy group at C6 and the other as part of the carbamate ring at C12. The activity of telithromycin demonstrated that the presence of the cladinose sugar is not essential for activity.

The synthesis of telithromycin involves the reaction of a primary amine with a macrolide containing an acyl

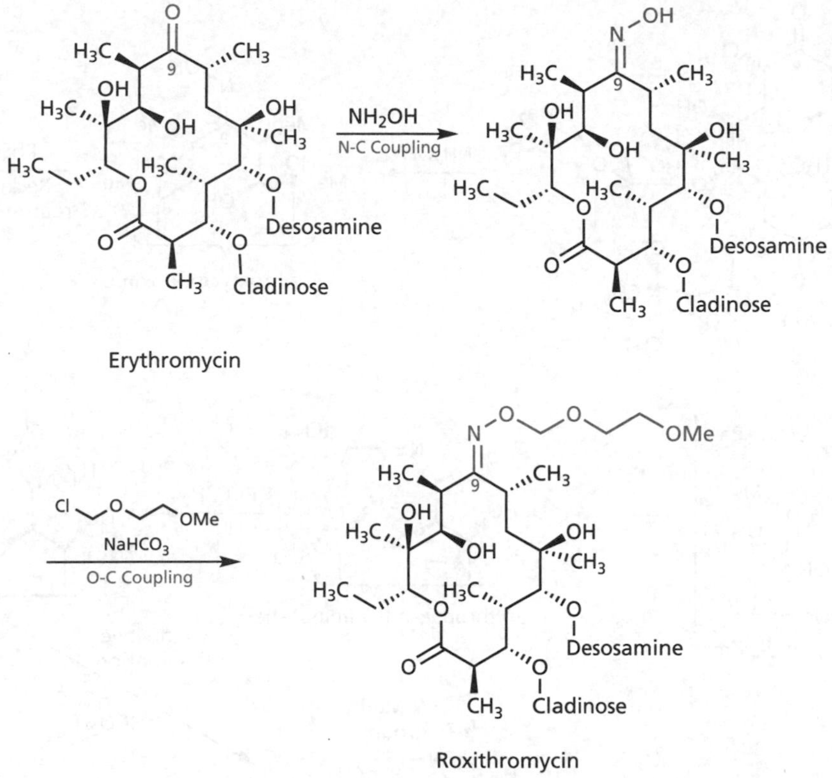

FIGURE 13.6 Semi-synthetic method of synthesizing roxithromycin.

imidazole group at position 12, resulting in an intermolecular cyclization (Fig. 13.8).

The acyl imidazole required for this reaction was synthesized from clarithromycin as shown in Figure 13.9. Treating clarithromycin with aqueous acid cleaves the cladinose sugar from position 3 to leave an alcohol group. The alcohol group on the remaining sugar is then protected with an acetyl group before the alcohol group at position 3 is oxidized to a ketone. The alcohol group at position 11 is then mesylated to make it a better leaving group for the subsequent elimination. Finally, the alcohol group is acylated with the imidazole carbonyl group.

13.4 Biosynthesis of erythromycin

A different approach has been used to obtain erythromycin analogues that are not possible to synthesize by synthetic or semi-synthetic methods. This involves the development of genetically modified cells where the biosynthesis of erythromycin has been modified to produce analogues with different functional groups and substituents around the macrocyclic ring. This is an approach known as **synthetic biology**. In order to appreciate how

this is achieved, it is necessary to understand how erythromycin is biosynthesized.

Erythromycin is derived from a biosynthetic intermediate called **6-deoxyerythronolide B** which is converted biosynthetically to erythromycin by the introduction of alcohol groups at C6 and C12 and the addition of two sugar molecules to the alcohol groups at C3 and C5 (Fig. 13.10). Clearly, 6-deoxyerythronolide B has much of the complexity of erythromycin, containing as it does a 14-membered macrocyclic ring which includes a lactone group, three alcohol groups, a ketone, one ethyl substituent, six methyl substituents, and ten chiral centres. Therefore, if we wish to explore the possibility of preparing analogues with different functional groups and substituents, we need to know how deoxyerythronolide is biosynthesized.

The biosynthesis of 6-deoxyerythronolide B involves a biosynthetic pathway called the **polyketide pathway**, which is commonly used by bacterial cells to build up the carbon skeletons of various natural products. You might imagine that the polyketide pathway must be extremely complex in order to create a molecule such as 6-deoxyerythronolide B. In fact, it is remarkably simple in concept (Fig. 13.11). Most of the biosynthesis involves the synthesis of a long acyclic carbon chain

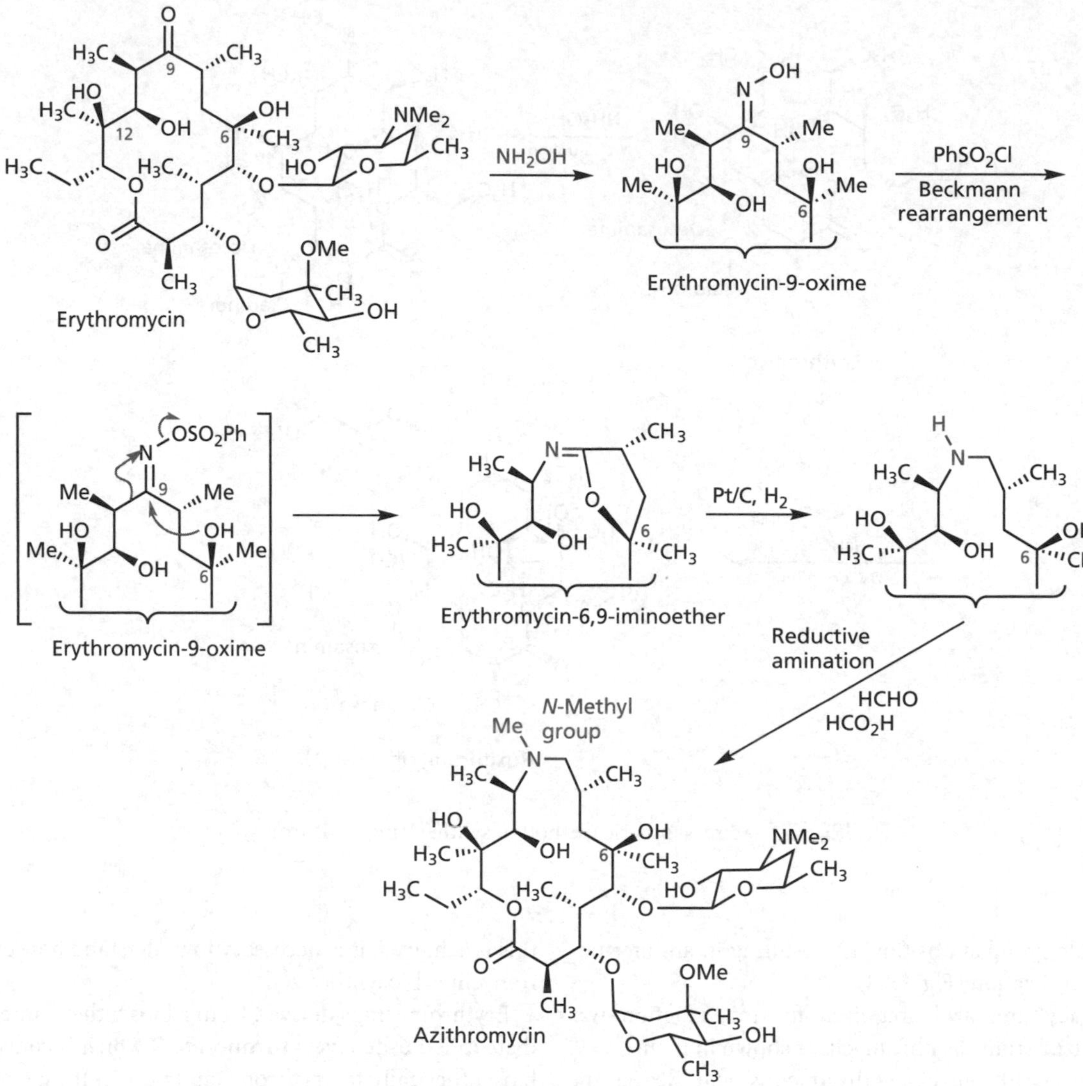

FIGURE 13.7 Semi-synthetic preparation of azithromycin.

where molecular building blocks are linked together one after the other to extend the carbon chain two carbon units at a time. After each linkage has taken place, there may be some functional group modification before the next 2C extension of the chain, and this is crucial to both the final complexity of the molecule and the manner in which the molecule will cyclize. It is only when the full chain has been completed that a cyclization reaction takes place to create the complex macrolactone observed in 6-deoxyerythronolide B. In the case of 6-deoxyerythronolide B, seven building blocks are involved to introduce the segments highlighted in colour in Figure 13.11.

Most of the building blocks used in the polyketide synthesis of 6-deoxyerythronolide B are provided by **methylmalonyl coenzyme A**, which acts as nature's equivalent of an activated ester. At each extension stage of the synthesis, a molecule of methylmalonyl coenzyme A acts like a carbanion to condense with a thioester at the end of the growing polyketide chain (Fig. 13.12). This reaction is equivalent to a Claisen condensation. The linking process is accompanied by loss of carbon dioxide which helps to drive the reaction, and a ketone group is formed from the original thioester. The chain has now been extended by two carbon atoms, and includes a methyl substituent.

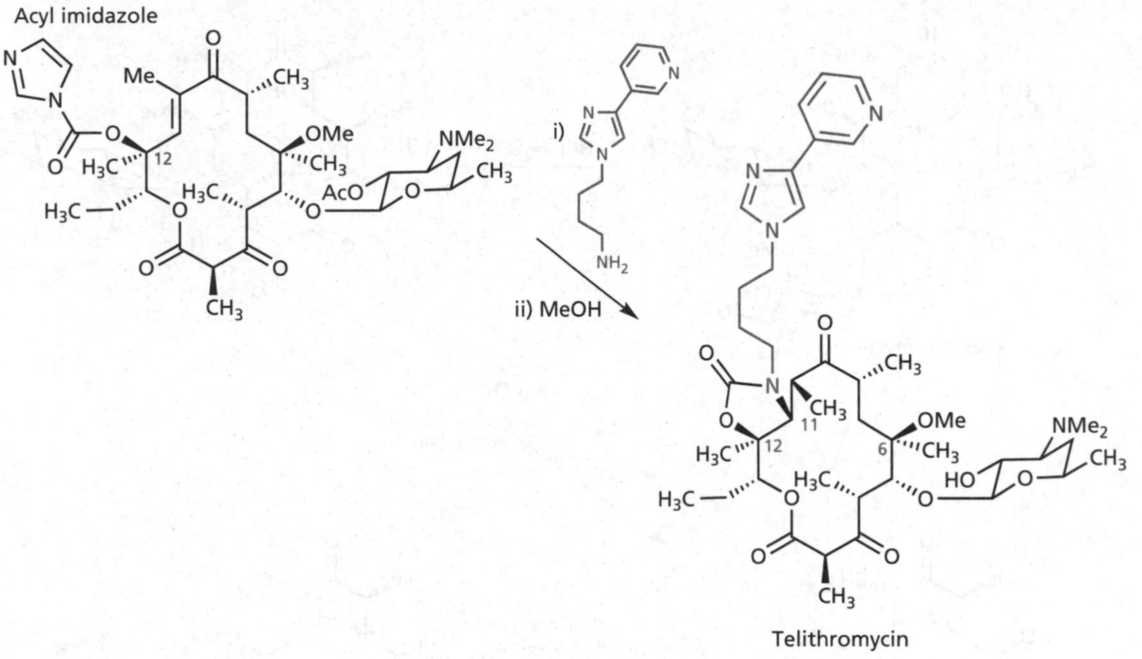

FIGURE 13.8 Synthesis of telithromycin.

Once each molecular building block has been added, a chemical modification of the newly formed ketone group may or may not take place. There are four possibilities (Fig. 13.13):

- no modification at all;
- reduction of the ketone to an alcohol;
- reduction of the ketone to an alcohol followed by dehydration to form an alkene;
- formation of an alkene as described above, followed by reduction to remove the functional group altogether—overall, this corresponds to a deoxygenation.

The polyketide that is created in the biosynthesis of 6-deoxyerythronolide B involves the linkage of seven building blocks. The first of these is propionyl coenzyme A and the remaining six are methylmalonyl coenzyme A. There are also four reductions and one deoxygenation in the synthetic route (Fig. 13.14). Once the full polyketide chain has been completed, a cyclization takes place involving the alcohol at one end of the chain and the thioester at the other end.

Although the concept is basically quite simple, there are some intriguing features about the biosynthesis. For example, the functional group modifications following each condensation are not consistent. Four of the ketone groups are reduced to an alcohol, one is deoxygenated, and one is not modified at all. How is

the pathway controlled such that the correct modifications occur in the right order after each condensation? Moreover, the chiral centre at the methyl substituent is *S* after three condensations, and *R* after the other three, yet the same building block is used for each condensation. Finally, the configurations of the alcohol groups formed by reduction are not identical either. Obviously, enzymes are responsible for catalysing the various stages, but how does that explain the different modifications and chiral centres that occur at each of the condensations?

The truth is that the process does not involve the growing polyketide chain floating through the cytoplasm of the cell to seek out each enzyme in turn. Instead, the whole process is catalysed from start to finish by a 'super-enzyme' containing all the different active sites required for the full polyketide synthesis. Moreover, the growing polyketide chain is covalently linked (or tethered) to this super-enzyme throughout the whole process and is transferred from one active site to the next like a car moving along an assembly line from one workstation to the next. This means that the growing polyketide chain is moving through a series of reactions in a precisely defined order and cannot escape from the super-enzyme until the whole process has been completed. This has been called **physical channelling** or **substrate channelling**.

The super-enzymes involved in polyketide synthesis are called polyketide synthases (PKSs) and have a

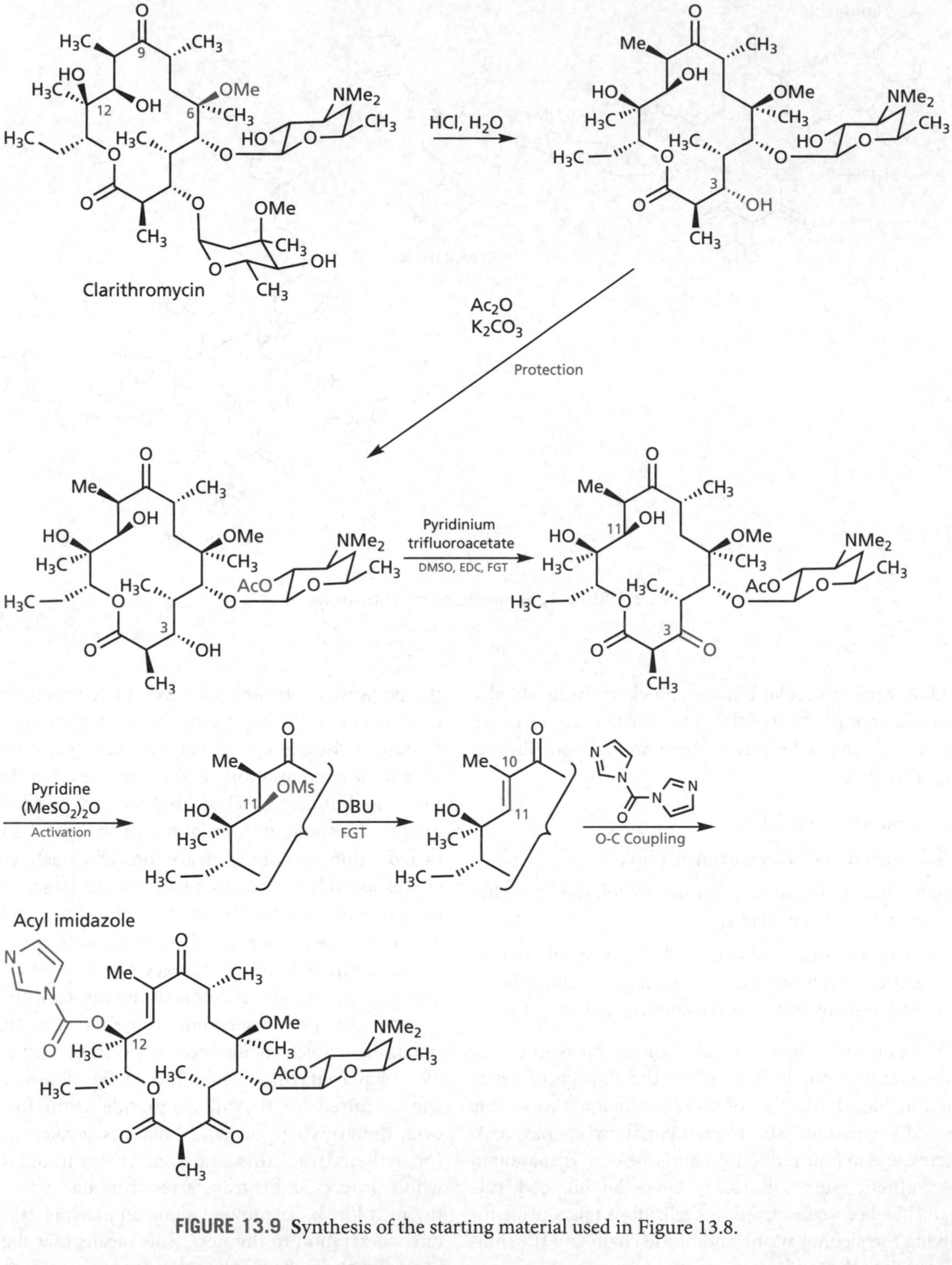

FIGURE 13.9 Synthesis of the starting material used in Figure 13.8.

molecular weight of over 900 000, as well as the capability of catalysing 28 reactions. There are different types of PKS responsible for the synthesis of different structural classes of natural product. The one involved in the synthesis of 6-deoxyerythronolide B is called **6-deoxyerythronolide B synthase** (**DEBS**) and is found in the microorganism *Saccharopolyspora erythraea*. DEBS consists of three large proteins known as **megasynthases**, with each protein being in the form of a dimer made up of more than 3000 amino acids. Each megasynthase is essentially a multifunctional polymer of different enzymes linked together by means of a single polypeptide chain. As a result, the full megasynthase protein contains all of the catalytic sites that are required to carry out its particular part of

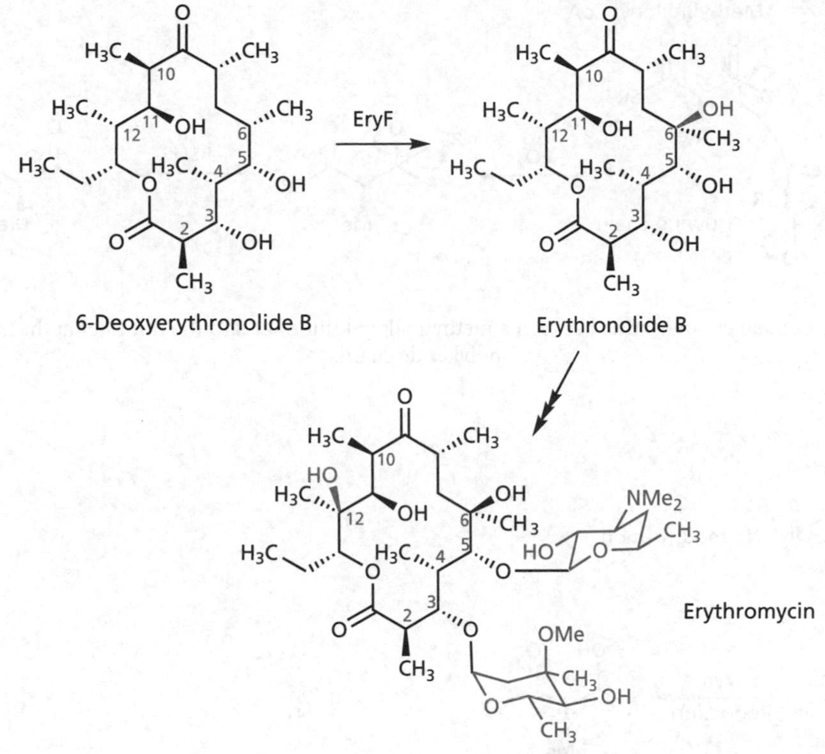

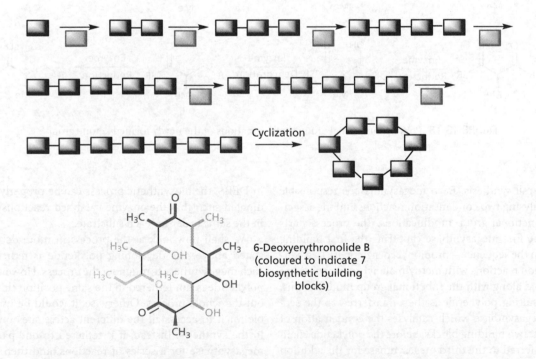

FIGURE 13.10 Biosynthetic conversion of 6-deoxyerythronolide B to erythromycin.

FIGURE 13.11 The polyketide pathway—linking building blocks and cyclization.

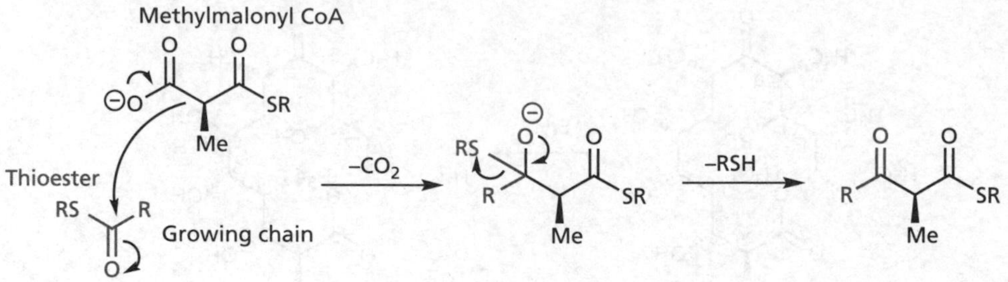

FIGURE 13.12 The Claisen condensation between a methylmalonyl thioester and the thioester at the end of the growing polyketide chain.

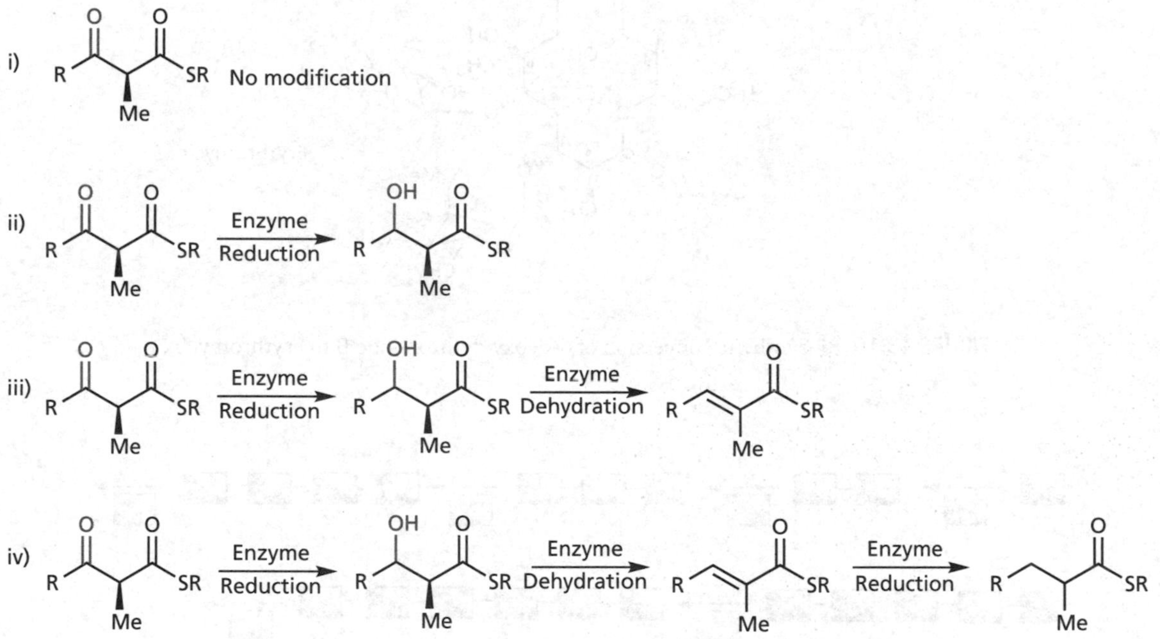

FIGURE 13.13 Possible functional group transformations of the newly formed ketone group.

the overall synthesis. Each megasynthase is responsible for catalysing two condensation reactions and the associated functional group modifications. The process starts with the first megasynthase capturing the first building block in the sequence—propionyl coenzyme A. Two condensation reactions with methylmalonyl coenzyme A are catalysed along with the functional group modifications. The resulting polyketide is then transferred to the second megasynthase which catalyses the condensation of the next two building blocks, before the polyketide chain is transferred to the third megasynthase for the addition of the final two building blocks. The complete polyketide chain is then cleaved from the third megasynthase and undergoes cyclization. Because all the enzyme-catalysed reactions can be carried out by the three megasynthases

in DEBS, the biosynthetic process can be properly coordinated such that the enzyme-catalysed reactions occur in the same order for each substrate.

We shall now look at the process in more detail. As stated above, the developing polyketide is tethered to each megasynthase throughout the process. However, the polyketide is not tethered to the same position throughout the whole process. Otherwise, it would be impossible for it to access all the different active sites involved in the synthesis. Instead, it is tethered to one part of a megasynthase for a series of reactions, and then transferred to a different part for the next series of reactions. In all, the polyketide is moved about the super-enzyme (DEBS) to seven different locations. Each location within each megasynthase is defined as an **acyl carrier protein**

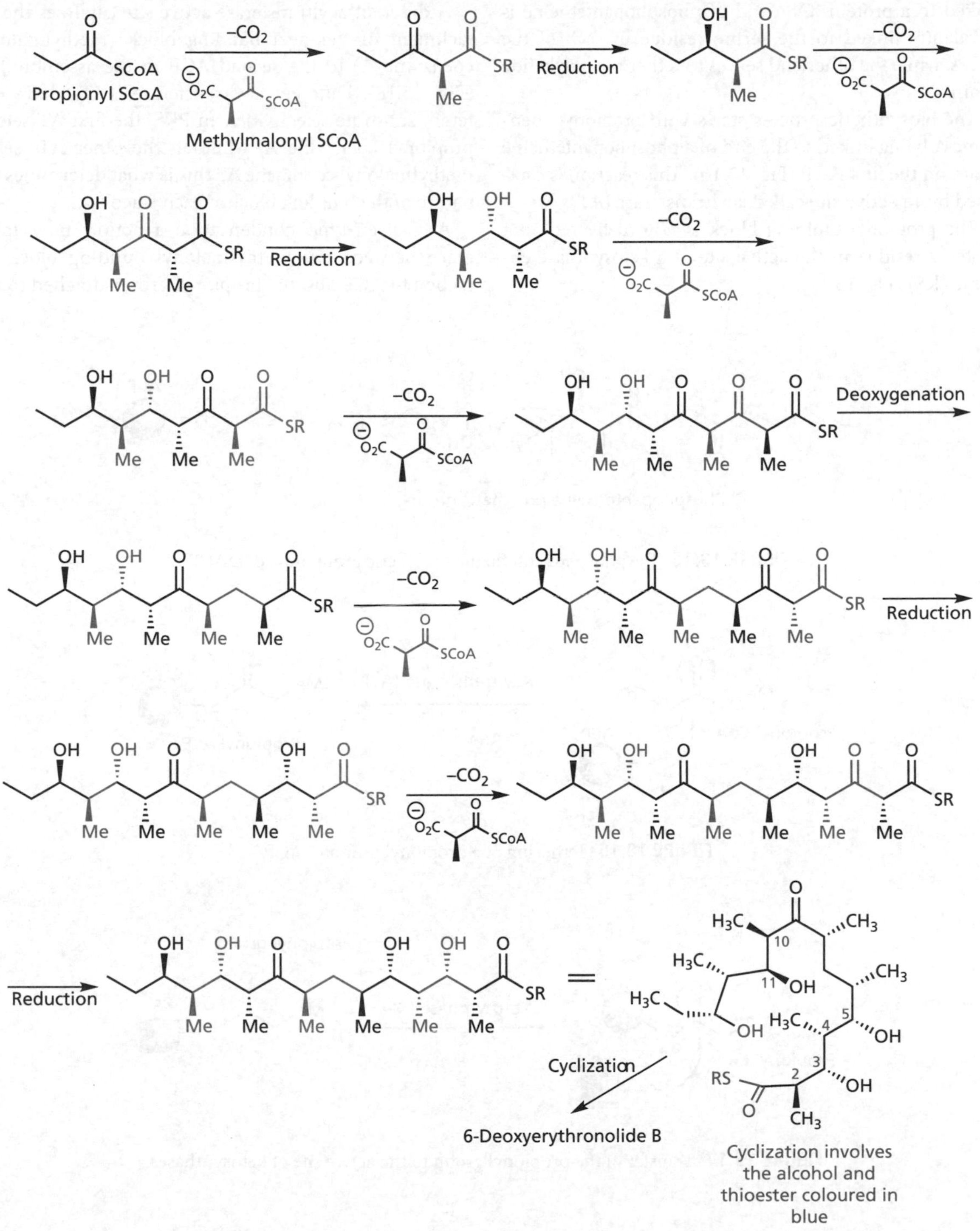

FIGURE 13.14 Polyketide synthesis involved in the biosynthesis of 6-deoxyerythromycin B.

(ACP), and each ACP contains a serine residue that is covalently linked to a long linear molecule called **phosphopantetheine** (Fig. 13.15). This molecule acts as a long flexible arm which serves not only to tether the polyketide chain, but also to provide the flexibility needed to manoeuvre the polyketide from one active site to the next in the correct sequence. Phosphopantetheine is not an integral part of the protein, but it is essential to the functioning of the whole catalytic process. Such a group is known as a **prosthetic group** when it is covalently linked to a protein. One end of phosphopantetheine is covalently linked to the serine residue in each of the ACPs, while the other end serves to tether the polyketide chain.

The biosynthetic process starts with propionyl coenzyme A being linked to the end of a phosphopantetheine chain on the first ACP (Fig. 13.16). This reaction is catalysed by an active site called acyltransferase (AT).

The propionyl building block is now 'delivered' to a cysteine residue in the active site of a ketosynthase enzyme (KS) (Fig. 13.17).

A different acyltransferase active site catalyses the attachment of the next building block (methylmalonyl coenzyme A) to the second ACP in the assembly line (Fig. 13.18). Different acyltransferases (ATs) have different substrate selectivities. In PKS, the first AT selects propionyl coenzyme A, while all the other ATs select methylmalonyl coenzyme A. This is what determines the nature of the building block at each stage.

A Claisen-type condensation reaction now takes place between the methylmalonyl building block attached to ACP and the propionyl group attached to the

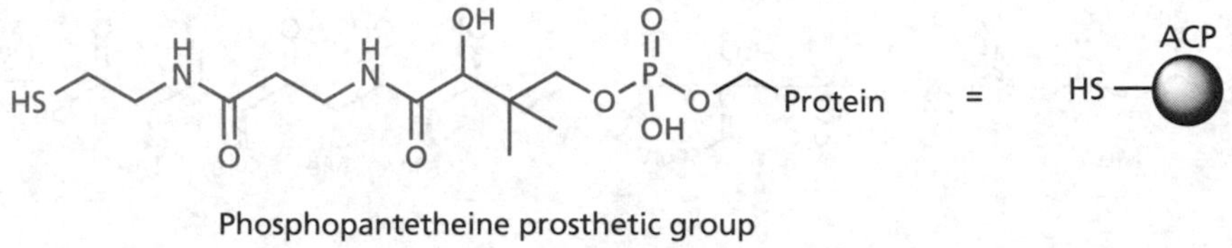

FIGURE 13.15 The phosphopantetheine prosthetic group linked to ACP.

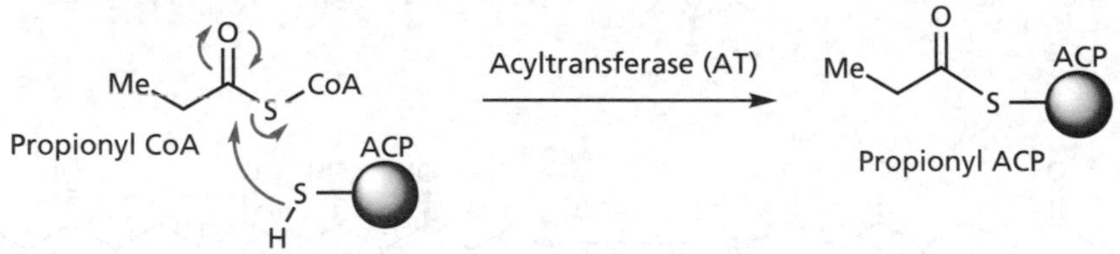

FIGURE 13.16 Tethering of a propionyl group to ACP.

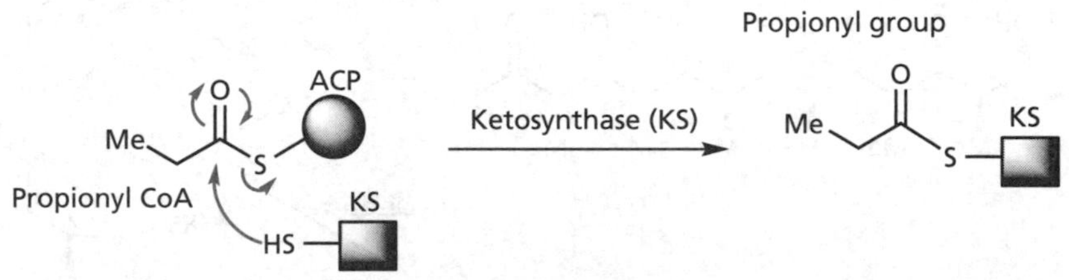

FIGURE 13.17 Transfer of the propionyl group to the active site of ketosynthase.

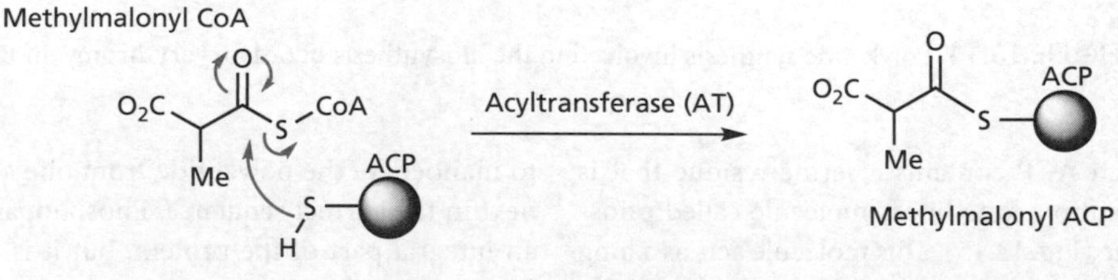

FIGURE 13.18 Tethering of a methylmalonyl group to the second ACP.

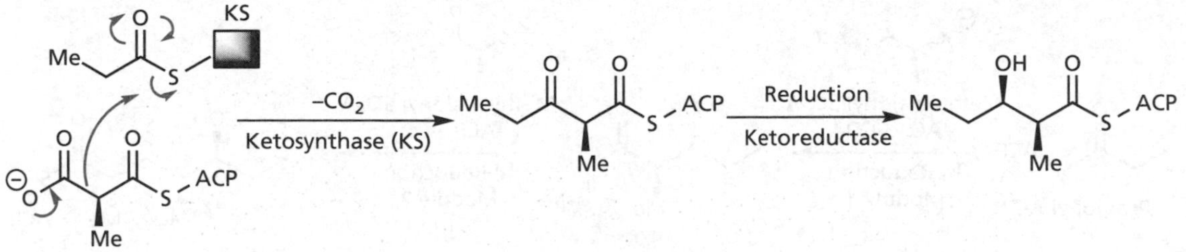

FIGURE 13.19 Claisen condensation followed by reduction.

active site of the ketosynthase enzyme. The two building blocks become linked and the propionyl group is removed from the ketosynthase active site (Fig. 13.19). The phosphopantetheine 'arm' now swings the newly formed diketide chain round to the next active site. This is a ketoreductase which catalyses the reduction of the ketone to an alcohol.

The modified diketide is now transferred to the active site of the second ketosynthase in the production line. Another molecule of methylmalonyl coenzyme A is linked to the third ACP in the sequence and condensation of the diketide chain with this building block sees the growing polyketide chain transferred again. And so the process continues until all the building blocks have been added to the growing polyketide chain.

Note that at each extension stage, a three-carbon building block is attached to the main chain, with the backbone being extended by two carbons. Because the growing polyketide chain is tethered to the megasynthases throughout the whole process, the process can be viewed as nature's equivalent of solid phase synthesis, with the important exception that the growing chain is transferred from one linker molecule (phosphopantetheine) to the next and is not fixed to the same linker throughout.

The catalytic regions responsible for all of these reactions are called **domains**, and the domain involved in each linkage and subsequent modification is called a **module**. In each module there are always three domains which are responsible for the linking process. These are a ketosynthase (KS), an acyltransferase (AT), and the acyl carrier protein (ACP). Each module also has a variable number of enzymes responsible for the modification reactions that occur following each condensation. These are a ketoreductase (KR), a dehydratase (DH), and an enoyl reductase (ER). The modification(s) that are carried out depend on which of these enzymes are present in the module.

Each of the three megasynthases in DEBS contains two modules. In other words, each megasynthase can carry out two condensations along with the subsequent

modifications. Therefore, the first module of the first megasynthase is responsible for the processes shown in Figures 13.16–13.19.

The product in Figure 13.19 is still attached to an ACP in module 1 and is now transferred to the active site of the ketosynthase in module 2. Another set of catalytic regions adds the next molecule of methylmalonyl CoA and the product is reduced by another ketoreductase (Fig. 13.20).

The second megasynthase protein containing modules 3 and 4 now comes into play. Another two molecules of methylmalonyl CoA are added to the polyketide chain, but there are differences in how the condensation products are modified. Module 3 does not modify the product at all and so the ketone group is retained, whereas module 4 removes the oxygen completely to leave a saturated centre. This is carried out by three domains—a ketoreductase (KR), a dehydratase (DH), and an enoyl reductase (ER) (Fig. 13.21).

The third megasynthase protein now comes into play with modules 5 and 6 to add the final two building blocks. After each condensation, the ketone is reduced to an alcohol. At the very end of the process, the polyketide chain is transferred to a serine residue in a thioesterase active site. The resulting ester link is then hydrolysed in a catalysed cyclization to cleave the final structure of 6-deoxyerythronolide B as a macrocyclic lactone.

This assembly line process involved in the biosynthesis is often portrayed as shown in Figure 13.22. This emphasizes the various domains and modules, and the order in which they appear. The nature of the growing polyketide chain at various stages is also shown.

Therefore, the pattern and nature of the building blocks and the substituents present in the final structure is dependent on the catalytic regions or domains within each module. The acyltransferases determine the nature of the building blocks used, while the presence or otherwise of the three domains (ketoreductase, dehydratase, and enoyl reductase) determines whether the ketone resulting from the condensation is modified.

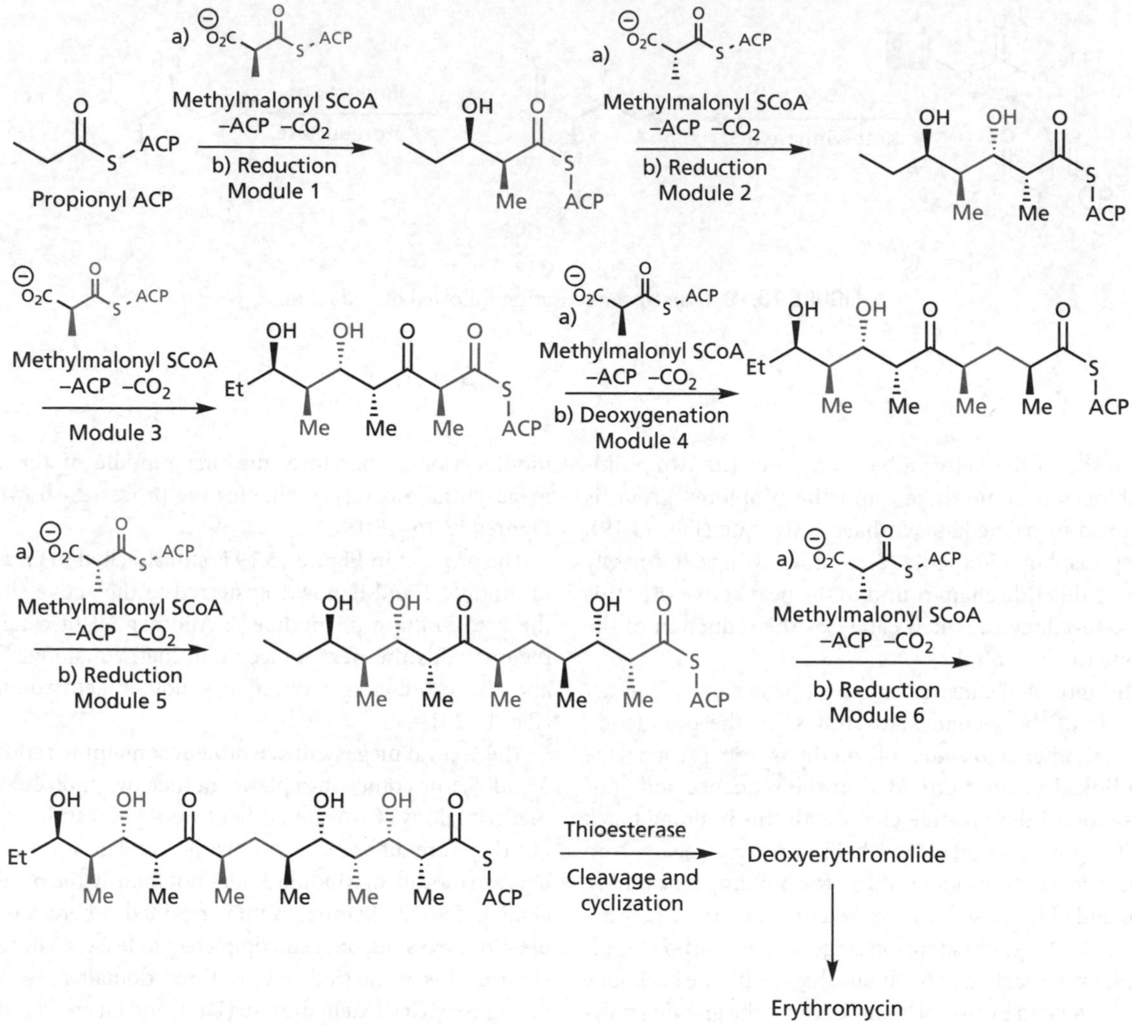

FIGURE 13.20 Biosynthesis of deoxyerythronolide and erythromycin.

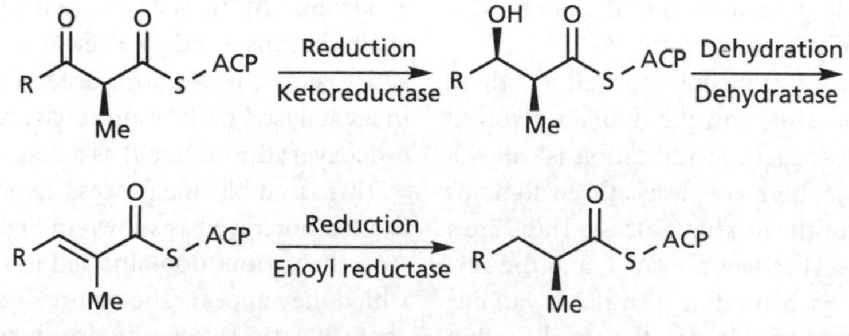

FIGURE 13.21 The deoxygenation process catalysed by three domains in module 4.

If no ketoreductase is present (as in module 3), the ketone group is retained. If a ketoreductase enzyme is present (as in modules 1, 2, 5, and 6) the ketone is reduced to an alcohol. It is interesting to note that the ketoreductases involved are not identical in the stereochemical outcome. The absolute configuration of the resulting alcohol is *R* in module 1, and *S* in modules 2, 5, and 6.

Module 4 contains a ketoreductase (KR), a dehydratase (DH), and an enoyl reductase (ER) domains, which catalyse a three-stage process to remove the ketone group completely (Fig. 13.23).

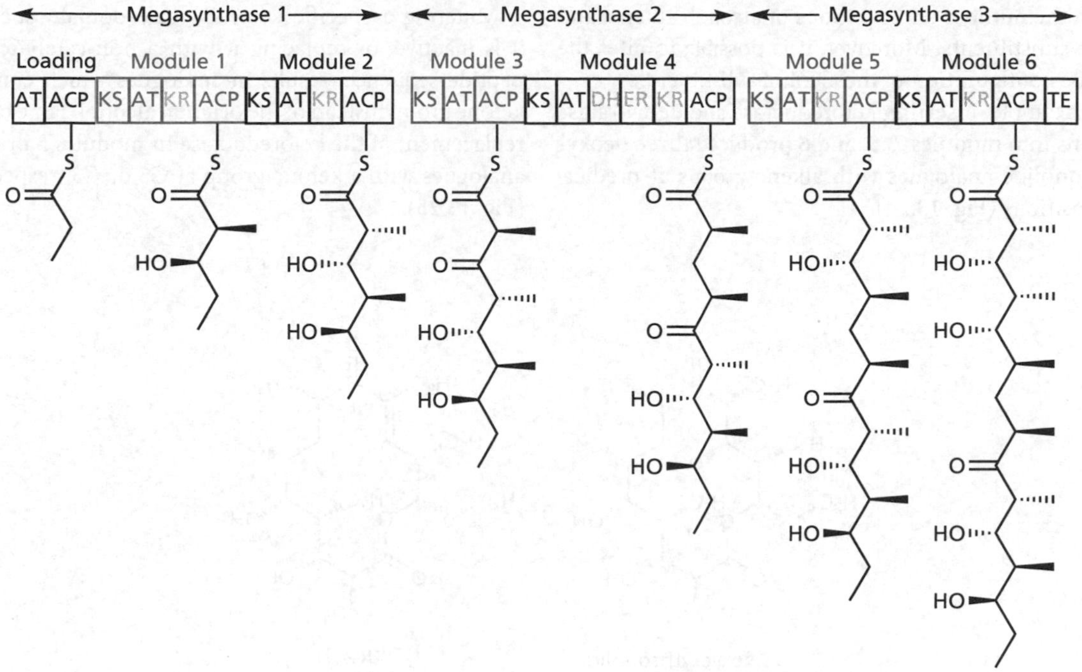

FIGURE 13.22 The assembly line process for the biosynthesis of 6-deoxyerythronolide B.

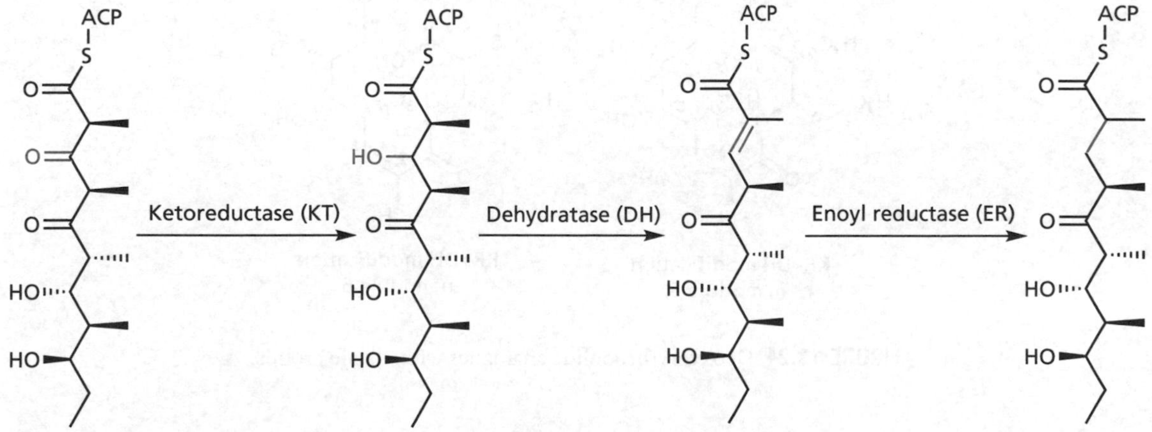

FIGURE 13.23 The deoxygenation of a ketone.

13.5 Precursor-directed biosynthesis as a means of synthesizing macrocyclic analogues

As described in section 13.4, the nature of the final polyketide produced by the polyketide synthase DEBS is determined by the order and the nature of the domains that are present within the super-enzyme. In essence, DEBS contains the blueprint or code for the final polyketide structure. The discovery of this highly ordered assembly line opens up the possibility of using genetic engineering to modify the various catalytic regions or domains that are present in the three megasynthases making up DEBS to generate analogues with different substituents and chiral centres. By altering the domains responsible for functional group modifications, deoxyerythronolide and erythromycin analogues would be synthesized which vary in the number and positions of alcohol, alkene, and ketone substituents. Moreover, it is possible to alter the absolute configuration of the alcohol substituents.

For example, inserting ketoreductase and dehydratase domains into modules 2, 5, and 6 produced three deoxy-erythronolide analogues with alkene groups at predictable positions (Fig. 13.24).

Mutating a specific ketoreductase domain such that it is inactive, or replacing it with a non-catalytic polypeptide sequence, results in analogues which contain a ketone group instead of the original alcohol. For example, replacement of the ketoreductase in modules 5 or 6 gave analogues with a ketone group at C5 or C3, respectively (Fig. 13.25).

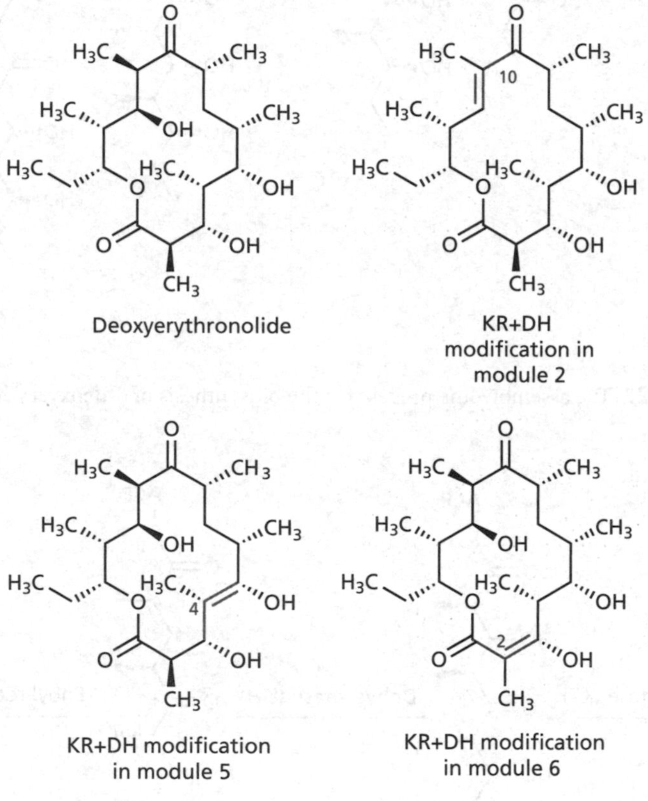

FIGURE 13.24 Deoxyerythronolide analogues with alkene groups.

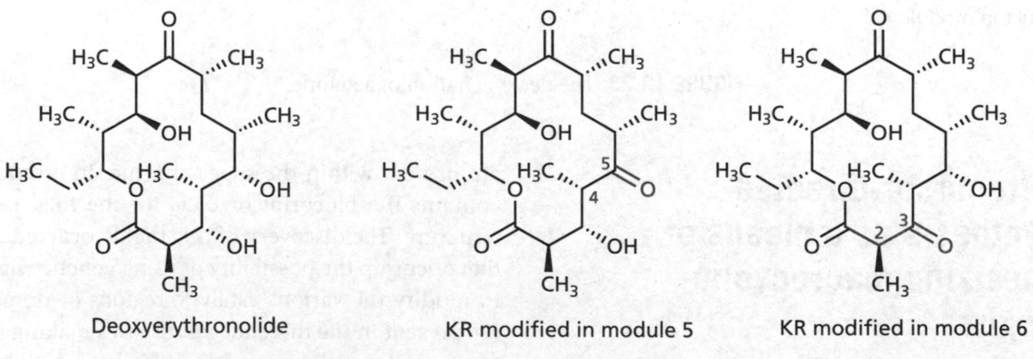

FIGURE 13.25 Deoxyerythronolide analogues arising from replacing a ketoreductase domain with a non-catalytic polypeptide.

Replacing the ketoreductase in modules 2 and 5 with a sequence made up of a dehydratase, an enol reductase, and a ketoreductase gave analogues where the functional group had been completely removed from positions 11 and 5, respectively (Fig. 13.26).

Acyltransferases (ATs) can also be modified. In DEBS, the first acyltransferase identifies propanoyl coenzyme A as the building block, whereas all subsequent acyltransferases identify methylmalonyl coenzyme A as the building block. However, there are acyltransferases in other polyketide synthases which identify different structures such as malonyl coenzyme A or ethylmalonyl coenzyme A. By replacing ATs in DEBS with different ATs, it is possible to prepare analogues which lack the methyl substituents at specific positions of the structure.

For example, replacing the native AT in module 2, 5, or 6 with an AT showing selectivity for malonyl coenzyme A gave analogues lacking methyl substituents at positions 10, 4, and 2, respectively (Fig. 13.27). It has also been

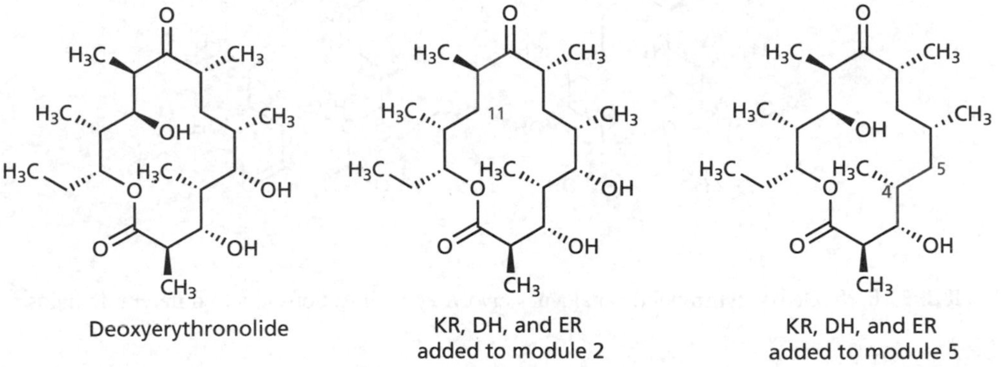

Deoxyerythronolide   KR, DH, and ER   KR, DH, and ER
                                  added to module 2   added to module 5

FIGURE 13.26 Deoxyerythronolide analogues resulting from adding ketoreductase, dehydratase, and enoyl reductase to specific modules.

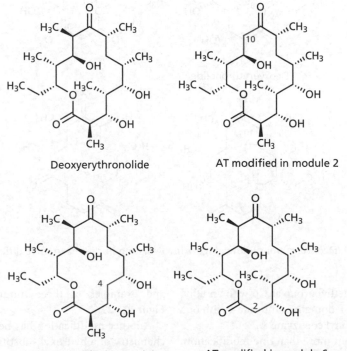

Deoxyerythronolide   AT modified in module 2

AT modified in module 5   AT modified in module 6

FIGURE 13.27 Deoxyerythronolide analogues obtained by replacing acyltransferases in modules 2, 5, and 6.

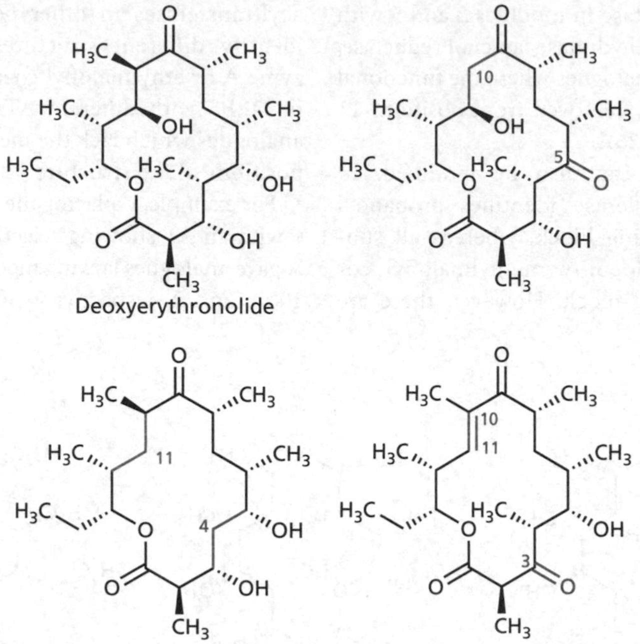

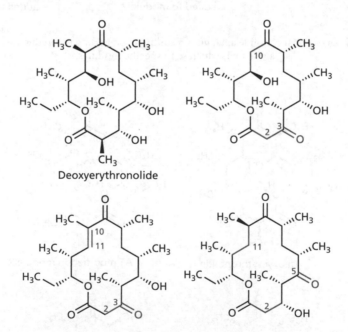

FIGURE 13.28 Deoxyerythronolide analogues involving modifications to two different domains.

FIGURE 13.29 Deoxyerythronolide analogues involving modifications to three different domains.

possible to replace the methyl group at C6 with an ethyl group by replacing the AT domain in module 4 with one that can accept ethylmalonyl coenzyme A.

It is feasible to carry out more than one modification, which increases the potential for generating large libraries of modified structures. Examples where two modifications have been carried out are shown in Figure 13.28,

and examples of three modifications are shown in Figure 13.29.

Another modification has been to change the stereochemistry of a hydroxyl substituent by replacing the native reductase with a reductase that creates the opposite stereochemistry (Fig. 13.30). However, this has only been successful with module 6.

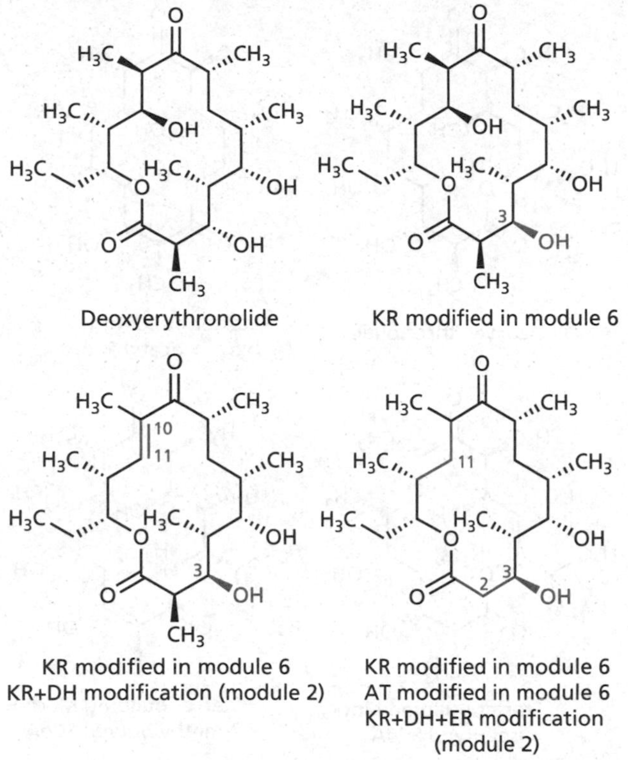

FIGURE 13.30 Deoxyerythronolide analogues which include a change in stereochemistry at a chiral centre.

The loading module for DEBS can also be modified. The native AT/ACP module shows selectivity for propionyl coenzyme A, but can also accept acetyl coenzyme A. Replacing the native module with a different AT/ACP module allows analogues to be produced where the ethyl substituent has been replaced with a methyl substituent or a branched substituent (Fig. 13.31).

Another approach has been to inactivate the KS1 domain such that DEBS cannot load the initial propionyl building block. Providing the cells with an unnatural substrate corresponding to a diketide resulted in the substrate being taken up by DEBS and shuttled through to the corresponding macrocyclic analogue (Fig. 13.32). This approach allows different analogues to be generated with variation at positions 12–15 of deoxyerythronolide, as well as ring expanded analogues.

This ability to genetically modify the polyketide synthase DEBS has led to the concept of **combinatorial biosynthesis** where genetic modifications are carried out to generate a library of polyketide synthases which differ in the number, sequence, and nature of the different domains within their structure. This, in turn, results in a library of novel analogues which can potentially be generated from fast-growing microbial cells or cell-free conditions.

Much of the research work carried out so far has focused on the generation of novel analogues of 6-deoxyerythronolide B. The antibacterial activity of this biosynthetic intermediate is less than that of erythromycin itself, but it is more convenient to isolate and test analogues of this structure than to convert all of the structures to erythromycin analogues. Any analogues of 6-deoxyerythronolide B that *do* show useful activity can be converted to the corresponding erythromycin analogues by feeding them to cultures of the microorganism and then extracting the final products.

Recently, novel analogues of **6-deoxyerythromycin D** have been produced in genetically modified cells of *Escherichia coli*, which contain plasmids carrying the enzymes required for the biosynthesis of this structure (Fig. 13.33). Two plasmids carry a modified version of DEBS where the initial loading domain and the first module have been omitted. Thus, the biosynthesis can only proceed if the system is fed with a suitable diketide precursor. Another two plasmids code for the enzymes required to link the two carbohydrate rings to produce 6-deoxyerthyromycin D or its analogues.

Some of these analogues have similar antibacterial activity to 6-deoxyerythromycin D itself, and one analogue

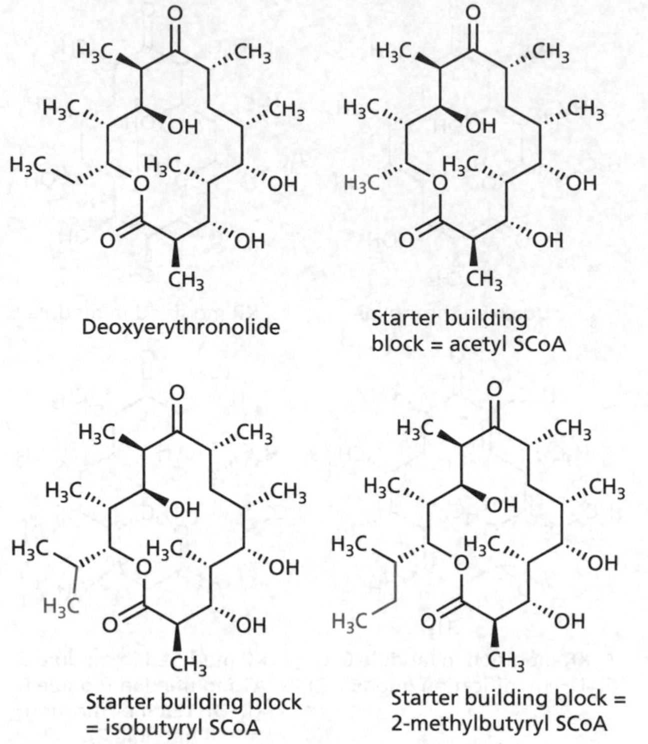

FIGURE 13.31 Deoxyerythronolide analogues obtained as a result of modifications to the loading module.

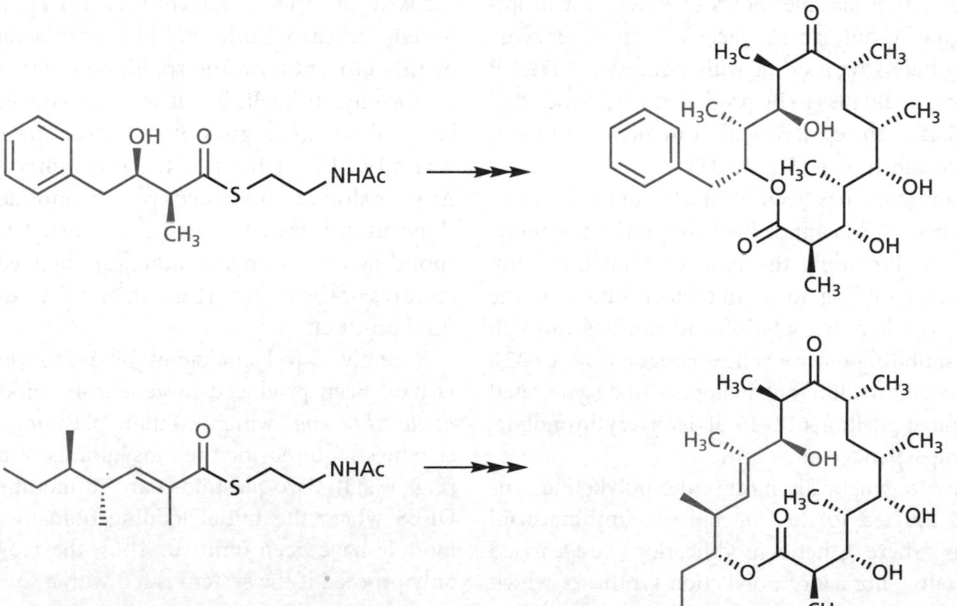

FIGURE 13.32 Deoxyerythronolide analogues obtained as a result of adding modified diketides to DEBS with an inactivate KS1 domain.

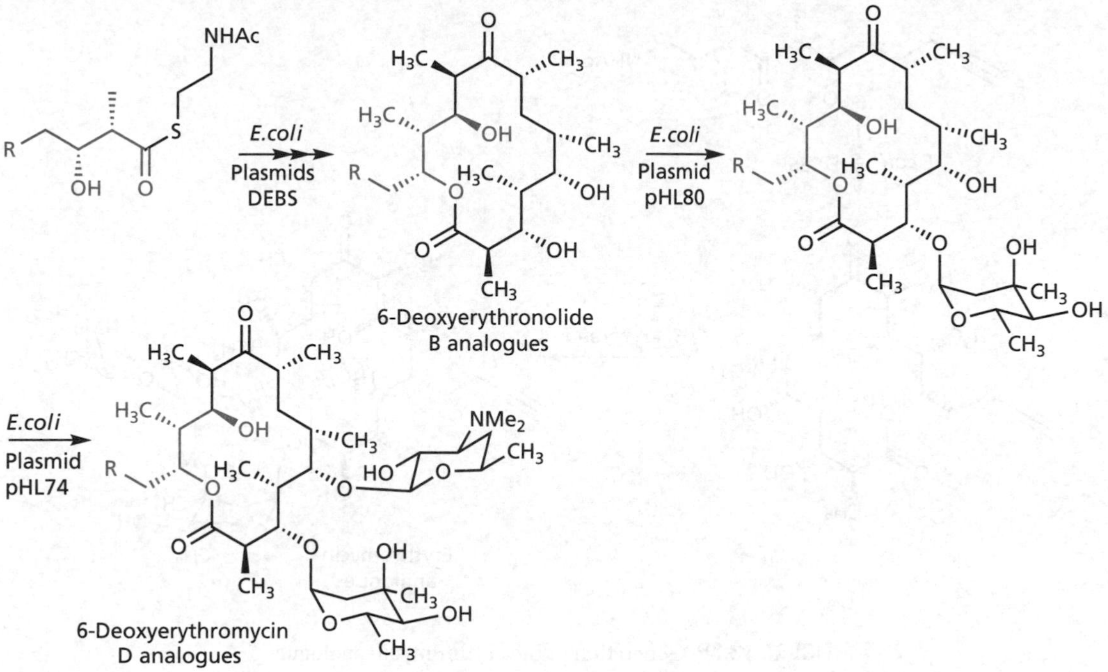

FIGURE 13.33 6-Deoxyerythromycin D analogues.

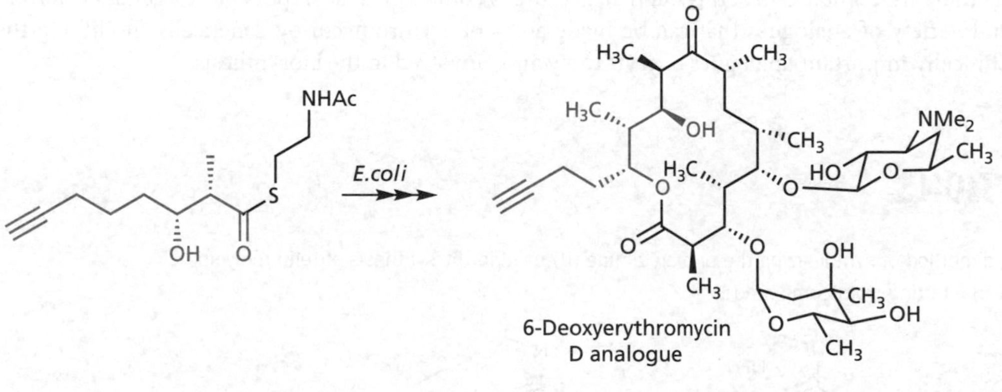

FIGURE 13.34 The most active 6-deoxyerythromycin D analogue.

containing an alkyne group in the side chain proved twice as active (Fig. 13.34).

Since 6-deoxyerythromycin D lacks the OH groups that are present at C6 and C12 in erythromycin, it was decided to generate the equivalent erythromycin analogue. The biosynthetic precursor containing the alkyne group was again fed to *E.coli*, but this time the microorganism lacked the plasmids responsible for carbohydrate biosynthesis and linkage. This allowed the

corresponding 6-deoxyerythronolide B analogue to be isolated and fed to a culture of *S.erythraea*, which completed the biosynthesis of the erythromycin analogue (Fig. 13.35).

The analogue proved to be similar in activity to erythromycin itself and five times more active than the 6-deoxyerythromycin analogue. The presence of the alkyne offers the potential for further elaboration of the structure.

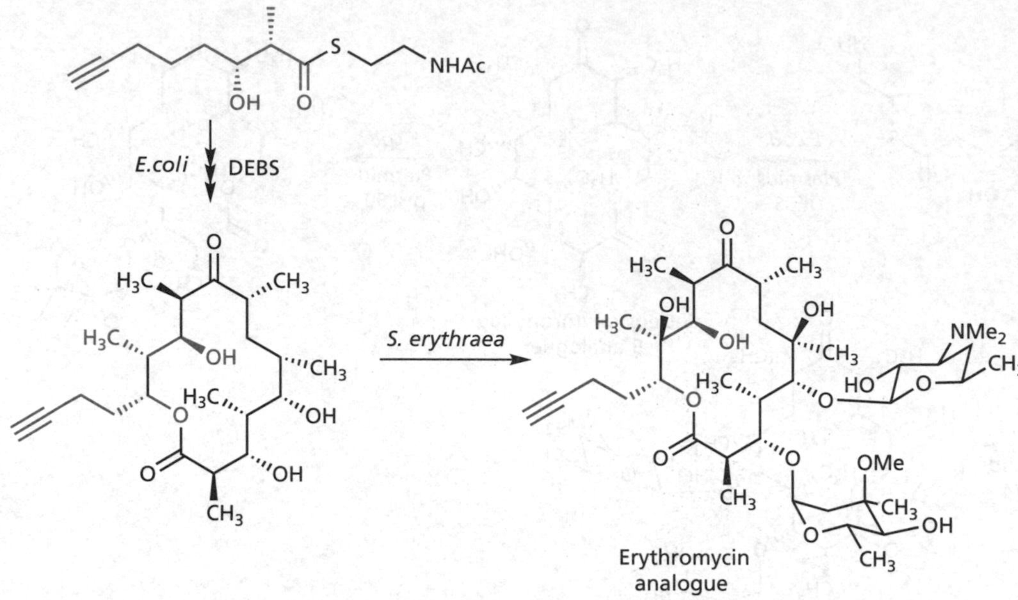

FIGURE 13.35 Generation of the erythromycin analogue.

13.6 **Conclusions**

Macrolides have been extremely useful antibacterial agents, but they are complex structures and hence the number and variety of analogues that can be made are limited. Clinically important analogues of erythromycin have been prepared by semi-synthetic methods. These analogues have been modified to prevent an acid-catalysed intramolecular reaction that leads to inactive compounds. It is possible to generate novel analogues of erythromycin by genetically modifying the enzymes involved in the biosynthesis.

QUESTIONS

1. Suggest a method of synthesizing the primary amine (II) used in the synthesis of telithromycin (Fig. 13.8), starting from compound I.

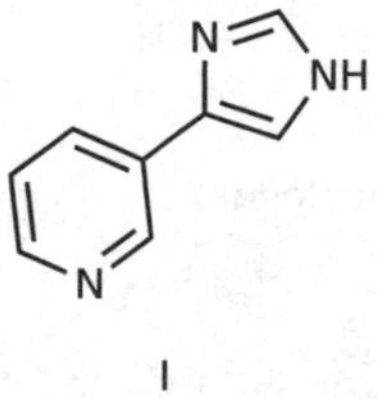

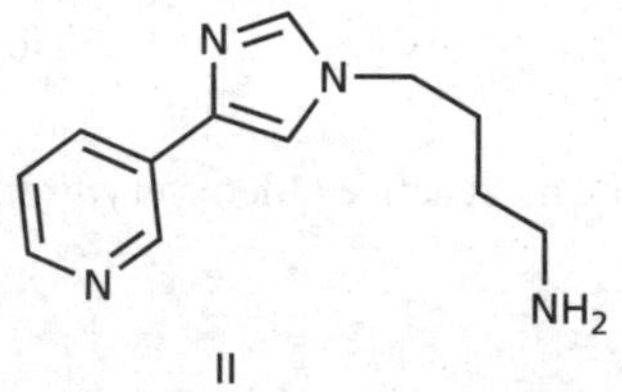

2. Propose a mechanism by which telithromycin is formed in the reaction shown in Figure 13.8.

FURTHER READING

Alekseyev, V.Y., et al. (2007) 'Solution structure and proposed domain–domain recognition interface of an acyl carrier protein domain from a modular polyketide synthase', *Protein Science*, **16**, 2093–107.

Bedford, D., et al. (1996) 'A functional chimeric modular polyketide synthase generated via domain replacement', *Chemistry and Biology*, **3**, 827–31.

Cane, D.E., Walsh, C.T., and Khosla, C. (1998) 'Harnessing the biosynthetic code: combinations, permutations and mutations', *Science*, **282**, 63–8.

Donadio, S., et al. (1991) 'Modular organization of genes required for complex polyketide biosynthesis', *Science*, **252**, 675–9.

Donadio, S., et al. (1993) 'An erythromycin analog produced by reprogramming of polyketide synthase', *Proceedings of the National Academy of Sciences of the USA*, **90**, 7119–23.

Harvey, C.J.B., et al. (2012) 'Precursor directed biosynthesis of an orthogonally functional erythromycin analogue: selectivity in the ribosome macrolide binding pocket', *Journal of the American Chemical Society*, **134**, 12 259–65.

Kapur, A., et al. (2010) 'Molecular recognition between ketosynthase and acyl carrier protein domains of the 6-deoxyerythronolide B synthase', *Proceedings of the National Academy of Sciences of the USA*, **107**, 22 066–71.

McDaniel, R., et al. (1999) 'Multiple genetic modifications of the erythromycin polyketide synthase to produce a library of novel 'unnatural' natural products', *Proceedings National Academy of Science*, **96**, 1846–51.

McDaniel, R., Welch, M., and Hutchinson, C.R. (2005) 'Genetic approaches to polyketide antibiotics: 1', *Chemical Reviews*, **105**, 543–58.

Marsden, F.A., et al. (1998) 'Engineering broader specificity into an antibiotic-producing polyketide synthase', *Science*, **279**, 199–202.

Oliynyk, M., et al. (1996) 'A hybrid modular polyketide synthase obtained by domain swapping', *Chemistry and Biology*, **3**, 833–9.

Ruan, X., et al. (1997) 'Acyltransferase domain substitutions in erythromycin polyketide synthase yield novel erythromycin derivatives', *Journal of Bacteriology*, **179**, 6416–25.

Stassi, D.L., et al. (1998) 'Ethyl-substituted erythromycin derivatives produced by directed metabolic engineering', *Proceedings of the National Academy of Sciences of the USA*, **95**, 7305–9.

Wong, F.T., et al. (2010) 'Protein–protein recognition between acyltransferases and acylcarrier proteins in multimodular polyketide synthases', *Biochemistry*, **49**, 95–102.

Woodward, R.B., et al. (1981) 'Asymmetric total synthesis of erythromycin. 3: Total synthesis of erythromycin', *Journal of the American Chemical Society*, **103**, 3215–17.

Wu, N., et al. (2001) 'Assessing the balance between protein–protein interactions and enzyme–substrate interactions in the channeling of intermediates between polyketide synthase modules', *Journal of the American Chemical Society*, **123**, 6465–74.

Xue, Q., et al. (1999) 'A multiplasmid approach to preparing large libraries of polyketides', *Proceedings of the National Academy of Sciences of the USA*, **96**, 11 740–5.

Quinolones and fluoroquinolones

14.1 Introduction

The quinolone and fluoroquinolone antibacterial agents are fully synthetic structures that are particularly useful in the treatment of urinary tract infections, as well as infections which prove resistant to the more established antibacterial agents.

Nalidixic acid (Fig. 14.1) was the first therapeutically useful agent in this class of compounds, and was synthesized in 1962. Various analogues were synthesized but offered no great advantage. However, a breakthrough was made in the 1980s with the development of **enoxacin** (Fig. 14.1), which showed improved broad-spectrum activity. The development of enoxacin was based on the discovery that a single fluorine atom at position 6 greatly increased both activity and cellular uptake. A basic substituent such as a piperazinyl ring at position 7 was also beneficial for a variety of pharmacokinetic reasons because of the ability of the basic substituent to form a zwitterion with the carboxylic acid group at position 3.

The introduction of a cyclopropyl substituent at position 1 further increased broad-spectrum activity, while replacement of the nitrogen at position 8 with carbon reduced adverse reactions and increased activity against *Staphylococcus aureus*. This led to **ciprofloxacin** (Fig. 14.1), the most active of the fluoroquinolones against Gram-negative bacteria.

14.2 Mechanism of action

The **quinolones** and **fluoroquinolones** inhibit the replication and transcription of bacterial DNA by stabilizing the complex formed between DNA and bacterial topoisomerases.

14.2.1 Function of topoisomerases

DNA is an extremely long molecule which is coiled into a more compact three-dimensional shape such that it can fit into the nucleus—a process known as **supercoiling**. This process requires the action of a family of enzymes called **topoisomerases**, which can catalyse the seemingly impossible act of passing one stretch of DNA helix across another stretch. They do this by cleaving one or both strands of the DNA helix to create a temporary gap, and then resealing the strand(s) once the crossover has taken place. Supercoiling allows the efficient storage of DNA, but the DNA has to be uncoiled again if replication and transcription are to take place. If uncoiling did not take place, the unwinding process that occurs during replication and transcription would lead to increased tension due to increased supercoiling of the remaining DNA double helix. You can demonstrate the principle of this by pulling apart the strands of a rope. The same topoisomerase enzymes are responsible for catalysing

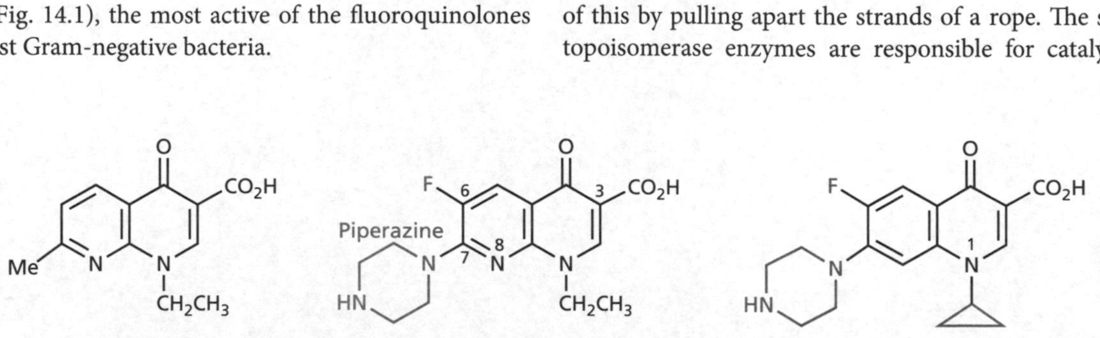

FIGURE 14.1 Quinolones and fluoroquinolones.

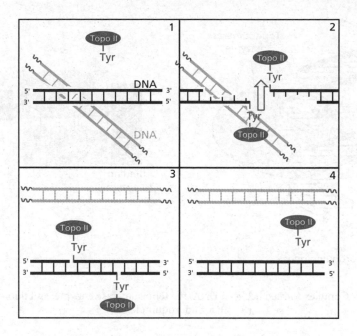

FIGURE 14.2 Method by which topoisomerase II catalyses the crossover of DNA strands. Note that the same enzyme bonds covalently to each DNA strand.

the uncoiling process, and so inhibition of these enzymes would effectively block transcription and replication.

Topoisomerase II is a mammalian enzyme that is crucial to the effective replication of DNA. The enzyme binds to parts of DNA where two regions of the double helix are in near proximity (Fig. 14.2). The enzyme binds to one of these DNA double helices, and tyrosine residues are used to nick both strands of the DNA (Fig. 14.3). This results in a temporary covalent bond between the enzyme and the resulting 5′ end of each strand, thus stabilizing the DNA. The strands are now pulled in opposite directions to form a gap through which the intact DNA region can be passed. The enzyme then reseals the strands and departs.

Topoisomerase I is similar to topoisomerase II in that it relieves the torsional stress of supercoiled DNA during replication, transcription, and the repair of DNA. The difference is that it cleaves only one strand of DNA, whereas topoisomerase II cleaves both strands. The enzyme catalyses a reversible transesterification reaction similar to that shown in Figure 14.3, but where the tyrosine residue of the enzyme is linked to the 3′ phosphate end of the DNA strand rather than the 5′ end. This creates a 'cleavable complex' with a single-strand break. Relaxation of torsional strain takes place either by allowing the intact strand to pass through the nick or by free rotation of the DNA about the uncleaved strand. Once the

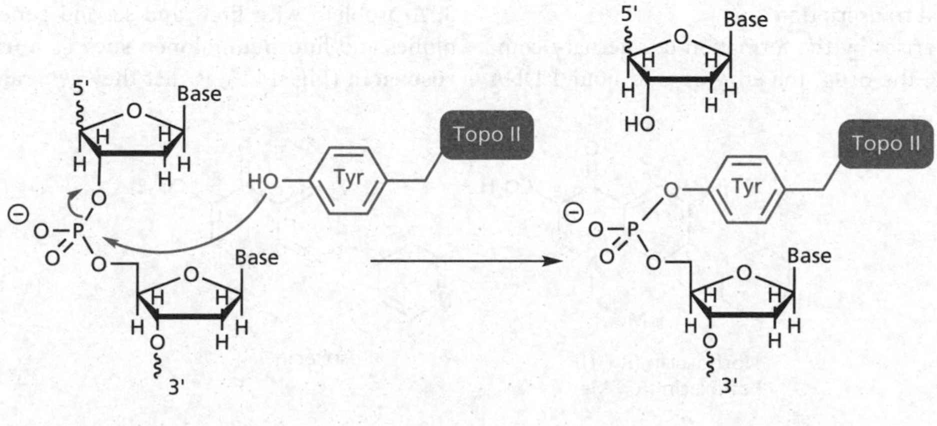

FIGURE 14.3 Mechanism by which topoisomerase II splits a DNA chain.

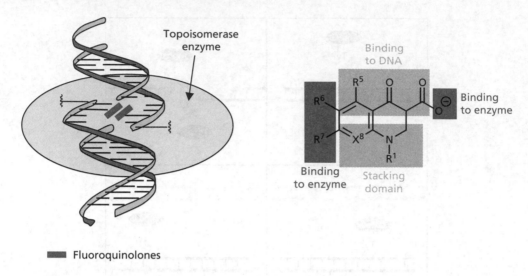

FIGURE 14.4 Complex formed between DNA, the topoisomerase enzyme, and fluoroquinolones. (R^6 = F for fluoroquinolones.)

torsional strain has been relieved, the enzyme rejoins the cleaved strand of DNA and departs.

Topoisomerase IV is a bacterial enzyme that carries out the same process as the mammalian enzyme topoisomerase II, and this is the target for the fluoroquinolone antibacterial agents (section 14.2.2).

14.2.2 Mechanism of inhibition by fluoroquinolones

In Gram-positive bacteria, fluoroquinolones bind to and stabilize the DNA–topoisomerase IV complex, with the drugs showing a 1000-fold selectivity for the bacterial enzyme over the corresponding enzyme in human cells. In Gram-negative bacteria, the main target for fluoroquinolones is the complex between DNA and a topoisomerase II enzyme called **DNA gyrase**. It has the same role as topoisomerase IV in reverse and is required when the DNA double helix is being supercoiled after replication and transcription.

Inhibition arises by the formation of a ternary complex involving the drug, the enzyme, and bound DNA

(Fig. 14.4). The binding site for the fluoroquinolones only appears once the enzyme has 'nicked' the DNA strands, and the strands are ready to be crossed over. At that point, four fluoroquinolone molecules are bound in a stacking arrangement such that their aromatic rings are coplanar. The carbonyl and carboxylate groups of the fluoroquinolones interact with DNA by hydrogen bonding, while the fluoro-substituent at position 6, the substituent at C-7, and the carboxylate ion are involved in binding interactions with the enzyme.

14.3 Properties and SAR

A large number of quinolones and fluoroquinolones have now been synthesized. Those agents having good activity all have a similar bicyclic ring system which includes a pyridone ring and a carboxylic acid at position 3. A problem with first- and second-generation quinolones and fluoroquinolones, such as **norfloxacin** and **rosoxacin** (Fig. 14.5), is that they generally show only

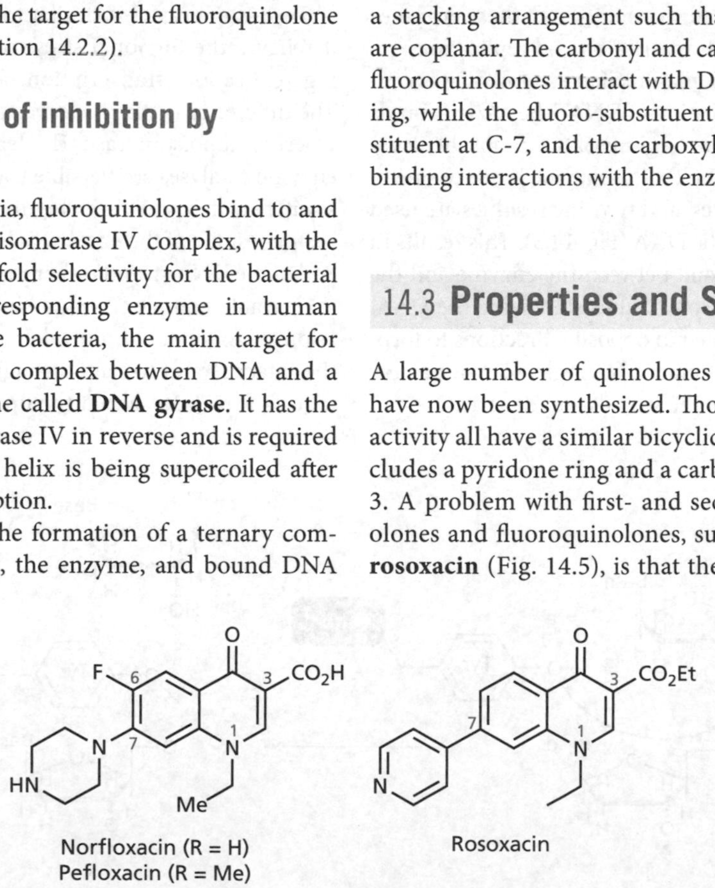

Norfloxacin (R = H)
Pefloxacin (R = Me)

Rosoxacin

FIGURE 14.5 First- and second-generation quinolones and fluoroquinolones.

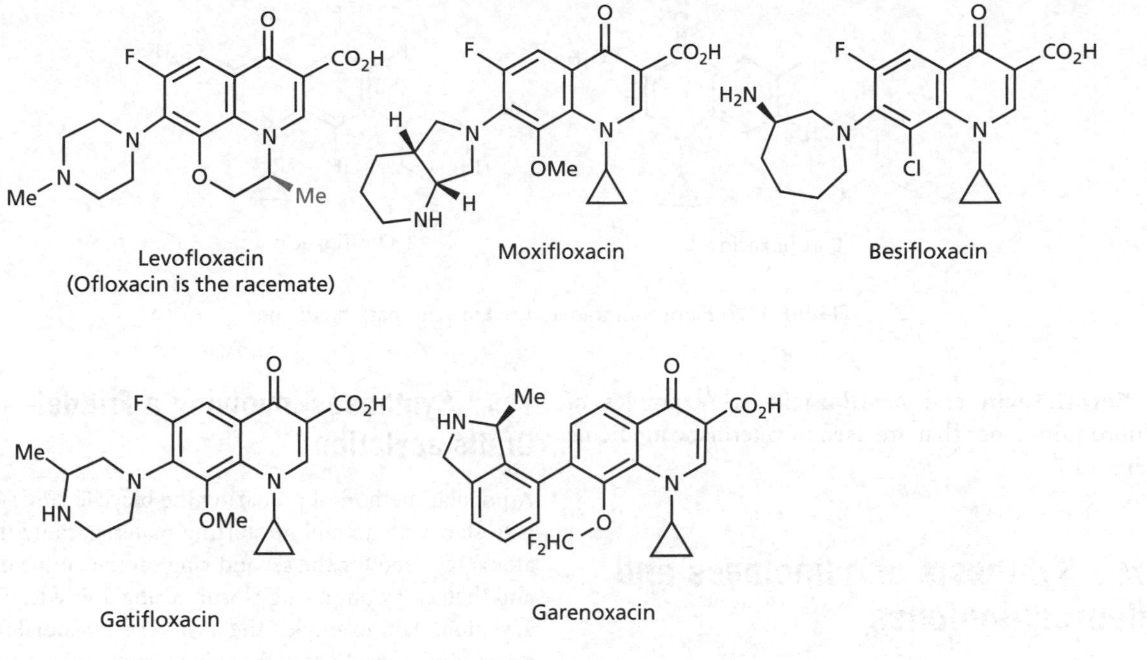

Levofloxacin
(Ofloxacin is the racemate)

Moxifloxacin

Besifloxacin

Gatifloxacin

Garenoxacin

FIGURE 14.6 Third- and fourth-generation fluoroquinolones.

moderate activity against *S.aureus*, followed by rapidly developing drug resistance. Furthermore, only marginal activity is shown against anaerobes and *Streptococcus pneumoniae*. Third- and fourth-generation fluoroquinolones such as **ofloxacin**, **levofloxacin**, **moxifloxacin**, **besifloxacin**, **garenoxacin**, and **gatifloxacin** (Fig. 14.6) began to be developed in the early 1990s to tackle these issues. Ofloxacin has an asymmetric centre and is sold as a racemic mixture of both enantiomers, one of which is active and one of which is not. Levofloxacin is the active enantiomer of oflaxacin and is twice as active as the racemate.

14.4 Clinical aspects of quinolones and fluoroquinolones

Nalidixic acid is active against Gram-negative bacteria and is useful in the short-term therapy of uncomplicated urinary tract infections. It can be taken orally but, unfortunately, bacteria can develop a rapid resistance to it. **Enoxacin** has a greatly increased spectrum of activity against Gram-negative and Gram-positive bacteria. It also shows improved oral absorption, tissue distribution, and metabolic stability, as well as an improvement in the level and spectrum of activity, particularly against Gram-negative bacteria such as *Pseudomonas aeruginosa*. **Ciprofloxacin** is used in the treatment of a large range of infections involving the urinary, respiratory, and

gastrointestinal tracts (e.g. travellers' diarrhoea), as well as infections of skin, bone, and joints. It is also used for gonorrhoea and septicaemia, and as part of a cocktail of drugs for anthrax. It has been claimed that ciprofloxacin may be the most active broad-spectrum antibacterial agent on the market.

In contrast with nalidixic acid, resistance to the fluoroquinolones is slow to appear, but when it does appear, it is mainly due to efflux mechanisms which pump the drug back out of the cell. Less common resistance mechanisms include mutations to the topoisomerase enzymes which reduce their affinity to the agents, and alteration of porins in the outer membrane of Gram-negative organisms which limit access of the agents into the cell.

Third-generation fluoroquinolones show improved activity against *S.pneumoniae*, while maintaining activity against enterobacteria. **Ofloxacin** is administered orally or by intravenous infusion to treat septicaemia, gonorrhoea, and infections of the urinary tract, lower respiratory tract, skin, and soft tissue. **Levofloxacin** has a greater activity against pneumococci than ciprofloxacin and is a second-line treatment for community-acquired pneumonia. It is also used for acute sinusitis, chronic bronchitis, urinary tract infections, skin infections, and soft tissue infections. **Moxifloxacin** also has greater activity against pneumococci than ciprofloxacin. It is used to treat sinusitis and is a second line treatment for community-acquired pneumonia. **Besifloxacin** is a fourth-generation fluoroquinolone which was approved in 2009.

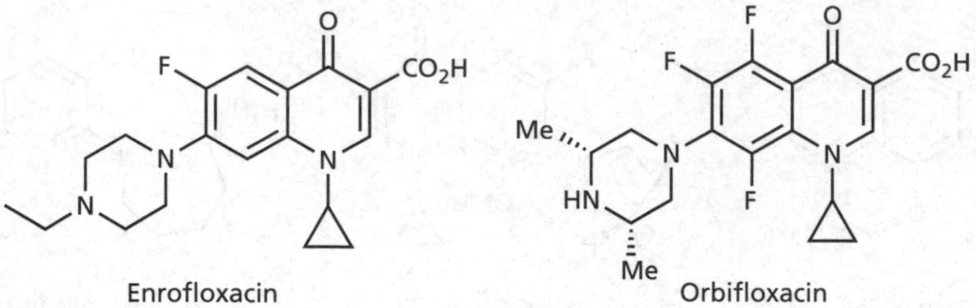

FIGURE 14.7 Fluoroquinolones used in veterinary medicine.

Enrofloxacin and **orbifloxacin** are examples of fluoroquinolones that are used in veterinary medicine (Fig. 14.7).

14.5 Synthesis of quinolones and fluoroquinolones

14.5.1 Introduction

The bicyclic ring system present in fluoroquinolones has been synthesized in a number of ways. A retrosynthetic analysis of the bicyclic ring system suggests two possible disconnections between the aromatic ring and the piperidinone ring (Fig. 14.8). One is between the aromatic ring and the ketone group. The reverse reaction of this disconnection would correspond to a Friedel–Crafts acylation. The second is between the aromatic ring and the nitrogen atom. The corresponding reaction is a nucleophilic substitution of an aryl halide with an amine. This is feasible if the aromatic ring contains electron-deficient substituents.

The syntheses used to prepare fluoroquinolones illustrate both these approaches.

14.5.2 Syntheses involving a Friedel–Crafts acylation

A popular method of preparing the bicyclic ring system is to start with an aniline starting material, build up the atoms required for the second ring on the amine group, and then carry out a ring closure using a Friedel–Crafts acylation. For example, the synthesis of **norfloxacin** starts from a disubstituted aniline structure that already contains the fluoro substituent required for position 6 of the final structure (Fig. 14.9). The chloro substituent is not in the final structure but it is crucial to the synthesis because it will eventually be substituted with the amine that is required at position 7.

The first stage in the sequence is to react the aniline starting material with diethyl 2-(ethoxymethylene)malonate, which provides the atoms required for the second ring. The reaction corresponds to a nucleophilic substitution of the vinyl ethoxy group. A Friedel–Crafts acylation can then be carried out in the presence of Dowtherm to create the required bicyclic ring system. The remaining ester substituent is hydrolysed to the carboxylic acid required at position 3, and then the nitrogen at position 1 is alkylated with iodoethane. The final stage is to substitute the chloro substituent at position 7 with piperazine. An excess of

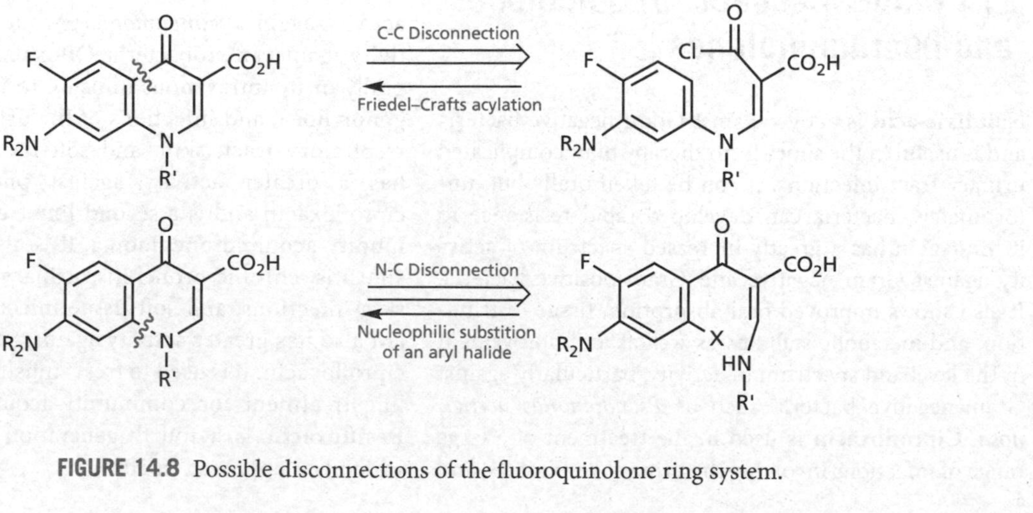

FIGURE 14.8 Possible disconnections of the fluoroquinolone ring system.

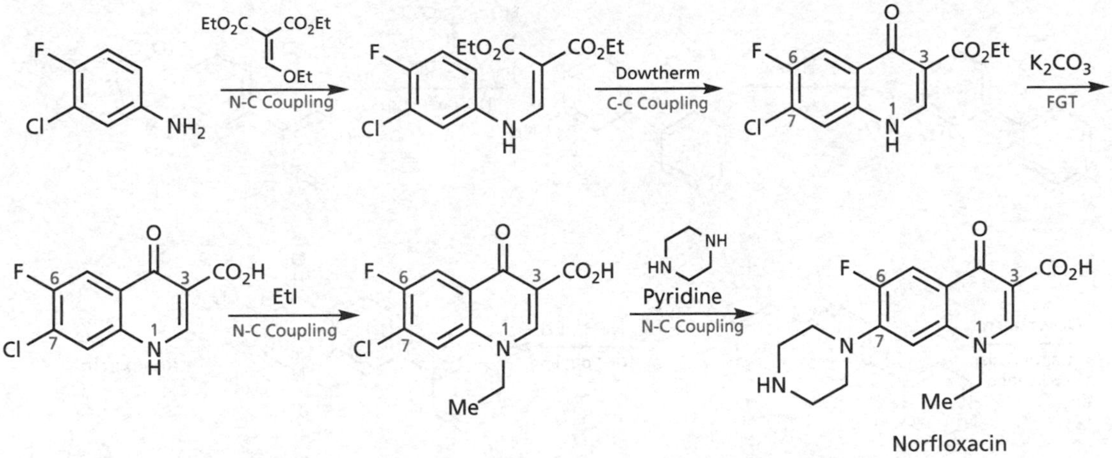

FIGURE 14.9 Synthesis of norfloxacin.

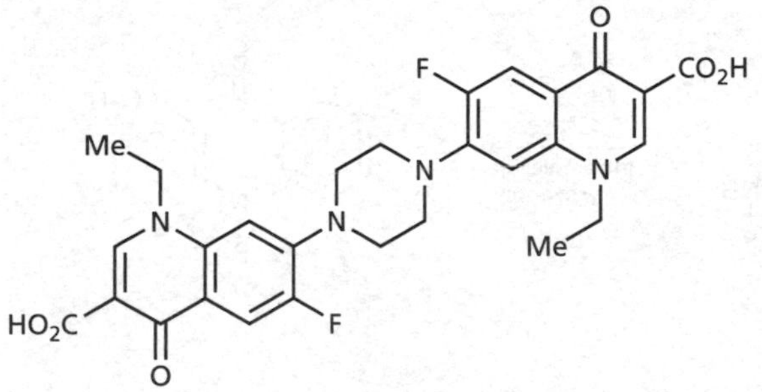

FIGURE 14.10 Possible side product from the final reaction step.

piperazine is used to ensure that the piperazine ring only reacts once. Otherwise, there is a risk of obtaining a di-substituted piperazine side product (Fig. 14.10).

Norfloxacin is a first-generation fluoroquinolone which was discovered in 1980 and is occasionally used for the treatment of urinary tract infections.

The same starting material as used in Figure 14.9 can be used to prepare other fluoroquinolone analogues with different substituents at positions 1 and 7. For example, **pefloxacin** (Fig. 14.11) is a third-generation fluoro-quinolone which has been used as a last resort for the

treatment of serious or life-threatening infections. If substituents are wanted at positions 5 and 8, then the same synthesis can be used but different starting materials will be required.

Quinolone structures lacking the fluoro substituent at position 6 can also be synthesized in this manner if a suitable starting material is used. For example, **rosoxacin** was synthesized from a biaryl starting material containing a nitro substituent (Fig. 14.12). The nitro group was first reduced to an amine and then the bicyclic ring system was prepared in the conventional manner, before alkylating at position 1. The substituent at position 7 is a pyridine ring, which was present in the original starting material.

14.5.3 Syntheses involving nucleophilic substitution of an aryl halide

A different approach to the synthesis of fluoroquinolones is to start with a substituted benzoic acid or benzoyl chloride and to use the carbonyl group to build up the atoms required for the second ring (Fig. 14.13). In other words, the second ring is grown in the 'opposite direction' to the synthesis described in section 14.5.2. The starting material must have a halo substituent (usually

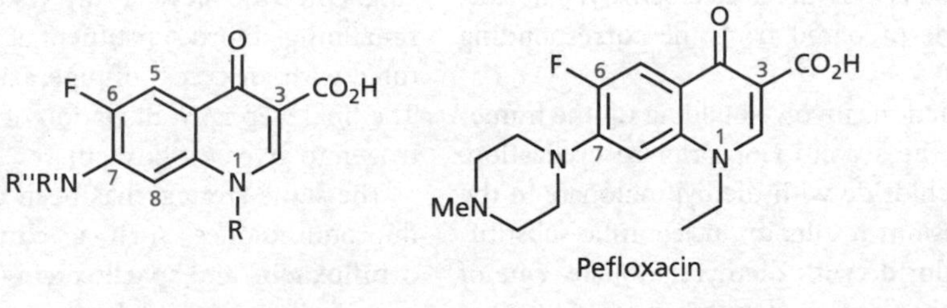

Pefloxacin

FIGURE 14.11 Types of analogue that are possible from the synthesis described in the text.

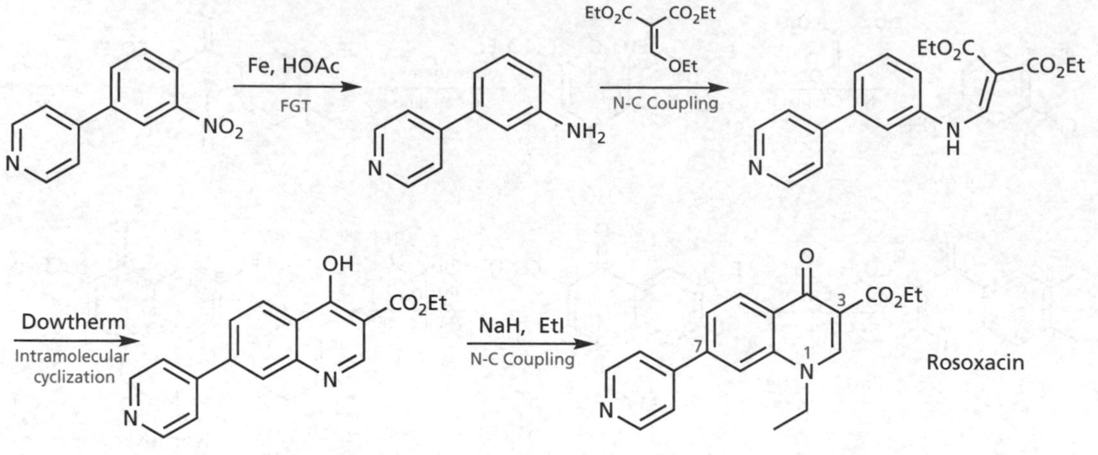

FIGURE 14.12 Synthesis of rosoxacin.

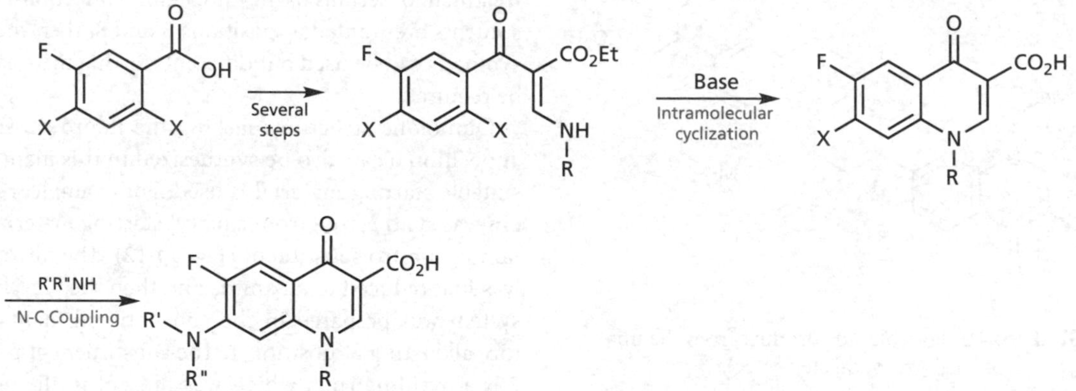

FIGURE 14.13 General approach used to synthesize fluoroquinolones from a benzoic acid starting material (X = halogen).

fluorine or chlorine) that is *ortho* to the carboxylic acid to allow an intramolecular cyclization by means of a nucleophilic substitution. Another halogen positioned at the *para* position to the carboxylic acid in the starting material will end up at position 7 of the bicyclic ring system. This will allow an amine substituent to be incorporated at a later stage.

An example of this approach is illustrated by the synthesis of **ciprofloxacin** and **enrofloxacin** (Fig. 14.14). The starting material is a trisubstituted benzoyl chloride, which can easily be prepared from the corresponding benzoic acid.

The first four reactions involve building up the framework required for the second ring prior to cyclization. Reacting the acid chloride with diethyl malonate in the presence of magnesium results in nucleophilic substitution of the acid chloride with diethyl malonate. One of the ester groups is then removed in the presence of *para*-toluenesulphonic acid (ptsa). The resulting keto ester is

then treated with acetic anhydride and triethyl orthoformate to create an ethoxy substituted alkene. The ethoxy substituent is then substituted with an amine to provide the framework needed for the subsequent intramolecular cyclization. Note that the final reaction in this sequence introduces the substituent that will be present at position 1 of the final structure.

The cyclization is now carried out such that the amine group substitutes the chloro substituent on the aromatic ring. Once the bicyclic ring system has been formed, the remaining chloro substituent at position 7 can be substituted with an excess of piperazine to give ciprofloxacin. The final step is *N*-alkylation of the other piperazine nitrogen to give enrofloxacin.

The same strategy has been used to synthesize other fluoroquinolones, such as **clinafloxacin, gatifloxacin, orbifloxacin**, and **sparfloxacin** (Fig. 14.15).

A similar approach has been used to synthesize quinolones such as **garenoxacin** which represents a new

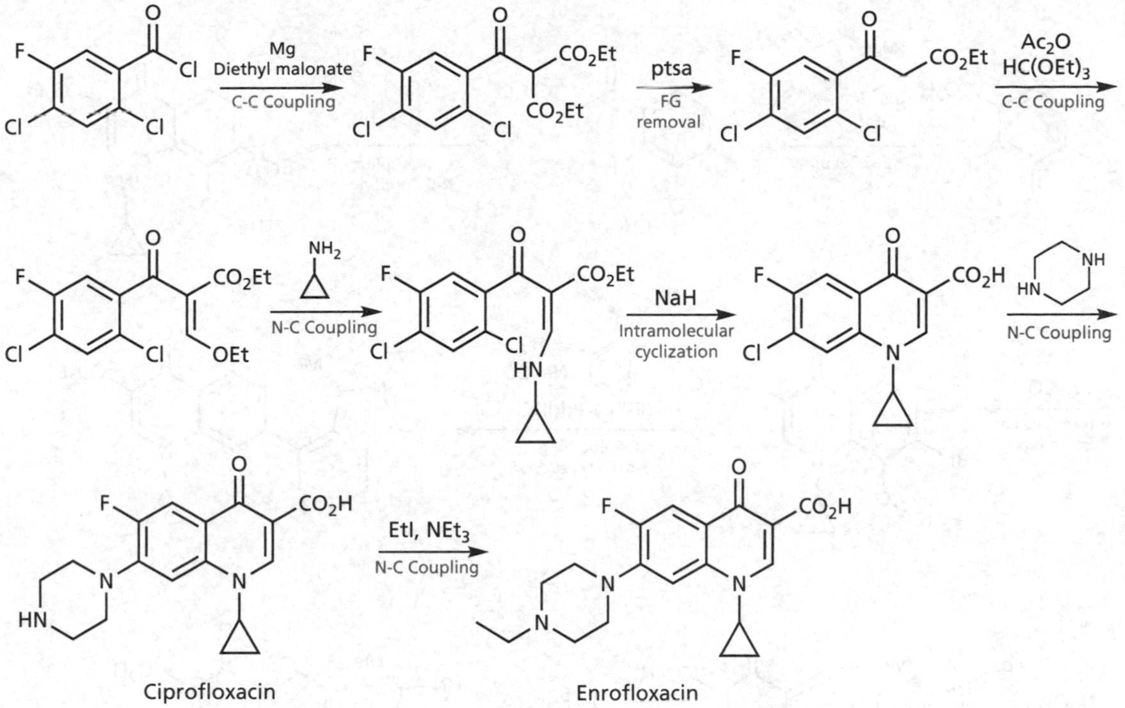

FIGURE 14.14 Synthesis of enrofloxacin.

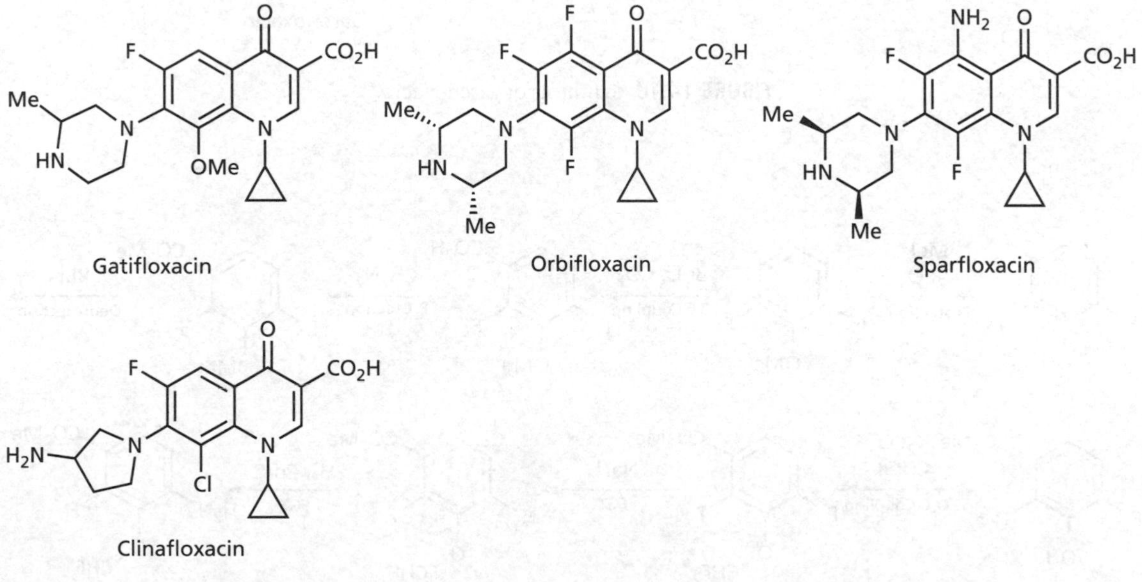

FIGURE 14.15 Gatifloxacin, orbifloxacin, sparfloxacin, and clinafloxacin.

generation of quinolone agents (Fig. 14.16). Note that the introduction of the substituent at position 7 involves a Suzuki coupling reaction between an aryl boronic acid and the aryl bromide (see also Case Study 2, section CS2.3).

The starting material required for the synthesis of garenoxacin was not commercially available and had to

be synthesized from a disubstituted phenol (Fig. 14.17). This illustrates that the number of steps required in a drug synthesis can expand quite dramatically if it is decided to incorporate unusual substituents, such as the $OCHF_2$ substituent at position 8. This is not the sort of analogue one would plan to synthesize when using a synthetic route for the first time, but it may well be

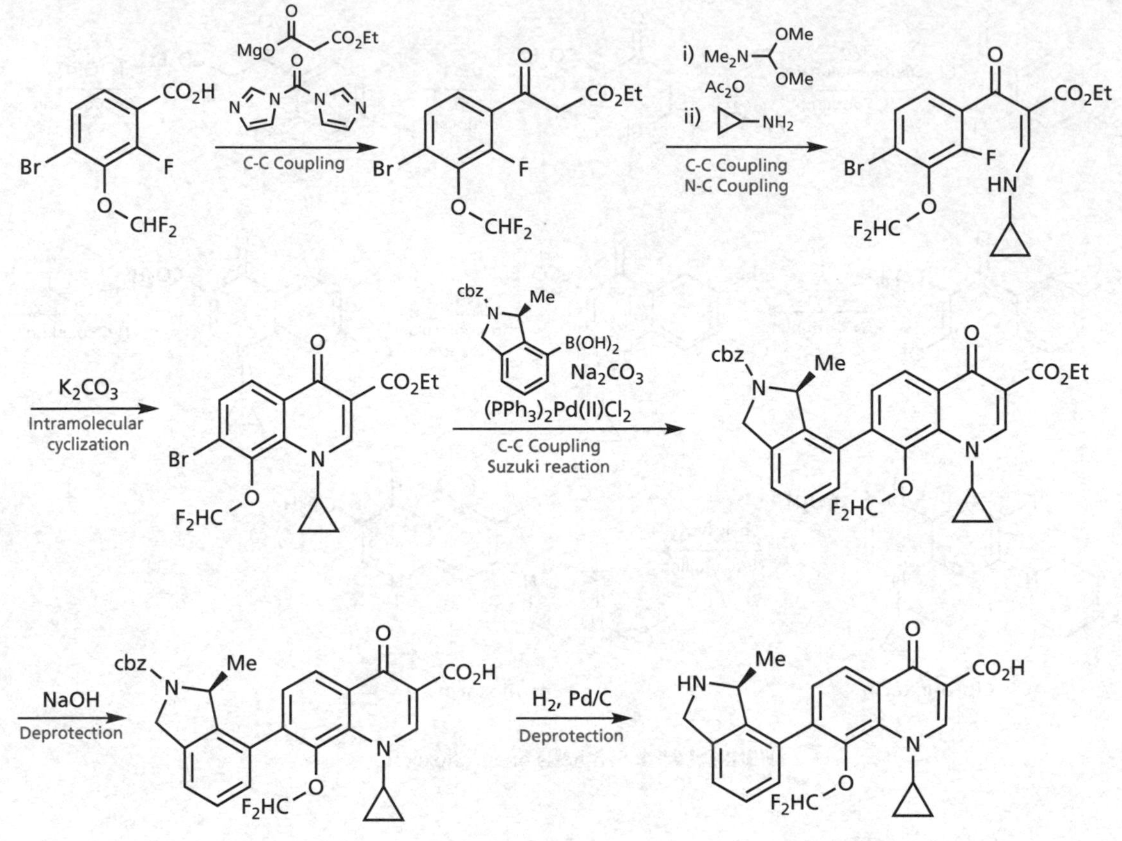

Garenoxacin

FIGURE 14.16 Synthesis of garenoxacin.

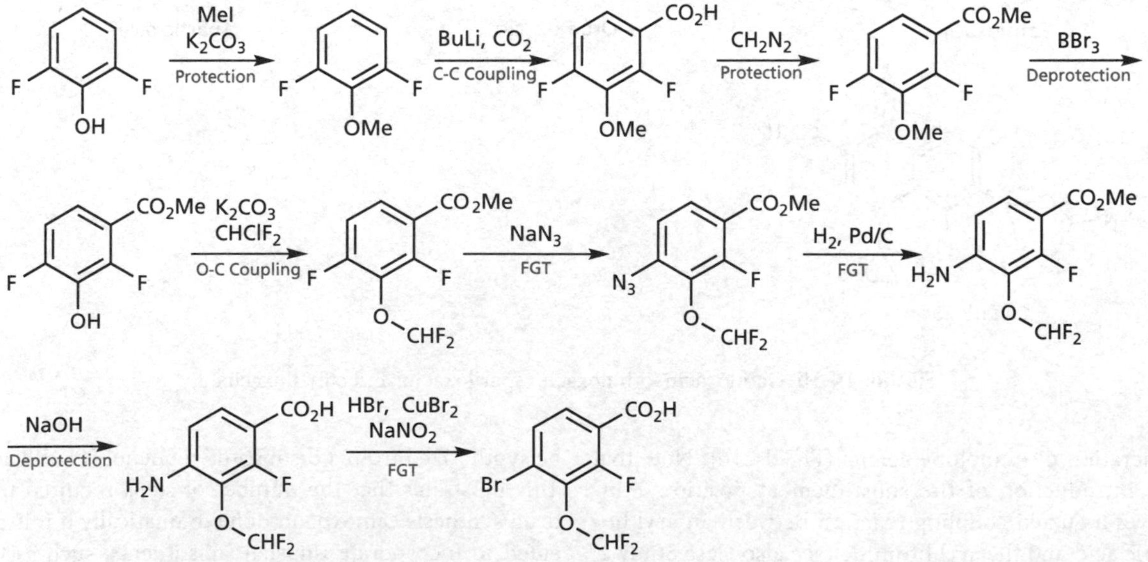

FIGURE 14.17 Synthesis of the starting material used for the synthesis of garenoxacin.

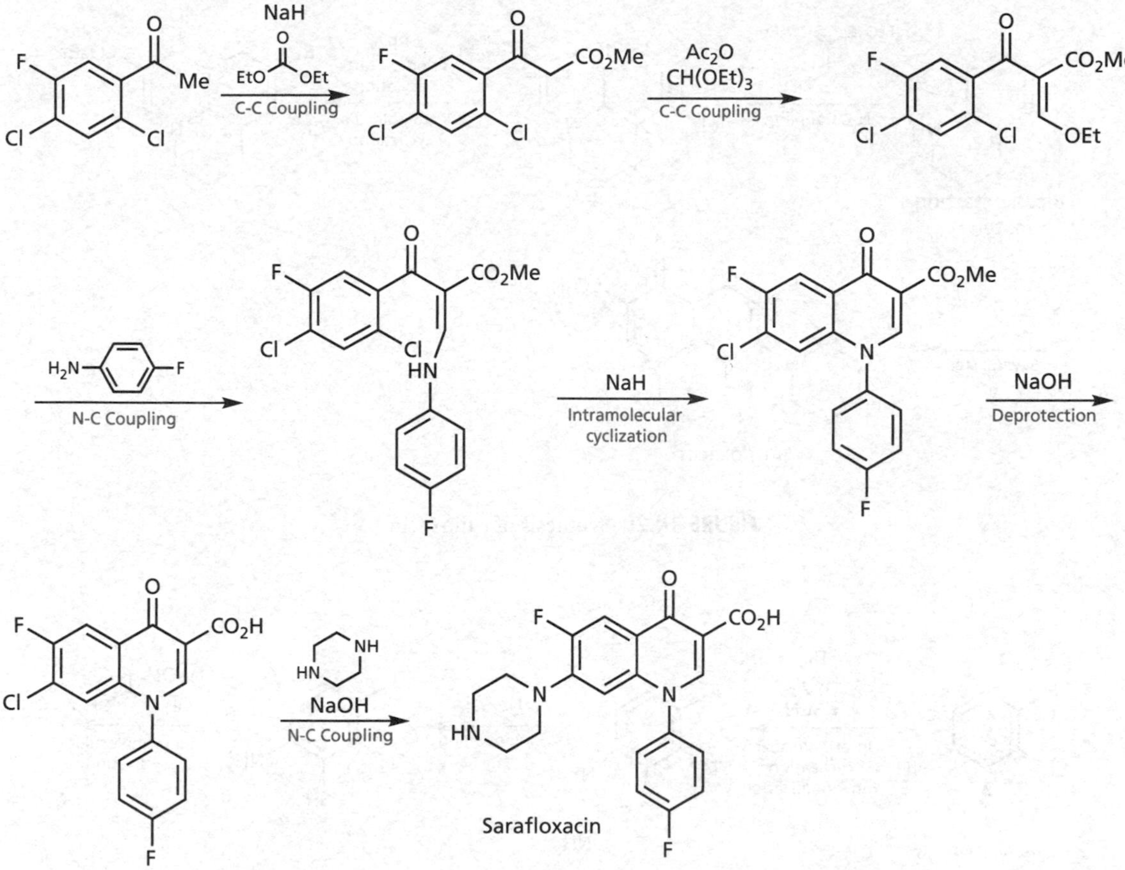

FIGURE 14.18 Synthesis of sarafloxacin.

one that becomes desirable at a later stage after having studied the structure–activity relationships of the initial analogues.

A variation of this approach is to start from an aceto-phenone structure—one containing a methyl ketone sub-stituent. This is illustrated in the synthesis of **sarafloxacin** (Fig. 14.18), which has now been discontinued. The pro-tons on the methyl group are slightly acidic as they are next to a carbonyl group. Therefore, treatment with base produces a carbanion, which can then be reacted with diethyl carbonate to give the required keto ester. The rest of the synthesis is similar to the previous examples.

Temafloxacin (Fig. 14.19) has been synthesized in a similar fashion. The drug was approved in 1992, but was withdrawn in the same year because of serious side effects.

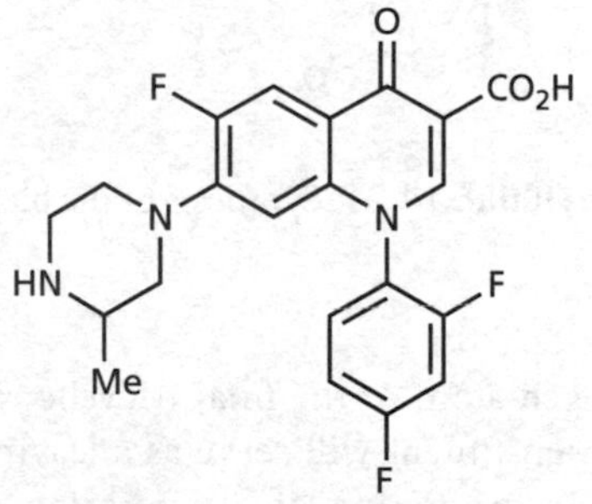

14.5.4 Tricyclic ring systems incorporating a quinolone ring system

A number of tricyclic ring systems which include the fluoroquinolone ring system have been prepared. As one would expect, the synthesis of these is more

FIGURE 14.19 Temafloxacin.

complex. One approach is to create the fluoroquin-olone moiety from a bicyclic starting material using the Friedel–Crafts acylation approach described in section 14.5.2. This is illustrated in the synthesis of **ru-floxacin** (Fig. 14.20).

The bicyclic structure required for the above synthesis was not commercially available and had to be synthe-sized from 3-chloro-4-fluoroaniline using a four-step process (Fig. 14.21). The starting material contains the fluoro substituent required in the final structure, as well

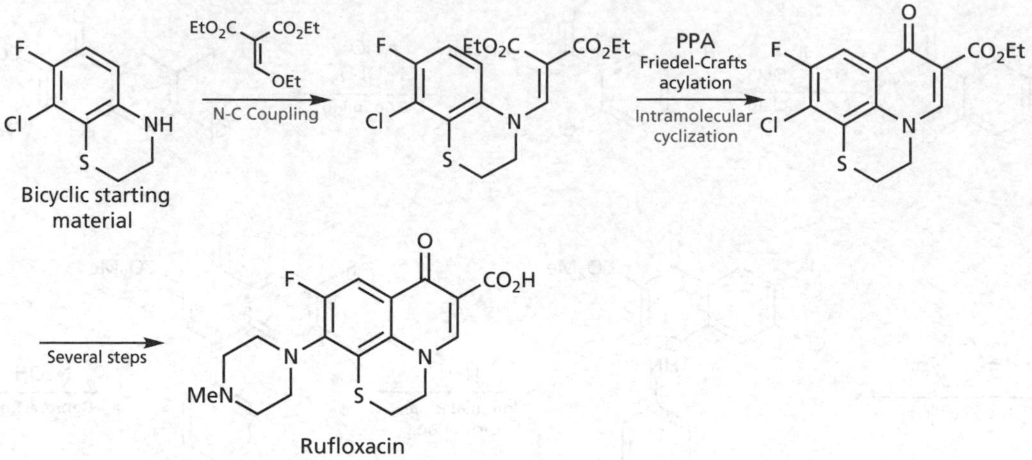

FIGURE 14.20 Synthesis of rufloxacin.

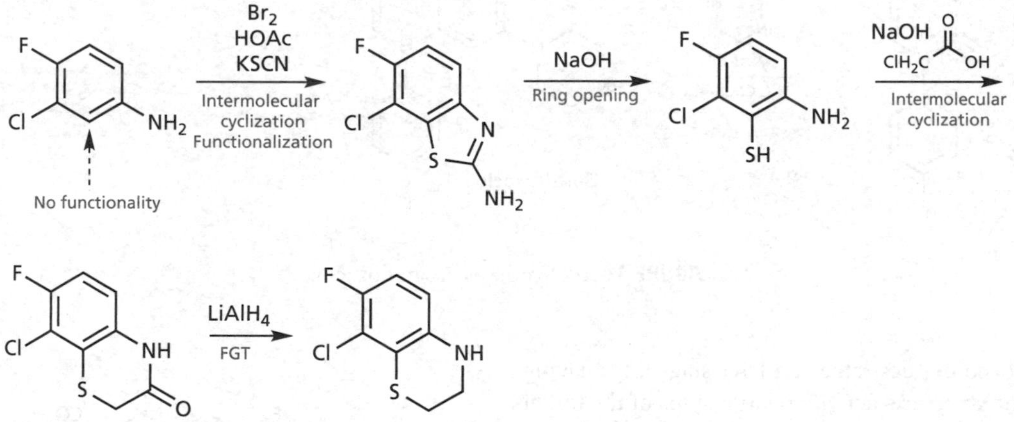

FIGURE 14.21 Synthesis of the bicyclic starting material required for the synthesis of rufloxacin.

as the nitrogen atom of the final tricyclic ring system. The chloro substituent will serve as a leaving group to allow later incorporation of a piperazine ring. However, the starting material lacks the sulphur substituent that will provide the sulphur atom required in the final product. Moreover, the relevant position is unsubstituted. Therefore, the first priority is to functionalize that part of the aromatic ring and introduce a sulphur substituent. This was achieved in a two-stage process, which involved an intermolecular cyclization followed by ring opening under basic conditions. Another intermolecular cyclization was then carried out involving the thiol and amino groups acting as nucleophiles with a dual-acting electrophile. The carbonyl group that resulted from this cyclization was reduced with lithium aluminium hydride.

The synthesis of rufloxacin was carried out in the conventional manner. However, the sulphur atom had to be protected as a sulphoxide before the piperazine ring was introduced (Fig. 14.22).

A similar approach was used for the synthesis of the tricyclic system in **ofloxacin**. The synthesis of ofloxacin required the bicyclic ring system starting material shown in Figure 14.23.

The required bicyclic ring system was obtained from 2,3,4-trifluoronitrobenzene (Fig. 14.24). The fluoro substituent that was *ortho* to the nitro group was substituted with a hydroxyl group in an aromatic nucleophilic substitution. This is aided by the presence of three other electron-withdrawing substituents. The resulting phenol group was alkylated, and then the nitro group was reduced with Raney

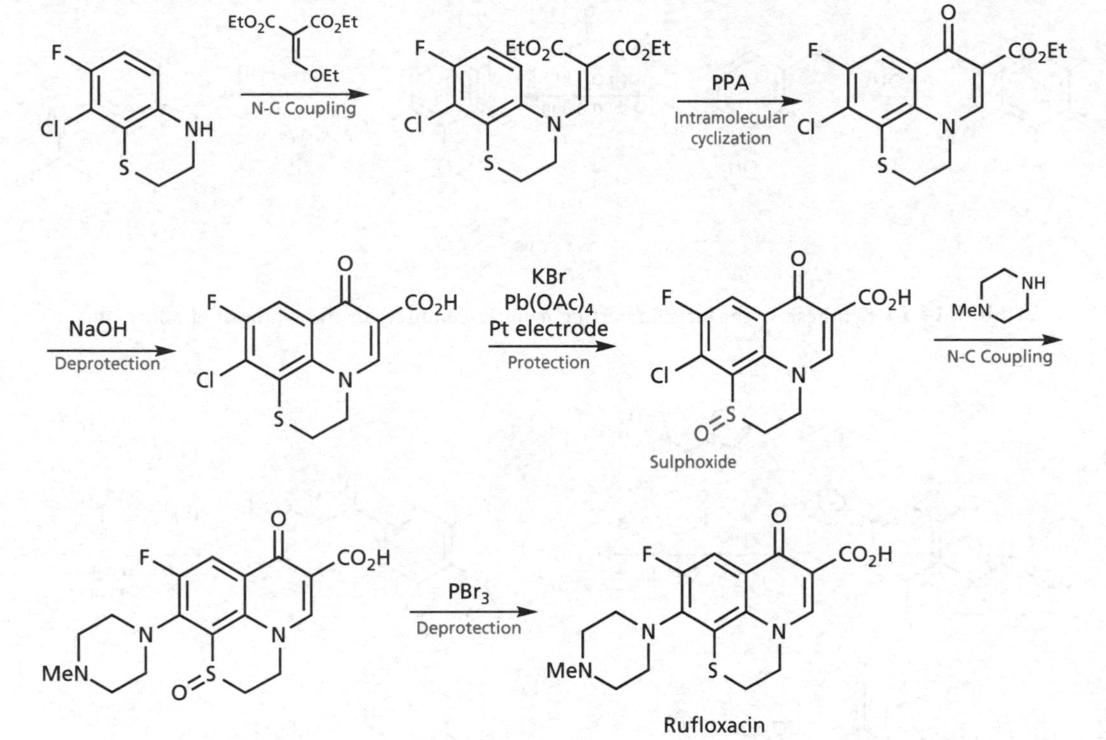

FIGURE 14.22 Full synthesis of the rufloxacin from the bicyclic starting material.

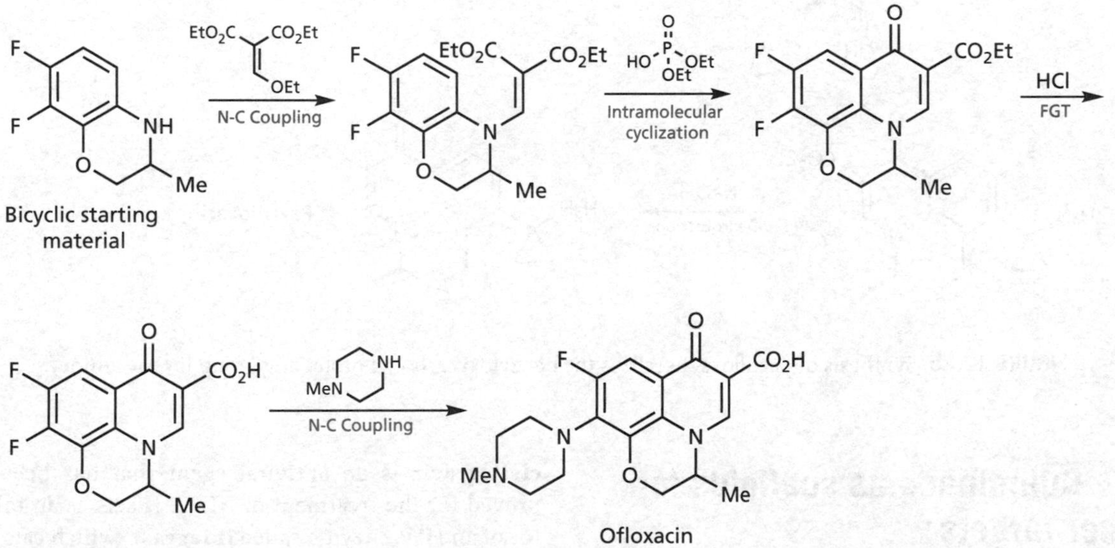

FIGURE 14.23 Synthesis of ofloxacin.

nickel to give a primary amine. Since a ketone group was present in the side chain, a spontaneous intramolecular cyclization took place to give the bicyclic ring system. This would normally have given an imine group (C=N), but this was reduced in the presence of Raney nickel.

The opposite strategy of starting from a benzoic acid structure is also feasible for the synthesis of tricyclic structures. This is illustrated in the synthesis of **pazufloxacin**, where two intramolecular cyclizations were carried out in one step in the presence of potassium carbonate (Fig. 14.25).

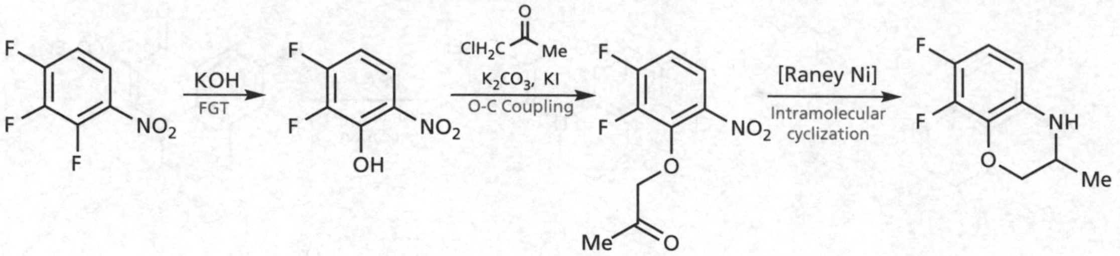

FIGURE 14.24 Synthesis of the bicyclic starting material required for the synthesis of ofloxacin.

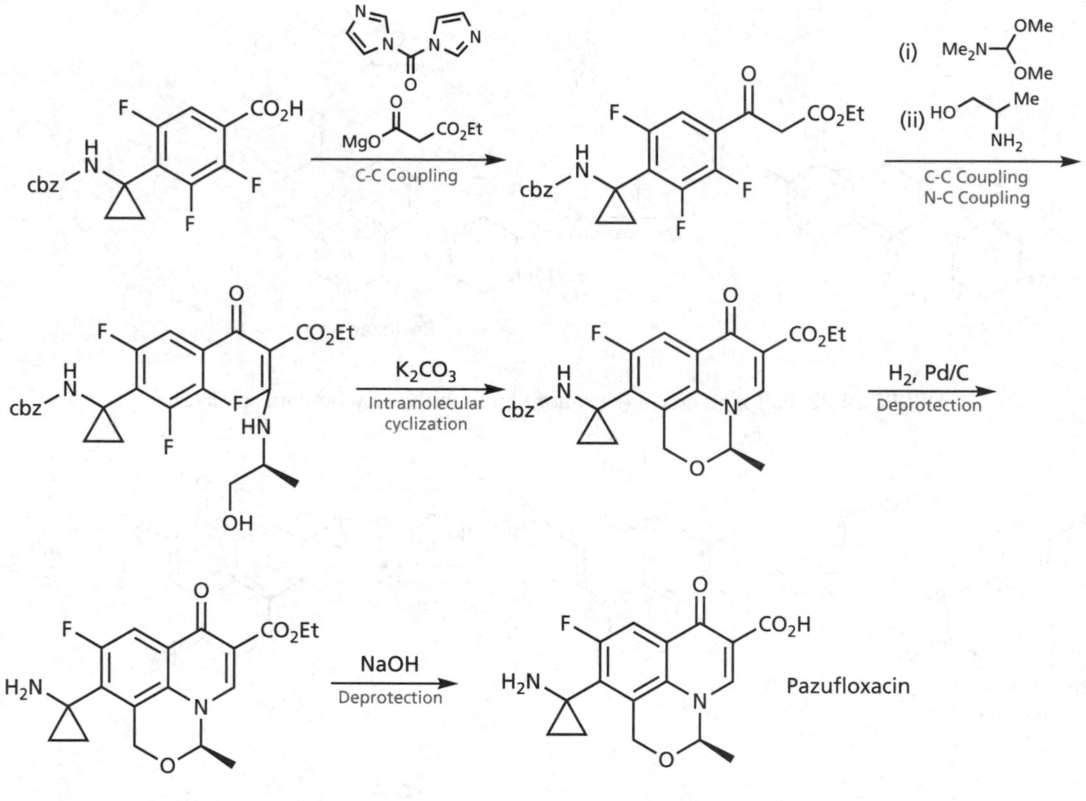

FIGURE 14.25 Synthesis of pazufloxacin (cbz is the benzyloxycarbonyl protecting group for the amine).

14.6 Quinolones as scaffolds for other targets

There is a risk of pigeon-holing particular structural classes of compound to specific fields of therapy. Quinolones and fluoroquinolones are best known for their antibacterial activity, but there is no reason why these systems cannot be used as scaffolds to create drugs that have different types of activity. Different substituents and substituent patterns around the ring system can result in quinolone or fluoroquinolone structures that can interact with different targets. For example,

elvitegravir is an antiviral agent that has been approved for the treatment of AIDS. It acts as an inhibitor of an HIV enzyme called **integrase,** which catalyses the splicing of viral DNA into host cell DNA. Inhibiting this process interrupts the viral lifecycle. There are many structural similarities between elvitegravir and the antibacterial fluoroquinolones, but there are significant differences in the substituents present at positions 1, 6, and 7. The structure has been synthesized from a benzoic acid starting material in the conventional manner with the key intermediates and cyclization reaction shown in Figure 14.26 (see also Case Study 2, section CS2.5).

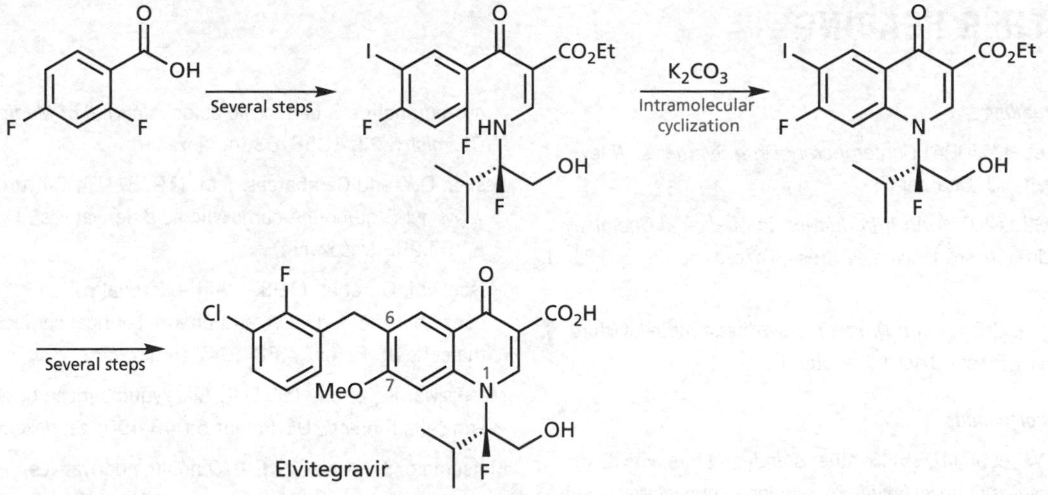

FIGURE 14.26 Synthesis of elvitegravir.

QUESTIONS

1. Show how the synthesis used to produce **enrofloxacin** can be modified to produce **gatifloxicin** instead. What is the added complication in this synthesis and how significant it is likely to be?

2. What starting material would be needed in order to synthesize **orbifloxacin** if you wished to use the same strategy involved in synthesizing **enrofloxacin**?

3. The final step of the synthesis leading to orbifloxacin involves an N–C coupling where a fluoro substituent is replaced at position 7. Why does the nucleophilic substitution occur selectively at this position and not at any of the other three positions bearing fluoro substituents?

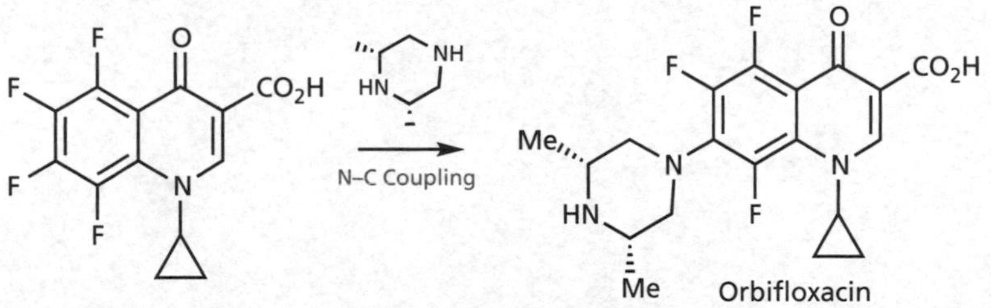

4. **Flumequine** is a first-generation fluoroquinolone which has now been discontinued, although it is still used in veterinary medicine. Propose a possible synthesis.

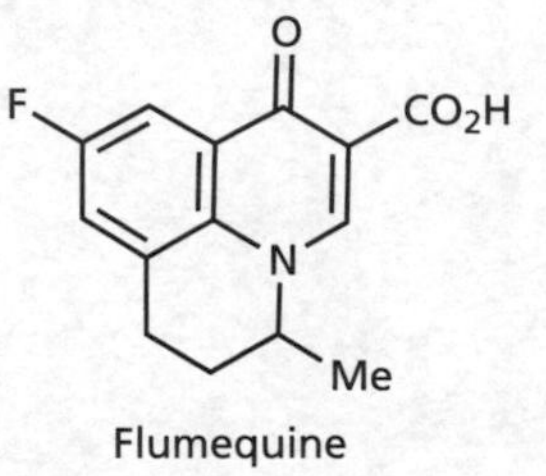

5. Propose mechanisms for the two cyclizations shown in Figure 14.21.

FURTHER READING

General reading

Li, J.-J., et al. (2004) *Contemporary Drug Synthesis.* Wiley, Hoboken, NJ, Chapter 7.

Patrick, G.L. (2013) *An Introduction to Medicinal Chemistry* (5th edn). Oxford University Press, Oxford, pp 74–6, 121–3, 457–9.

Saunders, J. (2000) *Top Drugs; top synthetic routes.* Oxford University Press, Oxford, Chapter 10.

Specific compounds

Chu, D.T.W., et al. (1985) 'Synthesis and structure-activity relationships of novel arylfluoroquinolone antibacterial agents', *Journal of Medicinal Chemistry*, **28**, 1558–64 (sarafloxacin).

Chu, D.T. (1988) 'Quinoline antibacterial compounds', US Patent 4,730,000 (temafloxacin).

Grohe, K., et al. (1987) '7-Amino-1-cyclopropyl-4-oxo-1, 4-dihydro-quinoline and naphthridine-3-carboxylic acids and antibacterial agents containing these compounds', US Patent 4,670,444 (ciprofloxacin and enrofloxacin).

Hayakawa, I., Hiramitsu, T., Tanaka, Y. (1984) 'Synthesis and antibacterial activities of substituted 7-oxo-2,3-dihydro-7H-pyrido[1,2,3-de][1,4]benzoxazine-6-carboxylic acids', *Chemical & Pharmaceutical Bulletin*, **32**, 4907–13 (oflaxacin).

Koga, H., et al. (1980) 'Structure-activity relationships of antibacterial 6,7- and 7,8-disubstituted 1-alkyl-1,4-dihydro-4-oxoquinoline-3-carboxylic acids', *Journal of Medicinal Chemistry*, **23**, 1358–63 (norfloxacin).

Lesher, G.Y. and Carabateas, P.M. (1973) '1,4-Dihydro-4-oxo-7-pyridyl-3-quinoline-carboxylic acid derivatives', US Patent 3,753,993 (rosoxacin).

Mascellani, G., et al. (1987) 'Anti-bacterial pyrido-benzothiazine derivative and pharmaceutical composition thereof', US Patent 4,684,647 (rufloxacin).

Masuzawa, K., et al. (1991) '8-Alkoxyquinolinecarboxylic acid and salts thereof', US Patent 5,043,450 (gatifloxacin).

Matsumoto, J., et al. (1989) 'Quinoline derivatives, pharmaceutical composition and method of use', US Patent 4,886,810 (orbifloxacin).

Matsumoto, J. et al. (1989) '5-Substituted-6,8-difluoroquinolines useful as antibacterial agents', US Patent 4,795,751 (sparfloxacin).

Narita, H., et al. (1991) 'Pyridone carboxylic acid derivatives and salts thereof, process for producing the same and antibacterial agents comprising the same', US Patent 4,990,508 (pazufloxacin).

Satoh, M., et al. (2005) '4-Oxoquinoline compounds and utilization thereof as HIV integrase inhibitors', US Patent 2005/0239819 (elvitegravir).

Todo, Y., et al. (2000) 'Quinolone carboxylic acid derivatives or salts thereof', US Patent 6025370 (garenoxacin).

Appendix 1

Functional group transformations

There are a large number of possible functional group transformations (FGTs) in organic synthesis. The following are examples of the most commonly used FGTs in drug synthesis. Further details on each reaction are available in the book's Online Resource Centre.

 http://oxfordtextbooks.co.uk/orc/patrick_synth/

1 Acetals and ketals

(a)

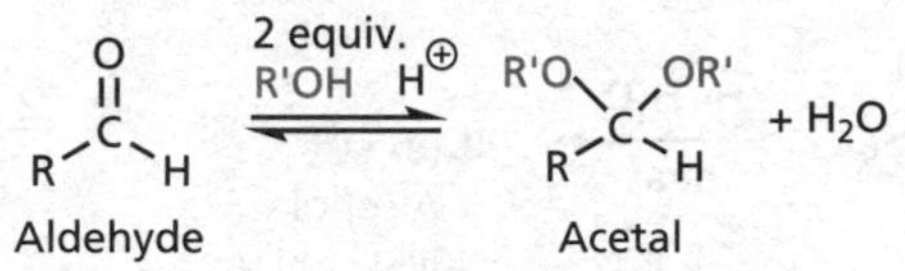

Aldehyde $\xrightarrow[\text{R'OH}\ \text{H}^{\oplus}]{\text{2 equiv.}}$ Acetal $+ H_2O$

(b)

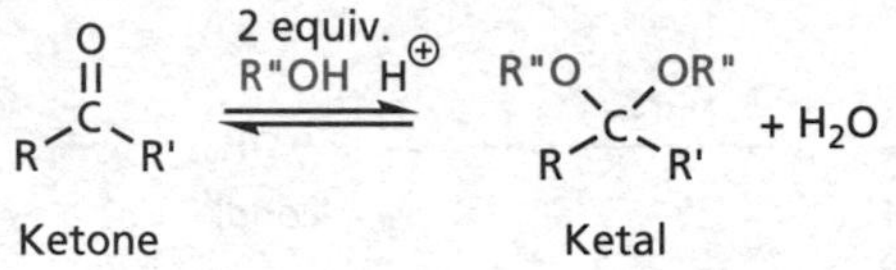

Ketone $\xrightarrow[\text{R"OH}\ \text{H}^{\oplus}]{\text{2 equiv.}}$ Ketal $+ H_2O$

2 Acid anhydrides

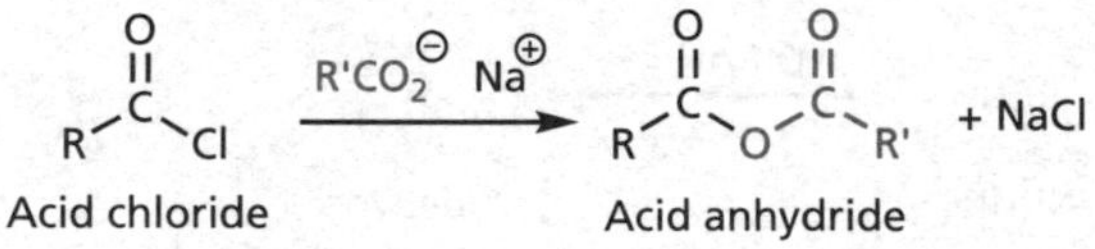

Acid chloride $\xrightarrow{\text{R'CO}_2^{\ominus}\ \text{Na}^{\oplus}}$ Acid anhydride $+ \text{NaCl}$

3 Acid chlorides

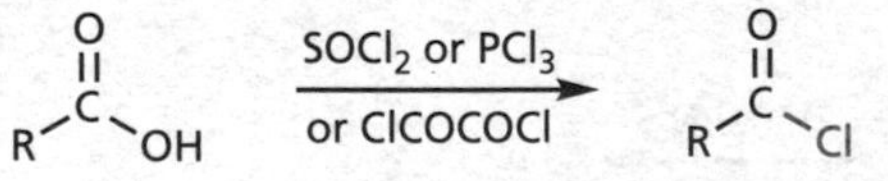

$\xrightarrow[\text{or ClCOCOCl}]{\text{SOCl}_2 \text{ or PCl}_3}$

4 Alcohols

(a)

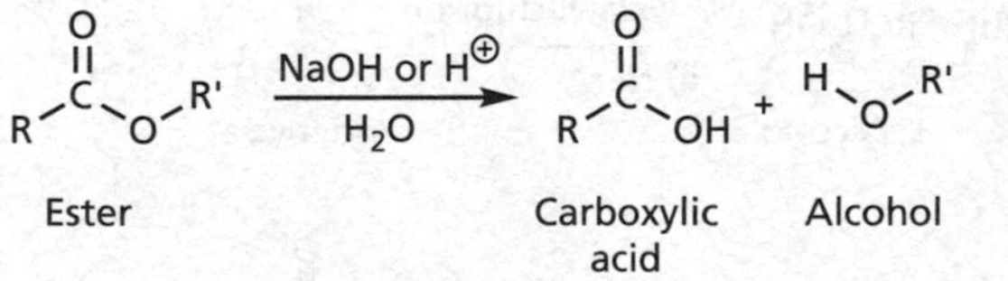

Ester $\xrightarrow[\text{H}_2\text{O}]{\text{NaOH or H}^{\oplus}}$ Carboxylic acid + Alcohol

(b)

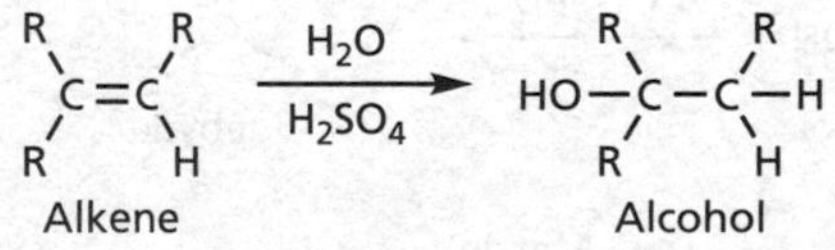

Alkene $\xrightarrow[\text{H}_2\text{SO}_4]{\text{H}_2\text{O}}$ Alcohol

(c)

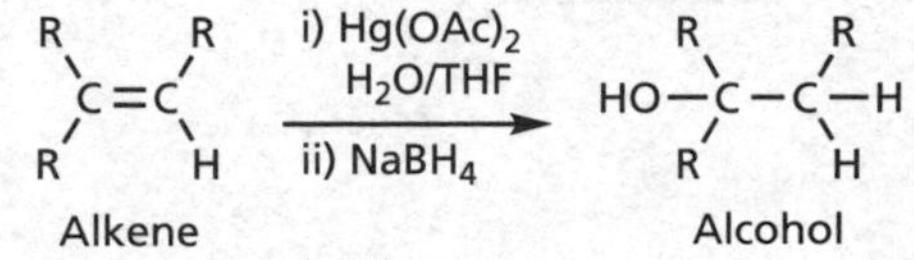

Alkene $\xrightarrow[\text{ii) NaBH}_4]{\text{i) Hg(OAc)}_2 \ \text{H}_2\text{O/THF}}$ Alcohol

(d)

Alkene $\xrightarrow[\text{ii) H}_2\text{O}_2/ \text{NaOH}]{\text{i) BH}_3}$ Alcohol

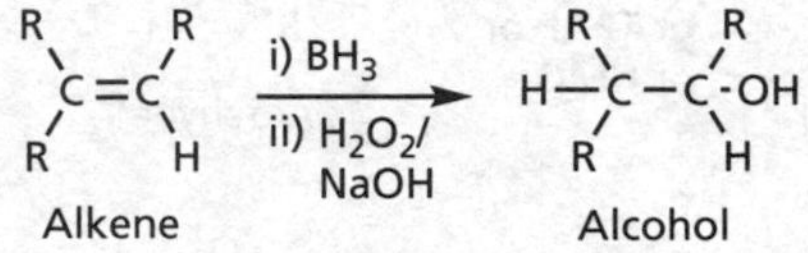

(e) Aldehyde $\xrightarrow[\text{ii) H}_3\text{O}^+]{\text{i) NaBH}_4}$ 1° Alcohol

(f) Ketone $\xrightarrow[\text{ii) H}_3\text{O}^+]{\text{i) NaBH}_4}$ 2° Alcohol

(g) Carboxylic acid $\xrightarrow[\text{ii) H}_3\text{O}^+]{\text{i) LiAlH}_4}$ 1° Alcohol

(h) Carboxylic acid $\xrightarrow[\text{or DIBAH}]{\text{B}_2\text{H}_6}$ 1° Alcohol

(i) Ester $\xrightarrow[\text{ii) H}_3\text{O}^+]{\text{i) LiAlH}_4}$ 1° Alcohol

(j) Ester $\xrightarrow{\text{DIBAH}}$ 1° Alcohol

(k) R–X Alkyl halide $\xrightarrow{\text{NaOH}}$ R–OH Alcohol

(l) R–O–R' Ether $\xrightarrow[\text{heat}]{\text{HI/H}_2\text{O}}$ RI + R'OH Alcohol + alkyl iodide

5 Aldehydes

(a) Acid chloride $\xrightarrow[\text{or H}_2, \text{Pd/C, EtN(}^i\text{Pr)}_2]{\text{LiAlH[O}^t\text{Bu]}_3}$ Aldehyde

(b) RCN Nitrile $\xrightarrow{\text{DIBAH}}$ Aldehyde

(c) Ester $\xrightarrow[\text{NaAlH}_2 \text{ (OCH}_2\text{CH}_2\text{OMe)}_2]{\text{DIBAH or}}$ Aldehyde

(d) ArCN Aromatic nitrile $\xrightarrow[\text{HCl, SnCl}_2]{\text{DIBAH or}}$ Aldehyde

(e) Amide $\xrightarrow[\text{LiAlH[O}^t\text{Bu]}_3]{\text{DIBAH or}}$ Aldehyde

(f) R≡ Terminal alkyne $\xrightarrow[\text{ii) H}_2\text{O}_2/\text{HO}^-]{\text{i) 9-BBN}}$ Aldehyde

(g) RCH$_2$OH 1° Alcohol $\xrightarrow{\text{PCC or PDC}}$ Aldehyde

(h) RCH$_2$OH 1° Alcohol $\xrightarrow[\text{ii) NEt}_3]{\text{i) DMSO Oxalyl chloride}}$ Aldehyde

(i) RCH$_2$OH 1° Alcohol $\xrightarrow[\text{or TPAP or or NMO}]{\text{Dess-Martin periodinane}}$ Aldehyde

(j) Alkene $\xrightarrow{\text{O}_3}$ Aldehyde

(k) 1,2-Diol $\xrightarrow[\text{–CH}_2\text{O}]{\text{NaIO}_4}$ Aldehyde

(l) RCH$_2$NO$_2$ 1° Aliphatic nitro group $\xrightarrow{\text{TiCl}_3}$ Aldehyde

6 Alkenes

(a) R—≡—R Alkyne → (Lindlar's catalyst or P-2 catalyst / H₂) → Z-Alkene

(b) R—≡—R Alkyne → i) Na or Li NH₃ ii) NH₄Cl → E-Alkene

(c) Alkyl halide → NaOMe / MeOH → Alkene

(d) Alcohol → H_2SO_4 or H_3PO_4 or $POCl_3$/pyridine → Alkene

(e) R—CH₂—NH₂ 1° Amine → i) excess CH_3I ii) Ag_2O/H_2O heat → Alkene

(f) ᵗBu—O—CH(CH₃)₂ 3° Ether → HI / H_2O → H_3C—C(CH₃)=CH₂ + HO—CH(CH₃)₂ Alkene + alcohol

7 Alkyl fluorides

(a) R—X Alkyl halide (X = Br, Cl) → KF → R—F Alkyl fluoride

(b) R—OH Alcohol → Et_2NSF_3 (DAST) → R—F Alkyl fluoride

(c) Aldehyde or ketone → Et_2NSF_3 (DAST) → Alkyl fluoride

(d) Carboxylic acid → Et_2NSF_3 (DAST) → Acyl fluoride

(e) Carboxylic acid → SF_4 → R—CF₃ Alkyl fluoride

(f) R—OTf Triflate → $Et_3N.3HF$ → R—F Alkyl fluoride

(g) R—OH Alcohol → $C_4F_9SO_2F$ / N-Ethyl piperidine → R—F Alkyl fluoride

(h) Ar—Me Aromatic methyl → i) Cl_2, PCl_5 ii) SbF_5 → Ar—CF₃ Alkyl fluoride

8 Alkyl halides

(a) Alkene → HX → Alkyl halide

(b) R—C(R)(R)—OH Alcohol → HX → R—C(R)(R)—X Alkyl halide

(c) RCH₂OH Alcohol → $SOCl_2$ / NEt_3 → RCH₂Cl Alkyl chloride

(d) RCH₂OH Alcohol → PBr_3 or PPh_3/CBr_4 → RCH₂Br Alkyl bromide

(e) RCH$_2$OMs $\xrightarrow{\text{NaI}}$ RCH$_2$I
Mesylate · Alkyl iodide

(f) RCH$_2$OTs $\xrightarrow{\text{NaI}}$ RCH$_2$I
Tosylate · Alkyl iodide

(g) R—X $\xrightarrow[\text{Acetone}]{\text{NaI}}$ R—I
Alkyl halide (X = Br, Cl) · Alkyl iodide

(h) R—OR' $\xrightarrow{\text{HI or HBr}}$ R—X + HOR'
Ether · Alkyl halide + alcohol

9 Alkynes

(a)

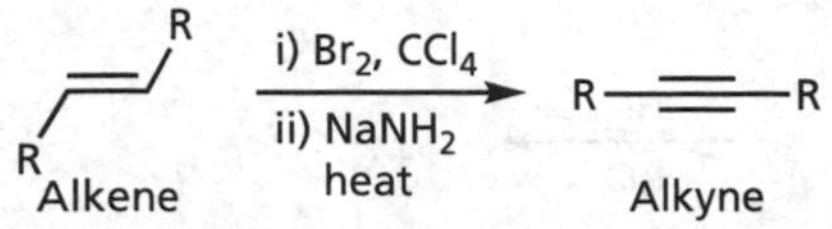

i) Br$_2$, CCl$_4$
ii) NaNH$_2$ heat

Alkene → Alkyne R—≡—R

(b)

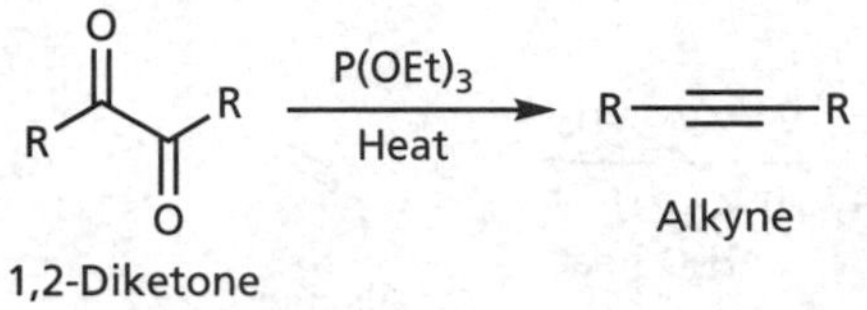

$\xrightarrow[\text{Heat}]{\text{P(OEt)}_3}$ R—≡—R

1,2-Diketone · Alkyne

(c)

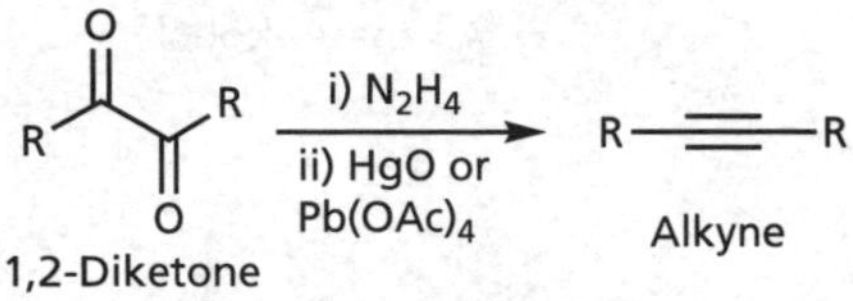

i) N$_2$H$_4$
ii) HgO or Pb(OAc)$_4$

1,2-Diketone → R—≡—R Alkyne

10 Allylic alcohols

(a)

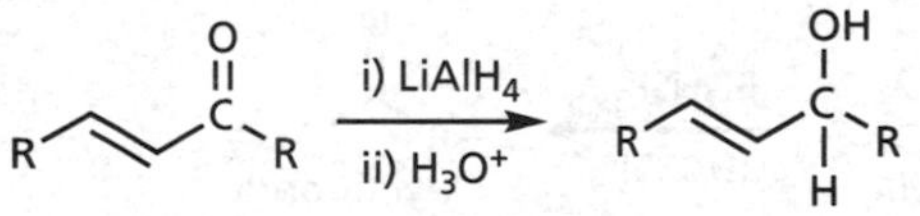

i) LiAlH$_4$
ii) H$_3$O$^+$

(b)

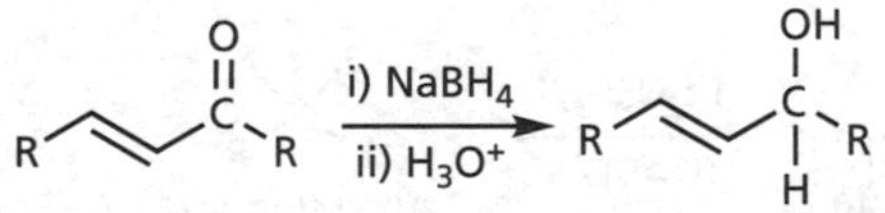

i) NaBH$_4$
ii) H$_3$O$^+$

(c)

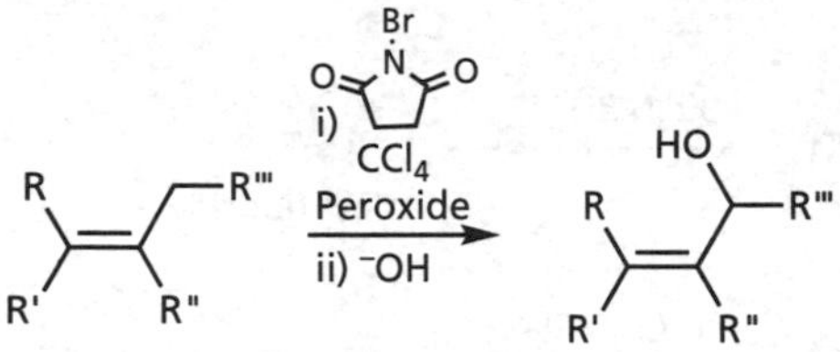

i)

CCl$_4$
Peroxide
ii) $^-$OH

(d)

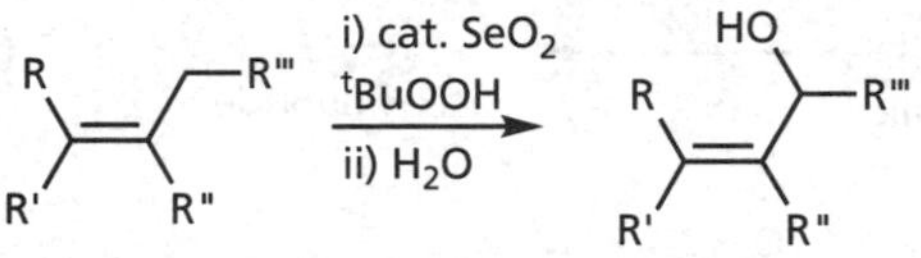

i) cat. SeO$_2$
^{t}BuOOH
ii) H$_2$O

(e)

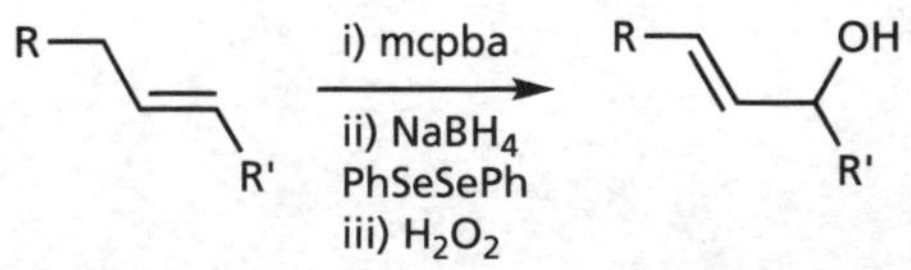

i) mcpba
ii) NaBH$_4$
PhSeSePh
iii) H$_2$O$_2$

(f)

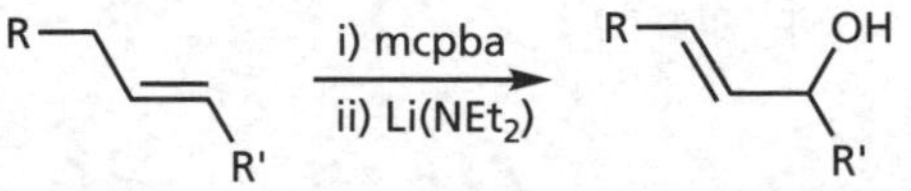

i) mcpba
ii) Li(NEt$_2$)

(g)

i) H$_2$O$_2$, NaOH
ii) NH$_2$NH$_2$

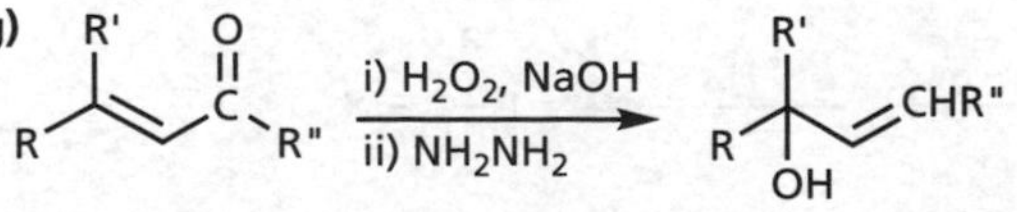

11 Amides

(a)
$$R-C(=O)-Cl \xrightarrow[\text{HNR'}_2]{\text{Pyridine or NaOH}} R-C(=O)-NR'_2$$
Acid chloride → Amide

(b)
$$R-C(=O)-OCOR \xrightarrow[\text{HNR'}_2]{\text{Pyridine}} R-C(=O)-NR'_2$$
Acid anhydride → Amide

(c)
$$R-C(=O)-OH \xrightarrow[\text{H}_2\text{NR'}]{\text{DCC or DIC or EDC}} R-C(=O)-NHR'$$
Carboxylic acid → Amide

(d)
$$R-C(=O)-OH \xrightarrow[\text{H}_2\text{NR'}]{\text{DCC, HOBt}} R-C(=O)-NHR'$$
Carboxylic acid → Amide

(e)
$$R-C(=O)-OH \xrightarrow[\text{H}_2\text{NR'}]{\text{PyBOP or PyBrBOP}} R-C(=O)-NHR'$$
Carboxylic acid → Amide

(f)
$$R-C(=O)-NHR' \xrightarrow[\text{ii) MeI}]{\text{i) NaH}} R-C(=O)-NR'Me$$
2° Amide → 3° Amide

12 Amines

(a)
$$R-C(=O)-NHR \xrightarrow[\text{H}_2\text{O}]{\text{H}^+} H_2N-R$$
Amide → Amine

(b)
$$R-C(=O)-NHR \xrightarrow{\text{LiAlH}_4} RCH_2NHR$$
Amide → Amine

(c)
$$RX \xrightarrow[\substack{\text{ii) LiAlH}_4 \text{ or} \\ \text{H}_2, \text{ cat. or} \\ \text{PPh}_3}]{\text{i) NaN}_3} RNH_2$$
Alkyl halide → 1° Amine

(d)
$$R-NO_2 \xrightarrow{\text{LiAlH}_4 \text{ or H}_2, \text{Pd/C}} RNH_2$$
Nitro → 1° Amine

(e)
$$RCN \xrightarrow[\substack{\text{B}_2\text{H}_6 \text{ or} \\ \text{H}_2, \text{PtO}_2}]{\text{LiAlH}_4 \text{ or}} RCH_2NH_2$$
Nitrile → 1° Amine

(f)
$$RX \xrightarrow[\text{ii) H}_2\text{O, HO}^-]{\substack{\text{i) Potassium} \\ \text{phthalimide}}} RNH_2$$
Alkyl halide → 1° Amine

(g)
$$R-C(=O)-NH_2 \xrightarrow[\text{Br}_2, \text{H}_2\text{O}]{4 \text{ NaOH,}} RNH_2$$
1° Amide → 1° Amine

(h)
$$R-C(=O)-N_3 \xrightarrow[\text{heat}]{\text{H}_2\text{O}} RNH_2$$
Acyl azide → 1° Amine

(i)
$$RNH_2 \xrightarrow[\text{ii) NaOH}]{\text{i) R''-X}} RNHR'' + RNR''$$
1° Amine → 2° and 3° Amines

(j)
$$RNHR' \xrightarrow[\text{ii) NaOH}]{\text{i) R''-X}} RNR'R''$$
2° Amine → 3° Amine

(k)
$$RNH_2 \xrightarrow[\substack{\text{ii) LiAlH}_4 \\ \text{iii) H}_3\text{O}^+}]{\text{i) Acid chloride}} RNHCH_2R'$$
1° Amine → 2° Amine

(l)
$$RNHR' \xrightarrow[\substack{\text{ii) LiAlH}_4 \\ \text{iii) H}_3\text{O}^+}]{\text{i) Acid chloride}} RR'NHCH_2R''$$
2° Amine → 3° Amine

(m)
$$R-C(=O)-H \xrightarrow[\text{NaBH}_3\text{CN}]{\text{R'NH}_2} RCH_2NHR'$$
Aldehyde → Amine

(n)
$$R-C(=O)-R \xrightarrow[\text{NaBH}_3\text{CN}]{\text{R'NH}_2} R_2CHNHR'$$
Ketone → Amine

(o)
$$R-C(=O)-R' \xrightarrow[\text{ii) LiAlH}_4]{\text{i) NH}_2\text{OH}} R-CH(NH_2)-R'$$
Aldehyde or ketone → 1° Amine

(p)
$$R_2NMe \xrightarrow[\text{ii) MeOH}]{\text{i) VOC-Cl}} R_2NH$$
3° Amine → 2° Amine

(q) 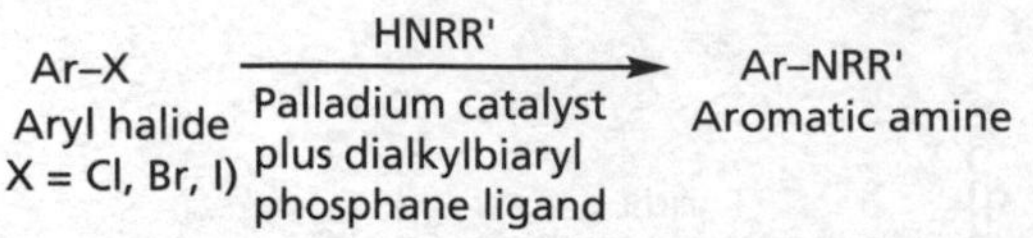 Ar–X
Aryl halide
X = Cl, Br, I)

$\xrightarrow[\text{Palladium catalyst plus dialkylbiaryl phosphane ligand}]{\text{HNRR'}}$

Ar–NRR'
Aromatic amine

(r) 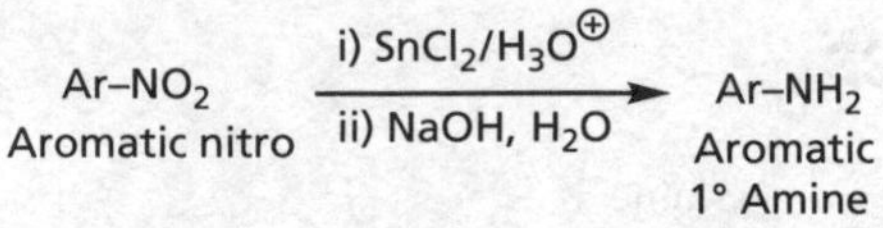 Ar–NO$_2$
Aromatic nitro

$\xrightarrow[\text{ii) NaOH, H}_2\text{O}]{\text{i) SnCl}_2/\text{H}_3\text{O}^\oplus}$

Ar–NH$_2$
Aromatic
1° Amine

13 Amino acids

(a) 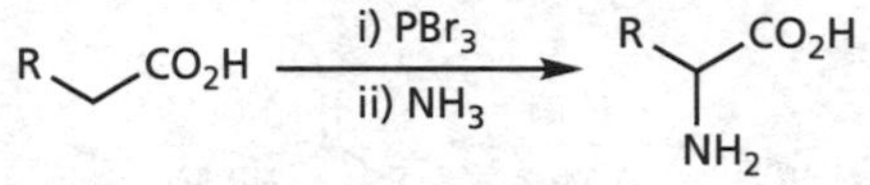 R⌒CO$_2$H $\xrightarrow[\text{ii) NH}_3]{\text{i) PBr}_3}$ R⌒CO$_2$H / NH$_2$

(b) 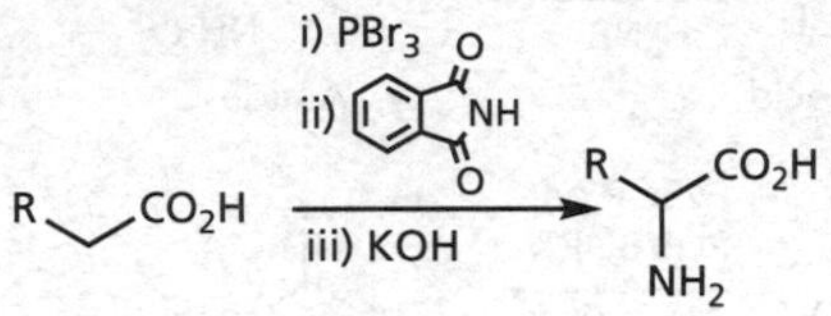 R⌒CO$_2$H $\xrightarrow[\text{iii) KOH}]{\text{i) PBr}_3 \text{ ii)}}$ R⌒CO$_2$H / NH$_2$

(c) 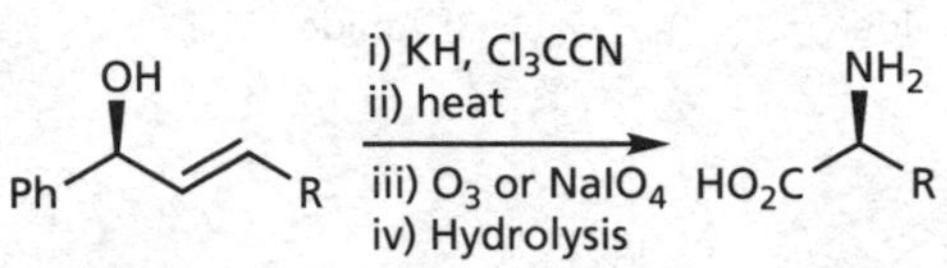 Ph—CH(OH)—CH=CH—R $\xrightarrow[\text{iv) Hydrolysis}]{\substack{\text{i) KH, Cl}_3\text{CCN} \\ \text{ii) heat} \\ \text{iii) O}_3 \text{ or NaIO}_4}}$ HO$_2$C—CH(NH$_2$)—R

(d) 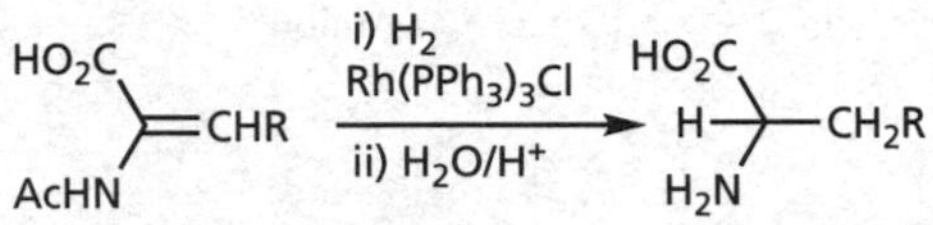 HO$_2$C—C(AcHN)=CHR $\xrightarrow[\text{ii) H}_2\text{O/H}^+]{\substack{\text{i) H}_2 \\ \text{Rh(PPh}_3)_3\text{Cl}}}$ HO$_2$C—CH(NH$_2$)—CH$_2$R

(e) 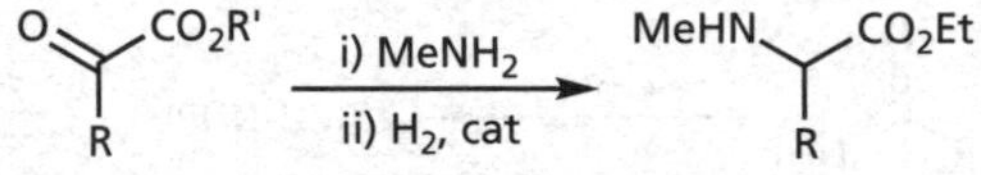 O=C(R)—CO$_2$R' $\xrightarrow[\text{ii) H}_2\text{, cat}]{\text{i) MeNH}_2}$ MeHN—CH(R)—CO$_2$Et

14 Amino alcohols

(a) 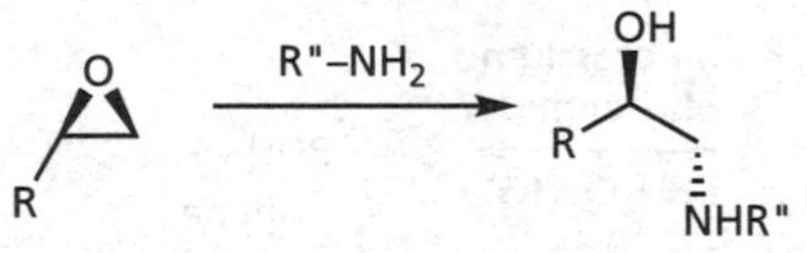 (epoxide) R $\xrightarrow{\text{R''–NH}_2}$ R—CH(OH)—CH(NHR'')

(b) 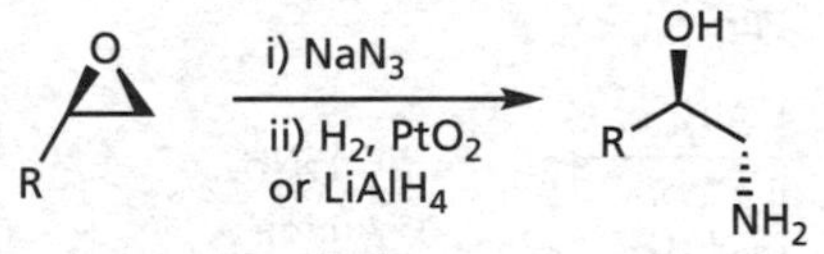 (epoxide) R $\xrightarrow[\text{or LiAlH}_4]{\substack{\text{i) NaN}_3 \\ \text{ii) H}_2\text{, PtO}_2}}$ R—CH(OH)—CH(NH$_2$)

(c) 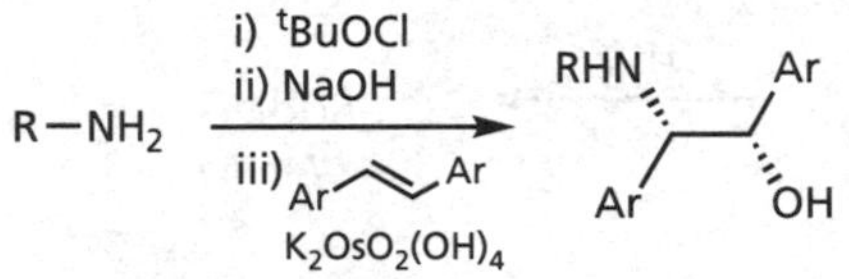 R—NH$_2$ $\xrightarrow[\substack{\text{iii) Ar—CH=CH—Ar} \\ \text{K}_2\text{OsO}_2(\text{OH})_4}]{\substack{\text{i) }^t\text{BuOCl} \\ \text{ii) NaOH}}}$ RHN—CH(Ar)—CH(Ar)—OH

(d) 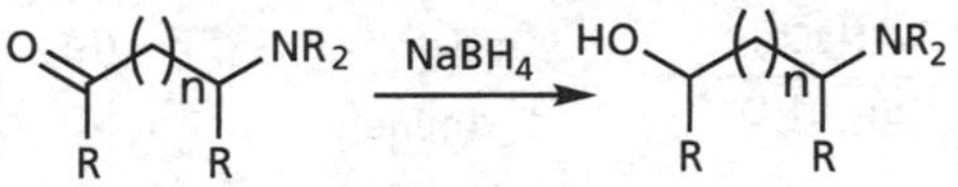 O=C(R)—(CH$_2$)$_n$—CR$_2$—NR$_2$ $\xrightarrow{\text{NaBH}_4}$ HO—CH(R)—(CH$_2$)$_n$—CR$_2$—NR$_2$

15 Aryl halides

(a) 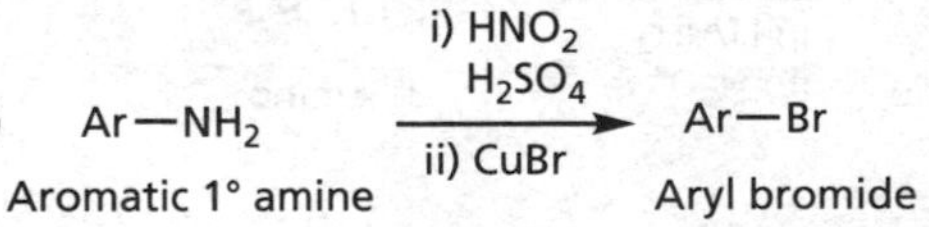 Ar—NH$_2$
Aromatic 1° amine
$\xrightarrow[\text{ii) CuBr}]{\substack{\text{i) HNO}_2 \\ \text{H}_2\text{SO}_4}}$
Ar—Br
Aryl bromide

(b) 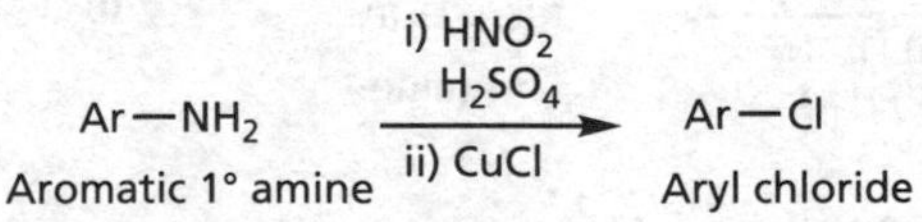 Ar—NH$_2$
Aromatic 1° amine
$\xrightarrow[\text{ii) CuCl}]{\substack{\text{i) HNO}_2 \\ \text{H}_2\text{SO}_4}}$
Ar—Cl
Aryl chloride

(c) 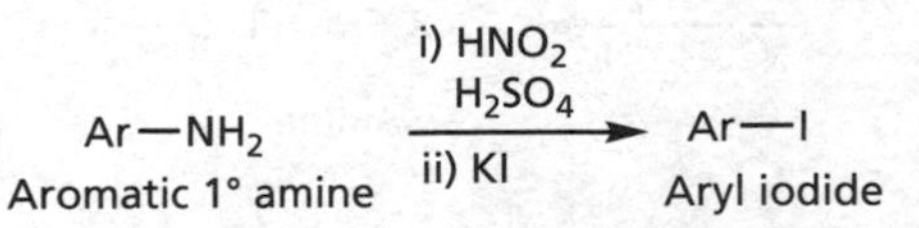 Ar—NH$_2$
Aromatic 1° amine
$\xrightarrow[\text{ii) KI}]{\substack{\text{i) HNO}_2 \\ \text{H}_2\text{SO}_4}}$
Ar—I
Aryl iodide

(d) 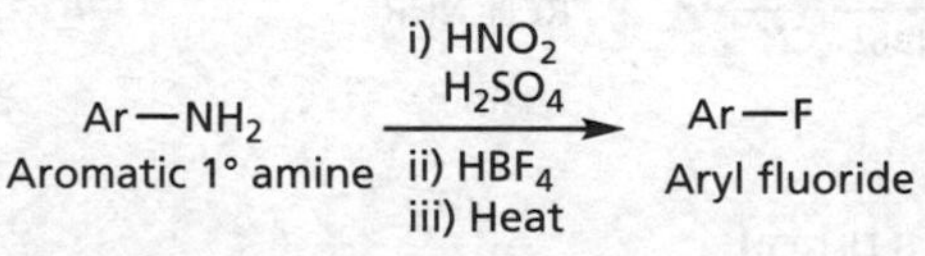 Ar—NH$_2$
Aromatic 1° amine
$\xrightarrow[\substack{\text{ii) HBF}_4 \\ \text{iii) Heat}}]{\substack{\text{i) HNO}_2 \\ \text{H}_2\text{SO}_4}}$
Ar—F
Aryl fluoride

16 Azides

R—X   →(NaN₃)→   R—N₃
Alkyl halide   Azide

17 Aziridines

(a) α-Haloamine →(Base)→ Aziridine

(b) Epoxide →(i) NH₃, ii) MeSO₂Cl)→ Aziridine

(c) Epoxide →(i) Azide, ii) PPh₃)→ Aziridine

(d) Alkene →(TsNClNa, PhNMe₃⁺Br⁻)→ Aziridine

(e) Alkene →(i) PhNMe₃⁺Br⁻, ᵗBuSO₂NClNa, ii) CF₃SO₃H, Anisole)→ Aziridine

18 Carboxylic acids

(a) Ester →(NaOH or H⁺, H_2O)→ $R-CO_2H$ Carboxylic acid (+ alcohol)

(b) Amide →(H⁺, H_2O)→ $R-CO_2H$ Carboxylic acid (+ amine)

(c) RCN Nitrile →(H_2O, acid or base)→ $R-CO_2H$ Carboxylic acid

(d) $R-CH_2OH$ 1° Alcohol →(CrO_3, H_2SO_4)→ $R-CO_2H$ Carboxylic acid

(e) Aldehyde →(CrO_3, H_2SO_4)→ $R-CO_2H$ Carboxylic acid

(f) Aldehyde →(AgO, $NH_3/H_2O/EtOH$)→ $R-CO_2H$ Carboxylic acid

(g) Aldehyde →(RCO_3H)→ $R-CO_2H$ Carboxylic acid

(h) Alkene →(i) $KMnO_4$, HO⁻, heat, ii) H⁺)→ 2 $R-CO_2H$ Carboxylic acid

(i) Alkene →(i) O_3, ii) H_2O_2)→ 2 $R-CO_2H$ Carboxylic acid

(j) Methyl ketone →(i) NaOH, I_2, ii) NaOH)→ $R-CO_2H$ Carboxylic acid

19 Cyanohydrins

(a) Aldehyde → i) HCN/KCN, ii) H_2O → Cyanohydrin

(b) Ketone → i) HCN/KCN, ii) H_2O → Cyanohydrin

20 Dienes (conjugated)

(a) → $LiAlH_4$ →

(b) → i) NBS, ii) KO^tBu →

(c) → i) $TsNHNH_2$, ii) 2RLi →

(d) OMe → Acid catalyst → OMe

21 Dienes (unconjugated)

(a) CO_2H → Li or Na, NH_3, EtOH → CO_2H

(b) OMe → Li or Na, NH_3, EtOH → OMe

(c) NMe_2 → Li or Na, NH_3, EtOH → NMe_2

22 Diketones

(a) → CrO_3 →

(b) → CrO_3 →

(c) Me — Ph → i) HNO_2, ii) H^+/H_2O → Me — Ph

(d) → SeO_2 →

(e) → O_3 →

(f) → aq. acid, Piperidine →

(g)

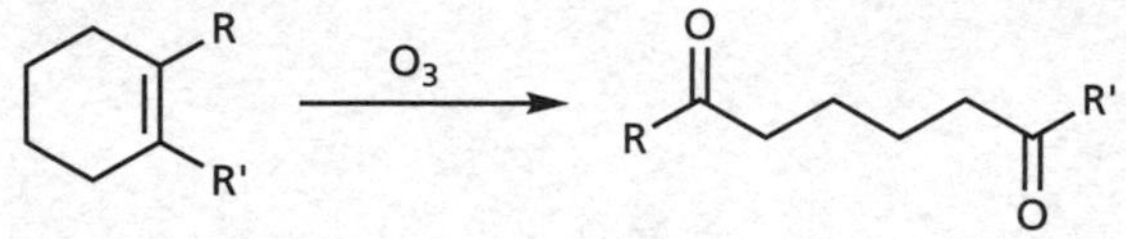

$$\text{cyclohexene} \xrightarrow{O_3} R\text{-CO-}(CH_2)_4\text{-CO-}R'$$

23 Diols

(a) alkene $\xrightarrow[\text{ii) NaHSO}_3]{\text{i) OsO}_4}$ diol

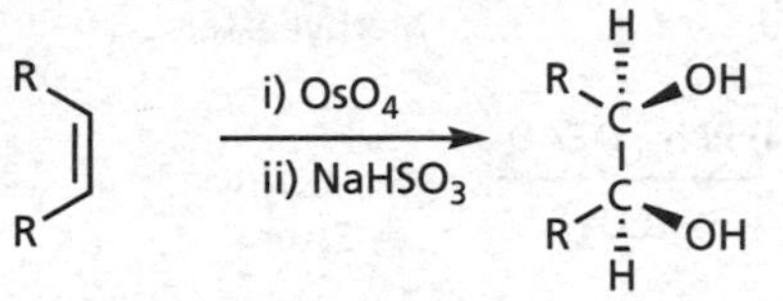

(b) alkene $\xrightarrow[\text{ii) H}_2\text{O, HO}^-]{\text{i) KMnO}_4}$ diol

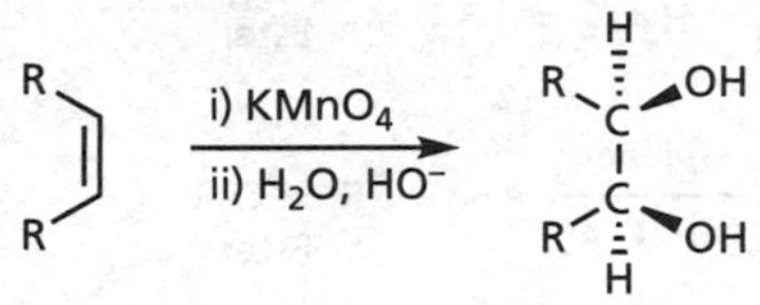

(c) alkene $\xrightarrow[\text{ii) NaOH or H}^+]{\text{i) mcpba}}$ diol

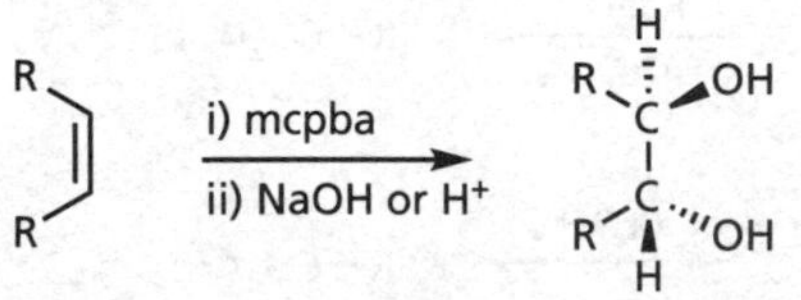

(d) $R\text{-CH=CH-CH=CH-}R' \xrightarrow[\text{iii) Reduction}]{\substack{\text{i) mcpba}\\\text{ii) (TiClCp}_2)_2}}$ R-CH(OH)-CH=CH-CH(OH)-R'

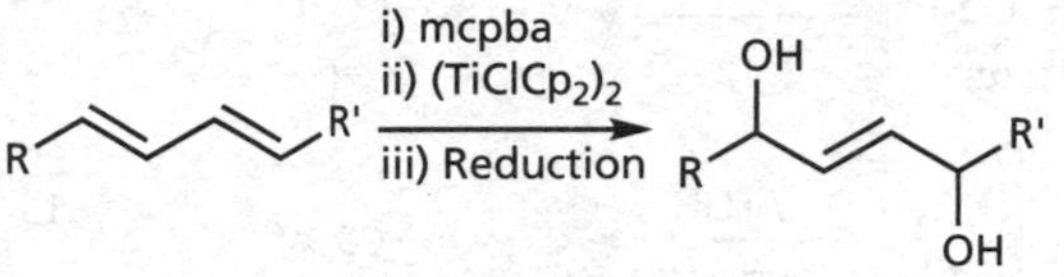

24 Disulphides

$$2\ R\text{—SH} \xrightarrow{\text{Br}_2 \text{ or I}_2} R\text{—S—S—}R$$
Thiol → Disulphide

25 Enamines

(a) Aldehyde $R\text{-CH}_2\text{-CHO} \xrightarrow[\text{NHR}^1\text{R}^2]{\text{H}^+}$ Enamine $R\text{-CH=CH-NR}^1\text{R}^2$

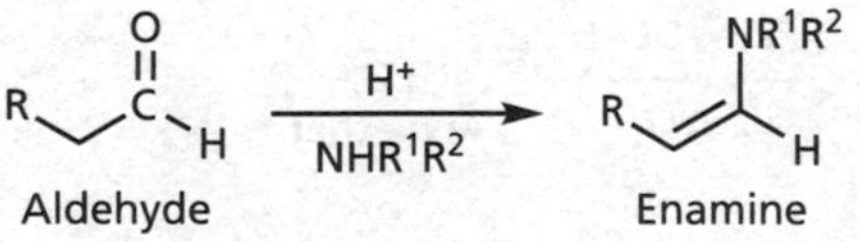

(b) Ketone $R\text{-CH}_2\text{-CO-}R' \xrightarrow[\text{NHR}^1\text{R}^2]{\text{H}^+}$ Enamine

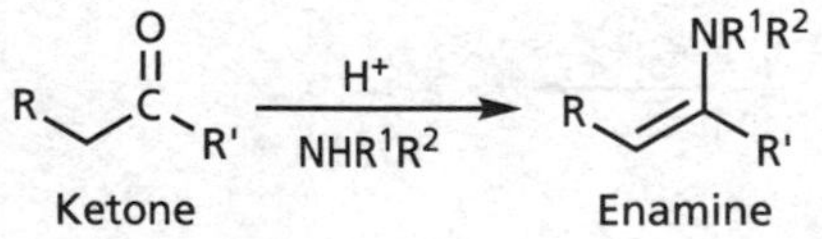

26 Epoxides

(a) Alkene $\xrightarrow{\text{mcpba}}$ Epoxide

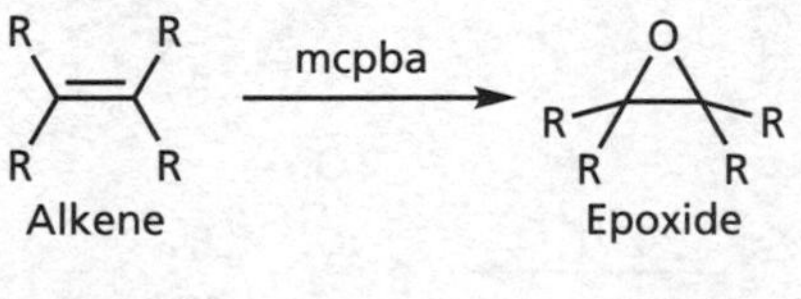

(b) Alkene $\xrightarrow[\text{Me-ReO}_3]{\substack{\text{aq. H}_2\text{O}_2\\\text{Pyridine or}\\\text{pyrazole}}}$ Epoxide

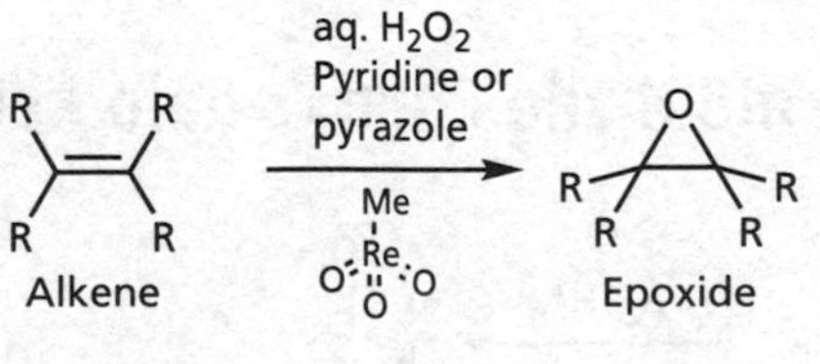

(c) Alkene $\xrightarrow[\text{ii) NaOH}]{\text{i) X}_2/\text{H}_2\text{O}}$ Epoxide

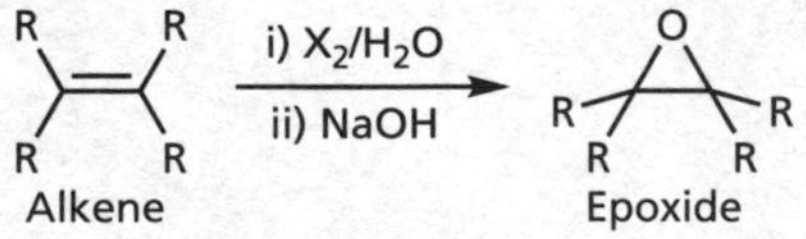

(d) Halohydrin $\xrightarrow{\text{NaOH}}$ Epoxide

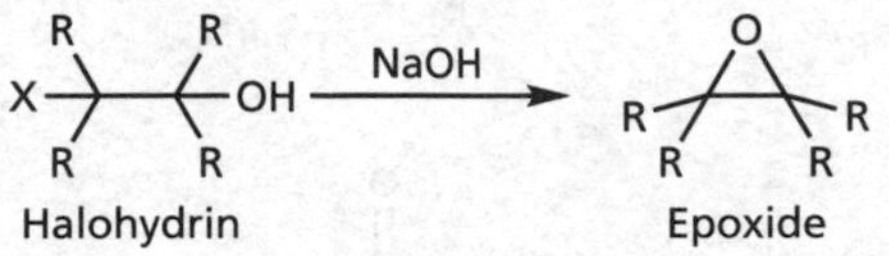

27 Esters

(a)
$$\underset{\text{Acid chloride}}{R-\overset{\overset{\displaystyle O}{\|}}{C}-Cl} \xrightarrow[\text{pyridine}]{R'OH} \underset{\text{Ester}}{R-CO_2R'}$$

(b)
$$\underset{\text{Alcohol}}{HO-R'} \xrightarrow[\text{NaOH}]{Ac_2O} \underset{\text{Acetate ester}}{Me-CO_2R'}$$

(c)
$$\underset{\text{Carboxylic acid}}{R-CO_2H} \xrightarrow[\text{H}_3O^+]{R'OH} \underset{\text{Ester}}{R-CO_2R'}$$

(d)
$$\underset{\text{Carboxylic acid}}{R-CO_2H} \xrightarrow{CH_2N_2} \underset{\text{Methyl ester}}{R-CO_2Me}$$

(e)
$$\underset{\text{Carboxylic acid}}{R-CO_2H} \xrightarrow[\text{ii) R'-X}]{\text{i) NaOH}} \underset{\text{Ester}}{R-CO_2R'}$$

(f)
$$\underset{\text{Alcohol}}{HO-R'} \xrightarrow[\text{ii) RCO}_2^-]{\text{i) PPh}_3,\text{ DEAD}} \underset{\text{Ester}}{R-CO_2R'}$$

(g)
$$\underset{\text{Ester}}{R-CO_2R'} \xrightarrow{R''OH/H^+} \underset{\text{Ester}}{R-CO_2R''}$$

(h)
$$\underset{\text{Nitrile}}{R-CN} \xrightarrow[\text{acid}]{R'OH} \underset{\text{Ester}}{R-CO_2R'}$$

(i)
$$\underset{\text{Acid chloride}}{R-\overset{\overset{\displaystyle O}{\|}}{C}-Cl} \xrightarrow[\text{pyridine}]{ArOH} \underset{\text{Ester}}{R-CO_2Ar}$$

(j)
$$\underset{\text{Acid anhydride}}{R-\overset{\overset{\displaystyle O}{\|}}{C}-O-\overset{\overset{\displaystyle O}{\|}}{C}-R} \xrightarrow{ArOH} \underset{\text{Ester}}{R-CO_2Ar}$$

(k)
$$\underset{\text{Ketone}}{R-\overset{\overset{\displaystyle O}{\|}}{C}-R'} \xrightarrow{RCO_3H} \underset{\text{Ester}}{R-CO_2R'}$$

28 Ethers

(a)
$$\underset{\text{Alcohol}}{R-OH} \xrightarrow[\text{ii) R'X}]{\text{i) NaH or Na}} \underset{\text{Ether}}{R-OR'}$$

(b)
$$\underset{\text{Alcohol}}{R-OH} \xrightarrow[\text{Ag}_2O]{R'-X} \underset{\text{Ether}}{R-OR'}$$

(c)
$$\underset{\text{Phenol}}{Ar-OH} \xrightarrow[\text{ii) R-X}]{\text{i) NaOH}} \underset{\text{Aryl ether}}{Ar-OR}$$

(d)
$$\underset{\text{Alcohol}}{R-OH} \xrightarrow[\text{ii) ArO}^-]{\text{i) PPh}_3,\text{ DEAD}} \underset{\text{Aryl ether}}{R-OAr}$$

(e)
$$\underset{\text{Alkene}}{\overset{R}{\underset{R}{}}C=C} \xrightarrow[\substack{\text{ii) R'OH/THF} \\ \text{iii) NaBH}_4,\text{ HO}^-}]{\text{i) Hg(O}_2\text{CCF}_3)_2} \underset{\text{Ether}}{}$$

29 α-Halo aldehydes and α-halo ketones

(a)
$$\underset{H}{\overset{O}{\|}}\text{...} \xrightarrow{X_2,\ H^+} \text{...}$$

(b)
$$R-\overset{\overset{\displaystyle O}{\|}}{C}\text{...} \xrightarrow{X_2,\ H^+} \text{...}$$

(c)
$$R-\overset{\overset{\displaystyle O}{\|}}{C}-CH_2-R' \xrightarrow[\substack{\text{ii)} \\ \text{Ph-S-N-S-Ph}}]{\text{i) LDA}} R-\overset{\overset{\displaystyle O}{\|}}{C}-\overset{F}{\underset{}{CH}}-R'$$

30 Halo amines

(a) Ar—N(H)—CH₂CH₂—OH →[POCl₃]→ Ar—N(H)—CH₂CH₂—Cl

31 α-Hydroxy ketones

(a) R—CH₂—C(=O)—R →[i) KHMDS; ii) PhSO₂—N(O)—Ph]→ R—CH(OH)—C(=O)—R

32 Imines

(a) R—C(=O)—H (Ketone) →[R'NH₂ / H⁺]→ R—C(=NR')—H (Imine)

(b) R—C(=O)—R" (Aldehyde) →[R'NH₂ / H⁺]→ R—C(=NR')—R" (Imine)

33 α-Keto acids and α-keto esters

R—C(=O)—CH=CH—R' →[O₃]→ R—C(=O)—CO₂H →[R'OH, H⁺]→ R—C(=O)—CO₂R'

34 Ketones

(a) R—CH(OH)—R' (2° Alcohol) →[CrO₃, H₂SO₄ or Na₂Cr₂O₇ H₂SO₄]→ R—C(=O)—R' (Ketone)

(b) R—CH(OH)—R' (2° Alcohol) →[Al(OR)₃ Acetone]→ R—C(=O)—R' (Ketone)

(c) R₂C=CR₂ (Alkene) →[i) KMnO₄, HO⁻; ii) H⁺]→ 2 × R—C(=O)—R (Ketone)

(d) R₂C=CR₂ (Alkene) →[O₃]→ 2 × R—C(=O)—R (Ketone)

(e) R—CH=CH—C(=O)—R' (α,β-Unsaturated ketone) →[H₂/Pd or Na/NH₃]→ R—CH₂CH₂—C(=O)—R' (Ketone)

(f) R—C(NO₂)—R' (2° Nitro) →[TiCl₃]→ R—C(=O)—R' (Ketone)

(g) R—C≡C—R (Alkyne) →[H⁺/H₂O HgSO₄]→ R—CH₂—C(=O)—R (Ketone)

(h) R—C≡C—H (Alkyne) →[H⁺/H₂O HgSO₄]→ R—C(=O)—Me (Methyl ketone)

(i) Ar—C≡C—R (Alkyne) →[H⁺/H₂O HgSO₄]→ Ar—C(=O)—CH₂—R (Aromatic ketone)

35 Lactones

(a)

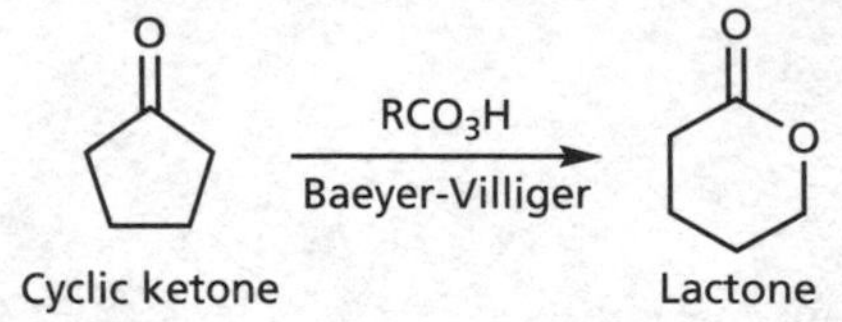

Cyclic ketone $\xrightarrow[\text{Baeyer-Villiger}]{\text{RCO}_3\text{H}}$ Lactone

(b)

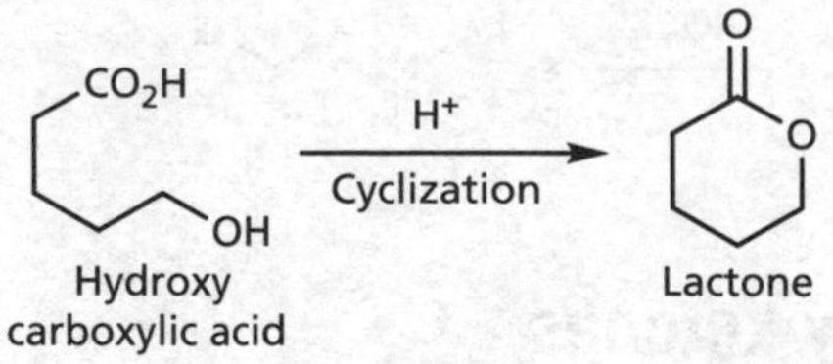

Hydroxy carboxylic acid $\xrightarrow[\text{Cyclization}]{\text{H}^+}$ Lactone

36 Nitriles

(a)   RCH$_2$X $\xrightarrow{\text{KCN}}$ RCH$_2$CN
1° Alkyl halide   Nitrile

(b) $\underset{\text{1° Amide}}{\text{R}-\overset{\overset{\text{O}}{\|}}{\text{C}}-\text{NH}_2}$ $\xrightarrow[\text{or P}_2\text{O}_5]{\text{SOCl}_2}$ R$-$CN
                                    Nitrile

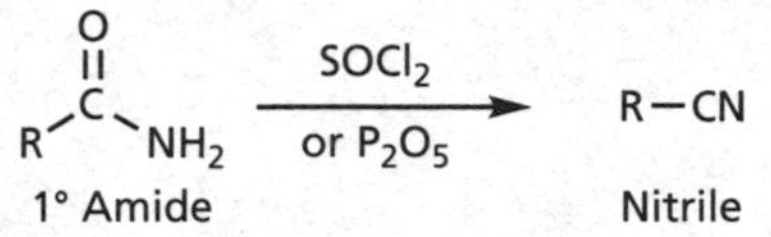

(c)   Ar$-$NH$_2$ $\xrightarrow[\text{ii) CuCN}]{\text{i) HNO}_2/\text{H}_2\text{SO}_4}$ Ar$-$CN
Aromatic   Aromatic nitrile
amine

37 Organometallic reagents

(a)   R$-$X $\xrightarrow[\text{ether}]{\text{Mg}}$ R$-$MgX
Alkyl halide   Grignard
                    reagent

(b)   R$-$X $\xrightarrow[\text{pentane}]{\text{Li}}$ R$-$Li
Alkyl halide   Organolithium
                    reagent

(c)   2 R$-$Li $\xrightarrow{\text{CuI}}$ Li$^+$(R$_2$Cu$^-$)
Organolithium   Organocuprate
reagent   reagent

(d)   R$-$Li $\xrightarrow{\text{CuI}}$ RCu
Organolithium   Organocopper (I)
reagent   reagent

(e)   R$-$MgX $\xrightarrow{^n\text{Bu}_3\text{SnCl}}$ ^{n}Bu$_3$SnR
Grignard   Organotin
reagent   reagent

(f)   Akene $\xrightarrow{\text{R'}_2\text{BH}}$ Organoboron reagent

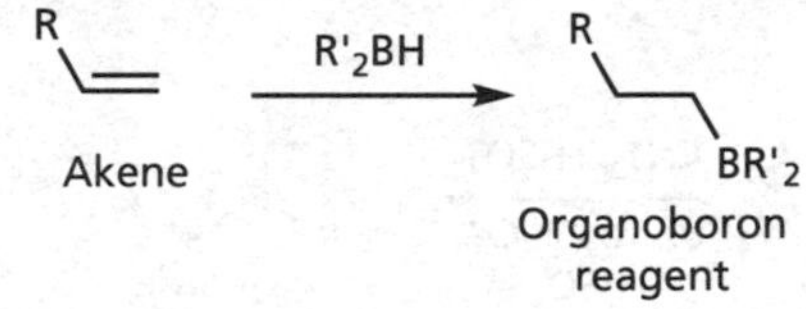

38 Oximes and oxime ethers

(a) $\underset{\substack{\text{Aldehyde}\\\text{or ketone}}}{\text{R}-\overset{\overset{\text{O}}{\|}}{\text{C}}-\text{R'}}$ $\xrightarrow{\text{NH}_2\text{OH}}$ $\underset{\text{Oxime}}{\text{R}-\overset{\overset{\text{NOH}}{\|}}{\text{C}}-\text{R'}}$

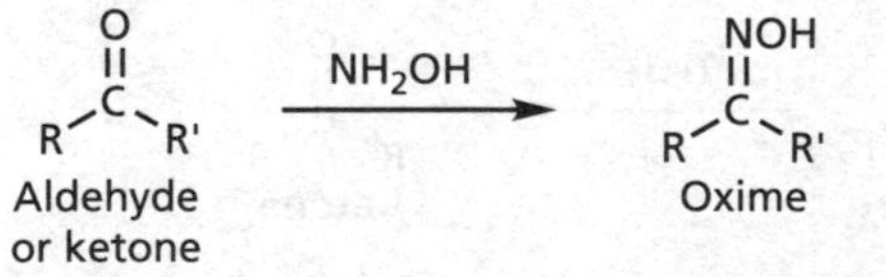

(b) $\underset{\text{Oxime}}{\text{R}-\overset{\overset{\text{NOH}}{\|}}{\text{C}}-\text{R'}}$ $\xrightarrow{\text{R''}-\text{X}}$ $\underset{\text{Oxime ether}}{\text{R}-\overset{\overset{\text{NOR''}}{\|}}{\text{C}}-\text{R'}}$

(c) $\underset{\text{Oxime}}{\text{R}-\overset{\overset{\text{NOH}}{\|}}{\text{C}}-\text{R'}}$ $\xrightarrow{\text{CH}_2\text{N}_2}$ $\underset{\text{Oxime methyl ether}}{\text{R}-\overset{\overset{\text{NOMe}}{\|}}{\text{C}}-\text{R'}}$

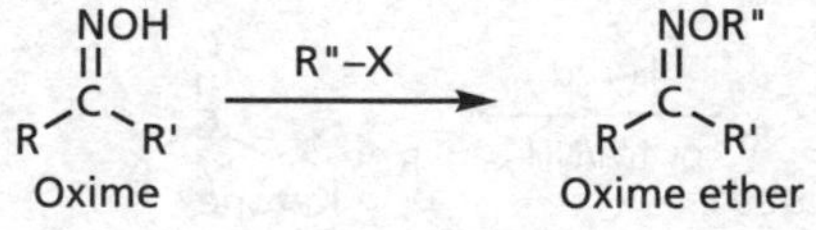

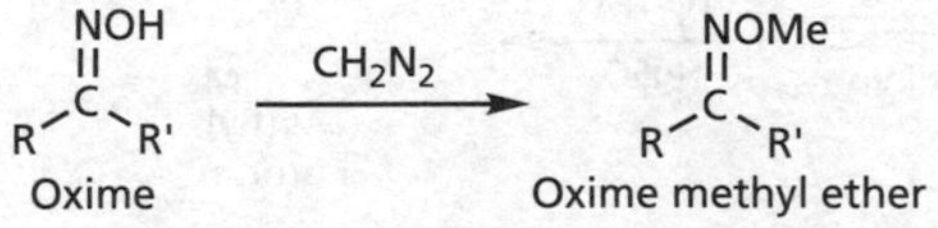

39 Phenols

(a) R—CO₂Ar $\xrightarrow[\text{or NaOH}]{H_3O^+}$ HO—Ar
Aryl ester   Phenol
(+ carboxylic acid)

(b) R—OAr $\xrightarrow[\text{heat}]{\text{Conc. HX}}$ HO—Ar
Aryl ether   Phenol
(+ alkyl halide)

(c) Ar—NH₂ $\xrightarrow[\text{ii) } H_3O^+]{\text{i) NaNO}_2/\text{HCl}}$ Ar—OH
Aromatic amine   Phenol

(d) Ar—SO₃H $\xrightarrow[\text{ii) } H_3O^+]{\text{i) NaOH}}$ Ar—OH
Aromatic   Phenol
sulphonic acid

40 Sulphonamides

Sulphonyl chloride $\xrightarrow[\text{Pyridine}]{\text{HNR'R''} \atop \text{Amine}}$ Sulphonamide

41 Sulphonates

(a) R—OH $\xrightarrow[\text{Pyridine}]{\text{Cl—Ts}}$ R—OTs
Alcohol   Tosylate

(b) R—OH $\xrightarrow[\text{CH}_2\text{Cl}_2]{\text{Cl—Ms} \atop \text{NEt}_3}$ R—OMs
Alcohol   Mesylate

(c) R—OH $\xrightarrow[\text{Pyridine}]{(\text{CF}_3\text{SO}_2)_2\text{O}}$ R—OTf
Alcohol   Triflate

42 Sulphoxides and sulphones

Thioether $\xrightarrow[\text{H}_2\text{O}]{\text{H}_2\text{O}_2}$ Sulphoxide $\xrightarrow{\text{CH}_3\text{CO}_3\text{H}}$ Sulphone

43 Thioethers

(a) R—X $\xrightarrow{\text{NaSR''}}$ R—S—R''
Alkyl halide   Thioether

(b) 2 × R—X $\xrightarrow[\text{H}_2\text{S}]{\text{KOH}}$ R—S—R
Alkyl halide   Thioether

44 Thiols

(a) R—X $\xrightarrow[\text{Excess} \atop \text{H}_2\text{S}]{\text{KOH}}$ R—SH
Alkyl halide   Thiol

(b) R—X $\xrightarrow[\text{ii) NaOH/H}_2\text{O}]{\text{i) } S=\!\!\!\!{{NH_2}\atop{NH_2}}}$ R—SH
Alkyl halide   Thiol

(c)

$$\text{Disulphide} \xrightarrow[\text{H}^+]{\text{Zn}} 2 \times \text{R-SH} \quad \text{Thiol}$$

45 α,β-Unsaturated aldehydes

$$\text{(OH)} \xrightarrow{\text{Dess-Martin periodinane}} \text{CHO}$$

46 α,β-Unsaturated ketones

(a) $\xrightarrow{\text{Br}_2/\text{HBr}}$

(b) $\xrightarrow[\text{iii) H}_2\text{O}_2]{\text{i) LDA/THF} \atop \text{ii) PhSeBr}}$

(c) $\xrightarrow[\text{Pd(OAc)}_2]{\text{i) NEt}_3 \atop \text{Me}_3\text{Si-Cl}}$

(d) $\xrightarrow[\text{CrO}_3/\text{pyridine}]{\text{SeO}_2 \text{ or}}$

(e) $\xrightarrow{\text{ptsa}}$

47 Ureas

(a) R-N=C=O Isocyanate $\xrightarrow{\text{H}_2\text{NR'}}$ RHN NHR' Urea

(b) RHN-CN Cyanamide $\xrightarrow[\text{HO-R'}]{\text{H}^+}$ RHN NHR Urea

48 Urethanes

(a) R-N=C=O Isocyanate $\xrightarrow{\text{HO-R'} \atop \text{Alcohol}}$ RHN OR' Urethane

(b) Cl OR' Chloroformate $\xrightarrow[\text{NaOH}]{\text{R-NH}_2 \atop \text{Amine}}$ RHN OR' Urethane

49 Vinyl esters and ethers

(a) R CHO Aldehyde $\xrightarrow[\text{ii) Acid chloride}]{\text{i) Base}}$ Vinyl ester

(b) R CHO Aldehyde $\xrightarrow[\text{ii) R}_3\text{SiCl}]{\text{i) Base}}$ Vinyl silyl ether

(c) R CHO Aldehyde $\xrightarrow[\text{ii) R'OTs}]{\text{i) Base}}$ Vinyl ether

Appendix 2

Functionalization

The following reactions are examples of how a functional group can be introduced at a position which has no such group. Further details on each reaction are available in the book's Online Resource Centre.

 http://oxfordtextbooks.co.uk/orc/patrick_synth/

1 Aldehydes

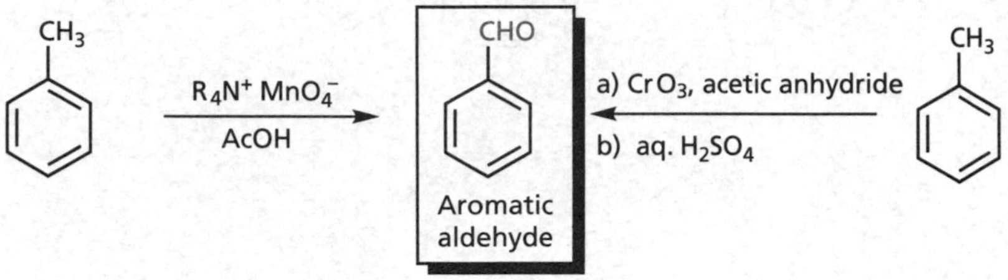

CH$_3$ → (R$_4$N$^+$ MnO$_4^-$ / AcOH) → CHO (Aromatic aldehyde) ← a) CrO$_3$, acetic anhydride b) aq. H$_2$SO$_4$ ← CH$_3$

2 Alkyl halides

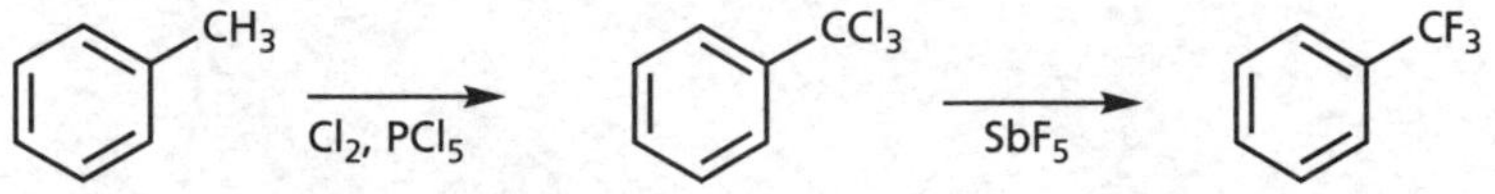

CH$_3$ → (Cl$_2$, PCl$_5$) → CCl$_3$ → (SbF$_5$) → CF$_3$

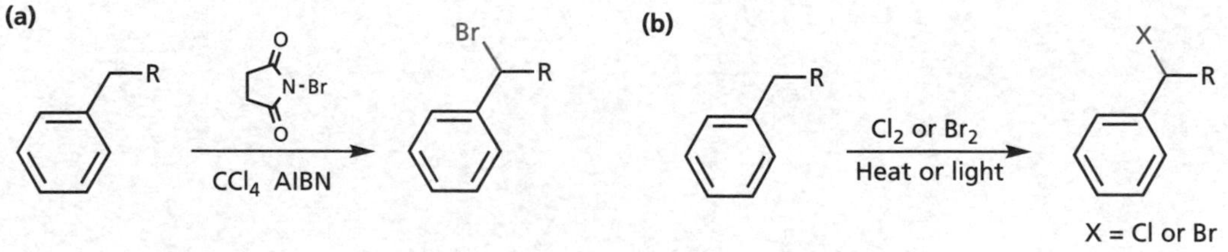

(a) R + (N-Br / CCl$_4$ AIBN) → Br—R

(b) R → (Cl$_2$ or Br$_2$ / Heat or light) → X—R   X = Cl or Br

3 Aryl halides

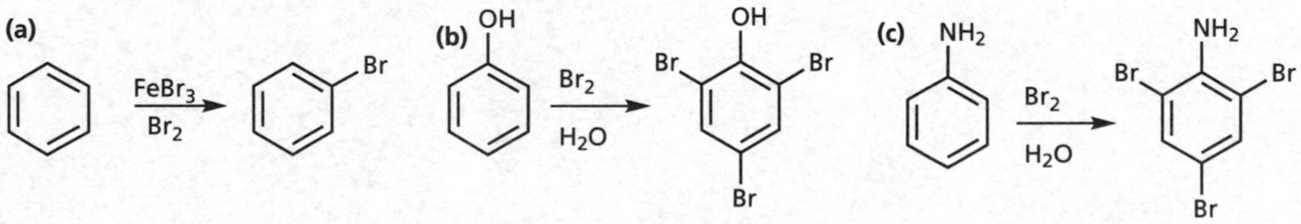

(a) → (FeBr$_3$ / Br$_2$) → Br

(b) OH → (Br$_2$ / H$_2$O) → Br, OH, Br, Br

(c) NH$_2$ → (Br$_2$ / H$_2$O) → Br, NH$_2$, Br, Br

4 Aryl sulphonic acids

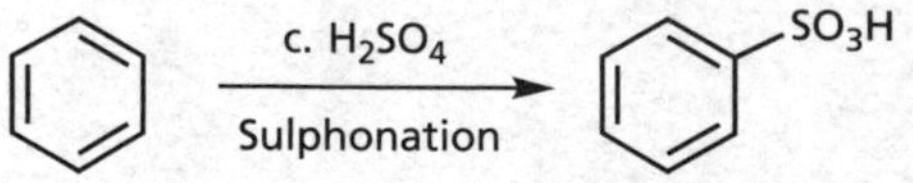

5 Carboxylic acids

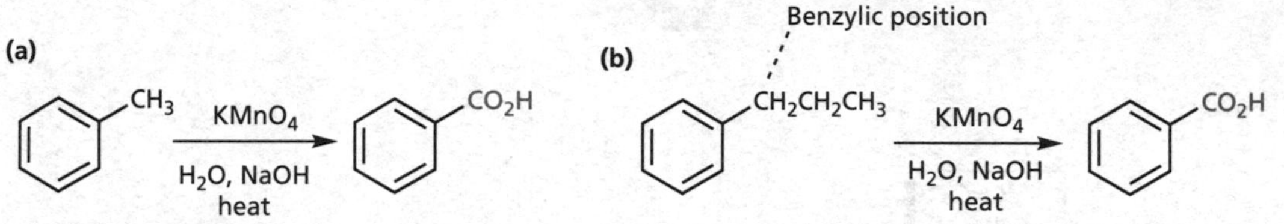

(a)

(b)

Benzylic position

6 Nitroaryls

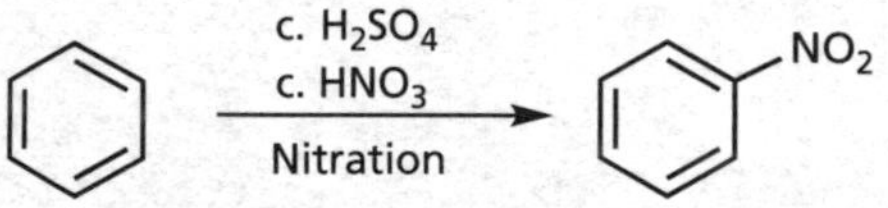

Appendix 3

Removal of functional groups

It is sometimes necessary to remove a functional group during a synthesis. The following summarize some of the methods of carrying out this operation. Further details on each reaction are available in the book's Online Resource Centre.

 http://oxfordtextbooks.co.uk/orc/patrick_synth/

1 Reduction of aldehydes and ketones

(a)

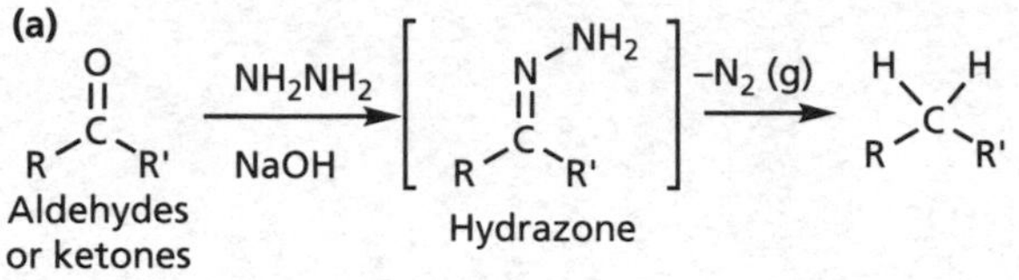

Aldehydes or ketones $\xrightarrow[\text{NaOH}]{\text{NH}_2\text{NH}_2}$ [Hydrazone] $\xrightarrow{-\text{N}_2\text{ (g)}}$ product

(b)

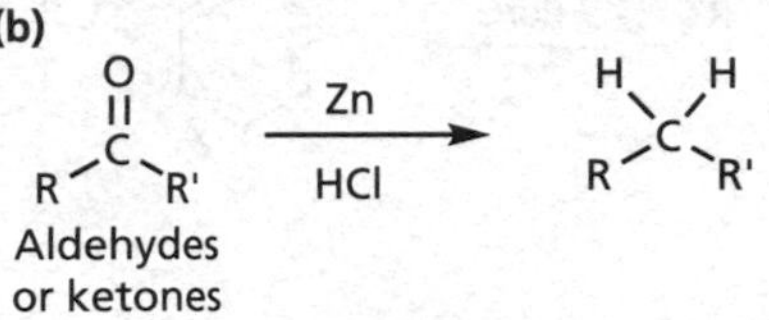

Aldehydes or ketones $\xrightarrow[\text{HCl}]{\text{Zn}}$ product

(c)

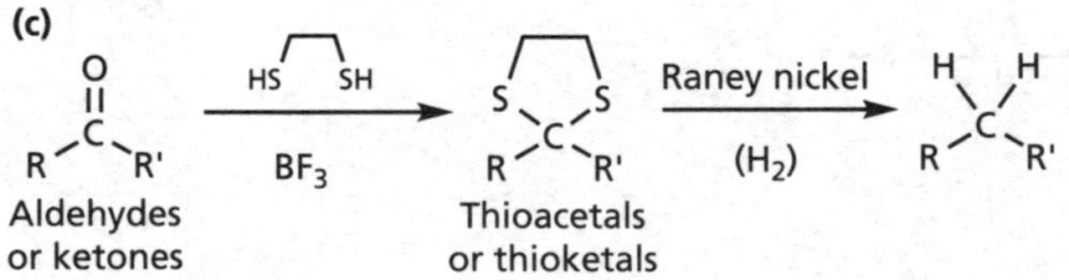

Aldehydes or ketones $\xrightarrow[\text{BF}_3]{\text{HS} \quad \text{SH}}$ Thioacetals or thioketals $\xrightarrow[\text{(H}_2\text{)}]{\text{Raney nickel}}$ product

(d)

Aldehydes or ketones $\xrightarrow[\text{Pd/C}]{\text{H}_2}$ product

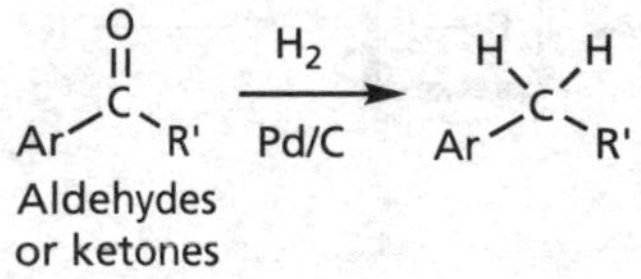

2 Reduction of alkenes, alkynes, and aromatic rings

(a)

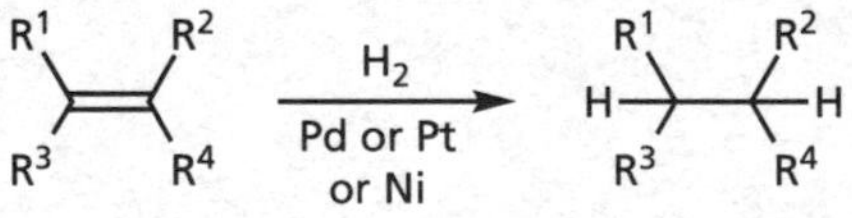

$\xrightarrow[\substack{\text{Pd or Pt} \\ \text{or Ni}}]{\text{H}_2}$

(b)

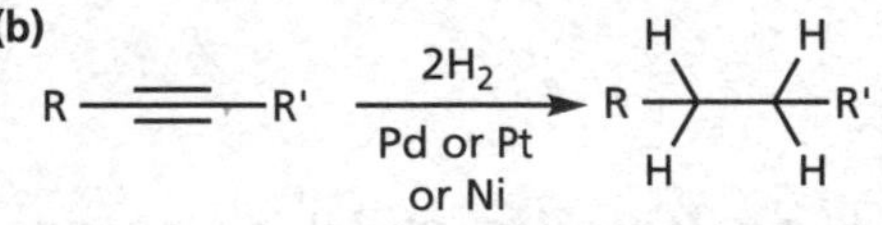

$R\text{—}\!\!\equiv\!\!\text{—}R' \xrightarrow[\substack{\text{Pd or Pt} \\ \text{or Ni}}]{2\text{H}_2}$

(c)

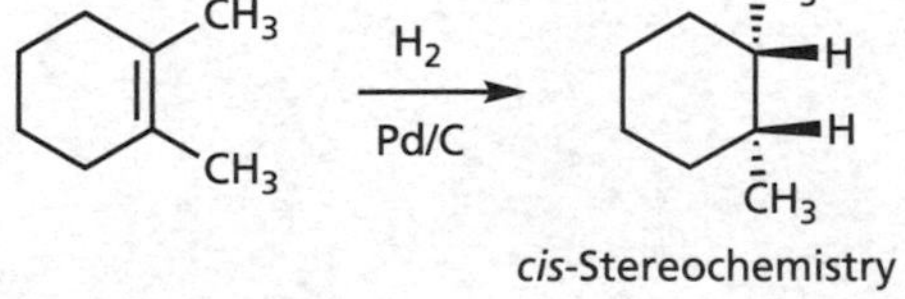

$\xrightarrow[\text{Pd/C}]{\text{H}_2}$

cis-Stereochemistry

(d)

$\xrightarrow[\text{metal catalyst}]{\text{H}_2}$

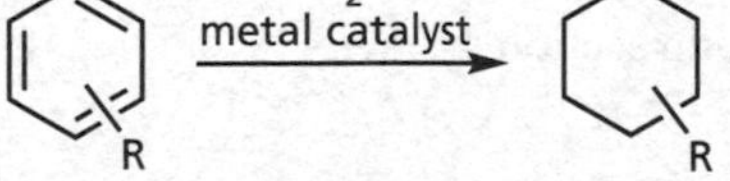

3 Removal of an alcohol group

(a)

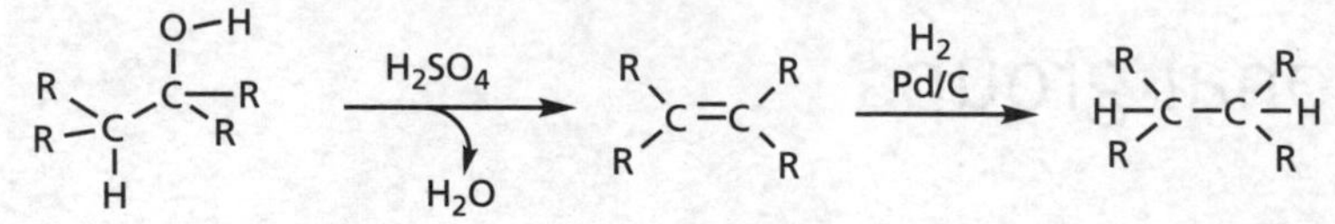

(b)

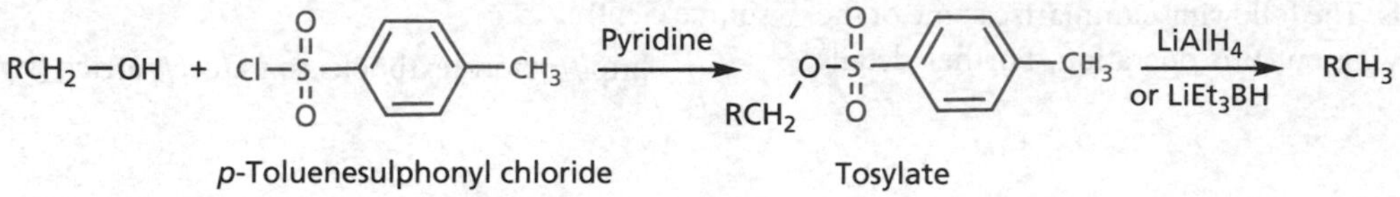

p-Toluenesulphonyl chloride   Tosylate

(c)

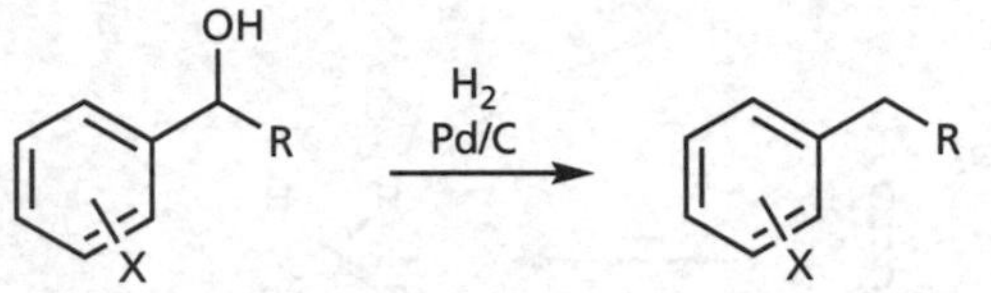

4 Removal of an alkyl halide—dehalogenation

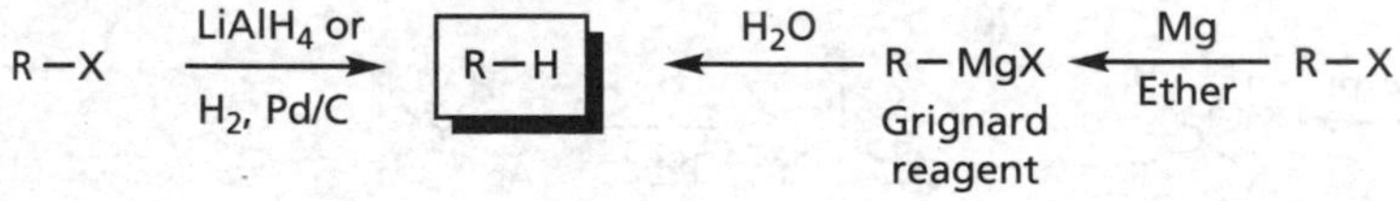

5 Removal of aromatic amines or nitro groups

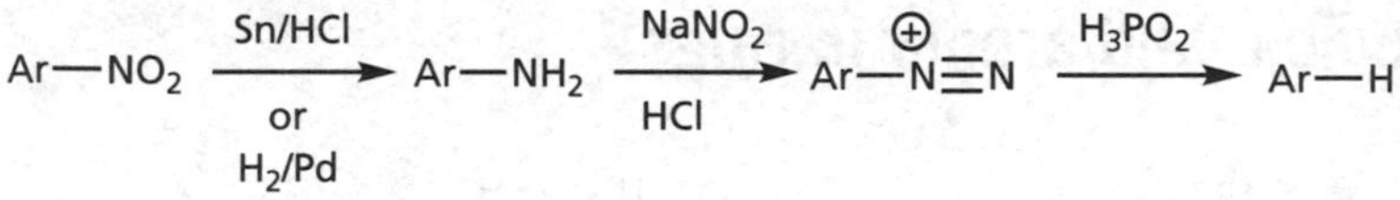

6 Removal of a carboxylic acid or ester

Carboxylic acids and esters can be reduced to alcohols,
then treated as described in section 4 above.

Appendix 4

Coupling reactions involving carbon–heteroatom bond formation

There are a large number of possible methods of coupling molecules through carbon–heteroatom bond formation. The following are examples of the most commonly used C–X couplings in drug synthesis. Further details on each reaction are available in the book's Online Resource Centre.

 http://oxfordtextbooks.co.uk/orc/patrick_synth/

1 Synthesis of amides

(a) R–CO–Cl (Acid chloride) → [Pyridine or NaOH, HNR'$_2$] → R–CO–NR'$_2$ (Amide)

(b) R–CO–OCOR (Acid anhydride) → [Pyridine, HNR'$_2$] → R–CO–NR'$_2$ (Amide)

(c) R–CO–OH (Carboxylic acid) → [DCC or DIC or EDC, H$_2$NR'] → R–CO–NHR' (Amide)

(d) R–CO–OH (Carboxylic acid) → [DCC, HOBt, H$_2$NR'] → R–CO–NHR' (Amide)

(e) R–CO–OH (Carboxylic acid) → [PyBOP or PyBrBOP, H$_2$NR'] → R–CO–NHR' (Amide)

(f) R–CO–NHR' (2° Amide) → [i) NaH, ii) MeI] → R–CO–NR'Me (3° Amide)

2 Synthesis of amines

(a) RX (Alkyl halide) → [i) NaN$_3$, ii) LiAlH$_4$ or H$_2$, cat. or PPh$_3$] → RNH$_2$ (1° Amine)

(b) RX (Alkyl halide) → [i) Potassium phthalimide, ii) H$_2$O, HO$^-$] → RNH$_2$ (1° Amine)

(c) R–CO–NH$_2$ (1° Amide) → [4 NaOH, Br$_2$, H$_2$O] → RNH$_2$ (1° Amine)

(d) R–CO–N$_3$ (Acyl azide) → [H$_2$O, heat] → RNH$_2$ (1° Amine)

(e) RNH$_2$ (1° Amine) → [i) R''–X, ii) NaOH] → RNHR'' + RNR'' (2° and 3° Amines)

(f) RNHR' (2° Amine) → [i) R''–X, ii) NaOH] → RNR'R'' (3° Amine)

(g) RNH$_2$ (1° Amine) → [i) Acid chloride, ii) LiAlH$_4$, iii) H$_3$O$^+$] → RNHCH$_2$R' (2° Amine)

(h) RNHR' (2° Amine) → [i) Acid chloride, ii) LiAlH$_4$, iii) H$_3$O$^+$] → RR'NHCH$_2$R'' (3° Amine)

16 Synthesis of esters

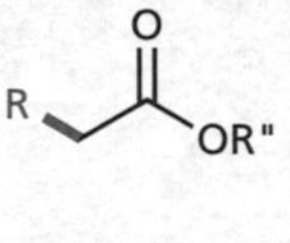

Alkylation of an ester or diethyl malonate

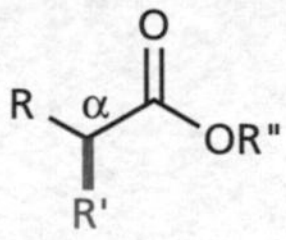

Alkylation of an ester

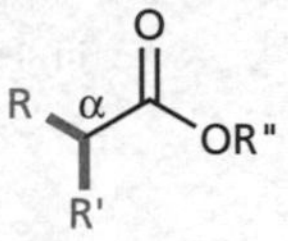

Alkylation of diethyl malonate

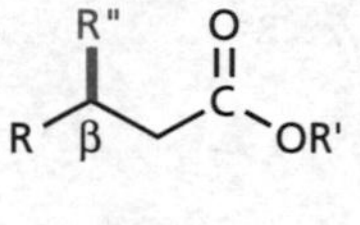

Alkylayion of an α,β–unsaturated ester

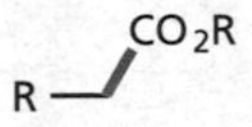

From carboxylic acids

(a) RCH_2-CO_2R $\xrightarrow[\text{ii) R'-X}]{\text{i) LDA/THF}}$ $\underset{R'}{RCH}-CO_2R$

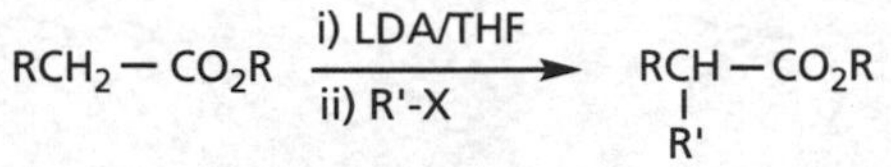

(b) $R-X$ $\xrightarrow[\substack{\text{ii) } CO_2 \\ \text{iii) R'OH, H}^+}]{\text{i) Mg}}$ $R-CO_2R'$

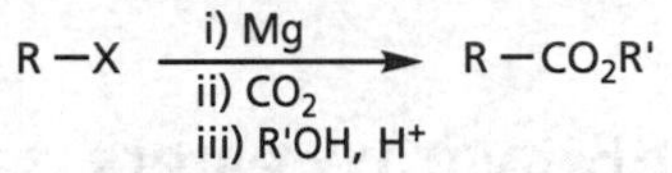

(c) $R-X$ $\xrightarrow[\substack{\text{ii) } H_3O^+ \\ \text{iii) R'OH, H}^+}]{\substack{\text{(i) Diethyl} \\ \text{malonate}}}$ $R-CH_2CO_2R'$

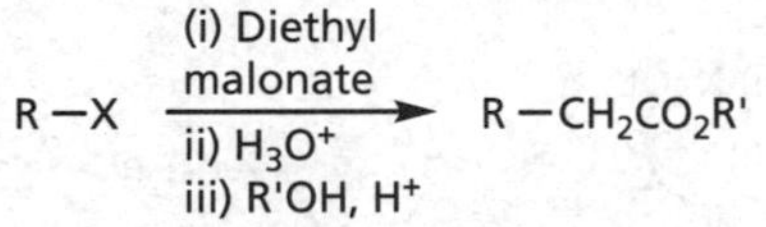

(d) $R-X$ $\xrightarrow[\substack{\text{ii) Base, R'X} \\ \text{iii) } H_3O^+ \\ \text{iv) R"OH, H}^+}]{\substack{\text{i) Diethyl} \\ \text{malonate}}}$ $R-\underset{R'}{\overset{H}{C}}-CO_2R"$

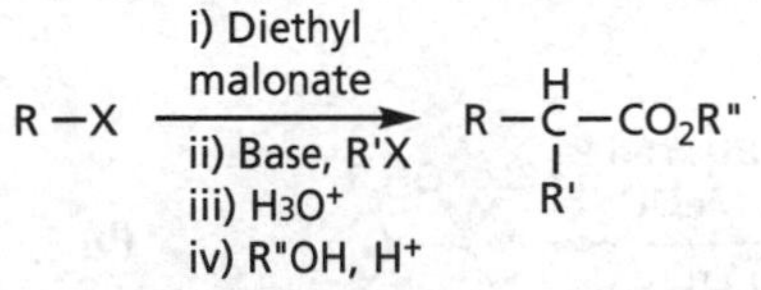

(e) $\xrightarrow[\text{ii) } H_3O^+]{\text{i) Li}^+ \text{(R"}_2\text{Cu}^-)}$

α,β-Unsaturated ester

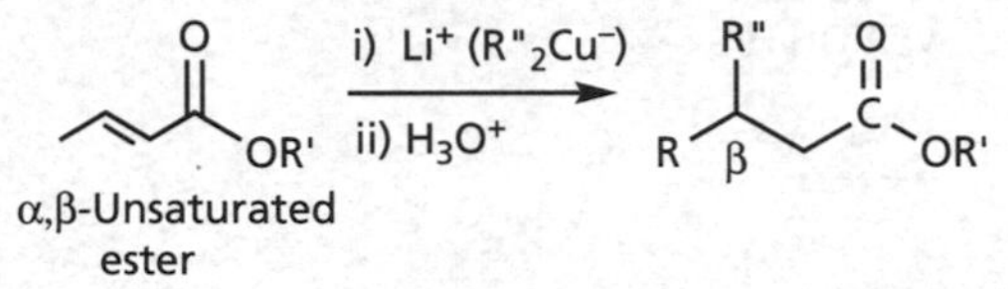

(f) $R-CO_2H$ $\xrightarrow[\substack{\text{ii) } CH_2N_2 \\ \text{iii) Ag}_2O, \text{R'OH}}]{\text{i) SOCl}_2}$ $R-CH_2CO_2R'$

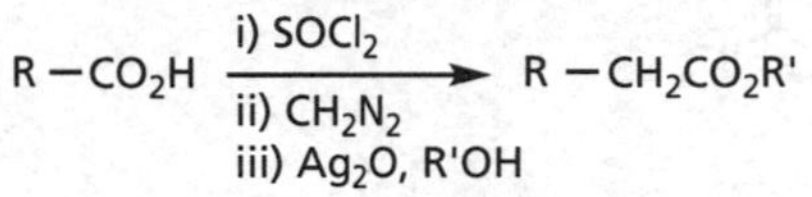

17 Synthesis of α-hydroxy aldehydes and ketones

(a) $2\,Ar-CHO$ $\xrightarrow{\text{NaCN}}$

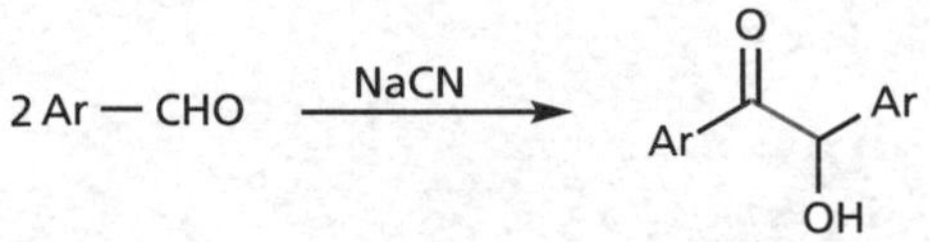

(b) $R-CHO$ $\xrightarrow[\substack{\text{ii) } H^+ \\ \text{iii) DIBAL}}]{\text{i) NaCN}}$

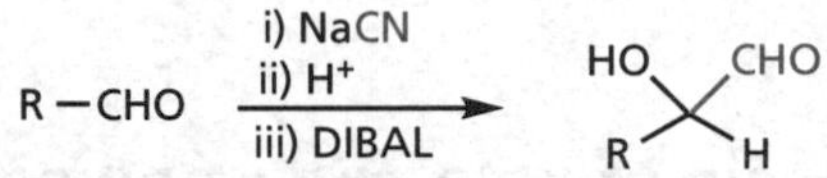

(c) $\xrightarrow[\substack{\text{ii) } H^+ \\ \text{iii) DIBAL}}]{\text{i) NaCN}}$

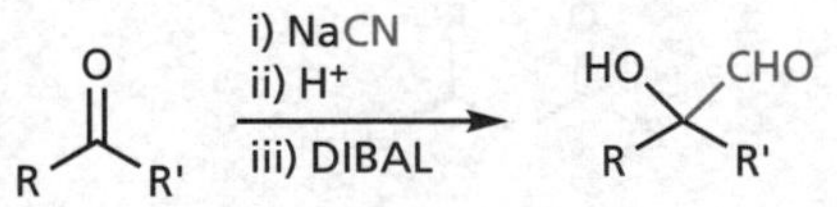

(d) $H-\!\!\equiv\!\!-H$ $\xrightarrow[\substack{\text{iii) HgO, H}_2SO_4}]{\substack{\text{i) NaNH}_2, \text{NH}_3 \\ \text{ii)}}}$

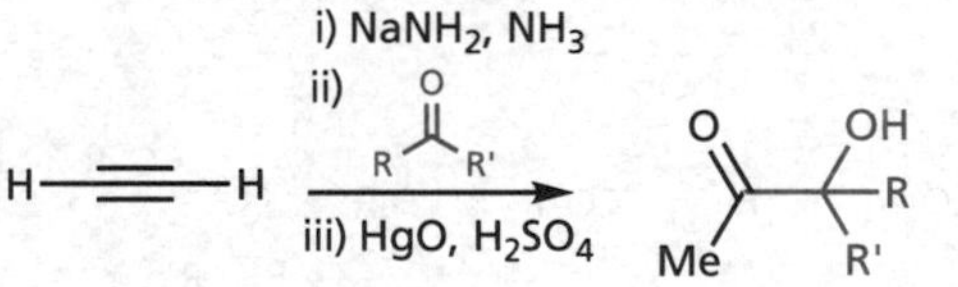

(e) $R-CHO$ $\xrightarrow[\substack{\text{ii) BuLi} \\ \text{iii) R'CHO} \\ \text{iv) Hydrolysis}}]{\text{i) HS}\frown\text{SH}}$

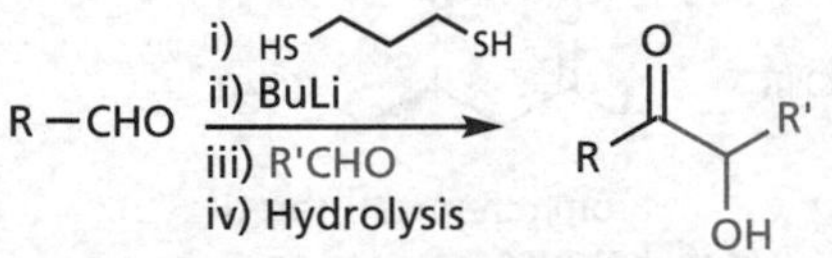

(f) $\xrightarrow[\text{ii) } H_2O, H^+]{\text{i) Na, Me}_3\text{SiCl}}$

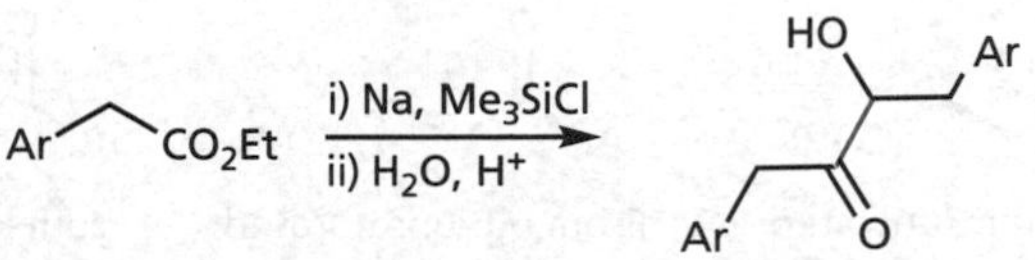

18 Synthesis of β-hydroxy aldehydes and ketones

(a) $\xrightarrow[\substack{\text{ii)} \\ \text{iii) } H_2O}]{\text{i) Base}}$

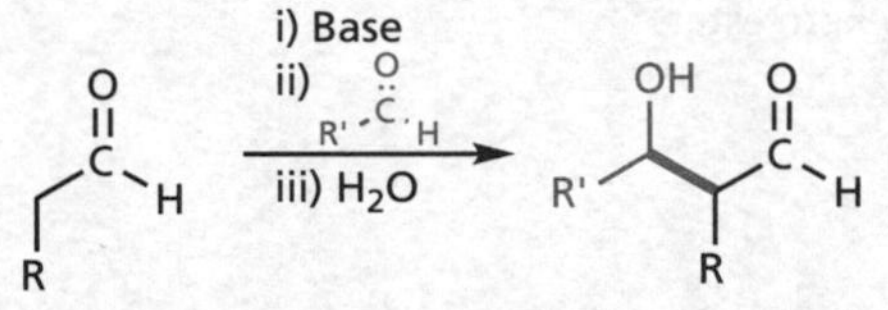

(b) $\xrightarrow[\substack{\text{ii)} \\ \text{iii) } H_2O}]{\text{i) Base}}$

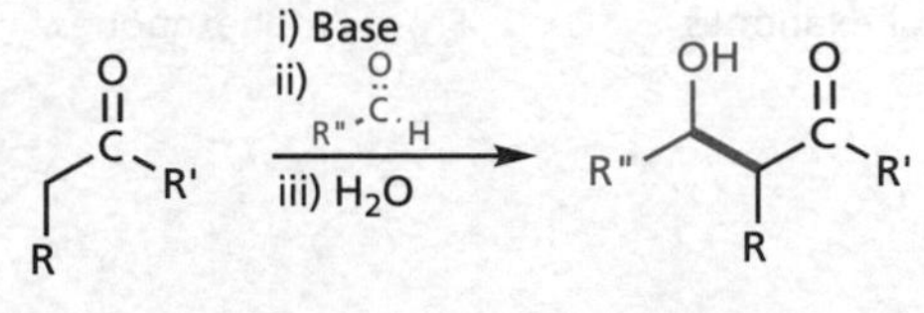

(c) R'CH₂C(=O)R → i) LDA, ii) R"CHO, iii) H₂O → R"CH(OH)CHR'C(=O)R

(d) RC(=O)CH₂R' → i) TMS-Cl, NEt₃, ii) R"C(=O)R''' / TiCl₄, iii) H₂O → RC(=O)CHR'C(OH)R"R'''

(e) CH₃CHO → i) RNH₂ or NH₂NHMe, ii) LDA, iii) PhC(=O)Me, iv) Hydrolysis → HC(=O)CH₂C(OH)(Me)(Ph)

19 Synthesis of β-hydroxy esters

(a) RCH₂CO₂H → i) Br₂, red P; ii) MeOH; iii) Zn; iv) R'C(=O)R" → R"R'C(OH)CHR'... CO₂Me

(b) CH₂(CO₂Et)₂ → i) NaOEt; ii) epoxide (R); iii) aq. Base; iv) aq. Acid; v) R'OH/H⁺ → HO–CHR–CH₂–CO₂R'

20 Synthesis of α-hydroxy ethers

(a) epoxide (R) → NaOR' → HO–CHR–CH₂–OR'

(b) epoxide (R) → R'OH / H⁺ → R'O–CHR–CH₂–OH

21 Synthesis of keto acids and esters

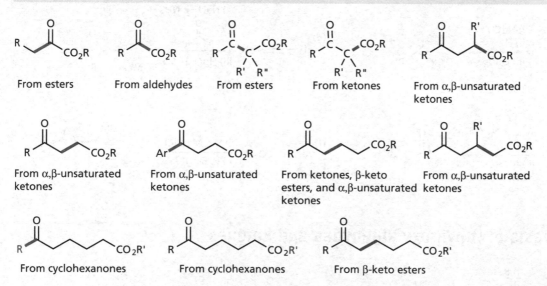

R–CH₂–C(=O)–CO₂R   From esters

R–C(=O)–CO₂R   From aldehydes

R–C(=O)–C(R')(R")–CO₂R   From esters

R–C(=O)–C(R')(R")–CO₂R   From ketones

R–C(=O)–CH₂–CH(R')–CO₂R   From α,β-unsaturated ketones

R–C(=O)–CH₂CH₂–CO₂R   From α,β-unsaturated ketones

Ar–C(=O)–CH₂CH₂–CO₂R   From α,β-unsaturated ketones

R–C(=O)–CH₂CH₂CH₂–CO₂R   From ketones, β-keto esters, and α,β-unsaturated ketones

R–C(=O)–CH₂CH(R')–CO₂R   From α,β-unsaturated ketones

R–C(=O)–(CH₂)₄–CO₂R'   From cyclohexanones

R–C(=O)–(CH₂)₄–CO₂R'   From cyclohexanones

R–C(=O)–(CH₂)₃–CO₂R'   From β-keto esters

(a) R–CH₂–CO₂Et → i) NaOEt, EtO₂C–CO₂Et; ii) Hydrolysis; iii) Decarboxylation → R–CH₂–C(=O)–CO₂H

(b) R–CHO → i) HS–SH, BF₃, Et₂O; ii) ⁿBuLi; iii) ClCO₂Et; iv) NBS → R–C(=O)–CO₂Et

(c) 2 RCH₂–C(=O)–OEt → i) NaOEt; ii) H₃O⁺ → RCH₂–C(=O)–CH(R)–C(=O)–OEt

(d) 2 R–CH(R')–CO₂Et → Strong base → R(R')CH–C(=O)–C(R)(R')–CO₂Et

(e) H₃C–C(=O)–OEt → i) LDA/THF, Ph–C(=O)–OEt; ii) H₃O⁺ → Ph–C(=O)–CH₂–C(=O)–OEt

(f) R–C(=O)–CH₂R' → i) NaOEt, EtO–C(=O)–OEt; ii) Base; iii) R"-X → R–C(=O)–C(R')(R")–CO₂Et

(g) R–C(=O)–CH=CH–R' → i) [CH₂=C(OMe)]₂CuLi; ii) O₃, MeOH → R–C(=O)–CH₂–CH(R')–CO₂Me

(h) R–C(=O)–CH=CH–R' → i) KCN; ii) HO⁻, H₂O; iii) ROH, H⁺ → R–C(=O)–CH₂–CH(R')–CO₂R

(i) R–C(=O)–CH₂–CO₂Et → i) Base; ii) Br–CH₂–CO₂Et; iii) HCl → R–C(=O)–CH₂–CH₂–CO₂Et

(j) R–C(=O)–CH₂R → i) Morpholine; ii) Br–CH₂–CO₂Et; iii) HCl → R–C(=O)–CH(R)–CH₂–CO₂Et

(k) Ar–H → i) AlCl₃, (succinic anhydride); ii) ROH, H⁺ → Ar–C(=O)–CH₂–CH₂–CO₂R

(l) R–C(=O)–CH₂–CO₂Et → i) cat. ᵗBuO⁻; ii) CH₂=CH–C(=O)–OMe; iii) Hydrolysis; iv) Decarboxylation → R–C(=O)–CH₂–CH₂–CH₂–CO₂H

(m) R–C(=O)–CH₂R → i) pyrrolidine; ii) CH₂=CH–C(=O)–OR; iii) H₃O⁺; iv) ROH, H⁺ → R–C(=O)–CH(R)–CH₂–CH₂–CO₂H

(n) CH₂(CO₂Et)₂ + R–C(=O)–CH=CH–R' → i) NaOEt; ii) Hydrolysis; iii) Decarboxylation; iv) ROH, H⁺ → R–C(=O)–CH₂–CH(R')–CH₂–CO₂R

(o) cyclohexanone → i) RMgBr; ii) H⁺; iii) O₃; iv) R'OH, H⁺ → R–C(=O)–(CH₂)₃–CH₂–CO₂R'

(p) 2-R-cyclohexanone → i) mcpba; ii) R'-Li → R–C(=O)–⋯–CO₂R'

(q) R–C(=O)–CH₂–CO₂Et → i) NaOEt, Br–CH₂CH₂CH₂–CO₂Et; ii) Hydrolysis; iii) Decarboxylation; iv) R'OH, H⁺ → R–C(=O)–CH₂CH₂CH₂CH₂–CO₂R'

22 Synthesis of ketones

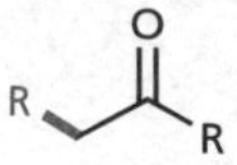

Alkylation of
β-keto esters

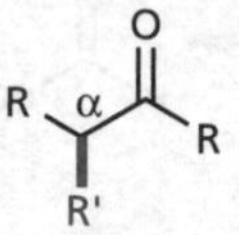

Alkylation of
aldehydes, ketones,
enamines or β-
keto esters

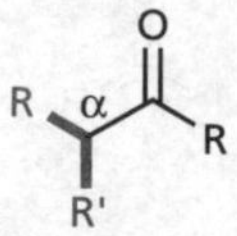

Dialkylation of
β-keto esters

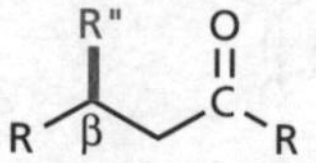

Alkylation of
α,β-unsaturated
ketones

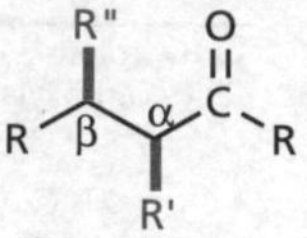

Alkylation of
α,β-unsaturated
ketones

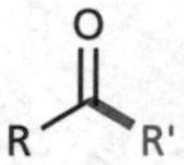

From acid chlorides or thioesters
From nitriles
From carboxylic acids
From ketones
From aldehydes

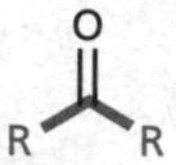

From organolithium
reagents and CO_2

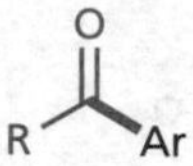

From Friedal–Crafts
acylation of aromatic
rings
From thioesters
From nitriles

(a) 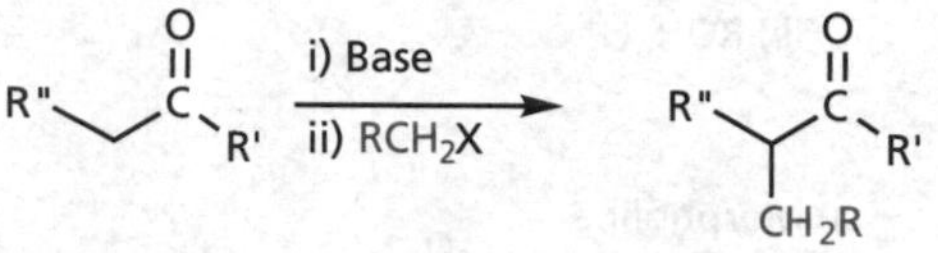 i) 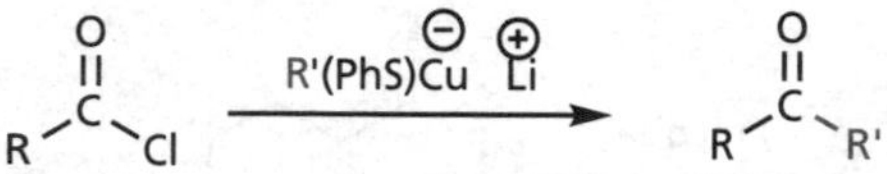 Base ii) RCH_2X

(b) 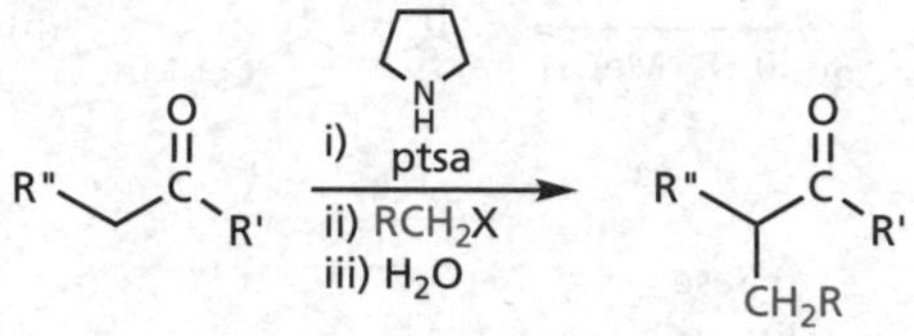 i) ptsa ii) RCH_2X iii) H_2O

(c) 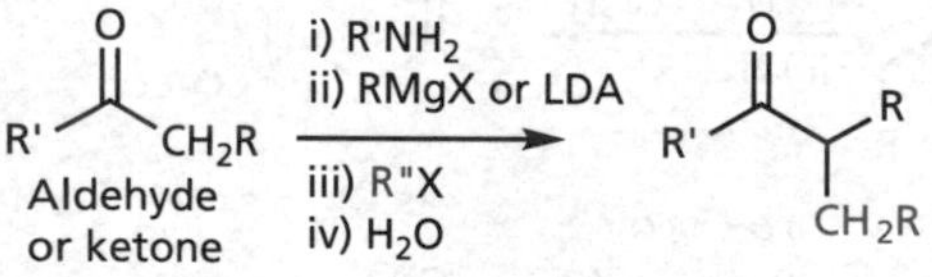 Aldehyde or ketone i) $R'NH_2$ ii) RMgX or LDA iii) $R''X$ iv) H_2O

(d) 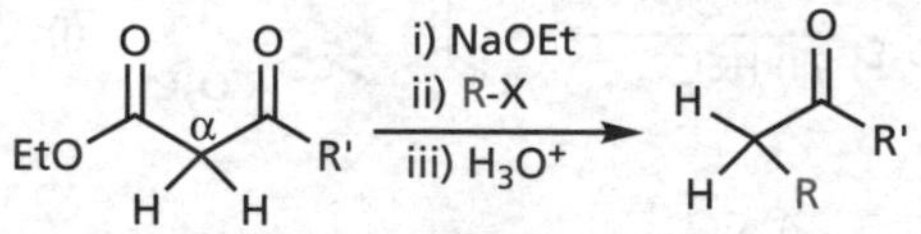 i) NaOEt ii) R-X iii) H_3O^+

(e) 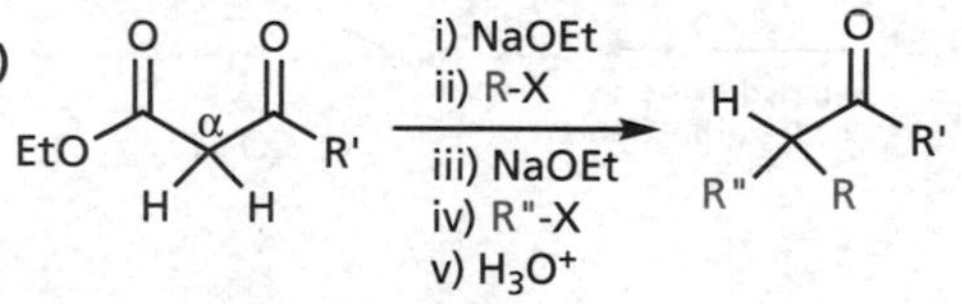 i) NaOEt ii) R-X iii) NaOEt iv) R''-X v) H_3O^+

(f) 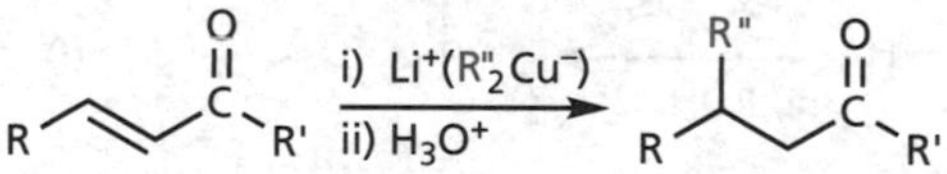 i) $Li^+(R''_2Cu^-)$ ii) H_3O^+

(g) 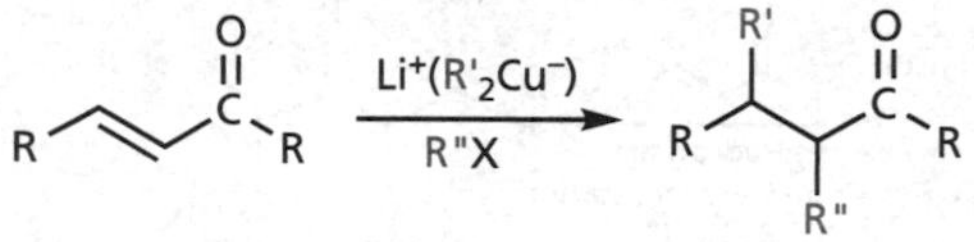 $Li^+(R'_2Cu^-)$ $R''X$

(h) 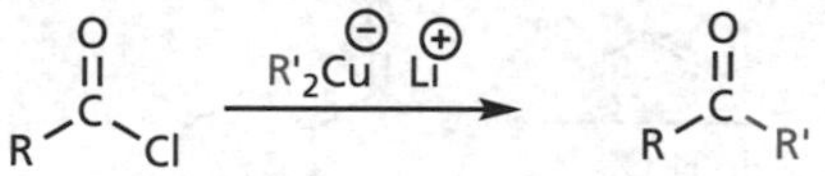 $R'_2Cu^- Li^+$

(i) $R'(PhS)Cu^- Li^+$

(j) 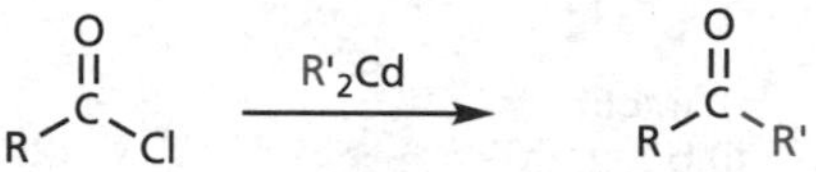 R'_2Cd

(k) 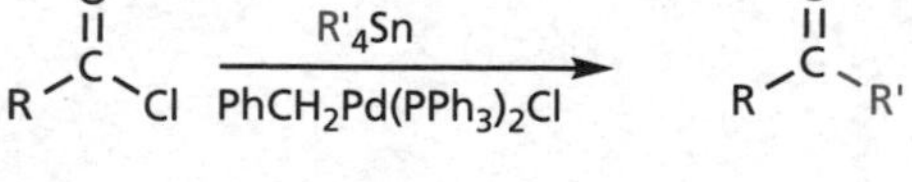 R'_4Sn $PhCH_2Pd(PPh_3)_2Cl$

(l) 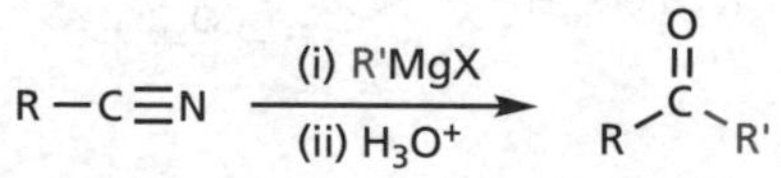 $R-C\equiv N$ (i) $R'MgX$ (ii) H_3O^+

(m) 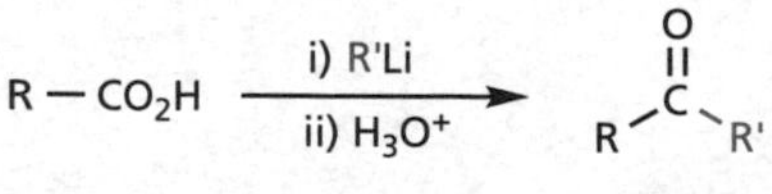 $R-CO_2H$ i) $R'Li$ ii) H_3O^+

(n) RLi i) CO_2 ii) RLi iii) H_3O^+

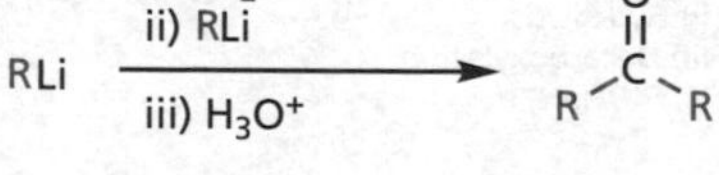

(o) 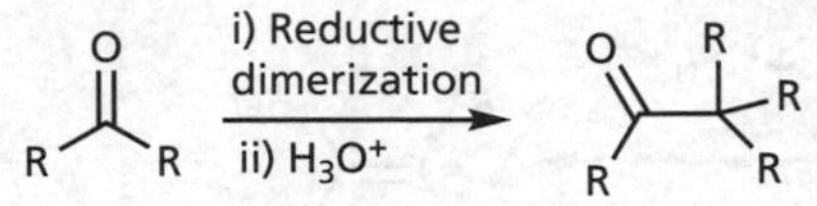 i) Reductive dimerization ii) H_3O^+

(p) 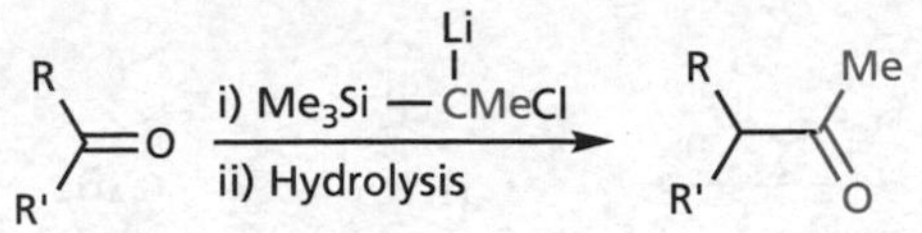 i) $Me_3Si—CMeCl$ (Li) ii) Hydrolysis

(q) 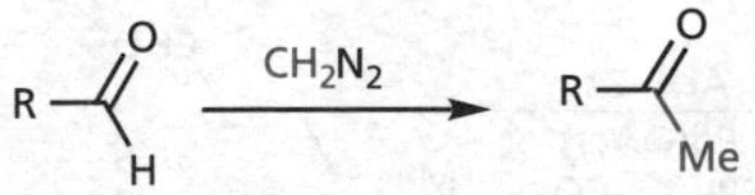 CH_2N_2

(r) 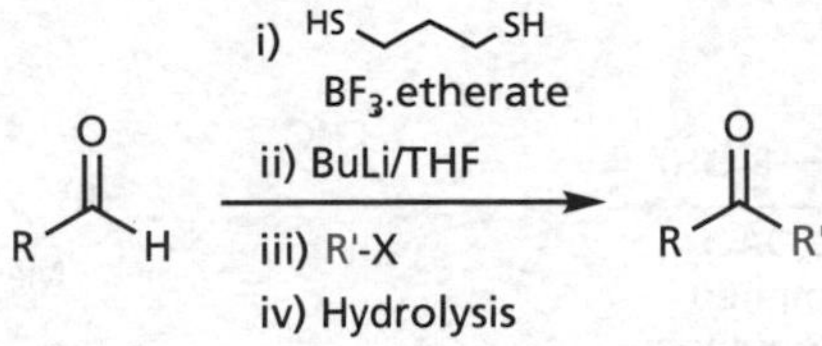 i) $HS{\sim}SH$ BF_3.etherate ii) BuLi/THF iii) R'-X iv) Hydrolysis

(s) 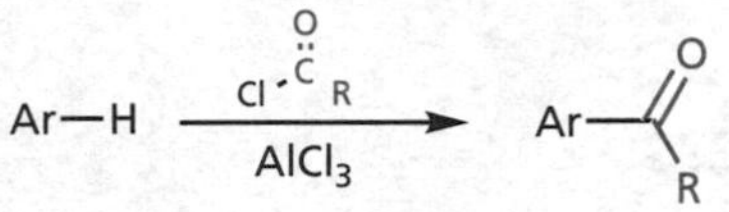 Ar—H $\xrightarrow[AlCl_3]{Cl—C(=O)R}$

23 Synthesis of nitriles

R—X $\xrightarrow{NaCN}$ R—CN

24 Synthesis of nitroalkanes

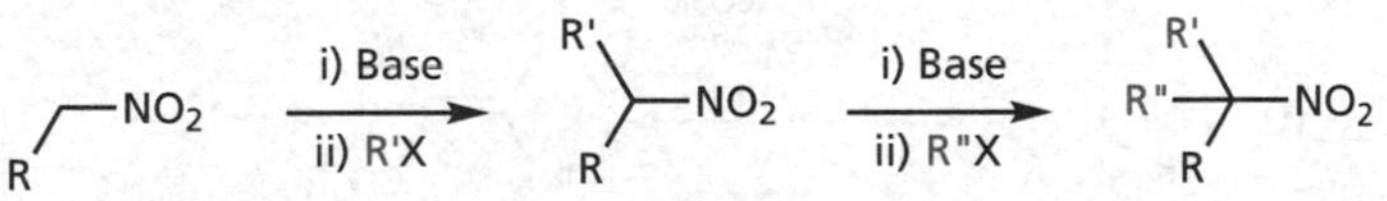

$\xrightarrow[\text{ii) R'X}]{\text{i) Base}}$ $\xrightarrow[\text{ii) R''X}]{\text{i) Base}}$

25 Synthesis of α,β-unsaturated aldehydes

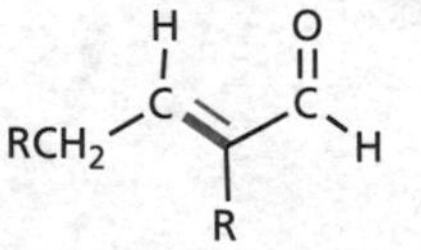

From aldehydes 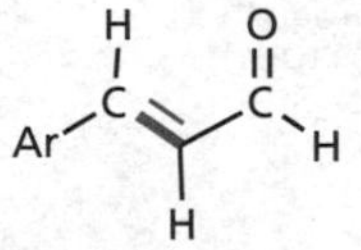   From aldehydes 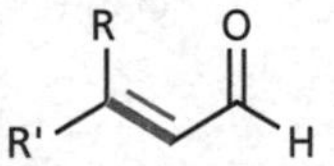   From ketones 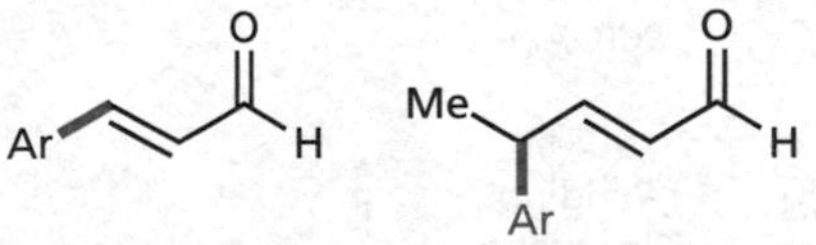   From α,β–unsaturated aldehydes

(a) 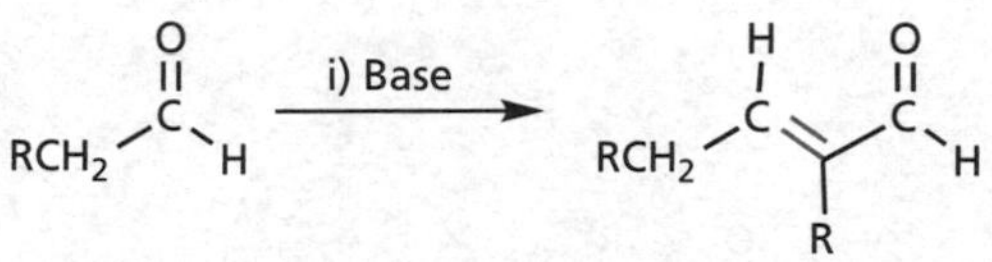 i) Base

(b) 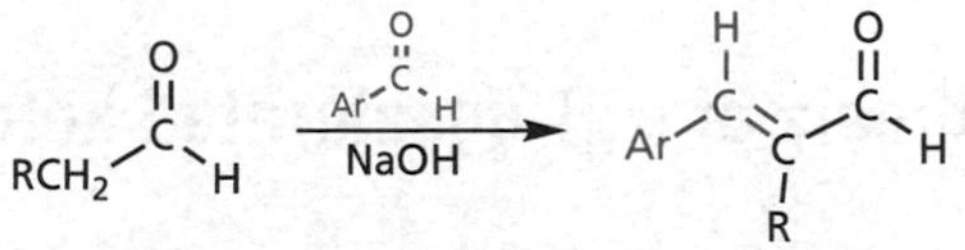 $\xrightarrow[NaOH]{ArCHO}$

(c) 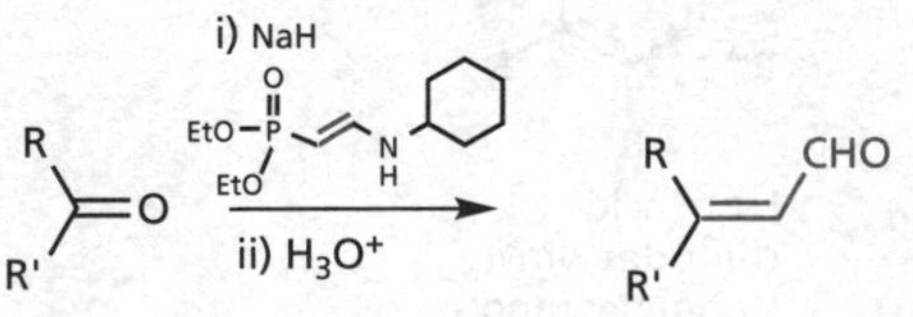 i) NaH ... ii) H_3O^+

(d)

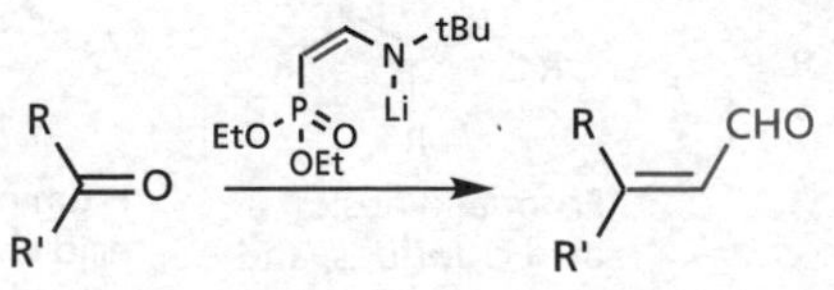

(e)

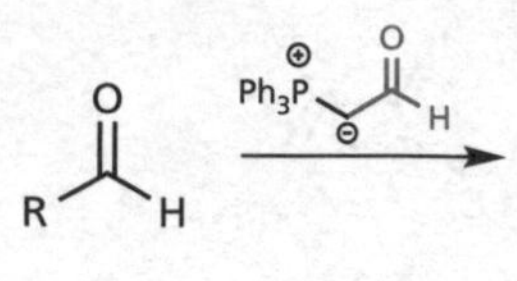

(f)

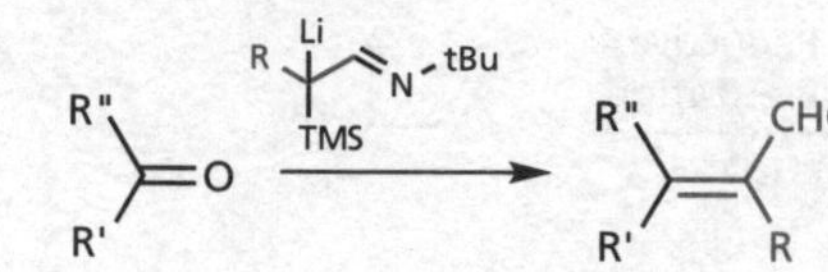

(g)

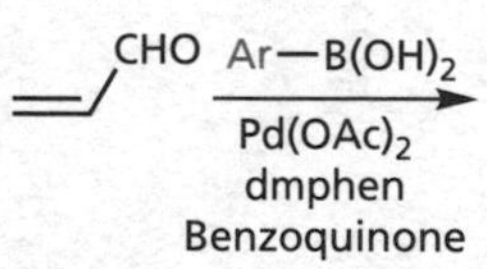

(h)

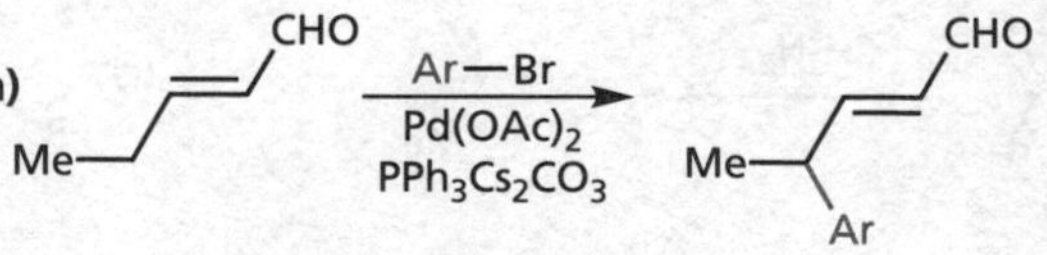

26 Synthesis of α,β-unsaturated acids and esters

(a) 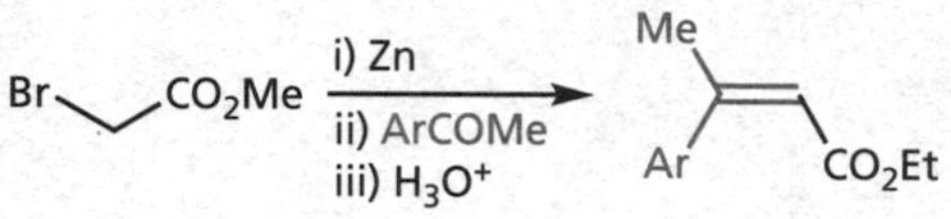 Br—CO$_2$Me
i) Zn
ii) ArCOMe
iii) H$_3$O$^+$

(b) 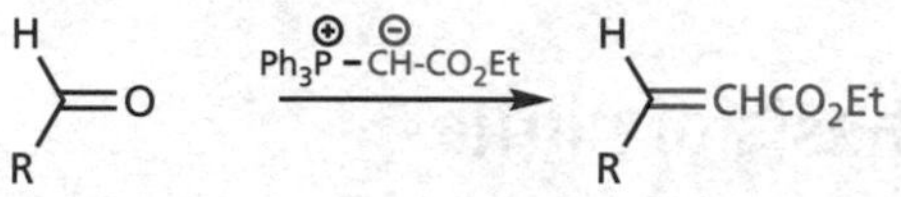
Ph$_3$P$^{\oplus}$—CH–CO$_2$Et

(c) 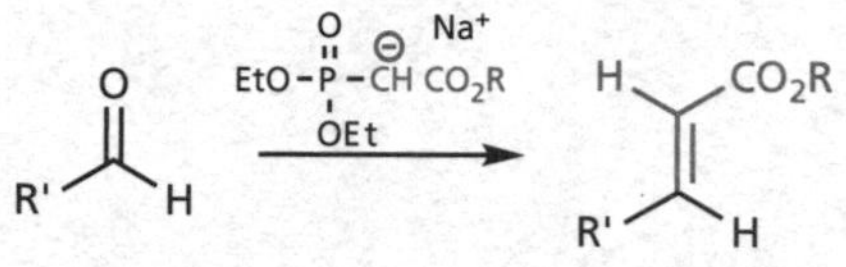
O
EtO–P$^{\ominus}$–CH CO$_2$R
OEt
Na$^+$

(d) 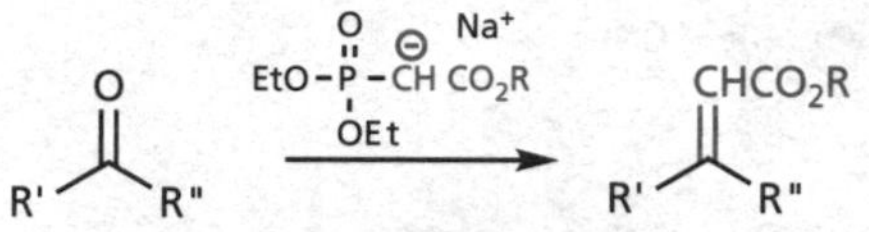
O
EtO–P$^{\ominus}$–CH CO$_2$R
OEt
Na$^+$

(e) 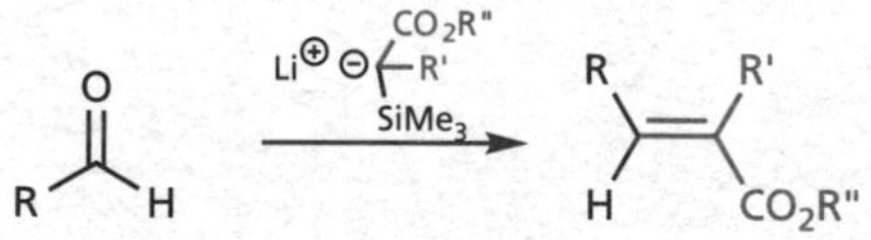
Li$^{\oplus}$ $^{\ominus}$O—R'
CO$_2$R"
SiMe$_3$

(f) 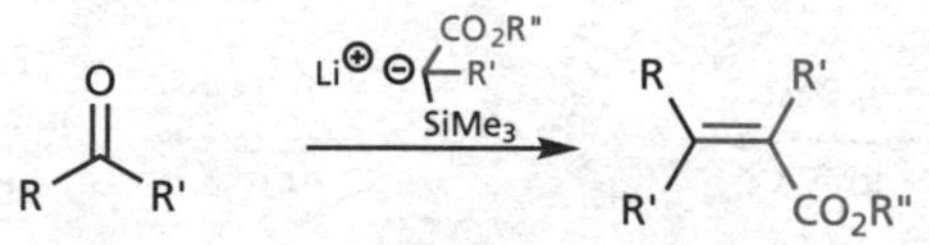
Li$^{\oplus}$ $^{\ominus}$O—R'
CO$_2$R"
SiMe$_3$

(g) 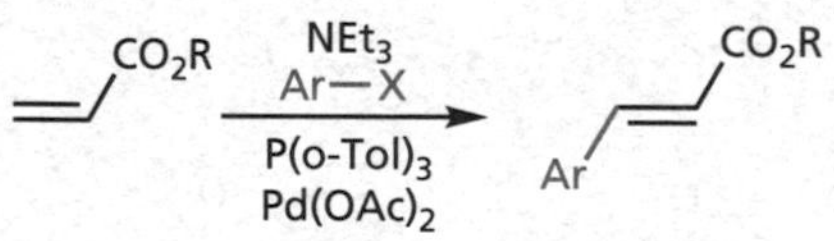
CO$_2$R
NEt$_3$
Ar—X
P(o-Tol)$_3$
Pd(OAc)$_2$

(h)
CO$_2$Et
CO$_2$Et
i) HOAc, R$_2$NH
ii)
H–R
iii) aq. HO$^-$
iv) aq. H$^+$.
v) EtOH/H$^+$

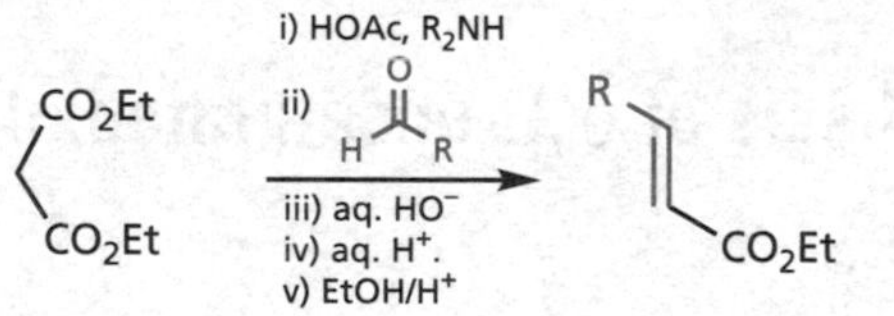

(i) 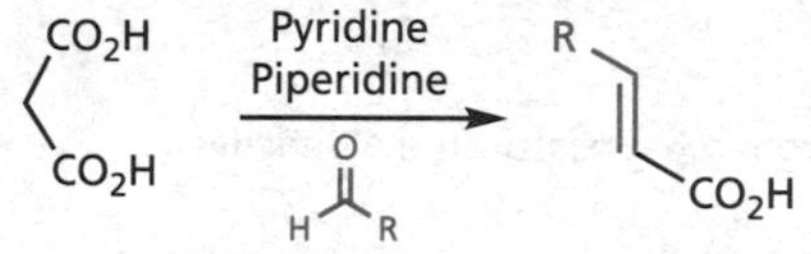
CO$_2$H
CO$_2$H
Pyridine
Piperidine
H R

(j) 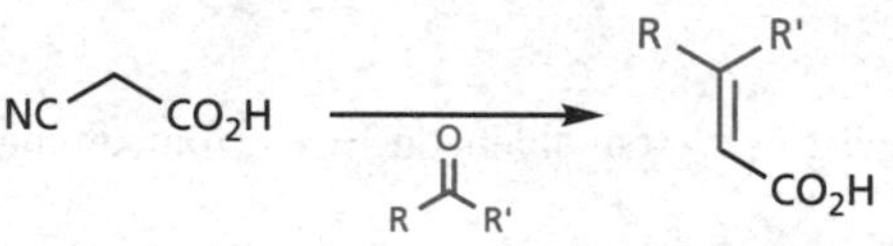
NC—CO$_2$H

27 Synthesis of α,β-unsaturated ketones

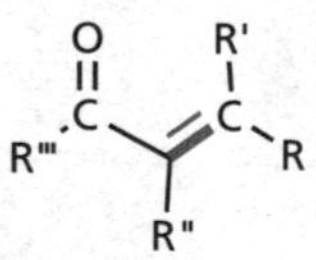

From ketones and aldehydes

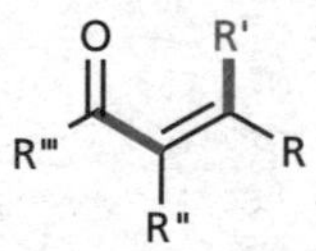

From alkynes, acid chlorides, and organocuprates

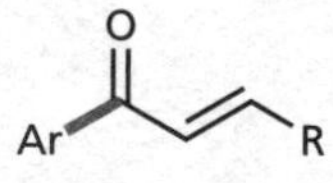

From α,β-unsaturated acid chlorides and lithium organocuprates

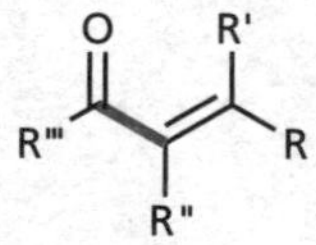

From acid chlorides, vinyl stannanes, and alkynes

(a)

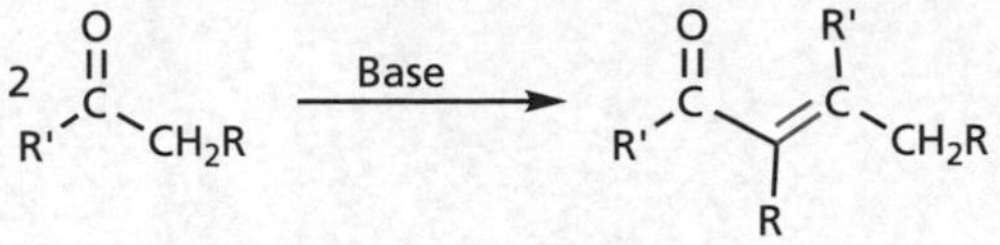

(b)

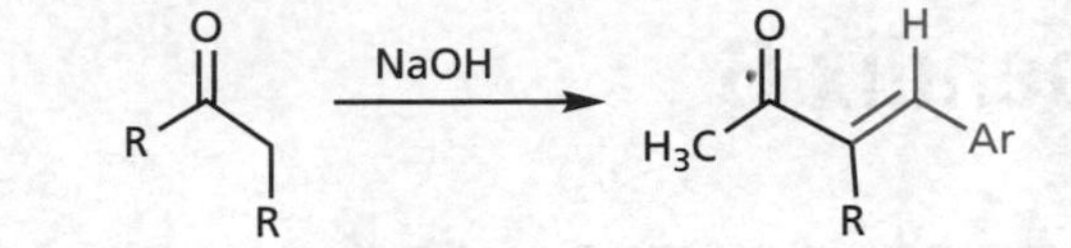

(c)

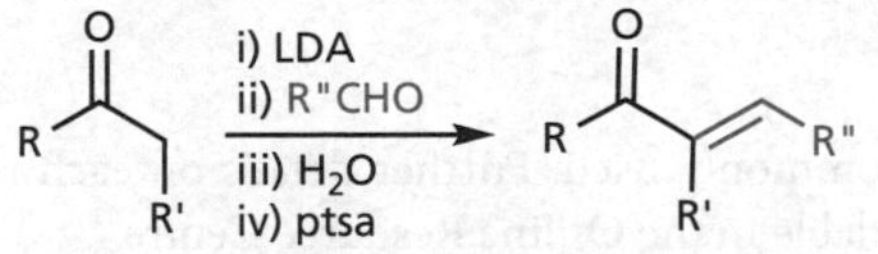

(d)

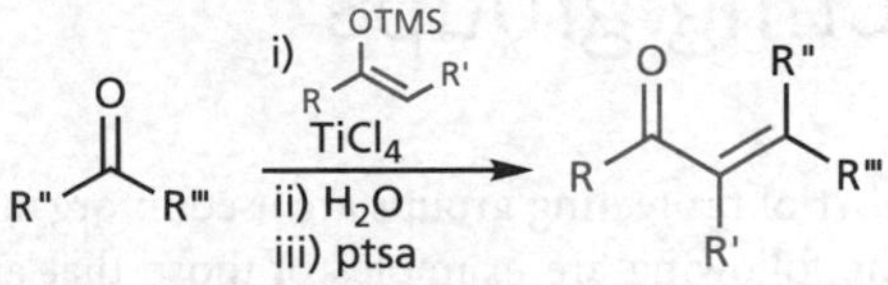

(e)

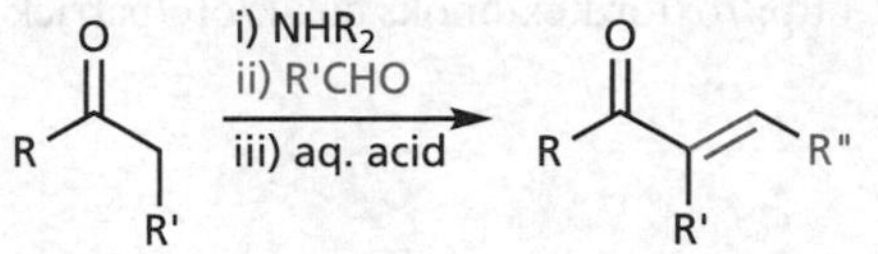

(f)

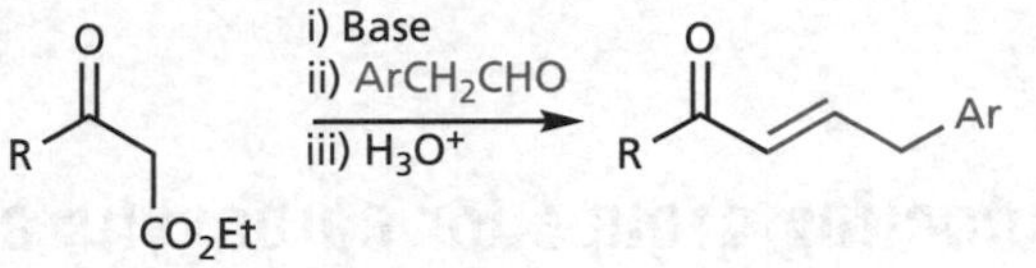

(g)

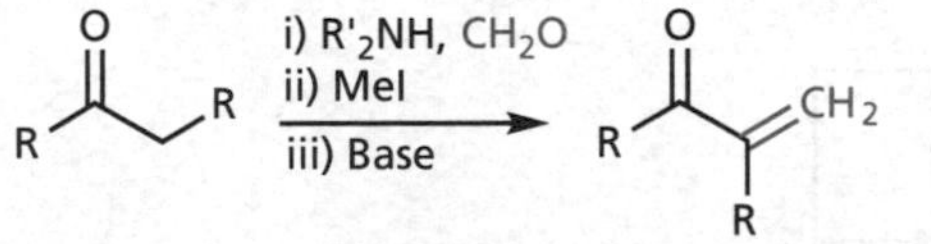

(h)

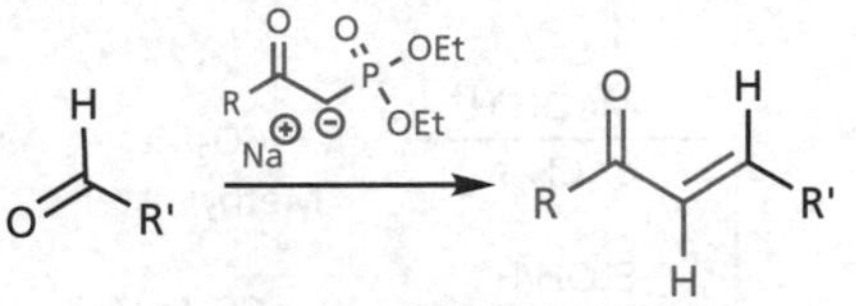

(i)

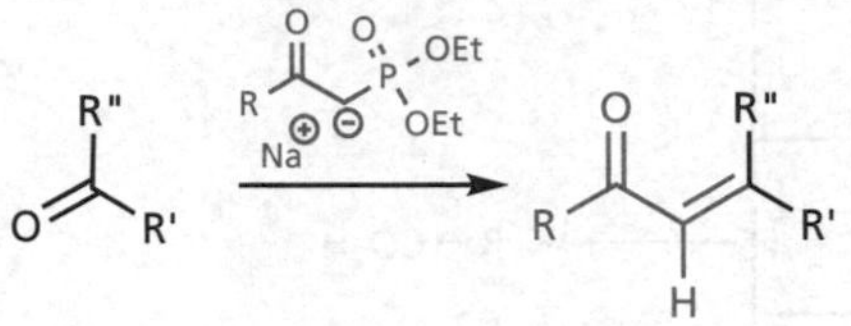

(j)

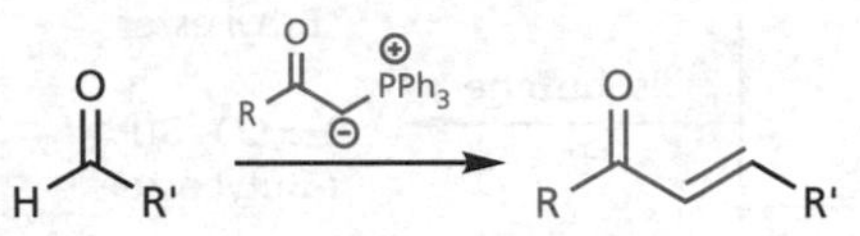

(k)

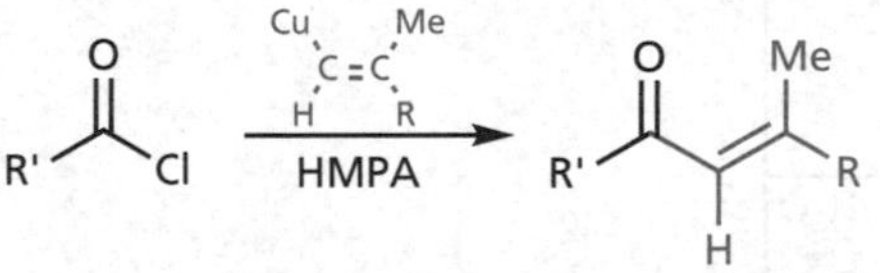

(l)

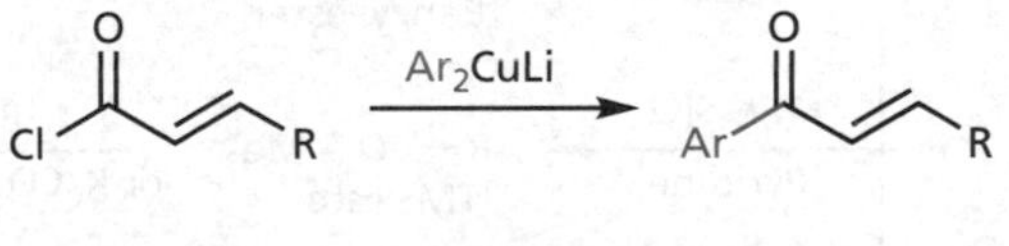

(m)

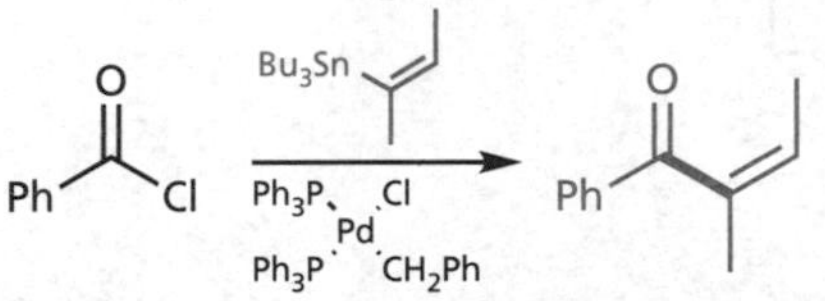

(n)

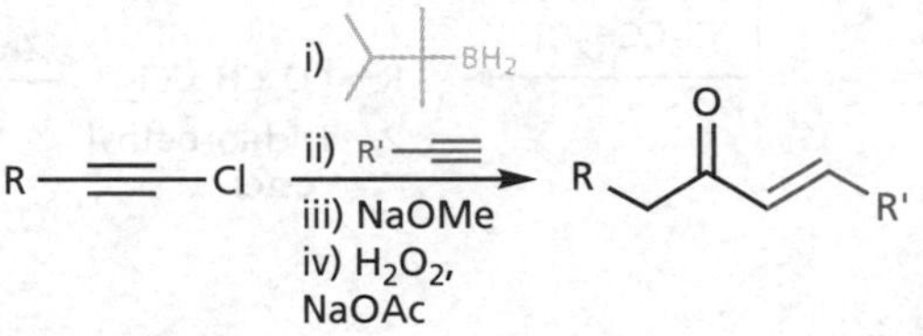

28 Synthesis of α,β-unsaturated nitriles

(a)

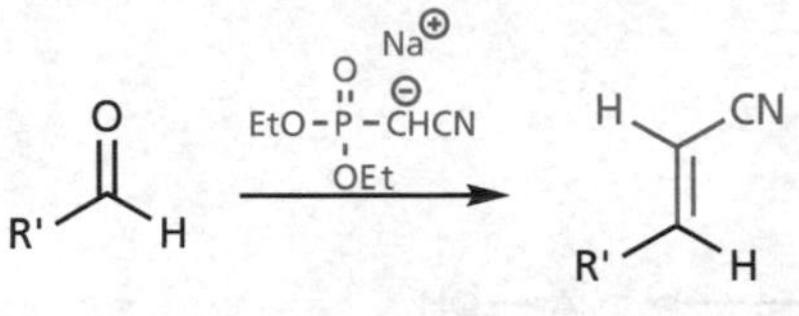

(b)

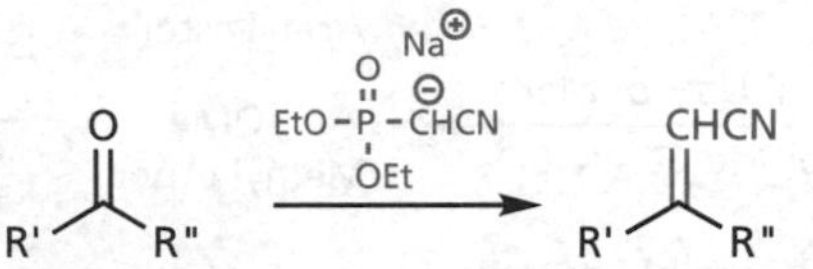

(c)

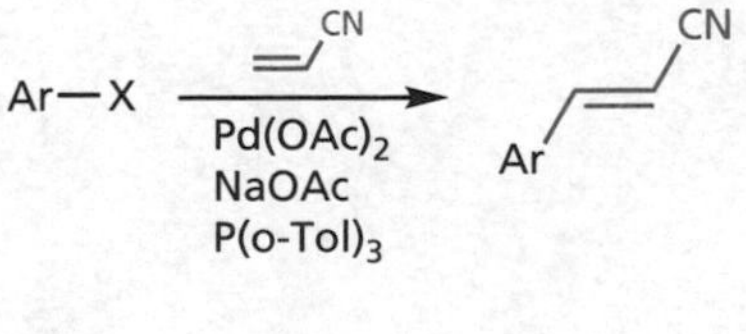

Appendix 6

Protecting groups

A large number of protecting groups are used in organic synthesis. The following are examples of those that are most commonly used. Further details on each reaction are available in the Online Resource Centre.

 http://oxfordtextbooks.co.uk/orc/patrick_synth/

1 Protecting groups for carboxylic acids

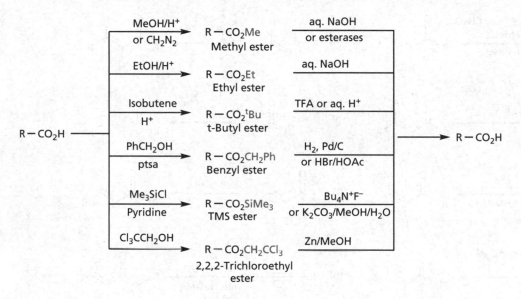

2 Protecting groups for phenols

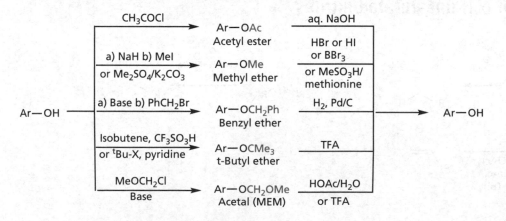

3 Protecting groups for alcohols

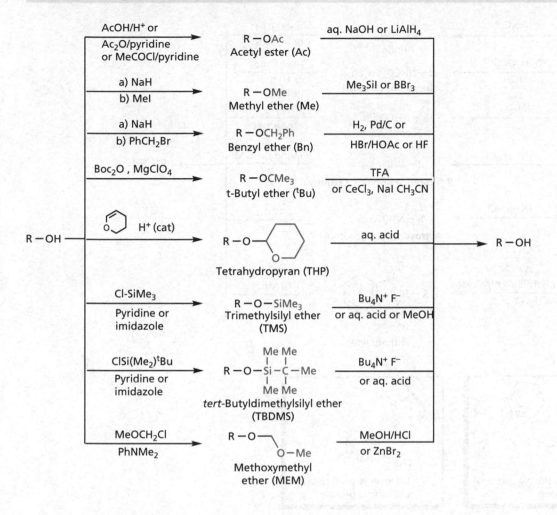

4 Protecting groups for amines

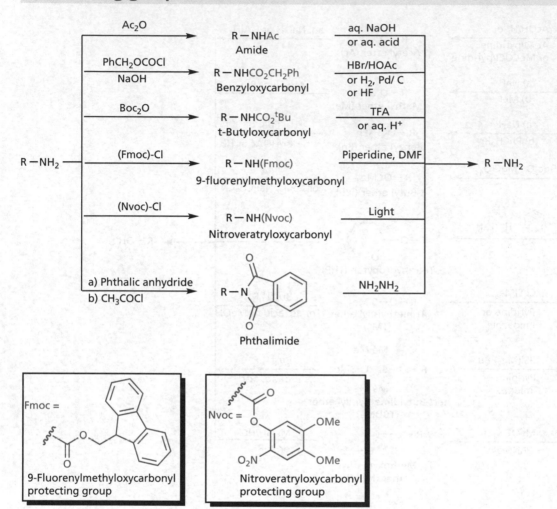

Fmoc =

9-Fluorenylmethyloxycarbonyl
protecting group

Nvoc =

Nitroveratryloxycarbonyl
protecting group

5 **Protecting groups for aldehydes and ketones**

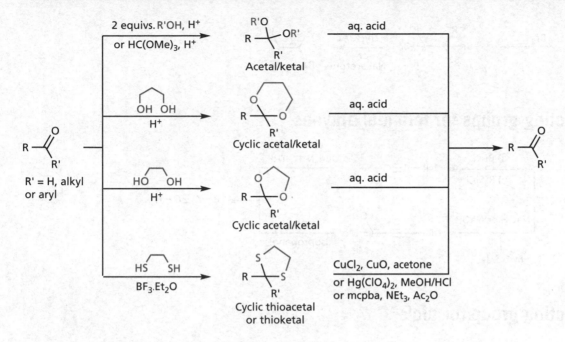

6 **Protecting groups for thiols**

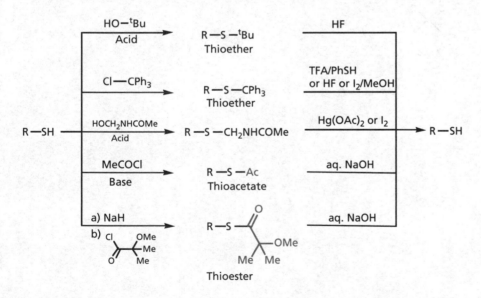

7 Protecting group for alkenes

8 Protecting groups for terminal alkynes

9 Protecting group for diols

Appendix 7

Structures of amino acids

The following amino acids are the essential amino acids that are present in mammalian proteins.

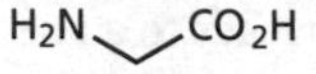

Glycine
(Gly or G)

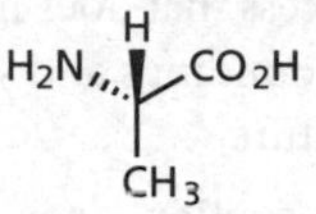

Alanine
(Ala or A)

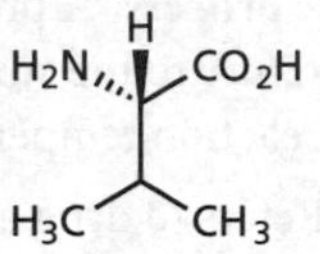

Valine
(Val or V)

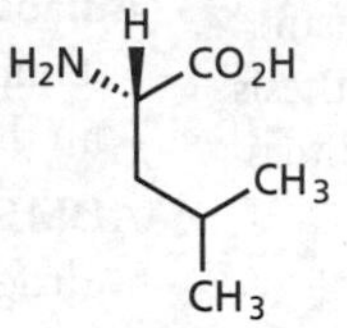

Leucine
(Leu or L)

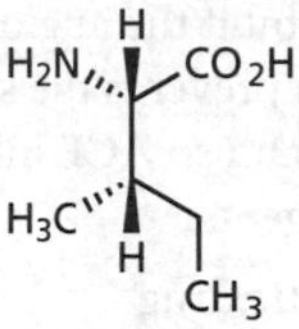

Isoleucine
(Ile or I)

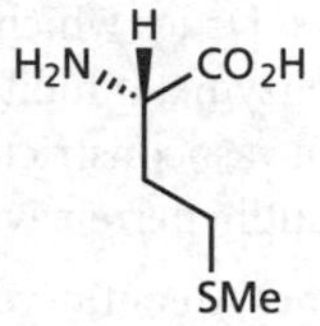

Methionine
(Met or M)

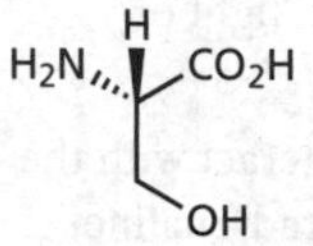

Serine
(Ser or S)

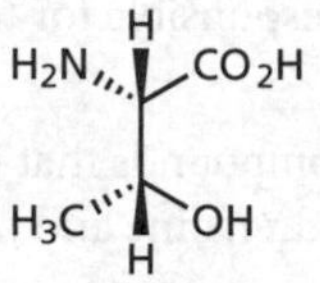

Threonine
(Thr or T)

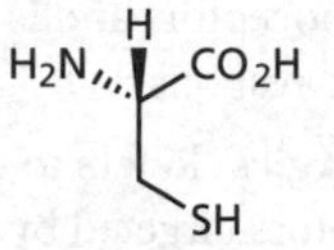

Cysteine
(Cys or C)

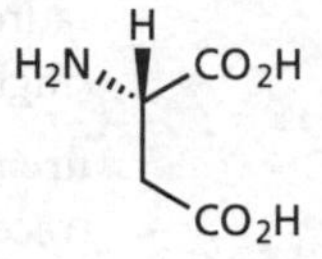

Aspartic acid
(Asp or D)

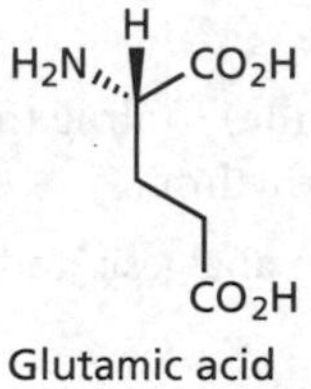

Glutamic acid
(Glu or E)

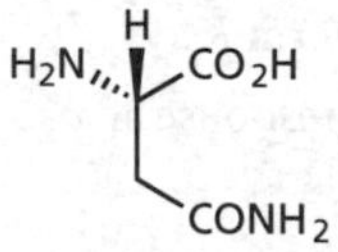

Asparagine
(Asn or N)

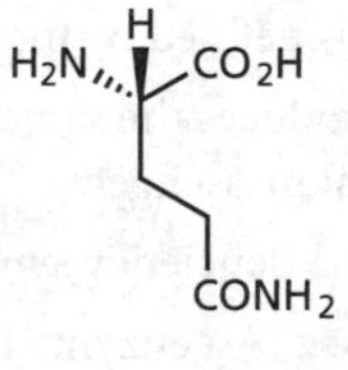

Glutamine
(Gln or Q)

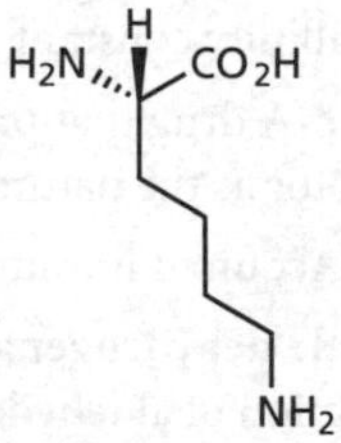

Lysine
(Lys or K)

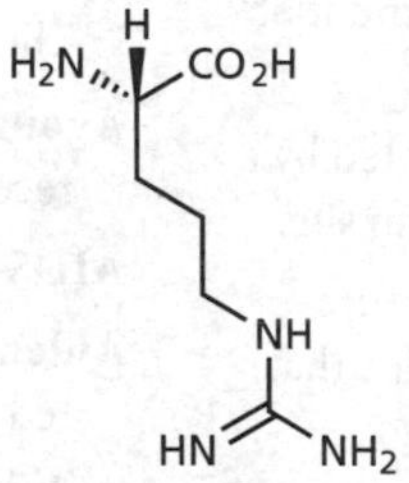

Arginine
(Arg or R)

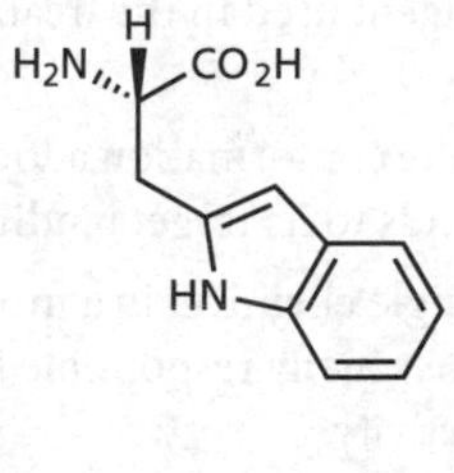

Tryptophan
(Trp or W)

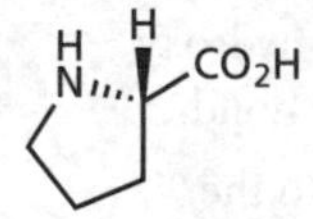

Proline
(Pr or P)

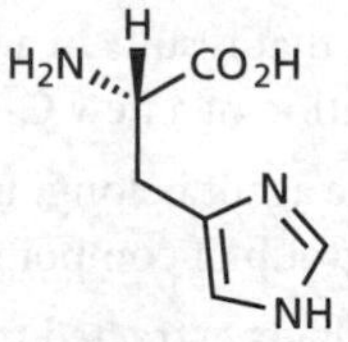

Histidine
(His or H)

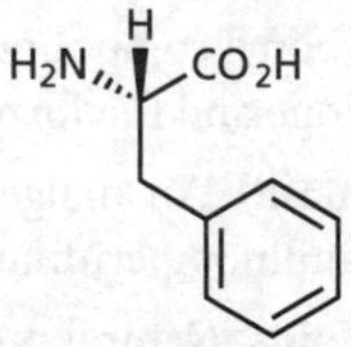

Phenylalanine
(Phe or F)

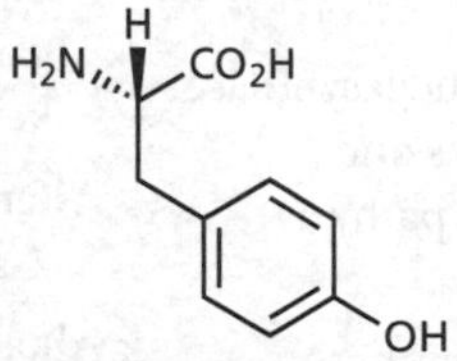

Tyrosine
(Tyr or Y)

Glossary

Abl kinase An enzyme that is the target for the anticancer agent imatinib.

Absolute configuration Defines the orientation of substituents at an asymmetric centre.

ACE inhibitors Drugs which inhibit the angiotensin converting enzyme. Inhibition prevents the synthesis of a powerful vasoconstrictor and so ACE inhibitors are used as antihypertensive agents.

Acetophenazine An antipsychotic drug.

Acetylcholine A neurotransmitter that activates nicotinic and muscarinic cholinergic receptors.

Acetylcholinesterase An enzyme that hydrolyses the neurotransmitter acetylcholine.

N-**Acetylprocainamide (acecainide)** A drug that is being considered as an anti-arrhythmic.

Acetylsalicylic acid (aspirin) An analgesic and anti-inflammatory agent.

Achiral An adjective that identifies a molecule as not being chiral.

Aciclovir An antiviral agent used in the treatment of herpes infections.

Active conformation The conformation adopted by a compound when it binds to its target binding site.

Active principle The single chemical in a mixture of compounds which is chiefly responsible for that mixture's biological activity.

Active site The binding site of an enzyme where a reaction is catalysed.

Acyclic Describes a structure or part of a structure that contains no rings.

Acyl carrier protein A location within a megasynthase enzyme which tethers the building blocks and intermediates involved in a biosynthetic pathway involving peptide or polyketide synthesis.

Acyltransferase An active site in a megasynthase enzyme that catalyses the tethering of a building block involved in polyketide or peptide synthesis to an acyl carrier protein.

Addiction Addiction can be defined as a habitual form of behaviour. It need not be harmful. For example, one can be addicted to eating chocolate or watching television without suffering more than a bad case of toothache or a surplus of soap operas.

Adenosine diphosphate A nucleotide that is produced naturally by the hydrolysis of adenosine triphosphate.

Adenosine triphosphate A nucleotide that is an important building block for RNA and DNA synthesis, as well as being involved as a cofactor in many enzyme-catalysed reactions.

Adiabatic process A process that does not give off temperature to the surroundings and so leads to a rise in the reaction temperature.

ADME Refers to drug absorption, drug distribution, drug metabolism, and drug excretion.

ADMET Refers to ADME plus the toxic properties of a drug.

Adrenaline A natural hormone that activates adrenoceptors and is responsible for the 'fight or flight' response.

Adrenergics Refers to compounds that interact with the receptors targeted by adrenaline and noradrenaline.

Adrenoceptors or adrenergic receptors Receptors that are activated by adrenaline and noradrenaline.

Affinity and affinity constant Affinity is a measure of how strongly a ligand binds to its target binding site. The affinity constant is defined by the term K_i.

Agonist A drug that produces the same response at a receptor as the natural messenger.

AIDS Acquired immune deficiency syndrome.

Aldehyde dehydrogenase An enzyme that catalyses the reduction of aldehydes to alcohols.

Aldol condensation A reaction that involves an aldol addition followed by dehydration to give an α,β-unsaturated ketone.

Aldol reaction A coupling reaction between two carbonyl compounds that results in a β-hydroxy ketone and the formation of a new C–C bond.

Alfentanil An analgesic agent belonging to the 4-anilinopiperidine group of compounds.

Alkaloids Natural products extracted from plants that contain an amine functional group.

Alkylating agents Agents which act as electrophiles and form irreversible covalent bonds with macromolecular targets. These agents are classed as cytotoxic and are used as anticancer agents.

Allosteric Refers to a protein binding site other than the one used by the normal ligand, which affects the activity of the protein. An allosteric inhibitor binding to an allosteric binding site induces a change of shape in the protein which disguises the normal binding site from its ligand.

Almotriptan An anti-migraine agent.

Alpha decay A form of radioactive decay that involves the release of an α-particle corresponding to the nucleus of a helium atom.

Alprazolam An anxiolytic.

Altinicline A cholinergic agonist acting at nicotinic receptors. It has entered clinical trials as a potential drug in the treatment of Parkinson's disease.

Aluminium chloride A Lewis acid.

Amadacycline An antibacterial agent that was synthesized from minocycline. It is an example of an aminomethylcycline.

Amino acids The molecular building blocks for peptides and proteins. They contain an amino group and a carboxylic acid, as well as a side chain.

γ-Aminobutyric acid The natural ligand for the benzodiazepine receptors.

7-Aminocephalosporinic acid A cephalosporin structure that lacks the side chain at position 7. It is an important starting material for the semi-synthetic preparation of cephalosporin analogues.

Aminoglycosides A group of antibacterial agents that contain sugar components and a basic amino function.

Aminomethylcyclines A group of tetracyclines with useful antibacterial activity.

6-Aminopenicillanic acid A penicillin structure that lacks the side chain at position 6. It is an important starting material for the semi-synthetic preparation of penicillin analogues.

Aminopeptidase enzyme An enzyme that hydrolyses the terminal amino acid from the N-terminus of a peptide.

Amlodipine A dihydropyridine that acts as a calcium ion channel blocker.

Amorphadiene synthase An enzyme that catalyses the cyclization of farnesyl diphosphate to form amorpha-4,11-diene.

Amoxapine A tricyclic antidepressant.

Amoxicillin A broad-spectrum penicillin which is orally active.

Amphetamine A stimulant. The word is also used as a general term for analogues of amphetamine.

Amphetaminil An amphetamine stimulant that was used to treat obesity, ADHD, and narcolepsy. It has now been withdrawn from the market.

Amphotericin An antifungal agent.

Ampicillin A broad-spectrum penicillin.

Anchor—*see* linker

Angiotensin II A hormone that plays an important role in narrowing blood vessels and increasing blood pressure. It is produced from another structure called angiotension I.

Angiotensin converting enzyme An enzyme that is responsible for catalysing the conversion of angiotensin I to angiotensin II.

4-Anilinopiperidines A group of compounds that act as analgesics. Fentanyl is one such structure.

Antagonist A drug which binds to a receptor without activating it, and which prevents an agonist or a natural messenger from binding.

Anthracyclines A group of antibiotics that are important in anticancer therapy.

Anti-arrhythmic agent An agent that is used to treat irregular heart rhythms.

Antibacterial agent A synthetic or naturally occurring agent that can kill or inhibit the growth of bacterial cells.

Antibiotic An antibacterial agent derived from a natural source.

Antibody A Y-shaped glycoprotein that is generated by the body's immune system to interact with an antigen present on a foreign molecule. It marks the foreign molecule for destruction.

Antibody–drug conjugates Refers to antibodies with drugs covalently linked to their structure.

Anticholinergic agent An agent that acts as an antagonist at cholinergic receptors and blocks the actions of acetylcholine.

Anticholinesterases Agents which inhibit the enzyme acetylcholinesterase.

Antigen A region of a molecule that is 'recognized' by the body's immune system and which will interact with antibodies targeted against it.

Antihistamine An agent that acts as an antagonist at histamine receptors and blocks the actions of histamine.

Antihypertensive agent An agent that lowers high blood pressure.

Anxiolytic An agent that relieves anxiety.

Apoptosis The process by which a cell commits suicide.

Aprindine An agent used in cardiovascular medicine as an anti-arrhythmic.

Aquaporins Membrane-bound proteins containing a pore that allows water to pass through the membrane.

Arachidonic acid An endogenous structure formed from prostaglandin H2 by the action of a cyclooxygenase enzyme.

Armodafinil The *R*-enantiomer of modafinil. It was approved in 2007 for the treatment of excessive sleepiness.

Arsphenamine (Salvarsan) An atypical antipsychotic agent.

Artemisinic acid A biosynthetic intermediate in the biosynthesis of artemisinin.

Artemisinin A natural product that is used as an antimalarial agent.

Arteries The major blood vessels responsible for taking blood away from the heart.

Arterioles Narrower blood vessels resulting from the branching of arteries.

Articaine (carticaine) A local anaesthetic used in dentistry.

Aspirin—*see* acetylsalicylic acid

Asymmetric centre An atom with four different substituents that frequently results in asymmetry for the whole molecule.

Asymmetric reaction A reaction which results in a new chiral centre and produces one configuration in preference to the other.

Asymmetric synthesis A synthesis of an asymmetric compound which shows selectivity for a particular enantiomer or diastereomer.

Asymmetry Lack of symmetry.

Atazanavir A protease inhibitor that is used as an antiviral agent in the treatment of AIDS.

Atorvastatin A cholesterol-lowering agent.

Atropine A natural product that acts as a cholinergic antagonist.

Attention deficit hyperactivity disorder (ADHD) A disorder associated with children—the name speaks for itself!

Aureomycin—*see* chlorotetracycline

Autoradiography A method of detecting radioactivity on chromatographic plates using photographic film.

Azaperone A sedative that is used in veterinary medicine and has occasionally been used as an antipsychotic agent.

Azeotrope A distillate that contains two different solvents.

Azithromycin A semi-synthetic macrolide antibacterial agent.

Baeyer–Villiger oxidation A reaction where a ketone is converted to an ester. Essentially, an oxygen atom supplied by a hydroperoxide reagent is inserted into the molecule.

Baldwin's rules A guide to predicting which types of intramolecular cyclization reaction are most likely to occur.

Barbiturates A series of synthetic compounds with sedative properties.

Bar code A method of identifying which structure is present on a resin bead by attaching molecular tags to the bead to indicate which reagents are used in a solid phase synthesis. The tags are released and 'read' by identifying the peaks in a gas chromatograph.

Barlos resin A resin containing a trityl linker that is used in solid phase synthesis.

Benzocaine A compound with weak local anaesthetic properties.

Benzodiazepine receptor A receptor protein that is targeted by the benzodiazepines. The natural ligand is γ-aminobutyric acid (GABA).

Benzodiazepines A group of cyclic structures that are generally used to treat psychological problems.

Benzotriazol-1-yloxytripyrrolidinophosphonium hexafluorophosphate (PyBOP) A coupling agent that is used in the solid phase synthesis of peptides.

(Benzotriazol-1-yloxy)tris(dimethylamino) phosphonium hexafluorophosphate (BOP) A peptide coupling agent that is now discouraged because of the formation of hexamethylphosphoramide, which is carcinogenic.

Benzyloxycarbonyl group A protecting group used to protect amines.

Besifloxacin A fourth-generation fluoroquinolone that was approved as an antibacterial agent in 2009.

Beta-blockers A series of compounds that block or antagonize β-adrenoceptors. Particularly useful in cardiovascular medicine.

Beta decay A form of radioactive decay that involves the release of an electron (β^- or e^-) or positron (β^+ or e^+) depending on the nucleus involved.

Beta-lactam ring A four-membered lactam ring that is present in penicillins and cephalosporins, and is crucial to the antibacterial mechanism of action.

Bexarotene An anticancer agent.

Binding affinity—*see* affinity

Binding group A term to identify a functional group or substituent that is present on a drug, and which interacts with a complementary binding region in a target binding site.

Binding region A region within a binding site that is capable of a specific intermolecular interaction with a drug or endogenous ligand.

Binding site The location where an endogenous molecule or drug binds to a macromolecule. It is normally a hollow or cleft in the surface of the macromolecule.

Bioassay A test that is carried out to determine the activity of a drug.

Bioavailability The fraction of drug that is available in the blood supply following administration.

Bioisostere A chemical group which can replace another chemical group without adversely affecting the desired activity of a drug.

Blood–brain barrier Blood vessels in the brain are less porous than blood vessels in the periphery. They also have a fatty coating. Drugs entering the brain have to be lipophilic in order to cross this barrier.

Boc A short hand term for the protecting group *t*-butyloxycarbonyl.

Bromazine An antihistamine.

Bromotripyrrolidinophosphonium hexafluorophosphate (PyBrOP) A coupling agent that is used in the solid phase synthesis of peptides.

Bronchodilator An agent which dilates the airways and can combat asthma.

Bupivacaine A local anaesthetic agent.

Buprenorphine A semi-synthetic orvinol that has been used as an analgesic and as a method of weaning addicts off heroin.

Burgess reagent A mild dehydrating reagent that is used to convert secondary and tertiary alcohols to alkenes.

Busulfan An anticancer agent that acts as an alkylating agent.

Butorphanol An opioid analgesic that is a member of the morphinan group of compounds.

***t*-Butyldimethylsilyl group** A bulky protecting group used to protect alcohols.

***t*-Butyloxycarbonyl group** A protecting group used to protect amines.

CAESA A program which evaluates how easily a structure can be synthesized by carrying out a retrosynthetic analysis to identify building blocks.

Cahn–Ingold–Prelog sequence rules The rules used to determine whether an asymmetric centre is defined as *R* or *S*.

Camphorsulphonyl chloride A chiral compound that can be used to derivatize an alcohol group. It is commonly used as a method of resolving racemic mixtures of chiral alcohols by forming diastereomeric derivatives.

Capillaries Small blood vessels.

Captopril An example of an ACE inhibitor used to lower high blood pressure.

Carbocyclic Describes a ring system where the scaffold is made up purely of carbon atoms.

Carbonic anhydrase An enzyme that catalyses the conversion of a bicarbonate ion to carbon dioxide and water.

Carcinogen A compound that induces cancers.

Carcinogenesis The birth of a cancer.

Carcinogenicity The potential of a compound to trigger carcinogenesis.

Carrier protein A protein in the membrane of a cell which is capable of transporting specific polar molecules across the membrane. The molecules transported are too polar to cross the membrane themselves and are crucial to the survival and functions of the cell.

Catalyst A compound or structure that lowers the activation energy of a reaction and is not altered itself.

Catch and release A method used in parallel synthesis where a reagent is attached to a solid support to 'capture' the starting materials and then release the product at a later stage.

C–C coupling A reaction that links two molecules together with the formation of one or more bonds between the two carbon atoms involved.

Celecoxib An anti-inflammatory agent that acts as a cyclooxygenase-2 selective inhibitor.

Cell membrane A phospholipid bilayer surrounding all cells that acts as a hydrophobic barrier.

Central nervous system The nerves that are present in the brain and the spinal column.

Cephalosporin C A naturally occurring cephalosporin that was the lead compound for further development.

Cephalosporins A group of β-lactam semi-synthetic antibacterial agents that target the bacterial transpeptidase enzymes.

Cerivastatin A cholesterol-lowering agent.

Chemical development The process by which a synthetic route is developed to produce a drug on a sufficiently large scale to allow preclinical and clinical trials.

Chemical space A term that is used to define all possible molecules that could be synthesized.

Chemoselectivity The selectivity of a reaction for one functional group over another.

Chiral auxiliary A chiral agent that is added to a reaction to introduce asymmetry to a reaction. Many chiral auxiliaries act as metal ligands to form chiral complexes.

Chirality The property of asymmetry where the mirror images of a molecule are non-superimposable.

Chiral switching The replacement of a racemic drug on the market with its more active enantiomer or stereoisomer.

meta-**Chloroperbenzoic acid** An oxidizing agent used to prepare epoxides from alkenes.

Chloropyramine An antihistamine used to treat allergies and asthma.

Chloroquine An antimalarial agent.

Chlorotetracycline The first naturally occurring tetracycline antibiotic to be discovered.

Chlorphenamine An antihistamine used to treat allergies and hay fever.

Chlorphenoxamine An agent that has antihistamine properties which is used to treat itching and Alzheimer's disease.

Cholinergic receptors Receptors that are activated by acetylcholine.

Cholinergics Compounds that interact with cholinergic receptors.

Chromophore A conjugated unsaturated feature within a molecule that is responsible for its colour.

Cimetidine An example of an anti-ulcer drug that acts as a histamine antagonist.

Ciprofloxacin A fluoroquinolone that acts as an antibacterial agent.

Claisen reaction A C–C coupling reaction between two esters that results in the formation of a 1,3-diketone.

Claisen rearrangement A reaction whereby an allyl vinyl ether undergoes a 3,3-sigmatropic rearrangement to form a γ,δ-unsaturated carbonyl compound.

Clarithromycin A semi-synthetic macrolide, synthesized from erythromycin, that is used as an antibacterial agent.

Click chemistry in situ A click reaction which takes place between two molecules that are bound to the binding site of a target protein.

Click chemistry/reaction A reaction that is thermodynamically favourable and proceeds under mild conditions in high yield to produce new C–X bonds.

Clinafloxacin A fluoroquinolone antibacterial agent.

Clinical trials Trials which are carried out to establish the efficacy, safety, and dosing regimes for a new drug.

ClogP A measure of a drug's hydrophobic character calculated by relevant software programs.

Cloning The process by which identical copies of a DNA molecule or a gene are obtained.

Clopidogrel An antithrombotic agent.

Co-crystal A co-crystal that contains two different components such as two different molecules. Different co-crystals of the same drug can result in different solubility characteristics.

Codeine An opioid structure obtained from opium that has analgesic activity.

Coenzyme A small organic molecule that acts as a cofactor.

Cofactor An ion or small organic molecule (other than the substrate) which is bound to the active site of an enzyme and takes part in the enzyme-catalysed reaction.

Combinatorial biosynthesis A process where genetic modifications are carried out to generate a library of super-enzymes, which differ in the number, sequence, and nature of the catalytic regions within their structure. This, in turn, results in a library of novel analogues which can potentially be generated from fast-growing microbial cells.

Combinatorial libraries A store of compounds that have been synthesized by combinatorial synthesis.

Combinatorial synthesis A method of synthesizing large quantities of compounds on small scale using automated or semi-automated processes. Normally carried out as solid phase syntheses.

Compound banks or libraries A store of synthetic compounds that have been produced by either traditional methods or combinatorial syntheses.

Configurational isomerism Non-interconvertible isomers that have similar atoms and bonds, but which are orientated differently. They include enantiomers, *cis* and *trans* disubstituted ring systems, and *Z* and *E* alkenes.

Conformational restriction A tactic used in drug design to rigidify a molecule such that it is less flexible and can adopt fewer conformations. If the active conformation is retained, the binding affinity to the target is likely to increase.

Conformational space The space that can be occupied by a molecule when all its possible conformations are considered.

Conformations Conformations are different three dimensional shapes for a compound, arising from single-bond rotations.

Conglomerates Compounds where the racemate normally crystallizes to give crystals made up solely of one enantiomer or the other.

Conjugated antibodies Antibodies that contain a radioisotope.

Conjugation In the chemical sense it refers to interacting systems of pi bonds. In the microbiological sense, it refers to the process by which bacterial cells pass genetic information directly between each other.

Conjugation reactions—*see* **phase II metabolism**

Continuous-flow system A process where reagents are mixed and reacted together in a continuous flow process.

Convergent synthesis A synthetic scheme where two halves of the target structure are synthesized separately, and then coupled towards the end of the synthesis.

Cope rearrangement The 3,3-sigmatropic rearrangement of 1,5-dienes.

Corey–Bikashi–Shibata reaction A reaction where a ketone group is reduced enantioselectively with a catecholborane in the presence of a chiral oxazaborolidine.

Coriolin An anticancer agent.

Cortisol A glucocorticoid steroid with anti-inflammatory activity.

Countercurrent extraction An extraction process where the two phases are in contact with each other, but flowing in opposite directions.

Coupling agents Reagents that are used to assist the coupling of two molecules under mild reaction conditions.

Coupling reactions Reactions that result in the coupling of two molecules with formation of a new bond.

Cryptand A cage-like molecule that has been used to enhance the solubility of inorganic salts such as potassium fluoride.

CrystalLEAD A screening method used in fragment-based design which uses electron density maps to identify compounds that bind to a target protein.

Crystal polymorphism The ability of a compound to crystallize in different crystal forms.

Curie A measure of radioactivity corresponding to 2.22×10^{12} disintegrations per minute.

Cyclic adenosine monophosphate A secondary messenger which plays an important part in several signal transduction processes.

Cyclobenzaprine A muscle relaxant that is used to treat muscle pain and injury.

Cyclooxygenases Enzymes that are important in the production of prostaglandins.

Cyclopentamine A vasoconstrictor that was used as a nasal decongestant, but has now been withdrawn.

Cyclo-release strategy A strategy used in solid phase synthesis where a structure is released from the solid phase and is set up to undergo spontaneous cyclization.

Cyclotron An instrument that is used to generate radioisotopes from a nuclear reaction.

CYP Abbreviated nomenclature for cytochrome P450 enzymes (e.g. CYP3A4).

Cytochrome P450 enzymes Enzymes which catalyse oxidation reactions and are extremely important in the metabolism of drugs.

Cytoplasm The contents of a cell.

Dapoxetine A selective serotonin reuptake inhibitor originally developed as an antidepressant. It has also been marketed as a treatment for premature ejaculation.

10-Deacetylbaccatin III A natural product that is used as a precursor for the semi-synthetic preparation of paclitaxel.

Decarboxylation A reaction that results in the loss of carbon dioxide from the starting material.

Deconvolution The isolation and identification of an active compound in a mixture of compounds obtained from a combinatorial synthesis.

Decoupling A method used in NMR spectroscopy to prevent coupling between different nuclei. It is used in carbon-13 NMR to prevent carbon signals being split by protons.

Degarelix acetate A decapeptide structure that was approved in Europe in 2009 for the treatment of prostate cancer.

Dehydratase An active site in a megasynthase enzyme that catalyses the dehydration of an alcohol to an alkene.

17β-Dehydroxysteroid dehydrogenase type 1 An enzyme responsible for catalysing the reduction of estrone to estradiol.

Denaturation Denaturation of a protein involves the protein adopting an unnatural tertiary structure resulting in loss of function.

De-novo **drug design** The use of specialized software programs to create ligands for a binding site based on the structure of that binding site.

6-Deoxyerythronolide B A biosynthetic intermediate of the macrolide erythromycin.

6-Deoxyerythronolide B synthase A super-enzyme responsible for catalysing the formation of 6-deoxyerythronolide B.

Deprotection The removal of a protecting group in order to restore a functional group.

Desolvation A process that involves the removal of surrounding water from molecules before they can interact with each other—for example, a drug with its binding site. Energy is required to break the intermolecular interactions involved.

Dexlansoprazole The *R*-enantiomer of the proton pump inhibitor lansoprazole.

Diamorphine An opioid analgesic with severe addiction and withdrawal side effects.

Diastereoselectivity The selectivity of a reaction in producing one diastereoisomer in preference to others.

Diastereotopic faces The two faces of a functional group containing a prochiral trigonal centre. The molecule already has an asymmetric centre and so a reaction showing selectivity for one face over the other will produce one diastereoisomer in preference to another.

Diastereotopic groups Two identical groups in a chiral molecule, which would result in an additional asymmetric centre if one of the groups underwent a reaction rather than the other. The two possible products would be diastereoisomers.

Diazepam A benzodiazepine structure that is used as an anxiolytic.

Dicyclohexylcarbodiimide A coupling agent used for the synthesis of amides and esters.

Dieckmann condensation The intramolecular cyclization of a diester to form a β-keto ester.

Diels–Alderase An unnatural enzyme that catalyses the Diels–Alder reaction.

Diels–Alder reaction A concerted cycloaddition reaction between two molecules resulting in the formation of two new C–C bonds. The reaction involves a diene and a dienophile.

Dienophile A structure that contains an electron-deficient alkene or alkyne that undergoes the Diels–Alder reaction with a diene.

Diethyl tartrate A structure containing two chiral centres that is used as a chiral auxiliary in the Sharpless epoxidation.

Digonal A linear system involving sp hybridized atoms.

Dihydrocodeine An opioid analgesic.

Dihydrocodeinone An opioid analgesic.

Dihydroetorphine An opioid analgesic used in China, which is also used as a treatment for opioid addiction.

Dihydropyridine calcium channel blockers A family of dihydropyridine structures that block calcium ion channels.

1,3-Diisopropylcarbodiimide (DIPC) A coupling agent that is used for the solid phase synthesis of peptides. The resulting urea (DIU) is soluble in dichloromethane and is easily washed from the solid phase.

Diltiazem A cardiovascular drug.

Dimazole An antifungal agent.

4-Dimethylaminopyridine An organic base that is commonly used in reactions.

N-[**(Dimethylamino)-1H-1,2,3-triazolo[4,5-b]pyridin-1-ylmethylene]-*N*-methylmethanaminium hexafluorophosphate (HATU)** A coupling agent that is used to synthesize amides.

Dimethylphenylsilyl group A group that can be used as a protecting group for alcohols.

Diosgenin A plant steroid that was used as the starting material of the semi-synthetic production of progesterone.

Dipeptidyl peptidase-4 An enzyme that has been targeted for the treatment of diabetes.

Dipole–dipole interactions Interactions between two separate dipoles. A dipole is a directional property and can be represented by an arrow between an electron-rich and an electron-deficient part of a molecule. Different dipoles align such that an electron-rich area interacts with an electron-deficient area.

Diprenorphine An opioid analgesic.

Directed evolution A term used in genetic engineering when mutating a protein to accept a wider range of substrates and to cope with harsher reaction conditions. The conditions are altered gradually and simple substrates are gradually made more complex.

Disconnection A term used in retrosynthesis to identify which bond in a target structure is disconnected to produce two synthons. It is not a chemical reaction in this context.

Displacement experiments Experiments that are carried out using a radioligand and an unlabelled drug to determine the affinity of the unlabelled drug.

Displacer A test compound that competes with a radioligand for the binding site of a receptor.

Dissociation binding constant Equivalent to the equilibrium constant for bound versus unbound ligand.

Distomer The less active enantiomer of a chiral drug.

Diversity-orientated synthesis The design of synthetic routes to obtain compounds with as diverse a nature as possible.

Diversity steps The steps in a synthesis where different reagents/building blocks can be used to produce different products.

DNA Deoxyribonucleic acid is a nucleic acid that contains the genetic code for proteins.

DNA gyrase A topoisomerase enzyme found in Gram-negative bacteria. It is a target for the fluoroquinolone antibacterial agents.

Dofetilide A cardiovascular drug used to treat irregular heart rhythms.

Domain A catalytic region within a megasynthase involved in the biosynthesis of polyketides and peptides.

L-Dopa A drug used in the treatment of Parkinson's disease.

Dopamine A neurotransmitter in the central nervous system.

Dosulepin A tricyclic antidepressant.

Double-blind studies Studies carried out during clinical trials such that neither the patient nor the doctor knows whether a drug or a placebo is being administered.

Double-cleavable linker A linker which allows a product to be released from a solid support in two stages.

Double-labelling experiments A biosynthetic experiment designed to test whether a radiolabelled structure is incorporated intact into the final structure. Typically, one part of the molecule is labelled with ^{14}C and the other with ^{3}H.

Doxorubicin An anthracycline that is used as an anticancer agent.

Doxycycline A semi-synthetic tetracycline agent that was prepared from methacycline. It is a clinically important antibacterial agent.

Dronedarone An agent used in the treatment of cardiac arrythmias.

Drug candidate A structure that is chosen to move forward from the research phase to preclinical trials.

Drug–drug interactions Related to the effect one drug has on the activity of another if both drugs are taken together.

Drug extension A drug design strategy where extra substituents or functional groups are added to an active compound in order to find additional binding interactions with a target binding site.

Drug metabolism The reactions undergone by a drug when it is in the body. Most metabolic reactions are catalysed by enzymes, especially in the liver.

Drug optimization The synthesis of analogues aimed at optimizing a number of predetermined properties in order to obtain an active compound with good selectivity and a minimum number of side effects.

Dual-acting agent A term used in medicinal chemistry to describe an agent that has been designed to interact with two different targets. In organic synthesis, the term can be used to describe a reagent that can undergo two separate reactions.

Dual-acting electrophile A reagent that can act as an electrophile in two consecutive reactions.

Dual-acting nucleophile A reagent that can act as a nucleophile in two consecutive reactions.

Dynamic combinatorial chemistry The generation of a mixture of products from a mixture of starting materials in the presence of a target. Products are in equilibrium with starting materials, and the equilibrium shifts to products binding to the target.

Dynamic structure–activity analysis The design of drugs based on which tautomer is preferred for activity.

Ebalzotan An antidepressant drug that acts as a selective serotonin antagonist.

Efficacy A measure of how effectively an agonist activates a receptor. It is possible for a drug to have high affinity for a receptor (i.e. strong binding interactions) but have low efficacy.

Efflux pumps Proteins that are responsible for transporting susceptible drugs out of bacterial cells—one of the factors that leads to drug resistance.

Electron capture A process where a proton in the nucleus 'captures' an inner-shell electron to produce a neutron.

Electrophile A structure containing an electron-deficient centre capable of reacting with a nucleophilic reagent.

Electrophilic centre An atom or region within a structure that is electron deficient and reacts with a nucleophilic reagent.

Electrostatic interaction—*see* **ionic interaction**

Eletriptan An anti-migraine agent.

Eltrombopag An agent used to treat low blood platelet counts.

Elvitegravir A fluoroquinolone structure that is used as an antiviral agent in the treatment of AIDS. It targets a viral integrase enzyme.

Emetic An agent that induces nausea and vomiting.

Enalapril A prodrug that is converted to enalaprilate, which acts as an ACE inhibitor. The compound is used in the treatment of hypertension.

Enantiomer A non-superimposable mirror image form of an asymmetric molecule.

Enantiomeric excess A measure of how enantioselective an asymmetric reaction has been in generating one enantiomer in preference to the other.

Enantioselectivity The selectivity of an asymmetric reaction to produce one enantiomer in preference to the other.

Enantiotopic faces The two faces of a planar functional group containing a prochiral trigonal centre. Selective reaction at one face over the other produces a new chiral centre with preference for one enantiomer over the other.

Enantiotopic groups Two identical molecules in an achiral molecule. If a reaction occurs at one group and not the other, a new asymmetric centre is formed, and one enantiomer is formed in preference to the other.

Encryption code—*see* **bar code**

Endogenous Endogenous compounds are chemicals which are naturally present in the body.

Endorphins Endogenous polypeptides which act as analgesics.

Enkephalins Endogenous peptides which act as analgesics.

Enoxacin An early member of the fluoroquinolone family of antibacterial agents.

Enoyl reductase An active site within a megasynthase protein that catalyses the reduction of the double bond in an enoyl group.

Enrichment process A process which is designed to increase the proportion of molecules containing a heavy isotope.

Enrofloxacin A fluoroquinolone that is used as an antibacterial agent in veterinary medicine.

Enterosoluble An adjective that is used to describe an object such as a tablet or capsule that is insoluble in the stomach but dissolves in the intestines.

Entropy A term used to describe the state of disorder of a defined system.

Enzyme A protein that acts as a catalyst for a reaction.

Enzyme inhibitor An agent that has been designed to bind to an enzyme and prevent the normal enzyme-catalysed reaction.

Epidermal growth factor A hormone that interacts with a membrane-bound receptor to trigger cell growth and division.

Epidermal growth factor receptor A membrane-bound kinase receptor that is activated by epidermal growth factor.

Epimerization The inversion of an asymmetric centre.

Epitopes Small molecules that bind to part of a binding site but do not produce a biological effect as a result of binding.

Epothilones Naturally occurring anticancer agents that inhibit tubulin depolymerization.

Eravacycline A fluorocycline antibacterial agent which has broad-spectrum antibacterial activity.

Ergosterol A fungal steroid that is an important constituent of fungal cell membranes.

Erythromycin A naturally occurring macrolide which is used as an antibiotic.

Eschenmoser's salt A strong dimethylamino methylating agent which reacts with enolates, enol silyl ethers, and some ketones.

Escherichia coli A bacterial strain that is commonly used in genetic engineering experiments.

Escitalopram An enantiomer of citalopram. It is used to treat depression, panic attacks, anxiety, and obsessive–compulsive disorder.

Eslicarbazepine acetate A prodrug of eslicarbazepine, which is a metabolite of the anti-epileptic drug oxcarbazepine.

Esomeprazole An enantiomer of the protein pump inhibitor omeprazole. It is used as an anti-ulcer agent.

Estradiol A mammalian female steroid hormone.

Estrone A female sex hormone with a steroid structure.

1-Ethyl-3-(3-dimethylaminopropyl)carbodiimide (EDC or EDCI) A water-soluble coupling agent used in peptide synthesis.

Ethynylestradiol A contraceptive.

Etidocaine A local anaesthetic used in surgical operations and childbirth.

Etoglucid An anticancer agent.

Etorphine A semi-synthetic opioid that is used as a sedative in veterinary medicine.

Eudismic ratio A measure of how a eutomer and a distomer differ in their activities.

European Agency for the Evaluation of Medicinal Projects (EMEA) The European regulatory authority for the testing and approval of drugs.

Eutomer The more active enantiomer of a chiral drug.

Exocytosis The process by which vesicles within a cell fuse with a cell membrane and release their contents from the cell.

Exotherm Refers to processes that are exothermic.

F A symbol used in pharmacokinetic equations to represent oral bioavailability.

Farnesyl pyrophosphate (FPP) pathway A biosynthetic pathway in bacterial cells which provides the necessary biosynthetic precursors for terpenes and sterols.

Farnesyl transferase An enzyme that attaches a farnesyl group to the Ras protein to allow membrane attachment.

Fast-tracking A method of pushing a drug through clinical trials and the regulatory process as quickly as possible. Applied to drugs that show distinct advantages over current drugs in the treatment of life-threatening diseases, or for drugs that can be used to treat diseases that have no current treatment.

Febuxostat An agent used for the treatment of gout.

Felkin–Anh model A model that is used to predict the diastereoselectivity of a nucleophilic addition to a carbonyl group when that carbonyl group is next to a pre-existing asymmetric centre.

Fentanyl An analgesic belonging to the 4-anilinopiperidine family of compounds.

Fenticonazole An antifungal agent.

Fexofenadine An antihistamine.

Fight or flight response The reaction of the body to situations of stress or danger, which involves the release of adrenaline and other chemical messengers that prepare the body for physical effort.

First-pass effect The extent to which an orally administered drug is metabolized during its first passage through the gut wall and the liver.

Fischer indole synthesis An intermolecular cyclization reaction resulting in the formation of an indole bicyclic ring.

FK506 binding protein A protein involved in the suppression of the immune response.

Florbetaben A compound used as a diagnostic tool in PET scans to study whether amyloid plaques are being broken down in the treatment of Alzheimer's disease.

Florbetapir An agent that is used as an imaging tool in PET scans for the diagnosis of Alzheimer's disease.

Fluanisone An antipsychotic agent.

Fluorenylmethyloxycarbonyl group A protecting group used for amines.

Fluorocyclines A group of tetracycline analogues that show potential as novel antibacterial agents.

[^{18}F]Fluorodeoxyglucose An agent that has been used in PET scans for medical imaging. It is the most commonly used PET radiopharmaceutical in the world.

6-[^{18}F]FluoroDOPA A radioactively labelled analogue of DOPA which is used in PET scans as a diagnostic tool to study dopamine biosynthesis and the degeneration of dopamine neurons in the brains of patients suffering from Parkinson's disease.

Fluoroquinolones A group of synthetic antibacterial agents.

Fluorous solid phase extraction A method of removing fluorinated molecules from solution by passing the solution through a silica column that already contains fluorinated molecules.

Flutemetamol A fluoro-analogue of Pittsburgh compound B. It has been used as a diagnostic tool for PET scans in the study of Alzheimer's disease.

Fluvastatin A cholesterol-lowering agent.

Food and Drugs Administration (FDA) The drugs regulatory authority for the United States of America.

Formate dehydrogenase An enzyme that catalyses the oxidation of formate to carbon dioxide using NAD as a cofactor.

Formulation Studies aimed at developing a preparation of a drug that is suitable for administration to patients.

Forskolin An agent that acts as a receptor agonist and results in the increased production of cyclic AMP. It is used in experiments that determine whether a drug acts as an agonist or antagonist.

Fragmentation pattern The molecular fragments produced when a compound is bombarded by electrons in mass spectrometry.

Fragment evolution A strategy used in fragment-based drug design where a lead compound is 'grown' *in silico* within a binding site from a single fragment.

Fragment self-assembly A form of dynamic combinatorial chemistry where fragments bind to different regions of a binding site, and then react with each other to form a linked molecule in situ.

Friedel–Crafts acylation A C–C coupling reaction between an aromatic or heteroaromatic ring and an acid chloride in the presence of a Lewis acid.

Friedel–Crafts alkylation A C–C coupling reaction between an aromatic or heteroaromatic ring and an alkyl halide in the presence of a Lewis acid.

Frondosin B A natural product which is a potential lead compound for the development of novel treatments for cancer, inflammatory diseases, and HIV.

Functional group A defined arrangement of atoms and bonds that determines the properties and reactions of that group.

Functional group activation The conversion of a functional group to one that is more reactive.

Functional group removal The removal of a functional group such that the position no longer has a functional group.

Functionalization A reaction that introduces a functional group at a position that did not previously contain one.

Furazolidone An antibacterial agent.

Gacyclidine An agent that is being studied for the treatment of tinnitus.

Galantamine A natural product that acts as an acetylcholinesterase inhibitor and has been used in the treatment of Alzheimer's disease.

Gamma decay A form of radioactive decay that involves the release of high energy electromagnetic radiation.

Garenoxacin A fluoroquinolone antibacterial agent.

Gastrointestinal tract Consists of the mouth, throat, stomach, and upper and lower intestines.

Gatifloxacin A fluoroquinolone antibacterial agent.

Gefitinib A protein kinase inhibitor used as an anticancer agent.

Genetic engineering The science by which modifications are made to the genes of a cell such that they code for proteins not normally produced by that cell.

Genome The genetic code for an organism.

Genomics The study of the genetic code for an organism.

Gliotoxin A fungal metabolite which shows immunosuppressant, anticancer, antibiotic, antifungal, and antiviral properties.

Glomerulus A knotted arrangement of blood vessels which fits into the opening of a nephron and from where water and small molecules are filtered into the nephron.

Glycylcyclines A group of semi-synthetic tetracyclines which have antibacterial activity.

Good Clinical Practice (GCP) Scientific codes of practice that apply to clinical trials and which are monitored by regulatory authorities.

Good Laboratory Practice (GLP) Scientific codes of practice that apply to a pharmaceutical company's research laboratories and which are monitored by regulatory authorities.

Good Manufacturing Practice (GMP) Scientific codes of practice that apply to a pharmaceutical company's production plants and which are monitored by regulatory authorities.

G-protein-coupled receptors Membrane-bound receptors that interact with G-proteins when they are activated by a ligand.

G proteins Membrane bound proteins consisting of three subunits. They are important in the signal transduction process from activated G-protein coupled receptors.

Gram negative and Gram positive A definition used for bacterial strains based on what colour they appear following a staining process with two different dyes. This is related to the type of cell wall that they have.

Grignard reaction A C–C coupling reaction between a Grignard reagent and an aldehyde, ketone, or ester that results in an alcohol.

Grignard reagent An organomagnesium reagent used in the Grignard reaction which is prepared from reacting magnesium with an alkyl or aryl halide.

Growth factors Hormones that activate membrane-bound receptors and trigger a signal transduction pathway leading to cell growth and division.

Grubbs catalyst A ruthenium catalyst ($RuCl_2(p\text{-}cymene)_2$) that is used in olefin metathesis.

Half-life The biological half-life is the time taken for the plasma concentration of a drug to fall by half. The radioactive half-life is the time taken for a radioactive isotope to decay to half of its initial radioactivity.

Haloperidol An antipsychotic agent used in the treatment of schizophrenia.

Hantzsch condensation An intramolecular cyclization procedure that results in dihydropyridine rings.

Heck reaction A palladium-catalysed reaction where an alkene can be coupled to an aromatic ring by reacting the alkene with an aryl halide.

Heteroaromatic ring An aromatic ring that includes one or more heteroatoms.

Heteroatom An atom other than carbon which is part of the molecular skeleton of a compound —normally oxygen, nitrogen, phosphorus, or sulphur.

Heterocycle A cyclic structure that contains one or more heteroatoms, such as oxygen, nitrogen, or sulphur.

Hexylcaine A local anaesthetic.

Highest occupied molecular orbital The molecular orbital of highest energy containing electrons.

High throughput screening An automated method of carrying out a large number of *in vitro* assays on a small scale.

Histamine and histamine receptors Histamine is a local hormone that is released from cells and interacts with histamine receptors in the cell membranes of other cells.

HIV protease A viral enzyme that catalyses the hydrolysis of viral polyproteins during the life cycle of HIV. It has proved a useful target for antiviral agents used to treat AIDS.

Hiyama reaction A palladium-catalysed coupling reaction between an organosilane and an aryl halide.

HMG-CoA reductase—*see* **3-Hydroxy-3-methylglutaryl-coenzyme A reductase**

Hock cleavage A C–C bond cleavage that occurs with hydroperoxides to produce two carbonyl groups.

Hofmann rearrangement A reaction where a primary amide is converted to a primary amine with one less carbon unit.

Homology models A term used in molecular modelling for the construction of a model protein binding site, based on the structure of known proteins or binding sites.

Hormones Endogenous chemicals that act as chemical messengers. They are typically released from glands and travel in the blood supply to reach their targets. Some hormones are local hormones and are released from cells to act in the immediate area around the cell.

Horner–Wadsworth–Emmons reaction A C–C coupling reaction between a stabilized phosphonate carbanion and an aldehyde or ketone to give an *E*-alkene. It is a useful method of synthesizing α,β-unsaturated carbonyl compounds.

Human Genome Project The sequencing of human DNA.

Huperzine A A natural product that acts as an anticholinesterase. It inhibits the acetylcholinesterase enzyme.

Hydantoinase Also called dihydropyrimidinase. It is a metalloenzyme that belongs to the hydrolase group of enzymes and catalyses a stage in the degradation of pyrimidines.

Hydrocodone An opioid analgesic.

Hydrocortisone—*see* **cortisol**

Hydrogenation A reaction where a compound is reduced by the addition of hydrogen.

Hydrogen bond A non-covalent bond which takes place between an electron-deficient hydrogen and an electron-rich atom, particularly oxygen and nitrogen.

Hydrogen bond acceptor A functional group which provides the electron-rich atom required to interact with a hydrogen in a hydrogen bond.

Hydrogen bond donor A functional group which provides the hydrogen required for a hydrogen bond.

Hydrogenolysis A reaction where a compound is reduced, along with bond cleavage.

Hydrolases Enzymes that catalyse hydrolysis reactions.

Hydromorphone An opioid analgesic.

Hydrophilic Refers to compounds that are polar and water-soluble. Literally means water-loving.

Hydrophobic Refers to compounds that are non-polar and water-insoluble. Literally means water-hating.

Hydrophobic interactions Refers to the stabilization that is gained when two hydrophobic regions of a molecule or molecules interact and shed the ordered water 'coat' surrounding them. The water molecules concerned become less ordered, resulting in an increase in entropy.

1-Hydroxybenzotriazole A reagent that is used in coupling reactions alongside dicyclohexylcarbodiimide to reduce the risk of epimerization.

3-Hydroxy-3-methylglutaryl-coenzyme A reductase An enzyme that catalyses the rate-determining step in the biosynthesis of cholesterol. It is the target for a group of drugs known as the statins.

Hygroscopic Describes a compound which absorbs water from the atmosphere.

Hypoglycaemia Lowered glucose levels in the blood.

Ibritumomab An antibody that contains the radioactive isotope ^{90}Y and is used as an anticancer agent. It recognizes a protein called CD20, which is located on the surface of B-lymphocytes.

Ibutilide A cardiovascular agent used to treat irregular heart rhythms.

IC$_{50}$ The concentration of an inhibitor required to inhibit an enzyme by 50%.

Imatinib A protein kinase inhibitor that is used as an anticancer agent.

Impurity profiling The identification of impurities in the final batch of a production process.

Indacaterol A β-adrenergic agonist used in the treatment of asthma.

Indinavir An antiviral agent used in the treatment of AIDS. It inhibits the HIV-protease enzyme.

Induced dipole interactions The situation where a charge or a dipole on one molecule induces a dipole in another molecule to allow an ion–dipole interaction or a dipole–dipole interaction, respectively. An induced dipole normally requires the presence of pi electrons.

Induced fit The alteration in shape that arises in a macromolecule such as a receptor or an enzyme when a ligand binds to its binding site.

Inhibitor An agent that binds to an enzyme and inhibits its activity.

Inhibitory constant—*see* **affinity and affinity constant**

In silico Refers to procedures that are carried out on a computer.

Insulin A peptide hormone produced in the pancreas that regulates carbohydrate and fat metabolism.

Integrase A viral enzyme that catalyses the splicing of viral DNA into host cell DNA. It plays an important role in the life cycle of HIV.

Intermediate trapping experiment A biosynthetic study where a radiolabelled starting material is fed to a culture, and a proposed intermediate is extracted to see whether it is radiolabelled.

Intermolecular bonds Bonding interactions that take place between two separate molecules.

Intermolecular cyclization A reaction between two or more molecules that results in the creation of a new ring.

Intramolecular bonds Bonding interactions other than covalent bonds that take place within the same molecule.

Intramolecular cyclization A cyclization reaction that occurs within a molecule to create a new ring.

***In vitro* tests** Tests that are carried out to measure a drug's binding affinity or pharmacological activity using isolated enzymes, cell membrane preparations, whole cells, or tissue samples.

***In vivo* bioassays** Tests that are carried out to measure a drug's activity using living organisms.

[^{123}I]Iobenguane An agent that targets adrenergic receptors and is used for imaging purposes.

[^{131}I]Iobenguane An agent that targets adrenergic receptors and has proved useful in targeting and eradicating tumours that have an affinity for noradrenaline.

Iodobenzene diacetate An oxidizing agent.

[^{123}I]Iofetamine An agent that has been used as a diagnostic tool to evaluate stroke and as an early diagnosis of Alzheimer's disease.

[^{123}I]Ioflupane A diagnostic agent that binds strongly to presynaptic dopamine transport proteins. It is used for the diagnosis of Parkinson's disease.

[^{11}C]Iomazenil A diagnostic agent that binds to benzodiazepine GABA$_A$ receptors. It is used for imaging the foci of epileptic seizures.

[^{123}I]Iomazenil A diagnostic agent that binds to benzodiazepine GABA$_A$ receptors. It is used for imaging the foci of epileptic seizures.

Ion channels Protein complexes in the cell membrane which allow the passage of specific ions across the cell membrane.

Ion–dipole interactions A non-covalent bonding interaction that takes place between a charged atom and a dipole moment, such as the interaction of a positive charge with the negative end of the dipole.

Ionic interaction A non-covalent bonding interaction between two molecular regions having opposite charges.

Irreversible inhibitor An enzyme inhibitor that binds so strongly to the enzyme that it cannot be displaced.

Isoelectric point The pH when a compound, capable of being positively or negatively charged, is neutral.

Isotope effects Effects that are observed when an isotope in a molecule is replaced with a different isotope of the same element. Bond strengths and reaction rates may differ between the labelled compound and the unlabelled compound.

Isotopes Elements which differ in the number of neutrons they have in their nucleus.

K_d The dissociation binding constant used to measure binding affinities.

K_i The inhibitory or affinity constant.

Karasch—Kumada—Corriu—Tamao reaction A coupling reaction involving an alkyl or aryl Grignard reagent reacting with an aryl halide in the presence of a palladium or a nickel catalyst.

Ketoreductase An active site within a megasynthase protein that catalyses the reduction of a ketone to alcohol.

Ketosynthase An active site within a megasynthase protein that catalyses the attachment of a building block to the end of a growing polyketide or peptide chain.

Kidneys The organ that is chiefly responsible for drug excretion.

Kinases Enzymes which catalyse the phosphorylation of alcoholic or phenolic groups present in a substrate. The substrate is normally a protein.

Kinetic isotope effect The effect that incorporating a different isotope of a particular element may have on reaction rates for a specific compound.

Kinetic resolution A resolution process that involves carrying out a reaction where only one of the enantiomers is affected.

Knoevenagel condensation A version of the aldol reaction that is used to synthesize α,β-unsaturated enones.

Koshland's theory of induced fit—*see* **induced fit**

Kosugi—Migita—Stille reaction—*see* **Stille coupling**

Lachrymatory Describes compounds which irritate the eyes, resulting in tears.

Lactate dehydrogenase An enzyme that catalyses the interconversion of pyruvate and lactate.

Lapatinib An anticancer agent that acts as a kinase enzyme inhibitor.

Latent group A relatively unreactive functional group that can be converted to a more reactive group when it is required.

Lawesson's reagent A thiation reagent that is used to convert ketones to thioketones.

Lead compound A compound showing a desired pharmacological property which can be used to initiate a medicinal chemistry project.

Lead discovery by NMR The use of NMR spectroscopy to identify whether small molecules bind to different parts of a target binding site, and the identification of lead compounds which link those molecules.

L-Leucine dehydrogenase An enzyme that catalyses the reductive amination of an α-keto acid to form the amino acid leucine.

Levetiracetam The *S*-enantiomer of etiracetam. It is an anticonvulsant agent used in the treatment of epilepsy.

Levofloxacin An enantiomer of the fluoroquinilone antibacterial agent ofloxacin.

Levorphanol An opioid analgesic.

Lewis acid A structure that acts as an acid but has no acidic protons.

Lidocaine A local anaesthetic used in dentistry.

Ligand A term used in medicinal chemistry for any molecule capable of binding to a binding site. In organic synthesis, it refers to molecules that bind to metal catalysts.

Linear synthesis A synthetic route where the target structure is built up stage by stage from a simple starting material.

Linezolid A synthetic antibacterial agent.

Linker A term used in combinatorial chemistry for a molecule that is covalently linked to a solid phase support and contains a functional group to which another molecule can be attached for the start of a synthesis.

Lipases Esterase enzymes that catalyse the hydrolysis of fats or lipids. They are important in digestion.

Lipinski's rule of five A set of rules that are obeyed by the majority of orally active drugs. The rules take into account the molecular weight, number of hydrogen bonding groups, and hydrophobic character of the drug.

Lipophilic Refers to compounds that are fatty and non polar in character. Literally means fat-loving.

Liquid scintillation counters Instruments that are used to measure the level of radioactivity present in liquid samples.

Lisinopril An antihypertensive agent.

Lithium aluminium hydride A reducing agent.

Lithium bis(trimethylsily)amide A strong non-nucleophilic base.

Liver The organ which is chiefly responsible for drug metabolism. It contains cytochrome P450 enzymes that play a major role in drug metabolism.

Log*P*—*see* **Partition coefficient**

Lollipop phase separator A method used in solution phase organic synthesis to remove an aqueous phase from an organic phase by freezing the aqueous phase on a cold pin or 'lollipop'.

Loratadine An antihistamine.

Lovastatin A cholesterol-lowering agent and a member of the statin family of drugs.

Lowest unoccupied molecular orbital The lowest energy molecular orbital that contains no electrons.

Loxapine An antipsychotic agent.

Lyconadin A A natural product which is of interest in the treatment of Alzheimer's disease.

Macrocycle A cyclic system involving a ring which contains nine or more members.

Macrolides Macrocyclic structures that act as antibacterial agents. Erythromycin is the best known example of this class of agents.

Macromolecule A molecule of high molecular weight such as a protein, carbohydrate, lipid, or nucleic acid.

Mannich reaction A reaction between a ketone, formaldehyde, and a primary or secondary amine. The ketone must have an acidic proton at the α-carbon and the product is a β-aminocarbonyl compound.

Marketing Authorization Application (MAA) A document provided to the EMEA in order to receive marketing approval for a new drug.

Mass spectrometry A method of identifying the molecular weight of a compound as well as its fragmentation pattern when it is broken up by bombardment with electrons.

Matrix metalloproteinases Enzymes that catalyse the hydrolysis of the proteins making up basement membranes. A target for new anticancer drugs called matrix metalloproteinase inhibitors.

Me-better drugs Drugs which have been modelled as variations of an existing drug and have better pharmacodynamic or pharmacokinetic properties.

Medicinal chemistry A branch of chemistry which focuses on the design and synthesis of novel pharmaceutical agents.

Megasynthase A multifunctional polymer of different enzymes linked together by means of a single polypeptide chain.

Melphalan An alkylating agent that is used as an anticancer agent.

Mephobarbital—*see* **methylphenobarbital**

Mepivacaine A local anaesthetic.

Meprylcaine A local anaesthetic.

Merrifield peptide synthesis A method of synthesizing peptides using solid phase synthesis.

Merrifield resin A resin used in solid phase peptide synthesis.

Messenger RNA (mRNA) Carries the genetic code required for the synthesis of a specific protein.

Mestranol A contraceptive agent.

Metabolic blockers Groups added to a drug to block metabolism at a particular part of the skeleton.

Metabolite A chemical structure that is formed in the cells of organisms or microorganisms—*see also* drug metabolism.

Metalloproteinases Enzymes that catalyse the hydrolysis of peptide bonds in protein substrates and which contain a metal ion as a cofactor in the active site.

Metazocine An opioid analgesic.

Methacycline A semi-synthetic tetracycline that was prepared from oxytetracycline.

Methadone An analgesic which is also used to treat opioid addicts.

Methisazone An antiviral agent that was used in the treatment of smallpox.

Methyldopa An adrenergic agonist used in the treatment of hypertension.

Methylmalonyl coenzyme A A building block used in the polyketide synthesis of 6-deoxyerythronolide B.

Methylphenobarbital An anticonvulsant barbiturate that also acts as a sedative and anxiolytic. It has generally been withdrawn from the market.

Metitepine An antipsychotic agent.

Me-too drugs Drugs which have been modelled as variations of an existing drug.

Metoprolol An adrenergic antagonist with selectivity for β_1-adrenoceptors. It is used in the treatment of hypertension.

Michael addition/reaction The 1,4-addition of a nucleophile to an α,β-unsaturated carbonyl group. The C=C double bond is lost as a result of the addition, but the carbonyl group is retained.

Microfluidics The manipulation of tiny volumes of liquids in a confined space.

Micromanipulation A term used in combinatorial synthesis where individual resin beads are isolated in order to identify and assay the attached product from a synthesis.

Microwave-assisted organic synthesis The use of microwave technology in organic synthesis in place of conventional heating.

Minocycline A semi-synthetic tetracycline with a broader range of activity than the naturally occurring tetracyclines, including activity against tetracycline-resistant staphylococci strains.

Minoxidil An agent that is used to counter baldness.

Mirabegron An adrenergic agonist that acts as a muscle relaxant and is used to treat an overactive bladder.

Mitsunobu reaction The conversion of an alcohol to an ester in the presence of triphenylphosphine, DEAD, and a carboxylic acid.

Mix and split The procedure involved when synthesizing mixtures of compounds by combinatorial synthesis.

Modafinil A CNS stimulant which is used to treat narcolepsy.

Module A module is made up of the domains involved in a megasynthase protein that catalyse each linkage and subsequent modification of the growing polyketide or peptide chain.

Molecular modelling The use of specialized software to study the shapes and properties of a molecule as well as the interactions between molecules.

Molecular 'signatures' An arrangement of functional groups and substituents that is indicative of a particular reaction.

Molindone An antipsychotic agent.

Montelukast An anti-asthmatic agent.

Morphine A natural product obtained from opium that acts as an analgesic.

Moxifloxacin A fluoroquinolone antibacterial agent.

Mupirocin A naturally occurring antibiotic.

Muscarinic receptor One of the two main types of cholinergic receptor. It is present in smooth muscle and cardiac muscle.

Mutation An alteration in the nucleic acid base sequence making up a gene. Results in a different amino acid in the resultant protein.

Myasthenia gravis A condition characterized by muscle weakness.

Nadolol A non-selective beta-blocker.

Nalbuphine An opioid analgesic.

Nalidixic acid A quinolone antibacterial agent.

Nalmefene An opioid antagonist.

Naloxone An opioid antagonist.

Naltrexone An opioid antagonist.

Naratriptan An anti-migraine agent.

Narcolepsy An excess tendency to sleep during the day.

Natural products Compounds that are isolated from organisms and microorganisms in the natural world.

N–C coupling A reaction that links two molecules together with formation of one or two bonds between the nitrogen and carbon atoms involved.

Negeshi coupling reaction A coupling reaction involving an alkyl or aryl zinc reagent reacted with an alkyl or aryl halide (or aryl triflate) in the presence of a palladium or nickel catalyst.

Nemonapride An antipsychotic agent.

Neostigmine An agent that acts as an acetylcholinesterase inhibitor and is used in the treatment of myasthenia gravis.

Nephrons Tubes that collect water and small molecules from the glomeruli and carry these towards the bladder. Much of the water, along with hydrophobic molecules, is reabsorbed into the blood supply from the nephrons and does not reach the bladder.

Nerve agents Agents that act as irreversible inhibitors of the acetylcholinesterase enzyme.

Neuromuscular blocking agents Agents that are used to block the transmission of cholinergic signals from nerves to skeletal muscle. They act as nicotinic antagonists.

Neuron A nerve cell.

Neurotransmission The process by which neurons communicate with other cells.

Neurotransmitter A chemical released by a neuron ending that acts as a chemical messenger by interacting with a receptor on a target cell.

Neutrino A neutral elementary subatomic particle with very tiny mass.

Nevirapine An antiviral agent that is used in the treatment of AIDS. It inhibits a viral enzyme called reverse transcriptase and is classed as a non-nucleoside reverse transcriptase inhibitor.

New chemical entity (NCE) A novel drug structure.

New Drug Application (NDA) A document provided to the FDA in order to receive marketing approval for a new drug.

New molecular entity (NME)—*see* **new chemical entity**

Nicotinamide adenine dinucleotide (phosphate) Enzyme cofactors that are important in enzyme-catalysed reactions involving reduction or oxidation.

Nicotinic receptors One of the two main types of cholinergic receptor.

Nifedipine A dihydropyridine structure that is used as a calcium ion channel blocker.

Nitrogen mustards Alkylating agents used in anticancer therapy.

Nitroveratryloxycarbonyl A protecting group used in peptide synthesis that can be removed in the presence of light.

NMR spectroscopy A spectroscopic method that is used to study the structures of molecules based on the chemical shifts of the atoms present.

Nomifensine An antidepressant.

Non-nucleoside reverse transcriptase inhibitors A group of antiviral agents that target an allosteric binding site on the viral enzyme reverse transcriptase.

Non-ribosomal peptide synthases A family of super-enzymes that are responsible for the biosynthesis of cyclic secondary metabolites in microbial cells.

Noradrenaline A neurotransmitter found in both the central nervous system and the peripheral regions of the body.

Norbinaltorphimine A highly potent and opioid antagonist that is selective for kappa opioid receptors.

Norfloxacin A fluoroquinolone that is used as an antibacterial agent.

Normorphine The *N*-demethylated analogue of morphine.

Nucleic acids RNA or DNA macromolecules made up of nucleotide units. Each nucleotide is made up of a nucleic acid base, a sugar, and a phosphate group.

Nucleophile A structure containing an electron-rich centre capable of reacting with an electrophilic reagent.

Nucleophilic centre An atom or region within a structure that is electron rich and reacts with an electrophilic reagent.

Nucleoside A building block for RNA or DNA that consists of a nucleic acid base linked to a sugar molecule.

Nucleoside reverse transcriptase inhibitors A group of antiviral agents that mimic nucleosides and target the viral enzyme reverse transcriptase.

Nucleotide A molecule consisting of a nucleoside linked to one, two, or three phosphate groups.

O–C coupling A reaction that links two molecules together with formation of a bond between the oxygen and the carbon atoms involved.

Oestradiol—*see* **estradiol**

Ofloxacin A fluoroquinolone that is used as an antibacterial agent. It contains a tricyclic ring system.

Olanzapine An antipsychotic agent.

Olefin metathesis A catalysed reaction involving two alkenes, where the alkenes are split and then re-formed to give hybrid alkenes.

Oligonucleotides A series of nucleotides linked together by phosphate bonds—smaller versions of nucleic acid.

Omeprazole An anti-ulcer agent that acts as a proton pump inhibitor.

Opioid receptors Receptors that act as the targets for opioid drugs such as morphine, as well as endogenous opioids such as the enkephalins. There are three main types—MOR, DOR, and KOR (mu, delta, and kappa opioid receptors).

Oppenauer oxidation Oxidation of a secondary alcohol to a ketone using an aluminium alkoxide and acetone.

Optical isomers Another term for enantiomers, where one enantiomer rotates plane polarized light clockwise and the other rotates plane polarized light anticlockwise.

Optical purity—*see* **enantiomeric excess**

Oral bioavailability The fraction of drug administered orally that circulates in the blood supply.

Orbifloxacin A fluoroquinolone that is used as an antibacterial agent in veterinary medicine.

Organometallic reagents Organic molecules that contain a metal and are used to carry out C–C coupling reactions. Examples are Grignard reagents, organolithium reagents and organocuprates.

Oripavines Complex multicyclic analogues of morphine which have powerful analgesic and sedative properties.

Orphan drugs Drugs that are effective against rare diseases. Special financial incentives are given to pharmaceutical industries to develop such drugs.

Orthogonality A term that is used in protecting group strategy when it is possible to remove one type of functional group without affecting another.

Orvinols Complex multicyclic analogues of morphine which have powerful analgesic and sedative properties.

Oseltamivir An antiviral drug used in the treatment of flu.

Osmium tetroxide An oxidizing agent.

Oxamniquine An anti-protozoal drug.

Oxazolidinones A group of synthetic antibacterial agents that act against protein synthesis.

Oxcarbazepine An anti-epileptic drug.

Oxidases Enzymes that catalyse oxidation reactions.

Oxycodone An opioid analgesic.

Oxymorphone An opioid analgesic.

Oxytetracycline A naturally occurring tetracycline with antibacterial activity.

Paclitaxel A natural product that is used as an anticancer agent.

Pagoclone A partial agonist at GABA$_A$ receptors. It has anxiolytic properties and has been considered as a treatment for stammering.

Paracetamol An analgesic and anti-inflammatory agent.

Parallel syntheses A method of synthesizing a large number of compounds on a small scale using automated procedures. Each reaction vial or well contains a single structure.

Parietal cells Cells lining the stomach which release hydrochloric acid into the stomach.

Paroxetine An antidepressant.

Partial agonist A drug which acts like an antagonist by blocking an agonist, but which retains some agonist activity of itself.

Partition coefficient (P) A measure of a drug's hydrophobic character—usually quoted as a log$P$ value.

Pazufloxacin A fluoroquinolone antibacterial agent which contains a tricyclic ring system.

Pefloxacin A third-generation fluoroquinolone that is an antibacterial agent of last resort for the treatment of life-threatening infections.

Pemetrexid An anticancer agent.

Penicillin acylase An enzyme that catalyses the formation of 6-aminopenicillanic acid from penicillin G or penicillin V.

Penicillin G The first penicillin to be discovered.

Penicillins Natural and semi-synthetic antibacterial agents that are bactericidal in nature.

Penicillin V An analogue of penicillin G with improved oral activity.

Pentacyclines Analogues of tetracycline that contain an additional fused ring.

Pentazocine An opioid analgesic.

Peptidases Enzymes which hydrolyse peptide bonds.

Peptide bonds The amide bonds that link amino acids in a peptide or protein structure.

Peptidomimetics Agents that have been developed from peptide lead compounds, such that their peptide nature is removed or disguised in order to improve their pharmacokinetic properties.

Peptoid A peptide which is partly or wholly made up of non-naturally occurring amino acids. As such, they may no longer be recognized as peptides by the body's protease enzymes.

Perphenazine An antipsychotic agent.

Peterson reaction The reaction of an α-silyl carbanion with an aldehyde or ketone to give an alkene.

Pethidine An analgesic compound that acts on opioid receptors.

Pfeiffer's rule The eudismic ratio for a series of chiral compounds increases with increasing potency of the eutomers.

Pharmacodynamics The study of how ligands interact with their target binding site.

Pharmacokinetics The study of drug absorption, drug distribution, drug metabolism, and drug excretion.

Pharmacology The study of how drugs interact with molecular targets.

Pharmacophore Defines the atoms and functional groups required for a specific pharmacological activity, and their relative positions in space.

Pharmacophore triangle A triangle connecting three of the important binding centres making up the overall pharmacophore of a molecule.

Phase I metabolism Reactions undergone by a drug which normally result in the introduction or unmasking of a polar functional group. Most phase I reactions are oxidations.

Phase II metabolism Conjugation reactions where a polar molecule is attached to a functional group that has often been introduced by a phase I reaction.

Phase separation columns Columns that are used to separate a dense chlorinated organic layer from an aqueous phase during solution phase parallel synthesis.

Phenobarbital A sedative belonging to the barbiturate class of drugs.

Phenylalanine dehydrogenase An enzyme that catalyses the conversion of a β-keto acid to phenylalanine by a transamination reaction.

Pholcodine An opioid analgesic.

Phosphatase An enzyme that catalyses the hydrolysis of phosphate bonds.

Phosphodiesterases Enzymes which are responsible for hydrolysing the secondary messengers cyclic AMP and cyclic GMP.

Phosphopantetheine A molecule that acts as a long flexible arm to tether a growing polyketide or peptide chain to a megasynthese protein and transfer it from one active site to another. It is an example of a prosthetic group.

Phosphorus pentachloride A chlorinating agent.

Photolithography A method of combinatorial synthesis involving the synthesis of products on a solid surface. Reactions only occur on those areas of the surface where photolabile protecting groups have been removed by exposure to light.

Photosensitizer A molecule that produces a chemical change in another molecule in a photochemical process.

Physical channelling—*see* **substrate channelling**

Physostigmine A natural compound which acts as an anticholinesterase agent.

Pi (π) bond A weak covalent bond resulting from the 'side on' overlap of p-orbitals. Only occurs when the atoms concerned are sp or sp² hybridized, and when the bond between the atoms is double or triple.

Pi-cation interactions The intermolecular interactions that take place between one molecule containing a positively charged group (e.g. an aminium ion) and another molecule containing a pi system (e.g. an aromatic or heteroaromatic ring).

Pictet–Spengler reaction The reaction of a β-arylethylamine with a ketone or an aldehyde to give a condensation product which undergoes intramolecular cyclization.

Piloty–Robinson synthesis A method of creating a pyrrole ring.

(-)-α-Pinene An optically active solvent used to crystallize one enantiomer of a racemate in preference to the other.

Pinocytosis A method by which molecules can enter cells without passing through cell membranes. The molecule is 'engulfed' by the cell membrane and taken into the cell in a membrane-bound vesicle.

Pirindol A muscle relaxant.

Pitavastatin A cholesterol-lowering agent.

Pittsburgh compound B An agent that is used in diagnostic tests to diagnose Alzheimer's disease, and to assess the effectiveness of drug therapy.

Pivampicillin A prodrug for penicillin.

pK_i The log of the inhibitory or affinity constant K_i.

Placebo and placebo effect A preparation that contains no active drug, but should look and taste as similar as possible to the preparation of the actual drug. Used to test for the placebo effect where patients improve because they believe that they have been given a useful drug.

Placental barrier Membranes that separate a mother's blood from the blood of her fetus. Some drugs can pass through the placental barrier.

Plasma proteins Proteins in the plasma of the blood. Drugs which bind to plasma proteins are unable to reach their target.

Platelet-activating factor A hormone released from inflammatory cells that promotes platelet aggregation.

Polar surface area A quantitative measure of the polarity associated with the surface area of a drug. It is calculated using software programs.

Polyketide pathway A pathway that is commonly used by bacterial cells to build up the carbon skeletons of various natural products.

Polyketide synthases Super-enzymes that catalyse the biosynthesis of multicyclic secondary metabolites in microbial cells.

Polymorphism—*see* **crystal polymorphism**

Polysaccharides Macromolecules consisting of carbohydrate (sugar) monomers.

Positional scanning libraries A method used in combinatorial synthesis where the same compounds are prepared with a different residue kept constant in each library.

Positron It is also called an anti-electron and is the antimatter equivalent of an electron. It has a charge of + 1, a spin of 1/2, and the same mass as an electron.

Positron emission tomography (PET) A diagnostic method used to image body tissues.

Potassium permanganate An oxidizing agent.

Potency Refers to the amount of drug required to achieve a defined biological effect.

Pralatrexate An anticancer agent.

Prasugrel An anticancer agent.

Preclinical trials Trials that are carried out on a drug candidate prior to clinical trials in order to study the pharmacological and toxicological properties of the compound. Formulation and stability tests are also carried out.

Precursor-directed biosynthesis A method of synthesizing analogues of natural products by the use of genetically modified super-enzymes.

Preferential crystallization The preferential crystallization of one enantiomer from a racemic solution by seeding with a pure enantiomer. The chiral compound has to be one that normally crystallizes to form conglomerates, where mixtures of crystals are formed with each crystal formed from a single enantiomer.

Prelog's rule A rule used to predict which enantiomer is more likely to be formed when a prochiral ketone is reduced to a chiral alcohol using baker's yeast.

Principle of chemotherapy The principle that a drug shows selective toxicity towards a target cell but not a normal cell.

Privileged scaffolds Scaffolds that are commonly present in established drugs with different pharmacological activities.

Procainamide An anti-arrhythmic agent.

Procaine A local anaesthetic.

Process development Studies aimed at developing a synthetic route in order to produce a drug in the production plant.

Prochirality and prochiral centres A molecule that has the potential to become chiral following a reaction. The centre that will become the new chiral centre is called the prochiral centre.

Prodrug A molecule which is inactive in itself, but is converted to the active drug in the body, normally by an enzymatic reaction. Used to avoid problems related to the pharmacokinetics of the active drug, or for drug targeting.

Progesterone A female sex hormone belonging to the progestogen class of steroids.

Prolintane A stimulant that is used for the treatment of senile dementia.

Promiscuous ligand A term used in medicinal chemistry to describe a compound that binds to a variety of different targets.

Promoter A chemical which can be added at a catalytic level in order to promote reactions on a large scale.

Prontosil A dye that led to the discovery of the sulphonamide antibacterial agents.

Proparacaine A local anaesthetic that is used topically in eye drops for ophthamology.

Propionyl coenzyme A The initial building block used in the biosynthesis of 6-deoxyerythronolide B.

Propranolol An example of a first-generation beta-blocker used to lower blood pressure.

Prostaglandin H2 The substrate for the cyclooxygenase enzyme, and a precursor to arachidonic acid.

Prostaglandins Endogenous chemicals that play an important role as chemical messengers. They have a large variety of biological functions.

Prosthetic group A cofactor that is covalently linked to an enzyme and plays a role in its function.

Protease inhibitors A group of antiviral agents which inhibit the HIV protease enzyme.

Proteases Enzymes which hydrolyse peptide bonds.

Protecting group A group that is added to a functional group to prevent it undergoing unwanted reactions.

Protection The process by which a protecting group is added to a functional group.

Protein A macromolecule made up of amino acid monomers. Includes enzymes, receptors, carrier proteins, ion channels, hormones, and structural proteins.

Protein kinases—*see* **kinases**

Protein tyrosine phosphatase An enzyme that hydrolyses phosphate groups from phosphorylated tyrosine residues on protein substrates. The enzyme reverses the effect of kinases.

Proteomics A study of the structure and function of novel proteins discovered from genomic studies.

Proton pump inhibitors A series of drugs which inhibit the proton pump responsible for releasing hydrochloric acid into the stomach.

Proxymetacaine—*see* **proparacaine**

Pseudomonic acid—*see* **mupirocin**

PS-isocyanate An electrophilic resin that contains isocyanate functional groups. It is used as a scavenging resin to remove excess amine from reaction solutions.

PS-morpholine A resin that contains attached morpholine. It is used to provide basic catalysis during a reaction.

PS-trisamine A nucleophilic resin that contains an amine. It is used as a scavenging resin to remove electrophilic reagents from reaction solutions.

Pulegone A naturally occurring plant monoterpene that is commercially available.

Pyridinium chlorochromate A mild oxidizing agent.

Pyrophoric Describes a compound that can ignite spontaneously in air.

Quetiapine An antipsychotic agent.

Quinine An antimalarial agent.

Quinolones A group of synthetic antibacterial agents, largely replaced by fluoroquinolones. In some texts, the term encompasses both quinolones and fluoroquinolones.

Racemate or racemic mixture A mixture of the various stereoisomers of a molecule. A molecule having one asymmetric centre would be present as both possible enantiomers.

Racemization A reaction which affects the absolute configuration of asymmetric centres to produce a racemic mixture.

Radafaxine A drug that was considered as an antidepressant and a treatment for restless leg syndrome.

Radiochemical purity A measure of how pure a radiolabelled product is, relative to any radiolabelled impurities that might be present.

Radiodilution analysis A means of identifying radiochemical purity. A sample of the radiolabelled product is diluted with a solution of the pure unlabelled product, then crystallized to constant specific activity.

Radioligand labelling The use of a radioactively labelled ligand in experiments designed to measure the affinity of unlabelled ligands.

Radiolysis Decomposition of a radiolabelled compound caused by free radicals that are formed as a result of β-particles emitted during radioactive decay.

Radiosynthesis The design of a synthesis that allows the incorporation of a radioactive isotope into the target structure.

Random peptide integrated discovery A process aimed at producing cyclic peptides *in vitro*, using the components of the translation process.

Ranitidine An anti-ulcer drug that acts as a histamine antagonist.

Receptor A protein with which a chemical messenger or drug can interact to produce a biological response.

Recursive deconvolution A method of identifying the constituents in a combinatorial synthetic mixture. The method requires the storage of intermediate mixtures.

Reductases Enzymes that catalyse reduction reactions.

Regioselectivity The selectivity of a reaction to occur at one position of a functional group rather than another.

Remifentanil An analgesic belonging to the structural family of 4-anilinopiperidines.

Renal Refers to the kidney.

Replication The process by which DNA produces a copy of itself.

Resolution The process by which the enantiomers of a racemic mixture are separated.

Retrosynthesis A method of planning synthetic routes by working backwards from the target structure to simple starting materials.

Reverse transcriptase A viral enzyme present in HIV that catalyses DNA from an RNA template.

Reverse transcriptase inhibitors A group of antiviral compounds that inhibit the viral enzyme reverse transcriptase.

Reversible inhibitors Enzyme inhibitors which compete with the substrate for the enzyme's active site and can be displaced by increasing the concentration of substrate.

Ribosomal nucleic acid (rRNA) Present in ribosomes as the major structural and catalytic component.

Ribosomes Structures consisting of rRNA and protein which bind mRNA and catalyse the synthesis of the protein coded by mRNA.

Rigidification strategies Strategies used to limit the number of conformations that a drug can adopt with the aim of retaining the active conformation.

Rink resin A resin used in solid phase synthesis.

Risperidone An antipsychotic agent.

Ritonavir An antiviral agent that acts as a protease inhibitor against HIV.

Rizatriptan An anti-migraine drug.

RNA Ribonucleic acid.

Rocuronium A neuromuscular blocking agent that acts as a cholinergic antagonist.

Rofecoxib A COX-2 inhibitor.

Rosoxacin A fluoroquinolone antibacterial agent.

Rosuvastatin A cholesterol-lowering agent that acts as an enzyme inhibitor.

Roxithromycin A semi-synthetic macrolide that is synthesized from erythromycin, and is used as an antibacterial agent.

Rufloxacin A fluoroquinolone antibacterial agent that includes a tricyclic ring system.

***Saccharomyces cerevisiae* A** A yeast that is commonly used in genetic engineering experiments.

Saccharopolyspora erythraea The fungus from which the antibiotic erythromycin was extracted.

Safety catch linker An example of a linker in combinatorial chemistry on which two molecules can be constructed, one the target molecule and the other a tagging molecule.

Salbutamol An adrenergic agonist that is used in the treatment of asthma.

Salmeterol An adrenergic agonist that is used in the treatment of asthma.

Sancycline A tetracycline structure.

Saquinavir An antiviral agent used in the treatment of AIDS. It acts as an inhibitor of the HIV protease enzyme.

Sarafloxacin A fluoroquinolone antibacterial agent which has now been discontinued.

SAR by NMR—*see* **lead discovery by NMR**

Saturation binding studies Studies involving radiolabelled ligands to determine the extent of binding to a target molecule.

Saxagliptin An enzyme inhibitor that acts as an oral hypoglycaemic and is used for the treatment of diabetes.

Scaffolds The molecular core of a drug onto which the important binding groups are attached as substituents.

Scatchard plot A plot used to measure the affinity of a drug for its binding site.

Scintillation proximate assay A visual method of detecting whether a ligand binds to a target by its ability to compete with a radiolabelled ligand that emits light in the presence of scintillant.

Screening A procedure by which compounds are tested for binding affinity or biological activity.

Secondary messenger A natural chemical which is produced by the cell as a result of receptor activation, and which carries the chemical message from the cell membrane to the cytoplasm.

Secondary metabolites Natural products that are not crucial to cell growth and division. Generally produced in mature cells.

Selective serotonin reuptake inhibitors Agents which bind to the transport proteins responsible for transporting serotonin from nerve synapses into the neurons that released it. As a result, they block the reuptake of serotonin and prolong its action.

Semi-synthetic product A product that has been synthesized from a naturally occurring compound.

Sequential release Refers to the sequential release of the products of a combinatorial synthesis from the solid support.

Serotonin and serotonin receptors A neurotransmitter that is present in the central nervous system. It activates serotonin receptors in the cell membranes of cells.

Sertraline An antidepressant.

Sharpless epoxidation An asymmetric reaction involving the enantioselective epoxidation of an allylic alcohol.

Sheppard's polyamide resin A resin that is used in solid phase synthesis. It is more polar than the Merrifield resin.

Shikimic acid A natural product that is an important intermediate in the biosynthesis of a large number of aromatic natural products.

Sigma (σ) bond A strong covalent bond taking place between two atoms. It involves strong overlap between two atomic orbitals whose lobes point towards each other.

Signal transduction The mechanism by which an activated receptor transmits a message into the cell, resulting in a cellular response.

Sildenafil The active component of Viagra, which is responsible for its anti-impotence properties.

Simmons–Smith reaction A reaction that results in the formation of cyclopropane rings from alkenes, α,β-unsaturated ketones, or allylic alcohols using diiodomethane with a zinc catalyst.

Simplification strategies The simplification of a drug to remove functional groups, asymmetric centres, and skeletal frameworks that are not required for activity.

Single photon emission computer tomography (SPECT) A diagnostic method used to image body tissues.

Sitagliptin A drug that is used for the treatment of diabetes.

Sodium borohydride A reducing agent.

Sodium cyanoborohydride A reducing agent.

Sodium triacetoxyborohydride A reducing agent.

Solabegron An agent that is undergoing clinical trials as a treatment for incontinence. It is a selective agonist of β_3 adrenoceptors, and relaxes the smooth muscle of the bladder.

Solid phase extraction The use of solid phase resins to remove impurities or excess reagents from a reaction solution.

Solution phase organic synthesis Procedures used to carry out parallel synthesis in solution instead of on solid phase.

Sonogashira reaction The palladium-catalysed coupling of a terminal alkyne to an aryl halide.

Sorafenib An anticancer drug that acts as a kinase inhibitor against a range of membrane-bound receptor kinases.

Sparfloxacin A fluoroquinolone antibacterial agent.

Specific activity A quantitative measure of radioactivity for a radiolabelled compound, which is measured in Ci/mmol or μCi/mmol.

Specifications The tests that have to be carried out on a manufactured drug, and the standards of purity required.

Spider scaffolds Scaffolds which have binding group substituents placed round the whole scaffold.

'Spring-loaded' functional group A functional group which reveals a second functional group following a reaction.

SPROUT A software program used in *de novo* drug design.

Squalene A biosynthetic precursor for the fungal steroid ergosterol.

Squalene synthase An enzyme that catalyses the formation of squalene.

Statins A family of structures sharing a similar pharmacophore which act as cholesterol-lowering agents.

Stereoselectivity　The selectivity of a reaction in terms of the geometry of the reaction mechanism and the resulting configuration of the product.

Steric shields　Bulky groups that are added to molecules to protect vulnerable groups by nature of their size.

Steroids　Hormones that share a tetracyclic ring structure. They are responsible for a wide range of effects and interact with intracellular steroid receptors.

Stille coupling　A coupling reaction between an arylstannane and an aryl halide (or triflate) in the presence of a palladium catalyst.

Strecker synthesis　A synthesis that is used to prepare amino acids.

Streptomyces coelicolor　A microbial strain that is commonly used in genetic engineering experiments.

Structure–activity relationships (SARs)　Studies carried out to determine those atoms or functional groups which are important to a drug's activity.

Structure-based drug design　The design of drugs based on a study of their target binding interactions with the aid of X-ray crystallography and molecular modelling.

Subsites　Often refers to enzymes that accept peptides or proteins as substrates. The subsites are binding pockets that accept amino acid residues on the substrate.

Substrate　A chemical which undergoes a reaction that is catalysed by an enzyme.

Substrate channelling　A description of how a growing polyketide or peptide chain moves through an ordered series of reactions catalysed by a super-enzyme. It is only released when the whole process has been completed.

Substrate walking　A term used in genetic engineering when modifying proteins such that they accept more complex substrates. A simpler substrate than the target substrate is used to find a modified enzyme that will accept it, and then the studies are continued with increasingly complex substrates.

Sufentanil　An analgesic having a 4-anilinopiperidine structure.

Suicide substrates　Enzyme inhibitors which have been designed to be activated by an enzyme catalysed reaction, and which will bind irreversibly to the active site as a result.

Sulfadiazine　A sulphonamide antibacterial agent.

Sulphonamides　In medicinal chemistry sulphonamides are synthetic antibacterial drugs that are bacteriostatic in nature. In chemistry, sulphonamide refers to a specific functional group which is present in the sulphonamide antibacterials.

Sumatriptan　An anti-migraine agent.

Super-additivity effect　An effect where the binding affinity of two linked fragments is much greater than might be expected from the binding affinities of the two independent fragments.

Supercoiling　The process by which DNA is coiled into a compact three-dimensional shape.

Super-enzyme　An enzyme complex containing proteins that have several active sites within their structure.

Suxamethonium　A neuromuscular blocking agent that acts as a cholinergic antagonist.

Suzuki coupling　A coupling reaction between an aryl halide and an aryl boronate (or aryl boronic acid) in the presence of a palladium catalyst.

Suzuki–Miyaura reaction—see Suzuki coupling

Swern oxidation　An oxidation reaction where a secondary alcohol is converted to a ketone in the presence of oxalyl chloride, DMSO, and triethylamine.

Synapse　The small gap between a neuron and a target cell across which a neurotransmitter has to travel in order to reach its receptor.

Synthetic biology　The genetic modification of a microbial cell which alters metabolic pathways, and results in the biosynthesis of products that are alien to the host cell.

Synthon　A molecular fragment that is obtained after a bond disconnection carried out as part of a retrosynthetic study. The synthon is not a real molecule, but can be used to identify suitable reagents that will represent it in a reaction.

Tadalafil　An agent that is used to treat erectile dysfunction.

Tadpole scaffold　A scaffold where substituents acting as binding groups are located at one region of the scaffold.

Tagging　A method of identifying what structures are being synthesized on a resin bead during a combinatorial synthesis. The tag is a peptide or nucleotide sequence which is constructed in parallel with the synthesis.

Tamoxifen　An antitumour drug.

Tandem reaction　A reaction where the product from one reaction is set up for a second reaction.

Tapentadol　An agent which acts as an agonist at opioid receptors, as well as an inhibitor of noradrenaline reuptake from nerve synapses.

Tazarotene　An agent which is used in the treatment of acne.

Telescoping A term used in chemical or process development where a series of reactions are carried out in the same reaction vessel by adding reagents in sequence and without the need for work-ups or the isolation of intermediates.

Telithromycin A semi-synthetic derivative of erythromycin that is used as an antibacterial agent.

Temafloxacin A fluoroquinolone antibacterial agent that was approved in 1992, but withdrawn the same year because of serious side effects.

Tentagel resin A resin used in solid phase synthesis which provides an environment similar to ether or tetrahydrofuran.

Teratogen A compound that produces abnormalities in a developing fetus.

Teratogenicity The property of producing abnormalities in a developing fetus.

Terramycin—*see* **oxytetracycline**

Testosterone A male sex hormone that has a steroid structure.

Tetracycline A naturally occurring compound with antibacterial activity.

Tetracyclines Tetracyclic antibiotics that bind to bacterial ribosomes and are bacteriostatic in their action.

Thalidomide A sedative that was withdrawn because of its teratogenic properties. It is now used to treat leprosy, and has useful anticancer activity.

Thebaine An opioid structure that is present in opium, and is used as the starting material for the preparation of thevinols and orvinols.

Therapeutic index or ratio A measure of how safe a drug is. The larger the therapeutic index, the safer the drug. The therapeutic index compares the dose level which leads to toxic effects in 50% of cases studied with the dose levels leading to maximum therapeutic effects in 50% of cases studied.

Thevinols Structures that are prepared from thevinone by the Grignard reaction, and can be converted to orvinols.

Thevinone A structure formed from the Diels–Alder reaction of thebaine with methyl vinyl ketone.

Thioesterase enzymes Enzymes that are present in super-enzymes and are responsible for catalysing the hydrolysis of a thioester group.

Thionyl chloride A chlorinating agent.

Thrombopoietin A hormone that is involved in blood platelet formation.

Ticarcillin An antibacterial penicillin.

Tigecycline A broad-spectrum tetracycline antibacterial agent that was approved in 2005. It can treat many infections that have developed resistance to the older tetracyclines. It is synthesized from minocycline.

Tinnitus A condition causing 'ringing' in the ears.

Tomoxetine An antidepressant.

Topoisomerases Enzymes that catalyse transient breaks in one or both strands of DNA to allow coiling and uncoiling of the molecule. These act as targets for several antibacterial and anticancer drugs.

Tositumomab An antibody that contains the radioactive isotope ^{131}I, and is used in the treatment of non-Hodgkin's lymphoma.

Toxicology Studies carried out to determine whether a drug produces toxic effects.

Transaminases Enzymes that catalyse the conversion of a ketone to an amine.

Transcription The process by which a segment of DNA is copied to mRNA.

Transdermal absorption The ability of a drug to be absorbed through the skin into the blood supply.

Transfer RNA (tRNA) An RNA molecule that bears an amino acid which is specific for a particular triplet of nucleic acid bases.

Transition state A high energy intermediate that must be formed during a reaction. The energy required to reach the transition state determines the rate of reaction. Catalysts and enzymes work by lowering the energy of transition states.

Translation The process by which proteins are synthesized based on the genetic code present in mRNA.

Transpeptidases Important bacterial enzymes that catalyse the final cross-linking of the bacterial cell wall. Targeted by penicillins and cephalosporins.

Transport proteins—*see* **carrier protein**

Tricyclic antidepressants A series of tricyclic compounds that have antidepressant activity by blocking the uptake of noradrenaline from nerve synapses back into the presynaptic neuron.

Triflate group A trifluorosulphonate group that is commonly used as a good leaving group.

bis(Trifluoroacetoxy)iodobenzene A reagent that is used in the Hofmann rearrangement reaction when it is carried out under acidic conditions. The reagent is also used to convert thioacetals to carbonyl compounds.

Trigonal Describes a planar system that is made up of sp² hybridized atoms.

Trihexylphenidyl A cholinergic antagonist that is used in the treatment of Parkinson's disease.

Triplet code Refers to the fact that the genetic code is read in sets of three nucleic acid bases at a time. Each triplet codes for a specific amino acid. The term has also been used for the bar coding method used to identify the reagents used in a combinatorial synthesis.

Triptans A family of compounds that are used in the treatment of migraine.

Tscherniac–Einhorn reaction. A form of Friedel–Crafts reaction.

Tubocurarine A natural product that acts as an antagonist at nicotinic receptors.

Tyrocidine A A naturally occurring cyclic decapeptide that has antibacterial activity.

Tyrocidine synthetase A super-enzyme that catalyses the biosynthesis of tyrocidine A.

Ugi coupling reaction A coupling reaction involving three reagents—an aldehyde, an isocyanide, and carboxylic acid.

Umpolung A synthon which has the opposite charge from the natural polarity of the structure.

Uramustine An anticancer agent.

Valsartan An antagonist of the angiotension II receptor. It is used as an antihypertensive agent to lower blood pressure.

Vancomycin A natural product that is used as an antibiotic.

van der Waals interactions Weak interactions that occur between two hydrophobic regions and which involve interactions between transient dipoles. The dipoles arise from uneven electron distributions with time.

Vardenafil An anti-impotence drug.

VARICOL A continuous chiral separation method which uses a chiral polysaccharide stationary phase.

Vasoconstriction The narrowing of blood vessels.

Vasoconstrictor Agents which promote the narrowing of blood vessels.

Veber's parameters A set of parameters that are used as a guideline to predict whether a drug is likely to be orally active. The parameters include the polar surface area, hydrogen bond donors, hydrogen bond acceptors, and the number of rotatable bonds.

Venlafaxine An antidepressant.

Vesicle A membrane-bound 'bubble' within the cell. Neurotransmitters are stored within vesicles prior to release.

Vinyloxycarbonyl chloride A reagent that is used to remove an *N*-methyl substituent from a tertiary amine.

Viruses Non-cellular infectious agents consisting of DNA or RNA wrapped in a protein coat. They require a host cell to multiply.

Voriconazole An antifungal agent.

Wang resin A resin used in solid phase synthesis.

Wilkinson's catalyst A rhodium catalyst used in the reduction of alkenes with hydrogen.

Withdrawal symptoms The symptoms that arise when a drug associated with physical dependence is no longer taken.

Wittig reaction A C–C coupling reaction that results in the formation of an alkene. The reaction is carried out between an aldehyde or ketone with a Wittig reagent.

Wittig reagent A reagent that is used in the Wittig reaction. It is an organophoshorus reagent formed from an alkyl halide and triphenylphosphine, followed by treatment with base.

Xanthine oxidase An enzyme that is the target for the enzyme inhibitor febuxostat in the treatment of gout.

X-ray crystallography A method of identifying the structure of a compound by studying the diffraction patterns produced by firing X-rays at a crystal of the compound.

Z A short hand term for the protecting group benzyloxycarbonyl.

Ziprasidone An antipsychotic agent.

Zolmitriptan An anti-migraine agent.

Index